Springer Series in Biophysics

Volume 27

The "Springer Series in Biophysics" spans all areas of modern biophysics, such as molecular, membrane, cellular or single molecule biophysics. More than that, it is one of the few series of its kind to present biophysical research material from a biological perspective.

All postgraduates, researchers and scientists working in biophysical research will benefit from the comprehensive and timely volumes of this well-structured series.

Alexander I. Zhmakin

Heat Transfer in Vivo

Springer

Alexander I. Zhmakin *(deceased)*

ISSN 0932-2353 ISSN 1868-2561 (electronic)
Springer Series in Biophysics
ISBN 978-3-031-95458-0 ISBN 978-3-031-95459-7 (eBook)
https://doi.org/10.1007/978-3-031-95459-7

This Springer imprint is published by the registered company Springer Nature Switzerland AG
The registered company address is: Gewerbestrasse 11, 6330 Cham, Switzerland

Preface

Art is a lie that makes us realize truth
Pablo Picasso

A model is a lie that makes us realize truth
Jeffrey J. Derby

Physical phenomena encountered in the heat transfer in the living tissues and the mathematical models used to simulate them are discussed. An attempt is made to consider all topics relevant to the heat transfer in humans, animals, and plants in nature and medicine from the whole body down to cellular and sub-cellular models.

Effects of the high or low temperature on the biological systems at different levels—cell, tissue, organism as well as the role of the blood circulation and of the structure of the vascular network on the heat transfer, are considered. The classic Pennes bioheat equation, a number of the non-Fourier heat transfer models (including the single-phase-lag and dual-phase-lag models), the porous media models, the models based on the fractional differential equations, and the discrete vascular models are analyzed.

Saint-Petersburg, Russia
November 2024

Alexander Zhmakin

Acknowledgements The author is grateful to Anton V. Chizhov, Evgeniy V. Galaktionov, Andrey M. Kuzmin, Dmitry Kh. Ofengeim, Vasiliy L. Preobrazhenskii, Georgiy G. Prokhorov, Alexander A. Schmidt, Valentine S. Yuferev, Igor A. Zhmakin, Tatiana P. Zhmakina for the useful discussions and to Marina N. Nemtseva for the help with the manuscript preparation.

Alexander Zhmakin: In Memoriam

Almost a century ago, Albert Einstein wrote that he was convinced that God is not playing at dice. I want to believe (in the absence of a verifiable reference) that Niels Bohr did, indeed, respond on that: "Don't tell God what to do with his dice". Even if he did not, he, most certainly, could have. This is what happens when determinism meets stoicism.

The necessary part of any development is relying on the previous experience, conclusions and observations. In this bulk of assumptions, many peacefully lose any relevance, rendering "common knowledge" rigid and often obsolete. Difficult to question it without getting into despair, denial or rebellious nihilism. When I think of my father's personality, this is what comes into my mind: he was always very open to alter his vision, no matter how hard was the path to it, or how fundamental something is. I never heard from him anything about grass being greener back in the days, unless there was a funny story related to the observation.

Extremely thorough (you can see it in this book), meticulous to exhausting details. Still, quick to generate bright ideas, bring a new perspective. A genuine artist of scientific simulation, no less. Hard to define what that special trick is, perhaps the ability to always look for adding dimensions to the space of the subject? Applicable way beyond the domain, no doubt.

In my own little professional breakthroughs I noticed, not once, that the way of thinking I applied, bears some semblance of what I might have picked up from him. More and more, I discover new things I happened to learn from him, in his subtle way where he, seemingly, just shared his thoughts. I dare to be sure that many of his colleagues would relate to this.

He had a unique strength to feel, have strong opinion, stand by it when matters, and yet not to judge. He always looked for the pure beauty of the universe, no matter where. Be it some paradigm, a system design nuance, or the way to describe it. When he worked on "Fundamentals of Cryobiology," he spent a lot of effort on epigraphs to emphasise the philosophical vibe of the chapters. Because this way the book gets prettier, and therefore no less important than completeness and precision of all references.

Unsure if he would agree, but for me he was always a true Master of the Glass Beads Game: no matter where the tough circumstances bring him, his actions seem to paint a consistent picture.

He finished this book and left. It's a loss which I will fail to describe.

Norway
February 2025

Ingvarr Zhmakin

Alexander Zhmakin: Biography

Education and Career

Alexander Igorevich Zhmakin was born on March 12, 1951, in Leningrad, USSR. In 1974 he graduated from the Leningrad Polytechnic Institute (Department of Physics and Mechanics) and joined the A.F. Ioffe Physical-Technical Institute (Ioffe Institute), where he worked throughout his career, eventually becoming a Leading Researcher of the Computational Physics Laboratory.

In 1980, he defended his Candidate of Physical-Mathematical Sciences (PhD equivalent) thesis on the numerical simulation of non-equilibrium shocked flows. Subsequently, he earned the Doctor of Physical-Mathematical Sciences degree (Dr.Sci.) in 1992 for research in numerical simulation of gas- and liquid-phase epitaxial crystal growth.

Beyond academia, Zhmakin had a longstanding collaboration with Soft-Impact Ltd., a company providing consulting services and developing simulation software for crystal growth processes and semiconductor device modelling. He served as Technical Director from 1997 to 2003 (soon after the company was founded), as Research Director from 2004 to 2008, and as Scientific Consultant from 2009 onward. He also collaborated with the St. Petersburg Branch of the Joint Supercomputing Center (JSCC) of the Russian Academy of Sciences.

He mentored numerous undergraduate and graduate students, researchers and engineers.

Research

Zhmakin's early studies, later presented in his Candidate thesis, focused on computational fluid dynamics of supersonic magnetohydrodynamic flows and problems of shock-wave interactions with obstacles.

In the mid-1980s, his scientific interests began to shift to low Mach number viscous gas flows and heat transfer and subsequently to crystal growth processes. During this period, Zhmakin developed and applied simulation algorithms for gas-, liquid-, and vapour-phase epitaxy, as well as for bulk and sublimation growth methods. He focused both on the overall processes and on specific aspects such as fluid dynamics and heat transfer. Furthermore, he co-authored several papers on fluid flows and crystal growth under microgravity.

From the 1990s onward, Zhmakin's research increasingly involved industrial collaborations, largely through his work in Soft-Impact, where he oversaw much of the simulation software development, actively contributing to architecture and research.

In the 2000s, his work expanded to include semiconductor nanostructures (A3B5 compound structure models) and light-emitting diodes based on complex heterostructures. Zhmakin explored interdisciplinary fields such as human motion modelling and cryobiology. His comprehensive monograph, "*Fundamentals of Cryobiology: Physical Phenomena and Mathematical Models*" (Springer, 2008) covered the state of the art in both theoretical and practical aspects, particularly in cryosurgery and cryoconservation. His interest in human motion modelling was first presented in a 2011 article, and later developed into the book "*Human Motion: from Animation to Simulation. A View from the Outside*" (Lap Lambert, 2021).

During his final years, Zhmakin increasingly focused on non-classical heat transfer phenomena. His 2023 book "*Non-Fourier Heat Conduction: From Phase-Lag Models to Relativistic and Quantum Transport*" (Springer) explored advanced theoretical and numerical aspects of heat conduction for scenarios where classical models prove inadequate—such as ultrafast heating and materials with non-homogeneous internal structure, including biological tissues. He later expanded upon this work in "*Heat Transfer in Vivo*" (2024), a comprehensive analysis of the properties of living organisms from the standpoint of thermal physics.

Contents

Part II Models of Heat Transfer

Acronyms

ADP	Adenosine DiPhosphate
AFM	Atomic Force Microscopy
AFP	Antifreeze Protein
AFGP	Antifreeze Glycoprotein
AI	Artificial Insemination
AIMD	*ab initio* MD
ALARA	As Low As Reasonably Achievable
AMF	Alternating Magnetic Field
AO	Adaptive Optics
AOX	Alternative Oxidase
APD	Absorbed Power Density
ART	Assisted Reproduction Technique
ATP	Adenosine TriPhosphate
BAT	Brown Adipose Tissue
BC	Ballistic-Conductive
BIE	Boundary Integral Equation
BHT	Bioheat Transfer
BHTE	Bioheat Transfer Equation
BMR	Basal Metabolic Rate
CAD	Computer Aided Design
CBF	Cerebral Blood Flow
CCR	Critical Cooling Rate
CDL	Cell Depleted Layer
CFD	Computational Fluid Dynamics
CFL	Cell Free Layer
CIT	Classical Irreversible Thermodynamics
CITT	Classical Integral Transform Technique
CHT	Computational Heat Transfer
CHTC	Convective Heat Transfer Coefficient
CMV	Critical Minimum cell Volume
CNT	Carbon Nanotube

CP	Cryopreservation
CPA	Cryoprotective Agent
CSF	Cerebrospinal Fluid
CSC	Cryogen Spray Cooling
CFD	Computational Fluid Dynamics
CHT	Computational Heat Transfer
CT	Computer Tomography
CVS	CardioVascular System
CVTT	Counter-current flow Vessels Transiting Tumor
CWR	Critical Warming Rate
DEP	DiElectroPhoretic
DFT	Density Functional Theory
DIVA	DIscrete VAsculature model
DMA	DiMethylAcetamide
DMSO	DiMethylSulfOxide
DNA	DeoxyriboNucleic Acid
DOM	Discrete Ordinate Method
DPD	Dissipative Particle Dynamics
DPL	Dual-Phase-Lag
DPM	Diffraction Phase Microscopy
DPPC	Dipalmitoyl Phosphatidycholine
DR	Deformation Ratio
DSC	Differential Scanning Calorimetry
DSM	Dynamic Scattering Microscopy
DTR	Dangerous Temperature Region
DWE	Damped Wave Equation
EC	Effective Connectivity
EF	Electric Field
EFGM	Element Free Galerkin Method
EFS	Mixture of Ethylene glycol, Ficoll and Sucrose
EG	Ethylene Glycol
EIT	Extended Irreversible Thermodynamics
ELF	Extremely Low Frequency
EMF	ElectroMagnetic Field
EPR	Entropy Production Rate
ESH	Effective Specific Heat
FC	Fractional Calculus
FDTD	Finite-Difference Time-Domain
FEM	Finite Element Method
FEWGM	Finite Element Wavelet Galerkin Method
FFT	Finite Fourier Transform
FGR	Flux-Gradient Relation
FHCE	Fourier Heat Conduction Equation
FDPL	Fractional Dual-Phase-Lag
FSPL	Fractional Single-Phase-Lag

FTPL	Fractional Three-Phase-Lag
FV	Finite Volume
II	Intracellular Ice
GBM	GlioBlastoma Multiform
GCE	Generalized Cattaneo Equation
GITT	Generalized Integral Transform Technique
GMP	Good Manufacturing Process
GNP	Gold Nano Particle
GPU	Graphical Processing Unit
GSBF	Glabrous Skin Blood Flow
HAT	Heterogeneously Anomalous Transport
HB	Hydrogen Bond
HSF	Heat Shock Factor
HSP	Heat Shock Protein
IBM	Immersed Boundary Method
IDQ	Integral of Difference-Quotient
IIF	Intracellular Ice Formation
ILPT	Interstitial Laser Photocoagulation Therapy
INA	Ice Nucleating Agent
IONP	Iron Oxide NanoParticle
IRT	InfraRed Thermography
ISF	Interrupted Slow Freezing
ITVOF	Infrared-Transparent Visible-Opaque Fabric
IVF	In Vitro Fertilization
IVM	In Vitro Maturation
HB	Hydrogen Bond
HCE	Human body, Clothing, Environment
HDA	High-Density Amorphous Ice
HIFU	High Intensity Focused US
HHCE	Hyperbolic Heat Conduction Equation
HSP	Heat Shock Protein
HU	Hounsfield Unit
HVAC	Heating, Ventilation and Air-Conditioning
LCG	Liquid Cooling Garment
LDA	Low-Density Amorphous Ice
LDF	Laser Doppler Flowmetry
LED	Light Emitting Diode
LFC	Local Fractional Calculus
LFE	LeidenFrost Effect
LHS	Latin Hypercube Sampling
LLLT	Low Level Laser Therapy
LTE	Local Thermal Equilibrium
LTM	Laplace Transform Method
LTK	Laser ThermoKeratoplasty
LTNE	Local Thermal Non-Equilibrium

MBE	Multiregion Bioheat Equation
MC	Monte Carlo
MCA	Microwave Coaxial Antenna
MCM	Metabolic Connectivity Mapping
MD	Molecular Dynamics
MIS	Minimally Invasive Surgery
MFH	Magnetic Fluid Hyperthermia
MNP	Magnetic Nanoparticle
MP	Machine Perfusion
MPI	Magnetic Particle Imaging
MPS	Moving Particle Semi-implicit
MRI	Magnetic Resonance Imaging
MW	MicroWave
MWA	MicroWave Ablation
NFCI	NonFreezing Cold Injury
NIR	Near-InfraRed
NMH	NanoMagnetic Hyperthermia
NMR	Nuclear Magnetic Resonance
NP	Nanoparticle
NSE	Navier-Stokes Equations
OCT	Optical Coherence Tomography
ODE	Ordinary Differential Equation
OHP	Oscillating motion Heat Pipe
OTSI	Objective Thermal Stress Index
PBHTE	Pennes Bioheat Transfer Equation
PD	Power Density
PDE	Partial Differential Equation
PEC	Perfect Electric Conductor
PEG	PolyEthylene Glycol
PES	Potential Energy Surface
PET	Positron Emission Tomography
PIRL	Picosecond Infrared Laser
PML	Perfectly Matched Layer
POLS	Pace-Of-Life Syndrome
POPE	Palmitoyl-Oleyoyl-Phosphatidyl-Ethanolamine
PPE	Personal Protective Equipment
PPTR	Pulsed Photothermal Radiometry
PPTT	Plasmonic Photo-Thermotherapy
PTT	Photo-Thermal Therapy
PWS	Port Wine Stain
RBC	Red Blood cell
RDF	Radial Distribution Function
RET	Rational Extended Thermodynamics
REV	Representative Elementary Volume
RF	Radio Frequency

RFA	Radio Frequency Ablation
RL	Riemann-Liouville
RNA	RiboNucleic Acid
ROI	Anatomical Region of Interest
ROS	Reactive Oxygen Species
RTE	Radiative Transfer Equation
SAR	Specific Absorption Rate
SAS	Solvent Accessible Surface
SATT	Single Artery Transiting Tumor
SAV	Supply Arteries and Veins
SAW	Self Avoiding Walk
SBC	Selective Brain Cooling
SCBA	Self Contained Breathing Apparatus
SCS	Static Cold Storage
SoCS	Sodium Cellulose Sulfate
SDM	Species Distribution Model
SDPD	Smoothed Dissipative Particle Dynamics
SEL	Stress Entropic Load
SIDQ	Singular Integral of Difference-Quotient
SHV	Submarine Hydrothermal Vents
SLP	Specific Loss Power
SOV	Separation of Variables
SPA	Specific Power Absorption
SPC	Simple Point Charge
SPL	Single-Phase-Lag
SPH	Smoothed Particle Hydrodynamics
SRS	Sample Random Sampling
STS	Selective Thermal Stimulation
TE	Transverse-Electric
TEE	Total Energy Expenditure
TGP	Thermo-reversible Gelation Polymer
TIP	Transferable Intermolecular Potential
THA	Thermal Hysteresis Activity
TM	Transverse-Magnetic
TNZ	ThermoNeutral Zone
TPL	Triple Phase Lag
TRTE	Transient Radiative Transfer Equation
TTF	Tumor Treating Field
TWMBT	Thermal Wave Model Bioheat Transfer
UHI	Urban Heat Island
US	UltraSound
UV	UltraViolet
VAT	Volume Averaging Theory
VMN	Vascular Magnetic Nanoconstruct
VOC	Volatile Organic Compound

VTPCSF	Poisson Conduction Shape Factors for Vessel-Tissue Heat Transfer
VVPCSF	Poisson Conduction Shape Factors for Vessel-Vessel Heat Transfer
YAG	Yttrium Aluminum Garnet
WBC	White Blood cell
WCI	Wind Chill Index

Chapter 1
Introduction

There are two reasons to study a response of the biological objects—cells, tissues, organs, organisms—to the action of the physical factors (the ultrasound [403, 2324], the electric [679, 835, 2027, 2769] or the magnetic field [293, 2800], the radiofrequency (RF) electromagnetic (EM) field [262, 727, 729, 3148], the ionizing radiation [415], the laser radiation [776, 1623, 2169, 2634, 3093, 3179], the extreme—high or low—temperatures [3510, 3513]):

- to assess the effects of the natural and the technogenic impacts on the organisms;
- to use the physical fields in the medicine for the diagnostics [612] and the therapy.

The mammals (including humans) are *homeothermic* (from the Greek *homoio*—like, resembling), i.e. they control the core body temperature within a narrow range independent of the environment temperature. E.g., the rorquals twice a year undergo the migration between the tropical and polar waters, involving a variation of no less than 25° C.

For humans the core temperature is the deep body temperature in the internal environment of the body, i.e., the abdominal, the thoracic, and the cranial cavities [1873] or the temperature of the venous blood returning to the heart [924, 2669]. The variation of the core temperature for the humans is about $\pm\,0.4°$ C [1365], the temperature in the human foetus is higher by 0.3° C–0.5° C [572].

The peripheral (shell) [1091] temperature, measured in the skin, the hands, and the feet, is approximately 4° C lower than the core temperature [28], usually the skin temperature is in the range 31.5° C–33.5° C [292].

Sometimes the mean skin temperature is used that is the composite of the body-surface temperatures measured at the chest (30%), arm (30%), thigh (20%) and leg (20%) [2571].

The skin temperature is much more variable than the core temperature—it is affected by the thermoregulatory reflexive autonomic responses [645] such as the skin blood flow and the sweat secretion, by the temperatures of the underlying tissues and by the environment.

A. I. Zhmakin, *Heat Transfer in Vivo*, Springer Series in Biophysics 27,
https://doi.org/10.1007/978-3-031-95459-7_1

As a rule, the homeotherms maintain the central nervous system and the internal organs at a more constant temperature, while allowing the periphery temperature to vary [2201, 3009]. The humans maintain a stable temperature for their vital organs despite the large fluctuations in the environmental conditions by carefully balancing the heat production and the heat loss [1247, 1248]. In the cold, the humans change the blood flow to allow the hands and feet to cool to conserve the internal heat.

A stability of the temperature of the human brain is about one tenth of a degree Centigrade [1610]. If more heat is lost than produced, the body temperature will drop. Vice versa, if more heat is produced than lost, the body temperature will increase. The control of the heat balance is referred to as the *thermoregulatioon* and is executed by the modulating behaviour (for example, by changing the clothes) and the physiological mechanisms (e.g., by the shivering, the sweating or the panting [221]).

Research shows that the rates of nearly all physiological functions are altered by 6–10% per a degree Centigrade [1488]. The heat can be added or removed during the therapeutic or the diagnostic procedures to produce or measure a targeted effect, e. g., to improve the patients' outcome after the brain injury [3525].

The core temperature will increase in the presence of the disease or the injury. The certain cells, when fighting the disease, release the agents into the blood which will, when they reach the brain, cause the core temperature to be raised. The changes in the core temperature give information about the progress of the disease. It is dangerous if the core temperature in the humans rises above 41° C or falls below 35° C [1896].

The tissues tolerate temperatures from just above the freezing to nearly 45° C; 45° C is also the temperature at which heating the skin causes pain (thermal pain) [3293, 3381].

Frequently it is assumed that the upper limit for normal (ectothermic) core temperature is about 40° C [1090, 1875], but the evidence derived from the cancer patients [416, 2447] and the long distance runners [418, 1809, 2524, 2553] showed the humans can tolerate the core temperature up to 42° C.

However, according to other data the brain cells could be damaged when the temperature is above 40° C [1648, 3493]—when it is above 42° C, the series of the cytotoxic events will occur, such as the protein denaturation, the aberrant oxidative phosphorylation and the impaired DNA synthesis, resulting in the multiorgan failure and the neuron non-reversible damage [1826]; in the cultured neurons, the spontaneous activity stopped irreversibly between 42° C and 43° C indicating the direct cell-level injury [1037].

Probably the first to report the cold injury was Baron Dominique Jean Larrey, Chief Surgeon of Napoleon's Grande Armée, in his "Surgical Memoirs of the Occupation of Russia, Germany and France". The coldest core temperature from which a person has been successfully rewarmed is 15° C [680].

Recognition that people can be killed by exposure to heat is documented in the Bible, for example,

- the death of the young son of a farmer from exposure to the midday heat during the harvest in his father's fields: he "went out to ... the reapers, said to his father, 'my head, my head,' " and died on his mother's lap (II Kings 4:18–20);
- Judith's husband, Manasseh, was out in the fields, supervising the binding of the barley sheaves: "he got sunstroke, and took to his bed and died" (Judith 8:3).

The pattern of the thermoregulation in which the body temperature depends on the high rate of the heat production (*tachymetabolic* from the Greek *takhus*—fast) is called *endothermy* (*endo* —inside) in contrast to *ectothermy* in the reptiles and other nonavarian and nonmammalian *poikilothermic* animals with *bradymetabolism* (*bradus*—slow) [317]—before about 200 million years ago most of the animals on the planet had a body temperature determined by the environment, primarily the ambient temperature, i.e. were ectothermic. Still, some fish lineages have evolved an ability to maintain parts of the body warm through the use of the countercurrent heat exchangers called retia (found in the locomotor muscles of the large tuna and some sharks). Some fish, such as the blue marlin (*Makaira nigricans*), use these circulatory adaptations, in combination with the modified muscles called the heater organs, to warm the regions of the central nervous system [2201, 2489].

The heat is produced by the catalysed by the enzymes aerobic and anaerobic metabolic transformation of the chemical energy by oxidation of the nutrients supplied to the living tissues $\dot{q}_{\mathrm{met}}$ and, as was stated for the first time by Hermann von Helmholtz [494], by the muscular activity.

Muscle plays a critical role in the thermal budget of the endotherms because the muscle is the most abundant tissue in the birds and mammals (in most mammals it makes up to 55% body mass [1442, 2429]) and it produces the considerable heat, even at rest. The locomotion enhances the rate of the muscle heat production. The exercise and the heat stress are characterised by an increased anaerobic contribution to meeting the metabolic demand of the skeletal muscle [1041].

Many birds and mammals can also use the skeletal muscle to generate the heat by the shivering (also shuddering)—a bodily function in response to the cold in the warm-blooded animals. When the core body temperature drops, the shivering reflex is triggered to maintain the homeostasis (*thermal* homeostasis [332]). The skeletal muscles begin to shake in the small movements, creating the warmth by expending the energy. The shivering is controlled by the motor neurons. The individual myofibres contract but the motor units are uncoordinated and the whole muscle undergoes no gross movement. This activity is fuelled by the lipid oxidation [2201].

The shivering is a strategy that works for short periods of the cold exposure, but it is not useful for the prolonged cold stress. If the shivering persists, or repeats frequently, the muscles are rapidly depleted of the nutrients and they become exhausted—the attenuation of the shivering capacity ("shivering exhaustion") [3289]. The shivering, in contrast to the sweating that is centrally controlled by the hypothalamus—when the body senses an increase temperature the hypothalamus increases nerve impulses to the sweat glands—is an involuntary response of the skeletal muscles when the passive body cooling exceeds the metabolic energy production.

The mammals also possess a unique way of generating heat in the deposits of the brown adipose tissue (BAT) [432] located near the back and shoulder region; this mechanism is most important for the thermogenesis in the skeletal muscles [2430] of the small mammals that is under the control of the sympathetic nervous system and is seen in the neonates and disappears in the adults [2249].

The BAT heat production called the non-shivering thermogenesis (NST) [686, 2201] (also "jitter-free thermogenesis") is related to the expression of a protein called thermogenin that stimulates the rate of the mitochondrial respiration and consequently the heat production [1445, 1867]—specialized proteins in the cell mitochondria of these tissues enable the cell to reduce the efficiency of adenosine triphosphate (ATP) production and release more heat [1637]. The increase in the heat production by the NST is around +5% to +20% of the normal heat production [3156]. The heat generation in the fatty layer of the hypodermis skin also occurs [2859].

The heat produced in the body is transported through the body by the conduction and convection by the blood flow (the study of the role of the blood flow in the heat transfer in the living tissues dates back to the experimental studies of Bernard in 1876 [492]).

The living tissue is a complex medium in which the highly branching blood vessels [290] form a fractal network of the fast heat transport, causing the heat propagation in it with anomalous properties [688, 1030] and an active medium that responds to the strong heating by the local growth of the blood perfusion rate [2907]. The experiments showed for certain tissues (viz., the cow haunch muscle) the significant apparent local anisotropy in the thermal conductivity that is caused by the formation of the long channels in the muscle tissue running along the fibres and the motion of the interstitial liquid due to the nonuniform heating of the tissue that affects remarkably the heat transfer [688].

Finally the heat is dissipated into the ambient mainly by the skin [1847, 3044, 3311]—the largest single organ of the body—(85–90%) via

- the convection (if the air is hotter than the skin, a reverse transfer of heat from the air to the body will occur),
- the radiation (the body continually loses heat via the electromagnetic waves whenever the body temperature is warmer than the environment $Q = \sigma_{SB}\varepsilon(T_{skin}^4 - T_{air}^4)$, where σ_{SB} is the Stefan–Boltzmann constant, ε is the emission coefficient of the skin surface; if the environment includes surfaces that are hotter than the skin, the flow of heat by radiation goes from the environment to the body—the human thermal radiation is nearly perfect blackbody radiation in the middle infrared band [1354, 1845], however, the effective radiating surface is smaller than the total body due the parts of the body surface that exchange heat by radiation with other parts of the body surface and depends on the posture),
- the evaporation (the heat loss at a rate of 2.43 KJ per gram of the evaporated sweat occurs as heat from the skin is used to evaporate the sweat; the average person has

2.6 million sweat glands [2523] of two types—eccrine and apocrine that differ by the sweat composition [3293]; the rate of sweat loss can range from about 1 to 3.5 L/h, depending on the body weight, the sweat gland volume and activity, the work intensity, the climate, and state of the heat acclimatization [109, 405, 2091]). It requires more energy to evaporate the water from the salty sweat—the sweat is made up of the water and the electrolytes such as sodium, chloride, and potassium—than from the pure water. The efficiency of the evaporative cooling also depends on the air humidity and the air velocity. The animals that do not produce sweat, e.g., kangaroos, lick the skin surfaces, which then cool as the saliva evaporates [2201].

The remaining 10% is released via the respiratory system [292]. The human respiratory airways having the largest surface area contact of the body with ambient air provide the gas exchange (supplying of the oxygen, removing of the carbon dioxide) and heat exchange. During passing the respiratory tract the inhaled air is warmed up to 37° C sequentially in the nasal vestibule, the nasal valve area, the anterior turbinate area, the nasopharinx, the nasal mucosa [955]. On the other hand, inhaling of the hot air could lead to the thermal injury—the "burned airways", first of all the nostrils, the vestibule, the nasal cavity [1123].

The experimental studies are limited for the obvious reasons, thus the convective heat transfer coefficients (CHTCs) inside the human respiratory tract could be determined using the computational fluid dynamics (CFD). As a rule, a low Reynolds $k - \varepsilon$ model of turbulence is used for the computations. The difficulties result from the complex anatomy and the very narrow nasal passageways. The geometry of problem is obtained by reconstruction of the body scans produced by the computed tomography (CT).

In humans, the evaporative heat loss accounts for up to 80% of the heat dissipation during the intense exercise, making it the primary channel of the heat dissipation [109, 1873]. Some of the heat dissipation occurs via hair or fur above skin surface [3376], but if the hair strands are dense that they can trap a layer of air and thus work as an insulation layer [1054].

The ectothermic animals can have essential distinctions from the endothermic ones. For example, many Antarctic fish have not the ability to express the functional oxygen-binding proteins, such as haemoglobin and myoglobin. These fish can survive without these oxygen carriers because they have low metabolic rates and the surrounding polar waters are rich in the oxygen [2201].

The endotherms rely on the futile cycling, in which the chemical potential energy is spent to generate heat. Most futile cycles involve cycling of the hydrolysis of the adenosine triphosphate (ATP) and the ATP synthesis—the heat is released in the ATP hydrolysis to the diphosphate ADP (also adenosine pyrophosphate) + phosphate, but a great deal more heat is produced when the cell uses the intermediary metabolism to regenerate the ATP [2201]. The endotherms can enhance the heat production either by increasing the rate of the ATP turnover or by reducing the efficiency of the ATP production; in both cases, most of the heat arises directly or indirectly from the mitochondrial oxidative phosphorylation.

The endotherms create the opportunities that cannot be exploited by the ectotherms. The ectotherms rely on the environmental conditions to heat up before they can perform the vital actions such as foraging or reproducing. The endotherms are less dependent on the environmental conditions and can therefore live in a wider range of climates, thereby reducing the competition for resources and increasing chances of the survival. In the cold or hot environments, relatively more resources have to be spent to regulate the body temperature than in the thermoneutral environments. Therefore, it might not be coincidental that the earliest civilizations developed in the climatic regions close to the thermoneutrality [1637].

The heat losses by the convection and the radiation depend on the environment and could vary from 30–50 W/m^2 to 110–140 W/m^2 due to the intensification of the sweat glands function; the heat production by the shivering could be as high as $3\dot{q}_{met}$ [291, 3149]. The heat transfer rates between the environment and the skin covered with the different textiles depends on the material, microstructure and design of textiles [1771, 2371, 2419, 2421].

Evidently, all of these mechanisms of the heat exchange with the environment are valid for the terrestrial animals: e.g., the whales can lose heat solely by the convection to the water flowing past it.

The skin of the endothermic species is not a passive insulator—it actively controls the outward passage of heat by the control of the through the skin [2631]—the important mechanisms of the thermoregulation are the shivering, the regulatory sweating (the sudomotor response) and the vasomotion [1487, 1744]—the vasoconstriction and the vasodilation (to maintain heat balance in a hot environment, vasodilation facilitates heat transfer from core to skin and sweat evaporates on skin to increase heat loss to the environment; to maintain the heat balance in a cold environment, the vasoconstriction [1535] minimizes the and ultimately heat transfer from core to skin to reduce heat loss from the skin to the environment) and/or opening/closing of the arteriovenous shunts (*Arterio-Venous Anastomoses, AVA*, located in the deep dermis close to the level of the sweat glands [3381]) that divert the blood from the skin at the low temperatures reducing the heat loss and allowing the blood to move through the s closer to the skin surface in the hot environment, enhancing the heat loss [2201]. The AVA rate is greater than the capillary rate [1540, 3006]. The AVA can divert the large amounts of the hot arterial blood from going through the high resistance path of the capillaries to a lower resistance path which connects the arterial and venous blood directly. The maximum vasodilatation happens when body temperature is 37.2° C (although it starts at the core temperature above 36.7° C) and the maximum vasoconstriction happens when the skin temperature reaches 27.8° C although it starts when the core temperature is below 33.7° C [1540].

Under the normothermic conditions, the blood vessels are under a basal sympathetic tone. Any increase in the sympathetic nerve activity induces the vasoconstriction, and decrease in the sympathetic nerve activity induces the passive vasodilation. The sympathetic adrenergic (vasoconstrictor) nerves innervate both the glabrous skin (for example, lips, forehead, palms and soles) and the non-glabrous skin. Therefore all skin regions are able to modulate the skin by vasoconstriction or the passive vasodilation [490]. On the other hand, the active vasodilation only occurs in the non-glabrous skin, this is because the sympathetic cholinergic (vasodilator) nerves innervate the non-glabrous but not glabrous skin. The experiments showed that acetylcholine released by the cholinergic nerves is broken down by the endothelium and the nitric oxide is formed which is a powerful vasodilator. Release of acetylcholine also stimulates the production of sweat from the sweat glands, thus, the onsets of sweating and the active vasodilation often coincide [1637].

The humans are exceptionally well (in comparison with animals) adapted to dissipate the excess heat: in addition to an ability to sweat, which in the humans can be effected over most of the body surface, the dynamic range of the rates in the skin is much higher than in other species [1613].

The malignant tissues are different in their structure, the cellular network, the metabolic processes, and the energy- and alkaline balances from their healthy counterparts that determine the reaction to heating [2993, 2995]. The deviations from the normal temperature (both of the core and the local temperature in the tumour) are indicators of the illness and are used for the medical diagnostics. The tumours generally have an increase in the blood supply and angiogenesis [2294, 2909] as well as an increased metabolic rate that generates an increase in temperature [474] that can be detected using the digital infrared imaging [112].

In spite of the progress of the measurements of the temperature of the biological objects (for example, the nanometre scale thermometry in a living cell [1716]), the *direct* measurement of the human *core* temperature is a problem—swallowing a sensor attached to a cord is inconvenient, thus, the alternative sites (ear, underarm, rectal, under tongue) are used. The non-invasive measurement of the human core temperature [924] is possible by a combination of the sensor system with the numerical simulation [2933]. Correlations between the human pathophysiology and the skin temperature is the basis of the infrared thermography (IRT) that allows one to diagnose a number of diseases via the noncontact measurements [1422, 1758, 2181].

The temperature variation could result from an accident (e.g., the heat flux applied at the skin surface due to the flush fire exposure, the inhalation of the heated gases, the resistance heating due to the electrical current flowing through the body [758]). The thermal effects that are present in the action of other physical fields (e.g., the electromagnetic near-field exposure effects due to the wireless electronic devices [990, 3299]) are both direct (the temperature rise) and indirect (the thermally induced membrane depolarization that could cause a temporary cessation in the firing of the pacemaker neurons [990]).

Medical conditions could lead to the variation of the local temperature of the area affected and thus temperature monitoring can be used for the diagnostic [715, 1224, 3068] of some diseases, for example, the arthritis of the joints [1590, 2607, 2608] or the breast cancer [322, 386].

There are a number of the approaches to the diagnostic that are based on the local heating of the living tissues, for example, by the ultrasound [752, 2773] or by the laser radiation [1965] (e.g., the pulsed photothermal radiometry, the pulsed photothermal radiometry (PPTR) [1428, 2143, 2144] or the infrared tomography [3031]).

The biological effect could be significant even for the weak action that plays a role of a trigger in this case [293]. E.g., the human visual system has an extremely high sensitivity: the energy of a nerve impulse is by several orders of magnitude larger than the energy of the quantum of light that triggered it; the additional energy is provided by the biology system itself, usually through the metabolic processes. Even a small (about 1° C) temperature increase can alter the production of the hormones [3299] and cause the certain brain disorders [661].

The human eye is one of the most vulnerable organs since the lens and the cornea are the avascular tissues with the low metabolism that have no thermal sensors and protective reflexes [2343, 2752, 3298].

Absorption of the ultrasound (US) waves during the diagnostic procedures results in the thermal effects that could be harmful to the cranial or the spinal tissues of an embryo [572], thus the US investigations in the pregnancy should follow ALARA ("as low as reasonably achievable") strategy [2347, 3036].

The extreme temperatures affect the biological systems at the different levels—the cell, the tissue, the organism; the effects of the low temperatures are more complex because of the ice formation. The extreme temperatures could be used to kill the malignant tissue by the low or high temperatures [758]—the minimally invasive surgery (MIS), the "keyhole" surgery—an alternative to the classical resection surgery—the hyperthermia [1202, 1203] (the cancer cells are more sensitive to heat) and the cryosurgery.

MIS usually provides the faster recovery, the lower cost, and frequently could be performed in an ambulatory surgery setting under the local anesthesia [2351]. However, cooling from the large blood vessels may affect the treatment efficiency and the lethal complications may develop if the vital s adjacent to the tumours are severely damaged [2395, 2901]. The treatment planning requires the determination of the energy absorption distribution in the tumour and the normal tissue and the temperature distributions. The heating patterns induced by the hyperthermia apparatus have to be studied to focus the energy on a given region of the body and provide a means for protecting the surrounding normal tissues. MIS has also advantage over the chemotherapy that is based on the spread cytotoxic drugs to destroy the abnormal growth of the cancer cells—the limitations of the chemotherapy are for the large solid tumours having high interstitial fluid pressure, which limits the penetration of drugs into infected organs, many anti-malignant drugs are unable to penetrate more than 40–50 μm in the tumour from the vasculature [1288].

The application of the thermal energy for the therapeutic purposes dates back thousands of years. Both Hippocrates (400 B.C.) and Galen (200 A.D.) recommended use of the hyperthermia for the accessible tumours [2770]. The thermal baths were used in the Roman Empire [1202]. The moxibustion—the local heating method using the combustion of the moxa (herb) sticks on the particular points on the body that produces the infrared radiation with a peak at 3.5 μm—is a traditional Chinese medicine [2813].

The first paper on the hyperthermia was published by W. Bush in 1886. The history of the hyperthermia had been reviewed by R. W. Y. Habash et al. [1202, 1203] and by D. Haemmerich and P. F. Laeseke [1206].

The microwave hyperthermia can be also used for diagnostics of diseases (for example, by the permittivity measurements and the attenuation measurements) [3210].

The hyperthermia and cryosurgery are also referred to as "energy-based surgical techniques" [850] as well as the electrosurgery—the application of a high-frequency (radio frequency) alternating electrical current to the biological tissue as a means to cut and desiccate the tissue [337, 1210, 2058]. M. C. L. Peek et al. [2412] published an extensive systematic review of the application of the minimally invasive ablative techniques (radiofrequency, high intensity focussed ultrasound, cryo, laser or microwave ablation) to treat the breast cancer considering over a thousand and a half patients. The authors analysed, among other issues, the completeness of the ablation, the complication rate and the tumour recurrence.

When the living tissue gains the heat from an outer heat source, the internal water evaporates [3400], the tissue shape is deformed and induced the thermal strain in the tissue [2811]. For example, F. Xu et al. [3376, 3378] investigated the thermomechanical response of the skin under the medical electromagnetic heating. The temperature increase is a direct consequence of the altered thermal balance resulting from the metabolic and externally applied heat source [2181]. Even the relatively small increases in the body temperature, as well as the thermal deformation, could lead to such effects as the altered production of the hormones, the suppressed immune response, and the protein denaturation [1787].

Evidently, in both the hyperthermia [960, 3143, 3157] (local, regional [1670] or whole body) and the cryosurgery the control of the intensity and location of the action is of the paramount importance—the tumour tissues should be *completely* destroyed to avoid the tumour recurrence (remember the Greek myth—the 2nd labour of Hercules—Iolaus had to burn the neck of the Lernaean Hydra the instant Hercules took a head off to prevent the stump from regenerating) while the neighbour healthy tissues should be kept intact. However, in the case of the excessive heat deposition the complications such as burns to the neighbouring organs or the bile duct could occur. Due to the restrictions of the real-time monitoring, the mathematical modelling is extremely important [81, 2161, 2507]. There are three stages of the analysis in silico:

1. The simulation of the evolution of the physical field and its absorption by the tissues to determine the heat source term [730, 1234]. These complex questions could include, e.g., the nonlinear effects in the high intensity focused ultrasound (HIFU) that could led to the regimes with the shock wave formation and boiling in the tissues [430] or the interaction of the laser light with the human cornea where the latter could be considered as the semiconductor [2044].
2. The heat transfer in the target and surrounding tissues.
3. The assessment of the thermal damage.

The numerical simulation is needed for the diagnostics as well, for example, to assist the prediction of the burn depth—while there is no difficulty to correctly classify the 1st and 3rd degree burns, the differentiation between the IIa and IIb burn wounds is problematic even for the experienced practitioners [232].

The difficulties of the mathematical description of the heat transfer in the vascular tissues [492, 494, 494, 758, 2840, 2946] stem from the complexity of the vasculature—the branching blood vessels' network [158, 493], the presence of the complex network of pairs of arteries and veins with the countercurrent flow [155] and variation of the perfusion intensity due to the change of the organism state (e.g., the blood perfusion rate in the active muscle tissue [2498] or due to the strong local heating [2907] is over an order of magnitude larger than that in the muscle at rest). The heat transfer is affected by the geometry, the local rates, and the thermal capacity of the blood. The rate and the size may change as a response to the local temperature, the local pH value, and the concentration of local O_2 and CO_2 levels [3525].

Somewhat simpler is the heat transfer in the avascular tissues, e. g.,

- in the human tooth: the laser assisted caries prevention, the cavity preparation etc.—the problem is to avoid the overheating of the enamel and the dentine [1889]);
- in the bone-cement-prothesis system [1239];
- in the knee cartilage under the cyclic mechanical loading [174] accounting for a contribution to the heat transfer of the synovial fluid flow within the intra-articular gap and inside the porous cartilage—even the moderate change of the temperature can have a considerable effect on the biochemical reactions such as the synthesis of the chondrocytes or the destruction of the articular cartilage collagen fibres [2167];
- in the bones during grinding with the high-speed abrasive tools and drilling in the surgery [681, 1513] and total knee joint replacement [2723]—there is a danger of the overheating (a bone necrosis occurs at about 50° C [3255]).

The whole body thermal models are needed to understand the behaviour of the body under the different environment conditions and the different activity levels, e.g., to asses a thermal comfort of the occupants in the vehicle [957, 1178], to assess the effect on the human of the military and civilian emergency situations, to assess the occurrence of the heat illnesses that will endanger the physical health and life safety during the fitness exercises and the training competitions in the hot environment [3536], to assess the possibilities of the human adaptation to the global warming [1238, 1873] or to develop the heating, ventilation and air-conditioning (HVAC) systems for the comfortable, healthy and safe indoor environments [3432].

One of the reasons to study the thermoregulation in the animals is to asses the possibility to model the human thermal physiology and the drug development. The whole body thermal models of the animals are also used to develop the species distribution models [2030, 3356] and assess the evolutionary adaptation to the climate change [1340, 2383]. The temperature is a powerful driver of the global biogeography and species distributions frequently reflect the temperature gradients in both aquatic and terrestrial habitats; many species adopt the thermal strategies (such as the thermoregulation or the acclimatization) that determine their thermal niche [638, 1382–1384].

Another problem is the validity of the Fourier law for the heat conduction in the living tissues—the Fourier law leads to the infinite speed of the signal propagation (thus the acusuality) and excludes the thermal waves that were anticipated by, among others, J. C. Maxwell, W. Nernst, L. Onsager and were observed in the form of the second sound in the helium II by V. Peshkov in 1944 [2435, 3261] (see also [901]).

The wave nature of the heat propagation becomes dominant in the ultrafast heating (e. g., the femtosecond heating of the metal films or the thin film of solid argon), the heat transfer at the nanoscale and in the biological tissues [1909, 1910, 2625, 2960, 3043, 3513, 3515]. The modern surgery such as the RF and the laser surgery involves the progressively smaller time scales and the higher energy fluxes [2034, 3096].

The tissues contain the cells, superstructures, liquids, solid/soft elements. Heating or cooling of the living tissues induce a series of the chemical, electrical and mechanical processes, for example, the diffusion, the electrical potential change and the osmosis across the cell membrane [48, 3290]; the cell membranes could store the energy [1898]. While in the homogeneous materials the relaxation time ranges from 10^{-8} to 10^{-14} s, its reported value in the biological objects varies up to 30 s. For the observations of the temperature oscillation in the living tissue [1898, 1908, 1909] see, e.g., the review by F. Xu et al. [3377]. The computations of the heat transfer processes which occur for time periods less than the relaxation time of the medium may yield the incorrect results if the Fourier model is used for applications such as, for example, the burning of the skin subjected to the instantaneous heating [1905] or the cryo-preservation of the skin [713]. The errors in the predicted temperature distribution in the case of the cryosurgery and the cryopreservation can manifest themselves in the thermal stress distribution [714, 2549, 2550] and estimation of the tissue fracture [1372, 2822] due to large the volumetric expansion [2820]; the thermal relaxation of tissue will delay the appearance of peak temperature during thermal treatments leading to a lower thermal dose level [2829].

In the book the physical phenomena encountered in the heat transfer in the living tissues and the mathematical models used to simulate them are discussed. The effects of the high or low temperature on the biological systems as well a role of the blood circulation on the heat transfer are considered.

The classic Pennes bioheat equation and its extensions, a number of the non-Fourier heat transfer models, the porous media models, the models based on the fractional differential equations, the discrete vascular models are analysed as well as the thermal properties of some human and animal tissues and some exact solutions of the heat transfer in vivo problems.

Part I
Heat Transfer in Nature

Chapter 2
Effects of Extreme Temperatures on Living Tissues

V. V. Gafiychuk and I. A. Lubashevsky [1030] listed three characteristics of the living tissue: "hierarchically organized", "active" and "heterogeneous" medium.

The living tissue consists of two subsystems:

- the cellular tissue treated as a uniform medium and
- a highly branching hierarchical vascular network involving the arterial and venous vessels.

The blood flow through the arterial bed supplies the cellular tissue with oxygen, nutritious products, etc. and controls the heat balance in the system. Through the venous bed blood flow withdraws products resulting from a life activity. The vascular network is embedded into the cellular tissue and in spite of its small relative volume mainly determines the heat and mass propagation due to the fast convective transport with the blood flow in the vessels. Such a feature makes the living tissue the highly heterogeneous media where the low-dimensional heterogeneities form the fast transport paths.

The living tissue differs significantly from such media the polycrystals and the crystals with dislocations where the grain boundaries or the dislocations form the fast diffusion paths in that vessels make up a highly branching network and blood flow in the vessels of one level are directly related to the blood flow in all other vessels [1030].

The living tissue is an active medium since the cellular tissue under the various conditions requires the different amount of oxygen, nutritious products, etc. Due to expansion of a single vessel leading to the blood flow redistribution over the whole vascular actually all vessels should take part in the vascular networks response to the local variations of the tissue state parameters—thus, the vascular network response is the cooperative action of all the vessels.

The thermal injury is the irreversible change in the living tissue caused by the temperature. The injury can occur when the tissue temperature exceeds the range between which the normal life processes exist. Both the high and low temperature

A. I. Zhmakin, *Heat Transfer in Vivo*, Springer Series in Biophysics 27,
https://doi.org/10.1007/978-3-031-95459-7_2

can cause the irreversible changes in the biomolecules, resulting in the injury, e.g., the burns and the frostbites. In the living organisms the main repositories of energy are the macromolecules, which store energy in the form of the covalent and non-covalent chemical bonds, and the unequal concentrations of solutes, principally ions, on opposite sides of the cell membrane.

Under some kinds of the moderate thermal stress that is subthreshold to the injury, the cells produce molecules (the heat shock proteins [2184], e.g., Hsp70 [521]) that render the temporary protection against levels of many types of the stress (thermal, mechanical, chemical, etc.) that would normally cause the injury. The heat tolerance is the cellular adaptation caused by a single sublethal heat exposure that allows the organism to survive a subsequent exposure to the lethal heat stress since the part of the thermal energy could be transformed into other types of energies, thus prevents tissue temperature from being increased quickly [1898], for example, by the protective effects of the heat shock proteins production on the cellular structures [2196]. At the temperature between 39° C and 40° C the heat-shock proteins are activated to induce the heat tolerance and enhance the cellular protection [1242].

The temperature affects the biochemical, physiological and reproductive processes of all living organisms. The mild increases in the temperature may simply slightly accelerate the cellular processes with no overall effect [572], the moderate temperature increases may arrest or retard the cell division [2088]. The effects of a moderate rise above normal physiological temperature can have the important consequences for developing the embryos or the foetuses—the actively dividing cells of the embryonic and foetal central nervous system are known to be highly susceptible to the changes in temperature [2835, 3275]. The core temperatures that drops below 31° C or rises above 43° C can be fatal [1866].

The most common type of the thermal injury is the accidental burn [373, 1475, 2948] encountered in the domestic, industrial and military settings; the degree of the injury depends on both the temperature (the damage to the human tissue is an exponential function of the temperature) and the duration of the heating [2186] ("burn duration" [86]). Most burns result from the propagation of the heat inward into the tissues as a result of the contact of the skin with a hot solid, liquid, or vapour, the exception is the electrical burn in which the tissue temperature is elevated owing to the dissipation of the electrical energy when a voltage is applied—the primary source of heating is internal since the impedance of the muscle is higher than that of the skin and fat.

2.1 The Thermoregulatory System

All homeothermic organisms are capable of maintaining the stable body temperature in the changing environment conditions using the various negative feedback mechanisms. The discovery of the homeostasis is usually attributed to Walter Cannon [431], however, Claude Bernard described the processes of the physiological control in 1878. Although the term homeostasis was originally used to refer to the processes within the living organisms, it is now used to describe the autonomous control in any system and is understood as a way of maintaining the adaptiveness of any system, either a living one or a machine [1212].

All livings systems need supply of the energy for the various processes such as the biosynthesis of molecules, the transport of signals in the form of molecules and ions across the cellular boundaries,and to perform the mechanical work for the cellular movement and the muscle contraction. The energy required for the mechanical and chemical activities are derived from the oxidation of food and light. The human digestive system breaks down the food into the smaller particles by the physical and chemical means, which are absorbed as individual molecules by the intestinal walls. The blood vessels distribute the corresponding molecules throughout the body, to be taken up by cells, according to the need such as the carbohydrate molecules for energy, the amino acid molecules for building the proteins and the lipid molecules for building the membranes and the hormones.

The blood flow through the complex vascular network of arteries, veins and capillaries regulates the local temperature of the tissues deep inside the body as well as close to the skin. The total number of the blood vessels in the human is about 10^{10} [830]. The total amount of the blood in humans varies with age, sex, weight, body type, and other factors, but an average figure for adults is about 60–65 ml per kg of the body weight [2627].

The blood flow through vessels acts as the local thermal regulator during the hypothermia, hyperthermia and the changing environmental conditions.

The metabolic heat generated in the tissues is either lost to the environment and/or transferred to other locations by combination of the several mechanisms: conduction, convection, radiation. For a body to be in the thermal equilibrium,the rate of the heat gain and the rate of the heat dissipation should be equal.

The principle of the homeostatic regulation of a biological parameter is that all responses act together in a highly coordinated fashion to adjust the bodily parameters, including the multiple negative feedback mechanisms. The human body is a thermal system that generates heat by the biochemical reactions at the cellular level and maintains its temperature with various thermoregulatory mechanisms. The overall thermal behaviour of the human body is the combined effect of the heat generation and the heat transfer at various scales: microscale (molecular or cellular level, mesoscale (tissue and organ level) and macroscale (thermal interaction of the complete body with the environment) [284].

There are two neuronal models of the temperature regulation (homeothermy) in the mammals are known:

- *the set-point model.* This model includes an adjustable set point and signals from the peripheral and/or central temperature-sensitive neurons that are integrated and compared with a set point at the level of the hypothalamus. The thermogenic responses can correct the core temperature toward the set point level [339, 1226]. The fever or the hypothermia result in a shift in the set point [420].

- *the null-zone model.* In this model body core temperature is defended around a "set level" or "null zone". The existence of this "null zone" has been demonstrated in several species, including humans [2112]. The null-zone model is based on the interaction of two variables rather than on the comparison of a variable to a constant set point. The reciprocal cross inhibition between a cold sensor and a heat production effector pathway and a warm sensor and a heat loss effector pathway, with the goal of defending a null zone of the core temperature, is the basis of this model [316, 2203].

The human body is exchanging the heat with the environment yet manages to maintain the fairly constant internal temperature, in spite of the changing environmental factors such as, for example, the air temperature, the air movement, the thermal radiation.

Temperature-measuring devices can fall into the two categories, invasive and noninvasive:

- the invasive temperature sensors—the thermocouples, the thermistor beads, the optical fibre sensors, etc.—have the advantages of the small size, the fast response time, the extreme sensitivity to the temperature changes, and the high stability, however, they involve a limited number of the measurement locations, the uncertainties about the anatomic placement of the thermometry devices and the periodic rather than continuous temperature monitoring; unlike a thermocouple or a thermistor bead, the fibre optic temperature probe does not interfere with an electromagnetic field, however, it is relatively big in size and has a lower temperature resolution [3525];
- The noninvasive temperature-measuring techniques include the magnetic resonance imaging (MRI) thermometry, the infrared thermography, etc. Because of the theoretical sensitivity of some of its parameters to temperature, the MRI has been considered to be a potential noninvasive method of mapping the temperature changes during therapies using the various forms of the hyperthermia. The MRI imaging has the advantage of producing the three dimensional anatomic images of any part of the body in any orientation. In clinical practice, MRI characteristic parameters such as the molecular diffusion coefficient of water, the proton spin-lattice relaxation time [2382] and the temperature-dependent proton resonance frequency shift have been used to estimate the in vivo temperature distribution in tissues [3525]. MRI provides good spatial localization and sufficient temperature sensitivity. However, this technique is limited due to the low thermal resolution [643, 1059, 2622]. Special attention should be given to the difference of the electromagnetic behaviours between the homogeneous phantom and the heterogeneous human tissues in the case when the implantable medical devices [3324] are present such as typical orthopaedic medical devices—the healthcare products applied to shoulder, humerus, hip, femur and tibia [1190].

The temperature strongly influences the biochemical reactions, the gene expression, the membrane ion transport, the protein-protein interactions and the overall cell functioning [2389, 2460, 2792]. The intracellular temperature is difficult to measure [1361]. However, advances in the thermometry (for example, P. Song et al. [2912] developed the transient microscopy method for study the heat transfer in single cells by measuring the cooling time of a heated gold nanoparticle with a temporal resolution of $\approx 5\,\mu s$ and a spatial resolution of $\approx 250\,nm$, R. Tanimoto et al. [3015] used a novel method for sensing temperature in the living cells based on the imaging technique of fluorescence of the quantum dots, K. Okabe et al. [2330] employed the fluorescent polymers and J. Donner et al. [788]—the green fluorescent proteins, G. Kusko et al. [1715] used the nanoscale thermometry that is based on the coherent manipulation of the electronic spin associated with the nitrogen–vacancy colour centres in the diamond that makes possible to detect the temperature variations as small as 1.8 mK) have allowed the high-resolution temperature mapping within the plant and animal cells [2094, 2331].

Results show that the nucleus and specific organelles can act as the local heat sources, leading to the localized variations in the intracellular temperature [1267, 3015, 3115]. For example, the temperatures in the nucleus, mitochondria, and centrosome are significantly higher than those of the cytoplasm and the cell membrane [1361]; the mitochondria are physiologically maintained at the temperature close to 50° C in the in mammalian cells [561] (and thus called cellular radiators [233]).

Such temperature variations depend on the biochemical reaction processes; the excess energy can be transformed into heat [3015].

In the nucleus, DNA replication, transcription, and RNA processing contribute to the local thermogenesis [1767]. Muscle is the primary thermogenic organ in most vertebrates, since the muscle contraction is coupled to the heat production [2651]. During the muscle contraction, heat is generated through both the myosin-mediated adenosine triphosphate hydrolysis and Ca^{2+} transport [2430].

In the mitochondria, the local heat generation is a result of the active energy metabolism (from Greek "metabolikon"—disposed to cause or suffer change—coined by the German biologist Theodor Schwann), with a potential contribution from the proton transfer through the mitochondrial inner membrane [85, 1267, 2884]. The hydrolysis of the ATP/GTP or the phosphorylation of the centrosomal proteins induces the local temperature increase at the centrosome [75, 796]; the calcium pump activity may also cause the local heating in the endoplasmic reticulum [165].

Compared to the measuring temperature, the measurement of the heat dissipation is more difficult: it requires breaking the thermal equilibrium, creating a heat transfer process at a subcellular level, and tracking the heat dissipation dynamics before the thermal equilibrium was rebuilt; the disturbance to the local thermal environment should be as low as possible to avoid the effects from temperature changes.

P. Song et al. [2912] explored the heat transfer in cells under the different environmental temperatures and found that the cells exhibit a self consistent regulation on their heat transfer properties responsive to the environmental temperature aimed at maintaining the intracellular temperature within a reasonable range, although the mechanisms were unclear. The authors hypothesized that such cellular level heat regulation might exist in all homoeothermic animals.

A. Murakami et al. [2229] used two distinct intracellular thermometers and revealed that D9-desaturase DESAT1 is involved in the cell-autonomous regulation of the intracellular temperature via the mitochondrial respiration.

Mitochondria are the organelles that have essential functions in a wide range of the biological processes, including the energy metabolism—the production of the adenosine triphosphate (ATP) via the mitochondrial respiration. The ATP provides the chemical energy needed a variety of biochemical process, for example, muscle contraction. The ATP is involved in the synthesis of the DNA and in the chemical communications between and within cells [1270].

The mitochondria contain two membranes: an outer mitochondrial membrane (OMM) and an inner mitochondrial membrane (IMM) [986]. At the IMM, the mitochondrial respiratory chain complexes I–IV pump protons from the mitochondrial matrix into the intermembrane space, forming a proton gradient across the IMM. Driven by this proton gradient, F1Fo-ATPase converts the ADP into the ATP. Mitochondria play an important role in cold adaptation, as the proton gradient can be uncoupled from the ATP synthesis to release energy as heat [432]. The enhancement of the proton leakage by uncoupling proteins is known to increase the mitochondrial heat production [932]. The maximal overall thermodynamic efficiency of the mitochondrial ATP synthesis is in the range of 40%, thus about 60% of the input energy is dissipated as heat [2275]. The mitochondrial tissue-nonspecific alkaline phosphatase production is found to control the thermogenesis by the hydrolysis of the phosphocreatine [2980].

Two main mechanisms (not mutually exclusive) can increase the mitochondrial heat production rate [233]:

- the reduction of the thermodynamic efficiency of the oxidation to ADP phosphorylation (OXPHOS), at a constant or increased rate of substrate oxidation, thus increasing the heat production at the expense of the ATP production;
- the increase in the ATP turnover and consequently the substrate oxidation rate, with an inherent loss of energy as heat.

A. Murakami et al. found that the mitochondrial respiratory activity is potentiated by the cold exposure. Thus, the production of the unsaturated fatty acid-containing phospholipids during the cold exposure is involved not only in the control of the physicochemical properties of the cell membranes but also in the enhancement of the mitochondrial respiration in response to the cold environments.

Heat is generated in the human body at a cellular level, mainly as a by-product of the breaking down of the proteins (on the protein metabolism relation to the nitrogen balance see, e.g., [3437]), the lipids and the carbohydrates. The major result of this process—the metabolism—is the production of the body's internal energy source, the ATP. The end products of the aerobic metabolism are the carbon dioxide and water. The production of the ATP, the main intracellular energy storage molecule, is accompanied by heat: the energy lost during the electron transport and the oxidative phosphorylation is largely converted into the heat. In the eukaryotic cells, mitochondria have a crucial role in the ATP and heat production and called "the powerhouses of eukaryotic cells" [233], over 90% of mammalian oxygen consumption is by mitochondria [2599].

The first attempt to quantify the basal metabolic rate of a living being was made in 1789 by Antoine Laurent Lavoisier and Armand Séguin who set up a device to measure the oxygen consumption of a live guinea pig and later of a resting human being and discovered that the physical activity, as well as the digestion of food, enhances the rate of oxygen uptake; they correctly concluded that respiration is a form of combustion. Later Eduard Pflüger discovered that the cells of the body are the sites of the biological oxidation [2307].

The production of the ATP is inefficient with nearly two thirds of the energy being "lost" as heat—only about 25–30% of metabolic heat produced is used by muscles to perform the mechanical work [806, 1661]. The remaining >70% of the metabolic heat is stored in the body that explains the increase in the body temperature when performing the physical work [930, 3446].

The metabolic rate depends on the tissue temperature as

$$\dot{q}_{\mathrm{met}} = \dot{q}^{0}_{\mathrm{met}} Q_{10}^{\frac{T_0 - T}{10}} \tag{2.1}$$

where $T_0 = 37^\circ$ C and Q_{10} is the temperature dependent coefficient usually taken as the constant from the range [2, 3]. n empirical unit of measurements is introduced equal of the "sitting-resting" metabolism of a man $1met = 58.15\,\mathrm{W\,m^{-2}} = 50kcalh^{-1}m^{-2}$ [317].

It is said that Enrico Fermi had estimated (and got a fairly accurate value) the rate of the metabolic heat production in humans assuming that the main ingredient of the human being is water and using the time needed for the corpse to become cool he got from the detective stories.

There are hormones—catecholamines and thyroxine—that can increase the metabolic rate (the catecholamines stimulate many enzyme systems, thus increasing the cellular metabolism; the thyroxine magnifies the metabolic response to the catecholamines and stimulates the oxidation in the mitochondria [3293]).

The conversion of the energy to heat could be delayed when the energy that was used to form glycogen or protein is released as the heat when the glycogen is converted back into the glucose, or the protein back into the amino acids.

All internal muscle work is eventually converted to the heat and there are other chemical reactions in the body which purely produce heat. The generated heat is distributed throughout the body by the vascular system which transfers heat from the sources of the heat production to areas of the body with the low metabolic rates. The body then losses heat to the environment at either the skin surface or via the respiratory tract. The excess metabolic heat in the muscle is transferred to the arterial blood flowing into the capillary bed [930] and stored in the venous blood flowing out of the muscle. Veins have wider diameters than arteries of the same length, thus, the blood pressure drop across a microcirculatory bed is mainly caused by the resistance of its arterial bed [1030]. Heat stored in the venous blood is circulated back to the heart and conserved within the body to cause the heat storage or to be transported to the skin surface for dissipation into the environment [2248].

The role of the blood flow in the heat transfer is critical. The heat transfer for the systemic thermoregulation is one of the three primary functions of the blood flow, the others are [3525]

- mass transfer in support of the body metabolisms, including the transports oxygen to the rest of the body and transport of the carbon dioxide and other waste from the cells;
- the regulation of the systemic blood pressure.

The theoretical study shows that if the heat produced in the central areas of the body could escape only by tissue conduction, the body temperature would not reach a steady state until it was about 80° C (a lethal temperature would be reached in only 3 hours), during exercise, the human body temperature would have typically risen 12° C in an hour if no heat was lost by the blood flow [3525]. The heat imbalance may cause an abnormal body temperature and even result in an irreversible physiological injury [349, 1359, 1826].

The temperature affects many biological processes through the specific effects on such functions as the electrical properties and fluidity of the cell membranes and through a general effect on most chemical reaction rates. The reaction rates vary approximately as an exponential function of the temperature within the physiological range, and increasing the temperature by 10 Centigrade degrees increases the reaction rate by a factor of 2–3. For any reaction, the ratio of the reaction rates at two temperatures 10 Centigrade degrees apart is called the Van't Hoff Q_{10} coefficient for that reaction, and the effect of temperature on the reaction rate is called the Q10 effect. This concept is often generalized to apply to a group of reactions that are involved in a physiological process, for example, the effect of the body temperature on the metabolic processes is clinically important in caring for patients with the high fevers who are receiving fluid and nutrition intravenously, and a rule states that each Centigrade degree of fever increases a patient's fluid and calorie needs by 13% [805, 3293].

When the body temperature drops, the thermoregulatory system responds by causing the shivering—the cold-induced thermogenesis attributable to the skeletal muscle contractile activity [2943], humans initiate this thermogenesis through the involuntary shivering or by the voluntarily modifying behaviour, i.e., increasing the physical activity. Shivering may start immediately, or after several minutes of the cold exposure, usually beginning in the torso muscles, then spreading to the limbs [240, 3053]. The intensity and extent of the shivering varies according to the severity of the cold stress. As the shivering intensity increases and more muscles are recruited to shiver, the whole body oxygen uptake increases.

When the body temperature rises, the body starts to sweat. The body also can to increase or decrease the flow of blood near the skin surface to increase or decrease the rate of heat loss from the skin.

The actual amount of the heat produced by the individual body depends on the activity level: the metabolic rate will be in the region of 30–50 W m^{-3} under the resting conditions and greater than 600 W m^{-3} when doing the exercise. The human resting energy expenditure is highest in the afternoon and evening and reduced by 10–30% at sleep [2739, 3539].

For example, P. O. Fanger [921] listed the metabolic rates for walking on the flat ground with the speed from 3 to 8 km/hr and up a grade from 5 to 25%; M. N. Sawka and K. B. Pandolf [2680] presented the metabolic rates for selected military tasks (e.g., lying on the ground, standing in the foxhole, sitting in the truck, driving the truck, walking on the hard surface, walking on the loose sand, lift and carry the 45-kg artillery shell, crawling with the full pack, the foxhole digging, the field assaults).

The metabolic heat during the high intensity physical exertion can overwhelm the heat dissipation mechanisms and promote the heat storage in the body [1090, 1873, 1875]. During the prolonged intense physical work the core temperature can increase to over 41° C [2553]. The muscles can increase the energy usage by 15–20-fold (and ten times the blood flow [3525]) from the base level; the amount of the heat being produced also depends on the type of work being done—in the exercise, the energy is required to do the external work, while in the shivering virtually all the energy is converted into the heat [1224].

J. Bienertová-Vašků et al. [289] introduced the stress entropic load (SEL)—a variable designed to facilitate the physical measurement of the stress load of a human body by monitoring the energy and the matter flows in the body, a time-dependent function which may be calculated from the different types of the heat, the temperature changes and the gases exchange. Their model could be considered as an extension of the model of the thermoregulation with the heating saturation by H. F. Nijhout et al. [2303] who assumed that the core body temperature T_C remains approximately constant in a wide range of the air temperatures T_A [2193]. Bienertová-Vašků et al. assumed that the core body temperature T_C is affected by the air temperature through an additional layer of skin. Hence, both the air temperature T_A and the core temperature T_C affect the skin temperature T_S. The only direct heat transfer between the body core and the ambient air takes places through breathing and is proportional to the volume of the inhaled air, and, thus to the volume of the oxygen uptake V_{O_2}.

The authors formulate the model of the mammal thermoregulation as

$$\frac{\mathrm{d}T_S}{\mathrm{d}t} = k_1(T_A - T_S) + k_2(T_C - T_S) \tag{2.2}$$

$$\frac{\mathrm{d}T_C}{\mathrm{d}t} = k_3(T_S - T_C) + f_S(T_C) + k_4 V_{O_2}(T_A - T_C) \tag{2.3}$$

where k_1 is the rate of the heat transfer between the skin and the ambient air, k_2 is the rate of the heat transfer between the skin and the core, k_3 is the rate of heat transfer between the core and the skin, k_4 is the rate of the heat transfer between the core and the ambient air through breathing, f_S is the saturating function that is smooth in T_C, has only one maximum and one minimum, e.g.

$$f_S(T_C) = p(T_C^{Set} - T_C)e^{-\upsilon(T_C^{Set} - T_C)^2} \tag{2.4}$$

where p and υ are the parameters of the saturating function (the authors used $p = 1$ and $\upsilon = 0.5$).

For the comparison of the saturating functions of the different shapes the authors defined the area below the saturation curve

$$E_S = \int_{T_C^{\vartheta_1}}^{T_C^{\vartheta_2}} |f_S(T_C)| \mathrm{d}T_C. \tag{2.5}$$

The number of parameters of model (2.2) and (2.3) by Bienertová-Vašků et al. could be reduced from nine to seven using the following transformation $\tau = k_1 t, a = k_2/k_1, b = k_3/k_1, c = p/k_1, d = k_4 V_{O_2}/k_1$:

$$\frac{\mathrm{d}T_S}{\mathrm{d}t} = (T_A - T_S) + a(T_C - T_S) \tag{2.6}$$

$$\frac{\mathrm{d}T_C}{\mathrm{d}t} = b(T_S - T_C) + c(T_C^{Set} - T_C)e^{-\upsilon(T_C^{Set} - T_C)^2} + V_{O_2}(T_A - T_C). \tag{2.7}$$

The authors analysed the equilibria of system (2.2) and (2.3) for sufficiently smooth function f_S. Setting the right-hand sides of Eqs. (2.2) and (2.3) equal to zero yields an equation for core body temperature equilibrium $T_C^\star$

$$T_A = T_C^\star - \frac{k_1 + k_2}{k_1 k_3 + k_4 V_{O_2}(k_1 + k_2)} f_S(T_C^\star). \tag{2.8}$$

If T_C^0 exists such that

$$\frac{\mathrm{d}f_S(T_C)}{\mathrm{d}T_C}|_{T_C = T_C^0} > \frac{k_1 k_3}{k_1 + k_2} k_4 V_{O_2} \tag{2.9}$$

then it is possible to establish T_A for which (2.8) has three solutions [289].

Depending on the system parameters, the two possible stable equilibria exist. We can distinguish between them using the core temperature T_C at the equilibrium point. The core temperature at the first equilibrium point is close to the core temperature set-point T_C^{Set}. The authors called this equilibrium homeostatic. The core temperature at the second equilibrium point is close to the ambient air temperature. The authors called this equilibrium non-homeostatic. The authors found that the humans may use two types of the adaptation strategies with respect to the thermoregulation. The first adaptation strategy type takes into account the adaptation costs measured by the area below the saturation curve f_S while the second is based on CO_2 production [289].

M. P. Castellani et al. [464] considered the human thermoregulatory model that simulates the thermal responses through a passive and active system. The passive system describes the heat balance within the body and between the skin surface and the environment using the Pennes equation [1264, 1561, 3372]. The active system that describes the thermoregulatory mechanisms—the vasodilation, the vasoconstriction, the sweating, and the shivering heat production—is simplified into the three subsystems:

- The *sensors* (the thermoreceptors) located in the hypothalamic region of the brain [2714], viz. in the preoptic area [2249], as well as in other brain areas [840, 1536] and in the peripheral regions. Each type of the thermoreceptor is activated within a specific range of temperature. When a thermoreceptor is subjected to an abrupt change in the temperature, it is first strongly stimulated (this is called the "overshoot"). The strong stimulation phase is short and the stimulation decreases asymptotically until it reaches a steady response rate [1343].

  The signals that activate the hypothalamic temperature regulating centres come from the two sources: the temperature-sensitive cells in the anterior hypothalamus and the cutaneous temperature receptors.

  The preoptic-anterior hypothalamus contains the neurons that [340, 341]

  – respond to the temperature in the brain and
  – receive input from the thermoreceptors from the skin and spinal cord.

  Note that not all neurons are uniformly activated because the brain does not cool uniformly that is confirmed by a significant nonuniformity in the temperatures in the various areas of the brain in the animals [1272].

  The cells in the anterior hypothalamus sense the temperature of the body core, viz, the temperature of the arterial blood that passes through the head. The certain neurons in the hypothalamus that respond to the cold stimuli also respond to the certain chemical changes (these nonthermal signals may explain the observations of humans complaining of being cold when they are actually dehydrated or hungry [2499]).

  The hypothalamus is part of the limbic system that serves as the common point of the consolidation for the efferent outputs to preserve the physiological homeostasis through the autonomic regulation [1873, 2004, 2194]. It has been shown that the spinal cord also participate in the thermoresponse to cold [2862, 2863].
- The *integrator* system serves as the processing centre of the thermoreceptor signals that are combined and transformed into the appropriate effector commands—the efferent responses include an increase in the heart rate, the peripheral vasoconstriction, the tensing of muscles, and the higher metabolic rate caused by the release of the various hormones;

- The *effector* system (for example, the cardiovascular system, the skeletal muscles, the sweat glands) responds to these commands and converts them into the actions.

In the human there are two sets of the sensors that measure the temperature. The skin contains the warm and cold receptors (the number of the cold receptors is 10-times larger [3293]). The temperature of the blood is constantly measured by a second set of the thermoreceptors which are found in the hypothalamus that also acts as the integrating centre, collecting the information from a variety of sensors and then initiating an appropriate response.

The main effector organ in the thermoregulation is the skin. The regulation of the skin blood flow and the non-shivering thermogenesis are stimulated by the sympathetic nervous system. The sympathetic nervous system is composed of the adrenergic and cholinergic branches. Each branch has the specific neurotransmitters and the co-transmitters that are released from their nerve endings. For the sympathetic adrenergic nerves, the main neurotransmitter is the noradrenaline and for the sympathetic cholinergic nerves the main neurotransmitter is the acetylcholine [1637]. The tissues that are innervated by these nerves contain the receptors that respond to the specific neurotransmitters. For example, the blood vessels in the skin are densely innervated by the sympathetic adrenergic nerves and contain many α-adrenergic receptors, which, when stimulated by the noradrenaline, cause the vasoconstriction.

The low-cost thermoregulatory responses are applied before more costly (in terms of energy, water, minerals) responses—is relatively inexpensive to decrease or increase the blood flow to skin, thus the vasoaction is used earlier than shivering and sweating [3149].

The skin has three ways of altering the heat gain/loss depending on the nerve impulses from the CNS:

- *Sweating*. The humans have the sweat glands spread over almost all the surface of the skin. The sweat is only produced when the body temperature is too high as the evaporation of the sweat from the surface of the skin leads to a cooler skin. The sweating is a powerful thermoregulatory mechanism: increase of the core temperature by 1° C can produce an increase in the sweat rate by a factor of 10–20 times [332].

  However, since the fluid in the sweat is derived from the plasma in the circulating blood [405, 1682, 2441], the prolonged excessive sweating without the adequate fluid replacement can lead to the depletion of the blood volume [17, 1120, 2720, 2722] and to a reduction in the cardiac output and the blood flow to the skin and muscle, which in turn compromise the heat dissipation [17, 944, 2721, 2722]. The blood circulation can be interrupted due to a mismatch between the venous return, the cardiac output and the demand for blood flow, which can result in the fainting [1119, 1120]—a self-limiting mechanism at the extreme of dehydration that prevents further depletion of the blood volume by forcing the cessation of the physical work and by lowering the metabolic heat production [1873].

Thus, the cost of the sweating and evaporative heat loss is borne by the cardiovascular system (CVS). The excessive sweating in the absence of the adequate fluid replacement leads to a decrease in the blood volume, which compounds the stress on the compensatory mechanisms in the CVS to preserve the cardiac output [2722]. At the extreme, the blood circulation can be interrupted due to a mismatch between the venous return, the cardiac output and the demand for the blood flow leading to the fainting [1119, 1120]. The fainting at the extreme of dehydration is a self-limiting mechanism that prevents further depletion of the blood volume by forcing the cessation of the physical work and by lowering the metabolic heat production. This mechanism under a state of the dehydration may contribute to the occurrence of the heat exhaustion and may also explain the absence of fatality due to the dehydration in the sport and the social settings.

- *Hairs*. The hairs on the skin play an important role in the thermoregulation in many mammals [1055]. If the body temperature drops, the CNS causes the hair erector muscles to contract and pull the hair to a more vertical position in the follicle, a thicker layer of air is trapped between them and so the body is better insulated against the heat loss. The unique properties distinguish the hair of the polar bear that have the air trapped inside; it should be considered as the fractal medium [3249].
- *Shifting patterns of blood flow in the skin*. This is the main effector mechanism in the human thermoregulation [903]. If the body is getting too cold, the pattern of the blood flow switches in the skin so less blood flows in the capillary beds near the surface of the skin composed of arterioles, arterial and vein capillaries, and venules (most of the microvasculature is contained in the papillary dermis within 1–2 μm below the epidermal surface) [3381] and more blood is retained deeper in the skin structure. This is achieved by narrowing the arterioles that supply the capillary beds near the surface. This narrowing of the arterioles is called the vasoconstriction.

The majority of the land-based mammals have the fur covering almost all of their skin. This keeps the animal warm in the cold temperatures at the expense of the lack of cooling in the hot temperatures. There are numerous explanations about why the humans lost most of their hair [946]. They include, for example, the increased dissipation of the body heat for living on the open savannah, the metabolically active brain, the improved heat dissipation needed for the transition to the bipedalism, leading to the faster running and the better success at the hunting [3309], use of the clothing [1745], reduction of the parasites (lice, fleas and ticks) that live in the hair, the discovery and use of the fire, the allometry—the study of the body size and physiology combined with the observation that the larger apes have the less dense hair [2748].

The thermoregulatory system exhibits some degree of the thermo-plasticity that is capable of the semi-permanent adaptation to the heat through the process of the heat acclimatization [101, 102, 1286, 3364]. The daily exposure to the moderate intensity work under the sub-lethal heat stress for up to 14 days enhances the heat dissipation mechanisms to result in the lower body temperature under the same workload and environmental condition [103, 1090, 2247].

These adaptations can be reversed with the prolonged absence of the heat and the exercise exposures [108, 1286]. The long term passive exposure to the warm and humid weather conditions also induces some degree the heat acclimatization [1238, 1874]. The highly adaptable attributes of the core temperature regulation has played the critical roles in allowing man to survive and function in the heat [1873]. There are two physiological systems that are involved in maintaining the core temperature during the intense exercise in the heat: the cardiovascular system (CVS) and the fluid balance mechanisms [1873]. The higher demand on these systems results mainly from the diversion of the venous blood to the skin for the heat dissipation and the loss of the plasma volume through sweating [1088, 2248].

A thermal stimulus (heat or cold) applied to human skin is one of the main causes of pain; solving how to relieve the thermally induced pain implies further study and development of the medical thermal methods [3376, 3381].

V. V. Gafiychuk and I. A. Lubashevsky [1030, 1031, 1972] considered the cooperative mechanisms of self-regulation in the large hierarchical systems—such systems occurring in nature are characterized by a great information flow that none of their elements can possess the whole information required of governing the system. The authors noted that the problem of the heat and mass transfer in the living tissue is closely connected with the turbulent transport problem in the hydrodynamics because the heat and mass propagation in the living tissue is also governed by the cooperative influence of the blood velocity field $\mathbf{v}(\mathbf{r}, t, l)$ in the vessels of all the lengths from L to l_{cap}.

The living tissue possesses a peculiarity that makes it distinct from physical and mechanical systems—the vascular network is active and responds to variations in the cellular tissue state (the temperature, the oxygen concentration, etc.)—the vessels will expand or contract, increasing or decreasing blood flow in order to supply the tissue with the blood.

However, a cooperative mechanism of the regulation could exist that involves the individual response of each element to the corresponding hierarchical piece of information and leads to ideal system response due to the self—processing of information.

The authors started with the equations describing the heat transfer in the living tissue at the mesoscopic level—considering the different vessels individually. Then, using the equivalence of the diffusion type process and random walks, they developed a procedure to average these mesoscopic equations over all scales of the hierarchical vascular network describing evolution of certain smoothed fields. The microscopic equations obtained in this way describe the living tissue in terms of an active medium with the continuously distributed self-regulation. A cooperative mechanism of self-regulation which involves individual response of each element to the corresponding hierarchical piece of information and leads to the ideal system functioning due to self-processing of information.

V. V. Gafiychuk and I. A. Lubashevsky [1030] explained why the total effect of the blood flow in the counter current pairs must be cooperative and the tissue phantom containing these vessel pairs can be described in terms of an effective medium using the random walkers description of the heat transfer: for a single pipe the blood flow effect may be treated in terms of the convective transport. For a single counter current pair the blood flow effect is diffusive in nature: a walker during its motion visits alternately vessels with the opposite blood flow directions.

Since the effective diffusion coefficient D_{eff} is the result of cooperative influence of all the vessels the authors represented the total value as the sum of the particular values of the effective diffusion coefficients induced by the blood flow in the vessels of the various groups

$$D_{eff} = D[1 + F_v(G) + F_c(G)] \tag{2.10}$$

where

$$G = \frac{4jl_0^2}{3\sqrt{3}\pi D} \ln\left(\frac{l_0}{a_0}\right),$$

the term $F_v(G)$ is caused by the blood flow in the second class vessels and $F_c(G)$ by the flow the capillary system.

2.1.1 Heat Transfer in the Brain

The brain is surrounded by a layer of dense bone, the skull, and is cushioned by the cerebral spinal fluid underneath. The arterial structure, the circle of Willis, provides a level of the circulatory redundancy so that if one feeding artery becomes obstructed, the downstream tissue does not become ischemic (starved of oxygen) [319].

The heat production caused by the cerebral metabolic activity—the power generated by the sodium metabolic pumps—may constrain the mammalian brain architecture. Since the energy expenditure efficiency is highly temperature-dependent [3444], the brain's thermal regulatory capacity may define its anatomical and physiological architecture and limits its processing capacity [3252].

J. Karbowski [1542] studied the physical limits on the brain wiring and changes in the brain temperature imposed by the thermodynamics of the heat balance determined by the Na+/K+-ATPase reactions, the cerebral blood flow (CBF), and the heat conduction.

The brain like any computational device [244, 1775] dissipates the energy [1362, 2855]—in 1892 William James, analysing Moritz Schiff's 1860s studies of the brain temperature in the awake dog, stated that "the brain activity seems accompanied by a local disengagement of heat" [3252].

The net reaction of the oxygen and the glucose generates the energy required for the cerebral metabolic activities. About one third of this energy is immediately released as heat, the rest is used to produce the ATP molecules to fuel a complex chain of the chemical reactions; the final ATP hydrolysis releases the energy to the biological system as the heat [2855]. The heat transfer from the neurons to the capillaries is fast due to the small distance (about 10 μm) [1978].

The human brain at rest has an estimated metabolic rate of 3–3.5 mL O_2 (100 g cerebral tissue)$^{-1}$ min^{-1} [1786, 2006]. On average, 0.66 J is released every minute per gram of the brain tissue [3391].

The experiments showing that the regional brain temperature rises more quickly and to a larger extent than do the arterial blood temperature prove that the local cerebral heat production is the primary cause of such functional brain hyperthermia [3252].

The measurements of the brain temperature in the animals and the humans demonstrate that the results vary with the method of the measurement (the thermistor, the infrared camera, the magnetic resonance, etc.), the type of stimulus, the animal preparation (anesthetized, restrained, freely moving, etc.), the species, the age and the size of the animal [3252]. The large animals have a positive "brain-body temperature gradient" (i.e., the brain temperature is greater than the body core temperature) that was demonstrated in monkeys, dogs, rabbits, cats, sheep, for example, the hottest parts of the brain, 0.5° C–0.6° C warmer than the blood [1272] or in baboons the temperatures measured directly in the hypothalamus exceeded the blood temperatures by 0.5° C [2028]. In contrast, the small anesthetized animals have a negative brain-body temperature gradient (the brain temperature is less than the body temperature)—the small animals have a high brain-surface-to-volume ratio to induce the significant thermal interactions with the environment [3252].

In animals an important factor is the relative position of the heart and the brain. For example, in the arboreal snakes the heart is placed closer to the head, in comparison with the aquatic snakes [1872]. This facilitates the blood perfusion to the brain.

The bony and soft coverings of the brain composed of head hair, scalp, skull help to maintain the brain temperature homeostasis by shielding the brain from external thermal challenges. Without these layers the surgically exposed cerebral cortex has the temperatures 5° C–10° C below the core body temperatures [1523]. The round shape of the head minimizes surface area to volume ratio and thus further protects the brain from environmental thermal challenges.

The *in vivo* experiments in the mammals demonstrate that the neuronal activity is a function of the temperature [1523]. The thermal perturbation influences such neurophysiological properties as the resting potential, the action potential, the nerve conduction velocity, the synaptic transmission [2636, 3088, 3203, 3252]. The temperature variations also have effect on the terminal transmitter release and the presynaptic uptake processes [1806, 3088].

The small temperature gradients generated in the axon terminals upon the activation of the brain uncoupling proteins (UCPs) can modulate the presynaptic and postsynaptic events since the neurotransmitter diffusion and convection depend on the temperature [85, 1020, 1352]. The UCPs as a family of the mitochondrial anion-carrier proteins are located in the inner membrane of mitochondria. Activation of the UCPs reduces the mitochondrial membrane potential by causing a proton leak back into the matrix. This process dissipates the proton motive force, uncouples the oxidative phosphorylation from the ATP synthesis, and thereby generates the local temperature gradients. The role of the UPCs is well understood in the context of the brown adipose tissue [432] where they generate heat associated with the vasodilation and increase the glucose consumption [2068]. The UPC variant, UPC-2, is expressed in the neural tissue and localized predominantly to axons and axon terminals [3252].

The cerebrospinal fluid (CSF) that circulates in the subarachnoid space in between the arachnoid and the pia layers and covers the cortex plays essential role in the brain heat transfer. The CSF provides a second fluid circulation system for the CNS, similar to the lymphatic system for the rest of the body [2131].

The CSF could be considered as the CNS thermal buffer—the contact surface between the CSF and the brain is extraordinarily large [3252]—in the adult humans the cerebrocortical surface is approximately 2300 cm^2 [853]; the fluid-brain contact surface at the capillary level through the perivascular spaces and the interstitial spaces is estimated at 250 cm^2/g of the tissue [2565]. The vast fluid-brain and fluid-vessel contact surface is essential for the CSF to stabilize and regulate the thermal environment of the CNS [3252]. For example, the CSF of the basal subarachnoid space in the primates was reported to have the temperatures at the level of the arterial blood temperature [1271]. The directly measured CSF temperature is on average 0.3–0.9° C above the body core temperature [1326, 2642].

The brain consumes the relatively more energy than other organs [37, 136, 1824]. The brain represents only from 2 to 3% of the human body weight but it uses 20 and 25% of the body's total consumption of the oxygen and the glucose, respectively [1362, 2203]. The energy metabolism in the brain is mainly aerobic; 95% of the glucose used by the brain undergoes oxidative metabolism, about 40% of the energy provided by the glucose is used to produce the ATP; the remainder is converted into the heat.

The cerebral metabolism changes linearly with the brain temperature, with 6–8% changes in the metabolism per degree Celsius of the temperature [1778, 2640]. For example, in the anesthetized dogs at 28° C, the cerebral metabolism represents only 50% of that at 37° C [2125, 2203].

The brain has a high average value of the Van't Hoff Q_{10} coefficient, on the order of 2.3 in the physiological temperature range [2124]. However, cells in some cerebral structures demonstrate even higher temperature sensitivity. For example, within the physiological temperature range (34° C–39° C), murine substantia nigra dopamine neurons in vitro [1176] increase their discharge rate with warming (Q_{10} = 3.7). Below this range (29° C–34° C), their discharge rate dramatically decreases (Q_{10} = 8.5). Temperature intrinsically affects brain metabolic rate as well as the affinity of haemoglobin for oxygen.

The potential thermal imbalance between the heat produced and removed could lead to the overheating and the improper functioning, since the neural properties are temperature dependent [1664]. Thus, the problem of the heat transfer could be the serious factor shaping the brain evolution, organization, and function [161, 916, 1647, 2566].

The neural membrane dynamics is determined by the Na^+/K^+ pump: it pumps out 3 Na^+ and pumps in 2 K^+ per 1 ATP molecule consumed [1529]. The equations governing the membrane and the pump under the assumption that the membrane is equipotential, i.e., the axons and the dendrites have on average the equal potentials, are written (neglecting the chloride current [549, 3276]) as [1542]

$$CS\frac{\mathrm{d}V}{\mathrm{d}t} = -g_{Na}S(V - V_{Na}) - g_K S(V - V_K) - g_L S(V - V_L) - I_p - I_s, \tag{2.11}$$

$$\frac{\mathrm{d}[Na]}{\mathrm{d}t} = -g_{Na}S(V - V_{Na}) - 3I_p - \alpha I_s, \tag{2.12}$$

$$\frac{\mathrm{d}[K]}{\mathrm{d}t} = -g_K S(V - V_K) + 2I_p - \beta I_s, \tag{2.13}$$

where is C is the membrane capacitance per unit area, V is the membrane potential, the symbol [...] denotes the intracellular ionic concentration, S is the neuron's membrane surface area, V_{Na} and V_K are Na^+ and K^+ the reversal potentials, g_{Na} and g_K are the voltage-dependent Na^+ and K^+ conductances per unit membrane area, g_L is the leak conductance, V_L is the reversal potential corresponding to the leak current. The current I_p is the Na^+/K^+ pump current, I_s is the synaptic current, α and β are the voltage dependent proportionality parameters.

The non-zero membrane potential is caused by the concentration gradients of the charged ions across the membrane, and the temporal changes in the potential are directly related to the temporal changes in the ionic concentrations.

In studying the metabolic and thermodynamic properties of the brain tissue the author used the following neuroanatomical relations [355]:

- the volume density of the synapses is the brain size independent;
- the fraction of the gray matter volume taken by the fibres (axons and dendrites) is approximately constant;
- the white matter volume scales with the gray matter volume [3478];
- the volumes of the axons and the dendrites are equal [550];
- the average values of the cortical axon and the dendrite diameters are roughly invariant with respect to brain size (i.e., approximately the same for the mouse and the human).

Since the metabolic rate in the white matter is 3–4 times lower than in the gray matter [2855], the white matter contribution to the total brain metabolic power was neglected. The average power dissipated by N neurons in the gray matter is given by

$$P_{ATP} = \frac{N}{\Delta t}\int_0^{\Delta}[-3I_p(V - V_{Na}) + 2I_p(V - V_K)]\mathrm{d}t \tag{2.14}$$

where the integral represents the electrical work performed by the Na^+/K^+-ATP pumps during the long time much larger than the average interspike interval. This work goes for removing 3 Na^+ ions and importing 2 K^+ ions against their electrochemical gradients, which cost 1 ATP molecule [1529].

The heat coming from other reactions such as the glycolysis (the conversion of the glucose to the ATP) is assumed to be less important and is neglected. In the deep brain regions the heat generated by the hydrolysis of the ATP molecules is removed by the conduction and the circulating cerebral blood flow, whereas at the surface (scalp) the heat is removed mostly by the convection/conduction and the radiation. The transfer of the heat to the circulating cerebral blood in the deep brain regions is possible because the incoming blood (artery) has a slightly lower temperature than the brain tissue in this region [1647, 2319].

When the neuron fires repeatedly, its intracellular sodium accumulates with every spike because the Na^+/K^+ pump is weak and cannot remove all Na+ promptly. The enhanced intracellular sodium fluctuations can effectively reduce the concentration gradients across the neuron's membrane to values close to zero, and this would have a devastating effect on the neuron's functionality if the high firing rates were maintained for a prolonged period of time.

The average firing rate in the mammalian brains can be determined indirectly from the cerebral glucose utilization rates, the power generated scales linearly with the glucose consumption rate.

In the superficial regions the brain temperature is always smaller than the blood temperature. This fact has important consequences for the heat transfer in the mammalian brains. In the deep brain regions the cerebral blood flow plays the role of a coolant, whereas in the superficial regions it serves as a strong brain heater. Consequently, the net effect of the blood flow is to warm up the brain tissue. The surgically exposed cerebral cortex can have the temperatures 5° C–10° C below the core body temperatures [1127, 1523].

The microcirculation in the human brain cortex was identified [828] that is similar to the local skin blood flow (the skin blood circulation could be 200–500 mL/min [1978]) autoregulated microcirculation [487]. It was demonstrated that the regional CBF changes depend on the cortical temperature alterations [3252].

There are two hypothesis of the local regulation of the cerebral blood flow and metabolism by the neuronal activity [737]:

1. *metabolic* hypothesis: changes in the neuronal activity drive changes in the energy metabolism, which *then* drive vasodilation/constriction and the associated changes in blood flow.
2. *neurogenic* hypothesis: the blood flow and the energy metabolism are driven in *parallel* by the neuronal activity—changes in the neuronal activity drive vasodilation and vasoconstriction by feed-forward mechanisms releasing neurotransmitter and neuropeptide molecules related to neuronal signalling.

A. Devor et al. [737] described the *in vivo* optical imaging of the cerebral blood flow and metabolism that uses the point optical fluorescence and the absorption measurements of the cellular metabolism and the haemoglobin concentrations. This mettod has the ability to descend to the single-cell and single capillary levels *in vivo* and observe firing of individual neurons, vasodilation, glucose uptake, and infusion of oxygen into the tissue.

A few investigations indicate the existence in the humans of a thermoregulation process called the “selective brain cooling” that keeps the brain cooler than the trunk core during the hyperthermia [419, 422]. The selective brain cooling in the humans depends on the venous blood that has been cooled by the evaporation of sweat on the skin of the head, and then drains into the cranium to exchange heat at several sites, particularly the cavernous sinus [419, 421].

M. Bommadevara and L. Zhu [330] developed a vascular heat transfer model to simulate the temperature decay along the carotid arteries in the humans and to determine the temperature differences between the body core and the arterial blood supplied to the brain. The authors considered the local blood perfusion rate, the blood vessel bifurcation in the neck, and the blood vessel pairs on both sides of the neck. The potential for cooling the blood in the carotid artery by the countercurrent heat exchange with the jugular veins and the byradial heat conduction to the neck surface was evaluated. It was also found that cooling along the common and internal carotid arteries is up to 0.87° C during the hyperthermia.

A 1.1° C temperature drop along the carotid arteries is possible when the neck surface is cooled to 0° C [330]. The clinically feasible brain cooling methods include a head hood or helmet with the chemical or water cooling, the head immersion in ice water. The temperature of the cooling fluid inside the helmet may be as low as 0° C. Under those circumstances, the radial heat conduction from the warm arterial blood to the cold neck surface may be greatly enhanced to achieve a reduction in the brain temperature. The centre of the human brain is from 0.5° C to 1° C warmer than the epidural space [2113]. The physiological fluctuations of 1° C–3° C in the brain temperature have been reported in the cats and other animal models [3252] (see, for example, refs. [1649, 2789]). The cerebral structures have their own basal temperatures, with the temperature differences between the various regions remaining relatively constant [1649, 3252]. For example, the temperature difference of 1.4° C between the hypothalamus and the cortex was reported in the free moving cats [2789].

The efficiency of the cerebral cooling— the selective brain cooling (SBC) [3468], the lowering of the brain temperature, either locally or as a whole, below the arterial blood temperature [368]—is species specific [161, 1458, 1458, 1484].

The SBC has been shown to occur in many animals, including mammals, due to the presence of a carotid rete (also known as a rete mirablia) [1458]. This is a dense mesh of arteries near the brain that are cooled by venous blood from facial extremities. Humans do not possess this feature and generally resort to other methods, e.g., the circulation of cerebral spinal fluid (CSF), the proximity of the upper airway passage to the brain, and the emissary veins in the skull.

In the felids, the arterial blood for the brain flows through the vascular network at the base of the skull, the carotid artery is very close to the cavernous sinus,which receives the cool blood from the mucosal surfaces of the nose; in the canids, the carotid rete is rudimentary, but the large surface of the cavernous sinus, which is in the close contact with the base of the brain, allows direct cooling of the rostral brain stem [2203].

In the humans, the face and the mucosal surfaces of the nose, which are sources of the cool venous blood, are small in relation to the mass of the brain. A specialized heat exchanger similar to the carotid rete does not exist in the humans, and a substantial fraction of the blood supply to the brain is provided by the vertebral arteries, which have no direct contact with the cool venous blood. The cool blood from the skin of the head can flow into the cranium and cool the brain via the emissary veins of the temporal and parietal bones [421]. The emissary veins connect the vessels in the scalp to the superior cerebral veins. These vessels are valveless which allows flow to occur in either direction [319].

There are a number of methods available to attempt to artificially induce the SBC in humans. The non-invasive methods do not to produce the significant results. P. Andrews, B. Harris B and G. Murray attempted to reduce the brain temperature using the airflow passed through the respiratory tract in the brain-injured patients [84]. L. Nybo, N. H. Secher and B. Nielsen showed that face fanning had no effect on the jugular venous blood temperature, an indication the brain temperature was unaffected [2319]. However, using the MRS measurements, B. Harris et al. [1252] found in a study of five healthy adults that a convective cooling head device reduced the average brain temperature by 0.45° C. It is possible to use the carotid arteries to deliver a local cooling, but due to the recirculation of the blood, this method would reduce the core body temperature as well. J. H. Choi et al. [554] demonstrated that introduction of the isotonic saline into an internal carotid artery reduced the jugular venous bulb temperature more than the systemic temperature, indicating the SBC. A cooling cap was shown to be able to reduce core brain temperatures selectively [319]. M. A. Neimark et al. [2279] showed that a cooling cap could maintain hypothermic temperatures that were reached through intracarotid cooling, but the scalp cooling alone is ineffectual. The low efficiency of the scalp cooling at inducing the hypothermia is the high levels of the perfusion present within the brain—it is called "temperature shielding effect" of the blood perfusion [319] and was demonstrated in the rat brains [3532, 3534].

The relative importance of the two major mechanisms of the heat transfer inside the brain is dependent on the brain size. For the very small brains, the heat cooling rate via the conduction is comparable to the heat warming rate via the blood flow. The reason for such a strong warming through the cerebral blood is that it must compensate the heat loss due to the scalp convection/conduction and radiation which are much larger than the metabolic rate. On the other hand, for the large brains, the cooling rate is almost twice as large as the warming rate, because the heat production due to the metabolic activity is more significant than for the small brains.

J. Karbowski [1542] estimated the thermal bounds on the fibre diameter that allow to maintain the safe temperature above which the molecular changes in the neurons and the synapses become critical and irreversible. The deep brain temperature depends significantly on the effective fibre diameter [2203].

The minimal effective fibre diameter was found to depend very weakly on the gray matter volume, thus, the brain size is not a critical factor determining the thermodynamic safety of the cerebral tissue: the thermal properties of the mouse and the elephant brains, differing by four orders of magnitude in the size, are similar. The minimal effective fibre diameter d_{min} depends stronger on the firing rate. The estimated average firing rates in the mammals are in the range from 1.7 Hz for the human to 6.2 Hz for the mouse, and they scale systematically with brain size [1542]. The too thin fibres are not beneficial for the thermal balance. For very small brains the blood flow is not a sufficient cooling mechanism deep in the brain.

The average value of the empirical fibre diameter in the gray matter is the brain size independent and is about 0.45 μm. This value is about 10 times larger than the largest value of d_{min}, thus, the mammalian brains operating under the normal physiological conditions are rather safe from the excessive overheating that would cause the brain damage for all ranges of the firing frequency [1542]. The margins of the thermal safety could be compromised in the pathological conditions, e.g., abnormalities associated with the strongly reduced cerebral blood flow. The thermal lower bound on the fibre diameter also determines the upper bound on the fibre length per neuron.

The brain cooling is not a classic problem of volume to surface ratio, as it is the case with natural objects. Instead, the brain cooling could be compared to the cooling of the combustion heat engine, which receives a liquid coolant—in the brain the role of the coolant is played by the cerebral blood , but only in the deep region because there blood has a slightly lower temperature than the brain tissue [1542]. In the superficial regions the brain tissue has a smaller temperature than the cerebral blood, and there blood warms up the brain.

The theoretical bound on the axon diameter implies a corresponding lower bound on the speed of the signal propagation in the gray matter.

A. L. Sukstanskii and D. A. Yablonskiy [2971] developed a model of the temperature changes during the arbitrary functional brain activity and found the theoretical boundaries on these temperature changes. The authors explained how these boundaries depend on the blood flow and metabolism and the environment.

The authors noted that there no agreement in the experimental data on the magnitude and even sign of the reported temperature changes in the brain temperature of humans and animals upon the functionally induced changes in the brain activity. They based their analysis on the theory of the brain temperature changes in the functionally activated regions developed by D. A. Yablonskiy et al. [3391] where, based on the human brain imaging with the positron-emission tomography data, these changes are related to the disproportional increases in the blood flow as compared with the oxygen consumption during changes in the brain functional activity. In the absence of the heat exchange with the surrounding nonactivated regions, the temperature within the activated region decreases as the cooling effect of the increased blood flow exceeds the heating effect of the increased oxygen metabolism.

The blood flow acts as a heat exchanger with pipes (the blood vessels) penetrating into all brain structures, equilibrating the brain and body temperature. As a result, the brain temperature has very little dependence on the brain metabolism and is primarily defined by the temperature of the incoming arterial blood.

While under the normal physiological conditions the tissue temperature in the deep brain regions is higher than the temperature of the arterial blood, the situation could be inverted near the brain surface due to the heat exchange with the environment. This phenomenon is regulated by the "temperature-shielding effect of blood flow" [3532]—the cerebral blood flow prevents the *extracranial cold* from penetrating the deep brain structures—the direct cerebral blood flow measurements with the microspheres in rats show that the characteristic temperature shielding length Δ is inversely proportional to the square root of the cerebral blood flow [3532]

$$\Delta = \left(\frac{\lambda}{\varrho \varrho_b c_b CBF} \right)^{1/2} \tag{2.15}$$

where λ is the tissue thermal conductivity, ϱ is the tissue density, ϱ_b and c_b are the density and specific heat of blood.

The estimates for the gray matter of the normal adult humans yield $\Delta \approx 3.6\,\text{mm}$ which is much smaller than the adult human brain size [3532].

V. Riedl et al. [2604] noted that the energy consumption is an essential in the neuronal communication—for all species, the greatest amount of energy metabolism in the brain is dedicated to signalling [1390]. Up to 75% of this energy is consumed postsynaptically, i.e., at the target neurons [58, 136, 1253]. In 1848, Herman von Helmholtz performed experiments to measure both the conduction velocity of the action potential and the heat produced by it in the frog motor neuron [3252]. The first reliable measurements of the heat production from a train of the neural impulses were performed by A. V. Hill in 1929 in the crustaceans and gave the average variation of $2\,\mu^\circ$ C per impulse [1319, 3252]. A. L. Hodgkin and A. F. Huxley noted the temperature dependence of the action potential [1336]. The introduction of the superior signal amplification equipment allowed B. C. Abbott et al. [4] in 1958 to obtain the first recordings of the heat production from a single neuronal impulse in the crab; the results showed the heat production is 80% greater than previously estimated [3252].

V. Riedl et al. proposed the metabolic connectivity mapping (MCM) method that combines the undirected functional effective connectivity (EC) across the brain states with the local energy metabolism from the functional magnetic resonance imaging (fMRI) and the positron emission tomography (PET). The method is based on the concept that most energy required for the neuronal communication is used by the target neurons. It could be used for the derivation of a voxel-wise, whole-brain mapping of the EC in humans. This approach can be used to determine the relations between the different regions of the cortex. While the early anatomic studies suggested a bottom-up hierarchy along the visual streams, the authors found the importance of feedback signalling onto the early visual areas.

2.1.2 Selective Thermal Stimulation

The primary site for thermoregulatory control is located in the brain (preoptic anterior hypothalamus). However, it has now been documented that parallel thermostat structures exist peripheral to the brain, most prominently along the spinal cord. This peripheral location provides an opportunity to apply safe thermal stimulation to manipulate the thermoregulatory function to therapeutic or prophylactic advantage for a number of the physiological applications [298].

In the method termed the selective thermal stimulation (STS) a small amount of energy is input to a very localized volume of the control tissue to produce a much larger overall thermoregulatory effect [762, 763].

The existence of an extracerebral thermoregulation control function associated with the spine has been recognized for more than century [298]. Manipulation of the spinal cord temperature as a means of controlling body temperature has been used for more than 50 years [1459], including experiments in numerous mammalian and avian species that maintain a constant core temperature, for example, ox [1463], goat [1464], sheep [1215], dog [1460–1462], pig [445], rat [1214], monkey [476], rabbit [1185], pigeon [2581], penguin [1227], goose [1289].

Virtually all of these experiments were conducted using an invasive procedure in which water-circulation tubing was implanted along the length of the animal's spine through or around the vertebral canal and, sometimes, surrounding the hypothalamus. Water was circulated at regulated temperatures to independently establish the temperatures of the spinal cord and hypothalamus. For example, in dogs the separate individual warming of both the hypothalamus and the spine caused the animals to upregulate their endogenous heat-rejection processes. This response presumably occurs because key thermosensory tissues are made locally warmer than the body in general, eliciting a heat rejection response, even though the region of overheating is limited to only the spine or hypothalamus. When both warming stimulations were applied simultaneously, the effect was additive. The threshold core temperatures for which heat rejection mechanisms are invoked were reduced from normal values. Furthermore, individual and simultaneous cooling of the spine and the hypothalamus produced the same types of results, except in the opposite direction of the thermoregulatory behaviour.The magnitude of the metabolic upregulation due to shivering was greater for simultaneous cooling stimuli than for cooling either the spine or hypothalamus individually [1460].

One of the most impressive demonstrations of the STS phenomenon was developed using an ox as a model subject [1463]. An ox had 10 U-shaped polyethylene water flow tubes inserted into the peridural space of the vertebral canal spaced from the lower lumbar to the upper cervical region. The animal rejected heat into the environment at a rate greater than basal metabolism that was in proportion to the heating temperature. Although only a very localized region of thermoregulatory tissue along the spine was being heated, the animal's response was to reject eight times as much body heat into the environment. The microwave heating of the hypothalamus of goats [83] resulted in a 10° C core temperature drop, indicating the existence of parallel thermoregulatory control functions in the hypothalamus and spine.

The human hypothalamic controller responsive to the dynamic alterations in a person's physiological state, to the level and nature of the physical activity and to the environmental thermal loading—cooling, neutral, or heating. If the STS acts additively to the hypothalamus, there may be circumstances in which the thermoregulatory function may be boosted by activation of the STS for the therapeutic benefit. One can either lower or raise the core temperature by a two-step process of

1. using the STS to enhance the AVA blood flow and
2. applying the heat exchangers to the glabrous skin surfaces to add or remove heat from the blood before it returns to the core.

This approach can extend the magnitude of operation of the thermoregulatory processes, providing access to the therapeutic states not otherwise available. A major advantage of the STS is that it presents an opportunity to proactively and effectively manipulate the body core temperature for the diverse medical purposes, involving cooling or heating [298]:

- adjusting the core temperature upward (warming the body core is of the critical clinical importance during the surgical procedures under the general anesthesia to defend against the occurrence of perioperative hypothermia);
- adjusting the core temperature downward (providing therapeutic hypothermia in response to events that cause major organ ischemia, especially of the brain—lowering the core temperature presents a striking physiological challenge because the body has effective defense mechanisms to guard against any significant drop in its internal temperature);
- depressing the action of vasoconstriction on AVAs to achieve a higher level of GSBF.

The first two involve the application of either cooling or warming heat exchanges to the palms and soles in conjunction with upregulating the GSBF by the STS. The third simply uses native heat exchange with environmental air.

2.2 High Temperatures

An increase (decrease) in the local temperature causes an increase (decrease) in the local blood flow. The smooth muscles surrounding the 40–200 mm diameter arterioles play a dominate role in controlling the local blood flow.

The different effects are observed in the tissues with increasing the temperature, e.g., inhibiting of the synthesis of the DNA, the RNA, and the proteins, the denaturation of the enzymes and the looseness of the cell membranes (40° C–45° C) [2708], the coagulation (the well-known example is the coagulation of the albumin in the white of a cooked egg) and the protein denaturation (60° C) [2610], including the enzymes, the intranuclear proteins and the DNAs [2961], the drying out (100° C), the carbonization (150° C), the pyrolysis (above 300° C). There is a rule of thumb that rate of the cell killing doubles (reduces) by a factor of four for every degree rise (respectively, drop) relative to 43° C; an empirical time—the temperature combinations for the thermal damage could not be extrapolated from the one tissue to another [743].

By increasing the temperature the state of the hyperthermia (40° C–50° C) [2244] is reached—some molecular bonds are destroyed, the membrane is altered—the membrane permeability increases [2649], the reduction in the enzyme activity is observed [90, 1313]. These effects are reversible.

The denaturation of the proteins and the collagen leads to a significant increase in the membrane permeability of the mitochondria [2861] and then to the coagulation [223] of the tissue and it can the necrotize cells. At higher temperature the equilibrium of the chemical concentration is destroyed as the permeability of the membrane of the cells increases. Temperature increase to 45° C–48° C—the so-called hyperthermia temperature interval [285]—can lead to an increment in the cells' sensitivity to injury [3216] and aggregation of proteins, inhibition of the repair of the DNA damage [1672].

The thermal denaturation of collagen—the major constituent of the dry skin—occurs when the temperature exceeds a certain threshold (about 43° C); during denaturation also the hydration of the collagen changes [3376]. Thus the thermal denaturation of a collagenous tissue can lead to remarkable changes in skin mechanical, thermal, electrical properties—the thermal and mechanical problems are fully coupled [512].

The vaporization of the water occurs at 100° C. Within the vaporization phase, the temperature of the tissue does not alter and the gas bubbles are formed. The propagation of these bubbles accompanied with the alteration of their volume causes the thermal decomposition of the tissue fragments [90]. If all water molecules are vaporized, the carbon atoms are released and the adjacent tissues are blackened. This stage is called carbonization [35]. The interactions of the high intensity laser irradiation on the dental hard tissues (the teeth are mainly composed by the hydroxyapatite that has the high heat capacity value and the low heat conduction value) could cause the surface carbonization or cracking [3469]. At the temperatures above 300° C melting might occur of, e.g., of the human teeth.

The most biological effects which occur at temperatures higher than 60° C are irreversible. Between 60° C and 100° C, protein coagulation occurs instantly with irreversible damage of key cytosolic and mitochondrial enzymes and nucleic acid-histone complexes [1798].

The irreversible thermal "damage" Ω is frequently assumed to accumulate according to the classic Arrhenius theory of the chemical reactions [741, 961, 1294, 1295, 2162].

The thermal damage process is considered as an irreversible first-order reaction that transitions from the state S_A to the state S_B

$$S_A \Longrightarrow S_B \tag{2.16}$$

where the reaction rate is temperature dependent

$$k \propto \exp\left(-\frac{E_a}{RT}\right). \tag{2.17}$$

The degree of the thermal damage can be quantified by the variable Ω

$$\frac{d\Omega}{dt} = At \exp\left(-\frac{E_a}{RT}\right), \tag{2.18}$$

and, thus

$$\Omega = \int_0^t A \exp(-E_a/RT)dt \tag{2.19}$$

where T is the temperature, E_a is the activation energy. The strong temperature dependence indicate that the zone of the tissue destruction is more compact than the zone of temperature rise. For example, C.-Y. Li et al. [1861] divided the thermal damage factor of skin tissue into three levels: $\Omega = 0.53$ (the first level); $\Omega = 1$ (the second level); $\Omega = 10^4$ (the third level).

While it captures the kinetic nature of the damage process, the Arrhenius damage model has a drawback: a recovery process was not taken into account in the model, which precludes representation of the so-called "shoulder region" observed at low temperatures in the early phase of hyperthermia [3518].

To improve this model, new models have been proposed. For example, Y. Feng et al. [947] formulated a two-state model based on statistical mechanics, which agrees with in vitro data and captures well the "shoulder region". The price is that one more parameter is introduced, and the model is not suitable for dynamic simulations where temperature changes. A three-state model was developed by D. P. O'Neill et al. [2340] to address this shortcoming.

It should be noted that the experimental data showing the "shoulder regions" were obtained at the hyperthermia-relevant temperature and the time scale, usually 40° C–50° C for minutes to hours. In the thermal ablation with rapid heating to high temperature (50° C for a few minutes) the release of the heat shock proteins and the repair mechanism are shut down before they are fully functional. Thus the recovery process is negligible, and one can use an Arrhenius damage model for the practical predictions.

The meaning of "damage" depends on the problem, e.g., for the skin it could be defined as the heat-induced damage of the collagen and the collageneous tissues [3380] via a concentration $C(t)$ of the undenaturated collagen. K. R. Diller and J. A. Pearce [764] pointed out that the dimensionless indicator of damage is the logarithm of the relative concentration of the or un-denatured collagen, in the collagen denaturation process, $\Omega(t) = \ln(C(0)/C(t))$ and the fraction of the denaturated collagen $Deg(t) = (C(0) - C(t))/C(0) = 1 - \exp(-\Omega(t))$.

The collagen in human dermis is mainly type I collagen that has a domain within the triple helix (composed of three polypeptide chains, each of approximately 300 nm in length [3343]) that is completely devoid of hydroxyproline. Since hydroxyproline readily forms hydrogen bonds that stabilize the molecule, its absence makes this domain particularly susceptible to thermal damage [2130]. There are two levels at which breakdown is thermodynamically significant [3435]:

- the collagen molecule itself, in which three peptide chains are twisted around each other to form a helical, rod-shaped molecule;
- the semi-crystalline fibril in which collagen molecules are assembled side-by-side in a staggered manner with the long axis of each molecule aligned with axial orientation of the fibril.

When collagen is heated, the heat-labile intramolecular crosslinks are broken and the collagen undergoes a transition from a highly organized crystalline structure to a random, gel-like state, which is the denaturation process [975]. Collagen shrinkage occurs through the cumulative effect of the unwinding of the triple helix, due to the destruction of the heat-labile intramolecular cross-links, and the residual tension of the heat-stable intermolecular cross-links [57, 110, 975].

The effects of heating on collagen can be reversible or irreversible and the precise heat-induced behaviour of collagenous tissue and shrinkage depend on several factors, including the collagen content [575], the maximum temperature reached and exposure time [57].

Different metrics have been used to characterize the thermal denaturation and the heating-induced damage of the collagen and the collagenous tissues [3376]:

- biological metrics—S. Bhowmick and J. C. Bischof [283] suggested to use the enzyme deactivation; D. M. Green and K. R. Diller [1154] proposed to exploit the extravasation of fluorescent tagged plasma proteins;
- thermal metrics—S. L. Jacques [1426] and C. A. Miles [2129] suggested to relay on the changes in enthalpy;
- mechanical metrics—numerous researches suggested to use the thermal shrinkage [511, 512, 1683, 1888] that could be detected, for example, by the second harmonic generation microscopy [1888];
- optical metrics—the thermally induced loss of birefringence [692, 3048].

During the denaturation, not only the structure, but also the hydration of the collagen changes, which may involve an initial liberation of, and subsequent absorption of, water via the water bridges; thus, the thermal denaturation of the collagenous tissue can result in the marked changes in the thermal, mechanical and optical properties [3376].

2.2.1 Protein Denaturation

The stability of the native protein conformation is determined by the difference of the Gibbs free energy of the native and the denaturated states [1741] that is a parabolic function of the temperature with a maximum in the range 15° C–25° C [426, 3012]; a hydrogen bonds (HBs) net in the system of the water molecules interacting with the protein plays a significant role in the process of the denaturation [1836]; for an additional information on the protein stability please consult a review by G. D. Rose and R. Wolfenden [2635].

A protein's biological activity depends on the location of the electrical charges in the molecule and on the overall configuration. The variation of the temperature can alter a protein's configuration and charge distribution, and thus change its activity, without affecting the sequence of the amino acids.

The thermal protein denaturation is usually irreversible [299, 2026] in contrast to the cold protein denaturation (the distortion of the secondary structure defined mainly by the hydrogen bonds [2587], although some disulfide bonding can also occur [299]), observed in the temperature range 0° C–20° C [3336] that could be either reversible or irreversible; in particularity, the freezing of the globular proteins frequently results in the reversible partial denaturation [3129]. During the denaturation the individual covalent bonds between the atoms of the polypeptide of the protein molecule are not broken; just the rearrangement of the hydrogen bonding from the ordered native state to the less ordered one occurs [299].

Frequently the process of the denaturation is considered as an abrupt cooperative transition [1270, 1869] (the transition of the type "all or nothing" [963]). This notion is an oversimplification (about the intermediate states of proteins see a review by R. L. Baldwin [168]), but the one-stage model which could be described by the irreversible reaction of the first order $N \xrightarrow{k} D$, where N and D denote the native and denaturated proteins is a good approximation for many problems. A fraction of the denaturated protein is expressed as

$$F_d = 1 - \exp\left(-\int_0^t k dt\right), \tag{2.20}$$

where the reaction rate dependence on the temperature is assumed to obey the Arrhenius model

$$k(T) = A \exp\left(-\frac{E_a}{RT}\right), \tag{2.21}$$

here E_a is the activation energy.

The application of the kinetic models of the higher order that explicitly consider the intermediate states of protein between the native and denaturated ones [299] to the process of the protein denaturation is certainly justified if a distinction is made between the states of the protein after the (reversible) unfolding U and the final denaturated state D that arises due the aggregation (either ordered one called gelation or the random aggregation—coagulation), the chemical modification of the amino acid residues or by other reasons [1996]. A two-stage model of Lamri and Eiring $N \Longleftrightarrow U \longrightarrow D$ should be used in this case. A description of more complex models of the denaturation could be found in the cited paper. The denaturated state of the protein could be non-unique [168].

The use of the lattice models of the macromolecules [1869] reduces the requirements to the computer resources due to the radical reduction of the number of the possible conformations of the macromolecule. However, these model introduce a priori the space anisotropy [2734]; thus lately the most attention is paid to the "off-lattice" models. Both the water molecules and the parts of the macromolecule are represented by the spheres of the different diameter; the interaction between the fragments i and j of the macromolecule that are not neighbours is described by the short range potential. Frequently, the cutoff Lennard-Jones potential

$$V_{ij}(r) = \begin{cases} \varepsilon\left[(\sigma_{ij}/r)^{12} - 2(\sigma_{ij}/r)^{6} + v_{ij}\right], & r \le R_{ij} \\ 0, & r > R_{ij} \end{cases} \tag{2.22}$$

is used. Here a shift v_{ij} provides the continuity of the potential at $r = R_{ij}$, while the cutoff radius depends on the properties of the elements i and j. To monitor the particles' position, the algorithm of the "random lattices" could be used [2199].

The main approaches to the construction of the hybrid models are

- A representation of the space fragment of the macromolecule (e.g., the phosphate or methyl group) as a single particle or a pair of particles with some effective properties.
- Combining the discrete (atomistic) description of the macromolecule with the continuum description of the solvent (water in our case) [1850]. The review of such "implicit" treatment of the water focused on the simulation of the structure and function of the biological membranes, including continuum and particle-based—tethered, non-tethered, non-tethered solvated, non-tethered implicit solvent—models has been published recently by G. Brannigan et al. [357].

In the HP model of the proteins [756] (as well as in its off-lattice generalization—the AB model [504, 2942]) the amino acid residues are reduced to the point entities that could be either polar (P, represented by the charge or dipole) or non-polar (H) [426]. The hydrophobicity is described as intention to reduce the area of the surface in the contact with water and is related to the reordering of the neighbour molecules into the energetically more attractive structure similar to the ice. This effect could be described formally as an attractive interaction of the non-polar amino acid residues. A hydrophobic molecule in solution displays the water molecules that at the low temperatures form an ice-like cage around the hydrophobic molecule, producing a structure that is energetically more favourable than bulk water. The system tries to minimize the free energy by maximizing the number cages—exposing as many hydrophobic molecules as possible. This is the cold denaturation of the proteins [426]. At the higher temperatures water molecules could not build cages; the disordered water molecules around the hydrophobic amino acid residues of the protein are energetically less favourable than bulk water; thus proteins try to hide their hydrophobic parts inside their core. P. De Los Rios and G. Caldarelli [426] model the protein as a self-avoiding walk (SAW) on a lattice. A variable with q states (labelled from 0 to $q-1$) is attributed to every lattice site non-occupied by the polymer. The state $q=0$ is associated to the cage configuration which is energetically favourable when the water molecule is in the contact with the polymer and the rest states $q=1,\ldots,q-1$ are associated to the disordered, unfavourable, configurations. The Hamiltonian of the system is written as

$$H=\sum_{i=1}^{N}\sum(-J\delta_{s_j,0}+K(1-\delta_{s_j,0})) \tag{2.23}$$

Here the first sum run over N monomers of the polymer while the second one—over the water occupied nearest neighbours of each monomer, J and K are the positive constants representing the energy of the cage and of the disordered configurations; there are no monomer-monomer interactions.

The partition function of the system is written as

$$Z_N=\sum_C Z_N(C), \tag{2.24}$$

where $Z_N(C)$ is the partition function associated to a single configuration C. Note that the maximum number of the water sites in the contact with the polymer is $M=2(d-1)N+2$ for the hypercube lattice in the d-dimensional space; the real number of the contacts is, of course, smaller than M. A particular water site could be counted more than once. The configuration partition function is expressed as

$$Z_N=q^{n_0(C)}\prod_{l=1}^{z}Y_l(C) \tag{2.25}$$

where

$$Y_l(C)=\left((q-1)e^{-\beta lK}+e^{-\beta lJ}\right)_l^n(C) \tag{2.26}$$

and $n_l(C)$ is the number of the water sites with l polymer contacts, z is the coordination number of the lattice, $\beta=1/k_BT$.

P. de los Rios and G. Caldarelli analysed the thermodynamic behaviour of the system on the two-dimensional lattice $d=2$ using the exact enumeration technique for polymers of length up to $N=25$. The so called Manhattan lattice was used (it is a two dimensional lattice on which rows (columns) are alternatively left/right (up/down) oriented). The authors state that simulations using the Manhatten lattices give results similar to those obtained on the regular Euclidian lattices. Only bimodal case $q = 2$ was considered by G. Calderelli and P. de los Rios [426].

The authors computed the energy of the monomer-monomer interaction ε using the partition functions for two configurations C and C' that differ in one monomer-monomer contact

$$\varepsilon = -\frac{1}{\beta} \ln \frac{Z(C')}{Z(C)} \tag{2.27}$$

and found the low and high temperature limits of this energy

$$\begin{gathered} \varepsilon \approx 2J, T \to 0 \\ \varepsilon \approx -2\left(K - \frac{J+K}{q}\right), T \to \infty \end{gathered} \tag{2.28}$$

Thus the energy of the effective monomer-monomer interaction is repulsive for low temperatures.

The explicit description of the water is necessary, in particular, for the correct simulation of the cold protein denaturation [414, 426].

The complete denaturation (cold unfolding) of the globular proteins upon decreasing temperature is similar to the abrupt swelling transition at low temperatures of the so called thermoreversible homopolymers. C. Buzano et al. [414] used the dynamic Monte Carlo simulations to study swelling of a hydrophobic polymer in the aqueous solutions. The water was described by a two dimensional model on a triangular lattice using the Mercedes-Benz-type model for the interaction potentials. The polymer was represented by a self-avoiding walk on the same lattice. The authors were able to observe the low-temperature-induced swelling transition at the high pressure only and formulate an important question: whether the lattice water models overestimate the stability of the ordered phase (ice)?

The simple model of the hydrophobic interaction does not reproduce the pressure effect on the cold protein denaturation; a correct simulation of this process at the high pressure requires the account of the water fluctuations [2051]. In the cited paper the non-polar heteropolymer is considered. Its interaction with the water (a partial ordering of the water molecules around the protein) is described by the Hamiltonian

$$\mathcal{H}_p = J_r n_{HB} \left(n_{max} - \sum_{i,j} n_i n_j \right) \tag{2.29}$$

where the parameter J_r defines the strength of the repulsive interaction, $n_{HB} \equiv N_{HB}/N_w$ is the numerical density of the hydrogen bonds, that are increasing the system volume, N_w is the number of water molecules, n_{max} is the maximal number of contacts of the amino acid residues with each other; $n_i = 1$ if ith node of the lattice is occupied by the residue and 0 otherwise. The expression in the braces is the measure of the protein compactness.

The Hamiltonian of the interactions between the water molecules is written as $\mathcal{H}_{HB} = -J \sum_{i,j} \delta_{\sigma_i,\sigma_j}$; the variable σ_i specify the orientation of the water molecule—the formation of the hydrogen bond between the neighbour molecules being feasible if $\sigma_i = \sigma_j$ [2051]. Enthalpy for the system "protein in the water bath" is $\mathcal{W} = \mathcal{H}_p + \mathcal{H}_{HB} + PV$. The authors used the multicanonic algorithm of the Monte-Carlo method to compute the two-parametric (the number of the hydrogen bonds and the number of the contacts of the amino acid residues) density of states. Then the dependence of the average number of the residue-residue contacts $\bar{N}_c$ on the temperature and on the pressure could be found

$$\bar{N}_c = \sum_{N_{HB}} \sum_{N_c} N_c g(N_{HB}, N_c) \frac{\exp - \left(\frac{W(N_{HB}, N_C)}{T} \right)}{Z}, \tag{2.30}$$

where Z is the partition function. The authors performed computations for the system of a protein with 17 nonpolar residues and 383 water molecules with periodic boundary conditions.

The results have been used to construct the phase diagram of the protein-water system in the $p - T$ plane. The condition of the denaturation was formulated as $\bar{N}_c / n_{max} \geq \alpha_d$ (it was assumed $\alpha_d = 0.96$ in the paper).

The authors stress that while they were able to reproduce the experimentally determined thermodynamics of the cold denaturation the key element of which is the pressure effect on the water density, the analysis of the kinetics of this process will require more detailed models accounting for the actual number of hydrogen bonds for each water molecule and not just the average number of bonds.

The free energy of the protein denaturation and its dependence on the pH of the solution was considered by P. J. Kundrotas and A. Karshikov [1741]. The authors, in contrast to many other models that deal with a single space distribution of the protein charges, consider the protein denaturated state where the titratable groups are distributed quasi randomly, obeying the protein sequence restrictions. The protein molecule was represented by a chain of N elements (the titratable sites) that are located on the sphere that comprises the material of denaturated protein. The allowed positions of elements are predetermined by a set of uniformly distributed points on the surface forming a spherical grid. The authors have also computed the desolvation energy and electrostatic interaction between proteins.

There are other approaches to account for the water explicitly, e.g., the model of the solvent-phantom, where the interaction of the water molecules with the macromolecules is considered while the interactions between the water molecules are neglected [2734]).

The protein denaturation can be the rate-limiting stage in the hyperthermic cell death that is confirmed by the experimentally measured correlation between the cell death and the protein denaturation [299]. The protein denaturation during the freezing or thawing can be also caused by the proteases leaking from the lyposomes that lost the membrane integrity [300].

Sometimes the Arrhenius thermal damage is considered that is expressed as

$$\Omega = \ln\left(\frac{C_0}{C_0 - C_d}\right) \tag{2.31}$$

where C_0 and C_d are the protein concentrations of the tumour before and after the treatment. The denaturation of the protein happens at $\Omega \geq 0.53$ [3287] and the permanent denaturation of the protein or the cell death happens at $\Omega \geq 1$ [2394].

2.3 Low Temperatures

Exposure to the cold stress involves mainly the heat conservation responses, including the increased heat production by the nonshivering and shivering thermogeneses, and the increased thermal insulation by constriction of the cutaneous arterioles and veins (the decreased skin blood flow), which also limit the heat conduction from the core to the periphery. The effects of the cold are first felt in the extremities (feet, hands).

The ability to maintain the core temperature is limited as, for example, the experiments with the volunteers immersed in the cold water bath showed [863]; there is a danger of a hypothermia for a patient in the perioperative period [1365]. In extreme situation such as heavy exercise or harsh thermal environment, the body temperature can shift to a high or low level from the normal range. Active control of body temperature is increasingly employed therapeutically in several clinical scenarios, most commonly to protect the brain from the consequences of either the primary such as the head trauma, stroke) or the secondary injury (for example, after cardiac arrest with brain hypoperfusion) [3525].

The mild to moderate hypothermia, during which the brain temperature is reduced to 30° C–35° C, has been studied as an adjunct treatment for protection from the cerebral ischemia[1] during the cardiac bypass injury [2318], the carotid endarterectomy [1438], the resection of aneurysms [3221]; it is also commonly employed in the massive stroke and the traumatic brain injury patients [2049]. Even mild reductions in the brain temperature as small as 1° C and importantly, the avoidance of any hyperthermia, can substantially reduce the ischemic cell damage and improve the outcome [2592].

An overloading or the failure of the thermoregulatory system could result in the heat stroke [336, 3419] and the heat syncope that characterized by a sudden and

[1] Ischemia—sudden reduction of blood flow results in failure of delivery of vital metabolites to the brain. Ischemic stroke [486]—a cerebrovascular accident (CVA), a sudden event due to pathology of the blood vessels in the brain—is one of the stroke types (the other is the hemorrhagic stroke).

sustained loss of the consciousness (the consciousness is usually lost at a body temperature of 28° C–30° C, but there are the isolated instances of the persons still being able to talk when the core temperature was as low as 24° C [2499]).

Heat stroke can be fatal if not appropriately managed [1934]. When an individual is exposed to the high ambient temperatures, the heat stroke can affect vital organs; one of the organes affected by the heatstroke is the gastrointestinal tract [2408]—studies have demonstrated damage to the small intestine following the heat stress [1934, 2490] including impairing the integrity of the intestinal barrier and the intestinal permeability [1854, 1934]. The reduction in gastrointestinal blood flow greatly impairs the epithelial cells of the small intestinal villi and induces excessive apoptosis. Studies in the multiple animal models have detected the heat stress-induced apoptosis in the small intestinal tissue and the intestinal epithelial cells [3424].

2.3.1 Adaptation to Cold in Nature

The adaptive reaction of the living organisms to cold includes the three components [2882]

1. to perceive (detect) the low temperature;
2. to transduce the signal (probably, by initiating the signalling cascade) to activate or repress the expression of the appropriate genes;
3. to utilize the genes to start the necessary changes of the biochemical processes.

The numerous types of the cells and the organisms satisfactorily endure the cooling to the temperatures higher than the crystallization temperature (for example, the human erythrocytes—up to one and a half months [2755]).

Some species not only survive in the cold environment but remain active at temperatures near 0° C or lower such as the snow fleas (Collembola) that could be seen hopping over the surfaces of the glaciers [723]. It is hoped that understanding the cold tolerance mechanisms can help in the development of the freeze resistant crop plants by the gene transfer [2233]. It is, however, difficult, since the cold resistance does not controlled by a single gene [225]—it is a complex of the interrelated processes including the regulation of the membrane fluidity, the synthesis of the low molecular weight and the high molecular weight cryoprotectants. At least 50 genes are identified now in the different species [369, 3064] that could be sorted into the following groups [369]:

- the genes that encode the structural proteins participating in the protection of the cell against the low temperature stress;
- the genes encoding the enzymes (the temperature changes can lead to the differential gene expression in order to create the different isozymes that are synthesised according to the temperature of the environment [2520]) involved in the synthesis of the different osmoprotectants (including the amino acid derived

ones), the desaturation of the lipid macromolecules and in the antioxidative response;
- the regulatory genes that control the overall mechanisms of the defence against the cold stress.

Although the interactions between the phospholipids are strong, the membrane must also remain fluid enough to allow the proteins to rotate and diffuse laterally within the membrane. The low temperatures cause the membrane lipids to solidify, which impairs the protein movement. Conversely, the high temperatures liquefy the membrane, which can compromise its integrity and reduce its effectiveness as a permeability barrier. The cells regulate the balance between the solid gel state and the liquid sol state. The ectothermic animals reduce the deleterious effects of the temperature by changing the composition of their membranes. In this process, called the homeoviscous adaptation, the cells remodel the membranes to preserve the fluidity by several mechanisms [2201]:

- *Fatty acid chain length:* The phospholipids with the short-chain fatty acids cannot form as many interactions with the adjacent fatty acids and therefore are highly mobile.
- *Saturation:* The double bonds create a kink in the fatty acid chain that prevents the effective bond formation with other fatty acids. With the fewer bonds between fatty acid chains, the membrane is more fluid.
- *Phospholipid classes:* The difference in the shape of the polar head groups alters the ability of the phospholipids to interact at the surface of the membrane. The phosphatidylcholine (PC) is more common in the membranes of the warm-acclimated cells, whereas the phosphatidylethanolamine (PE) is more common in the cold-acclimated cells.
- *Cholesterol content:* A pure phospholipid bilayer is mostly fluid at the high temperature and mostly solid at the low temperature. The cholesterol in a fluid membrane tends to make it more stable by plugging spaces in between the phospholipids, buffering against a loss of the integrity at high temperatures. When the cholesterol is in a fluid membrane that is cooled, it disrupts the structure in a way that prevents the membrane from solidifying.

When the biological material is subjected to the subzero temperatures for a sufficiently long time, the solidification of a part of the system into the ice phase will occur [758, 3511, 3513]. The biological systems as the chemically impure systems do not experience the phase change at a singular temperature—they solidify progressively over a range of the temperatures [758]; there are three macroscopic domains: the unfrozen, the partially frozen (mushy) and the frozen regions [1531].

In the absence of a seed surface the ice formation in the supercooled solution is initiated by the homogeneous or the heterogeneous nucleation. The heterogeneous nucleation dominates if the following inequality ("wetting criterium" [694]) is valid: $\sigma^{vs} - \sigma^{vl} < \sigma^{ls}$, where σ^{ij} is the surface tension on the interface between phases i and j, the superscripts "v", "l" and "s" denote the vapour, liquid and solid phases, respectively [1573]. The nucleation is observed at the cell membrane [1003] or at the objects contained in the cytoplasm; the importance of the nucleation on the subcellular objects increases as the temperature decreases [1380]; the heterogeneous nucleation has not been observed in the cells of some hardwood plants [2406] and the co-existence of ice and the supercooled water is possible [2923].

Two strategies for the adaptation of the organism to the low temperatures are observed in the nature [389, 1166, 1804]:

1. Avoiding the crystallization: an accumulation during the acclimation period [2882] of the oligosaccharides [470, 2953], other solvable carbonhydartes [393], the glycerol [1793], the free amino acids (especially the proline, the glycine and the betaine [1871]) [2406] that prevent the ice nucleation [2046] by increasing the water binding capacity and stabilize the protein structure; the multicomponent nature of the resulting cryoprotective mixture reduces the possible toxic effects associated with the concentration required for the single component to produce the same level of the cryoprotection [170]. Achieving this goal is also assisted by the antifreeze proteins, AFPs (the thermal hysteresis proteins, THPs) that reduce the crystallization temperature while keeping the melting temperature unaltered [723]; an amplitude of the hysteresis depends on the AFP concentration [684]; lowering of the freezing point by AFPs action is the non-colligative non-equilibrium. Sometimes the solutes accumulated by the cells at the low temperatures are referred to by the unifying name the "compatible solutes" [2882]. The cold acclimation is also accompanied by the changes in the gene expression [366, 1323, 2950]; some of the corresponding proteins, especially the protectin, stabilize the cell membranes during freezing [1323], other are engaged in providing the anoxia tolerance and the antioxidant defence [2950].

   The importance of the acclimation period could be illustrated by a couple of examples:

   - G. Bryant and J. Wolfe observed the significant differences in the survival rates (correlated with the morphological changes in the cellular membranes) during freezing for the protoplasts extracted from the leaves of the cold acclimated and unacclimated rye seedlings [391];
   - the needles of the Central European Scots pine *Pinus sylvestris L.* are lethally damaged by the exposition to the temperature $-10°$ C during the summer while in the winter they endure the exposure to $-80°$ C [225].

   It should be noted that examples of the very fast reaction of the organism to cooling is also observed in the nature: the alpine flowering plant in the family Campanulaceae *Lobelia telekki* growing only on the slopes of Mount Kenya, Mount Elgon, and the Aberdare Mountains of East Africa is able to cope, mainly by the ice nucleating agents (INA) production, with the temperature fluctuations over the range $-10°$ C $+10°$ C during a one day period [1003].
2. The regulation of the crystallization

   - *in time* (to be more exact, *in the temperature range*)— e.g., by the minimization of the supercooling ΔT and thus limiting the nucleation rate that strongly depends on the supercooling $J \propto \exp(B\tau)$, where $\tau \propto (\Delta T)^{-2}$,

ΔT is the supercooling [1003]); after the ice nucleation and the initial crystal growth, the temperature rises to the melting point and remains approximately constant during the subsequent crystallization of water—the so called "latent heat plateau" is formed
  and
- *in space* (the ice formation in the extracellular space): modification of the ice formation process by the synthesis of the INA—*nucleator* proteins that assist in the formation of the nuclei of the critical size[2] at the desired temperature (as a rule, at higher temperature than would occur in their absence, usually $\geq -10°$ C [1353, 1983]) and in those parts of the organism where the presence of ice is not lethal (the extracellular space where the nucleator proteins have been found in the living organisms). The freeze-tolerant animals endure the presence of ice in all extracellular spaces (e.g., in the abdominal cavity, bladder, brain vetricles, eye lens) and survive the temporary interruption of all vital functions (breathing, blood circulation, nerve and muscle activity), regaining them upon thawing.

Many animals including both the inverterbrate (the numerous terrestrial insects, several species of the intertidal marine mollusks) and the vertebrate (some species of the terrestrially hibernating amphibians and reptiles—several species of the frogs such as the wood frog *Rana sylvatica*, the gray tree frog *Hyla versicolor*, the spring peeper *Pseudacris crucifer*, the box turtles *Terrapene carolina* and *Terrapene orusta*, the Siberian salamander *Salamander keyserlingi* and the European common lizard *Lacerta vivipara* endure a partial crystallization of the body water [3511].

The largest of the freeze-tolerant animals are the box turtles *Terrapene carolina* and *Terrapene orusta* whose mass is about 0.5 kg. Most freeze-tolerant animals are from the North America; exceptions are the Siberian salamander and the European common lizard. Note, however, that freezing injuries still could occur in the surviving freeze-tolerant animals [632].

Some animals restore all the functions after freezing of about 50% (according to Layne and Lee [1795]—up to 70%) of the total body water (e.g., the wood frog *Rana sylvatica*—65% for up to 2 weeks; the conversion of up to 80% of body water into the ice was reported for the barnacles [1]) [2950, 2952, 2953]; there are cells (such as the isolated fat body cells from insects [1805, 2692], e.g., from the freeze-tolearant larva *Eurosta solidaginis* [170]) and organisms that survive the formation of the intracellular ice (the Antarctic nematode *Panagrolaimus davidi*[3308], the Pacific oyster *Crassotrea gigas* [2688]), the

[2] The ability of the organic molecules having a linear scale of their structure comparable to the lattice constant of the ice to serve as the crystallization nuclei for the vapour phase growth is well known: e.g., a monolayer of $C_nH_{2n+1}OH$ increases the crystallization temperature up to $-8°$ C for even $n \geq 22$ and up to $-1°$ C for $n = 31$; the size of the critical nucleus turns out to be only about 50 Å that approximately corresponds to the fifty water molecules and is comparable to the size of the critical nucleus in the homogeneous nucleation at $-39°$ C [3131].

insects survive the temperatures as low as − 80° C [389, 2135], some woody plants—down to − 50° C [369, 470], the twigs of the woody plants—the liquid nitrogen temperature [2681], the invertebrates and the dried seeds—down to the temperature close to the absolute zero [1966].

J. P. Sparks, G. S. Campbell and R. A. Black used the time domain reflectometry (TDR) to measure the liquid water content of otherwise frozen material [2923]. The authors considered the different types of wood (black locust *Robina pseudoacacia*, black cottonwood or California poplar *Populus trichocarpa*, lodgepole pine, shore pine or twisted pine *Pinus contoria*, western larch *Larix occidentalis*) and found that more than 25% of the water in wood of all species was liquid at temperature of −15–50° C. There are numerous examples of the bacteria being revived after the long-term storage in the Siberian [2821] and the Antarctic [2278] permafrost.

Different species "prefer" different cryoprotectors. For example, the glucose is the only cryoprotective agent (CPA) [2116] produced by *Rana sylvatica* (the reactions of synthesis of the glucose from the glycogen by the frog liver are the same as those used by all vertebrates) while the grey tree frogs and the Siberian salamander use glycerol as a primary CPA [2953]. The relatively low level of the colligative CPAs in the reptiles correlate with their generally lower freeze tolerance compared with the amphibians. The overwintering insects may contain as much as 25% of the glycerol [3197]. Glucose and glycerol are each highly efficient in preventing the hemolysis of the wood frog erythrocytes, while glucose turn out to be a poor CPA for the human erythrocytes—red blood cells (RBCs, also *erythroid* cells, and rarely *haematids*).

The controlled ice growth allows, *inter alia*, to provide the dehydration of the important organs [1793] and to avoid their injury; the direct evidence of the relationship between maintenance of the critical minimum cell volume (CMV) and the freezing survival was obtained by K. B. Storey, J. Bischof and B. Rubinsky in the experiments on freezing in the liver of *Rana sylvatica* [2954]. The cell shrinkage during the extracellular ice formation is minimized by the colligative cryoprotectants. Rapid distribution of the cryoprotectant is ensured by a high density of the membrane transport proteins. The AFPs are found in the organisms of this group along with the nucleator proteins; their task in these animals is to prevent the recrystallization of ice during thawing [170].

The survival of the freeze-tolerant animals also depends on the freezing rate, since the low rate of the temperature change provide enough time for the implementation of the cryoprotective measures such as the distribution of the cryoprotectant throughout the body. The control of the nucleation start allows these animals to avoid the dangerous fast ice formation that occurs if supercooling is too large. E.g., the typical crystallization rate in the wood frogs is less than 5% of the total body water per hour [2953].

In contrast to freezing which is a directional process with ice propagating inwards from the periphery to the core of the body, the thawing occurs relatively uniform throughout the body. The internal organs such as the heart or the liver melt quickly being still surrounded by the extra-organ ice since the high concentration of the CPA results in a low melting point. In fact, the heart beat is observed to resume before all ice in the body is melted [2953].

The important factors of the freeze-tolerance are the mechanisms of the anoxia tolerance for sustaining the cellular metabolism in the frozen state; sometimes the cold adaptation involves a conversion from an aerobic to a partially anaerobic metabolism at the low temperatures [574, 2951]. The oxidative stress at the low temperatures in the plants is usual due to the reduction of the photosynthesis rate [369].

The change of the adaptation strategy depending on the state of the environment is also observed [3202]; the European common lizard *Lacerta vivipara* could serve as an example of species that use both strategies [631].

The freeze avoidance is considered to be less favourable than the freeze tolerance since the former is associated with the significant energy loss [1794].

The ability of the plants to endure cold to a great extent depends on the environment. D. J. Beerling et al. [231] studied the effects of the long-term global climate changes (CO_2 enrichment and the increased exposure to the ultraviolet (UV) radiation) on the ice nucleation temperatures of some evergreen and deciduos woody shrubs and concluded that these changes will lead to the significant increase of the foliage damage of the subarctic plants.

2.3.2 *Antifreeze Proteins (AFPs)*

The antifreeze proteins were first described by J. Ramsay [2572] in 1964; later the AFPs and the antifreeze glycoproteins (AFGPs) were found in the species of different taxonomic groups including the Arctic and the Antarctic fishes such as the Antarctic cod, the Antarctic and the Arctic eel pout, the Arctic polar cod, the Winter flounder [738, 740, 974, 1660], over 40 species of insects (mainly beetles) [78, 1145–1147, 1712, 2693, 3102], the bacteria [1075], the fungi [2618] and the plants [1164, 1166, 2883, 3340] including the grass [2851]. The antifreeze proteins perform the dual function

- preventing freezing from external ice;
- inhibiting internal ice growth.

The important action for the fishes is suppression the ice growth, not the nucleation: some fishes were found to contain the ice crystals that do not grow even under the supercooling [1659].

The insect AFPs are considerably more active than those from the fishes [684]; the thermal hysteresis activity (THA) of the plant AFPs is low compared to that of the fishes and the insects (the levels of the THA for the plant AFPs are in the range $0.2 - (-0.6)°C$ [1165]), their inhibition of the ice recrystallization is, however, comparable or greater [2883, 3340]. There is a simple explanation of this fact:

supercoolings involved in the recrystallization are very small and thus a moderate hysteresis activity is sufficient to suppress the process [1659].

The different activity of the fish and insect AFPs is attributed not to the different affinity of the AFP to ice, but to the different morphology of the ice crystals that grow in the presence of the AFP [684]: all the fish AFPs shape ice into a hexagonal bipyramid that bursts out of the tips along the c-axis while the AFP from the spruce budworm *Choristoneura fumiferana* (CfAFP) having roughly four times the thermal hysteresis of the type I fish AFP produces the ice crystal that bursts along the a-axis.

The thermal hysteresis activity is the effect that is easiest to quantify and thus it is usually used to characterize the strength of the AFPs [1659]. While the THA of the fish AFPs (approximately 0.7–1.5° C) is considerably lower than that of the AFPs from the terrestrial arthropods ($3-(-6)°$ C) [3232], this level is sufficient to depress the freezing point of the body fluids below that of the seawater thus protecting the freeze-avoiding Arctic and Antarctic fishes. The measured THA of the certain insect AFPs is shown to inversely depend on the size of the ice crystal present in the sample [3457].

The fish AFPs are the most characterized and usually are divided into several groups according to their structure [505]. The first one contains the AFGPs and the type I AFPs containing the linear helical structure with the periodically repeating the ice-binding elements. The type I AFPs are the alanine-rich α-helical proteins with the molecular mass ca. 3300–4500 D [1496]. The second group comprises the globular proteins without any apparent repetitive substructures; it includes the type II (cysteine-rich globular proteins containing five disulfide groups) and the type III (globular proteins having the molecular mass about 6000 D) AFPs. The more lately discovered type IV glutamate- and glutamine-rich protein is considered as being intermediate to these groups [505] and is assumed to have an α-helix bundle structure [721]. The AFP molecules, as could be seen, are rather large; the smallest one discovered so far has the molecular weight about 1600 [1659].

In fact, the two different AFPs were found in the fish. The first one was largely confined to the liver and secreted into the blood to be distributed throughout the extracellular spaces of the organism. The second, *skin-type AFP*, is expressed in a wide range of the tissues, including the gills, most abundantly in the external epithelia—the tissue that come into the intimate contact with the ice-laden environment [2237].

The antifreeze glycoproteins have a molecular weight from 2600 to 34,000 while the antifreeze proteins—between 3200 and 14,000. In the solution these compounds usually form the helical rods. They are amphipathic molecules, with one side of the rod being composed of the hydrophobic residues and the other side being composed of the hydrophilic residues. On the hydrophilic side, there is a 4.5 Å separation between repeating threonine and aspartate residues that bind the protein to an ice lattice. There are great differences in both the primary sequences of the AFPs and their 3D structure [684, 945]. Still, there are some common features such as shaping of both the fish and the insect AFPs in such a way that the binding sites through which the AFP can dock to ice are relatively flat (the area of the binding surface is about a few hundred square angstroms) and constitute a significant part of the protein surface [684]. It was suggested that the threonine and aspartic acid residues could form the hydrogen bonds to the oxygen atoms on the primary prism plane of the ice [739, 1894].

There are a number of contributions to the energy of binding, in addition to the hydrogen bonds, viz., the enthalpic contributions from the van der Waals forces that come from the ideal surface-surface complimentary and probably the entropic contribution that follows from the fact that there is no need to solvate the hydrophobic ice-binding surface when it docks to ice [2916].

The THA of the AFPs depends, in addition to their specific activity and concentration, on the presence of the so called *enhancers* [813] that may be the low molecular mass substances such as the glycerol and the citrate [812, 1848] or other proteins. For example, other AFPs or the thuamatine-like proteins could serve as the antifreeze protein enhancers forming complexes that block a larger surface area of the potential seed ice crystal [3232]. The reported enhancement of the THA by the low molecular mass solutes varies from threefold (glycerol, sorbotol, alanine, ammonium bicarbonate) and fourfold (succinate, aspartate, glutamate) to up sixfold (citrate) [1848]. These enhancers could also inhibit the action of the INAs [812].

The genes encoding the AFPs cloned from some insects that are among the most active known natural AFPs that are considered for introduction into the plants to increase their cold resistance.

In contrast to the cryoprotectants such as glycerol, sugars and the dimethylsulfoxide (DMSO) that act colligatively and thus need to be at rather high concentration to produce the required effect, the AFPs' action is of the non-colligative nature and therefore are preferable for the use in the food industry since the extremely low concentrations of these substances (hundreds times lower than the corresponding colligative CPAs) will hardly have the adverse effects on the taste, the texture or the toxicity of the frozen food products [369]. S.R. Payne and O. A. Young reported the beneficial effects of the meat treatment with the AFGPs before freezing: the samples after thawing exhibit less drip with no detrimental effect on the texture, the tenderness or the juiciness [2403]. Addition of the AFPs also reduces the microbial activity within the food products [369, 1163].

The mechanisms of the ice growth inhibition by the AFPs and the AFGPs are difficult objects for the experimental study. In contrast to the enzymatic proteins that can often crystallized along with their ligands, there is no direct way to study the molecular interactions between the antifreeze proteins and the surface of the growing ice crystals [505]. Even more difficult is the investigation of the antifreeze protein effect on the ice nucleation [803]. Some hopes are related to new sophisticated techniques such as the "double oil layer micro-sized ice crystallization technique" that allows to minimize the influence of the container on the measurements. This approach provides a way to measure the value of the free energy barriers associated with the different dynamic steps of nucleation [803], including the incorporation of the water molecules onto the surface of ice nuclei at the kink sites and the inhibition of this process as a result of the adsorption of the additive or the impurity. The authors of the cited paper were able for the first time show that the mechanism of the nucleation inhibition on the ice crystal and on the surface of the foreign dust particles demonstrate some common features. The authors stress that it is important to distinguish between the nucleation induction time t_{nucl} and the time t_{g} needed for ice crystals to grow from the critical size to the observable size in the experimental investigation of the nucleation processes. The often used induction time is the sum of these two characteristic times and is close to the nucleation time only if $t_{\text{g}} \gg t_{\text{nucl}}$. Of course, this relation is often valid since the energy barrier of the nucleation is higher than that of the growth, but the exceptions are possible. N. Du et al. [803] also report that similarly to the surfactant molecules, the type III AFP molecules accumulate and self-organize on the surface of water that is attributed to the presence of both the hydrophobic and hydrophilic parts.

The free energy $\Delta G_{AFP-ice}$ of the AFP-ice binding could be estimated from the experimental data on the antifreeze activity. This energy proves to be rather small, being the difference of the two large quantities—the energy of interaction between the AFP and the ice binding surfaces in vacuum and the energy of the hydration of these surfaces [1496]. The authors of the just cited paper warn that it is difficult to accurately obtain this value in the computations for the two reasons:

- the uncertainty in the interaction potentials and
- the need to include a very large number of water molecules (thousands) into the system of the AFP and ice to ensure complete hydration.

The reversible process of the AFP binding could be formally written as $AFP + ice \Leftrightarrow AFP : ice$ with k_+ and k_- being the rates of the direct and inverse reactions related through the fraction ϑ of the ice surface covered by the AFP

$$k_+(1-\vartheta)C_{AFP} = k_-\vartheta \tag{2.32}$$

and whose ratio gives the association constant $K_a \equiv k_+/k_-$. The surface coverage could be thus expressed as

$$\vartheta = \frac{K_a C_{AFP}}{1 + K_a C_{AFP}} = \frac{C_{AFP}}{K_d + C_{AFP}}, \tag{2.33}$$

where the dissociation constant $K_d \equiv 1/K_a$ was introduced. Defining the fractional hysteresis activity ΔT_F as the ratio of the thermal hysteresis temperature at the given concentration of the AFP to the maximal thermal hysteresis temperature $\Delta T_F \equiv \Delta T_C/\Delta T_{max}$ and assuming that it is a function of the coverage $\Delta T_F = \psi(\vartheta)$, the authors note that experimental data could be reasonably fit for the choice $\psi(x) = x$. To get the needed estimate for $\Delta G_{AFP-ice}$ it is remains to assume that it is related to K_d by the Arrhenius equation $\Delta G_{AFP-ice} = RT \ln K_d$.

It should be, however stressed, that not the energy of the AFP binding to the ice matters, but the *difference* between its interaction energy with the ice and the liquid water [1659].

Still, up to date the numerical simulation is the main source of the information on the AFPs and their function.

A. Jorov et al. [1496] used the Monte Carlo (MC) method for the minimization of the energy of the complexes the AFP-ice in vacuum accounting for the nonbounded interactions, hydrogen bonds and the hydration potential for proteins. They computed MC trajectories from many different positions of the AFP at different ice samples. If the AFP helix is considered as a rigid body, its position and orientation are characterized by the six variables corresponding to the three translational and the three rotational degrees of freedom all of which were varied in the MC minimization. The mutual disposition of the AFP in the bounding state and ice when the helix axis of the AFP is approximately parallel to the ice surface could be described by the three parameters—the distance from this axis to the surface and the two angles (between the helical axis and the **a** axis on the basal plane of ice and the angle, characterizing rotation of the AFP around the helical axis). Variation of the latter angle will cause the different AFP faces to approach the ice surface. The authors found the three different binding configurations with very close values of energy and concluded that geometric complimentary alone does not explain the specificity of the AFP-ice association.

Consideration of the AFP-ice complex in water using the explicit water models requires the huge execution time; Jorov et al. refer to the computations made by J.D. Madura et al. of binding of the winter flounder AFP to ice that took months at the supercomputer [1496]. The MC minimization turned out to be considerably less computationally demanding. The authors found that convergence of the minimization process can be greatly improved by increasing the dielectric constant with only the minor effect on the optimal orientation of the AFP on ice.

The MD method was used to study the state of the antifreeze protein extracted from the winter flounder *Pseudopleuronectes americanus* in vacuum, in the aqueous solution and on the surface of ice [524]. The studied protein has 37 residues with 11 residue repeats. The AMBER all-atom force field for the proteins [3282] and the TIP3P potential for water were used. The computations were performed at the constant temperature 300 K. This choice rather than temperature 273 K was motivated by the computation complexity: it is easier to provide appropriate sampling of the phase space at the higher temperature.

The most favourable for the inhibition action of the AFP crystallographic planes of ice were established; the relative role of the hydrogen bonds and the Van der Waals forces as well as mutation effects (removing of one of the peptide groups) were considered. The authors identified the four distinct ice-binding regions on the AFP surface that could bond to ice simultaneously and determined the number of the hydrogen bonds each of these regions could form with the ice surface.

This approach was also used along with the energy minimization to study the system "ice - AFP" in vacuum and in the water. The aim was to find out what peptide groups are responsible for the suppression of the ice growth. Contrary to the existing opinion, the significant role of the hydrophobic amino acid residues was demonstrated [147, 505, 1496]. The effects of the solvent in the system "water - ice - AFP" were investigated by L. Kai et al. [1520].

G. Chen and Z. Jia [505] used a combination of the numerical approaches (the molecular docking, the energy minimization, the molecular dynamics) to find the best (i.e. the most energetically favourable) ice-binding surface of the fish type III AFP, exploiting no prior knowledge of the known ice-binding residues. They considered the two arrangements of ice: the "prism" ice representing the hexagonal ice I_h and the "random" ice, in which the water molecules are randomly positioned, still providing the same average density. The fish AFP considered consists of 66 residues; its structure is known from the X-ray crystallography and the NMR studies. The molecular dynamics simulation were performed after the energy minimization for the AFP-ice complex; no constraints were imposed at the MD stage except holding the ice oxygen atom position fixed. The authors determined the patch of the protein surface that is likely to bind to ice; the ice-binding site of Type III AFP turned to be larger than previously supported. Participating of the peripheral hydrophobic residues support the opinion that such interactions play an important role in the AFP binding to ice [146, 1496].

Jorov et al. presented a simple qualitative arguments while it should be so: the "rigid" water molecules incorporated into ice hardly can provide contacts with the AFP that are more optimal in terms of the van der Waals, the electrostatic and the hydrogen bond energy in comparison with the "soft" molecules of the bulk water that leaves the hydrophobic interactions as the most probably driving force of the AFP-ice association [1496]. The water molecules around the hydrophobic groups maximize the number of the hydrogen bonds by forming the enthropically unfavourable cage-like structures. The hydrophobic groups are getting together to minimize the total surface exposed to water. Interacting with ice, the AFP hydrophobic groups dip into the grooves on the surface of the ice crystal thus releasing the caged water to the bulk phase.

The analysis of the structure–activity relationships of the different AFPs from both the experimental data (mainly for the fish AFPs) and the numerical simulations allowed to formulate a number of the necessary conditions of the antifreeze activity [1496]:

- the AFP should be water-soluble;
- the AFP ice-binding surface should have the hydrophobic groups;
- the AFP should have the rigid backbone conformation to prevent the hydrophobic collapse of these groups in water;
- the AFP ice-binding surface should be complimentary to the ice surface;
- the AFP hydrophobic face should not be the self-complimentary to prevent the self-aggregation.

The above criteria are in a sense static ones and do not cover the important issue of the AFP molecule behaviour in respect to the crystallization front when the latter is propagating. Two points of view have been declared: (1) the adsorbed AFP molecules are in dynamical equilibrium with the free ones and (2) the absorption is permanent. C.A. Knight and A. Wierzbicki [1659] presented strong arguments for the latter option. If a dynamic equilibrium occurs, every time when the AFP molecule is (temporary) detached, the ice front could irreversibly advance some distance into the supercooled water, since there is no energetic barrier to the new layer nucleation, except the case of growth on the basal plane [1659]. To fulfil its function, the permanently absorbed AFP molecule should resist both an engulfment by the propagating ice crystallization front and pushing forward from the interface.

C.A. Knight and A. Wierzbicki suggested a quantitative measure of the AFP molecule suitability for the task, assuming it can be approximated by a spherical particle and characterized by its interfacial energies with water and ice γ_{pw} and γ_{pi} and introducing notions of "icephobic" and "icephillic" properties that are determined by the sign of the difference $(\gamma_{pw} - \gamma_{pi})$. The authors, however, warned that this construction is based on the sharp water/ice interface that in reality could represent a transition zone of finite width. The unacceptance of the nonpermanent absorption is probably the reason why the AFP molecules are rather large: other circumstances being equal, the molecule with a smaller binding site to ice will be more easily detached from the surface by the thermal fluctuations.

2.3.3 Ice Nucleating Agents (INAs)

The first INA was found in the bacteria isolated from the frost-damaged plants; it is assumed that this nucleator induces the damage to the host thus giving the bacteria access to the nutrients from the plant [1983]. The ice nucleators have been found in many other organisms including the lichen, the fungi, the insects [3456], the mollusks, the vertebrates. The INAs are thought to be active when they are attached to the surface of the cellular membrane and to keep their properties (the ice-nucleation activity) even for 1 : 1 000 000 dilutions [1983]. The extracellular INAs actually prevent the formation of the intracellular ice by initiating crystallization in the extracellular space at the high sub-zero temperatures.

In the plants both the extrinsic INAs such as the ice-nucleating bacteria (for example, a rod-shaped, Gram-negative bacterium *Pseudomonas syringae*) and the intrinsic nucleating agents synthesized by the plants [121, 390, 1173] are acting [3325]. These nucleators induce the stems to freeze at the warm, subzero temperatures. The ice propagation into the lateral appendages such as the buds or the newly extended primary tissues is prevented by the hydrophobic barriers that are the most effective if the initial freezing occurs at the relatively warm temperature [3325, 3341].

A new class of the heterogeneous ice nucleators was found in the freeze-tolerant larvae of the gall fly *Eurosta solidaginis*—probably, the most popular insect in the cryobiology [2204]. The large crystalloid spheres present within the Malpighian tubules of the larvae turned out to be a hydrate of tribasic calcium phosphate. The spherule size usually varies from 100 to 300 μm in the diameter, some large spherules over 500 μm were also observed. Unlike many known heterogeneous nucleators that have a hexagonal physical structure similar to that of ice lattice (such as some organic compounds or the silver iodide well known to the atmospheric scientists), the amorphous calcium phosphate spherules could induce the ice nucleation via cracks or other structural defects. The body fat cells seem to be another source of the ice nucleators.

The ice-nucleating activity of the spherules is less than that of the ice-nucleating proteins and lipoproteins. Still, their activity, along with the contribution from the body fat cells, provide a sufficiently high shift of the whole-body nucleation temperature needed for the survival of the goldenrod gall fly *Eurosta solidaginis*.

The interaction of the AFP or the AFGP with the ice surface is a specific example of the general problem of the protein interaction with the solid surfaces that is of great interest for the bio- and nanotechnology [1153] (the integration of an implanted device or material with the tissue, the hybrid biological/inorganic devices, the bio-nano-assembly technologies [3041]).

The most important distinction of the freeze-tolerant organisms is the difference between the freezing temperature and the lethal temperature [170].

Often the cold tolerance of the animal is expressed in terms of the freezing temperature (the supercooling point (SCP)) alone. J. S. Bale [169, 170] notes that the greatest threat to the survival is the cumulatative effect of the prolong exposure to the low temperatures rather than freezing per se and thus the dichotomy "freeze-tolerant—freeze-avoiding" is too coarse to describe the variety of the organism's reaction to the cold observed in the nature. Bale proposed to extend the classification by adding three more groups of the organisms:

- *chill tolerant*: the species with the low SCP (usually from $-20°$ C to $-30°$ C) that differ from the freeze-avoiding group by the occurrence of some mortality above the SCP that is increasing with the exposure time;
- *chill susceptible*: the species that die after the brief exposures (minutes or hours) to the temperatures significantly higher than the SCP;
- *opportunistic survival*: the species that are unable to survive below the threshold temperature for development.

The lowering of the temperature of the organism initiates the heat shock response (HSR) [2288] that is a mechanism to cope with the proteotoxic stress by inducing the expression of molecular chaperones and other heat shock response genes—the expression of the heat shock proteins (HSP) (also called the heat stress proteins [3189] or the cold shock proteins [1003, 2046]) [299, 931, 1711, 2373, 2443, 2673].

The unfortunate name reflects the history of the discovery of these proteins (by Ferruccio Ritossa in 1962, who observed that a transient increase in the temperature induced puffing patterns in the chromosomes in salivary glands of *Drosophila melanogaster* larvae [2611]): in fact, they are expressed as a reaction to the physiological stress of different nature such as heating, cooling, anoxia, ischemia, the presence of toxic compounds; their universality is also in covering practically all living organisms—from a bacteria to human [1568, 1711].

An objection [2373] against the more correct term the "stress" proteins [723]—these proteins are synthesized at the normal conditions as well—could equally be applied to the term 'heat shock proteins'.

About the specific and non-specific reaction of the plant cells to the three forms of the abiotic stress—the cold, the water depletion, the increased salinity etc. see [226].

It should be noted that the non-specific reaction to a variety of stresses on the organism level is treated by the well-known theory of the general adaptation syndrome formulated in 1936 by H. Selye [2781].

The HSPs assist in the keeping the cell homeostasis [251] (for example, by the compensation of the reduction of the metabolic reaction [2406]—the temperature effect is described by the Arrhenius equation $k = Ae^{-E_a/RT}$, that is the exponential decrease of the reaction rate with the temperature; frequently the mediators are the cold adaptive, or the psychrophillic enzymes, both intracellular and extracellular, that are distinguished by the reduced activation energy E_a and the enhanced catalytic activity [667, 3018]).

The HSPs belong to a class of proteins called the *molecular chaperones* that cope with the stress-induced denaturation of other proteins [931] and produce the adaptive responses such as the nuclear DNA repair [2921]. The genes encoding are highly conserved and are present even in the smallest of genomes. In addition, the HSPs help to sequester severely damaged proteins for degradation [2918]. For example, the heat shock protein Hsp70 plays a critical role in the thermotolerance of mammalian cells due to its chaperone activity. E. A. A. Nollen et al. [2313] investigated the chaperone activity and cellular heat resistance when the overexpression of the Hsp70 was transiently induced by means of the tetracycline regulated gene expression system using firefly luciferase as a reporter protein. The chaperone activity was found to strictly correlate to the level of the Hsp70 expression.

Being expresses in response to almost every stress if sufficiently intense, the HSPs are also considered as very sensitive biomarkers that could indicate the presence of pollutants or toxins in the environment [931]. The problem, however, is how to attribute the changes in the HSP expression to a particular environment stress since organisms often undergo the multiple stresses simultaneously.

The heat shock protein uses the energy of the ATP to catalyse the protein folding after the translation. The chaperones can also help refold the proteins that have become denatured as a result of the thermal stress. Many cells exposed to the extreme temperatures undergo a heat shock response, which leads to a dramatic increase in the levels of the specific proteins that help repair the damaged proteins. During a heat shock, the cell undertakes a rapid increase in the synthesis of the several critical heat shock proteins. The cell can halt the transcription and translation of other genes, sparing the biosynthetic resources for the HSP synthesis. It stimulates the expression of the HSP genes by activating a heat shock factor (HSF), a transcription factor that binds to the heat shock elements in the promoters of genes for the heat shock proteins [2201].

The initial stage of the adaptation to cold is the detection of the low temperatures by the organisms and the transduction of the signals in order to activate the biochemical changes, including the gene response [869], that provide the defence against cold [342, 3188]. The plasma membrane is considered as the principal sensor of the low temperatures since the membrane rigidification [3187] causes the cytoskeleton rearrangement and changes in the ions influx to the cell [3193]. The direct proof of the effect of the membrane properties on the gene expression was obtained by B. L. Orvar et al. [2354] in the experiments on the fluidization of the membrane with the benzyl alcohol and the membrane rigidification with the dimethylsolfoxide. The membrane rigidification could, in turn, trigger the reorganization of the cytoskeleton [2354, 2705]. In addition to the direct effect on the membrane fluidity, the low temperatures influence the stability of the secondary structures of the ribonucleic acid (RNA) and the DNA as well as the activity of the enzymes, including those involved in the transcription and the translation as well as in the metabolic reactions [2882].

Studies of the signal transduction in the thale cress (mouse-ear cress or arabidopsis) *Arabidopis thaliana* [1658] and other plants [2467] showed that the intracellular calcium plays a significant role in the process of the cold adaptation: all cell types respond to the cold by the elevating calcium (probably, by a direct activation of the mechanosensitive calcium channels via the changes in the mechanical state of the membranes [768]); in fact, the elevation of calcium could be caused by the different abiotic stresses [1658], including salinity [1913]. Increase of the membrane viscosity affects the Ca^{2+} channels also causing an increase of the cytosolic calcium concentration [2175].

A number of components are considered as the probable intermediate agents in the transduction of the calcium signal: the calmodulin, the calcium-sensitive protein phosphtase, the calcium-dependent protein kinase [2882, 2886].

The increase of the intracellular concentration of calcium, in turn, activates the expression of the transcription factors that control the subsequent transcription of several cold-regulated genes [1430, 1658, 3187].

Many genes are regulated by the different biotic and abiotic factors [2882] (one speaks about the *cross-talk* between the different stress signalling pathways [2834] in this case), but some of them are specifically regulated by the low temperatures [1385, 3064].

The cold sensing in the plants also could be related to the temperature effects on the metabolic reactions and the imbalances between the absorption of the light energy and its metabolic consumption [858].

The observations indicate that the cells detect the rate of the temperature change dT/dt rather than the temperature itself [2147, 2467, 2882]. The more detailed information on the molecular and gene processes of the cold adaptation of the psychrophilic organisms could be found, e.g., in [667, 1323]. For the detailed discussion of the temperature sensing, the signal transduction and the gene regulation in the plants during the cold acclimation please see refs. [2258, 3193, 3194] and the references cited therein.

Notwithstanding the aforesaid, the cooling could result in the cell death [736, 2655]: the tertiary protein structure as well as the polypeptides contained in the protein are changed and the denaturation of the latter is possible [723, 2202]. The cooling could also cause the mechanical damage to the cytoskeleton which structure is determined by the bonds between the membrane proteins and the cell scaffold since lowering the temperature weakens these bonds. The cooling can cause the marked effect on the microtubules[3] including the depolymerization and disappearance of the microtubule organizing centres. Such behaviour is observed in the unfertilized oocytes [269]. The low temperatures can also act on other main components of the cytoskeleton, e.g., on the microfilaments composed of the polymerized actin (F-actin) that are important for the different cell processes including cytokinesis [269]. The action of the low temperatures on the cytoskeleton could be mediated by the increased level of the intracellular calcium that activate the calcium dependent proteases and phospholipoidases [2656].

The hypothermic damage of the cells was extensively studied in respect to the long-term RBC storage and even a special term *hypothermic storage lesion* was introduced. It could include the adenosine triphosphate (ATP) depletion which is critical for the maintenance of the electrolyte balance and powering the sodium and potassium ionic pumps as well as the membrane injury that is observed as the lipid loss, the macroaggregate formation and the reduction of the ratio of the surface area to the volume [2755].

The changes of the cell state caused by the low temperatures could manifest themselves in the behaviour of the larger biological structures comprising these cells. Thus, the elevation of the intracellular Ca^{2+} [2525] in the cardiomyocytes [1042, 1901, 3240] contribute to the posthypothermic myocardial failure [1685].

Some organisms, however, dye from the cooling to the moderate temperatures before the freezing occurs [169]. For a human patient being held at 15° C for the open-heart surgery, a maximal duration of the operation is about 2 hours [2413].

The hypothermia is dangerous for the newborns since they are at a greater risk of the heat loss for a number of reasons [144, 214, 1040, 1204]:

- the large surface area-to-body mass ratio;
- the decreased subcutaneous fat;
- the greater body water content;
- the immature skin leading to the increased evaporative water and the heat losses;
- the poorly developed metabolic mechanism for responding to the thermal stress (e.g., no shivering);
- the altered skin blood-flow (e.g., the peripheral cyanosis).

There are three major categories of the hypothermia (hypothermia is usually defined as a 2° C decrease in the core temperature) [2499]:

- *primary* hypothermia refers to the condition in which the casualty has the normal thermal regulatory responses, but these are ineffective against the environment;
- *secondary* hypothermia is caused by the impaired thermoregulation;

[3] Microtubules are structures consisting of the polymerized tubulin in equilibrium with the free tubulin pool in the cytoplasm; they play an important role in the organization of the spindle in the oocytes and the process of the cell division. The microtubules also, as R. Penrose supposed, play a crucial role in the brain function [2267, 2423].

- *clinically induced* (iatrogenic) hypothermia is defined as a decrease in the core temperature that is induced for the various surgical procedures such as the coronary bypass.

The accidental profound hypothermia should be distinguished from the anapyrexia (from Greek *ana*—reverse and *pyretos*—fever) that is a pathological condition in which there is a regulated decrease in the core temperature [317]; it is a compensatory response to such factors as the hypoxia; the importance of the anapyrexia is in its reducing the oxygen consumption and increasing the affinity of the haemoglobin to oxygen [2937]) is frequently lethal: the reported mortality related to the progressive reduction of the cardiac output and a sudden fall in the arterial blood pressure ("re-warming shock") is 52–80%, depending on the re-warming method [1685]. The low temperatures decrease the myofilamental Ca^{2+} sensitivity and increase the myocardial contractility [1255]; the significant changes are already observed for the moderate temperature decrease down to 30° C [1747].

Sometimes the cold detection is performed by the special cold receptors (such as the so called bulbs of Kraus residing in the dermis of the skin of humans [2498]). In nature, however, another mechanism is more important—the cellular level of the detection of the low temperatures by organisms and the transduction of the signals in order to activate the biochemical changes that provide the defence against cold [342]. The plasma membrane frequently is considered as the principal sensor of low temperatures and the membrane rigidification [3187] causes the cytoskeleton rearrangement and changes in the ions influx to the cell [3193]. The studies of the pathways of the "cold" signal transduction in *Arabidopis thaliana* showed that the intracellular calcium plays a significant role in the process: all cell types respond to cold by the elevating calcium; in fact, the elevation of calcium could be caused by the different abiotic stresses [1658], including salinity [1913]. The symptoms of the frost injury in plants are similar to those caused by the desiccation [3127]. The increase of the intracellular concentration of calcium, in turn, activates the expression of the transcription factors that control the subsequent transcription of several cold-regulated genes [1430, 1658, 3187]. The effects of the divalent cations of magnium and calcium on the freezing tolerance are similar, but Ca^{2+} is five times more effective than Mg^{2+} [2235].

2.3.4 Membrane Behaviour

The eukaryotic cells are considered below that in contrast to prokaryotes have the flexible cell walls and form both the unicellular and multicellular organisms. The eukaryotic cells present an extensive variety of shapes, as an adaptation to their specific function and location within the different tissues. The cortical cytoskeleton and the plasma membrane are the two main elements responsible for the cell shape and mechanical response [3080]. The different organelles occupy an important portion of the cell volume, and their presence implies that the cell must accommodate them.

The old eukaryote-centric ("five kingdom") classification of life contained three kingdoms of the multicellular organisms (*plantae, animalia, fungi*) that are clustered at the end of one branch of the new prokaryote-centric ("three kingdom") scheme based on the advances of the molecular taxonomy and phylogeny, primary on the study of the slowly evolving molecule 16SrRNA that is common to nearly all life on the Earth [2278].

The cell membrane is a bilayer of the amphiphilic (having both the hydrophobic and hydrophilic parts—from Greek *αμπις amphi*, on both ends, and *φιλια philia*, love, friendship) molecules—the lipids that include the phospholipids, glycolipids and steroids; the membrane could contain up to a hundred of the different lipid molecules [328] and a number of the proteins—both peripheral and integral. Lipids represent up to 50% of the total mass of the membrane in the mammalian cells [3154, 3417].

The lipids have a polar head (which prefers to contact and interact with other polar molecules, such as water) and a tail formed by two hydrocarbon chains which present a strong hydrophobicity and avoid the interacting with water.

If lipids are immersed in water, they tend to self-assemble to avoid the hydrophobic interactions with the surrounding water and can aggregate to form two basic structures [3080]

- *micelles*, a closed structure with all the tails in the inner, water-free region, and the lipid heads oriented to the exterior, in contact with water;
- *bilayers* and *vesicles* with two lipid monolayers fold in the opposite directions, so that the heads form two parallel sheets whereas the tails are trapped in the intermediate region, without contact with the aqueous environment (cholesterol—a small molecule with a polar hydroxyl group and a short hydrocarbon chain—occupies the space between the phospholipid tails in the inner region of the bilayer, with its head oriented close to the phospholipid head; the mammalian cell membranes are rich in cholesterol, which plays an important role in the control of the bilayer fluidity [2075].

For example, in the human erythrocytes—the red blood cells (RBC)—the membrane lipids (mostly the unestherified cholesterol and phospholipids, with small amounts of the free fatty acids and glycolipids) compromise up to 50% of the total mass of the membrane in the mammalian cells [3154, 3417] while the membrane proteins—about 52% of the mass [2172, 2755]. The lipids behave as a fluid in the bilayer plane, a property with important implications for the cell activity. The membrane is connected with the inner cytoskeleton, a three-dimensional mesh formed by actin filaments which provides compactness and structural ordering, and determines the cell shape, which in turn depends on the type of cell and its function; in some cells, an exterior cytoskeleton also exists, and it connects with neighbouring cells in order to facilitate a coordinate response of the tissue. The actin filaments that occupy most of the inner cytosolic volume and connects the different organelles and microstructures of the cell. The presence of the ATP is crucial for maintaining the cytoskeleton properties—when this molecule is depleted, the cell experiences drastic shape changes [3080]. It is assumed that in this case the network is subjected to continuous remodelling, which allows the relaxation of the cytoskeleton tensions [1849].

The membrane equilibrium is controlled by a number of processes, including a flip-flop rearrangement of the different lipid species of the bilayer, remodelling of the cytoskeleton, or the balance of lipid densities during vesiculation processes, which are able to occur due to the existence of lipid reservoirs in the interior of the cell [3080].

The lipid bilayers are selectively permeable to exchange of the polar molecules and host a high density of transmembrane proteins, which essentially define the specific membrane functionality. Since the proteins are much larger than the lipids, their concentration corresponds to a protein per about 50–100 lipids. The transmembrane proteins are also amphiphilic and orient their polar groups to the aqueous environment (cytosol and exterior of the cell), whereas the hydrophobic groups interact with the lipid tails [3080].

The membrane lipids have a small polar head and a long non-polar (hydrocarbon) tail. Most lipids are nearly insoluble in water (the solubility values are between 10^{-10} and 10^{-6} moles per litre [475]) and tend to form the aggregates spontaneously. The bilayers consist of two opposing hydrocarbon monolayers, separated from the surrounding water by two layers of the polar lipid headgroups.

The cell membrane forms the boundary between the intracellular and extracellular compartments and mediates the transport into and out of the cell. The voltage-gated sodium channels exhibit the non-Arrhenius behaviour at the temperatures below 10° C [610, 1273].

The membrane also provides the mechanical coupling between the cell and its environment—the neighbour cells and the extracellular matrix. The cell membrane is a composite structure described as the two dimensional "fluid mosaic" with inclusion of the various membrane-associated proteins [2864]. Among the membrane proteins are the ion channels, the receptors and the signalling macromolecules that are diffusing in the lipid bilayer or linked—to the cytoskeleton or to the extracellular matrix. In the erythrocytes, for instance, the peripheral proteins—the spectrin, the ankyrin, the actin, the adducin—found on the cytoplasmic side of the membrane form a mesh-like network of the microfilaments that provide the mechanical strength of the membrane, in particular, the cellular deformability [2755].

The interaction of the molecules that constitute the membrane with the water molecules determines both the membrane structure and stability; some water molecules—the hydration shells around the polar parts of the lipid molecules and proteins as well as the singular water molecules in the hydrocarbon zone of the membrane—are osmotically inactive. The surface of the membrane is usually charged (in the average) negatively [3332]. Depending on the water content in the system and the electrolyte concentration [2208], the different mesophase structures could be observed; the biological membrane in the normal state is a lyotropic liquid crystal [2914].

The classic fluid mosaic model of Singer and Nicolson [2864] assumes that only one type of the lipid molecules is present in the membrane; its generalization to the case of two different lipids [391, 2296, 3219] extends significantly the class of the problems that could be considered rigorously, in particularity, it becomes possible to correctly describe a deformation of the membrane [2296] and the phase separation [1729, 3219]. The heterogeneity and the dynamic character of the membrane are essential for its function [3189].

In case of the membrane consisting of the incompatible amphiphiles A and B, the phenomenological parameter φ is introduced defined as the fraction of the surface area occupied by the component A, i.e. $\varphi = area(A)/(area(A)+area(B))$. For the phase-separated membranes under the weak tension, the total free energy is written as the sum of the contributions from the bending elasticity F_b, the membrane surface tension F_λ and the line tension between phases F_σ [2296]: $F = F_b + F_\lambda + F_\sigma$, and these contributions are defined as [2772]:

$$F_b = \sum_{i=A,B} \iint_{S_i} \frac{k_i^c}{2}(2H - c_0)^2 dS, \quad F_\lambda = \sum_{i=A,B} \lambda \iint_{S_i} dS, \qquad F_\sigma = \sigma \int_{\partial S} dl, \tag{2.34}$$

where $i = A, B$ denotes the domains of the membrane occupied by the corresponding components, k_i^c is the bending rigidity, λ and σ are the surface and the line tension, H is the mean membrane curvature, c_0 is the spontaneous curvature.

The bending rigidity could be measured by different experimental techniques [767], for example, micropipette [877], AFM [2728] and optical tweezers [276]. The shear modulus of the bilayer is negligible due to its fluidic nature in the membrane plane, given that any shear stress is instantaneously relaxed by the rapid lateral rearrangement of lipids.

However, such experimental methods as the pipette aspiration [772], the electric field deformation [860], and optical tweezers [674] provide quantitative information about the shear and bending moduli of the RBC membranes in the static conditions only. Thus, the dynamic, frequency-dependent knowledge of the RBC mechanical response is very limited [2527]. M. S. Amin et al. [67] suggested a new optical technique, referred to as dynamic scattering microscopy (DSM), to quantify the microrheological properties of live cell membranes—the dynamics light scattering is performed using microscopy platform. Using DSM, the authors measure the time-fluctuating irradiance scattered at a certain angle by the micro beads conjugated to the RBC membranes. By employing the fluctuation-dissipation theorem and generalized Stokes-Einstein relationship, this dynamic scattering signal is translated into an effective frequency-dependent 3D complex shear modulus.

D. Ni et al. [2296] considered the case of the surface tension λ, the spontaneous curvature c_0 and the Gaussian curvature rigidity being the same for the both components (the only differences are in the bending rigidity k_i^c) and found several stable patterns of the phase arrangements including a linear array of the A-phase stripes and a hexagonal array of the circular A-phase caplets in the matrix of B-phase with the shape of the domains being strongly influenced by the composition φ. The authors also found that the different bending rigidities of the lipid components could induce the first-order transition between the phases. The case of the different signs of the spontaneous curvature of the components of the two-phase membrane was considered by J.L. Harden et al. [1245].

P. B. Sunil Kumar et al. [1729] studied the modulated phases in the multicomponent fluid membranes; the two-component (A, B) bilayer and the three-component (A, B, C) monolayer membranes were considered. For the former the authors found that two square and three hexagonal phase separation pattern are possible, for the latter in addition two stripe phases and the homogeneous A-rich, B-rich and C-rich disordered phases were observed. In the certain restricted regions of the parameter space, the behaviour of the three-component monolayer is similar to the behaviour of the two-component bilayer. The authors note, that the two phase coexistence between the gel and the fluid phases is more difficult for the analysis in comparison with the fluid-fluid coexistence, since the shear elastic properties of the gel domains should be taken into account.

The membrane reacts to the temperature variation (the *thermotropic mesomorphism* [328]—the order of the membrane constituents changes with the temperature): the heating results in the formation of the hexagonal structures in the lipid bilayer, the cooling leads to the transition of the membrane lipids into the gel phase ("the liquid-solid transition", also called *the main phase transition* [652, 1610]). The low temperatures also induce alterations in the lipid composition of the membrane (the degree of non-saturation) [2882]. The phase separation is a result of the difference in the phase transition temperatures of the different lipid molecules present in the membrane; often it leads to the formation of the gel-phase rich domains (the "aparticulate domains") [2591], in which the integral proteins have been excluded from the lipid layer [393]; the domains have also been detected on the whole cells [836].

The temperature of the main phase transition depends on a number of the factors including the composition of the membrane, the degree of the hydration, the presence of the impurities in the lipid bilayer and the composition of the suspending solution [393].

The transition of the membrane into the gel phase is accompanied by the change of the sizes of the elements of its structure [1172], an appearance of the defects between the membrane proteins and the lipid bilayer [297, 2597], increasing of the membrane permeability for the ions [93] and the possible leakage of the cytoplasmic solutes. The critical temperature of the transition depends on the water content in the system and attains the minimal value when it exceeds the water content that could be bond by the lipid structures [328] as well as on the membrane composition (the fraction of the saturated and unsaturated phospholipids; the well-known example of the composition effect is the variation of the fraction of the unsaturated phospholipids in the cells of the north deer *Rangifer tarandus* leg from the hooves upward in accordance with the variation—by tens of degrees Centigrade—of the characteristic temperature).

The cell dehydration influence the membrane behaviour: it increases the main phase transition temperature of the lipids. As a result, the phase separation could occur at the higher temperatures [2591]. The experiments on freezing of the prostate tumour cells showed that the onset of the liquid crystalline to the gel transition coincides with the ice nucleation temperature [3336]. It is unclear whether this relation is valid for other types of cells.

The freeze-induced dehydration was shown to result in the alteration of the permeability parameters of the protoplast membrane—the protoplast became osmotically unresponsive [2938]. The electron microscopy showed that the membrane defects such as the "fracture-jump lesion" related to the existence of the aparticulate domains are formed.

The cooling reduces the potential of the muscle membranes and the excitability of the muscle fibres that is explained by the loss of the integrity of membranes [723] and the changes in their composition, mainly removing of the hydrophilic lipid components [477]). The membrane fragmentation could be preceded by the loss of the membrane proteins [2406]. The reduction of the mechanical strength of the membranes is confirmed by the results of the centrifugation of the cooled cells [1959].

J. Chakrabarty et al. [477] investigated the variation of the membrane composition during the cryopreservation and found a significant decrease in the content of the neutral lipids, glycolipids and phospholipids. As a result the cholesterol:phospholipid ratio increases as well as the amount of the hydrocarbons. In another words, increase of the hydrophobicity of the membrane is observed caused by the shedding off the preferentially specific hydrophilic constituents of the cell membrane. The authors note that such change of the membrane hydrophobicity aim to counteract the cold stress could be not totally reversible upon the thawing.

The temperature lowering also results in the cytoskeleton depolymerization [2368] and reduces the ability of the integral proteins [1168] to control the intracellular electrolyte concentration that could finally cause the protein denaturation.

2.3.5 Cell Dehydration

When a cell is cooled slowly, its destruction is caused by the cell dehydration (*exosmosis*—the water is leaving the cell both by the agency of the diffusion across the bilipid cell membrane and by the hydrodynamic flow through the pores in the membrane [660]) that results in the growth of the intracellular concentration of the solutes and the cell shrinking. The diffusion of water molecules across the bilipid layer consists of three steps:

1. partitioning of the water molecules from the aqueous phase into the lipid phase;
2. diffusion of the water molecules through the lipid layer;
3. return of the water molecules from the lipid to the aqueous phase.

All these stages could be associated with some activation energy.

Considering the intracellular solution, it is necessary to distinguish the mean salt concentration $\bar{c}^i_{\text{salt}}$

$$\bar{c}^i_{\text{salt}} = \frac{\text{moles of salt}}{\text{total cell volume}} = \frac{N_{\text{salt}}}{V} \tag{2.35}$$

and the actual concentration c^i_{salt} [210]

$$c^i_{\text{salt}} = \frac{\text{moles of salt}}{\text{cytosol or osmotically active volume}} = \frac{N_{\text{salt}}}{V - V_b}, \tag{2.36}$$

where V_b is the osmotically inactive volume. Introducing the fraction of the osmotically inactive cell volume $\varphi = V_b/V$, the relation between these concentrations could be written as

$$\bar{c}^i_{\text{salt}} = c^i_{\text{salt}}(1 - \varphi). \tag{2.37}$$

If a semipermeable solute is present in the solution, its mean cellular concentration will be $\bar{c}^i_{\text{s}} = N_{\text{s}}/V$. To determine the actual solute concentration, it is necessary to use the Henry law of absorption with the coefficient k [210]:

$$c^i_{\text{s}}(V - V_b) + c^i_{\text{s}} k s_m + c^i_{\text{s}} K V_b = N_{\text{s}} \tag{2.38}$$

or

$$\bar{c}^i_{\text{s}} = c^i_{\text{s}}(1 - \varphi + k\alpha + K\varphi) \tag{2.39}$$

where $\alpha = s_m/V$ is the specific surface area of the organelle membranes and K is the partition coefficient into the organelles.

For a continuous volumetric density (for example, the mass density or the solute concentration) $\psi(\mathbf{r}, t)$, its total amount within the moving control volume $\mathcal{V}(t)$ is

$$\Psi(t) \equiv \int_{\mathcal{V}(t)} \psi(\mathbf{r}, t) \mathrm{d}V \tag{2.40}$$

and its variation in time, accounting for the fact that the cellular membrane could be considered as the non-material interface as [210]

$$\frac{\mathrm{d}\Psi}{\mathrm{d}t} = \int_{\mathcal{V}(t)} \left(\frac{\partial \psi}{\partial t} + \mathbf{v} \cdot \nabla \psi \right) \mathrm{d}V + \int_{\partial \mathcal{V}} \mathrm{d}\mathbf{S} \cdot (\mathbf{u} - \mathbf{v}) \psi. \tag{2.41}$$

where $\partial \mathcal{V}$ is the volume boundary moving with the velocity $\mathbf{u}$.

For the spherical cell these relations are simplified to

$$\Psi(t) \equiv \int_{0}^{R(t)} \psi(\mathbf{r}, t) r^2 \mathrm{d}r, \qquad \frac{\mathrm{d}\Psi}{\mathrm{d}t} = 4\pi \int_{0}^{R(t)} \frac{\partial \psi}{\partial t} r^2 \mathrm{d}r + 4\pi R(t)^2 u(t) \psi(R(t), t)). \tag{2.42}$$

When the salt or other solute concentration is chosen as ψ, the above relations give the rate of change of the corresponding substance in the cell

$$\frac{\mathrm{d}N}{\mathrm{d}t} = 4\pi R^2 \left(\bar{D} \frac{\partial \bar{c}}{\mathrm{d}r} + uc \right) |_{r=R}. \tag{2.43}$$

The quantity in the parentheses in the above expression is the flux (per unit area) of the substance which is zero for the salt while in case of the semipermeable solute it is related to the its distribution coefficient at the membrane—the measure of its ability for permeation.

Assuming that the intracellular solution is ideal, i.e. the difference in the chemical potentials of the cytoplasm and the water is proportional to $\ln(1 - \chi)$, where χ is the molar concentration of solutes, a depression of the crystallization temperature $\Delta T \propto \chi$ could be expressed as [1179]:

$$\Delta T = \frac{RT_f}{L} \chi = -K\chi \tag{2.44}$$

where L is the latent heat of crystallization, T_f is the freezing temperature of the pure liquid and K is the so called cryoscopic constant.

This effect is the colligative one, i.e. determined by the *quantity* of the solutes and does not depend on their nature; an additional effect observed in the case of the macromolecules is related to the properties of their surface [3335]. The effect of the solutes on the structure of the hydrogen bonds network and the crystallization temperature has been investigated experimentally by M.B. Baker and M. Baker for the small water drops and found in many respects to be similar to the effect of the higher pressure; it is even possible to express the solute content in terms of an approximate osmotic pressure that is proportional to the logarithm of the water activity in the solution [162].

There is a linear relationship between the equilibrium melting point and the homogeneous nucleation temperature of the various aqueous solutions [1636] $\Delta T_{f,hom} = \lambda \Delta T_m$, where $\Delta T_{f,hom} = T_f^0 - T_f$ and $\Delta T_m = T_m^0 - T_m$, T_f^0 and T_m^0 are the homogeneous nucleation temperature and the equilibrium melting point of the pure water, respectively, T_f and T_m are the homogeneous nucleation temperature and the equilibrium melting point of the solution.

The coefficient λ could be considered as the measure of the solute-specific supercooling capacity. N. Kimizuka and T. Suzuki [1635] measured the value of λ for the solutions of the small molecules such as salts, alcohols, sugars and found a linear relationship between the coefficient λ and the logarithmic value of the solute self-diffusion coefficient D_0. In the later paper N. Kimizuka et al. [1636] suggested an explanation of the observed correlation, involving the expression for the nucleation rate J

$$J = \frac{k\Gamma_z}{\eta(T)} \exp\left(\frac{-\Delta G^{nucl}}{kT}\right) \quad (2.45)$$

where ΔG^{nucl} is the activation energy for the nucleation, Γ_z is the Zeldovich constant and η is the solution viscosity. The homogeneous nucleation temperature could be found from the condition that the number of the nuclei per unit volume becomes equal to unity and thus evidently depends on the nucleation rate. Assuming further that the Stokes-Einstein relation is valid in the supercooled fluid, the authors get the needed relationship between the λ and D_0 through the viscosity of the solution $\eta(T)$.

The value of λ for polymer solution is rather high and increases with the molecular weight, indicating that these solutions allow the deep supercooling [1636].

The concentration of the extracellular solution increases due to the solute rejection from the ice crystallization front—the so called constitutional supercooling arises—until the eutectic temperature is reached at which the remaining solvent and solute crystallize together. In the case of sodium chloride it is $-21.1°$ C and the eutectic mixture contains 31% NaCl [2413]. The additional depression of the freezing point for the cytoplasm is related to the interaction of the ions in the solution with the charged sites of the proteins present inside the cell and bringing the proteins into the solution.

P. Mazur et al. reported in a series of papers [1180, 2084, 2086, 2087] the results of their search for the correlations between the temperature at which the intracellular ice formation (IIF) occurs ("flashing temperature") and the state of the external solution. The authors studied the IIF in the mouse oocytes and in the stage I and II of the African clawed frog *Xenopus laevis* oocytes in the solutions of glycerol or ethylen glycol. They found that the flashing temperatures in the *Xenopus* are significantly higher than those observed in the mouse oocytes that is attributed to the much larger size of the former [1180]. The IIF in the mouse oocytes is found to occur in the narrow range of the unfrozen fractions of the external solution that allowed the authors to state that these entities are related and to introduce the notion of the critical unfrozen fraction. The results for the *Xenopus* in the EG solutions do not show such clear trend while the IIF in this case could be correlated with the narrow range of the solution concentration (the solution osmolality). P. Mazur et al. supposed that in the case of the *Xenopus* oocytes in EG solution the dominant damage mechanism is the chemical damage of the membrane in the highly concentrated solutions [2084].

The specific mechanism that could be responsible for the correlation of the unfrozen fraction of the external solution with the IIF (as well as for the existence of the critical value of the unfrozen fraction) remains obscure. It is assumed that this mechanism could involve the contact of the external ice with the existing pores in the membrane and the membrane defects or the severe deformation of the membrane surface attempting to conform its shape to the available small extracellular space that is still in the liquid phase [2087].

Preventing of the ice formation inside the cell is by all means favourable for the cell survival [470], however, the growth of the solute concentration results in the "solution effect injury": the proteins denaturation, the lysis (a breaking of the cellular membrane), injuries to the mitochondria and the nucleus [2190, 2516]. Some biological objects are relatively resistant to the solute effects, e.g., the eight-cell embryos [2741]. The membrane behaviour is sensitive to the temperature of the solution crystallization [3336].

The two-compartments model (the compartment models are rather popular in the physiology, for example, they are used to study the volume flow across the amphibian skin, where the number of the compartments—the regions containing the well mixed solutions—could be up to four the in the modern models [1187]) of the cell's dehydration and shrinking (to be more exact, the cell volume change) in the freezing solution [2080] suggested by P. Mazur over forty years ago still forms the basis of most modern models. This model exploits the Kedem-Kachalsky equations for the non-electrolyte transport through the permeable membrane that divides two compartments containing the well-mixed ideal dilute solutions [1583]:

$$J_v = L_p \Delta P - L_p \sigma \Delta \Pi, \qquad j_s = \omega \Delta \Pi + (1 - \sigma) \bar{c} J_v, \tag{2.46}$$

where J_v is the volume flux, j_s is the solute flux, L_p is the hydraulic permeability (filtration coefficient), σ and ω are the reflection and transmission coefficients, ΔP and $\Delta \Pi$ are the drops of the hydraulic and osmotic pressures over the membrane, $\bar{c}$ is the average concentration.

The Eq. (2.46) were derived in the frame of the linear thermodynamics theory of the irreversible processes. The "mechanistic" derivation of the equations describing the transport through the membrane proposed by M. Kargol and A. Kargol [1544–1546] provides a simple interpretation of the coefficients entering into the equations; the membranes having pores of the different radii were also considered in the cited papers.

The osmotic transport of water out of the cell (exmosis) is limited by the membrane permeability that depends on the temperature

$$L_p = L_{p0} \exp\left(-\frac{E_a}{R}\left(\frac{1}{T_0} - \frac{1}{T}\right)\right) \tag{2.47}$$

and serves as a basic parameter of the Mazur's model, here E_a is the activation energy.

The water transport parameters of the cell membrane—the hydraulic permeability and its activation energy are shown to alter due to the cold shock and, to a lesser extent, due to the osmotic shock [733].

Extension of the model to the case of the solution containing an intracellular (permeating) cryoprotective agent adds two more parameters: the membrane permeability for the cryoprotectant and a parameter describing the solute interaction with the water molecules. Generalization of the model to the non-dilute solutions was performed by C.V. Studholm [2967].

Probably, the most general formulation of the problem for the cell in the multicomponent solution given by G. A. Ateshian et al. [132] takes into account the deviation of the segregation coefficients of the solutes from the unity and the mechanical stress in the membrane. It was assumed that the cell is spherical and the mechanical reaction of the membrane has a threshold nature (the surface tension is considered to become zero when the membrane surface area falls down below the specified critical value).

2.3.5.1 Membrane-limited vs. Diffusion-limited Transport

The described zero-dimensional model neglects the diffusion processes in the cell cytoplasm assuming the cytoplasm to be a well-mixed solution. This assumption is valid, for example, for the leucocytes and the lymphocytes, but is not reasonable for the erythrocytes [2529]. A one-dimensional model that incorporate the intracellular diffusion developed by K.A. Carnevale [447] allows to distinguish the dehydration regimes limited by the membrane transport and by the diffusion processes in the cell.

Diffusion inside the cell is govern by the usual parabolic equation for the concentration which for the spherical cell is expresses as

$$\frac{\partial c_{\mathrm{W}}(r,t)}{\partial t} = \frac{1}{r^2}\frac{\partial}{\partial r}\left(D(r,t)r^2\frac{\partial c_{\mathrm{W}}(r,t)}{\partial r}\right) \tag{2.48}$$

The water flux J_w through the cell membrane should be provided by the diffusive flux from the cell's interior. These fluxes, however, differ due to the cell membrane movements. Continuity requirements yield the following boundary condition for the concentration at the membrane surface $r = R$:

$$-D\frac{\partial c_{\mathrm{W}}(r,t)}{\partial r}|_{r=R} = J_w + c_{\mathrm{W}}\frac{\mathrm{d}R}{\mathrm{d}t}. \tag{2.49}$$

The zero diffusive flux is specified at the cell's centre

$$\frac{\partial c_{\mathrm{W}}(r,t)}{\partial r}|_{r=0} = 0. \tag{2.50}$$

The equation for the membrane velocity follows from the incompressibility of the intracellular fluid

$$\frac{\mathrm{d}R}{\mathrm{d}t} = -J_w v_w, \tag{2.51}$$

where ν_w is molar specific volume of water, $J_w = (L_p/\nu_w^2)\Delta\mu$, $\Delta\mu$ is the difference of the chemical potentials across the membrane $\Delta\mu = \mu_{\mathrm{ice}} - \mu_{\mathrm{w}}$ [447].

The incompressibility of the cytosol and conditions under which only water transport across the membrane occurs lead to the on-to-one correspondence between the cell volume and the amount of water contained in the cell

$$c_{\mathrm{w}} = \frac{(V_0 - V_b)c_{\mathrm{w}0}\nu_{\mathrm{w}} - (V_0 - V)}{(V - V_b)\nu_{\mathrm{w}}}, \tag{2.52}$$

where $c_{\mathrm{w}0}$ and V_0 are the initial values of the cell water concentration and the cell volume, respectively, and V_b is the osmotically inactive volume.

In computations K.A. Carnevale set the diffusion coefficient to be constant and estimated its value according to the Stokes-Einstein relation. The apparent hydrodynamic radius of the water molecule equals to $a = 1.4 \cdot 10^{-10}m$, for temperature dependence of the water viscosity Vogel-Fulcher-Tammann-type dependence $\eta \propto \exp(E/(T - T_0))$ was used and the viscosity of the suspension was considered as that of a suspension of the rigid spheres and estimated as

$$\eta = \eta_w \exp\left(\frac{k_E \varphi_{\mathrm{sph}}}{1 - \lambda\varphi_{\mathrm{sph}}}\right), \tag{2.53}$$

where $k_E = 2.5$ is the Einstein shape factor for the spheres, $\lambda = 0.609375$ is the hydrodynamic interaction parameter and φ_{sph} is the volume fraction of spheres. The latter was calculated under the assumption that the diffusing rigid sphere is the hydrated salt ion, thus

$$\varphi_{\mathrm{sph}} = c_s(\nu_s + h\nu_w), \tag{2.54}$$

where c_s is the salt concentration, ν_s is the molar specific volume of salt and h is the effective number of the water molecules in the hydration shell.

The chemical potential of the extracellular water is determined by the Clausius-Clapeyron relationship as

$$\mu_{\mathrm{ice}} = \mu_0 + L\left(\frac{T}{T_0} - 1\right), \tag{2.55}$$

where T_0 is the equilibrium freezing temperature of water and μ_0 is the chemical potential of water at T_0.

Under the assumption of the validity of the Raoult law (i.e. considering the cytosol as the ideal solution) one can write for the chemical potential of the intracellular water

$$\mu_{\mathrm{w}} = \mu_0 + R_g T \ln(\chi_{\mathrm{w}}), \tag{2.56}$$

where the mole fraction of the intracellular water χ_{w}

$$\chi_{\mathrm{w}} = \frac{c_{\mathrm{w}}}{c_{\mathrm{w}} + \sigma_{\mathrm{s}} c_{\mathrm{s}}}, \tag{2.57}$$

where $\sigma_{\mathrm{s}} = 2$ is the salt dissociation constant.

A few characteristic times could be defined for the problem of the cell dehydration. The first one characterizes the duration of the whole process $\delta t_B = T_0$, where $B = -\mathrm{d}T/\mathrm{d}t$ is the externally specified cooling rate.

The characteristic time of the intracellular diffusion is $\delta t_D = R^2/D_0$ where D_0 is the characteristic diffusion coefficient (e.g., at the cell centre), while that of the membrane transport is $\delta t_M = R\nu_{\mathrm{w}}/L_p R_g T$.

Since the cell dehydration involves both the membrane transport and the intracell diffusion, the water loss will be insignificant and the probability of the IIF high if either of two characteristic times δt_D or δt_M will be large: $\delta t_D \gg 1$ or $\delta t_M \gg 1$.

2.3.6 *Intracellular Ice*

When the cooling is fast, the water has not enough time to leave the cell and formation of the intracellular ice causes the cell destruction [2087] mechanically disrupting the cell membrane and the cytoskeleton [3042]; for a more detailed account of the mechanical cell injury see [183]. The intracellular ice formation observed in the experiments is often referred to as the "flashing" or "darkening" of the cells [14, 1201, 1691, 1816]. The lethal character of the IIF is known for a long time [478].

The temperature at which at which the intracellular ice formation occurs and the amount of water within the cell at that moment determine the total amount of the intracellular ice. The conversion of more than 5–10% of the normal intracellular water into ice is considered lethal to most cells while the lesser amounts of ice are often tolerated [3336]. The effect of the nucleation temperature on the thermotropic behaviour of the membrane is usually attributed to the differences in the extent of the cell dehydration.

The probability of the spontaneous nucleation depends on the cytoplasm volume, the degree of supercooling, the solute concentration and the fluid viscosity. Recently, as was already mentioned, P. Mazur et al. [2087] have studied the relation of the amount of the intracellular ice to the fraction of the unfrozen water.

A thermodynamic model of the IIF accounting for the heterogeneous nucleation at the cellular membrane and at the particles contained in the cytoplasm was developed by M. Toner et al. [3065]. The authors starting from the classic nucleation theory[4] as a sequence of bimolecular reactions $\alpha + \beta_m \Longleftrightarrow \beta_{m+1}$ found the rate of homogeneous nucleation I^s as a function of two parameters—thermodynamic Ω and kinetic $\varkappa$ ones:

$$I^s = \Omega \exp\left(-\varkappa/(\Delta T^2 T^3)\right), \tag{2.58}$$

where

$$\Omega = 2(\upsilon^\beta)^{1/3} \frac{(\sigma^{\alpha\beta} kT)^{1/2}}{(\eta \upsilon^\alpha) N_\upsilon}, \qquad \varkappa = \frac{16\pi (\sigma^{\alpha\beta})^3 T_f^4}{(3kL^2)}, \tag{2.59}$$

ΔT is the supercooling, υ^α is the molecular volume of the phase α, $\sigma^{\alpha\beta}$ is the free energy of the ice-water interface, L is the latent heat, η is the cytoplasm viscosity. The latter is approximated as

$$\frac{\eta}{\eta_{\rm w}} = \exp\left(\frac{5}{2} \frac{\varphi_m}{(1 - Q\varphi_m)}\right), \tag{2.60}$$

where $\eta_{\rm w}$ is the viscosity of the water, φ_m is the total fraction of the solutes (e.g., salts and proteins), Q is the interaction parameter. To approximate the viscosity dependence on the temperature either the Vogel-Fulcher law or the polynomial approximation is used.

---

[4] In this theory [2364] the nucleation rate is $J = J_0 \exp(-W^\star/kT)$, where the nucleation prefactor J_0 is related to the molecular mobility and the excess free energy of the critical cluster $W^\star$ is calculated assuming the spherical shape of the cluster and the constant density with a sharp interface (the "droplet" model) as $E = -4/3\pi R^3 \Delta F + 4\pi R^2 \sigma^{\alpha\beta}$, where ΔF is the difference of the free energy per unit volume and $\sigma^{\alpha\beta}$ is the specific surface energy.

The classical theory is based on the assumption that macrscopic materials properties could be used for the energy per unit volume of the new phase and the interfacial energy of the surface that is considered to be flat. The accuracy of the model drops for the very small critical embryos and polar molecules. In these cases, the sharp surface curvature has a significant effect on the interfacial energy.

A more rigorous model based on the density functional approach [2363] takes into account free energy dependence on the spherical density distribution versus radius $\varrho(r)$ rather than just the radius [1149]. The density of the new phase at the centre of the nucleus could differ from that in the bulk of the new phase. The critical density profile is the saddle point that separates small clusters that tend to disintegrate from large clusters that tend to grow.

L. Gránásy [1149] used the density functional approach, viz., single order parameter Cahn-Hillard-type model, to study water nucleation. The previous model of this type based on the piecewise parabolic free energy dependence on the order parameter reproduce some experimental results, but give a discontinuity in the chemical potential, and the results are sensitive to the way of elimination of this discontinuity. Gránásy used a quartic dependence for the free energy and had adjusted the three model parameters using the experimental data for the Gibbs free energy differences between phases as well as computational data on the interface thickness. He was able to reproduce experimental nucleation rates down to 233 K. As the author notes there are two weakness in his theory: it does account neither for the $LDL \Leftrightarrow HDL$ transition with the density change nor the volume change upon freezing.

The interfacial free energy for the ice/water system could be measured by the different methods most of which exploit the Gibbs-Thompson relationship for the depression of the melting temperature due to the curvature of the crystallization front; the recent measurements of the supercooling needed for the freezing in the porous media by W. B. Hillig [1321] gave the value $31 \pm 2.7\,\mathrm{mJ\,m^{-2}}$.

M. Toner et al. [3065] also proposed a modification of the parameters Ω and $\varkappa$ for the case of the heterogeneous nucleation that consists in the introduction into the Eq. ((2.3.6)) factors $(f(\vartheta_0))^{1/6}$ and $f(\vartheta_0)$, respectively, $f(\vartheta) = (1 - \cos\vartheta)^2(2 + \cos\vartheta)/4$, and ϑ_0 is the contact angle between the ice cluster and the inner membrane surface which depend linearly on the solution concentration. The proposed model is highly sensitive to the values of $\varkappa$ and η. Later this model was combined with the model of the transport through the membrane and used to study the cells behaviour in the mixture "water - NaCl - glycerol".

The ice crystals at the end of freezing could be too small to injury the cellular organelles but the recrystallization during the thawing leads to the fatal increase of the crystal size [3329]. At the low temperatures the passive ion transport is essentially unaltered, but the efficiency of the active ion transport through the membrane that is based on the metabolic reactions drops [837]. The cell could be destructed also in the case of the eutectic crystallization [1228].

C. Koshimoto and P. Mazur [1697] drew attention to a significant (eightfold) discrepancy between the cooling rate predictions of the IIF using the model of the kinetic cell dehydration and the results derived from the experimental survival curves versus the cooling rate for the mouse spermatozoa. The somewhat smaller discrepancy between the cooling rates needed to produce the IIF in the mouse spermatozoa and those inferred from the survival curves was reported by R. V. Devireddy et al. [733]. M. R. Curry, J. D. Millar and P. F. Watson found that the calculated optimal cooling rates fail to conform with the observation for the ram and human spermatozoa, with a difference as high as four orders of magnitude in the critical cooling rate for the ram spermatozoa [660].

P. Mazur and C. Koshimoto [2085] list the possible causes for this discrepancy. The first one is the range of the reasonable values for the poorly known critical parameters—the hydraulic permeability of the cell membrane L_p and its activation energy E_a—are large enough to account for the observed discrepancy. The spermatozoa, being the extremely permeable cells with the very low activation energy, are known to be difficult subjects for the L_p measurement [660]. These parameters are also extremely sensitive to the cryoprotective solution composition: e.g., in the mammalian embryos, the hydraulic permeability reduces by a factor of 2 or more in the presence of the intracellular CPA. G. A. Gilmore et al. [1081] have also reported the significant increase of the activation energy for the hydraulic permeability at the low temperatures when the glycerol, the DMSO, the ethylen glycol or propilen glycol is present in the solution. The greater changes are possible at the lower temperatures, but it is difficult to measure the membrane parameters below the freezing point. There are two routes for the water passage into the cell—by the diffusion through the lipid bilayer and through the pores in the membrane.

The porous membranes in comparison with the pure phospholipid membranes have the higher osmotic permeability and the lower activation energy.

M. R. Curry et al. [660] stress that the nucleation temperature needed for the calculations is usually chosen somewhat arbitrarily. The authors also note that the injury could result from the internal freezing in the cell organelles such as the mitochondria rather than the IIF in the whole cytoplasm. If so, the parameters of the organelle's membrane rather than those of the cell membrane should be used. Recently G. J. Morris reported the results of the study of the IIF in the human spermatozoa in the wide range of the cooling rates from 0.3° C/min to 3000° C/min [2190]. Using the cryo scanning electron microscopy (CryoSEM), the author did not find the evidence for the IIF within the resolution limits of the technique. The author asserts that in earlier studies where no direct evidence for the IIF was presented, the drop of the survival curve versus the cooling rate could be wrongly interpreted as the IIF while the cell death was caused by a long period of exposure to the hypotonic solution during warming.

No intracellular ice was observed by F. Fonseca et al. [978] who studied the cryoconservation of the *Lactobacillus delibrueckii* subsp. *bulgaricus*—the lactic acid bacteria widely used in the milk industry—in the glycerol solutions using three freezing protocols and two storage temperatures. The authors' aim was to separate the effects of freezing kinetics and the storage temperature. The authors note that due to the high viscosity of the water-glycerol mixture [2192] the growth of the ice becomes diffusion-limited and the amount of ice formed becomes dependent on the cooling rate.

The cells' damage occurs during warming that includes the crystallization of the residual water that was kinetically arrested during cooling. F. Fonseca et al. found that the recrystallization during warming is especially harmful if the samples were stored at the temperature above the glass transition temperature [978].

The role of the solution effect injury was studied by J. E. Lovelock [1959] and S. F. Mullen et al. [2214] using an imitation of the cell's environment evolution during the cryoaction: increasing of the solution concentration (the "crystallization of the solution", such process—immersion of the cellular material in the hypertonic solution—known as the osmotic dehydration is routinely used in the food industry as a pre-treatment prior to drying of the fruits [2, 1607, 3071]) followed by the solution dilution ("thawing"). The substitution of the lithium salt that causes the more strong lyotropic effect instead of the NaCl results in the more severe cell damage [1959].

J. Wolfe and P. L. Steponkus in their investigation of the mechanical properties of the membrane of the isolated plant protoplasts [3333] also used transfer of the sample to the hypertonic solution ("dehydration") and return to the isotonic solution ("rehydration") to get the cell's volume and the area excursions similar to those occurring during the freeze/thaw cycle. Over short periods (less than a second) the membrane behaves as an *elastic* two dimensional medium. A different behaviour with the resting tension being independent of the area was observed over longer time intervals. Stressing that such mechanical properties are not universal since the protoplast's membrane differs greatly from the membranes of the RBCs and some other cell types, the authors suggested an explanation of the long time membrane behaviour based on the slow exchange of the membrane material with some *reservoir* that provides an "extensive" response of the membrane to the tension variation caused by the osmotic manipulations. As a candidate for the role of such reservoir folds of the membrane or the vesicles are considered [3333].

In fact the biological cells are the *viscoelastic* medium [847, 1560, 2927, 3081]. The viscosity is a measure of a fluid's resistance to flow when subjected to the stress; the elasticity is a measure of a body's ability to return to original dimensions after the application and removal of the stress. A viscoelastic material is one that resists flow while also capable of partially returning to the original dimensions after the removal of the applied stress; the viscoelasticity is the exhibition of both the viscous and elastic properties through the simultaneous dissipation and storage of the mechanical energy [1761, 3196].

In a nonlinear viscoelastic material, the stress is $\sigma = F(\varepsilon(t), t)$ where ε is the time-dependent strain and F is a functional indicating both the properties of the material and the convolving conditions of measurement such as temperature and molecular perturbations; in the linear regime, the stress and time response can be separated so that a general expression for a linear viscoelastic material is $\sigma = \varepsilon G(t)$ where $G(t)$ is the time-dependent modulus of the material. [3196].

The viscoelastic properties of the eukaryotic cell are determined the cytoskeleton consisting of three main polymers: the actin filaments (or microfilaments), the microtubules and the intermediate filaments, all organized into networks whose architecture is controlled by several classes of the regulatory proteins [973]. A. Vo and A. Ekpenyong [3196] found that while the extraction of the model parameters from the viscoelasticity data of many cell types can be done successfully using the integer order mechanical and the power-law viscoelastic models, for some cell types (for example, for the macrophages, which are among the white blood cells that function in the immune system) the fractional order viscoelastic models such as, e.g., the fractional order Kelvin-Voigt model (in the classical integer order Kelvin-Voigt model the total shear stress is represented as the sum of viscous and elastic contributions [879]), are needed.

The cell's counteraction to the volume and the cytoplasm composition changes at the molecular level manifests themselves in the activation of the ion channels and the transport mechanisms or in the regulation of the protein expression [226, 2444].

There is an optimal value of the cooling rate for the cell survival [2082]. The dependence of the vital cells after the freezing versus the cooling rate is frequently called the inverse U-shaped curve or the "cell survival signature". This appearance of the survival signature is typical for the most types of cells; however, there are also exceptions when either the low-cooling-rate or the high-cooling-rate branch of the curve is not distinct such as the yeast cells *Saccharomyces cervisiae* and the *Candida utilis* or the human leukemia K562 cells [816].

The optimal for the survival value is determined by the characteristic time of the removal of the excess heat due to the crystallization τ_T and the characteristic time of the water diffusion through the membrane τ_D, thus the two most important cell parameters are its size (or the ratio of the volume to the surface for the cells which shape is far from the spherical one) and the hydraulic permeability of the membrane

[3013]. If two cells A and B have similar shape and dimension, but the hydraulic permeability of the first is smaller $L_p(A) < L_p(b)$, the optimal cooling rate for the cell survival will be smaller for the cell A.

For the erythrocytes having the disk-like shape and the high permeability the optimal cooling rate is much higher (up to 10^3 times [1530]) then for other types of cells: e.g., the optimal cooling rate for the human erythrocytes is circa 3000° C/min while for the bull spermatozoa it is 1° C/min [2413] and for the mouse embryo—0.5° C/min [2083].

One of the most difficult types of cells for the cryopreservation is the mammalian unfertilized egg (oocyte). The mammalian oocytes are characterized by the large volume, the large amount of cytoplasmic lipids and presence of the MII spindle [2214]. The oocytes are highly sensitive to the cooling and can be injured even at the room temperature due to the delicacy of the miotic spindle apparatus (the oocytes are locked in the Metaphase II of the cell division cycle [2083]) and to the high lipid content of the cells [73]. The MII oocytes are more resistant to freezing injury than the GV-stage oocytes that is attributed to the differences in the properties of the cytoskeleton elements, viz., in the configuration of the microtubules and microfilaments. These elements in the GV-stage are straight and rigid while the microtubules and microfilaments in the MII-stage oocytes are flexible [59]. Thus, hardening of the lipids in the cell membrane during freezing is more readily results in the permanent damage of the rigid GV-oocyte cytoskeleton.

The uniqueness of the oocytes is in that the DNA is held suspended in the cytoplasm on the meiotic spindle and not within the protective confines of the nuclear membrane. The chilling injury may also manifests itself as the thermotropic phase transition in the membrane of the oocyte. The oocytes also are very sensitive to osmotic swelling [2348].

The cryoconservation of the human oocytes is more difficult then conservation of other mammalian oocytes due to the species-dependent effect of the cryoprotectors: the temperature at which the IIF occurs in the human oocytes turn out to be much less depressed by the addition of the cryoprotectants (about 6.5 degrees of Centigrade in comparison with 23 degrees for the mouse oocytes [3076]) indicating, firstly, the very efficient heterogeneous nucleation for these cells, and, secondly, the limitedness of the animal models in the development of the cryoconservation procedures for the human cells. Increasing the seeding temperature was found to provide a narrow window for the human oocytes to dehydrate without incurring the lethal IIF [3076].

The sensitivity of the oocytes and the embryos at the different development stages (oocytes, zigotes, 2-, 4-, and 8-cell embryous) to the cryopreservation was studied by R. T. Pfaff et al. [2449]. The authors measured the water and the DMSO membrane permeabilities and found the significant differences for the 8-cell embryos only. The also found that the presence of the DMSO increases the hydrauvlic pearmeability of the membrane. This effect was attributed to the amphiphilic properties of this cryoprotectant. To summarize the cooling rate effect on the cell survival, the following cases are possible

- $\tau_D < \tau_T$: the excessive dehydration of the cell, the "solution effect";
- $\tau_D \approx \tau_T$: an intra/extra cellular equilibrium is established that is optimal for the cell survival;
- $\tau_D > \tau_T$: the intracellular ice formation;
- $\tau_D \gg \tau_T$: the vitrification.

Cells also differ in the sensitivity to the thawing rate [2079], and this dependence is more sharp in the presence of the cryoprotective agent [2922]; as a rule, the more fast thawing is preferable. The osmotic volume excursions—the shrinkage of cells during the freezing and their re-expansion during the thawing, or during the addition/removing of the CPAs—could result in the damage of some cells' types [2083].

2.3.6.1 Imitation Model

The imitation models are aimed at the simulation of functioning of the large systems. Usually a model includes a description of the possible system states along with some elementary processes that perform a transition of the system from one state to another. A model conserves both the logical structure of such discrete events as well as their casual relationships and, thus, the time sequence of events.

The imitation model does not try to reproduce the real transition rates and thus its predictions are qualitative only. However, such models allow one to gain a general understanding of system working, interaction of its components, and of the relative importance of the different parameters that govern the behaviour of the system. The imitation models, as a rule, are rather fast; frequently, an analytical solution is available for some limiting cases.

A. Zhang et al. [3481] considered a model of the tightly packed cells that comprises of several concentric ring layers of the small malignant cells surrounded by the extracellular shell. The layers are numbered from the outside inwards. The authors stated that four cell layers correspond to the pattern observed experimentally. Due to the smallness of this structures (radius is about 80μm), the temperature was assumed to be spatially uniform, while its variation in time was determined by the solution of the one-dimensional Pennes equation. The evolution of the volume of the first (outer) cell layer is determined by the water transport:

$$\frac{\mathrm{d}V_1}{\mathrm{d}r} = -\frac{L_p A_1 R_g T}{v_\mathrm{w}}\left[\ln x_1 - \frac{L v_\mathrm{w} \varrho}{R}\left(\frac{1}{T_R} - \frac{1}{T}\right)\right] - \frac{L_{p,c} A_2 R T}{v_\mathrm{w}} \ln\frac{x_1}{x_2}, \quad (2.61)$$

where V_1 is the first layer volume, A_1 and A_2 are the permeation surfaces, associated with the first and second layers, v_w and ϱ are the water molar volume and density, L is the latent heat, T_R is the reference temperature, R_g is the universal gas constant. It is assumed that for the cells in the close contact the hydraulic permeability $L_{p,c} = 0.5L_p$, x_i is the mole fraction of the intracellular water in the ith cell layer

$$x_i = \frac{V_{0,i} - V_{b,i}}{(V_{0,i} - V_{b,i}) + \gamma_s n_{s,i} v_w}, \tag{2.62}$$

where V_0 is the initial cell volume, V_b is the osmotically inactive cell volume, γ_s is the dissociation constant for salt, $n_s = C(V_{0,i} - V_{b,i})$ is the mole fraction of the salt ions, C is the cell osmolarity.

The water transport for other cell layers is driven by the concentration differences

$$\frac{dV_i}{dt} = -\frac{L_{p,c} A_i RT}{v_w} \ln \frac{x_i}{x_{i-1}} - \frac{L_{p,c} A_{i+1} RT}{v_w} \ln \frac{x_i}{x_{i+1}} \tag{2.63}$$

for the intermediate layers ($i = 2, \ldots, n-1$) and

$$\frac{dV_n}{dt} = -\frac{L_{p,c} A_n RT}{v_w} \ln \frac{x_n}{x_{n-1}} \tag{2.64}$$

for the central cell region. The authors used the model by M. Toner and E. G. Cravalho [3065] to write the probability of the intracellular ice formation for the innermost (central) cell layer $i = n$ as $P_{IIF} = 1 - \exp(-\tau)$, where

$$\tau = \int_{T_{\text{seed}}}^{T} A\Omega_0 \left(\frac{T}{T_f}\right)^{1/2} \frac{\eta}{\eta_0} \frac{A}{A_0} \exp\left(\frac{-k_0 (T_f/T_{f0})^{1/4}}{(T - T_f)^2 T^3}\right) dt, \tag{2.65}$$

where Ω_0 and k_0 are the cell-type dependent parameters of homogeneous nucleation, A is the surface area, T_f is the freezing temperature of the intracellular fluid whose dependence on the solute mole fraction could be expressed as [3481]

$$T_f = \frac{T_R L}{L + T_R((1 - x_i) - (1 - x_{i,0}))} \tag{2.66}$$

and η is the viscosity defined as [1833]

$$\eta = \eta(T, \varphi_m) = 0.139 \cdot 10^{-3} \left[\frac{T}{225} - 1\right]^{-1.664} \exp\left(\frac{2.5\varphi_m}{1 - 0.609375\varphi_m}\right), \tag{2.67}$$

the volume fraction of the solute φ_m could be approximated by $V_{b,i}/V_i$ and the subscript "0" denotes the isotonic state.

A. Zhang et al. [3481] assumed the propagation of the intracellular ice in the tightly packed cluster of four concentric spherical layers to be instantaneous. Then for the probability of the IIF for the ith cell layer the authors write

$$P_{IIF}(i,t) = (1 - \exp(-2\tau))P_{IIF}(i-1,t) + (1 - \exp(-\tau))(1 - P_{IIF}(i-1,t)) \tag{2.68}$$

The heat transfer problem was considered under the following assumptions:

- the phase transition occurs between 0° C and − 10° C;
- the density of the media does not change upon the phase transition;
- the thermal conductivity is constant, but different in the frozen and unfrozen regions;
- the blood perfusion rate and the metabolic heat generation reduce linearly with temperature down to zero at the crystallization front;
- the specific heat is defined as an effective property incorporating the latent heat;
- the thermophysical properties of the healthy tissues and of the tumour are the same.

In fact, the thermophysical properties of the healthy tissues and of the tumour could greatly differ. For example, in the normal brain tissues changes of the specific heat capacity are minimal as the tissue is heated from 37° C to 61° C. Conversely, in the brain tumour tissues, changes of the specific heat capacity between 37° C and 43° C are substantial, and the difference in the rate of change of the specific heat capacity between brain tumour tissues and normal brain tissues is significant [2707].

The computations by A. Zhang et al. showed that only cells in the first layer suffer the considerable dehydration while for the cells in the inner layers the IIF is more important. In accordance with the observations of J. Hong and B. Rubinsly [1347], the IIF occurs at the lower cooling rates than are typical for the cells in suspension.

Later O.M. Karlsson showed, however, that the assumption about the infinite rate of the intracellular ice propagation is not correct and the finite speed of the intracellular ice propagation should be taken into account [1549].

Even if the propagation is instantaneous, the probability of the IIF depending on the layer number does not make sense. As O.M. Karlsson righteously noted, under this assumption the intracellular ice will appear in all cells simultaneously upon the first nucleation event in any cell belonging to the cluster and the total nucleation rate in some cell in the cluster will be given by the sum of the IIF probabilities in the individual cells

$$J(t) = \sum_{j=1}^{n} \sum_{i=1}^{m_j} J_{i,j}(t), \tag{2.69}$$

where it has been assumed that the number of cells in the jth layer is m_j. If the nucleation could be described as the Poisson process with an average rate $J(t)$, the cumulative probability of the ice nucleation in the cluster will be expressed as

$$P_{\text{cluster}}^{IIF} = 1 - \exp\left[-\int_0^t J(t)\mathrm{d}t\right] \tag{2.70}$$

D. Irimia and O. M. Karlsson [1405] considered a pair of the contacting cells as a system with a finite number of states, viz., the three-state system:

(0) there is no ice in the either of cells;
(1) there is ice in one of the cells;
(2) there is ice in both cells.

The state of the ensemble of cells could be described by the two parameters, e.g., by the probability of "unfrozen" $P_0 = N_0/N$ and the "singlet" $P_1 = N_1/N$ states (clearly, the probability of the "duplet" state $P_2 = 1 - P_1 - P_2$). An evolution of the cell ensemble is governed by the two independent stochastic processes—a spontaneous nucleation of the intracellular ice J_s and the ice propagation through the intercellular contact J_p. Evidently,

$$\frac{\mathrm{d}N_0}{dt} = -2J_s N_0 \tag{2.71}$$

and

$$\frac{\mathrm{d}N_2}{dt} = (J_s + J_p)N_1 \tag{2.72}$$

Introducing a dimensionless time

$$\tau = \int\limits_0^t J_s dt \tag{2.73}$$

and a dimensionless rate of ice propagation $\alpha = J_p/J_s$, Eqs. (2.3.6.1), (2.72) could be written as

$$\frac{\mathrm{d}}{\mathrm{d}\tau}\begin{bmatrix} P_0 \\ P_1 \end{bmatrix} = \begin{bmatrix} -2, & 0 \\ 2, & -(1+\alpha) \end{bmatrix}\begin{bmatrix} P_0 \\ P_1 \end{bmatrix} \tag{2.74}$$

It can easily be seen that $P_0(\tau) = \exp(-2\tau)$. In the case of constant α an analytical solution for P_1 is written as [1405]

$$P_1(\tau) = \begin{cases} \frac{2}{1-\alpha}(\exp(-(1+\alpha)\tau - \exp(-2\tau)), & \alpha \neq 1 \\ 2\tau \exp(-2\tau), & \alpha = 1. \end{cases} \tag{2.75}$$

Maximal probability of the singlet state

$$P_1^\star = \left(\frac{1+\alpha}{2}\right)^{(1+\alpha)/(1-\alpha)} \tag{2.76}$$

is achieved at

$$\tau^\star = \frac{1}{1-\alpha}\ln\left[\frac{2}{1+\alpha}\right] \tag{2.77}$$

The ice formation and propagation in the aggregates of cells was considered under assumption that the rate of nucleation in the current cell is defined as

$$J = J_s + mJ_p, \tag{2.78}$$

where m is the number of the neighbour cells [1406].

The total number of states M in the system containing N cells comprising all distinct combinations of states (either frozen or unfrozen) of the individual cells

$$M = 2^{N-1} + 2^{(N-1)\div 2}. \tag{2.79}$$

In the ensamble of the identical cell aggregates an evolution of the vector $\mathbf{P}$ of the probabilities P_ξ of each state ($\xi = 0, \ldots, M-1$) is described as

$$\frac{d\mathbf{P}}{d\tau} = \mathbf{Q}(\alpha)\mathbf{P}, \tag{2.80}$$

where the matrix $\mathbf{Q}$ contains all the rates of transitions between the different states. This equation is sometimes is referred to as the Kolmogorov differential equation [1406].

An analytical solution for the case $N = 4$ was presented in the cited paper.

The formation of the intracellular ice for large N was simulated using the Monte-Carlo method. The tissue sample is described by a regular lattice in which each lattice site represent a cell that could be in one of two states (frozen/unfrozen). The ice propagation occurs between the immediate neighbours. Assuming that the ice propagation could be described by a Poisson process with a rate given by (2.78), D. Irimia and J. O. M. Karlsson write out the probability that an unfrozen cell j will freeze within a time interval Δt as

$$P_j^{IIF}(\Delta t) = 1 - \exp\left[-\int_t^{t=\Delta t}[J_j + m_j(t')J_p(t')]\mathrm{d}t'\right], \tag{2.81}$$

where m_j is the number of the nearest neigbours of the cell j that are frozen.

If one further assumes that both α and m_j are constant during this time interval Δt, one get a *homogeneous* Poisson process

$$P_j^{IIF}(\Delta t) = 1 - \mathrm{e}^{-(1+m_j\alpha)\Delta t}. \tag{2.82}$$

To enforce the validity of the assumption on the constancy of m_j during the time interval Δt, the authors required the probability of the IIF occurring anywhere in the sample during this time interval to be bounded from above by some small probability ε leading to the constraint on the time step

$$\Delta t \leq \left\{ \frac{-\ln(1-\varepsilon)}{\sum_j (1+m_j\alpha)} \right.$$

where the sum is taken over all unfrozen cells in the sample.

For time intervals $\Delta\tau$ satisfying this equations, due to the smallness of the probability of the simultaneous spontaneous nucleation (s) and the ice propagation from the neighbour cell (p), these mechanisms are considered as independent to yield

$$P_j^s(\Delta\tau) = 1 - \mathrm{e}^{-\Delta\tau} \approx \Delta\tau \tag{2.83}$$

$$P_j^p(\Delta\tau) = 1 - \mathrm{e}^{-m_j\alpha\Delta\tau} \approx m_j\alpha\Delta\tau, \tag{2.84}$$

respectively, and the total probability of the IIF in the cell j is approximately

$$P_j^{IIF}(\Delta\tau) \approx P_j^s + P_j^p \approx (1 + m_j\alpha)\Delta\tau. \tag{2.85}$$

The authors thus choose the time step for the Monte Carlo computations as

$$\Delta\tau = \frac{\varepsilon}{(1 + n\alpha)N}, \tag{2.86}$$

where the maximum number of the nearest neighbours m is $m = 2$ for the linear cell arrays and $m = 4$ or $m = 6$ for the two-dimensional cell aggregates, depending on the cell shape and packing. In computations either one or two independent quasi-random numbers were used, depending on whether the distinction between different nucleation mechanisms is to be monitored.

The one dimensional array of 100 cells has been considered by Irimia and Karlsson while Stott and Sumper [2957, 2975] solved the two-dimensional problems for the number of neighbours $m = 6$ and $m = 4$.

When the domain contain both pathological ("t") and healthy ("h") tissues, the nucleation rate is written as [2957] $J = J_s + m^t J_p^t + m^h J_p^h$. A new parameter that characterize the ratio of the ice propagation speeds in two kinds of tissue arises $\beta = J_p^t / J_p^h$. Computations for a wide range of the ice propagation speed in the healthy tissue revealed three growth regimes differing by the dependence of the pathological tissue destruction on the parameter β [2957].

Large scale Monte Carlo computations were performed by S. L. Stott et al. [2958] accounting for the characteristic length scale of the thermal gradient in the vicinity of the cryoprobe for a two-dimensional tissue sample containing 10^4 cells.

A sample was represented by a lattice of 99 × 99 cells with each interior cell interacting with four neighbours. The computational domain was divided into two regions corresponding to the tumour and healthy tissue. The centrally located tumour occupied one half of the domain. Similarly to the previous analysis, the dimensionless rates of the ice propagation were introduced

$$\alpha^h = \frac{J_p^h}{J_s^h}, \qquad \alpha^t = \frac{J_p^t}{J_s^h}, \qquad \gamma = \frac{J_s^t}{J_s^h}, \tag{2.87}$$

the latter being the dimensionless rate of the spontaneous II nucleation. The use of the same denominator in the definition of αs allowed one to define the relative amplitude of the ice propagation rates in the healthy and tumour tissues β (see above) that can also written in terms of the dimensionless propagation rates α^h and α^t as $\beta = \alpha^t/\alpha^h$.

For the heterotypic cell contact the author used the slowest propagation rate $J_p^{jk} = \min(J_p^h, J_p^t)$.

To imitate the effect of the spatial temperature non-uniformity around the cryoprobe, the authors have assumed that both the spontaneous nucleation rates and the ice propagation rates decay exponentially with the distance from the cryoprobe, i.e.

$$J_s^{jk} = J_s^h e^{-\frac{r_{jk}}{\lambda}} \qquad \text{and} \qquad J_s^{jk} = J_s^t e^{-\frac{r_{jk}}{\lambda}} \tag{2.88}$$

for the cell jk in the healthy and tumour regions, respectively, and similarly

$$J_p^{jk} = J_p^h e^{-\frac{r_{jk}}{\lambda}} \qquad \text{and} \qquad J_p^{jk} = J_p^t e^{-\frac{r_{jk}}{\lambda}} \tag{2.89}$$

for the homotypic interactions within healthy or tumour regions and

$$J_p^{jk} = \min(J_p^h, J_p^t) e^{-\frac{r_{jk}}{\lambda}} \tag{2.90}$$

for the boundary between the tumour and healthy tissues. Here the dimensionalless distance of the cell jk from the surface of the cryoprobe r_{jk} is related to the effective tumour radius R

$$r_{jk} \equiv \frac{\sqrt{x_{jk}^2 + y_{jk}^2}}{R}, \tag{2.91}$$

and λ is a parameter that specifies the rate of the decay of the nucleation and the propagation rates away from the cryoprobe (to the e^{-1} at the distance λ from the probe).

The above described Monte Carlo algorithm is based on the discrete nucleation events. It is possible to simulate the "continuous" nucleation exploiting the approach developed by D. T. Gillespie for the general chemical kinetics [1078]. To accomplish it, a joint probability function on the discrete variable j specifying the cell where the next IIF event will occur and on the continuous variable $\delta\tau$, representing the time interval until the occurrence of this IIF [2975]. To generate a pair of these random variables, two random numbers r_1 and r_2 uniformly distributed over the unit interval should be draught to permit the estimation of $\delta\tau$ as

$$\delta\tau = \frac{1}{a_0} \ln\left(\frac{1}{r_1}\right), \tag{2.92}$$

where a_0 is the specified parameter of the probability density function—the overall rate of the intracellular ice formation in the tissue—while the cell number j where the IIF will take place is determined by the following inequalities [2975]

$$\frac{1}{a_0}\sum_{m=1}^{j-1} h_m c_m < r_2 < \frac{1}{a_0}\sum_{m=1}^{j} h_m c_m, \tag{2.93}$$

where h_m is the state parameter in the Gillespie algorithm and c_m is the rate of conversion of the mth cell from the unfrozen to frozen state.

Sumpter [2975] notes that typical size of the cluster that is practically reasonable to simulated by the described Monte Carlo algorithm is about 10^5 cells that correspond to the tissue size 1–10 mm^2 in the biological sample and suggested a modification of the above mentioned continuous Johnson-Mel-Avrami-Kolmogorov model that describes the kinetic of the phase transformation by the evolution of the transformed volume fraction $X(t)$ in time as $X(t) = 1 - \exp(-kt^n)$ where k and n depend on the geometric factors and on the mechanisms of the phase transition.

Chapter 3
Whole-body Models

3.1 Heat Transfer in Evolution

The ability to sense, respond and adapt to the external environmental threats supported the evolution of life, from a single protocell unit to the vertebrates; this innate property of the thermo-plasticity forms the underlying physiological functions that allow to live under extremes of the environmental temperature [1089, 1873].

The heat provided the energy that drives the organization and formation of the protocells from the polypeptides, which evolve into the unicellular and multicellular organisms over time [995, 2215].

The role of the thermal gradients above the submarine hydrothermal vents (SHV) (a hydrothermal vent is a fissure—a fissure vent, also known as the volcanic fissure, the eruption fissure—on the seabed from which the geothermally heated water discharges; the seafloor hydrothermal vents are found by searching the bottom waters for the hydrothermal plumes drifting up and away from the vents, and by imaging the seafloor, using the deep-towed or the on-bottom sonars and cameras) [197] in the life evolution was analysed by A. W. J. Muller [2218, 2219] who considered the major stages of the evolution as the appearance of

- unicellular prokaryotes,
- unicellular eukaryotes,
- multicellular eukaryotes,
- metazoans (the multicellular, mitochondrial eukaryotes—the animals with the differentiated tissues, including the nerves and muscles; the metazoans are the heterotrophic animals—they consume other organisms or the products of other organisms since they cannot produce their own energy).

A. I. Zhmakin, *Heat Transfer in Vivo*, Springer Series in Biophysics 27,
https://doi.org/10.1007/978-3-031-95459-7_3

The author proposed the general conjecture that several major transitions of the evolution were effected by the thermal processes:

- the first organisms used the heat engines during the thermosynthesis in the convection cells;
- the first eukaryotic cells used the microtubules during the thermosynthesis in the thermal gradient above the SHV;
- the first metazoans used the transport of water or in water during the thermosynthesis above the SHV under an ice-covered ocean;
- the first mammalian brain used a thermal machinery based on the thermal gradients in or across the cortex.

The emergence of the photosynthesis on the early Earth (for example, the appearance of the plants) is related to the thermosynthesis that involves the biological heat engines working on the thermal cycling; the required thermal cycling is the effect of suspension in the convecting water such as present in the volcanic hot springs [2217, 2220, 2221].

The thermosynthesis is based on the chemiosmosis—the mechanism of the ATP synthesis [646, 2154, 2155, 3319]—that uses the biological heat engines working on the thermal cycling: during both the photosynthesis and the respiration protons are pumped across a membrane, which stores the free energy generated; the ATP synthase enzyme (this enzyme contains two moieties, the membrane bound F_0 part and the water-immersed F_1 part) transduces the free energy of this proton motive force into the free energy of ATP. During the respiration, the electrons move from a reductant to an oxidant while simultaneously protons are pumped across a membrane; during the photosynthesis, the electrons excited by the light also drive the proton pumping [2218].

In the thermal gradient above a hydrothermal vent, a thermometer performs a relaxation oscillation: while warm it folds, but since the relaxed the folded and relaxed unfolded protein have different shapes, the new conformation is under strain. It cannot instantaneously relax to the new equilibrium shape.

A native protein is folded—in the core of a protein, apolar amino acid side chains interdigitate and are tightly packed, forming rigid contacts. In contrast, the unfolded state of the protein is more flexible and fluid-like—the bonds in the amino acid side chains rotate relatively freely in an unfolded protein, and in the ideal case all amino acid side chains are completely exposed to the solvent.

After relaxation by expansion, the thermometer comes in contact with the cold ocean water and cools. The cold denaturation changes the conformation. The protein is again first in the strain, and then relaxes by the contraction. An attached “First Protein” that works on the cold denaturation permits the synthesis of the ATP from the thermal cycling caused by the oscillating thermometer. The proposed mechanism of the first protein is a thermal variation of the binding change mechanism of the ATP synthase [2216] and is called the ‘protein-associated thermosynthesis’.

A. W. J. Muller suggested that the heat engines—the systems where the heat transport down a thermal gradient is accompanied by the generation of the high free energy substance such as the ATP [2218]—using the more complex mechanism for the thermal cycling are invoked to explain the origin of the animals as well [2218]. The author noted that almost any temperature-dependent material characteristic is applicable in a heat engine. The convection cell, in which the rotating fluid does the mechanical work, is an example of the heat engine can even be self-organizing—the convection sustains the weather in the atmosphere, the ocean currents in the hydrosphere, the plate tectonics in the lithosphere, and probably the geodynamo in the Earth's core that generates the Earth's magnetic field [1341, 2615, 3477].

The thermosynthesis could be based on the relaxation oscillations in the thermal gradient between the submarine hydrothermal vent and the ocean resulted from large convection cells:

- the convection cells in the Earth's mantle that drive plate tectonics;
- the ocean water flow resulted from the thermohaline circulation in the ocean—the ocean currents move along the wind-driven path [3358] along the ocean's surface from the tropics to the poles where the water cools, sinks, and returns along the ocean floor–passing by the hydrothermal vents—to the tropics.

For example, Muller described the evolution of the flagellum-based heat engines that worked on a thermal gradient—in a bacterium carried along by the convection above the hydrothermal vent, the proton pump is driven by the thermal cycling. The resulting proton gradient drove the ATP synthase; this mechanism could work during the global glaciation since it does not require the sunlight [2215].

The emergence of the muscle can be explained similar to the emergence of the bacterial flagellar motor. A global glaciation again yielded the long evolutionary time required for the emergence of a thermosynthesis driven process where the ATP-consuming first protein evolved into the myosin and the thermometer into the actin [2218].

For the understanding the coordination of the cell division timing the cell temperature is an important question in the field of the developmental biology [557]. The temperature is the control parameter of the cell cycle duration since it can accelerate or decelerate the rate of biochemical reactions. However, the experiments at the cellular scale are difficult due to the limited availability of the biocompatible temperature sensors and the lack of the practical methods to control the local temperatures and the cellular dynamics. J. Choi et al. [557] suggested a method to probe and control the cell division timing using a combination of the local laser heating that produces a temperature gradient across the embryo and *in vivo* nanoscale thermometry using the quantum defects in the nanodiamonds.

The authors used a non-invasive temperature control method to study and manipulate the early embryonic development of the free-living transparent nematode about 1 mm in length that lives in temperate soil environments *Ceanorhabditis elegans*. The resulting temperature distributions were monitored in real time using the nanoscale quantum thermometers [1785, 2160, 3353] in which the nitrogen-vacancy centres in the biocompatible nanodiamonds are used to measure the local temperature with the high spatial resolution (down to 200 nm [1716]). The optical stability of the nitrogen-vacancy centres circumvents the photo-bleaching drawbacks of the existing fluorescence-based measurement techniques, allowing the continuous monitoring of the temperature gradients over the extended periods of time.

In order to explain the measured cell cycle times, the authors introduced two competing hypotheses: does the early embryonic cell

- follow the local, nucleus temperature at the heating spot or
- follow the mean temperature averaged over the cell volume?

They found that the cell cycle time is not solely determined by the DNA replication occurring in the cell nucleus, but is also dependent on the kinetics of cyclins and other proteins throughout the cytoplasm. The authors also stated that the cell cycle timings are controlled in a cell autonomous manner, with no significant contribution from the cell-to-cell communication.

As life evolved from the multicellular organisms into the biological organs and systems, the ability to regulate temperature within a physiological range remains a key requirement for the survival. The thermoregulatory system plays the vital role in supporting the human life [1269].

3.2 Humans

The human thermophysiological response is investigated at the different levels. Probably the first model was developed by A. C. Burton [407] in 1934 who considered the human as a single homogeneous cylinder with uniform metabolic heat generation. The most modern models consider the whole body as a combination of the multiple compartments [3525].

The whole body model by L. Zhu et al. [3533] uses the representation of the Pennes perfusion source term to assess the overall thermal interaction between the tissue and blood in the human body. The study is motivated the need to develop safe approaches for the body temperature reduction. In clinical practice, the most effective way to control the temperature in the critically ill patients is to directly cool the blood in the major veins using the intravascular cooling catheters [663]. Cooling rates as high as 5° C/hour can be achieved depending on the cooling capacity of the device and the patient's cooling response [794, 1399].

Based on the Pennes equation, the authors express the rate of the total heat loss from the blood to tissue as

$$Q_{blood-tissue} = \int\limits_{V_{body}} \varrho_{\mathrm{b}} c_{\mathrm{b}} \omega_b (T_{\mathrm{a}} - T_t) \mathrm{d}V = \varrho_{\mathrm{b}} c_{\mathrm{b}} \bar{\omega}_b (T_{\mathrm{a}} - \bar{T}_t) V_{body}$$

where $\bar{\omega}_b$ is the volumetric average blood perfusion rate defined as

$$\bar{\omega}_b = \frac{1}{V_{body}} \int\limits_{V_{body}} \omega_b \mathrm{d}V,$$

$\bar{T}_t$ is the weighted average tissue temperature that may not be equal to the volumetric average tissue temperature if the perfusion rate is not uniform in the body tissue.

The blood perfusion rate is defined as the amount of the blood supplied to the certain tissue region per minute per 100 g tissue weight. In the humans, there are several tissue regions, such as the kidney, the heart, and the choriocapillaries in the eye (that provide the transport of oxygen and nutrients to the retina and stabilize the temperature environment of the retina and retinal pigment epithelium [139]), possessing the high blood perfusion rate: the measured blood perfusion rate in the kidney is approximately 500 mL/min/100 g tissue, in the heart is around 300 mL/min/100 g, in the choriocapillaries the blood perfusion rate can be as much as 8000 mL/min/100 g tissue [3525].

The measurement of the blood flow is an important physiologic study of the humans. Many methods are utilized in the measuring the tissue blood flow, the most often one used is based on the injection of the radioactively labelled microspheres due to the shortcomings of the alternative methods of blood flow determination. The Doppler ultrasound is used to provide measurements of the average flow velocity in the large to medium-size s if the diameter is known. It is based on the average frequency shift of the Doppler spectrum that is proportional to the average particulate velocity over the cross-sectional area of the sample. The laser Doppler flowmetry (LDF) detects the frequency shift of the laser light imparted by moving the red blood cells in the blood stream—the incident light is carried to the tissue by the fibre optic cables, where it is scattered by the red blood cells; sampling the reflected light, it is possible to calculate the flux of the red blood cells. The variants of this technique are bidirectional laser Doppler velocimetry and Fourier domain Doppler optical coherence tomography. The Doppler shift indicates the speed of the optical scatterers, but applying the Doppler approach to small s is challenging since there are far fewer blood cells at the capillary level.

A. Joseph et al. [1499] performed the imaging of the single-cell blood flow in the smallest to largest s in the living retina. The retinal neurons have exceptionally high metabolic activity and demand timely and adequate nutrition and waste removal [2493], thus, the retinal circulation supplies the bulk of the inner retina with a stratified network of s, while the choroidal circulation supplies the photoreceptors. The the inner retina is typically fed by a single central retinal artery while return is mediated by the central retinal vein in the holangiotic retina. Understanding perfusion kinetics at this scale is critical to understanding the entire connected network, since blood flow dynamics are tightly linked to physiologically relevant neural loads [1499].

The functional MRI is one of the most popular imaging techniques for measuring blood flow in the brain, but it can only achieve a resolution of around 1 millimetre, and the tiniest capillaries measure less than a hundredth of a millimetre across. These s can be examined surgically, or by injecting dyes into the bloodstream, but these techniques carry a risk [1499]. The retina is part of the brain and, because the eye is transparent, its blood supply is much easier to see. Joseph et al. used 15 kHz adaptive optics (AO) imaging to noninvasively measure single-cell blood flow and measured flow across the full spectrum of retinal diameters (3.2–45.8 mm). The authors used the living mouse to benchmark the automation of blood velocity data—the challenge of imaging mouse retinal blood flow is attributed to the difficulties of imaging its rather small eye, with even the largest s being only a quarter the size of the largest human retinal .

The model developed by E. H. Wissler sixty years ago was based on the mapping the temperatures all over the body [3326]. This model was based on the presentation of the human body by a multi-segmental model—the six cylinders and included the simulations of the oxygen and the carbon dioxide mass exchange, the lactate production and distribution along the body. The model took into account the heat loss through the respiratory system and the countercurrent heat exchange between the large arteries and the large veins. The vascular system was composed of the arteries, veins and capillaries and the blood temperatures of the arteries and veins were assumed to be uniform. The blood circulated from the heart (in the upper part of the trunk) through the arteries to the arterial pool of each segment and further into the capillaries of each segment. The connection between the elements was only through the macrocirculation. The passive system of the Wissler's model was very detailed compared to the previous models.

The rather sophisticated thermal model with the temperature grid in the body (1 cm step for the trunk, 0.5 cm for other parts) represented by 400,000 points was developed by J. Werner [3296].

The human thermal models could be used to evaluate [2374] the critical exposure to the high or low temperatures, the unpleasant submersions in the cold or hot fluids, the whole-body discomfort, the exposure time period limits, the sweat rate and the dehydration level, the overheat process, the individual heat tolerance limits etc. The advantage of the simulation is the remove of the need to submit people to the unpleasant and/or unsafe conditions; the models are low cost comparatively to the experimental research.

Several thermal indexes [237, 867, 1178, 2180] are introduced to characterize the physiological stress that refers to the sum of physiological demands for an adaptation to maintain homeostasis and to protect the survival of the host [2782]. The heat stress is defined as the total heat load that the thermoregulatory system has to cope with to maintain physiological homeostasis and is indicated by the peak core temperature. The "heat" term is also used to described the organism states that only indirectly related to the temperature distribution (e.g., the *heat cramps*—the painful spasms of the muscles due to the profuse sweating are caused by the salt deficiency [317]). For example, the heat stress issues are important in the mining industry of many countries [1522, 3354].

The heat stress index developed by H. S. Belding and T. F. Hatch [237] and modified by T. F. Hatch [1259] and B. F. Hertig and H. S. Belding [1308] is the ratio of the total evaporative heat loss from the skin surface required for thermal equilibrium (the sum of metabolism and dry heat load) to the maximum evaporative heat loss possible for the environment, multiplied by 100, for steady-state conditions and with skin temperature held at 35° C.

The wind chill index (the wind chill index (WCI) is an empirical index developed from cooling measurements obtained in Antarctica on a cylindrical flask partly filled with water; it describes the rate of heat loss from the cylinder by radiation and convection for a surface temperature of 33° C, as a function of ambient temperature and wind velocity).

The physiological effective temperature was developed in 1920s as an empirical sensory index, combining into a single value the effect of the temperature and the humidity on the thermal sensations, with an adjustment for the effects of the air movement [2069, 2095]. Later a new effective temperature scale was developed defined as follows [332]: $ET^{\star} = T_0 + wi_m LR(p_a - p_{ET^{\star},s})$ where T_0 is the operative temperature (an average of the mean radiant and ambient temperatures weighted by their respective heat transfer coefficients), w is the skin wettedness, i_m is the moisture permeability index, LR is the Lewis ratio—the evaporative heat transfer coefficient, P_a is the ambient partial vapour pressure.

The $ET^{\star}$ is the temperature of the environment at 50% relative humidity that results in the same total skin heat loss as in the actual environment. This index combines the effect of three parameters (the mean radiant temperature, the dry-bulb or air temperature and the ambient partial vapour pressure) into a single index.

PMV (Predicted Mean Vote) method that combines four physical variables (the air temperature, the air velocity, the mean radiant temperature, and the relative humidity), and two personal variables (clothing insulation and activity level) used to establish the ASHRAE thermal sensation scale [117]: hot, warm, slightly warm, neutral, slightly cool, cool, cold. H. Zhang extended this scale, adding "very hot" and "very cold" [3486, 3487].

The heat strain and the heat stress are used interchangeably sometimes to describe a state of the positive heat storage in the body, although these are different indexes of the thermoregulation [1873]. From a physiological perspective, a "strain" refers to a stretch or departure from an original state of existence. Therefore, the heat strain is defined as the magnitude of departure from the resting core temperature, i.e., the difference between the peak and the resting core temperature.

The human thermal comfort—the function of both the psychological and the physiological factors—is influenced by the environmental and the physical factors such as the air temperature, the humidity, the air movement, the physical activity, and clothing; the physiological measures include the blood pressure, the heart rate, and the respiration rate. E.g., in the design, production and use of the helmets, thermal comfort is one of the important factors that must be considered [324, 326, 1066, 1216, 3117, 3405]—the good thermal comfort is a necessary condition for the wearer to work properly [1285]. However, due to the differences in the individual physiological attributes [325, 326, 1066, 1216], the thermal sensation varies from person to person [977, 1993, 2689, 3523]. The studies [1264, 1981, 2291] showed that the thermal sensation and the thermal comfort of the human body are strongly related to the internal heat balance—how the bioheat of metabolism is produced in the living tissue and transferred to the skin.

The respiratory heat losses are related to the minute ventilation and to the meteorological parameters such as the atmospheric temperature, the atmospheric pressure and the relative humidity and are mainly determined by the temperature difference between the air in the upper airways and the environmental air [291, 2845]. The total heat losses through the respiration of the human body demonstrate the considerable variations seasonally depending on the climate zones—the highest heat losses are determined in the region of the continental climate while the lowest in the region of the Mediterranean climate [292].

The human thermal comfort is defined as "the condition of mind which expresses satisfaction with the thermal environment" [921] or as the state where the thermoregulatory effort is minimized [1987].

However, there is no accurate method to quantitatively determine the effect of the thermal environmental condition on the comfort feeling in a human being, thus, there is a need to develop the objective measure of the thermal stress that would quantitatively provide an indication of the thermal comfort in the form of a single number.

S. C. Boregowda [332, 333], following the research by I. Aoki [98, 99] who was the first to use the concept of entropy in the human thermal physiology calculating the entropy production in the human body under the basal conditions and during the exercise, introduced an Objective Thermal Stress Index (OTSI) based on application of the second law of thermodynamics to the human thermal system to quantify the thermal stress or the discomfort in the human body. The stress has long been the subject of the psychological and the physiological interpretation. In fact the word "stress" itself is not well-defined and has been overused, meaning the different things to different people; in the engineering, stress is considered as a reaction to what is perceived to be a threat to either the individual's security or his/her accomplishment of the assigned task [332].

A real human adapts into changing conditions in the environment by means of the thermoregulatory system that displays some inertia. There is an old experiment described by John Locke in 1690: "A person places one hand in a basin of warm water and the other in a basin of cool water. After a short time, both hands are placed together in a third basin of water, which is at intermediate temperature. The hand previously in warm water feels cool and the hand previously in cool water feels warm even though they are actually at the same temperature" [1343].

The human beings have two distinct subsystems to regulate the body temperature [3293]:

- the physiological thermoregulation that is capable of the precise adjustments of the heat balance but is effective only within a narrow range of the environmental temperatures;
- the behavioural thermoregulation, through the use of the shelter, the space heating and the clothing, enables the humans to live in the most extreme climates on the earth, but it does not provide the fine control of the body heat balance.

The knowledge of the human thermoregulatory system is needed to develop the protective "microclimates", i.e. the garments designed for work in the hostile environments, for example, the garments of the fire fighters should protect them from the fire but prevent the overheating that could result in the heart attacks [494, 1342] or the infrared-transparent visible-opaque fabric (ITVOF) which provides passive cooling via the transmission of the thermal radiation emitted by the human body directly to the environment [3067, 3410]. The diving suits must reduce the heat loss to the cold water. Additionally, the excessive heat loss to the breathing gas must be avoided especially in the case of the helium-oxygen mixtures used in the deep diving due to the high thermal conductivity of the helium.

When the body temperature rises, the heat balance of the body is restored by the increased blood flow to the skin (under thermal stressing, the skin blood flow can rise to levels more than 50-fold its value [1507]; the amount of blood in the deep dermal plexuses may exceed that of the papillary capillaries by a factor of 30 [617]) and by the sweating. These responses increase the work of the heart and cause loss of the salt and water from the body and can overload the heart and cause the hemoconcentration, which can lead to the coronary and the cerebral thrombosis [1202]. These effects of the thermoregulatory adjustments occur with even mild heat loads and account for a great majority of the incidences of the heat-related illness and death [786].

An effect of the sweating on the skin temperature is illustrated by D. C. Sanyal and N. K. Maji [2709] who found that the temperature at an environment temperature of 15° C without sweating is higher than that in the 25° C environment with a sweating rate of 10^{-3}g/cm^{2}s due to the latent heat of evaporation.

In order to prevent heat stress and extend tolerance time, a cooling garment needs to be worn under the protective clothing. Cooling garments use circulating liquid and/or gas, or a phase changing substance. Liquid cooling garments (LCGs) can also be used in medical applications to control the core and skin temperatures, the cutaneous blood flow, and the sweat rates during certain procedures [3306].

An alternative classification of the environment temperature effect on the humans could be based on the clinical definitions of the heat (the heat stress—the physiological strain and the perceived discomfort; the heat exhaustion—an illness, of the mild to the moderate severity, where the core body temperature is abnormal but below 40° C; the hyperthermia—the heat dissipating mechanisms are impaired with a rise in the core body temperature above the hypothalamic set point over approximately 40° C; the heat stroke [1028]—a severe illness event caused by exposure to the heat and/or during the physical activity where the core body temperature has exceeded 40° C and the noticeable impact on the central nervous system, e.g., delirious, convulsions), including the fatal form of the heat injury to the central nervous system [630, 2803]—and cold (the frostbite, also called the freezing cold injury [2142], the damage may be temporary or permanent, gangrene and loss of digits or limbs may result [2391]—the temperature is below the point at which the skin tissue begins to freeze (while 0° C is considered the freezing point of water, the freezing point of the skin is marginally less due to the electrolytes; the observed freezing points range from − 4.8° C to − 0.6° C [1582, 3322]); nonfreezing cold injuries (NFCI)—an array of the injury events where tissue freezing has not occurred but the damage occurs [463] (the hypothermia injuries [2491]).

The whole body models are needed to understand the thermal behaviour of the body under the different environment conditions and the different activity levels, for example,

- to asses a thermal comfort of the occupants in the vehicle [957, 1178],
- to assess the effect on the human in the military and civilian emergency situations, such as, fo example, the disabled submarine scenarios that often involve the prolonged exposures to the warm and the very humid environments [532, 2731],
- to assess the occurrence of the heat illnesses that will endanger the physical health and life safety during the fitness exercises and the training competitions in the hot environment [3364, 3536],
- to estimate the effects of heat on the military operations (in about 400 BC, Herodotus was the first to mention this issue [1110]),
- to assess the possibilities of the human adaptation to the global warming [1238, 1873], especially in cities that are classified as the urban heat islands (UHIs) [1860] due to the higher environment temperature than in the rural areas due to heat radiated from the building and the road surfaces [2756, 2807],
- to develop the heating, ventilation and air-conditioning (HVAC) systems for the comfortable, the healthy and safe indoor environments [3432].

The whole body models could be useful in the analysis of the behaviour of the energy-efficient buildings, and particularly in the interaction between the indoor air quality, comfort, ventilation and the energy consumption including the increased insulation levels causing the over-heating problems during the summer and leading to installing an electricity-consuming cooling system [1343, 2785, 3262].

To operate the HVAC systems effectively, the feedback signals—the human thermal physiology signals and thermal comfort/discomfort status—are needed [3432]. The traditional sensor based the environmental parameters' measurements were not accurate enough because the sensors are normally located in one location and the indoor environmental parameters are spatially non-uniform. B. Yang et al. [3408, 3409] reviewed the new non-invasive (non-contact) sensing technologies (the infrared camera, the Euler video magnification technology, the cross-validation of the infrared camera, RGB camera and wearable devices, the skeleton keypoints model aided normal camera). The infrared imaging technology can also be used to identify the thermal state of animals, which is a common measuring method in the veterinary medicine, biology and other related fields.

The human thermoregulation models are classified by the following two types:

1. the unit model for the entire body;
2. the multinodes model with body parts [3433].

The first model was formulated in 1911 by Lefevre who considered the human body as a sphere, where the core produced heat and the shell transferred the heat to the environment [1343].

The model by A. P. Gagge et al. [1035] calculates the thermal response by means of the two energy balance equations (one for the core node and one for the skin node) and takes into account the effects of the heat accumulation, the physical work done, the conductive and the convective heat transfer by the blood flow between the core and shell, the sensible and latent heat loss caused by the respiration, the metabolic heat generated during the exercise and the shivering, the heat exchange processes between the skin and its surroundings by means of the convection, radiation and evaporation of moisture.

Yu. I. Luchakov et al. [1978] used similar model and found that the convective mechanism of heat transfer due to the blood flow is realized in the tissues of the thermal core of the organism. In the tissues of the thermal envelope there is basically conductive heat transfer due to the temperature gradient. Thus, the temperature in the core remains fairly high and uniform, and the shell is low and changing.

The model by A. P. Gagge et al. was improved [1036] by adding the control functions for the blood flow rate, the sweat rate, the shivering metabolic rate and the fraction of the total mass assigned to skin layer. The thermal control functions were functions of the core and skin temperature signals. When the core and skin temperatures exceeded their thermoneutral values their signals were warm, when the temperatures were lower than the thermoneutral values, the signals were cold. The warm core signal controlled vasodilation, the warm core and skin signals controlled the sweat rate, the cold skin signal controlled vasoconstriction and the cold skin and core signals controlled the shivering metabolic rate.

The empirical models are developed by exposing humans to a range of the thermal environments and "fitting" the mathematical models to the human response data collected [2384]. One such model predicting the core temperatures over the time developed by B. Givoni and R. F. Goldman [1092, 1093] was based on the phenomenological data and the statistical adjustments. The model by K. B. Pandolf et al. [2374] predicts the rectal temperature, the heart rate and the sweat loss as function of the work intensity and the clothing ensemble.

The whole body models are needed to estimate the personal protective equipment (PPE) designed to protect individuals from chemical, physical, and biological hazards that may be encountered during hazardous materials operations. The personal protective ensembles are required when the working environment poses a threat to the wearer's health. Extreme environments encountered in the space travel and the military activities necessitate the provision of the sophisticated garments to the astronauts and the military personnel for the thermal protection [3376].

There are four levels of the personal protective equipment [2523]:

1. *Level A* protection is required when the greatest potential for exposure to the hazards exists, and when the greatest level of the skin, the respiratory, and the eye protection is required; the examples include the positive-pressure, the full face-piece self contained breathing apparatus (SCBA) or the positive pressure supplied air respirator with the escape SCBA, the totally encapsulated chemical- and vapour-protective suit.

2. *Level B* protection is required under the circumstances where the highest level of the respiratory protection is needed with the lesser level of the skin protection; examples include the positive-pressure, full face-piece self contained breathing apparatus (SCBA) or the positive pressure supplied air respirator with escape SCBA, the chemical-resistant gloves, the face shield, the hooded chemical resistant clothing.
3. *Level C* protection is required when the concentration and type of the airborne substances is known; examples include the full-face air purifying respirators, the chemical-resistant gloves, the hard hat, the escape mask, and the chemical-resistant outer boots.
4. *Level D* protection is the minimum protection required. Level D protection may be sufficient when no contaminants are present or the work operations preclude the splashes, the immersion, or the potential for unexpected inhalation or contact with the hazardous levels of the chemicals.

When any material covers the body, the microenvironment is formed between the clothing and the skin—usually becomes warmer and more humid than the ambient environment because of the production of the metabolic heat and the sweat by person. For example, the body temperature of the pilots wearing the anti-G garments becomes uncomfortably or even dangerously high in the moderate to hot environments [2535].

The protective suits usually contain the multiple layers with the different functions, including the thermal barrier that limits the heat loss from the body. The insulation is provided by the fabrics themselves and the layers of air trapped between the skin and clothing and in the clothing layers. The insulation of the clothing is mainly determined by the air trapped in and between them.

Sometimes the heat control is applied locally. For example, the meibomian gland dysfunction (MGD) is one of the common cause of evaporative dry eye [335, 383, 993]. Regardless of the driving factor for MGD, ultimately it leads to an inadequate lipid layer, which increases the evaporation of the tear film. The warm compresses are regarded as a primary therapy [383, 1056] aiming to provide the thicker lipid layer heating the eyelid surface to have the appropriate temperature reach the meibomian glands.

Development of the simplified "lumped" human models have been started in the middle of the twentieth Century by A. P. Gagge et al. [775, 1998, 2689].

Usually the human body or its segments are approximated by the cylinders having a number of the layers (such as the bone, muscle, fat, skin [775], sometimes an additional "core" layer is introduced [1759]). For example, the human breast tissue is modelled with six layers: epidermis, papillary dermis, reticular dermis, fat, gland and muscle [1263].

The skin itself has several layers [3374] and frequently is considered together with the subcutaneous tissues. Note that the very important thermal effect of the dermal blood flow is significantly underestimated by treating skin as a single region [116]. The subcutaneous tissue consists of a fine, collagenous and fibrous septa network enveloping the clusters of the adipocyte cells; the density and orientation of septa, varies among patients and correlates with the cellulite grading [1121]. The thermal properties of skin vary across different layers (epidermis, dermis and subcutaneous tissue); even within the same layer, there exists large nonhomogeneity and anisotropy due to the presence of blood vessels [3376]. For example, D. C. Sanyal and N. K. Maji [2709] found that in the middle layer of the skin, the blood mass flow, the thermal conductivity, and the metabolic heat generation are position dependent because the density of blood varies within the skin. The structure of the human skin is different from site to site and, at a particular area on the human body, it depends on its functions; the thickness of different skin layers varies depending on the location of the skin [3381].

The 15-segments model by M. S. Fereira and J. I. Yanagihara is based on the more realistic elliptic cylinders and includes the head, the neck, the trunk, the arms, the forearms, the thighs, the legs and the feet [953, 1178].

In addition to the elliptical cylinders, the truncated circular cones and the elliptical cylinders with four different radii are used for some segments in order to improve the geometrical depiction of the body regions [953, 1532, 1759].

E. D. Yildrim and B. Ozerdam [3426] developed the multi-segmental multi layers model of the human body exposed to a combination of the convection, evaporation and radiation that are taken into account as the boundary conditions.

K. Muhsin and S. Gökhan [2206] computed the airflow, the heat transfer and the moisture transport around the standing human body using the combined model of the manikin and the room. Sorensen and Voigt [2920] performed modelling of the flow and the heat transfer around the seated female body by the computational fluid dynamics.

S. Murakami et al. [2230] used the CFD methods to analyse the wind effect on the human body under the various wind conditions and found that the wind effect depends not only on the air velocity but also on the turbulence intensity.

The multi-segmental model containing 15 spherical or cylindrical body elements (head, face,neck, shoulders, arms, hands, thorax, abdomen, legs, and feet) was used by D. Fiala et al. [956]. The model includes the metabolic heat produced within the body, respective heat distributed over the body regions by the blood circulation and heat conducted to the body surface. Insulated by clothing, the heat is lost from the body surface to the surroundings by the convection, radiation, and evaporation. The heat exchange by the respiratory tract is also included. To achieve the more realistic results all the phenomena of the heat diffusion account for the geometric and anatomic characteristics of the human body and consider the thermophysical and the basal physiological properties of the tissue materials. The multilayer model consists of the annular concentric tissue layers and uses seven different tissue materials: brain,[1] lung, bone, muscle, viscera, fat and skin. The skin had two layers: the inner and the outer layer. The metabolic heat is generated in the inner skin layer.

[1] The brain itself is composed of the scalp, the skull, the subarachnoid space (filled with the cerebrospinal fluid), the gray matter, the white matter and many blood s [2913].

The outer skin layer simulates the vapour barrier for the moisture diffusion through the skin.

The blood circulatory system consists of three components: the central blood pool, the counter current heat exchanges, and pathways to the individual tissue nodes. The blood flows from the central pool through the major arteries to the body elements. The blood is cooled by the heat lost to the counter current bloodstreams in the adjacent veins. The arterial blood exchanges heat in the capillary beds. The venous blood collects in the major veins and is rewarmed by heat from the adjacent arteries as it flows back to the central blood pool. It is assumed that heat is exchanged with the tissue in the capillary bed and no heat storage occurs in the bloodstream. The counter-current effect becomes more significant with decreasing blood flow rate [2156].

The criteria for division consists in separating the layers whenever a significant change in the tissue properties occurs. This model is composed by 19 compartments; later D. Fiala et al. [958] simplified it to the UTCI (universal thermal climate index) model reducing the 19 compartments to 12. J. Xu et al. [3388] modified the overall and local convective heat transfer coefficients of the standing and seated human bodies of the Fiala model. J. Westin [3306] adopted the Fiala thermoregulatory model for use in the LCG environments. The author also identified a problem of the Fiala model with the dynamic response of the central blood pool temperature: when there is a change in metabolic rate, there is an abrupt change in the central blood pool temperature. As a result, the model predicts an excessive change in core temperature. In reality, the cardiovascular system does not instantaneously respond to changing requirements. J. Westin incorporated an exponential type of increase or decrease of muscle blood flow after a change in the activity level.

The model developed by C. Huizenga et al. [1386] includes the counter flow heat exchange and the perfusion from s to the tissues. The model also calculates the heat transfer using the angle factors in the radiation heat flux model. The model can simulate an arbitrary number of the body segments each consisting of four layers (core, muscle, fat, and skin tissues) and the clothing layer. A blood flow model includes the counter flow heat exchange in the limbs, and the modified blood perfusion from s to the tissue.

J. P. Rugh et al. [2663] suggested a three-dimensional transient finite element model that consists of a human tissue system and a thermoregulatory system. The thermoregulatory system controls vasomotor action, sweating, and shivering. The human tissue system represents the human body, including the physiological and thermal properties of the tissues. The overall model consists of approximately 40,000 nodes and elements.

M. Salloum et al. [2689] developed a model for predicting the thermal and the regulatory responses of the nude human that contained 15 cylindrical segments divided into four nodes (core, skin, artery blood, vein blood). In any body element, the blood exiting the arteries and flowing into the capillaries is divided into the blood flowing in the core and into the skin layer. The model calculates the blood circulation flow rates based on the physiological data and the real anatomic positions of the arteries in the body.

The wall of large arteries is composed of three layers,the intima, the media and the adventitia [92]. The adventitia,the outer layer, is mainly composed by connective tissue,mostly collagen fibres with a longitudinal orientation, and provides mechanical support. The media is the thickest layer, composed by smooth muscle cells and connective tissue, mainly collagen and elastin, with both cells and matrix fibers oriented circumferentially, and its role is to provide vasomotility and elastic compliance . In normal subjects ,the intima is composed almost exclusively by a monolayer of endothelial cells on a basement membrane,and is responsible for the antithrombogenic behaviour of the vascular wall, for vasoregulation.

The networks can be reconstructed, for example, using the dynamic contrast enhanced multi-slice CT imaging (a technique for imaging cross-sections of a subject using series of X-ray measurements taken at different angles around the subject) [1666], the magnetic resonance(MR) imaging (based on the measurement of relaxation times of the net magnetization vectors induced in tissues when a magnetic field at a given frequency is applied) or the ultrasound imaging (based on the generation of ultrasounds from a piezoelectric transducer which is then used as a receiver for the waves reflected at the interfaces between two tissues with different acoustic impedance). The patient-specific vascular model reconstruction techniques rely on a set of techniques which imply low-level operator interaction, such as identification of lumen on the 2D images, reconstruction of the 3D model from the set of 2D sections by means of CAD software. The description of the 3D geometry of the reconstructed models,such as measurement of axis and diameters,is not always univoque, and requires the operator to take arbitrary decisions [92].

E. Foda et al. [977] have modified the Gagge model [1036] using their measurements of the local skin temperature. The modification was based on manipulating the local core set points, using a modified calculation procedure for the convective heat transfer coefficients, and adjusting the heat transfer term from core to skin using the common blood temperature along with the local core temperatures.

M. P. Castellani et al. [464] proposed the finite element human thermoregulatory model using an anatomically and geometrically correct human body presentation. The finite element body model includes 13 organs and tissues: skin, muscles, fat, bones, heart, lungs, brain, bladder, intestines, stomach, kidneys, liver, and eyes. The model simulates thermal responses through the passive and active system. The passive system describes the heat balance within the body and between the skin surface and the environment using the Pennes equation [1264, 1561, 3372]. The active system that describes the thermoregulatory mechanisms—vasodilation, vasoconstriction, sweating, and shivering heat production.

I. V. Khromova et al. [535–537, 1617] investigated the human thermoregulation system under the low temperature conditions in the model "human body – thermal protection – environment" system accounting for passive and active thermal protection, sex, age and an organism's health state as well as heat transfer with a blood flow, internal heat sources, thermophysical and geometric properties of calculated elements. The authors suggested the heat-hydraulic scheme of the human thermoregulation system which based on calculation of the mass flow rate of the heat carrier (blood) for the "core" and "shell". The materials in the range of thermal resistances from 0 to 1 $m^{2\circ}$ C/W under different environmental conditions were examined.

Most models were developed for a uniform thermal environment. D. Lai and Q. Chen [1759] developed a 12-segment model for the transient and non-uniform surrounding conditions by considering the two-dimensional heat transfer in each segment of a human body. This model could be considered by the extension of the Fiala one with a few improvements: an account for the conduction in the both radial and the angular directions and for the arterial blood temperature drop along the extremities as well as the method to quantify the clothing insulation at the local segments. The heat transfer included the convection, radiation, and evaporation on the bare skin and the skin covered with the non-uniform clothing insulation across the different body segments. Each segment consisted of the four or five concentric layers, which were a combination of the following seven tissue materials: brain, lung, viscera, bone, muscle, fat, skin. The heat transfer between the body segments was estimated from the blood circulation assuming the counter-current heat exchange.

To account for the heat exchange between the body segments the heat balance equation of each segment of the body—the inner core, the muscle layer, the fat layer, the skin layer, the central blood pool—that include, e.g., the blood circulation heat conduction between the body nodes, the heat conduction between adjacent layers of the same segment, the respiratory heat loss, the heat generated by the exercise work, the heat generated by the muscle tremor, the dry heat loss between the skin surface and the environment, the moist heat exchange on the skin surface due to the evaporative heat loss [3535] should be exploited [3536].

Frequently a simplified blood circulation model is applied that assumes that the heat transfer is ineffective in the central segments of the body (the head, the neck, the thorax, the abdomen) due to the relatively large blood flow and the short lengths and the blood temperature is equal to that in the central blood pool [1759]. The blood temperature decreases along the direction of the arterial flow in the extremity segments (the shoulders, the arms, the hands, the legs, the feet) due to the countercurrent heat exchange between the arteries and the veins. The heat loss to the environment includes the radiation, the convection, the evaporation; the heat transfer coefficients for the individual segments of the human body are measured in a wind tunnel [693]. The two-node Gagge's model was combined with the Pennes equation to produce the simplified thermoregulatory model that incorporate a number of the empirical relations (for example, the DuBois formula for the body surface area, the heat production by the shivering, the heat transfer coefficient at the skin surface etc.) [3541].

A. E. Iyoho et al. [1420] developed the human thermal model that accounts for the asymmetric environments and includes the extremities. The model incorporates the two dimensional (the radial and circumferential) heat transfer along with the arterial and the venous countercurrent blood flow. The digits (the fingers and the toes) were modelled as the arteriovenous anastomoses. There are about 20–25 of AVA structures per cm^2 in the finger pad and are located about 1–1.5 mm below the surface of the skin [2915]. The AVAs are the shunts that directly connect the arteries and venous plexuses. For the heat exchange, most of this vasculature is located in the hands, feet, ears and hairless regions of the face. When the AVAs are open, the large amounts of blood can pass through to the venous plexuses. It is estimated that blood flow into the subcutaneous venous plexuses can be as much as 30% of the total cardiac output [1148]. The AVAs play a key role in regards to the thermal comfort of the extremities. The authors stressed that the digits, especially the fingers, respond to changes in the thermal environment quicker than all other locations of the body. Thus, a change in a digit temperature is a good indication of the thermal imbalance in the body. The blood flow to the arteriovenous anastomoses in the glabrous skin (glabrous skin blood flow, GSBF) of the palms and soles constitutes the primary heat transfer portal between the body core and the environment [3016, 3029].

The heating of the selected areas of the skin surface can open the AVA flow when it is not already established, thereby creating a convenient and effective pathway to induce the heat flow between the body core and environment. Upregulation of the AVA flow is produced by the mild heating superficial to the spinal cord, particularly in the cervical region [298].

The human thermal model used 14 cylindrical elements to represent the body with each element having the four concentric regions: the core, muscle, fat and skin layers; each element contained the six equally spaced angular sectors. The vasomotor actions either increase or decrease the blood flow rate to the skin. During vasodilation, the blood flow rate to the skin is increased. environment. During vasoconstriction, the blood flow rate to the skin is decreased. Vasodilation occurs when the simulated subject is too warm, whereas vasoconstriction occurs when the simulated subject is cold. The state of the blood vessels could also be changed by different substances, e. g., substances called vasoconstrictors can reduce the size of blood vessels, thereby increasing blood pressure. Vasodilators (such as nitroglycerin) increase the size of blood s, thereby decreasing arterial pressure.

A. E. Iyoho et al. [1421] modelled the gender differences in the thermoregulation.

N. Jia et al. [1468] developed the thermophysiology comfort prediction model that includes the human body, clothing, and environment (HCE) as well as the human body physiological properties—a nonlinear heart rate regulation model was incorporated. The model was based on the mathematical description of the heat and moisture transfer processes in the clothing according to the Henry model [1296] and in the external environment.

L. G. Berglund [260] simulated several option of the micro-cooling

- the uniform whole body cooling under clothing;
- cooling of the upper torso with a water cooled vest;
- cooling of the upper torso with an air cooled vest.

The author found that the upper torso air and water cooling systems could overcool the skin causing the local vasoconstriction that reduces the micro cooling efficiency and registered the reduction of the sweat rates with the micro cooling.

M. Rida et al. [2603] used the 25 segments model to study the possibility to reduce or delay the state of heat stress by decreasing quickly the core temperature using the immersion of the hand and forearm in the cold water [1720, 2104]—D. W. DeGroot et al. [706] reported that the arm immersion up to the elbow results in the greater heat lost compared to the hand only immersion and may reduce the cardiovascular strain by lowering the heart rate by 10–25 beats/min (the water is known to conduct the heat from the body at about 20 to 26 times faster than the air at the same temperature [2885]).

Rida et al. used the model that includes the head, the upper trunk, the lower trunk, the upper arms, the lower arms, the palms and the fingers, the thighs, the calves, and the feet. Each segment was composed of the core node, the skin node, the artery node, the vein node, and when applicable a superficial vein node. The modelling of the blood flow was based on the multi-branched arterial tree model of A. P. Avolio. The authors found the fast reduction of the body core by 1° C in less than 20 minutes when the hands and forearms are immersed in the water at 15° C.

R. Holopainen [1343] developed the multi-node Human Thermal Model implemented in a dynamic building simulation environment that includes

- a passive system: the body tissues, the internal organs, the circulatory and respiratory systems;
- a control system that controls the physiological responses of the body;
- a thermal sensation and thermal comfort calculation system.

The model divides the body into 16 parts, each being further sub-divided into the four realistic tissue layers (bone, muscle, fat, and skin) by the concentric cylinders (the exception is the head, which is approximated by the concentric spheres). The model incorporates the presence of clothing which insulating effect is based on the still air layers between the skin and the different clothing layers, which reduce conduction losses to the environment, resists the convective losses by preventing convection and also reduces radiant heat loss by serving as a thermal radiation barrier and restricting the evaporation of sweat produced by the body. The author determined the heat exchange coefficients using a manikin that is electrically heated internally to simulate the human skin temperature distribution.

The circulation model is divided into the macrocirculation (the large s) and the microcirculation subsystems (the circulation through the small s such as arterioles, venules and the capillary bed). The capillary blood flows to the different tissues of the body part represents the microcirculation and the blood flow through the body part into the following body part represents macrocirculation. The multi-branched model of the human arterial system with 128 segments including the central s and the major peripheral arteries suggested by A. P. Avolio [141] was used, the arteries and s were located in the muscle tissue (thus the convective heat transfer from the arteries and s is taken into account only in the heat balance of the muscle tissue).

The microcirculation flow is characterized by a low Reynolds number $Re = \eta v D/\rho < 1$, where v is the velocity of the flow, D is the diameter of the micro, η is the dynamic viscosity of the fluid, ρ is the fluid density. The Reynolds number is defined as the ratio between the inertial and viscous forces, and in the micros it ranges between $0.001 < Re < 0.1$. Thus, the viscous effects are more significant than the inertial effects, and the flow is laminar. In micros over 200 μ diameter, it can be assumed that blood is a homogeneous continuous fluid. This assumption is not true for smaller s and here the individual motion of the RBCs becomes important [3080].

The transport of the thermal energy by the blood flow in the microcircular system is more important to the heat transfer throughout the tissue than the transport of the thermal energy in the macrocirculation: the capillary blood flow is the main factor in the exchange of energy between the blood and the tissue. It is assumed that the blood leaves the tissue at the tissue temperature. The skin blood flow is controlled by the mechanisms of vasodilation and vasoconstriction that result in a higher or lower skin blood flow. The correlations depend on the mean skin temperature and on the core temperature. If the vasodilation response is not sufficient to keep the core temperature from rising further, the core temperature will reach the sweating threshold, where sweating begins. Beyond this threshold, the sweat rate increases linearly with increasing core temperature. If vasoconstriction is not able to keep heat in the body, the metabolic thermoregulatory function takes place and the body will start shivering in order to heat the body.

Holopainen [1343] used the developed model to compare the effect of the alternative heating distribution systems on the thermal comfort index. The author found that the clothing level and the activity level have the greatest influence on the thermal sensation and comfort. The differences between the different heating distribution systems are relatively small in all cases. The highest thermal comfort is achieved with the floor heating system, which has the largest heating area. Holopainen also stressed that the commonly used the PMV method for the thermal sensation and thermal comfort calculation takes into account the average operative temperature and does not pay attention to the fact that the surface temperatures vary inside of a room, e.g. during the winter the heating devices are warmer and the window surfaces colder than other surfaces. Thus, the uneven surface temperature distribution can cause the thermal discomfort even though the average surface temperature might be acceptable.

3.3 Animals

One of the reasons to study the thermoregulation in the animals is to asses the possibility to model the human thermal physiology and the drug development [1584]. For example, V. Ŝkop et al. [2874] investigated the mouse thermal physiology in comparison with the human physiology in spite of the 3000-fold difference in the body weight and differences in the thermal behaviour—the mass-specific metabolic rate is sevenfold higher and the surface area/volume ratio is 14-fold higher than in humans [2875].

Explaining the emergence and maintenance of the phenotypic variation is an important topic in the evolutionary biology. A review by P. A. Biro and J. A. Stamps [296] indicated that nearly 3/4 of the studies had demonstrated a positive relationship between some measure of the metabolic rate (basal, resting or standard) and the behavioural trait (for example, aggressiveness, boldness, and activity). For example, individuals that exhibit the slow exploratory behaviour have the similarly low basal metabolic rate measurements [1139].

These finding were later confirmed in investigations by I. Krams et al. [1709] and by K. J. Mathot et al. [2063]). However, the negative relationships or no relationship between the metabolism and the behaviour have also been observed, suggesting that the linkage may be more complex involving the additional factors such as ontogeny, the reproductive status, and the environmental conditions, see, for example, [442, 443, 1073, 1360].

One of the models that integrates the variation in the life history, physiology, and the behaviour [296] is the *pace-of-life syndrome (POLS)* hypothesis [2602, 2652] according to which individuals with

- the *slow* pace-of-life individuals exhibit the slower metabolic rates, the slower stress reactivity, and the higher immune response and are predicted to minimize the risk-taking and behave more cautiously when facing the threats from predators and has the long life spans; the lower the metabolic rate of a species, the longer its life expectancy [5,6];
- the *fast* pace-of-life individuals are expected to adopt the "live fast—die young" strategy and are less risk averse due to the higher metabolic rates, the low immune response, the low hypothalamus-pituitary-adrenal responsiveness [1139] and the lower stress reactivity.

The endothermic mammals and birds with the smaller mass generally have a longer maximum life span and the higher mass-specific metabolic rate, but they also sleep longer [2719]: for example, the mice sleep about four times longer than the elephants [2599].

C. T. Goulet et al. [1139] proposed to extend the POLS hypothesis by adding a new dimension, the cold-hot axis. The authors tested the hypothesis by measuring the thermal traits of the heliothermic lizards (the delicate skinks, *Lampropholis delicata*) and linking them to their behaviour.

The increase of the energy expenditure in the cold environment was established by Antoine-Laurent de Lavoisier in 1790 [1989] and in the hot environment by Christian August Voigt in 1876 [2660]. The ambient temperatures where the metabolic rate is at a minimum and constant is the usual definition of the thermoneutral zone (TNZ) [433, 1128, 2744]. Another definition of the TNZ [1637]: the range of the ambient temperature at which the temperature regulation is achieved without the regulatory changes in the metabolic heat production or the evaporative heat loss. The TNZ has mainly been studied in the animals [60, 595, 2183, 2632], yet some studies focused on the human TNZ [644, 1246, 1311, 2718].

Below the lower critical temperature (LCT) of the TNZ the heat balance can be achieved by upregulation of the metabolic heat production (the non-shivering thermogenesis and/or shivering). Above the upper critical temperature (UCT) the heat balance can be achieved by increases in the evaporation. In the humans, the onset of the sweating is accompanied with the active vasodilation, which also contributes to the maintenance of the heat balance by increasing the heat transport from the body to the skin [1588]. Above the UCT the heat production also increases due to the increased blood circulation, the activity of the sweat glands and the increased body temperature [620].

The metabolic rate measured in the awake, resting, and postabsorptive state is the basal metabolic rate (BMR)—the BMR represents the integration of minimal activity of all the tissues in the body, or "basal metabolism is the minimal rate of energy expenditure compatible with life" [1297]. The BMR of mammals varies with body mass, with the same allometric exponent as field metabolic rate and with many physiological and biochemical rates. In general, metabolic rate scales with body size and temperature. The resting metabolic rate per gram of body weight of a mouse is about sevenfold greater than in the humans; about two-thirds of the resting metabolic rate is dedicated to homeothermy, i.e. to maintaining a stable core temperature [2599].

The membrane pacemaker theory proposes that the fatty acid composition of the membrane bilayers is an important determinant of a species BMR: in both mammals and birds, the membrane polyunsaturation decreases and monounsaturation increases with increasing the body mass [1387].

The heat loss in the TNZ is controlled chiefly by regulating the blood flow [433, 1129, 1638, 2632].

The mice at the typical housing temperatures (20° C–22° C) live below the thermoneutrality, and about half of their total energy expenditure (TEE) is devoted to maintaining the core body temperature (T_b). In contrast, the humans generally live in the thermoneutral micro-environment and their T_b is supported by the "waste" heat byproduct of the metabolic processes—the human thermal biology is more organized around the heat dissipation, rather than the generation or conservation [1045, 1129, 2593, 2874]. Ŝkop et al. found no ambient temperature zone where both the metabolic rate and body temperature were constant in mice.

The whole body models of the animals are used to solve the practical problems. For example, the dairy cows are sensitive to the environment temperature: even a 0.5° C change in the core-body temperature reduces the cow's performance, increases lameness; the extreme thermal stress could lead to death. The three dimensional model of a cow lying on a cooled surface (a truncated cylinder) was used to study an efficiency of the cooling [471].

3.3.1 The Square–Cube Law

The whole-body models are used to justify the empirical power laws of the form $y = x^a$ where a is the scaling exponent in the allometry [2661]—the study of the relationship of the body size to shape, anatomy, physiology, first outlined by Otto Snell in the end of the nineteenth century.

One of such empirical relationships is the so called the "index" surface rule [317] (or the square–cube law or cube–square law first described in 1638 by Galileo

Galilei in his "Two New Sciences"): "...ratio of two volumes is greater than the ratio of their surfaces".

This law applied to the physiology states that the basal metabolic rate (BMR) heat production in an organism in a rested, awake and the thermoneutral state is proportional to the surface and thus to 2/3 power of the body weight [1749]; the reasoning leading to this law is similar to that of E. Fermi: the metabolism rate is balanced by the losses to the ambient.

The ratio of the surface area to the volume influences all aspects of the heat exchange. This parameter comes into play when comparing the animals of the different body dimensions or the body mass. For example, an arctic wolf is about one-tenth the mass of a grizzly bear, but it has twice the ratio of surface area to volume. Thus, living in the similar climate, the wolf incurs greater thermoregulatory costs because of its size.

An animals can regulate the heat exchange by altering the posture of the body to minimize or maximize the exposed surface area. For example, the dogs stretch out in the hot environment to maximize the conductive heat loss to the ground, but roll up when cold to minimize the conductive heat loss to the air; when the python rolls into a ball, its externally exposed surface area decreases by about 85%, greatly reducing heat loss.

There are, however, doubts in the validity of the "2/3" law since the experiments for the mammals and the birds, in contrast to the cold-blooded vertebrates, show the power index to be in the interval [2/3,3/4] [2661]. An alternative model resulted in the 3/4 power relation was developed by A. Bejan [235]. The author stated that for the cold-blooded animals the surface determines the heat exchange with environment and thus arises the 2/3 law while for the mammalian and ovarian species the skin acts as an insulator; the heat transfer in the body is provided by the blood flow and heat loss to the ambient is determined by the structure of the blood s network.

The 3/4 exponent is valid at the limit where heat loss is impeded by mostly convective resistance within the body; the heat loss through the outer layer of skin to the ambient environment is mostly conductive with radiant and evaporative losses or resistance due to surface coatings also playing roles [2661]. Thus the allometric scaling when applied to small animals tends to scale with an exponent that can be less than 3/4 (i.e. with more surface area to volume both convective and conductive/surface factors influence empirical estimates).

3.3.2 Kleiber's Law

There is a related law—the Kleiber's law of the metabolism discovered when analysing a large number of the studies on the respiration within the individual species—which states that the metabolic rate of the homeotherms is positively correlated to the body size [1652, 2307]. Kleiber expected to find an exponent of 2/3—the bodies lose heat passively via their surface, but produce heat metabolically throughout their mass, the metabolic rate must scale in such a way as to counteract the square–cube law—but found an exponent of 3/4, i.e. the metabolic rate of an animal scales to the 3/4 power of the mass. The scaling of the body mass with respect to the metabolic activity is applied equally to the plants as well as to the animals [2307].

A model by G. B. West et al. [3301] suggests that the 3/4-scaling arises because of efficiency in the nutrient distribution and transport throughout an organism. In most organisms, the metabolism is supported by a circulatory system featuring the branching tubules (for example, the plant vascular systems, the insect tracheae, or the human cardiovascular system) and is assumed to be proportional to the total fluid and, thus, to the total number of the smallest tubules that scales as 3/4 exponent of the total fluid volume [876].

It is assumed that the observed 3/4 power scaling by Kleiber an artefact of the inclusion of the large herbivores [1097, 2661, 3312, 3313]. The experimental data show that the Kleiber's law does not hold over a wide variety of scales: the metabolic rates for the smaller animals (birds under 10 kg, or insects) typically fit to the 2/3 dependence much better than 3/4; for the larger animals, the reverse holds [780].

Nevertheless, when appropriate adjustments are made (for example, normalization to a constant temperature) the metabolic rates can be scaled from single cells to large plants [2304–2306] and animals [3300–3303] using weight to the 3/4 power relationships [2661].

Cellular transporters associated with cytoskeletal elements provide for the vectorial movements of fuels and a network for the convective heat exchange. In cells the specific transport and organized movement are best achieved by elements (proteins and transporters) that have a high affinity or avidity for given substances, but with fast rates of exchange so that movement can occur; these same 'transport' networks should allow for convective heat flow as well. It would appear that by design many types of organisms are similar as it relates to heat exchange [2661].

The substantial local heating within the cell introduces the possibility of the temperature gradient-driven convective circulation (the cytoplasmic streaming [1113]) that under some conditions could dominate over the random thermal diffusive motion and active transport by the molecular motor proteins at the scale of a single eukaryotic cell [1361] in contrast to the purely thermal diffusive motion that is typically the primary source of the substance transport in the prokaryotic cells lacking organelles and compartmentalization [2926, 3180]. The directional flow of the cytosol has been observed in the cells of the plants, the fungi, the animals and the human cells [1112, 1596, 2817, 2832].

R. Howard et al. [1361] performed the three-dimensional simulations of the cytosolic fluid motion in the spherical cell with the off-centre nucleus. The authors used the local temperature gradients that lie within the range of those observed experimentally [1267, 2330, 3015].

The steady state Navier-Stokes equations (NSE) [1945] and energy equation for the incompressible cytosol in the dimensionless form are written as

$$Ga(1 - \beta\vartheta T)(\mathbf{u} \cdot \nabla)\mathbf{u} = -\nabla p + \nabla^2\mathbf{u} + \frac{1}{3}\nabla(\nabla \cdot \mathbf{u}) - (1 - \beta\vartheta T)\mathbf{z},$$

$$\nabla \cdot ((1 - \beta\vartheta T)\mathbf{u}) = 0,$$

$$Ra(\mathbf{u} \cdot \nabla)T = \nabla^2 T$$

where $Ga = gL^3/\nu^2$ is the Galilei number, $Ra = UL/\varkappa$ is the Rayleigh number, $U = \varrho_0 g L^3/\mu$, ν is the kinematic viscosity, g is the gravity, $\varkappa$ is the thermal diffusivity, β is the thermal expansion coefficient, ϑ is a typical small deviation from the reference temperature, the equation of state becomes $\varrho = 1 - \beta\vartheta T$.

While various organelles and particles are suspended within the cytosol, around 70% of the cellular volume is water [1977], thus the authors used the thermophysical properties of the pure water.

In the considered configuration of the heated off-centre nucleus in the spherical cell the natural convective flow will necessarily occur [1039]. The computations showed the formation of the pattern of the motion corresponding to warm, less dense fluid rising around the heated nucleus, being horizontally displaced as it reaches the upper cell boundary, then cooling and sinking to create a "fountain" effect, i.e. forming the closed convective circulation.

G. B. West et al. [3303] calculated the smallest weight a mammal may have and maintain a flow network, i.e. where pulsative currents and blood flow can still occur. The mass of the mammal should be not less than 2 g or about the size of a very small shrew, the smallest known mammal.

The metabolism scales with the temperature as a function of the Arrhenius equation, but it is determined by the resource availability. However, the metabolic rate of the brain tissue is nine times that of the average of the metabolic rate of the rest of the body.

3.3.3 Species Distribution Models

The whole body models of the animals are also used to develop the species distribution models (SDMs) [2030, 3356] and assess and the evolutionary adaptation to the climate change [1340, 2383]. The temperature is a powerful driver of the global biogeography and the species distributions frequently reflect the temperature gradients in both the aquatic and terrestrial habitats; many species adopt the thermal strategies (such as the thermoregulation or the acclimation) that determine their thermal niche [638, 1382–1384]. For example, S. J. S. Wuitchik et al. [3356] investigated how the evolution of the thermal physiology alters the projected species distributions of the threespine stickleback (*Gasterosteus aculeatus*) that exhibits a wide thermal tolerance range bounded by a mean $T_{min} = 1.89°$ C and a mean

$T_{max} = 30.1°$ C under the climate change. J. C. Chato [494] mentions the study by R. C. Birkebak who successfully predicted the wintering places of the migratory birds by considering the energy balance requirements.

However, as M. W. Sears and M. J. Angilletta [2758] noted, the niche-based methods [354, 400, 401, 654, 1344, 1580] to predict how the species will shift their ranges fail to account for the environmental heterogeneity at the scale of an organism. Typically, the topographical features are considered constant throughout an area no less than 1 km^2, consequently, all individuals within that area are predicted to be either active or inactive at a particular time [2758].

Since the 1990s, the researchers have used an index of the thermal quality, q^e, that equals the absolute difference between the operative temperature of the environment and the preferred temperature of the species [1309]. Usually, the samples of q^e are averaged over space to obtain a simple descriptor of the thermal quality (e.g., [318]) but this descriptor carries no information about the heterogeneity or the spatial structure of the temperatures in the landscape. However, this heterogeneity and spatial structure determine whether an organism can maintain its preferred temperature. Thus, a theory of the thermoregulation must consider the abundance and distribution of the thermal resources in space [88].

Sears and Angilletta used an individual-based model to quantify the effects of the operative environmental temperatures, as well as their heterogeneity and the spatial structure, on the thermoregulation, movement, and energetics of the ectothermic animals. The individual-based modelling provides a framework for analysing the behaviours of organisms in the spatially explicit landscapes. Given the thermal landscape associated with the particular topography and the climate, one can model the temperatures of the individuals. The thermoregulatory performance in the landscape determines the activity and energetics, which can used to calculate the demographics of a population [400, 401].

The authors simulated the thermoregulatory behaviours of the animals in the two-dimensional landscapes (a tiled grid of cells) that differed in the availability of the preferred temperatures (availability), the spatial arrangement of the temperature (dispersion), and the thermal variance of the temperature (variance). Each cell was classified as either preferred or nonpreferred with respect to the temperature. The authors showed that the heterogeneity and the spatial structure of a thermal landscape are as important as its mean temperature.

There are a few rules of the "biogeography" ("patterns of the ecotypic variation" [2141])—"purely empirical generalizations describing parallelism between morphological variation and physiogeographical features" [2078]—valid for most of the mammals and birds [2201]. The ecogeographical rules are used for the studies of many species of the mammals, birds, other vertebrates, and even some invertebrates [2141]. There are three main patterns of the geographical phenotypic differentiation between the populations within the species:

- variation in the colour;
- variation in the size of the appendages;
- variation in the body size.

Several other less important "rules" are also mentioned—the "hair-rule" (the mammals have the longer and thicker hairs in the colder areas), the "rule of the shape of the wing (the birds have more slender and pointed wings in the colder areas), the "egg-rule" in the birds and the of the shape of the wing, "rule of the litter-size" in the mammals (the increased clutch or litter size in the colder areas) [2141], "the island rule" (also known as the " Foster's rule—a general tendency for small insular mammals to evolve towards the larger size and larger species towards the smaller size) [1946].

3.3.4 *Bergmann's Rule*

It was formulated by the German biologist Carl Bergmann in 1847: the animals living in the cold environments tend to be larger than animals in the warmer environments—organisms at the higher altitudes should be larger and thicker than those present near the equator, for example, the population of the white-tailed deer is larger in Canada than in Florida. Bergmann assumed that the surface area of an animal determines the rate of the heat dissipation while the heat production is determined by the volume. The larger animals have the smaller surface area to the volume ratio than the smaller animals, thus, they are able to keep themselves warm in the colder climate.

The Bergmann's rule is usually applied to the mammals and birds which are endotherms, but some researchers have also found evidence for the rule in studies of the ectothermic species [2335, 3055] such as, e.g., the ant in the subfamily Myrmicinae *Leptothorax acervorum*. The Bergmann's rule holds for most—between 62 and 83%—of the vertebrate species, with the exception of the squamates and the fishes, where fewer than 30% of the species conform to this rule [2141]. S. Meiri and T. Dayan [2111] investigated the occurrence of the Bergmann's rule in the birds (99 species) and in the mammals (149 species), using only studies where the statistical significance of the results was tested.

There is a variant of the Bergmann's rule: the organisms that are present deeper in the sea are larger in size than those present in the shallow water. According to the Bergmann's rule, the size of the sea creatures increases with the decrease in the temperature. The larger sea animals are found in the colder areas than the smaller animals. The low temperatures lead to an increase in the size of their cell and their lifespan. For example, the colossal squid lives 7200 feet below the sea. At such a depth, it is able to increase in the weight and height incredibly.

3.3.4.1 Exceptions

There are exceptions to the Bergmann's rule [118, 2111, 2141]. For example, the increasing size of the birds in California contradicts the Bergmann's rule: they should be getting smaller in size as a result of the global warming. However, the birds in California have increased in weight and wingspan by 2–5 per cent over the last 40 years.

For the humans, we now know that the body size and shape are regulated more by the social-economic-political-emotional factors as well as the nutrition-infection interactions—the temperature has virtually no effect [327].

3.3.5 Allen's Rule

The *Allen' s* rule—the ecogeographical rule formulated by Joel Asaph Allen in 1877—reflects the biological observation that strong positive correlations exist among the latitude, the ambient temperature, and the limb length in mammals and states that the animals in the colder climates tend to have shorter extremities than animals in the warmer climates: the body surface-area-to-volume ratio for homeothermic animals varies with the average temperature of the habitat to which they are adapted (i.e. the ratio is low in cold climates and high in hot climates).

According to the Allen's rule, the appendages size is reduced in the cold climates—it is based on the simple reasoning that the less surface area an organism has relative to its body mass, the more thermally efficient it will be. One way to reduce the heat loss is through the reduction of the appendages such as ears, tails or limbs [2743]. For example, the polar bear has the stocky limbs and the very short ears.

The Allen's rule has been found within and across the various species of the mammals and the birds. For example, the relative ear size increases in the fox from the artic fox (*Alopex lagopus*) through the red fox (*Vulpes vulpes*) to the desert fox (*Fennecus zerda*) [2141].

The usual explanation for temperature-growth effects in the skeletal extremities is an altered supply of the essential nutrients via the increased or decreased blood flow that results from the temperature-induced vasoconstriction/vasodilation [3082, 3086, 3272]. However, the blood not only provides developing limbs with a reservoir of essential growth factors, nutrients, and oxygen—it is also an important source of heat [2790]. The haemoglobin plays a key role in an animal's utilization of plant energy, transporting from lungs (or gills) to cells through-out the body the oxygen needed to burn plant "fuel." The energy of the organic molecules is released in the animals in a series of the reactions in which glucose, fats, and other organic compounds are oxidized (burned) to carbon dioxide, water and the heat.

Cartilage growth, which largely determines bone growth rate via endochondral ossification, is the process that may be sensitive to ambient temperature. Therefore, the amount of blood arriving at an extremity might modulate its temperature and thus the rate of bone elongation. Indeed, many arctic mammals rely on counter current vascular heat exchange mechanisms to prevent excessive heat loss in their extremities. Thus, there is another mechanism leading to the Allen's rule in vertebrates may be that the growth of cartilage—a resilient and smooth type of connective tissue—is at least partly dependent on temperature. Temperature can directly affect the growth of cartilage, providing a biological explanation for this rule.

Experiments by M. A. Serrat et al. [2790] who raised mice at different temperatures showed that the tails and ears were significantly shorter in the mice raised in the cold in comparison to the mice raised at warmer temperatures, even though their overall body weights were the same.

The Allen's rule can be applied to the humans: the differences in limb lengths have been observed when the different portions of a given human population reside at the different altitudes, for example, in Peru, individuals who lived at higher elevations (lower ambient temperature) tended to have shorter limbs, whereas those from the same population who inhabited the more low-lying coastal areas generally had longer limbs [3283]. Betti et al. [275] performed the measurements of the limb bones, including the first metatarsal and metacarpal, for almost four hundreds of the individuals from ten globally distributed human populations. The relationship between intra-limb indices and minimum temperature was tested using generalized least squares regression, correcting for spatial autocorrelation.

3.3.6 Gloger's Rule

According to this rule established by Constantin Wilhelm Lambert Gloger in 1983 the individuals that live in the warm and humid areas are darker in colour than those living in the cold and dry areas. The underlying mechanism for Gloger's rule is that the individuals with more cryptic colouration are under the reduced predation pressure. The Gloger's rule ia also related to the thermoregulation: the individuals with the darker pigmentation are found in the tropics, while the individuals with the lighter pigmentation predominate towards the poles. This rule was first observed in the birds but also applies to the mammals. A classic example is provided by the hares which have white fur in the northern latitudes (*Lepus arcticus*) and a brown fur in the temperate habitats (*Lepus alleni*).

3.4 Plants

The heterotrophic organisms (animals, fungi) gain the energy by consuming parts or the whole of other organisms. The autotrophic organisms (plants, some bacteria, and some protists) gain the energy by converting inorganic (non-carbon containing) chemicals (chemoautotrophs) or sunlight (photoautotrophs) into the useable forms of energy.

Chemoautotrophs are thought to have been the earliest organisms on earth, and are present today (for example, sulfur-oxidizing bacteria). Photoautotrophs capture the energy in sunlight and use it to power the production of the ATP and organic molecules like glucose. The process of photosynthesis is a complex series of enzyme-mediated steps, and the exact process differs between organisms. The organelle responsible for photosynthesis in plants is the chloroplast that consist of a double external membrane which houses internal membrane structures called thylakoids. Each thylakoid consists of a phospholipid bilayer membrane with proteins and pigment molecules suspended within it. Inside the thylakoid is an open area called the thylakoid space. Outside the thylakoids is a semi-liquid matrix called the stroma.

The plants have the ability to regulate their body temperature, a process known as the thermoregulation, which is achieved through the anatomical and physiological mechanisms that take lace in their leaves, stems, flowers, and fruits [3142].

While the plants lack movement in comparison to the animals, they, in fact, use a wide range of strategies to regulate their temperature. Some species, e.g., use the heliotropism and/or paraheliotropism, the processes that involves the adjustment of the leaf orientation in response to the sunlight to facilitate the interception of the light for photosynthesis, or to reduce the excessive light absorption and thus, the heating stress.

Plants do not perform breathing in its literal sense. Unlike animals, they do not possess any specialized structures like lungs for breathing. Stomata, the minute opening in leaves, and lenticels found in woody stems helps plants in gas exchange. The opening and closing of stomatal pores during the exchange of gases are regulated by the guard cells—the bean-shaped cells found in the epidermis of leaves and young stems surrounding a tiny gas exchange opening called a stoma.

Similar to other living organisms, respiration occurs in plants to fulfil their energy requirements. Respiration is a set of processes that break down nutrient molecules in living cells to release energy in usable form. In plants, the glucose produced during photosynthesis is broken down in a series of enzyme-facilitated reactions to release energy in the usable form of the ATP, producing carbon dioxide and water as byproducts.

The northern regions comprise a high number of the plant species characterised by the presence of the hollow stems, in fact, the highest prevalence in comparison to any other place in the planet—this feature plays a critical role underlining the plant thermoregulation. The analysis of the four species of the herbaceous plants—the oats, the dandelion, the thistle, and the sneezewort—revealed that the temperatures within the stems of all species differ by the several degrees from the ambient air temperature. The structures and features, such as the stem and fruit hollowness and the shape and colour of flowers, play important roles in the accumulation of heat that increases the chances of the reproduction, the dispersal, and the survival [1600]. The size and shape of the flowers influence their ability to capture the heat which, in conjunction with the microgreenhouse effects, contribute to the regulation of the body temperature; the flower colouration influences the absorption of radiant energy. Those processes also influence the presence, the behaviour, and the survival of other organisms, such as the pollinators, as the elevated temperatures elevate their nutritional reward.

The photosynthesis (the term was coined by botanist Charles Reid Barnes in 1893) is a process used by the green plants and other organisms to convert the light energy into the chemical energy that can later be released to fuel the organism's activities. Some of this chemical energy is stored in the carbohydrate molecules, such as the sugars and starches, which are synthesized from the carbon dioxide and water. Most plants[2], algae, and the cyanobacteria perform the photosynthesis; such organisms are called the photoautotrophs (i.e., the organisms that use the light energy and the inorganic carbon to produce the organic materials; for example, the eukaryotic photoautotrophs absorb energy through the chlorophyll molecules in their chloroplasts while the prokaryotic photoautotrophs—for example, the cyanobacteria, the purple bacteria and the green sulfur bacteria—use the chlorophylls and the bacteriochlorophylls present in the free-floating thylakoids in their cytoplasm).

The photosynthesis reaction involves a series of the electron transfer steps to convert the sunlight energy into the chemical energy [370, 1356] and uses the carbon dioxide and water to produce the oxygen and the glucose

$$6CO_2 + 6H_2O \rightarrow C_6H_{12}O_6 + 6O_2$$

and glucose monomers in the long stringy molecules called cellulose (cellulose is the most abundant organic compound on the Earth and the repository of over half of all the carbon of the biosphere [1270]).

The certain bacteria that perform the photosynthesis do not produce oxygen as the by-product of the photosynthesis—the anoxygenic photosynthetic bacteria (the bacteria that produce oxygen are called the oxygenic photosynthetic bacteria).

The photosynthesis takes place in two stages: the light-dependent reactions and the light independent reactions (the dark reactions, the Calvin cycle).

The light harvesting mediated by the Fenna-Matthews-Olson (FMO) photosynthetic complex demonstrate a remarkably high efficiency (over 95%). R. Alicki and W. Miklaszewski [53] stated that the popular picture of the excitation moving stepwise from the highest to the lowest excitonic state, by means of a diffusive hopping is not necessary correct and suggested a resonance mechanism of this phenomenon employing the exciton coherent motion and a coupling to the vibrational environment that is a necessary component of this mechanism—the resonant, coherent energy exchange between the donor site and the acceptor one accompanied by an irreversible energy transfer to the sink (the reaction centre). The role of the

[2] The leaves contain the microscopic cellular organelles known as the chloroplasts that contain four types of pigments—chlorophyll "a" $C_{55}H_{72}N_4Mg$, chlorophyll "b" $C_{55}H_{70}O_6N_4Mg$, carotenoids—carotene $C_{40}H_{56}$ and xanthophyll $C_{40}H_{56}O_2$.

Chlorophyll "a" is found in all photosynthetic organisms; chlorophyll "b" is produced in the vascular plants [1270]. Chlorophyll atoms are arranged in a circle with the help of four pyrrole rings (C_4H_5N), with Mg atom at the centre. For example, chlorophyll absorb energy from the blue and red wavelengths of light but reflect the green wavelengths. Therefore, our eyes perceive it as green.

Other photosynthetic pigments are phycocyanin $C_{34}H_{44}O_8N_4$ and phycoerythrin $C_{34}H_{46}O_8N_4$ [1270].

relatively strong coupling to the energy sink in achieving the resonance condition and the absence of heating of the vibrational environment are emphasized. The authors used the standard tight-binding model of the energy transport in the quantum networks in the single-exciton approximation which is valid under the assumption that the exciton lifetime is much longer than any other relevant time scale in this model.

B.-X. Wang et al. [3257] performed an experimental quantum simulation of the photosynthetic energy transfer using the nuclear magnetic resonance (NMR). The authors showed that the N-chromophore photosynthetic complex, with the arbitrary structure and the bath spectral density, can be effectively simulated by a system with $log_2 N$ qubits. The computational cost of simulating such a system with a theoretical tool, like the hierarchical equation of motion (HEOM), which is exponential in N, can be potentially reduced to requiring a just polynomial number of qubits N using NMR quantum simulation.

Plants continuously interact with their environment by the heat and mass exchange. An important component of the plant heat exchange is the convective heat transfer, the exchange of heat between the leaf surface and environment. The heat penetrates the boundary layer of a leaf that is characterized by the gradients of temperature, the gas concentrations, and the air velocities determining the boundary layer conductance [2578, 2745, 2746, 3198]. Because transpiration, i.e., heat loss via the water vapour, affects the leaf temperature T_L [1051] and T_L also affects the convective heat transfer [777], the convection and transpiration are coupled to each other.

The boundary layer conductance cannot be measured directly and it is often approximated by the heat transfer coefficients that relate the heat flux densities per unit leaf area to the difference between the leaf temperature T_L and the ambient air temperature $(T_L - T_a)$ [2578, 2746]. The heat transfer coefficients depend largely on the air speed (the forced convection) or the temperature differences (the free convection), which can be further approximated with the dimensionless numbers.

The radiation is an important factor influencing the photosynthesis and transpiration, variation of the thermal conductivity is often neglected in the quantitative analyses because the thermal conductivity of the plant leaves is generally low [1447]. The low thermal conductivity of the plant leaves is a result of a relatively high content of water resulting in a high specific heat capacity; water is the main component of leaves and considerably affects the leaf heat capacity. Therefore, the leaf water content per unit area (LWC), the leaf succulence, affects the dynamics of T_L response to heat absorption.

The thermal responsiveness of a leaf depends on the leaf heat capacity and the leaf heat transfer coefficient h_L which both can be estimated by the time constant τ of T_L dynamics. The use of τ as a measure of the thermal responsiveness of the leaves was proposed in the early studies in the 60's using the so-called "cooling curve technique"—the temperature of the leaves is transiently increased by a short heat pulse and the following cooling curve is recorded. This temperature cooling kinetics provides the value of τ which is the slope of the exponential decay curve [46].

The studies using the cooling curve technique were mainly performed using the thermocouples or the radiometers for T_L measurements. Using the thermocouples is not non-invasive, because the thermocouples are attached to the leaf surface and may affect the leaf heat transfer. The radiometers may be preferable being non-invasive but they do not provide the spatial information of T_L. H. Albrecht et al. [46] performed measurements using an active thermography protocol applying the short infra-red radiation pulses and tested the thermal responsiveness of the contrasting leaf types (i.e., the leaf structure, the vascular tissue) of the spring barley (*Hordeum vulgare*) and the common bean (*Phaseolus vulagris*).

The steady state leaf energy balance model assumes that all the absorbed heat that originates from the absorption of the solar and thermal radiation (φ_{in}) is dissipated by the heat flux densities that include the long-wave radiative heat flux density (LW), the convective heat flux density (H) and the evapotranspiration (λE) [1891, 2310]

$$0 = \varphi_{in} - LW - H - \lambda E$$

The authors found that τ and h_{leaf} decrease in response to the increased wind-speed, which affects the boundary layer thickness. The barley leaves showed higher h_{leaf} and lower τ values compared to the bean leaves. The smaller and narrower leaves have a thinner and the more homogeneous boundary layer. However, the response to the wind was quite similar for the barley and bean leaves. If the wind is present, the leaf boundary layer thickness will be reduced and heat will be increasingly transported away from the leaf surface with the air movement.

M. B. Dickinson [749] investigated the vascular cambium cell survival after the surface fire that cause a rise and fall of the vascular cambium temperatures in the bole, i.e., the trunk [750, 905, 3192]. The wick fires were used to generate the vascular cambium temperature regimes for 30 trees from 12 Bolivian, seasonal-forest tree species [2462]. The temperature regimes were measured with a thin thermocouple inserted along the vascular cambium. The thinner the bark, the higher the maximum temperature and more rapid the temperature rise and fall.

There are two approaches to predict the tissue response to the elevated temperatures:

- the tissue necrosis occurs if the *threshold* temperature (60° C for the plant tissue [3158]) is reached—the drawback of this approach is that the longer exposures to the temperatures below the threshold value also could result in the necrosis;
- an exponential relationship between temperature and the time required for the tissue necrosis at the constant elevated temperature [2114].

K. A. Finnikov et al. [965] investigated the heat transfer in the system that includes the plants aggregation, the air layer and the ground layer in one-dimensional formulation. It was found that influence of the plants in the aggregation on the heat exchange with the grows with the plants size and the plant number per ground area.

3.4.1 Thermogenic Plants

Plants are usually considered to be ectotherms, i.e. to derive the heat they require from the environment, but there are the thermogenic plants (also "heat-producing plants") that display endothermy, i.e. have the ability to raise their temperature above that of the surrounding air [1150, 1413, 1769, 2252, 2476]. Currently 14 such plant families (e.g., *Annonaceae, Araceae, Arecaceae, Cycadaceae, Nymphaceae, Rafflesiaceae* [220]) are known.

For example, R. S. Seymour and P. Schultze-Motel [2794] reported the values as high as 35° C above air temperature and noted that the rate of the heat production increases as air temperature decreases—a large drop in ambient temperature causes a smaller drop in the flower temperature which causes an increase in the rate of the heat production by the futile involvement of the cytochrome and the alternative oxidase respiratory pathways [2792]. K. A. Nagy, D. K. Odell and S. Seymour stated that the Philodendron selloum spadices can even heat themselves to 38o C–45° C when the ambient temperature is near freezing [2252].

In some plant species (e.g., the Arum lily or the Titan Arum *Amorphophallus titanum* [1695]) the thermogenesis is associated with the production of specific odours—the elevation in the floral temperature helps the release of the volatile organic compounds (VOCs) [3259].

An increase of the plant temperature could serve, for example, to produce a warm environment that attracts insects, particularly the pollinators [2795]—for example, the bees are known to be able to detect the within-flower temperature differences greater than the 2° C [1251], to aid in the seed dispersal; it allows the plants to thrive in environments with the low ambient temperatures, such as the cool forests or the shady areas.

The heat is generated in the plant mitochondria [273, 362, 1424, 2648] that have a unique enzyme called the alternative oxidase (AOX) [1150, 1414, 2629] that is also acts as a part of the electron transport chain [3268] in a process of the cellular respiration [1194]—the set of the metabolic reactions and processes that take place in the cells of organisms to convert the biochemical energy from the nutrients into the ATP, and then release the waste products [1429][3]—the process by which

[3] The cellular respiration (or the oxidative metabolism) requires fuel (glucose or another similar sugar) and an electron acceptor which drives the chemical process of turning energy into a useable form and can be either aerobic, requiring oxygen—the most powerful electron acceptor in the nature, or anaerobic.

The aerobic respiration is an extremely efficient process allows the eukaryotes to have the complicated life functions—they perform the cellular respiration in their mitochondria—organelles that are designed to break down sugars and produce the ATP very efficiently. However, it also means that they require a constant supply of oxygen. The aerobic respiration could occur via the glycolysis, the Krebs cycle, and the electron transport chain [2852].

The chemical formula for the aerobic respiration via the glycolysis is:

$$C_6H_{12}O_6 + 6O_2 \rightarrow 6CO_2 + 6H_2O + energy(heat + 36ATP)$$

biological fuels are oxidized in the presence of an inorganic electron acceptor, such as oxygen, to drive the bulk production of the ATP—called the thermogenesis.

The AOX is an integral membrane protein that is present as a homodimer in the inner mitochondrial membrane of all plants [3161]. The electron transport to oxygen via the AOX branches from the main mitochondrial electron transport chain and is largely uncoupled from the ATP production, with energy being released as heat [3268].

The distribution of this oxidase is widespread—it has been identified in all higher plants studied as well as in many algae, fungi, yeasts and protozoa [23]. The molecular biology of the alternative oxidase was considered, for example, by L. McIntosh [2105], A. M. Wagner and A. L. Moore [3220] and by C. Hiser and L. McIntosh [1331].

The alternative oxidase and the uncoupling proteins in the plants [618, 1764, 2337, 3176] are similar to those found in the mammals.

Examples of the thermogenic plants include the dead-horse arum *Helicodiceros muscivorus*, the eastern skunk cabbage [1415, 1416, 1662, 2881] (named for the

---

During the cellular respiration, the proteins called enzymes (enzymes act as catalysts, substances that speed up chemical reactions; enzymes catalyse reactions by reducing the activation energy for a specific reaction to occur and yet are neither destroyed nor altered during this process) bind with substrate molecules—the sugar or the glucose—and help to break the molecules apart. In doing so, the cells harvest the energy stored within chemical bonds. Through several enzyme-mediated chemical reactions, this energy is transferred and stored in molecules of the ATP. When cells need energy to do any form of work, the ATP is broken down and the stored energy is released and used to fuel the chemical reactions.

The prokaryotic organisms such as bacteria and archaebacteria can use other forms of the respiration, which are less efficient. This allows them to live in the environments where the eukaryotic organisms could not, because they do not require oxygen.

The different types of anaerobic respiration, which are performed by different species of bacteria and archaebacteria, and by some eukaryotic cells in the absence of oxygen are called fermentation that involves the incomplete oxidation of sugars; there are some types of fermentation:

- *Alcoholic fermentation*—this type of fermentation, performed by yeast cells and some other cells, metabolizes sugar and produces alcohol and carbon dioxide as byproducts;
- *Lactic acid fermentation*—this type of fermentation is performed by the human muscle cells in the absence of oxygen, and by some bacteria;
- *Proprionic acid fermentation*—this type of fermentation is performed by some bacteria, and is used to make swiss cheese;
- *Acetogenesis*—is a type of fermentation performed by bacteria, which produces acetic acid as its byproduct.

The methanogenesis is a special type of the anaerobic respiration that can only be performed by he archaebacteria—a fuel source carbohydrate is broken down to produce carbon dioxide and methane.

One can't measure cellular respiration directly [1684], but can measure the its results. E. g., one can measure the amount of heat given off by an organism, the rate of consumption of chemical reactants, e.g., oxygen [2380], or rate of production of chemical products during cellular respiration. The rate of consumption or production of these materials is directly proportional to an organism's metabolic rate.

smell emitted from bruised leaves) *Symplocarpus foetidus* (this plant blooms in the early spring in the North America, is capable of reaching temperatures as high as 22° C and its inflorescence sometimes melts a hole through the surrounding ice or snow [2794] and emerge earlier than other plants—the skunk cabbage blooms and maintains the temperature of the female spadix, the spike-like flowering stalk, at close to 20° C for about a week, even when the ambient air temperature drops below the freezing point of water [1270]), the elephant foot yam *Amorphophallus paeoniifolius*, the elephant ear *Philodendron selloum*, the lords-and-ladies *Arum maculatum*, the voodoo lily *Sauromatum venosum* [2107], *Sauromattum guttatum* [2879] that produces heat to mimic the body temperature of a decomposing animal (reaching 32° C above the environment temperature [2880]) attracting thus carrion beetles, which serve as pollinators for the plant, the aquatic lotus (also the Indian lotus flower) *Nelumbo nucifera* [1768], the Amazon water lily *Victoria amazonica*, the taro (*Colocasia esculenta*) [1417]. The *Symplocarpus foetidus* is poisonous for humans and in addition to the North America could be found in Russia (Sakhalin), Japan, Korea.

A small number of the thermogenic plants are capable of the thermoregulation, that is, sensing external temperature changes and regulating heat production at the cellular level to maintain tissue temperature within a narrow range [3268].

The *Nelumbo nucifera* presents a good example of the thermoregulation in plants: the temperature is kept constant in the range between 30◦C and 35° C, regardless of the ambient temperature. Mechanisms of the thermoregulation in the *Nelumbo nucifera* and in *Symplocarpus renifolius* wee investigated by J. R. Watling et al. [3268] and by L. I. Terry et al. [3040]. The authors considered both the alternative oxidase and cytochrome C as the terminal electron acceptor that is responsible for heat generation using a combination of stable oxygen isotope and standard protein quantification techniques.

The thermogenic flowers are usually large, some (for example, the titan arum) huge ones because small flowers with high surface to volume ratios are unable to retain enough heat to raise flower temperature noticeably. The *Nelumbo nucifera* receptacles, petals and stamens were all found to heat independently of one another: the receptacle heated most above non-thermogenic tissue temperature (8.1◦C ± 1.9° C) [1150].

Another general characteristic of such flowers is that they are protogynous, that is, the female parts of the flower mature first and pollination takes place before the male parts release pollen; the major episode of the heat production in these plants corresponds to the period when the female flower parts are most receptive to pollination and when the flower's scent is strongest [2794]. N. M. Grant [1150] found that heating in the fertile male florets was characterised by a pattern similar to sterile male florets but with a lower mean peak temperature compared with the sterile male peak temperature. The sterile and fertile male florets continued to heat more than 10° C above laboratory temperature for up to 30 hours following removal from the plant. The pattern of heating continued despite removal from the plant indicates that all resources for the thermogenic phase are within the florets themselves.

Since the flowers lack the complex nervous thermoregulatory features of animals, it is apparent that regulation must occur at a cellular level and be triggered directly by changes in the flower temperature [2793].

The thermogenesis is related to the photomorphogenesis that is the process by which plants respond to light and regulate their growth and development—the plants undergo a series of changes in their morphology, physiology, and gene expression. These changes include alterations in the leaf shape, stem elongation, chlorophyll synthesis, and flowering time. The responses are mediated by the specific photoreceptors, such as phytochromes, cryptochromes, and phototropins, which are sensitive to the different wavelengths of light.

The relation between the thermogenesis and the photomorphogenesis becomes evident when considering the role of light in the regulating thermogenic processes—light acts as a major environmental factor that influences the timing and intensity of the thermogenic processes in plants. E.g., certain thermogenic plants exhibit the increased heat production during the specific times of the day or under the specific light conditions.

Part II
Models of Heat Transfer

Chapter 4
Haemodynamics

Frequently the flow of the blood is simulated assuming that

- the blood is the viscous incompressible Newtonian fluid and
- the vessels walls are rigid.

In most cases the blood flow is considered in the cylindrical pipes with the constant cross-section area.

In fact, the walls of the blood vessels are the non-linearly viscoelastic tubes that could stretch [2152] while the cross-section area of the vessels varies due to the arterial obstruction (i.e. the stenosis) or the aneurysms [198, 450, 1126, 1948] and the blood is the suspension of the wide variety of cells of different sizes and elastic properties [1813] (the red blood cells—the erythrocytes, RBCs, the white blood cells—the leukocytes, WBCs, that are subdivided into the granulocytes—the neutrophils, the eosinophils or the basophils, the lymphocytes, macrophages, phagocytes, the monocytes and the natural killer cells, the platelets, PLTs [449]) in a fluid called the blood plasma.

The leukocytes vary in size and properties, for example, a typical inactivated neutrophil cell is the approximately spherical in shape with a diameter of $8\,\mu$. Neutrophils live for about 5 days and make up 60–70% of the total WBC count [2070].

The WBCs consist of a plasma membrane, an underlying cortical layer whose primary component is the actin, the three-dimensional cytoskeleton as part of the cytoplasm, and a nucleus; the nucleus is significantly stiffer than the rest of the cell [2487].

The rheological properties of the WBCs have been described by the viscoelastic models and by a model of a drop comprised of a viscous nucleus enveloped by two layers of the less viscous fluid representing the cytoplasm and the membrane layer [3077]. The WBCs interact with other cells including the endothelial cells via their adhesion molecules and the cell surface receptors.

A. I. Zhmakin, *Heat Transfer in Vivo*, Springer Series in Biophysics 27,
https://doi.org/10.1007/978-3-031-95459-7_4

The platelets (also known as the thrombocytes) are not cells, but rather the non-nucleated disk shaped cell fragments with the characteristic size of about 2–3 μ that are mainly involved in the blood coagulation; the platelets have the numerous surface receptors and the adhesion molecules that mediate the platelet-platelet interactions as well as the interactions with the endothelium and the plasma components [2487].

The blood cells are continually produced by the bone marrow over a human's life, for example, the erythrocytes have an average lifetime of 120 days [274], thus, the new cells are produced at the rate of nearly one billion per minute [1205][1]. Other blood cells have different life spans, for example, basophils about 8 hours, thrombocytes 10 days [2282].

During their lifespan, the red blood cells withstand the tremendous stresses and deformations as they travel hundreds of kilometres through the vascular system. Their distinctive biconcave discocyte shape has evolved to meet the requirements for the large, repetitive and reversible deformations [966, 2589]. The RBC's non-sphericity permits shape changes at the constant volume and surface area. The shape of red blood cells is coupled to the ambient plasma, leading to a variety of the RBC morphologies in the blood circulation [1537].

The half a century ago P. B. Canham [428] established that the reason for a red blood cell to have a biconcave shape is the minimum of the membrane bending energy.

The RBC is a container for haemoglobin by which alone oxygen necessary for the life of all tissue is transported. The tetramer structure allows haemoglobin to bind oxygen cooperatively, meaning that the binding of oxygen to a single heme increases the likelihood that the remaining chains will bind oxygen (this is due to the allostery of haemoglobin—haemoglobin is an example of a cooperative protein because the binding of one oxygen molecule increases the affinity for subsequent oxygen molecules); the conformation of haemoglobin changes from the oxygenated state to the deoxygenated state [1320].

The haemolysis is defined as the release of haemoglobin into the plasma due to a mechanical damage of the RBC membrane. The RBC deforms in the shear flow, while preserving the volume and surface area. Above a critical shear stress, the membrane is forced to stretch. The haemoglobin may be released via the complete rupture of the cell [2573] or through the pores appearing at the high stress in the

---

[1] In fact, formation and regulation of the red blood cells in the body (*haematopoiesis*) has several stages [2528]: they originate from the haematopoietic stem cells, at the early stages are located in the yolk sacs also called blood islands, then move to the spleen, liver and lymph nodes; after birth, they are formed only in the bone marrow mostly in vertebra, pelvis and sternum.

Haematopoietic growth factors are local signalling molecules called cytokines which are supplemented by two hormones. One of them is erythropoietin (EPO), balancing the rate of erythropoiesis to the prevailing need in the circulation. RBC production is accelerated due to, e.g., instance haemorrhage, donation of blood, and chronic hypoxia. Secretion of EPO is stimulated by a fall in tissue oxygenation, and its concentration in plasma is inversely related to the partial pressure of oxygen in the arterial blood [2070].

viscoelastic membrane [3504]. The shear stress-exposure time is a key factor in the haemolysis [1830]. Below a critical shear stress threshold, the surface-effects dominate, whereas above, the shear-stress effects dominate.

The human red blood cells allow description by a relatively simple mechanical model system [877, 878] due to their structural simplicity [3080]: the mammalian RBCs lack a nucleus and any internal structure such as the intracellular organelles [1689], so that their unique components are the plasma membrane with its underlying cytoskeleton [2171] and the shape of the RBC can be directly understood as the result of its membrane properties. However, the red cells of the lower vertebrates (for example, birds) have a nucleus. The red cells vary markedly in size among mammals, e.g., those of the goat are much smaller than those of humans, but the goat compensates by having many more red cells per unit volume of the blood; the concentration of the haemoglobin inside the red cell varies little between the species [2627].

The need for the RBCs—"*why should the haemoglobin be carried in a cell rather than in solution in the blood plasma?*"—was explained by J. Barcroft [186]: the proteins of the blood plasma cannot pass through the wall of a capillary, thus they exert an osmotic pressure, tending to draw water into and hold it within the lumen of the capillaries—the haemoglobin would have a high osmotic pressure and would draw water from the tissues to make impossible the circulatory conditions. Moreover, the red blood cell eliminates the iron toxicity [377] by containing the iron within a cell membrane and increases oxygen delivery rates by maximizing the surface area to volume ratio by adopting a biconcave disc shape [1320]. Finally, the comparison of the binding parameters for haemoglobin in the red blood cells shows that encapsulation within the RBC results in significantly lowered co-operativity and higher oxygen affinity than in solution. In some simple animals with the low oxygen needs, including the small worms and mollusks, the transported oxygen is merely dissolved in the plasma [2627].

The erythrocyte functions as a single unit separate from all other cells; as it utilizes no oxygen, it has no metabolism of its own; it is always in motion. The red blood cell was the first cell seen by the human eye ("Micrographia" published by Robert Hooke in 1665), the first quantitative information on the RBC ("size, the number, the volume, the surface, and the colour of the blood corpuscles of man and animals") was presented by H. Welcher in 1863 [1205].

The blood plasma is usually regarded as a Newtonian (the shear stress is proportional to the rate of shear and the viscosity is the constant of proportionality) incompressible fluid containing water with the inorganic constituents and proteins that make up about 7% of the plasma by weight [2627] (globulin, serum albumin, fibrinogen), hormones, nutrients, and gases. Under some circumstances the effect of these solutes can become significant and lead to the non-Newtonian effects (the non-linear relationship between the shear rate and the shear stress [49]) such as the plasma hardening [3163].

Usually, the rheological properties of fluids are experimentally determined by means of the rotational viscometers—these devices, characterized by a cylinder-in cylinder or the cone-plate geometry, operate maintaining a constant shear rate or a constant stress (fixed by the rotational speed) and thus the effective viscosity is extracted from the response of the fluid in terms of the measured shear stress or the plate velocity [1796]. However, the behaviour of the complex fluids is sensitive to the flow properties, and hence the fluid response in the pressure driven flows in the narrow channels may differ from that observed in the rotational viscometers at larger scales. Thus, to study the rheological behaviour of fluids below the microscale, the alternative procedures are used which consists of using the micro-PIV techniques, in which the small beads are added to the flow, and the tracking of these beads allows the reconstruction of the velocity field, the shear stress and rate can be then computed.

An alternative measure of the viscosity—termed the effective (or apparent) viscosity—is obtained from the ratio of the pressure drop to the flow rate in a channel. The chip viscometers based on this idea have been increasingly used in microfluidics due to their simplicity, the low-cost and fast measurement process.

Thus, the rheological models are needed to correctly describe the haemodynamics in the smaller arterioles and capillaries. The apparent blood viscosity reaches its minimum at the tube diameter of about 7 μ and shows a significant increase with respect to this minimum as the tube diameter decreases or increases [937].

However, the Newtonian law is still a good approximation for the high shear rate blood flow [3116] or in tubes with diameters larger than 500–600 μm [937]. Note that the shear rate is zero at the centre of the vessel and increases towards the vascular wall, thus the local viscosity values vary across the radius of the vessel [49].

In contrast to the ordinary fluids, such as water, where the molecular time scale (the molecular rotation, and vibration) is in the range of 10^{-12} seconds, which is much smaller than the macroscopic flow scales, for a complex fluid like blood, the red blood cells move and deform over time scales comparable to the global flow, so that the macroscopic constitutive law should carry information on the dynamics of the suspended entities [3195]. Probably the first theoretical analyses of the RBC motion in the capillaries was performed half a century ago simultaneously by A. C. L. Barnard et al. [195] and by M. J. Lighthill [1870]. In both of these papers, the cell shapes were considered to have the axial symmetry and the lubrication theory was used to describe the motion of the suspending blood plasma in the spaces between the RBCs and the capillary walls. Still, it is possible in many problems to consider the blood as the uniform non-Newtonian fluid [790, 1534] that reveals the viscoelasticity, the thixotropic and shear thinning characteristics [530, 3052]. The elastic characteristics of the RBCs are the major contributor to the viscoelasticity of the blood [1534, 1796]; the mutual interaction and dynamics of the RBCs are also responsible for the two non-Newtonian properties of the blood, the shear thinning and the yield stress [1534, 3051]. The RBCs are the most numerous of the cells in the blood with a normal volume fraction of the haematocrit of 45% in adults.

The RBCs have a biconcave shape [428] that has a large surface to volume ratio that facilitates the exchange of oxygen by increasing the reaction surface. A conserved and characteristic feature of the human and animal RBCs is their flattened shape [2600]. The diameter of the undeformed RBCs is 6–8 μ with a 2–3 μ thickness, the average volume of an erythrocyte is 90 μ^3 [449].

The RBCs could be described as the deformable closed shells with a membrane characterized by the bending rigidity and the stretching elasticity [939]. Their deformability originates from the structure that consist of a lipid bilayer that can dynamically attach and detach from the underlying supporting cytoskeleton formed by the elastic spectrin protein connected to each other via the actin filaments [771, 1140]—an elastic network of spectrin which has a triangular structure (and most of these triangles form the hexagons) in the network [3175]. This geometry, as well as the intrinsic elastic properties of the spectrin, allows the RBC to be highly deformable and elastic [2825]. The RBC membrane has a phospholipid bilayer that is responsible for near-conservation of the membrane surface area. The bilayer and cytoskeleton are tightly bound through interactions between the transmembrane proteins within the bilayer and proteins of the cytoskeleton [2691]. The cytoskeleton facilitates the stretch deformation through the folding and unfolding of the spectrin proteins, which are the long molecules that bind with actin at their ends to form a triangulated network beneath the bilayer [1930]. At the low strain the membrane behaves similarly to the elastic solids, however, at the high shear deformation it behaves more like a fluid [820].

The leukocytes are much larger than the erythrocytes (their diameter ranges between 6 and 17 μ [274]) and are roughly spherical, but they exist in a smaller number in blood. The white blood cells as well as platelets could demonstrate the margination—the migration of a particle or cell in blood flow towards the periphery of the blood vessel [440]. An in vivo study by B. Woldhuis et al. [3331] showed a striking difference between the platelet distribution in the arterioles and venules, with significantly more margination occurring on the arteriolar side of the vascular system.

W. Marth et al. [2052] numerically investigated the margination of the white blood cells and its dependency on the haematocrit, the deformability of the cells and the Reynolds number. Most of the papers consider the margination in the spatially constant geometries such as the shear flow, the pipes with the cylindrical or rectangular cross-section or the plane-Couette systems. C. Bächer et al. [148] investigated the margination of microparticles/platelets in blood flow through the complex geometries typical for the vessel networks: a vessel confluence and a bifurcation using the three-dimensional Lattice-Boltzmann simulations [36, 513]. The typical distance between the two bifurcations lies within 0.4 μ up to 1 μ in the microvascular system [1118, 2511].

The authors confirmed that behind the confluence of two vessels a cell-free layer devoid of the red blood cells develops in the channel centre due to a cross-stream migration of the RBCs in the flow governed by cell—wall hydrodynamic interactions. Despite its small size of roughly one micrometre, this central cell-free layer persists for up to 100 μ after the confluence.

The two-phase model for flow of blood in narrow tubes [2200, 2804] considers a flow core consisting mainly of the RBCs and a cell-depleted layer next to the tube wall devoid or depleted of cells. The CFL has a lower viscosity in comparison with the RBC core and serves as a lubrication layer reducing the effective blood viscosity. The cross-stream migration of RBCs in tube flow is governed by cell-wall hydrodynamic interactions, which drive the cells away from the wall, and by cell–cell hydrodynamic interactions, which tend to disperse RBCs.

The cell-free layer also is found to contain a significant amount of the anti-marginated microparticles/platelets [148].

At bifurcations, however, a similar effect does not appear and the margination is largely unaffected by the geometry. In the mixed suspension of the red blood cells and the stiff particles flowing through a vessel confluence as well as a vessel bifurcation the stiff particles can be regarded as models for the synthetic drug delivery agents or the naturally occurring stiff cells such as platelets.

The study of the blood flow near the vessel bifurcation is important since the atherosclerotic plaque formation mainly takes place near bifurcations and bendings has led to the hypothesis that irregular hemodynamic conditions may play a role in initiation and progression of atherosclerotic lesions [92, 2024].

The intracellular fluid, the cytosol, has a five times higher viscosity compared to the blood plasma. While the blood plasma is over 92% water, due to plasma proteins, its viscosity at, e.g., 37° C is around 1.7 times the viscosity of the water at the same temperature [3080].

The healthy RBCs are extremely deformable and able to easily flow through the narrow vessels. However, the RBC deformability is reduced in many pathological conditions and during the blood storage [831]. A measurable decrease in the deformability of the RBCs is the common indicator of the rheological dysfunction—decreased deformability is associated with cellular stress and can impede the transit of these cells through the microvasculature [1189]. M. Musielak [2240] reviewed the methods to measure the red blood cell deformability.

Q. Guo et al. [1189] suggested to measure the RBC cortical tension as an indicator of the RBC deformability based on the critical pressure required for the RBC transit through the microscale funnel constrictions. The authors assumed that the RBC behaves as a Newtonian liquid drop when constricted through a microscale pore. For an intracellular pressure P_C P_1 and P_2 are the pressures applied from the leading and trailing edges are according to the Laplace Law

$$P_1 - P_C = T_C \left(\frac{1}{R_{c1}} + \frac{1}{R_{c2}} \right), \quad P_2 - P_C = T_C \left(\frac{1}{R_{a1}} + \frac{1}{R_{a2}} \right) \tag{4.1}$$

where R_{c1} and R_{c2}, R_{a1} and R_{a2} are the radii of the leading and trailing edges constricted by the width and height of the funnel, T_C is the cortical tension of a cell membrane [1189], a property that contributes to the deformability of the cell, the applied pressure on the entire cell

$$P_2 - P_1 = T_C \left(\frac{1}{R_{a1}} + \frac{1}{R_{c1}} - \frac{1}{R_{a2}} - \frac{1}{R_{c2}} \right). \tag{4.2}$$

Since R_{c1} and R_{c2} are both constrained by the funnel height, these two radii are approximately equal and thus

$$P_{threshold} = T_C \left(\frac{1}{R_{a1}} - \frac{1}{R_{a2}} \right). \tag{4.3}$$

The alteration of the biomechanical properties of the red blood cells is often correlated with the abnormal cellular function and sometimes with the serious diseases [3530]. For example, the malaria infected the red blood cells could be much more rigid in comparison with the healthy ones [2810, 3256]. In contrast to the healthy cells, the sick RBC tend to a more swinging and tumbling behaviour when exposed to an external flow [1001].

The humans RBCs exhibit a huge intraindividual variability, with strong correlation with sex and age: cells of men are up to 20% larger than in women, and men also present higher hematocrit in the circulatory system. Aging affects to the RBC membrane rigidity, so that old individuals present more rigid cells [3080].

Three types of the RBCs motions in a linear shear flow are reported: tank treading, tumbling, vacillating breathing (the major axis of the RBC oscillates around the shear direction accompanied with the breathing like motion) [673].

The behaviour of the RBCs in the homogeneous shear is characterized with three regimes:

1. At the low shear rates the RBCs display the rigid-body-like flipping (tumbling motion [287, 819, 967, 2736] or the wheel-like rotation (rolling motion retaining the original biconcave shape [820].
2. The increase of shear leads to a swinging motion regime.
3. At the high shear rates the membrane deforms, (partially) detaches from the cytoskeletal structure, and starts to rotate around like the tread of a tank [2877] (the tank-treading motion [7, 967, 969, 2736]—the motion of the membrane around the cell perimeter [2768]). In such flows the RBCs tend to line up and align with the flow, which in turn leads to the drop in the bulk viscosity. This shear-thinning behaviour was observed in the pure shear flow (far from walls) in the Chien experiments [539]. These experiments also showed a steep increase in the viscosity as the shear rate falls below $10\,\mathrm{s}^{-1}$ that is caused by the aggregation of the RBCs into the column-like structures (rouleaux) [215, 2696]. The RBCs finally adopt the highly deformed polylobed shapes as the shear rate is gradually increased when the viscosity contrast between the cytosol and the blood plasma is large enough [2071].

   The tank tread (TT) at high enough shear stresses adopts a nearly stationary orientation in shear flow and its membrane performs rotating motion [2071]. The transition between the tumble and tank-treading motions is due to the existence of a minimum of the elastic energy when the membrane is in static equilibrium, which is referred to as shape memory [967].

M. Levant and V. Steinberg [1829] investigated the dynamics of a single rabbit red blood cell in a planar linear flow as a function of a shear stress σ and the dynamic viscosity of the outer fluid η_0. The authors observed three RBC dynamical regimes: tumbling (TU), intermediate (INT), and swinging (or tank-treading) (TT) and found that

- the RBC dynamical states and phase diagram in the case of the RBC in-shear-plane orientation in a planar linear flow is in a good agreement with that obtained in early experiments in a shear flow for human RBCs;
- the value of the critical shear stress of the TU-TT transition surprisingly coincides with that found in early experiments in spite of a significant difference in the degree of the RBC shape deformations;
- the INT regime, which is stationary, is characterized by the strong RBC shape deformations and is observed in a wide range of the shear stresses.

The similar motions are observed for the motion of the vesicles: the tank-treading in that the vesicle orients steadily itself with a certain angle with respect to the flow direction; the tumbling and vacillating-breathing where the vesicle long axis oscillates around the flow direction, whereas its shapes undergoes a breathing motion [3195]. The difference between these two objects is the a membrane around the drop that can either be an elastic solid (capsule, for example, the RBC) or a fluid lipid bi-layer (vesicle). The lipid molecules that comprise the lipid bilayer can slide past each other relatively easily, but resist being pulled apart. The vesicles are the simple structures formed by the enclosed lipid bilayers, that share with the RBCs the capability to deform, but differ in other aspects—shape and size[1796].

J. Mauer [2070] found essential differences for the shapes of RBCs at the different haematocrits. First, the single cells are more spherical than cells in a more concentrated solution. Second, the higher the haematocrit, the more the cells are stretched in the flow direction. For both the low haematocrit and the low shear rates, cells experience only small forces between the walls and thus, can extend in the gradient direction. For both the low haematocrit and the high shear rates, cells are stretched in the vorticity direction.

J. Mauer et al. [2071] constructed a state diagram of the RBC shapes and dynamics in shear flow as a function of the shear rate and the viscosity contrast using two different simulation approaches:

- a mesoscopic particle-based approach—smoothed dissipative particle dynamics—for modelling fluid flow with the RBC membrane represented by a triangulated network of springs whose vertices are coupled to the fluid via the frictional forces;
- a finite-volume solver for the incompressible Navier-Stokes equations on unstructured grids with fluid-structure coupling implemented using an immersed boundary method.

The RBC state diagrams with various shapes and dynamics for a wide range of the flow rates and the cell confinements were also obtained in the simulation studies in two dimensions by R. Wells and H. Schmid-Schönbein [3292] and by B. Kaoli et al. [1538] and three dimensions by D. A. Fedosov et al. [940].

F. Reichel et al. [2589] noted that there is no single well-defined RBC state for the fixed flow conditions, but rather a broad distribution of states. This result is attributed to the inherent variability in the RBC mechanical properties, which is confirmed by a model that takes the variation in the RBC shear elasticity into account.

The shear-thinning happens in the small size vessels or in the regions of the stable recirculation, like in the venous system and parts of the arterial vasculature where geometry has been altered and the RBC aggregates become more stable, like downstream a stenosis or inside a saccular aneurysm [274].

The tendency of the RBCs to migrate away from the vessel walls [2761] leads to a different variety of the shear-thinning. As the blood flows through the smaller vessels, the apparent viscosity drops, as denoted by the Fahraeus-Lindqvist effect.

As the RBCs vacate the vicinity of the vessel wall, they create a plasma rich, red cell free layer (CFL) [1630, 3256] (also the cell depleted layer, CDL [2818]), which effectively acts as a lubrication layer [601, 937, 2487]. The CFL thickness is on the order of the RBC diameter [1629]. The CFL occupies a larger relative volume in smaller blood vessels.

Due to the low cell concentration, the viscosity in the CDL is reduced compared to the cellrich core. As we go towards the smaller vessels the relative effect of this layer grow, leading to an overall decreasing viscosity [906, 2514]. The cell free layer has been measured and investigated numerically the in complex geometries, such as contractions and bifurcations [33, 184, 3427].

T. W. Secomb [2759] suggested a 'modified axial train' model of blood flow to describe a layer of plasma surrounding a column of flowing RBCs—in a simple twophase model, a cylindrical central 'core' region containing RBCs, has viscosity μ_c, is surrounded by a layer of plasma with viscosity μ_p, where $\mu_c > \mu_p$. The apparent viscosity can then be obtained as:

$$\mu = \frac{1}{1 - \lambda^4 \left(1 - \frac{\mu_p}{\mu_c}\right)} \tag{4.4}$$

where λ is the ratio of the radius of the core region to the vessel radius, i.e., $\lambda = 1 - \delta/r$ where δ is the width of the cell-free plasma layer and r is the tube radius. As δ increases, λ decreases from 1.

Due to the fourth-power dependence, the factor $(1 - \lambda^4)$ is sensitively dependent on λ and relatively narrow plasma layer can have a substantial impact on apparent viscosity. The presence of a plasma layer decreases the local viscosity in the region near the wall where viscous energy dissipation would otherwise be concentrated.

D. Alizadehrad et al. [55] numerically investigated the deformation of red blood cells as a function of radial position—the shape transition in response to local shear stress and the wall effect, the radial variation of the red blood cell deformation.

To quantify the deformation of the RBCs, the gyration tensor of an RBC is calculated as

$$\hat{G} = \frac{1}{M} \sum_{m=1}^{M} (\mathbf{x}_m - \mathbf{x}_g) \otimes (\mathbf{x}_m - \mathbf{x}_g) \tag{4.5}$$

where $\mathbf{x}_m$ is the position vector of the m-th membrane particle of the RBC, and $\mathbf{x}_g$ is the center of gravity of the RBC [2106, 2312, 2372].

The gyration tensor is a symmetric 3×3 matrix with three eigenvalues, $\lambda_1 \geq \lambda_2 \geq \lambda_3$.

Assuming that the shape of a stretched RBC is an ellipsoid, the eigenvalues correspond to the diameters, and the stretching ratios can be obtained from [55]

$$s_i = \frac{\sqrt{\lambda_i} - \sqrt{\lambda_i^0}}{\sqrt{\lambda_i^0}} \tag{4.6}$$

where λ_i^0 is the i-th eigenvalue at the reference state, and $\lambda_1^0 = \lambda_2^0 > \lambda_3^0$ for the initial biconcave shape.

The authors found that the red blood cell deformation in relatively large microvessels could classified into three different regions: near-center, middle, and near-wall regions; the effects of the local shear stress and wall vary with the vessel diameter, the hematocrit, and the shear rate. The authors considered the blood flow in the circular channels for diameters of 8–50 mm, Hcts of 20–45%, and shear rates of 20–150 s^{-1}.

The width of the CDL is affected by haematocrit, the RBC deformability, vessel diameter, flow rate, the presence (or absence) of the endothelial glycocalyx layer, and the amount of time allowed for the CDL to form [2818].

J. Lauri et al. [1790] showed that the quantification of the CFL thickness and spatial distribution of the RBCs in microcirculatory blood flow could be performed by the high-speed optical coherence tomography (OCT) with 1 μ axial resolution.

When the RBCs move towards the centre of the vessel, they create a highly non-homogeneous haematocrit distribution that peaks in the middle of the channel and goes to zero next to the wall (CFL). This results in an uneven viscosity distribution and causes a departure from the Poiseuille-profile that characterizes stationary Newtonian channel flows [441].

An analysis of the viscoelastic flows requires introduction in addition to the usual dimensionless criteria of the Newtonian flow—the Reynolds and the Prandtl numbers—the Weissenberg number (Wi) and/or the Deborah number (De). The Weissenberg number named after Karl Weissenberg compares the elastic forces to the viscous forces. It can be variously defined, e.g., by the relation of the stress relaxation time of the fluid and a specific process time. The Deborah number is defined as the ratio of the time it takes for a material to adjust to the applied stresses and the characteristic time scale of an experiment probing the response of the material (the Deborah number owes its name to the Prophetess Deborah who, in the Book of Judges, proclaimed *the mountains flowed before the Lord*).

Conditions under which the Weissenberg number is equivalent to the Deborah number were analysed by R. J. Poole [2483].

4.1 Non-Newtonian Fluid Models of the Blood

The non-Newtonian behaviour of the human blood depends on many aspects such as the shear rate and the vessel diameter [352, 2764]. The human vascular system is a complex network of the different blood vessels majorly categorized as arteries, capillaries, veins—all these vessels are different from each other in terms of their materialistic properties, the physiological behaviour, and the structure [2526].Due to the complexities in the structure of the vascular network and blood composition, there is no universally agreed rheological model for blood that can be applied to any blood vessel [1534].

The non-Newtonian fluid models of the blood are divided into two groups—the generalized Newtonian models that do not account for history-dependent elastic or thixotropic effects (thixotropy—from the Greek words "thixis" (touch) and "trepein" (to turn) meaning the change or transition due to mechanical load—is the dependence of the material properties on the time over which shear has been applied [274]) and the yield stress models [1534]:

- Generalized Newtonian models
  - *Power Law Model*
    The model is also known as the Ostwald de Waele equation $\mu = k(\dot{\Upsilon})^{n-1}$ where k the consistency index, n is the power-law index (for shear-thinning fluid $0 < n < 1$). The model is valid only for a limited range of the shear rates, since it predicts the zero viscosity when $\dot{\Upsilon} \to \infty$.
  - *Cross Model*
    M. M. Cross [651] proposed a four-parameter equation that predicts the effective viscosity variation for the complete range of the shear stress $\mu = \mu_\infty + (\mu_0 - \mu_\infty)/(1 + \lambda\dot{\Upsilon})^n$ where λ is the natural time (inverse of the shear rate) at which the fluid behaviour changes from the Newtonian to power law.
  - *Carreau Mode*
    The model is an extension of the Cross model
    $\mu = \mu_\infty + (\mu_0 - \mu_\infty)/[1 + (\lambda\dot{\Upsilon})^2]^{\frac{n-1}{2}}$.

- *Walburn and Schneck Model*
  In this model [3226] , apart from shear rate, the viscosity is also depends on the haematocrit (volume percentage of RBC in blood) and total protein minus albumin (TPMA): $\mu = C_1 e^{C_2 H} \left[e^{C_4} \left(\frac{TPMA}{H^2} \right) \right] (\dot{\Upsilon})^{-C_3 H}$.

- Yield stress models

  - *Herchel-Bulkley (HB) Model*
    This three-parameter model can approximate Newtonian and a variety of time-independent non-Newtonian fluids $\mu = k(\dot{\Upsilon})^{n-1} + \frac{\tau_y}{\dot{\Upsilon}}$ where τ_y is the yield stress at which fluid starts flowing.
  - *Casson Model*
    The blood is considered as non-Newtonian yield stress medium (Casson fluid) with a constitutive equation [1369]

$$\sqrt{\tau_s} = \sqrt{\tau_y} + \mu \frac{\partial u}{\partial y} \tag{4.7}$$

    where τ_s is the shear stress, τ_y is the yield stress, μ is viscosity, u is the axial blood velocity, y is the direction normal to the vessel wall.
  - *KL Model*
    This model formulated by X. Y. Luo and Z. B. Kuang aimed to capture shear thinning behaviour of blood at both low and high shear rates $\mu = \frac{1}{\dot{\Upsilon}} \left[\tau_y + \mu_c (\alpha_2 \sqrt{\dot{\Upsilon}} + \alpha_1 \dot{\Upsilon}) \right]$
    where τ_y is the function of haematocrit.
  - *Quemada model*
    The model is also known as the modified Casson model $\mu = \left(\sqrt{\mu_c} + \frac{\sqrt{\tau_y}}{\sqrt{\lambda} + \sqrt{\dot{\Upsilon}}} \right)^2$
    When $\lambda \to 0$, this model reduces to the Casson model.

4.2 Blood Properties

The human blood is a dense suspension consisting of 55% the slightly viscoelastic fluid—blood plasma and 45% blood cells, with over 98% of the blood cells being the red blood cells (RBCs) [3005] (the volume fraction of the RBCs is also called the *haematocrit* [1780]), the inorganic constituents such as ions, mainly Na^+, K^+, Ca^{2+}, Cl^-, proteins. The blood could be considered as a two phase fluid [2818]. The complete set of the blood components is usually referred to as the whole blood [3080].

Evidently, the hydrodynamic interactions of the individual RBCs that depend on the viscosity ratio between the cytoplasm and the blood plasma and the capillary number[2]

$$Ca = \frac{\mu_0 U}{G_s} \tag{4.9}$$

where G_s is the surface shear elastic modulus, are of the fundamental importance for the haemorheology.

The blood viscosity decreases with the shear stress [769, 3291]; this shear-thinning is related to the dynamics and mutual interactions of the red blood cells [1780] due to their aggregability and deformability [539, 2873].

The effect of the deformability of the RBCs on the blood viscosity is especially strong at the high shear rates and haematocrits [1780]. "Deformability" means that cells can change their shape in response to the mechanical stresses—both the out-of-plane and in-plane deformations of the membrane elements are observed.

The macromolecules dispersed in the plasma, for example, the fibrinogen, induce an attractive force between the RBCs leading to aggregation [2115, 3222]. The disaggregation is determined mainly by the mechanical shear forces—it occurs when a shear or elongational force applied to the aggregated cells exceeds the binding force.

A. I. Maslianitsyna et al. [2056] studied the dependence of the aggregation parameters of the human red blood cells on their deformability using the laser diffractometry, the diffuse light scattering and the laser tweezers for in vitro measurements. The different osmolarity of plasma and concentrations of glutaraldehyde were used to change the deformability of the healthy red blood cells in vitro. The authors found that with the cells becoming more rigid some aggregation parameters (for example, the fraction of aggregated cells) decrease, while some of them (e.g., the hydrodynamic strength of the aggregate) stay unchanged.

---

[2] The capillary number Ca, which is the ratio of the viscous and the elastic forces, characterises the deformability—the low Ca values correspond to configurations dominated by the elastic forces where particles easily recover the equilibrium shape and deformations are small.

The capillary number could be also defined [1796] as the ratio of the characteristic elastic relaxation time (related to the cell membrane bending modulus) to the viscous time (that is the inverse of the mean shear rate along the channel)

$$Ca = \frac{\tau_k}{\tau_\eta} = \frac{\eta_0 \bar{v}_z d^2}{\kappa} \tag{4.8}$$

where $\bar{v}_z$ is the mean velocity in the direction of the flow, η_0 is the viscosity of the surrounding fluid, d is the diameter of the RBC, b is a geometrical parameter which accounts for the confinement and κ is the membrane bending modulus. For the human RBCs, the typical values of the bending modulus are $\kappa \approx 50 k_B T = 2 \cdot 10^{-19}$ J [3080].

The ratio d/b is a parameter that relates the RBC size to the size of the system to establish the confinement. The dynamics of the cell membrane are controlled by the viscosity ratio between the inner and outer regions of the cell, and the capillary number.

Under the physiological conditions, the RBCs do not interact with the endothelium via the specific receptors; however, the receptor-ligand interactions are present the under pathological conditions [2487, 2841].

The haematocrit is the primary determinant of the blood viscosity, especially at the lowest shear rates: doubling of the haematocrit results in a three to fourfold increase in the blood viscosity at the high shear rates (predominantly a mass effect), while it prompts an almost 10-fold increase within the lowest shear rate range (the mass effect plus the aggregation/rouleaux formation) [49].

The viscosity of the blood is greatly influenced by the exceptional ability of the RBC to adjust their morphology in response to the local shear forces [2071]. The decreased RBC deformability occurs when the membrane lipid bilayer loses fluidity, if the internal viscosity of the cell increases, or when the cell's size and/or the hydration status is impacted, compromising its exceptional surface area-to-volume ratio [49]. Decreased deformability promotes an increase in the blood viscosity at the high shear rates.

Because the RBCs have a biconcave discoid (disk-like) shape at rest, they form the long floppy rouleaux structures which can reversibly and continuously break down to the single flowing discocytes for increasing the shear rates [938, 2533]; this change in the microstructure could be accompanied by a tenfold viscosity drop [2737].

C. Uzoigwe [3122] assumed that the disk-like shape of the RBCs has a low inertial momentum, so that it does not rotate when flowing in the main arteries, minimizing the formation of the turbulent flows.

However, L. Lanotte et al. [1780] stated that the main reason for the blood continuous shear-thinning is not the aggregation effect but the rich dynamical behaviour of the RBCs population, which is based primarily on the inability of the RBCs to tank-tread under the physiological conditions. The lack of the membrane fluidity for the high viscosity contrast between the inner and outer fluids is the key feature which controls the RBCs behaviour.

The non-physiological blood flow conditions occur in the mechanical system that are devices that manage the blood pumping function of the heart and/or the oxygenating function of the lungs, such as the Extra Corporeal Membrane Oxygenation (ECMO) systems—the blood clot formation can be initiated in the regions of the high shear and long residence time, such as the flow separation and stagnation [256]. The complications are linked to the local flow environment where regions of the low shear rates ($< 100^{-1}$) and the pathologically high shear rates ($> 10{,}000\,s^{-1}$).

The initiation of the platelet activation leading to the blood clot formation has been reported to occur in regions of the high shear and stagnation, but also in the recirculation zones where the time during which the platelets are exposed to shear becomes an important factor [1714]. Another flow characteristic identified as a potential platelet activator is the vortical flow structures that allows blood clots to develop [320]. Risk of haemolysis on the other hand is increased during events when the pump is exposed to the negative inlet pressures, elevating the risk of cavitation [1890, 2411]. Due to the design of the pumps combined with them operating at high rotational speeds, the above mentioned flow characteristics will appear, especially regions of the high shear rate.

G. K. Batchelor [207] half a century ago suggested the rheological description of the suspensions of the force-free particles and showed that the stress of the force-free particles due to the presence of particles could be evaluated using a particle stress tensor expressed as a sum of the stresslets in a domain. C. Pozrikidis [2500] analytically derived the effective stresslet of a deformable capsule consisting of an internal fluid enclosed by a thin elastic membrane.

There are two qualitatively different microstructures in the flow [2753]:

- an extended network of red blood cell aggregates;
- a tightly packed core of blood cells surrounded by a plasma layer.

R. Lima et al. [1877] experimentally showed that the thickness of the plasma layer tends to increase by decreasing the haematocrit. J. M. Sherwood et al. [2818] noted that the cell depleted layer which forms at the walls of microvessels has been observed to be enhanced by aggregation.

The aggregate formation is relatively universal within the mammalian kingdom; the species from the broad divisions such as the monotremes (e.g., echidna), marsupials (e.g., koala), and the placental mammals (e.g., humans) all exhibit varying degrees of this phenomenon [204].

J. L. McWhirter et al. [2106] studied the aggregation of the several RBCs and identified three phases

- the disordered-discocyte, in which the cells maintain the discocyte shape with random orientation— this phase is stable for rather high capillary numbers and respond to the screening effect between cells, which attenuates the effect of the flow perturbation on each cell;
- the zigzag slipper, where cells cluster in two stripes in the centre of the channel, accommodating their shape to the neighbouring cells;
- the aligned parachute, formed by an array of centred parachutes.

Two models of the red blood cell aggregation were suggested [49, 2488]:

- the RBCs aggregates form via the fibrinogen bridging, whereby the RBCs are attracted to one another through the attachment to the macromolecules, such as the fibrinogen, in a process that interlocks cells [541];
- the fibrinogen did not necessarily cause the RBCs bridging, but due to its immense size and physical characteristics attracted fluid via the local oncotic pressure gradient [2286].

By introducing a non-linear scaling parameter, the adhesion scaling number, the effects of aging on the RBC aggregation was quantified [3080]

$$A = \frac{\eta_0 \dot{\gamma} d^3}{k_\alpha E_0}. \tag{4.10}$$

Here η_0 is the viscosity of plasma, d is the average diameter of the RBC, E_0 is the energy scale associated with the aggregation energy between the RBCs [122], k_α is a scaling factor that accounts for the relative increase in the adhesion energy when the blood ages. This parameter can be interpreted as the ratio between the characteristic viscous energy scale and the aggregation energy.

The RBCs also aggregate under the low shear conditions in the presence of the long chain macromolecules such as Dextran [2818].

The usual haemoglobin[3] concentration in the RBCs varies from 27 to 37 g/dL that corresponds to the internal fluid viscosity $\mu_1 = 5 - 15\ cP$ [2172], the normal plasma viscosity is $\mu_0 = 1.1 - 1.3\ cP$ at 37° C [1250].

4.3 Blood Flow in Small Vessels

The dynamics of the red blood cells determine the microvascular blood flow by adapting their shape to the flow conditions in the narrow vessels. In the smallest vessels of the microvascular network, the RBCs flow in a single file (the linearly stable train [394, 395]) through the microvessels with a diameter comparable to their size [1221, 2584]: for example, the RBCs have to squeeze through 0.5 μ-thick endothelial slits in the spleen and they the experience elastic reversible.

At rest, the healthy RBCs have a biconcave, discocyte shape with a diameter of roughly 8 μm and a thickness of 2 μm, minimum thickness of 1 μm. The typical cell area and volume of a healthy cell are 140 μm^2 and 90 μm^3 [3463].

Probably Van Leeuwenhoek was the first to observe the ability of the red blood cells to traverse the capillaries smaller than themselves [2856].

The ability of the RBCs to repeatedly pass through narrow blood vessels makes the deformability a key characteristic for the RBCs to efficiently deliver dioxygen to the bodily tissues and to remove the waste products [540, 2172]. Less deformability may lead to the impaired perfusion, the higher blood viscosity, ischemia, and the occlusion in the microvessels [2070].

---

[3] About 95% of the dry weight of the RBC consists of haemoglobin. Haemoglobin occupies a quarter of the human RBC volume [2991].

Haemoglobin is a protein; a molecule contains four polypeptide chains (a tetramer), each chain consisting of more than 140 amino acids. To each chain is attached a chemical structure composed of a ringlike organic compound known as a porphyrin, to which an iron atom is attached. It is the iron atom that reversibly binds oxygen as the blood travels between the lungs and the tissues. There are four iron atoms in each molecule of haemoglobin, which can bind four atoms of oxygen. The complex porphyrin and protein structure provides the proper environment for the iron atom so that it binds and releases oxygen appropriately under physiological conditions. The affinity of haemoglobin for oxygen is so great that at the oxygen pressure in the lungs about 95% of the haemoglobin is saturated with oxygen. Haemoglobin has a much higher affinity for carbon monoxide than for oxygen. Carbon monoxide produces its lethal effects by binding to haemoglobin and preventing oxygen transport.

The deformations of the mammalian red cells should satisfy two significant constraints [1221]:

- the cells must deform at constant volume, because the contents of the cell are incompressible;
- the cells must deform at nearly constant surface area, because the red cell membrane strongly resists dilation.

D. Halpern and T. W. Secomb [1221] stated that there exists a minimal tube diameter below which passage of intact cells is not possible. A cell in a tube with this diameter has its critical shape : a cylinder with the hemispherical ends [429]—for the normal human cells the length of the cylindrical part is 12.26 μm, the critical radius is 1.42 μm; when the tube diameter is slightly larger than the minimal value, the cell shape is close to its shape in the critical case, but the rear end of the cell becomes flattened and then concave with a relatively small further increase in the diameter.

4.3.1 Mechanical Properties of Red Blood Cells

The flow behaviour of the RBC at the micrometer scale has been studied both experimentally [359, 1027, 1177, 2504, 2986, 3063] and numerically [100, 935, 1177, 1537, 2312, 2765, 2768, 3001, 3413, 3416]. The RBCs flowing in the small vessels can exhibit a plethora of the shapes depending on the channel confinement, the flow velocity, the viscosity of the surrounding fluid, and the mechanical properties of the RBC itself [1184] in both the suspension-level and single-cell considerations [9]. The most frequently observed shapes are the parachute and the slipper shape. While the nearly axisymmetric parachute-shaped RBCs flow in the channel centre, the nonaxisymmetric slippers flow at an off-centred position closer to the vessel walls [2584].

The parachute shape of the RBCs occurs at the strong confinements and the high shear rates, whereas at the weak confinements, an unstable slipper shape occurs. The slipper may experience the strong deformations depending on the shear rate. The slippers all rotate and deform continuously, and can be subdivided into two types: one with rather tumbling motion and the other one with rather tank-treading motion.

There are snaking cells at the low shear rates. All slippers are unstable and their motion could be called as the stationary rotation-deformation [2070].

The axisymmetric cell shapes can be described in terms of r(s), the distance from the axis $r(s)$ and the angle between the normal to the membrane and the axis $\theta(s)$, where r, θ, ϕ are the cylindrical polar coordinates and s is arc length, measured from the nose of the cell [2760]. The membrane strain is expressed in terms of the stretch ratios or extensions $\lambda_s = ds/ds_0$ and $\lambda_{phi} = r/r_0$ relative to the unstressed shape. Subscript s denotes components in a plane containing the axis, and ϕ the azimuthal components.

The components of in-plane membrane curvature are $k_s(s)$ and $k_\phi(s)$, the bending moments are $m_s(s)$ and $m_\phi(s)$, and the components of membrane tension are $t_s(s)$ and $t_\phi(s)$. The stresses in the membrane was evaluated using constitutive relations proposed by E. A. Evans and R. Skalak [881]. The bending moment was assumed to be isotropic and proportional to the increase in total curvature of the surface:

$$m_s(s) = m_\phi(s) = B[k_s(s) + k_\phi(s) - (k_s + k_\phi)_0] \tag{4.11}$$

where B is the bending modulus, viscous resistance to membrane bending was assumed to be negligible. The membrane tensions can be represented in terms of mean (isotropic) and deviatoric (shear) components:

$$t_s = t_m + t_d, \qquad t_\phi = t_m - t_d. \tag{4.12}$$

The membrane is nearly incompressible in a two-dimensional sense, so the deformation is locally area preserving. The mean tension cannot be expressed in terms of membrane deformation—it depends on the forces acting on the membrane and is analogous to the hydrostatic pressure in incompressible fluid flow [2760].

L. Shi et al. [2825] used the spring model to simulate the skeleton structure of the red blood cell membrane and to study the lateral migration properties of the cells in two-dimensional Poiseuille flows. The simulation results showed that the lateral migration toward the centre of the channel depends on the swelling ratio, the deformability of the cells, and the height of the channel.

T. Wang et al. [3256] used the two-dimensional micro-scale simulation to investigate the red blood cell passage through both symmetric and asymmetric bifurcated vessels. The authors studied the hydrodynamic behaviour of the multiple red blood cells in a horizontal channel filled with the Newtonian fluid with a bifurcation at the central part of the vessel. The red blood cells were modelled as membrane particles connected by springs with stretch and compression resistance and bending rigidity to account for the stiffness of the cell membrane [3090, 3246]. The authors investigated the effects of the cell deformability (by increasing the cell membrane stiffness by a factor of ten), of the initial position and shape, of the haematocrit.

T. Wang et al. [3256] applied the elastic membrane model. The elastic energy stores in the spring due to the change of the length l of the spring with respected to its reference length l_0 and the change in angle θ between two neighbouring springs. The total elastic energy of the red blood cell membrane, $E = E_l + E_b$ is the sum of the total elastic energy for stretch and compression and the total elastic energy for bending

$$E_l = \frac{k_l}{2}\sum_{i=1}^{N}\left(\frac{l_i - l_0}{l_0}\right)^2, \qquad E_b = \frac{k_b}{2}\sum_{i=1}^{N}\tan^2\left(\frac{\theta_i}{2}\right) \tag{4.13}$$

where N is the total number of the spring elements, k_l and k_b are the spring constants for changes in length and bending angle, respectively, θ_i is the angle formed by the ith particle and its two adjoined springs. The bending constant is closely related to the rigidity of the membrane. A higher value of k_b results in a less deformable cell. Thus deformability of normal and hardened red blood cells can be modelled by varying spring constants for these two resistances.

The shape of the red blood cell is obtained using the minimum-energy principle. Initially, the cell is assumed to be a circle of radius r_0. The circle is discretized into N membrane particles so that N springs are formed by connecting the neighbouring particles. The shape change is stimulated by reducing the total area of the circle s_0 through a penalty function

$$\Gamma_s = \frac{k_s}{2}\left(\frac{s - s_e}{s_e}\right)^2 \tag{4.14}$$

where s and s_e are the time-dependent area and the equilibrium area of the red blood cell. The total elastic spring energy is modified as quasi-energy $E_q = E + \Gamma_s$ and the force acting on the ith membrane particle is $\mathbf{F}_i = -\partial E_q/\partial \mathbf{r}_i$.

Each membrane particle moves according to the following equation of motion: $m\ddot{\mathbf{r}}_i + \gamma\dot{\mathbf{r}}_i = \mathbf{F}$.

Here, dot denotes the time derivative; m represents the mass of the membrane; γ is a friction coefficient for numerical calculation. The quasi energy decreases with the time elapse. The final shape of the red blood cell is obtained as the quasi energy is minimized.

T. Wang et al. found that the red blood cells migrated towards the centreline of the blood vessel downstream of the bifurcation even when the upstream position of the cells was coaxial with the microchannel. The local increase of haematocrit in the region around the apex of the diverging bifurcation caused the lower velocities in this region. Flow propagated almost symmetrically or totally asymmetrically in the two daughter vessels depended on the geometry of the bifurcation. The distribution of red blood cells in the two daughter vessels was not uniform because the cells have a higher tendency to enter the nearest branch or the branch with the higher flow rate. The results also demonstrate that red blood cells in microvessels experienced lateral migration in the parent channel and blunted velocity profiles in both straight section and daughter branches, and this effect was influenced by the shape and the initial position of the cells, the haematocrit, and the membrane deformability. The variation in haematocrit may cause fluctuation in blood effective viscosity which is a measure of blood flow resistance and a crucial factor of friction against the vessel walls.

H. Noguchi [2311] investigated the dynamics of the red blood cells in the simple shear flow using a theoretical approach based on three variables: (1) a shape parameter, (2) the inclination angle θ, and (3) phase angle ϕ of the membrane rotation. The author used the model of the motion of a tank-treading particle in a shear flow developed by S. R. Keller and P. Skalak [1587]. The particle was assumed to have a fixed ellipsoidal shape

$$\left(\frac{x_1}{a_1}\right)^2 + \left(\frac{x_2}{a_2}\right)^2 + \left(\frac{x_3}{a_3}\right)^2 = 1 \tag{4.15}$$

where a_i are the semi-axes of the ellipsoid, and the coordinate axes x_i point along its principal directions. The x_1 and x_2 axes, with $a_1 > a_2$, are on the vorticity (xy) plane, and the x_3 axis is in the vorticity (z) direction.

The velocity field on the membrane was assumed to be

$$\mathbf{v}^m = \omega\mathbf{u}^m\omega\left(-\frac{a_1}{a_2}x_2, \frac{a_2}{a_1}x_1, 0\right). \tag{4.16}$$

The energy supplied from the external fluid has to be balanced with the energy dissipated inside the vesicle and on the membrane. The motion of the vesicle is derived from this energy balance. The motions of the inclination angle θ and phase angle ϕ are governed by the equations [2311]

$$\frac{d\theta}{\dot{\gamma}dt} = \frac{1}{2}[-1 + f_0 f_1 \cos(2\theta)] - \frac{f_0 d\phi}{\dot{\gamma}dt}, \tag{4.17}$$

$$\frac{d\phi}{\dot{\gamma}dt} = -\frac{(c_0/\dot{\gamma}^\star V^\star)\frac{\partial F^\star}{\partial p} + \cos(2\theta)}{2f_1[1 + f_2(\eta^\star_{in} - 1) + f_2 f_3 \eta^\star_{mb}]} \tag{4.18}$$

where $f_0 = 2/(a_1/a_2 + a_2/a_1)$, $f_1 = 0.5(a_1/a_2 - a_2/a_1)$, $f_2 = 0.5g(a_1^2 + a_2^2)$, $f_3 = 0.5E_s R_0/(f_1^2 V)$, $g = \int_0^\infty(\alpha_1 + s)^{-3/2}(\alpha_2 + s)^{-3/2}(\alpha_3 + s)^{-1/2}ds$, $\alpha_i = a_i/(a_1 a_2 a_3)^{1/3}$, $E_s = \oint \tilde{e}_{ij}\tilde{e}_{ij}dS$, $\tilde{e}_{ij} = e_{ij} - 0.5\Theta P_{ij}$, $e_{ij} = 0.5P_{ik}P_{jl}(\partial u_k^m/\partial x_l + \partial u_l^m/\partial x_k)$, $\Theta = P_{ij}\partial u_i^m/\partial x_j$, $P_{ij} = \delta_{ij} - n_i n_j$, $\eta^\star$ is the viscosity, $\dot{\gamma}$ is the shear rate, and E_s is an integral over the membrane surface, n is the normal vector of the surface.

The author found that the phase diagram of the RBCs motions divided into three regions: (1) tank treading, (2) tumbling, and (3) intermediate regions. In the intermediate regions, the RBCs exhibit intermittent or synchronized rotations of the inclination angle and phase angle.

The shearing deformation and area dilation were described using the strain energy function by R. Skalak et al. (4.49) where G_S is the membrane shear elastic modulus, CG_S is the area dilation modulus, $I_1 = \epsilon_1^2 + \epsilon_2^2 - 2$, $I_2 = \epsilon_1^2 + \epsilon_2^2 - 1$ are the strain invariants of the Green strain tensor, ϵ_1 and ϵ_2 are the principal stretch ratios. The bending resistance is modelled following the Helfrich's formulation for the bending energy [3517]

$$W_B = \frac{E_B}{2}\int_S (2k - c_0)^2 dS \tag{4.19}$$

where E_B is the bending modulus, k is the mean curvature, c_0 is the spontaneous curvature, and S is the surface area. The spontaneous curvature is non-zero since the RBC membrane is made of the different lipids on both sides (e.g., the major lipids are phosphatidylethanolamine and phosphatidylserine of the inner side, sphingomyelin and phosphatidylcholine of the outer side [2070]).

The lipid membranes are similar to the liquid crystals, composed of the axial molecules that influence each other in their orientation. In the liquid crystalline phases, an equilibrium order of the molecules exists, which can be altered by performing the work (distortion).

W. Helfrich applied the general theory of elasticity to the particular characteristics of membranes, assuming that the cell membrane can be described as a two-dimensional sheet, based on its small thickness compared to the cell length. Helfrich also proposed that, from the main types of deformations that a layer can undergo—shear, tilt, stretch and bending—only the latter plays an essential role in the membrane elasticity to characterize the shape of the RBC.

P. Balogh and P. Bagchi discretized the computational domain using a rectangular, fixed mesh of about $8 \cdot 10^7$ points. The surface of each cell is discretized using 5120 Delaunay triangles. A finite-element method is used to compute the elastic tension at each vertex of the surface triangles. The membrane forces were coupled to the bulk fluid by introducing a source term in the Stokes equation.

The authors found that within the same vessel, the haemodynamic quantities can greatly vary over time with amplitudes several factors and even orders-of-magnitude higher than the average values. This high degree of temporal heterogeneity was absent in blood flow in the straight tubes.

G. R. Lázaro et al. [1796] analysed the membrane elasticity using the elastic energy of a lipid bilayer described by the Helfrich bending energy (4.19). The authors considered a symmetric membrane (the spontaneous curvature c_0 is zero) exploiting the phase-field method [1116, 1552, 2521]—a technique for simulating the microstructural evolution that is based on expressing the total free energy functional of the heterogeneous material in terms of the free energy of its constituent phases and the interfaces—to track the membrane evolution. The authors studied the rheological and morphological properties of the RBC in a Poiseuille flow at the high confinement, restricted the investigation to the viscous regime (i.e. zero Reynolds number) and concentrated on the regime in which the interaction with the walls is dominant, characteristic of the strong confinement.

The authors performed two-dimensional simulations that qualitatively capture the actual RBC behaviour—and sometimes recover many of the 3D results [1537], and enable a more extensive parameter range.

D. A. Fedosov et al. [937] written the shear modulus of the membrane as

$$\mu_0 = \frac{\sqrt{3}k_B T}{4pl_m x_0}\left(\frac{x_0}{2(1-x_0)^3} - \frac{1}{4(1-x_0)^2} + \frac{1}{4}\right) + \frac{\sqrt{3}k_p(n+1)}{4l_0^{n+1}} \tag{4.20}$$

where l_0 is the equilibrium spring length, $x0 = l_0/l_m$.

The area-compression K and Young's Y moduli are equal to $2\mu_0 + k_a + k_d$ and $4K\mu_0/(K + \mu_0)$.

The membrane shear viscosity η_m is related to the dissipative parameters γ^T, γ^C as $\eta_m = \sqrt{3}(\gamma^T + \gamma^C/4)$ [937].

The lipid bilayers are characterized by a high surface incompressibility due to the presence of strong bonds between lipid molecules; the membrane is impermeable to water exchange. For a closed membrane such as a cell, these features imply that both the cell surface and volume are conserved at normal conditions (i.e. pores are absent and there is no vesiculation). In order to fulfil these constraints the authors introduced two Lagrange multipliers γ and ξ

$$\mathcal{F}_{mem} = \frac{k}{2} \int\limits_{\partial\Omega} c^2 \mathrm{d}S + \int\limits_{\partial\Omega} \gamma \mathrm{d}S + \int\limits_{\Omega} \xi \mathrm{d}V \tag{4.21}$$

where

- g accounting for the surface conservation acts as an effective surface tension;
- x—the multiplier for the volume conservation—acts as an effective pressure difference between the inner and outer regions.

The phase-field method is based on an order parameter, f, which varies between -1 and 1 to represent the membrane. The membrane is identified from an auxiliary scalar field defined in the entire space, and the method details the dynamics of the field, instead of specifically dealing with the evolution of the interface; this formulation avoids the problem of defining the boundary conditions at the membrane surface [427, 804, 2410, 2639].

The phase-field free energy which accounts for the bending elasticity of the membrane is written as [1796]

$$\mathcal{F}^{\phi}_{mem} = \frac{k}{2} \int\limits_{\Omega} (-\phi + \phi^3 - \varepsilon^2 \nabla^2 \phi)^2 \mathrm{d}V + \int\limits_{\Omega} \gamma |\nabla \phi|^2 \mathrm{d}V. \tag{4.22}$$

Far from the interface, the free energy has two equilibrium solutions $f_{eq} = 1$ (the inner region of the RBC) and $f_{eq} = -1$ (the external fluid).

The chemical potential could be defined as

$$\mu = \frac{\delta \mathcal{F}^{\phi}_{mem}}{\delta \phi}. \tag{4.23}$$

For a flat interface in equilibrium $\mu = 0$ the equilibrium profile in the normal direction to the interface $f_{eq} = \tanh(x/\varepsilon\sqrt{2})$; the transition between the bulk phases is therefore smooth.

G. R. Lázaro et al. deride expressions for the pressure tensor

$$P^{m}_{\alpha\beta} = -\left(\mathcal{F}^{\phi}_{mem} - \phi\frac{\delta\mathcal{F}^{\phi}_{mem}}{\delta\phi}\right)\delta_{\alpha\beta} + \nabla_{\alpha}\phi\frac{\partial\mathcal{F}^{\phi}_{mem}}{\partial(\nabla_{\beta}\phi)} - \nabla_{\alpha}\phi\nabla_{\beta}\frac{\partial\mathcal{F}^{\phi}_{mem}}{\partial(\nabla^{2}\phi)}$$
$$+ \nabla_{\alpha}\nabla_{\beta}\phi\frac{\partial\mathcal{F}^{\phi}_{mem}}{\partial(\nabla^{2}\phi)}. \tag{4.24}$$

The force density exerted by the RBC membrane is obtained from the pressure tensor

$$\mathbf{f}_{mem} = -\nabla\cdot P^{mem}. \tag{4.25}$$

To study the evolution of the order parameter the authors used an advection–diffusion equation

$$\frac{\partial\phi}{\partial t} + \mathbf{v}\cdot\nabla\phi = M\nabla^{2}\mu \tag{4.26}$$

where M is the mobility of the order parameter, $\mathbf{v}$ is the velocity field of the fluid, the advective term $\mathbf{v}\cdot\nabla\phi$ describes how the fluid pushes the cell membrane.

The authors considered a two-dimensional channel of width $L_x = b$ with non-slip conditions at the walls and length L_z sufficiently large to recover a Poiseuille flow $v_z(x) = (f_0/2\eta_0)(x^2 - (b/2)^2)z$. The viscosity contrast (the ratio between the internal over external fluid viscosities) was neglected with the reference as the justification to the paper by N. Tahiri et al. [3001] whose authors have shown that the dynamics of the RBC does not differ substantially for viscosity ratios $\eta_i/\eta_e = 1-5$.

C. Trejo-Soto et al. [3080] used different equation to describe the dynamic of the membrane

$$\frac{\partial\phi}{\partial t} = \bar{\kappa}\nabla^{2}\left\{(3\phi^{2}-1)\Phi[\phi] - \varepsilon^{2}\nabla^{2}\Phi[\phi] + \varepsilon^{2}\bar{\sigma}\nabla^{2}\phi + \varepsilon^{2}\nabla\bar{\sigma}\cdot\nabla\phi\right\} \tag{4.27}$$

where $\bar{\kappa} = 3\sqrt{2}\kappa/(4\varepsilon^{3})$ is the mean bending modulus, ε is the interfacial width, $\Phi[\phi] = -\phi + \phi^{3} - \varepsilon^{2}\nabla^{2}\phi$, $\bar{\sigma}$ is the mean surface tension of the membrane.

To model the interaction of the membrane with the surrounding fluid, C. Trejo-Soto et al. used the (4.26) and the Navier–Stokes equation

$$\rho\left[\frac{\partial\mathbf{v}}{\partial t} + \mathbf{v}\cdot\nabla\mathbf{v}\right] = -\nabla P - \phi\nabla\mu + \eta\nabla^{2}\mathbf{v} + \mathbf{f}_{ext} \tag{4.28}$$

where $\mathbf{v}$ is the velocity of the fluid, ρ is the density of the suspension, P is the pressure exerted on the fluid and $\mathbf{f}_{ext}$ are the external forces applied to the fluid.

The minimization of the Helfrich free energy for an ellipsoidic shape leads to the biconcave discocyte of the RBC as the equilibrium shape. However, to obtain an accurate model of the RBC membranes, the cytoskeleton's elastic properties must be considered. The cytoskeleton presents a low resistance to bend, with a bending modulus at least two orders of magnitude lower than that of the bilayer; it does present, however, resistance to shear and compression in the membrane layer, and it is known to play a fundamental role in inhibiting budding and vesiculation processes [3080]. Different models have been used to model the cytoskeleton's elasticity

- representing it as a spring mesh;
- consider the finite strain theory [2166].

The elastic properties of the RBC membrane are highly dependent on the specific bilayer lipid composition, the ATP concentration, age of the cell, and the temperature [2170].

The RBC membrane is a composite of two main components

- an outer lipid bilayer composed of phospholipids and cholesterol with embedded proteins, and
- a spectrin-based cytoskeleton beneath.

The bilayer and cytoskeleton are tightly bound through interactions between transmembrane proteins within the bilayer and proteins of the cytoskeleton. From a mechanical perspective, the bilayer provides resistance to bending and thus restrains membrane curvature. The bilayer also resists changes in its surface area. The cytoskeleton facilitates stretch deformation through the folding and unfolding of spectrin proteins, which are long molecules that bind by short filaments of actin at their ends to form a triangulated network beneath the bilayer [935, 1930, 2277]. The cytoskeleton can rearrange according to an external mechanical force in the flow [72].

S. Barns et al. [196] investigated the RBC mechanical properties using the atomic force microscopy (AFM) with the aim to develop a coarse-grained particle method model to study the RBC indentation in both the 2D and the 3D. S. Sen et al. [2783] developed a model of the indentation and adhesive probing of a cell membrane with the AFM for the solid-to-solid contact of the elastic, isotropic materials where the size of the contact region was negligible compared to the bodies themselves. However, the model is also limited in its application only for the RBCs which have formed perfectly symmetrical dome shapes [196].

The Hertz-based models have been widely used for analysing the experimental RBC force indentation data in order to estimate the stiffness of the membrane [1086, 1770, 2002] due to their simplicity which enables a standard equation to be routinely fitted to the experimental curves to extract an effective Young's modulus of the RBC membrane. However, there are significant limitations in reasoning the assumptions of the solid mechanics contact for samples the Hertz-based equations were developed specifically for solid-to-solid contact of elastic, isotropic materials where the size of the contact region was negligible compared to the bodies themselves [196].

The force–deformation results from the AFM are sensitive enough to detect a difference in the effective Young's modulus between the normal RBCs and those from patients suffering from some diseases [196]. Thus, the cell indentation was proposed as a possible diagnostic tool as mechanical properties of neoplastic cells have been shown to differ significantly from the healthy cells.

An advantage of the AFM over other experimental methods is that force and deformation can be measured at the different locations over the cell membrane with the high accuracy: the AFM provides control over the indentation point, meaning that force acting on the membrane can be quantified locally. S. Barns et al. found that the membrane's bending stiffness was most influential in controlling the RBC physical behaviour for indentations of up to 200 nm. The AFM can be performed while cells are submerged in the liquid, meaning the experimental conditions can replicate aspects of the physiological environment [1085]. A significant aspect of applying the AFM to the RBCs is managing the elasticity of the membrane during imaging and indentation, while protecting its natural organisation [811, 2002].

Probe shape is an important consideration, with most AFM studies of the RBCs using conical and pyramidal tips [1086, 1770, 2002]. These sharp tips can push the membrane beyond physiological limits leading to penetration and rupture. To overcome these risks, spherical probes have been considered [367].

D. A. Fedosov et al. [940] listed the features a model must have to capture the realistic behaviour of the RBCs

- the membrane viscoelasticity with a viscous contribution of the lipid bilayer and with an elastic contribution of the spectrin network;
- the membrane bending resistance;
- the separate external/internal fluids with distinct viscosities.

S. Barns et al. used the coarse-grained particle method (CGPM) to develop the numerical RBC model. Initially N particles are evenly distributed around the cell—a circle in 2D [3090] or sphere in 3D [3089]. For example, in the 3D case this is done with the aid of a spherical surface mesh composed of N vertices, N_e edges and N_t triangles. Each vertex represents the location of a membrane particle, each edge represents a linear interaction between adjoining particles, and bending interactions are present between the adjoining triangular surfaces. On average, six triangles are formed around each vertex, aligning with the cytoskeleton's junctional complex structure [1930].

The total energy stored in the membrane is the sum of energy stored through the stretch resistance of the cytoskeleton, the surface area incompressibility of the bilayer, the bending resistance of the bilayer, and the volumetric incompressibility of the internal fluid. S. Barns et al. noted that the 2D model only has three components for the total energy compared to four for the 3D since in the 2D the areal incompressibility of the bilayer is effectively combined with the spectrin's stretch resistance to oppose the relative movement of the adjacent particles—in the 3D model, stretch resistance and surface area incompressibility are separated [196]. The bilayer is responsible for the resisting bending of the membrane, thus, S. Barns et al. introduced between adjoining particles in the 2D and triangular surfaces in the 3D model.

To validate the model, the authors reproduced the experimental conditions relating to substrate adherence and indentation—both the AFM and the confocal imaging showed that the RBCs formed dome-shapes when adhered. To incorporate this into the model, a constraint was introduced for specific membrane particles to be in contact with the substrate.

The authors found that at small indentation depths, indentation force is mainly used to modify the curvature of the membrane, impacting energy stored through the bending mechanism.

With further indentation, the probe causes the membrane to curve beneath and at the same time, the cytoplasm needs to redistribute to avoid being compressed. Thus, the cytoplasm applies pressure on the membrane, which in turn develops tension and in addition to a change in the bending energy, there is also an increase in energy stored through membrane tension.

It is expected that for deeper indentations, tension on the membrane would become dominant over bending stiffness, as the cytoskeleton is stretched.

4.3.1.1 Föppl–von Kármán Equations

A set of nonlinear partial differential equations describing the large deflections of thin flat plates [472] was independently derived by August Föppl [983] and Theodore von Kármán [3208]

$$\frac{Eh^3}{12(1-\nu^2)}\nabla^4 w - h\frac{\partial}{\partial x_\beta}\left(\sigma_{\alpha\beta}\frac{\partial w}{\partial x_\alpha}\right) = P \tag{4.29}$$

$$\frac{\partial \sigma_{\alpha\beta}}{\partial x_\beta} = 0 \tag{4.30}$$

where E is the Young's modulus of the material that is assumed to be homogeneous and isotropic), ν is the Poisson's ratio, h is the thickness of the plate, w is the out–of–plane deflection of the plate, P is the external normal force per unit area of the plate, $\sigma_{\alpha\beta}$ is the Cauchy stress tensor, and α, β are indices that take values of 1 and 2 (the two orthogonal in-plane directions). A good example is the membrane of the RBC since the cell size is a few micrometres while the membrane thickness (evolutionarily constrained by the slow, haemoglobin-restricted diffusion in cytoplasm [2600]) is a few nanometres [1287].

The 2-dimensional biharmonic operator is defined as

$$\nabla^4 w = \frac{\partial^2}{\partial x_\alpha \partial x_\alpha}\left[\frac{\partial^2 w}{\partial x_\beta \partial x_\beta}\right]. \tag{4.31}$$

The main assumption (also known as the *Kirchhoff hypothesis*) is that surface normals to the plane of the plate remain perpendicular to the plate after deformation.

It is also assumed that the in-plane (membrane) displacements are small and the change in thickness of the plate is negligible.

J. L. van Hemmen and M. A. Peterson [3147] suggested the generalization of the Föppl–von Kármán equations to the case of the initially precurved sheet where the geometrically computed moment of strain replaces the notion of bending moment; as the curvature approaches zero, i.e., the sheet becomes flat, the new equations reduce to the classic ones.

The three Föppl–von Kármán equations can be reduced by introducing the Airy stress function ϕ

$$\sigma_{11} = \frac{\partial^2 \phi}{\partial x_2^2}, \quad \sigma_{22} = \frac{\partial^2 \phi}{\partial x_1^2}, \quad \sigma_{12} = -\frac{\partial^2 \phi}{\partial x_1 \partial x_2} \tag{4.32}$$

to get

$$\frac{Eh^3}{12(1-\nu^2)} \Delta^2 w - h \left(\frac{\partial^2 \phi}{\partial x_2^2} \frac{\partial^2 w}{\partial x_1^2} + \frac{\partial^2 \phi}{\partial x_1^2} \frac{\partial^2 w}{\partial x_2^2} - 2 \frac{\partial^2 \phi}{\partial x_1 \partial x_2} \frac{\partial^2 w}{\partial x_1 \partial x_2} \right) = P \tag{4.33}$$

while the Airy function satisfies the force balance (4.30).

An equation for the Airy function is derived enforcing the representation of the strain as a function of the stress [1772]

$$\Delta^2 \phi + E \left\{ \frac{\partial^2 w}{\partial x_1^2} \frac{\partial^2 w}{\partial x_2^2} - \left(\frac{\partial^2 w}{\partial x_1 \partial x_2} \right)^2 \right\}. \tag{4.34}$$

In the finite strain theory (also called large strain theory, or large deformation theory) the components of the three-dimensional Lagrangian Green strain tensor are defined as

$$E_{ij} = \frac{1}{2} \left[\frac{\partial u_i}{\partial x_j} + \frac{\partial u_j}{\partial x_i} + \frac{\partial u_k}{\partial x_i} \frac{\partial u_k}{\partial x_j} \right]. \tag{4.35}$$

Neglecting the higher order terms, and enforcing the requirement that the plate does not change its thickness, the strain tensor components reduce to the *von Kármán strains*

$$E_{11} = \frac{\partial v_1}{\partial x_1} + \frac{1}{2} \left(\frac{\partial w}{\partial x_1} \right)^2 - x_3 \frac{\partial^2 w}{\partial x_1^2}, \quad E_{22} = \frac{\partial v_2}{\partial x_2} + \frac{1}{2} \left(\frac{\partial w}{\partial x_2} \right)^2 - x_3 \frac{\partial^2 w}{\partial x_2^2}, \tag{4.36}$$

$$E_{12} = \frac{1}{2} \left(\frac{\partial v_1}{\partial x_2} + \frac{\partial v_2}{\partial x_1} \right) + \frac{1}{2} \frac{\partial w}{\partial x_1} \frac{\partial w}{\partial x_2} - x_3 \frac{\partial^2 w}{\partial x_1 \partial x_2}, \quad E_{13} = E_{23} = E_{33} = 0. \tag{4.37}$$

The first terms are the usual small-strains, for the mid-surface. The second terms, involving squares of the displacement gradients, are non-linear, and need to be considered when the plate bending is fairly large. These first two terms together are called the *membrane strains*. The last terms, involving second derivatives, are the flexural (bending) strains that involve the curvatures. These zero terms are due to the assumptions of the classical plate theory, which assume elements normal to the mid-plane remain inextensible and line elements perpendicular to the mid-plane remain normal to the mid-plane after deformation.

4.3.2 Red Blood Cell Passage Through Microvessel

Due to their high deformability, the RBCs can accept the various shapes when flowing through the microvessels [152, 2871]—e.g., the axisymmetric parachute-like shapes or the asymmetric crepe-like shapes [2487]—and even pass through the capillaries smaller than their own size [1006, 1895, 2487, 2685]. C. Pozrikidis [2503] investigated the parachute shape of the RBCs in the capillaries with the different haematocrit values and the apparent blood viscosity using the boundary-integral method with both Mooney–Rivlin and Skalak models of the bending resistance for the RBC membrane—it was found that the bending resistance must be included in order to simulate the complex shaped RBCs when they deform in the shear flow. K. Tsubota et al. [3090] studied the parachute shape of RBCs between two parallel plates for various Hct in two dimensions for various haematocrit using a semi-implicit particle method combined with a spring model.

D. A. Fedosov et al. [933, 934, 940] used the mesoscale hydrodynamic simulations (the smoothed dissipative particle dynamics method for the blood plasma) to predict the phase diagram of shapes and dynamics of the RBCs in the cylindrical microchannels—the idealized microvessels—for a wide range of the channel confinements and the flow rates. The RBC membrane was modelled by a triangulated network of springs [935, 936, 2312] which includes the elastic, bending, and viscous properties. The model also contains the area- and volume conservation constraints that provides the area compression resistance to mimic incompressibility of the lipid bilayer and cell's cytosol [2589].

A RBC is represented by a collection of N_v particles connected by $N_s = 3(N_v - 2)$ springs forming N_t triangles. The potential energy of the system could be defined as [935]

$$U(\{x_i\}) = U_{in-plane} + U_{ending} + U_{area} + U_{volume}. \tag{4.38}$$

The potential that mimics the elastic spectrin network is written as

$$U_{in-plane} = \sum_j \left[\frac{k_B T l_m (3x_j^2 - 2_j^3)}{4\xi(1 - x_j)} + \frac{k_p}{l_j} \right] \tag{4.39}$$

where l_j is the length of the jth spring, l_m is the maximum spring extension, $x_j = l_j/l_m$, ξ is the persistence length, k_p is the spring constant. In addition to the elastic contribution each spring may also have the dissipative and random force terms [935] to incorporate the membrane viscosity.

The spring forces in membrane are a combination of conservative elastic forces that can be expressed in terms of the energy potential above, and dissipative forces The first term in Eq. (4.39) is the attractive wormlike chain (WLC) potential, and the second term defines a repulsive force for n > 0 often called the power force (POW), so that we abbreviate this spring model as named WLC-POW [935]. A nonzero equilibrium spring length is defined by the balance of these two forces.

D. A. Fedosov et al. modified the spring definition by adding the viscous contribution through dissipative and random forces—such terms fit naturally in the DPD method, where interparticle dissipative interactions are an intrinsic part of the method. D. A. Fedosov et al. followed the general framework of the fluid particle model [873] to define dissipative FD and random forces for each spring as

$$\mathbf{F}_{ij}^{D} = -\gamma^{T}\mathbf{v}_{ij} - \gamma^{C}(\mathbf{v}_{ij} \times \mathbf{e}_{ij})\mathbf{e}_{ij}, \tag{4.40}$$

$$\mathbf{F}_{ij}^{R} = \sqrt{k_T}\left(\sqrt{2\gamma^{T}}d\bar{\mathbf{W}}_{ij}^{S} + \sqrt{3\gamma^{C} - \gamma^{T}}\frac{tr\left[d\mathbf{W}_{ij}\right]}{3}\mathbf{1}\right) \times \mathbf{e}_{ij} \tag{4.41}$$

where γ^T and γ^C are dissipative parameters, $\mathbf{v}_{ij}$ is the relative velocity of spring ends, $tr\left[d\mathbf{W}_{ij}\right]$ is the trace of a random matrix of independent Wiener increments $d\mathbf{W}_{ij}$.

The simulations were performed using the same viscosity for both interior cellular (cytosol) and exterior (plasma) fluids; however, the authors discussed how the viscosity contrast could affect the shape diagram.

D. A. Fedosov et al. [940] obtained different shapes of the healthy RBC for the different channel confinements and the flow rates. For example, the RBC shape is similar to the biconcave discocyte shape for the slow flow. For higher flow rates a slipper shape may be observed, which is characterized here by an off-center position within the tube so that the membrane displays a tank-treading motion due to local shear gradients resembling the tank-treading in shear flow. In the tank-treading motion, a band of membrane around the cell is alternately lengthened and shortened. This continuous deformation results in viscous energy dissipation in the membrane—the internal viscous elements represent resistance to this motion [2760].

At the highest flow rate the parachute shape is obtained, where the cell is at the channel centre and the membrane is practically not moving in the lab frame. The parachute RBC shape is mainly observed in the region of the strong confinements and high flow rates, where large he flow forces are able to strongly deform the RBC. At the weak confinements, the authors found the off-centre slippers with tank-treading motion for the higher flow rates and discocytes with tumbling motion for the lower flow rates. At even higher confinements than those considered by the

RBCs are reported to attain the bullet shape which resemble an elongated cylindrical shape with a semi-spherical cap at the front end and a semi-spherical dip at the rear end [3063].

The numerical simulations have shown that the occurrence of the stable RBC shapes under the steady-state conditions is affected by the channel geometry, the initial RBC shape and orientation, and the initial cell position in the channel cross-section at the beginning of the microfluidic channel [1177].

In addition to the stable shapes, the various dynamical states such as tumbling, swinging, and tank-treading motions of the RBCs have been reported in the numerical [100, 781, 2071, 2126, 2877, 3001] and the experimental investigations [8, 820, 967–969, 1782, 3075].

S. M. Recktenwald et al. [2584] developed the microfluidic technique for the monitoring of the RBC motion that allows to track the objects such as the RBCs in the comoving frame over several centimetres along the channel flow direction: during the tracking loop, the position of the cell is determined in the every frame and a consistency check is performed based on the position of the cell in the previous frame; the every second frame, the stage velocity is adjusted using a non-linear integral controller to keep the cell at a given position in the field of view. The procedure is repeated until the cell exits the channel.

S. H. Bryngelson et al. [396] performed the long-time simulations of the flow of the red blood cells within the cylindrical vessels in which the RBCs flow 10^5 vessel diameters. The authors considered the three-dimensional Stokes flow of the elastic capsules confined in a rigid tube.

The interior cellular (cytosol) and exterior (plasma) fluids were assumed to have the same viscosity. While the actual red blood cells have an elevated cytosol viscosity by about a factor of 5 [769] the matched-viscosity models are simpler and have been shown to be sufficiently accurate to reproduce the microcirculatory phenomenology [1008]. The red blood cell elastic membrane is described with the Skalak constitutive model, with independent shear, dilatation, and bending moduli.

The authors solved the Stokes equations

$$-\nabla p + \mu \nabla^2 u = 0 \tag{4.42}$$

with the incompressibility constraint $\nabla \cdot u = 0$ enforced by the pressure p using the standard boundary integral formulation [2500, 2568] for which the velocity is

$$u_i(\mathbf{x}_0) = u_i^{\infty}(\mathbf{x}_0) - \frac{1}{8\pi\mu} \int\limits_{\Omega} G_{ij}(\mathbf{x} - \mathbf{x}_0) \Delta\sigma_j \mathrm{d}S(\mathbf{x}) \tag{4.43}$$

where $\mathbf{x}_0$ is a point on a cell surface, Ω is the union of all surfaces, $\Delta\sigma$ is the surface traction on the fluid, G is the periodic Stokes Green's function.

S. H. Bryngelson et al. used the spherical harmonics for the cell membrane description since a relatively small number of spherical harmonic modes are required to accurately describe the cell shape.

The computational results—the space–time correlations (a space–time separation map is used to show the dynamic correlations, patterns, and coherent structures in the data [2522]) and Fourier power spectra are used to search for any recurrent behaviours and frequencies—showed that the flow has neither classifiable recurrent features nor a dominant frequency. The flow kinematics were found to be sensitive to the initial flow configuration in a way consistent with chaos and Lagrangian turbulence [396].

The phase-space reconstructions showed that a low-dimensional attractor does not exist, so the observed long-time dynamics are effectively stochastic. The authors introduced a Markov chain model for the dynamics that reproduces the statistics of the cell positions.

I. G. Gould et al. [1138] combined the advanced microscopy data with the large scale haemodynamic simulations to quantify the effect of the angioarchitecture on the cerebral microcirculation. The authors found that the wide variation of haemodynamic states in the tortuous, randomly organized capillary bed is responsible for relatively uniform cortical tissue perfusion and oxygenation; computational analysis of microcirculatory blood flow and pressure drops indicates that the capillary bed, including capillaries adjacent to feeding arterioles, are the largest contributors to hydraulic resistance.

The simulations showed that even capillaries with identical blood flow, calibre, and path length may convey the RBCs at the different velocities. Though the capillary diameter and path length contribute to blood flow resistance, the variability of the RBC velocity within the capillary bed depends on the haematocrit fluctuations.

The RBC flow in microcapillaries has been widely studied in vitro, usually under steady-state conditions. The transient RBC deformation has been mostly investigated in unbounded flow geometries, such as cell shape relaxation upon exiting a micropipette [865]. The parameter measured from relaxation experiments is the RBC time constant, defined as the time that a deformed red cell takes to recover its original biconcave shape and given by the ratio of the elastic shear modulus, that governs uniaxial deformation at constant surface area, and the surface viscosity, that governs the rate of cell shape recovering [1335]. Thus, the time constant can be considered as a measure of cell deformability. For the healthy RBCs it is of order of 0.1 s [203, 1335].

F. Guglietta et al. [1182] investigated the time-dependent response of an erythrocyte membrane to external mechanical loads via ab-initio mesoscale numerical simulations, with a focus on the detailed characterisation of the RBC relaxation time following the arrest of the external mechanical load. The authors exploited a hybrid Immersed Boundary-Lattice Boltzmann Method (IB-LBM) coupled with the Standard Linear Solid model (SLS) to account for the RBC membrane viscosity.

LBM is a numerical approach which is grounded on the dynamical evolution of fluid molecules distribution functions $f_i(\mathbf{x}, t)$ which represent the density of fluid molecules with velocity $\mathbf{c}_i$ at position $\mathbf{x}$ and time t

$$f_i(\mathbf{x} + \mathbf{c}_i \Delta t, t + \Delta t) - f_i(\mathbf{x}, t) = -\frac{\Delta t}{\tau} \left(f_i(\mathbf{x}, t) - f_i^{eq}(\mathbf{x}, t) \right) + f_i^{(F)} \tag{4.44}$$

where τ is the relaxation time, $f_i^{(F)}$ is the force density, $f_i^{eq}(\mathbf{x}, t)$ is the equilibrium distribution function

$$f_i^{eq}(\mathbf{x}, t) = w_i \rho \left(1 + \frac{\mathbf{c}_i \cdot \mathbf{u}}{c_s^2} + \frac{(\mathbf{c}_i \cdot \mathbf{u}))^2}{c_s^4} - \frac{\mathbf{u} \cdot \mathbf{u}}{c_s^2} \right) \tag{4.45}$$

where ρ is the fluid density, c_s is the speed of sound.

Since two different fluids (one outside and one inside the membrane of a capsule) are considered, the relaxation time is not constant but it depends on the position, i.e. $\tau = \tau(\mathbf{x})$.

The the hydrodynamics quantities (the density and the velocity of the fluid) are written as

$$\rho(\mathbf{x}, t) = \sum_i f_i(\mathbf{x}, t), \quad \rho\mathbf{u}(\mathbf{x}, t) = \sum_i \mathbf{c}_i f_i(\mathbf{x}, t). \tag{4.46}$$

The properties of the membrane are defined through its free energy, whose main contributions are the bending and the strain energies. F. Guglietta et al. [1182] used the Helfrich's formulation for bending energy (4.19) and the Skalak constitutive law [2872] for the strain energy function.

F. Guglietta et al. used the viscosity of the membrane to relates the 2D viscous stress $\hat{\tau}^{\nu}$ to the strain rate tensor $\dot{\hat{E}}$, i.e. the time derivative of the strain tensor. The authors considered two contributions to the strain rate tensor—the shear and the dilatational ones. They exploited the standard linear solid (SLS) model to describe the viscoelastic behaviour of the membrane that can be considered as the linear combination of the Kelvin-Voigt model (in which a dashpot is connected in parallel with a spring) and the Maxwell model (in which the dashpot and an artificial spring are connected in series). The Kelvin-Voigt model can successfully describe membrane creep, while the Maxwell model accurately characterises the stress relaxation; their combination, the SLS model, correctly accounts for both phenomena.

G. Tomaiuolo and S. Guido [3062] studied the transient behaviour of the RBC shape in the confined Poiseuille flow in vitro using the high-speed video microscopy imaging. It is found that in the start-up flow, the RBCs undergo a complex transition from the biconcave shape at rest and reaching an asymptotic steady state configuration, for example, a parachute-like shape, through the membrane folding and cytoplasm reorganization. The steady state configuration was found to depend on the cell velocity, the cell orientation in the capillary, and the cell size.

An initial inwards bending of the rear portion of the rim leads to a folding onto the central thinner region of the cell body. Such folding can be attributed to the shearing action of the external flow which tends to bend the thicker rim around the inner region; at the same time, the cell leading edge becomes more pointed in the flow direction and bulges outwards due to a cytoplasmic redistribution [3062].

In order to provide a detailed description of the transient shape evolution, G. Tomaiuolo and S. Guido [3062] used the image analysis techniques to measure three shape parameters as referred to the two-dimensional shape projection, i.e., extension, dispersion, and orientation, based on the zero, first, and second order area moments [3412].

N. Takeishi et al. [3005] numerically investigated the effect of a wide range of viscosity ratios $\mu_1/\mu_0 = 0.1 - 10$, the non-dimensional shear rates and volume fractions ϕ on the behaviour of the RBCs at the various capillary numbers in the wall-bounded shear flow from the dilute ($\phi = 6 \cdot 10^{-4}$) to the dense suspensions ($\phi = 0.41$). The contribution of the individual deformed RBC modelled as a biconcave capsule to the bulk suspension rheology is quantified by the stresslet tensor [207]. The authors considered a cellular flow consisting of the plasma and the RBCs with radius a in a rectangular box with a resolution of 8 fluid lattices per radius of the RBC that was modelled as a biconcave capsule, or a Newtonian fluid enclosed by a thin elastic membrane, with a major diameter $2a$ and the maximum thickness $a/2$.

The membrane was considered as an isotropic and hyperelastic material and the surface deformation gradient tensor $\hat{F}_s$ is defined by $d\mathbf{x}_m = \hat{F}_s \cdot d\mathbf{X}_m$ where $\mathbf{X}_m$ and $\mathbf{x}_m$ are the membrane material points of the reference and deformed states, respectively. The local deformation of the membrane is measured by the Green-Lagrange strain tensor

$$\hat{E} = \frac{1}{2}(\hat{C} - \hat{I}_s) \tag{4.47}$$

where $\hat{I}_s$ is the tangential projection operator.

The Cauchy tension $\hat{T}$ is related to the elastic strain energy per unit area $w_s(I_1, I_2)$

$$\hat{T} = \frac{1}{J_s}\hat{F}_s \cdot \frac{\partial w_s(I_1, I_2)}{\partial \hat{E}} \cdot \hat{F}_s^T \tag{4.48}$$

where $I_1 = \lambda_1^2 + \lambda_2^2 - 2$ and $I_2 = \lambda_1\lambda_2 - 1$ are the two invariants of the in-plane strain tensor $\hat{E}$, λ_1 and λ_2 are the principal extension ratios, the Jacobian $J_s = \lambda_1\lambda_2$ expresses the ratio of the deformed to the reference surface areas.

The strain energy function satisfies the Skalak constitutive law [2872]

$$w_s^S = \frac{G_s}{4}(I_1^2 + 2I_1 - 2I_2 + CI_2^2) \tag{4.49}$$

where C is a coefficient representing the area incompressibility [3005].

The behaviour of the RBCs subjected to the shear flow is quantified by two different orientation angles θ and ψ where θ is the angle between the major axis of the deformed RBC and the shear direction, and ψ is the angle between the vortex axis and the normal vector at the initial concave node point [3005].

The contribution of the suspended particles to the bulk viscosity is expressed in terms of the particle stress tensor [207]

$$\hat{\Sigma}^{p} = \frac{1}{V}\sum_{i=1}^{N} \hat{S}_i \tag{4.50}$$

where V is the volume of the domain and $\hat{S}_i$ the stresslet of the i-th particle.

C. Pozrikidis [2499] derived the stresslet of a deformable capsule for an arbitrary viscosity ratio:

$$\hat{S}_i = \int \left[\frac{1}{2}(\mathbf{x}\otimes\mathbf{q}+\mathbf{q}\otimes\mathbf{x}) - \mu_0(1-\lambda)(\mathbf{v}\otimes\mathbf{n}+\mathbf{n}\otimes\mathbf{v})\right] \mathrm{d}A_i \tag{4.51}$$

where $\mathbf{x}$ is the membrane position relative to the centre of the RBC, $\mathbf{q}$ is the load acting on the membrane including a contribution of bending rigidity, $\mathbf{n}$ is the surface normal vector, μ_0 the outer fluid (plasma) viscosity, $\mathbf{v}$ the interfacial velocity of membrane, and A_i is the membrane surface area of the i-th RBC.

4.3.3 Red Blood Cell Passage Through Bifurcated Vessels

The red blood cells tend to slow down and aggregate around the apex of the diverging bifurcation. The geometry of the microchannel, the haematocrit percentage and the deformability of the cell membrane influence the formation and the size of the depleted zones.

The RBCs exhibit complex behaviour under the transient flow conditions when entering the small vessels from a larger reservoir and at the bifurcations [9, 2504]. G. Tomaiuolo and S. Guido [3062] investigated the start-up of the capillary flow in the microfluidic experiments and found that the RBCs transit from the initial biconcave shape at rest to a final parachute-like configuration and the time scale of this shape evolution from a shape at rest to a cell shape in the flow was roughly 0.1 s.

Microvascular system is a network of microvessels connected by irregular short bifurcated segments. In the microcirculation, the red blood cells flowing through bifurcations may deform considerably due to combination of the different phenomena that happen at the micro-scale level, such as: the attraction effect, the high shear, and the extensional stress, all of which may influence the rheological properties and flow behaviour of blood.

The local blood flow behaviour at the vicinity of the apexes of the diverging and converging bifurcation significantly affects the haemodynamics in microcirculation, such as the heterogeneous distribution of red blood cells throughout the microvasculature, the effective viscosity, and blood flow velocity [2513, 3256]. The blood flow behaviour in the bifurcations presents difficulties in analysis since it comprises the motion of extremely deformable cells in geometrically complex regions.

The blood flow in microcirculation is affected by many factors. One of the most crucial factors is the haematocrit—the volumetric percentage of the red blood cells in the whole blood. Due to the Fahraeus-Lindqvist effect,[4] the tube haematocrit in smaller branches can be as low as 10–20%, much lower than in the main vessels [3256].

The variation in haematocrit may cause fluctuation in blood effective viscosity which is a primary measure of blood flow resistance and a crucial factor of friction against the vessel walls [3256]. Another important factor that affects the microscopic blood flow is the deformability of the red blood cell—the red blood cells deform into various shapes to compensate the flow resistance [3238, 3246, 3251].

The partitioning of the RBC fluxes between the branches of a diverging microvessel bifurcation is generally not proportional to the flow rates, as RBCs preferentially enter the higher-flow branch leading to a higher haematocrit in that branch and a lower haematocrit in the low-flow branch. This nonuniform partitioning behaviour (also known as the phase separation effect [2630] or sometimes called the Zweifach-Fung effect [1018, 2988]) results in a highly heterogeneous distribution of the haematocrit among microvessels. The main cause of unequal partitioning of the haematocrit in the bifurcating microvessels is the presence of a cell-free or cell-depleted layer near the wall of the upstream mother vessel [185, 2768]. The fraction of the RBCs entering each branch in a bifurcation depends on the fraction of bulk blood flow entering the branch, the ratio of the two daughter vessel diameters, the size of the mother vessel, and the upstream haematocrit distribution [454, 709, 859, 948, 2510]. The lower-flow branch tends to receive a higher fraction of flow from this peripheral region, resulting in a lower haematocrit in that branch. Under standard physiological circumstances, a branch receiving typically one fourth of the blood inflow will see its hematocrit (volume fraction of red blood cells) drop down to zero, which will have obvious physiological consequences [797]. Under

[4] The Fahraeus-Lindqvist effect describes how the viscosity of the blood changes with the diameter of the tube it travels through: there is a 'decrease in viscosity as the tube's diameter decreases' because erythrocytes move over to the centre of the vessel, leaving only plasma near the wall of the vessel. The existence of this effect is proved not experimentally only but in the numerical simulations as well, e.g., of the blood flow as of a suspension of red blood cells using the dissipative particle dynamics method in microtubes ranging from 10 to 40 μm in diameter is simulated in three dimensions for values of hematocrit in the range of 0.15–0.45 by D. A. Fedosov et al. [937]. The velocity profiles for different hematocrit values show an increase in bluntness with an increase in hematocrit.

In the range between roughly 300 and 10 μ of the tube diameter, the effective viscosity decreases up to 4–5 times [3080]. P. Gaehtgens [1026] demonstrated that this reduction continues down to the capillaries with diameters of 4–5 μm. However, the apparent viscosity increases for vessels with diameters less than this, and tends to infinity as a critical diameter is reached [1221].

Because the size of the red blood cells is non-negligible comparing to the diameter of the blood vessels, the distribution of the cells is not homogeneous in the arteriole scale.

Due to their deformability, the red blood cells migrate away from the vessel walls, resulting in a region of the reduced cell concentration close to the walls, namely—a cell-poor, plasma-rich low viscosity layer [49]—"cell free layer". The cell free layer has been measured and investigated numerically [33, 184, 3427].

certain circumstances even an inversion of the Zweifach-Fung effect may occur [2814].

S. Roman et al. [2630] investigated the partitioning of the RBC fluxes in vitro for channels with diameters $< 20\,\mu$ for the intermediate haematocrits, that is physiological conditions under which the effect is significant, using microfluidic devices that model microvascular bifurcations. The position of the RBC flow separating line between both daughter branches was determined. The phase separation effect was found to be greater than previously described in simulations by V. Doyeux et al. [798].

T. W. Secomb et al. [2768], T. W. Secomb [2760] and J. O. Barber et al. [185] applied the two-dimensional model[5] to simulate the motion and deformation of the mammalian red blood cells in microvessels and to investigate the effects of the cell–cell mechanical interactions on the RBC partitioning in the bifurcations.

Due to their high concentration, the RBCs exert a strong influence on the flow properties of blood, both in bulk flow and in microvessels. For the steady laminar flow of a Newtonian fluid in a cylindrical tube, the volume flow rate Q is given by the Poiseuille's Law:

$$Q = \frac{\pi}{128}\frac{\Delta p D^4}{L\mu} \tag{4.52}$$

where L is the tube length, D is the diameter and μ is the plasma viscosity.

For the blood flowing in the microvessels frequently the apparent viscosity is defined as [2760]:

$$\mu_{app} = \frac{\pi}{128}\frac{\Delta p D^4}{Q\mu} \tag{4.53}$$

and the relative apparent viscosity as $\mu_{rel} = \mu_{app}/\mu$.

The RBC membrane and cytoplasm are represented by sets of the viscoelastic elements (the element represents the cross-sectional shape of a three-dimensional cell in a plane through the centre of the cell—the "model cell plane"; the cell membrane lying in this plane is represented as a chain of straight elastic elements, hinged at the nodal points) immersed in a low Reynolds number flow. The fluid flow around the model cell results in loadings on the external elements. The longitudinal (tension) force $t_i(s)$, the transverse (shear) force $q_i(s)$ and the bending moment $m_i(s)$ acting in external element i are consequently functions of distance s along the element. The equations of mechanical equilibrium for this system are given by Secomb et al. [2768].

---

[5] Reducing the dimensionality simplifies the problem, but at the cost of potentially introducing unrealistic features. Three-dimensional simulations of the motion of the deformable RBCs including real membrane properties have so far considered only a single cell [842, 2502] while the motions of multiple interacting cells have been simulated in 3D using a simplified representation of the RBC mechanics [830].

The coupling of the membrane and fluid forces is represented by the following equations:

$$\frac{dt_i}{ds} = -g_i, \quad \frac{dq_i}{ds} = -f_i, \quad \frac{dm_i}{ds} = q_i, \tag{4.54}$$

where s is distance from node i along the ith external element, $t_i(s)$, $q_i(s)$, and $m_i(s)$ are the membrane tension, the shear force, and the bending moment per unit length, respectively, and g_i and f_i are the tangential and normal forces acting on the membrane resulting from the external fluid forces and the internal pressure.

By integrating these equations, the forces that each external element exerts at a node can be expressed in terms of the fluid loadings acting on the element and the mean forces in the element. No fluid loading acts on the internal elements.

The internal elements represent both the internal viscosity of the cell and the cell membrane's resistance to three-dimensional out-of plane deformations. Membrane bending elasticity is represented by an elastic resistance at the nodes.

The effects of viscous resistance of the membrane to shear deformation that it experiences in 3D deformations and the viscosity of the cell interior, the resistance of the membrane to bending is represented by making the hinge points between the membrane elements elastic [2768]—the two effects, the viscous resistance of the membrane to deformation out of the model cell plane and the internal viscosity of the cell, are lumped together in the two-dimensional model.

With the proper choice of the parameters of the RBC mechanical properties, the tank-treading frequency and cell elongation agree with observations over a range of shear rates. In simulations of the RBC motion in the capillary-sized channels, initially circular cell shapes rapidly approach shapes seen experimentally in the capillaries, convex in front and concave at the rear. An isolated RBC entering an 8 μ capillary close to the wall is predicted to migrate in the lateral direction as it traverses the capillary, achieving a position near the centreline after travelling a distance of about 60 μ [2768].

The authors identified several types of the two-cell interactions that can affect partitioning

- 'trade-off': a cell entering one branch causes a following cell to enter the other branch;
- 'herding': the leading cell is caused to enter the same branch as the following cell;
- 'following': the trailing cell is caused to enter the same branch as the leading cell.

With increasing the haematocrit, the frequency of interactions increases, and more uniform partitioning results [185].

The distribution of the haematocrit in the bifurcations influences the functional behaviour of microvascular networks in several ways:

1. The effective viscosity of the blood in the microvessels depends on the haematocrit [2515], thus, the partition of the haematocrit influences the vessel flow rates.
2. The flow-dependence of haematocrit partitioning induces a correlation between flow rate and haematocrit in networks, which in turns leads to a reduction in the average network haematocrit [185]—is the so called generalized Fahraeus principle [2512].
3. Oxygen is carried by blood in the form of oxyhaemoglobin within the RBCs thus it is influenced by haematocrit distribution.
4. The dependence of haematocrit partition on flow rates, which themselves depend on haematocrit, indicates that cell–cell interactions are important [773, 2510] and results in a nonlinear coupled system that shows complex dynamics including oscillatory behaviour [455].

J. O. Barber et al. used a two-dimensional model to simulate the motion of a single RBC passing through small diverging bifurcations, with diameters 6–13 μ and found that there are two effects that can significantly alter the RBC partitioning [185]:

- cell migration: a cell drifts towards the centre-line of the mother vessel upstream of the bifurcation;
- cell obstruction: a cell that passes near the entrance of a daughter branch tends to enter that branch, even when its centre of mass lies on a background fluid streamline that enters the other branch.

J. O. Barber et al. investigated the distribution of the haematocrit in the bifurcations that has important implications for the functional behaviour of the microvascular networks and analysed different types of interactions leading to a switch of the branch where the cells eventually enters.

The non-spherical particles can migrate across the streamlines [137, 539] and this may have a significant effect on the RBC partitioning in the small bifurcations.

In the computation of the fluid flow, the fictitious domain method was utilized to solve the Navier-Stokes equations. In this method the irregular-shaped domain is extended to regular-shaped so that a simple structured mesh instead of unstructured mesh can be used, which substantially reduces computational complexity of the algorithm [1102, 1103]. The flow region Ω_f is embedded in a rectangular domain Ω. Then the fluid flow containing red blood cells was solved in the bigger domain Ω and the no-flow condition in the solid region was treated as constraints $\mathbf{u} = 0$ in $\Omega \,/\, \Omega_f$.

The immersed boundary method developed by C. S. Peskin [2436, 2437] was adopted to couple the cell deformation with the fluid. The force located at the immersed boundary node $X = (X_1, X_2)$ affects the nearby fluid mesh nodes $x = x_1, x_2$ through a discrete function:

$$\mathbf{F}(\mathbf{x}) = \sum_{x} \mathbf{F}(\mathbf{X}) D_h(\mathbf{X} - \mathbf{x}), \quad \mathbf{X} - \mathbf{x} \leq h, \tag{4.55}$$

where $D_h(\mathbf{X} - \mathbf{x}) = \delta_h(X_1 - x_1)\delta_h(X_2 - x_2)$ with the one-dimensional discrete δ_h functions [3256]

$$\delta_h(z) = \begin{cases} \frac{1}{4h}\left(1 + \cos\left(\frac{\pi z}{2h}\right)\right), & |z| \leq 2h, \\ 0, & |z| > 2h. \end{cases} \tag{4.56}$$

The movement of the immersed boundary node is affected by all the nearby fluid mesh nodes through the same discrete function

$$\mathbf{U}(\mathbf{X}) = \sum h^2 \mathbf{u}(\mathbf{x}) \cdot D_h(\mathbf{X} - \mathbf{x}), \quad \mathbf{X} - \mathbf{x} \leq h \tag{4.57}$$

and the position of the immersed boundary node is updated by after each time step Δt

$$\mathbf{X}_{t+\Delta t} = \mathbf{X}_t + \Delta t \mathbf{U}(\mathbf{X_t}). \tag{4.58}$$

The current models describe the red blood cell as Newtonian fluid enclosed by the cell membrane. There are two groups of the models

- *elastic membrane models* focus on the elastic property of the membrane and try to reproduce it through the strain energy functions [151, 1926];
- *spring models* represent cell membrane with a network of springs [935, 936, 2312, 3090, 3246].

V. Leble et al. [1799] used a micro-PTV system to investigate the flow of the RBCs through diverging and converging bifurcations with dimensions comparable to the arteriole-sized vessels using two media:

- the human red blood cells suspended in dextran 40 with about 14% of the hematocrit level;
- the pure water seeded with the fluorescent trace particles.

All the measurements were performed in the centre plane of rectangular microchannels using a constant flow rate of about $3.0 \cdot 10^{-12}\ \mathrm{m^3/s}$.

The authors found that the behaviour of the RBCs is asymmetric at the divergent and convergent side of the geometry while the velocities of tracer particles suspended in the pure water are symmetric. The formation of a RBC-depleted zone immediately downstream of the apex of the converging bifurcation is a result the enhanced adhesion of the RBCs to the wall surface in this region.

The authors also studied the RBCs motions in the high-haematocrit blood flowing through a stenosed microchannel [1016] and found that the trajectories of the healthy RBCs become asymmetric before and after the stenosis, though trajectories of the tracer particles in the pure water were almost symmetric between the upstream and the downstream of the stenosis; the asymmetry was larger in a 10% Hct case than in a 20% Hct case. The authors claimed that the deformability is the key factor in the asymmetry of the trajectories and the thickness of cell-free layer that was confirmed by experiments with the RBCs hardened by the glutarardehyde treatment. The RBCs are deformed by the traction forces generated by the inner- and outer-fluid of the membrane, which is dependent on the history of deformation along the trajectories. Thus, the deformation of the RBC becomes asymmetric before and after the stenosis.

H. Polwaththe Gallage et al. [2480] investigated the motion and deformation of a single two-dimensional RBC through a stenosed capillary using the elastic spring network to model the RBC membrane, while the RBC's inside fluid and outside fluid were treated as smoothed particle hydrodynamics (SPH) particles. The elastic energy stored in the springs due to stretching/compression was written as

$$E_l = \frac{1}{2}K_l \sum_{i=1}^{N} \left(\frac{l_i - l_0}{l_0} \right)^2 \tag{4.59}$$

and bending energy as

$$E_b = \frac{1}{2}K_b \sum_{i=1}^{N} \tan^2 \left(\frac{\theta_i}{2} \right) \tag{4.60}$$

where l_i, l_0 and θ_i are the length of the ith spring, original length of the same spring and the angle between two consecutive springs, K_l and K_b are the spring constants for stretching/compression and bending, N is the number of spring.

The authors studied the effect of the RBC's membrane stiffness, the inlet pressure and the geometrical parameters of the capillary on the motion and deformation of the RBC. During its motion through the stenosed capillary it exhibits the maximum deformation when it flows through the stenosed area; once it leaves the stenosed section, it recovers its typical deformed parachute shape. The mean velocity of the RBC also reaches its maximum value, when the cell passes through the stenosed area. The authors found that the energy related with the bending deformation is a direct measurement of RBC's bending deformation and shows a maximum value when the RBC passes though the stenosed area.

The deformability of the RBCs enhance the displacement perpendicular to the wall when they flow through a stenosis induced by the interaction between the RBCs and the stenosed wall. If one puts two rigid spheres in a simple shear flow, they interact hydrodynamically. The trajectories of two spheres, however, are symmetric before and after the interaction, as was shown by Batchelor and Green [208]. In contrast, if the particles are deformable, such as droplets and capsules, the trajectories become asymmetric. The displacement perpendicular to the flow, before and after the interaction, has been investigated numerically by E. Lac et al. [1752].

The effects of cell aggregation have generally been studied in straight vessels or capillary tubes in order to simplify the analysis. At microvascular bifurcations, the non-uniform the haematocrit distribution can lead to the branch with a lower proportion of the flow receiving disproportionately fewer cells, in an effect known as plasma skimming [1810].

The bulk measurements of haematocrit and flow rate allowed to quantify the plasma skimming phenomenon and to identify the most important parameters—the parent vessel haematocrit, the tube diameter and flow ratio (the proportion of the flow from the parent branch entering a given daughter branch) [2818].

X. Yang et al. [3402] investigated the white blood cell motion in the asymmetric bifurcations experimentally while C. Sun and L. L. Munn [2976] studied the interaction of six red blood cells flowing behind a white blood cell in the vicinity of vessel junctions.

4.3.4 Behaviour of Red Blood Cells in a Wall-bounded Shear Flow

The effect of viscosity ratio $\lambda = \mu_1/\mu_0$ on the steady motion of the single cell in the shear flow has been well investigated D. Cordasco and P. Bagchiet [629] and by J. Mauer et al. [2071].

The effective viscosity η of dilute suspension of hard spheres grows linearly with the volume fraction of particulate π according to:

$$\eta = \eta_0 \left(1 + \frac{5}{2}\phi \right) \tag{4.61}$$

where η_0 is the solvent viscosity.

M. Scott [2753] extended this result to the case when the particle is a drop of the incompressible Newtonian fluid with viscosity $\tilde{\eta}_0$ assuming that the shape of the particle remains spherical, and that departures from symmetry have negligible effect

$$\eta = \eta_0 \left(1 + \frac{\eta_0 + \frac{5}{2}\tilde{\eta}_0}{\eta_0 + \tilde{\eta}_0}\phi \right). \tag{4.62}$$

The behaviour of the deformable particles in the suspension was numerically studied by J. R. Clausen et al. [597, 2583], A. Kumar et al. [1731] and by D. Matsunaga et al. [2067] as the neo-Hookean spherical capsules and by M. E. Rosti et al. as the viscoelastic materials [2646, 2647].

M. E. Rosti and L. Brandt [2646] considered the deformable particles with a viscous hyper-elastic behaviour in the Couette flow. These are a special class of elastic materials where the work done by the stresses during a deformation process is dependent only on the initial and final configurations, the behaviour of the material is path independent and the stored strain energy function or the elastic potential can be defined.

The human blood is a dense suspension consisting of 45% blood cells, however, in narrow microvessels the concentration of cells reduces down to $\approx 10\%$ [1019]. The interactions between the RBCs consequently decrease and the dynamics of the cells is dominated by the interactions with the walls—this effect requires tubes of the similar width to that of the cell, roughly 10 mm, which includes a large proportion of the arterioles and capillaries that compound the circulatory system, as well as many of the typical widths of the channels currently used in the microfluidic devices [1796].

4.3.4.1 Behaviour of the Single RBC

N. Takeishi et al. [3005] studied the RBC placed perpendicular to the shear direction that keeps flipping along the vortex angle for low values of the capillary number (the tumbling motion) or randomly placed RBCs that tends to orient parallel to the shear plane.

The numerical results showed that the stable rolling RBCs has higher intrinsic viscosity than the tumbling RBCs. When Ca increases, the rolling motion becomes unstable. For example, RBCs subjected to the highest value of Ca investigated fluctuates showing a multilobeslike shape [1781].

4.3.4.2 Behaviour of the RBCs in the Semi-dilute and Dense Suspension

N. Takeishi et al. studied the individual RBCs in semi-dilute (the volume fraction $\phi = 0.05$) and dense (the volume fraction $\phi = 0.41$) suspensions.

It was found that in the semi-dilute suspensions, the RBCs subjected to the low Ca = 0.05 show small deformations, where both the rolling and tumbling motions coexist. In the dense suspensions, however, due to the high packing, the RBCs are forced to exhibit only the swinging motion resulting in the large elongations even for low Ca = 0.05 [3005].

The authors stated that the single RBC subjected to low Ca tends to orient to the shear plane and exhibits the rolling motion as a stable mode associated to higher intrinsic viscosity than the tumbling motion. As Ca increases, the mode shifts from the rolling to the swinging motion, and the intrinsic viscosity decreases. The hydrodynamic interactions (the higher volume fraction) also allows the RBCs to exhibit the tumbling (the presence of the walls partially inhibits the tumbling motion [1796]) or swinging motions resulting in a decrease of the intrinsic viscosity for the dilute and semi-dilute suspensions. Thus, a simple polynomial equation of the volume fraction for the relative viscosity cannot be applied to the RBC suspensions at the low volume fractions. The relative viscosity for the high volume fractions, however, can be well described as a function of an effective volume fraction defined by the volume of the spheres of the radius equal to the semi-middle axis of the deformed RBC.

R. Lima et al. [1876] investigated the radial dispersion the labelled RBCs flowing in the concentrated suspensions of the RBCs (the haematocrit was varied from 2 to 35%). The RBC radial dispersion was found to strongly depend on the haematocrit and to decrease with a decrease in the diameter of the glass capillary: the authors assumed that such behaviour is related to the Fahraeus-Lindqvist effect. The radial dispersion coefficient is defined as [1409, 1877]

$$D_{yy}(t) = \frac{1}{N}\sum_{i=1}^{N}\frac{\langle (R_i(t) - R_i(0))^2\rangle}{2t} \tag{4.63}$$

where $R_i(t)$ is the radial displacement, N is the number of RBCs.

M. E. Rosti and L. Brandt [2646] considered the flow of a suspension of the deformable viscous hyperelastic particles in an incompressible Newtonian viscous fluid in a channel with the moving walls—the plane Couette geometry.

The kinematic and dynamic interactions between the fluid and solid phases are determined by the continuity of the velocity and traction force at the interface between the phases

$$u_i^f = u_i^s, \qquad \sigma_{ij}^f n_j = \sigma_{ij}^s n_j$$

where n_j is the normal vector at the interface.

For the Newtonian fluid

$$\sigma_{ij}^f = -p\delta_{ij} + 2\mu^f D_{ij} \tag{4.64}$$

where p is the pressure, μ^f is the fluid dynamic viscosity, and D_{ij} is the strain rate tensor

$$D_{ij} = \frac{1}{2}\left(\frac{\partial u_i}{\partial x_j} + \frac{\partial u_j}{\partial x_i}\right). \tag{4.65}$$

For the incompressible viscous hyper-elastic solid material undergoing only the isochoric motion

$$\sigma_{ij}^s = -p\delta_{ij} + 2\mu^s D_{ij} + G_{ij} \tag{4.66}$$

where μ^s is the solid dynamic viscosity. Rosti and Brandt modelled the hyper-elastic contribution as the neo-Hookean material that satisfies the incompressible Mooney-Rivlin law, G is the modulus of transverse elasticity and B_{ij} is the left Cauchy-Green deformation tensor [2646]

$$B_{ij} = F_{ik}F_{jk}, \qquad F_{ij} = \frac{\partial x_i}{\partial X_j}, \tag{4.67}$$

x and X are the current and reference coordinates.

To solve the fluid-structure interaction problem, the authors used the "one-continuum formulation", where only one set of equations is solved over the whole domain with the additional transport equation for the solid volume fraction

$$\frac{\partial \phi}{\partial t} + \frac{\partial u_k \phi}{\partial x_k} = 0. \tag{4.68}$$

The wall-normal derivative of the stremwise velocity at the wall was used to estimate the effective viscosity of the non-Newtonian fluid made by the suspension of particles in the Newtonian fluid. The authors studied the variation of the effective viscosity as a function of the total volume fraction Φ and of the capillary number Ca and found that it is a monotonic non-linear function of both variables; the deformable cases have the lower effective viscosity then the rigid ones at the same volume fraction, and the difference is enhanced with the capillary number. The cell shape is strongly affected by Ca and weakly by Φ. In particular, the particles, originally spheres, progressively become elongated ellipsoids as Ca increases. To quantify this effect, the authors used the so-called Taylor parameter

$$\mathcal{T} = \frac{a - b}{a + b} \tag{4.69}$$

where a and b are the semi-major and semi-minor axis of the inscribed ellipse passing through the particle centre in the x-y plane. The Taylor parameter increases with both the capillary number and the volume fraction.

4.3.5 Motion of the RBC Through Microvessels with Irregular Cross-sections

The red blood cells passing through microvessels undergo large deformations—the deformability of the red cells is a key aspect of blood flow in the microcirculation. Usually the motion of the RBC through microchannels of the constant cross-section is considered [399, 1222, 1370]. T. W. Secomb and R. Hsu [2762] investigated the passing of the RBC through microchannel with variable diameter that has a smoothly defined profile, without sharp corners

$$a = \begin{cases} a_0 + \alpha a_0(\cosh(\lambda Z) - 1) & if \quad Z < 0 \\ a_0 & if \quad 0 \le Z \le Z \\ a_0 + \alpha a_0(\cosh(\lambda(Z - L) - 1) & if \quad Z > 0. \end{cases} \tag{4.70}$$

where L is the length of the cylindrical part of the pore, Z is axial distance measured from its entrance and a_o is the radius. The parameters α and λ determine the shape of the entrance and exit regions, for example, the values $\alpha = 1.8$ and $\lambda = 0.2\mu^{-1}$ results in constrictions with smooth but quite rapid transitions similar to the pores in a nickel filter described by T. Nakamura et al. [2254].

D. Bento et al. [250] studied the deformation of the RBCs flowing through the hyperbolic, smooth, and sudden-contraction microchannels and compared with the deformations of the air microbubbles and the droplets flowing through a microfluidic constriction. The authors showed that the RBCs flowing through a hyperbolic contraction experience a strong extensional flow with a region of the homogeneous strain rate along the centreline, thus, the hyperbolic-shaped microchannels have the potential to precisely control and detect small changes in the RBC deformability in pathological situations.

T. W. Secomb and R. Hsu used the Kelvin solid model with the axial and the circumferential components of the membrane tension (4.11) given by E. A. Evans and R. Skalak [881].

The cells were assumed to have axial symmetry and to conserve surface area during deformation. The effects of the membrane shear viscosity and elasticity are included, but bending resistance is neglected. A time-dependent lubrication equation describing the motion of the suspending fluid is solved, together with the equations for membrane equilibrium, using a finite difference method.

The equations for equilibrium of the normal and tangential forces on the membrane are:

$$p_0 - p = \frac{\partial \theta}{\partial s} t_s + \frac{1}{r} \sin\theta t_\phi \tag{4.71}$$

and

$$\frac{1}{r}\frac{\partial (r t_s)}{\partial \sigma} = \frac{1}{r}\cos\theta t_\phi - \tau_c \tag{4.72}$$

where p_o is the constant pressure in the cell interior, p is the pressure in the gap, a material coordinate σ is defined as the arc length measured along the cell from the origin in an axisymmetric reference shape, $s(\sigma, t)$ is arc length measured along the cell from the origin, $\theta(\sigma, t)$ is the angle between the normal to the membrane and the axis, τ_c is the fluid shear stress on the membrane.

The authors used the axisymmetric formulation of the lubrication theory [972]. The hydrostatic pressure within the gap is assumed to be constant across the gap, but to vary along it, and the transverse component of velocity is neglected. The axial velocity profile in the gap is calculated in terms of the pressure gradient, and the flow rate and shear stress on the membrane are deduced. The spatial derivative of the flow rate is related to the time derivative of the gap width by conservation of volume. Thus a time-dependent lubrication equation is

$$-\frac{1}{16\mu}\frac{\partial}{\partial \sigma}\left[(a^2 - r^2)\left(a^2 + r^2 - \frac{a^2 - r^2}{\ln(a/r)}\right)\frac{\partial p/\partial \sigma}{\partial z/\partial \sigma}\right] + \frac{1}{4}\frac{\partial^2 z}{\partial \sigma \partial t}\left(\frac{a^2 - r^2}{\ln(a/r)} - 2r^2\right) \tag{4.73}$$

$$+\frac{1}{4}\left(\frac{\partial z}{\partial t} - u_0\right)\frac{\partial}{\partial \sigma}\frac{a^2 - r^2}{\ln(a/r)} - \frac{\partial z}{\partial \sigma}\frac{\partial r}{\partial t} = 0 \tag{4.74}$$

The fluid shear stress on the membrane is written as [2762]

$$\tau_c = \frac{1}{4}\frac{\partial p/\partial \sigma}{\partial z/\partial \sigma}\left(2r - \frac{a^2 - r^2}{r\ln(a/r)}\right) - \frac{\mu}{r\ln(a/r)}\left(\frac{\partial z}{\partial t} - u_0\right). \tag{4.75}$$

T. W. Secomb and R. Hsu found that the time taken for cells to enter pores represents more than one-half of the transit time. The transit time increases with the increasing membrane viscosity and with the increasing cell volume and is relatively insensitive to changes in the internal viscosity and to changes in the membrane elasticity except in the narrowest pores at the low driving pressures.

4.3.6 Behaviour of the White Blood Cells Near a Wall in the Shear Flow

Rolling adhesion of the white blood cells (leukocytes) along the vascular endothelium plays a key role in the immune response and is an example for the interplay between transport and specific adhesion in biological systems. The white bloods cells circulate the body with the blood flow and perform receptor-mediated binding to the ligand-coated walls. C. B. Korn and U. S. Schwarz [1692, 1693] investigated how the white blood cells move in linear shear flow above a wall to which they can adhere via the specific receptor-ligand bonds.

The authors performed the computer simulations based on a Langevin equation accounting for hydrodynamic interactions, thermal fluctuations and adhesive interactions. In contrast to earlier approaches, their model not only includes stochastic rules for the formation and rupture of bonds, but also fully resolves both receptor and ligand positions. The authors identified five different dynamic states of motion in regard to the translational and angular velocities of the cell.

As a simple model system for a cell a rigid sphere is considered. For objects as small as white blood cells (with diameters $< 10\,\mu m$) the typical value for the Reynolds number is much less than one and inertia can be neglected. Thus, the flow around the cell is laminar and well described by the linear Stokes equation.

A six-dimensional state vector $\mathbf{X}$ which includes both the three translational and the three rotational degrees of freedom is introduced. The first three components of $\mathbf{X}$ denote the Cartesian coordinates, the last three components describe the rotation that maps a coordinate system fixed to the centre and the orientation of the sphere to the flow chamber fixed coordinate system. Similar compact notations are introduced for velocities and forces. The symbol $\mathbf{U}$ denotes a six-dimensional velocity with the first three components being translational velocities and the last three being angular velocities. $\mathbf{F}$ is the combined force and torque vector. The Langevin equation describing the motion of the cell is written as [1692]

$$\dot{\mathbf{X}} = \mathbf{U}^{\infty} + M(\mathbf{F}^S + \mathbf{F}^D) + k_B \nabla M + g_t^I \tag{4.76}$$

where M is the 6×6 mobility matrix. $\mathbf{U}^{\infty} \propto \dot{\gamma}$ is the velocity of the unperturbed shear flow at the position of the centre of the sphere, $\mathbf{F}^{S} \propto \dot{\gamma}$ is the shear force, which results from the hydrodynamic interaction between the cell and the wall, $\mathbf{F}^{D}$) denotes all direct forces (torques) acting on the sphere, e.g., the bond forces. The terms $k_B \nabla M$ and g_t^I describe the effect of thermal noise. g_t^I is the Gaussian white noise with

$$\langle g_t^I \rangle = 0, \quad \langle g_{t'}^I g_t^I \rangle = 2k_B T_a M \delta(t - t') \tag{4.77}$$

T_a is the ambient temperature, k_B Boltzmann's constant. The mobility matrix for a sphere above a wall depends the distance between sphere and wall.

4.4 Effect of Temperature on the RBC Motion in Microchannel

D. Pinho et al. [2463] investigated the effect of the temperature on the red blood cell motion at the microscale level. The authors noted that the random-like transverse motion and rotation of the RBCs in the shear flow play an important role in thrombogenesis [2100].

The authors used a confocal micro-PTV system to study the RBC's trajectories and to investigate the best approximation of the trajectories through a nonlinear optimization for the in vitro human blood pumped through a 100 mm glass circular microchannel at the different temperatures. The suspension of Dextran 40 with 12±2% (12Hct) of the human RBCs was studied.

The authors observed and analysed he motion of forty cells flowing in the middle of the microchannel: twenty at temperature 25° C and twenty cells at the temperature 37° C. For each cell i the data sets (x_j, y_j), $j = 1, ..., k_i$ were obtained.

The approximation for the data was obtained using nonlinear optimization The authors considered three different functions

- polynomials

$$f_1(x, p) = \sum_{i=0}^{9} p_i x^i; \tag{4.78}$$

- sum of trigonometric functions

$$f_2(x, a, b) = a_0 + \sum_{i=1}^{8} p_i x^i a_i \cos(ix) + b_i \sin(ix); \tag{4.79}$$

- sum of exponential functions

$$f_3(x, c, d, g) = \sum_{i=1}^{8} g_i exp\left(-\left(\frac{x - c_i}{d_i}\right)^2\right) \tag{4.80}$$

where $p \in \mathbb{R}^{10}$, $a \in \mathbb{R}^9$, $b, c, d, g \in \mathbb{R}^8$ are the function parameters and the vector $x \in \mathbb{R}^{k_i}$, i represents the cell number.

The authors investigated the errors of nonlinear least squares approximation of the selected cells

$$F_i = \sum_{j=1}^{k_i} (y_j - f_i(x_j)^2) \tag{4.81}$$

and found that a better approximation to the data is based on the trigonometric functions.

4.5 Red Blood Cell Deformation Under Thermal Fluid Flow

The temperature plays an important role in the mechanical properties of the RBCs [3269]. G. M. Artmann et al. [115] and C. Kelemen et al. [1585] showed that the red blood cells can easily pass though a micropipette with a small inner diameter of 1.3 μ if the temperature reaches 36.4° C [981]. J. J. Foo et al. [979] showed that a thermal effect modulates the viscous drag of a vesicle composed of phospholipid, which is the main constituent of biological membrane, entrapped by an optical tweezer (the device that uses the forces exerted by a focused laser beam to entrap and move micron and submicron particles [250, 1686]) when it is subjected to a low Reynolds number flow.

J. J. Foo et al. [981] studied the RBC deformation while heating through 37.0° C to 42.0° C. Each temperature was maintained for at least 20 min before each flow experiment was started, the cell was kept under an optical trap for 1–3 min at each flow velocity.

The Taylor deformation parameter

$$D_{12} = \frac{L - D}{L + D} \tag{4.82}$$

is utilized to illustrate the mechanical behaviour or the morphological changes of a *spherical* (or elliptic) particle or cell where L and D are two axes of an ellipse [200, 980, 2501].

J. J. Foo et al. [981] used the deformation parameter

$$D^\star = \frac{L/D - L_0/D_0}{L/D + L_0/D_0} \tag{4.83}$$

to characterize the degree of deformation of a trapped RBC under hydrodynamic flow, L_0 and D_0 represent the original cell length and width of the RBC.

An important question is the choice of the laser for the optical tweezers. As was shown by König et al. [1687, 1688] the laser beam with a wavelength of 700–800 nm in the optical tweezers adversely affects the cell metabolism and causes cell death within seconds. J. J. Foo et al. [981] used the optical tweezers powered by a laser with a wavelength of 1064 nm that have no apparent effect on the functions of the living cells.

The authors also computed the low Reynolds number hydrodynamic flow around a deformable red blood cell is modelled as a steady-state, incompressible, laminar and viscous flow with temperature-dependent fluid properties, including the density and viscosity of water. It has been experimentally illustrated that the biconcave RBC is transformed into a more axisymmetry and stream-wise shape with its cross-section parallel to the focal plane of the highly focused laser spot under flow. Thus, the authors assumed that the RBCs have an idealized elliptic boundary configuration.

It was found that the RBCs are aligned, deformed, elongated and, finally, transformed to a stream-wise geometry due to the viscous drag induced by the fluid shear stress on the cell surface at all temperatures, however, the extent of deformation for the RBC at body temperature (37.0° C is significantly higher than that at room temperature.

The temperature has a strong effect on the rate and extent of the haemoglobin gelation, the elastometric transition of the membrane spectrin and the prephase transition of the lipid phospholipid bilayer at 36.4° C, 28.3° C and 35° C–38° C, respectively [115, 979]. The temperature has been also found to play an important role in the local or global remodeling of the RBCs' cytoskeletons, which mainly contribute to the shear deformation of the protein-based network spanning underneath the plasma membrane of the phospholipids bilayer [2171].

The characteristic deformation parameter $D^\star$ was best fitted by a power-law with a nonlinear least-squares routine in order to elucidate the thermal effect on the flow-induced deformation of the trapped cell from its initial condition, that is $D^\star$ equals zero. $D^\star$ was positively correlated with the flow velocity at all temperatures. The authors stated that the higher the surrounding temperature, the larger the cell deformation during the heating process.

4.5.1 Thermal Fluctuations in Red Blood Cells

The thermal fluctuations of membranes affect their mechanical characteristics. Whereas these fluctuations are well understood for flat membranes, curved shells show anomalous behaviour due to the geometric coupling between in-plane elasticity and out-of-plane bending.

The mechanics of the shell are controlled by two elastic constants B and S that represent the bending and stretching moduli of the shell. For thin shells treated as elastic continua, they are related to the Young's modulus Y, Poisson's ratio ν, and thickness t [884]: $B \propto Yt^3$, $S \propto Yt$. Thus, the sufficiently thin sheets are more compliant to bending than to stretching. The relative importance of bending to stretching energies on a membrane with local radius of curvature R can be described by the dimensionless Föppl–von Kármán number $\gamma \equiv SR^2/B \propto (R/t)^2$.

A. A. Evans et al. [884] studied the role of the geometry in determining the spatial distribution of their surface undulations of the membranes of the RBCs. The effect of geometry on biological membranes is particularly interesting, because these structures are typically soft enough to support large undulations in thermal equilibrium.

The RBC membrane is made up of a lipid bilayer containing transmembrane proteins linked into the 2D triangular network on the cytosolic side of the membrane by the spectrin proteins. However, on scales much larger than either the thickness of the membrane or the lattice constant of the spectrin network [379, 994], the composite membrane may be treated as an elastic shell.

Based on this simplified elastic description and the assumption of flat membranes, a basic theory for the RBC undulatory dynamics was proposed by F. Brochard and J. F. Lennon [379]. Experimental exploration of the RBC membrane elasticity using micropipette aspiration, electric field-induced deformation, optical tweezers, and microrheology showed an unexpectedly complex spatial distribution of the observed thermal membrane undulations [884].

A. A. Evans et al. based their analysis on the data collected from diffraction phase microscopy (DPM) measurements of the RBCs that produce high-resolution images of the RBC flicker maps, which show a spatial distribution of membrane undulations. The authors demonstrated that this distribution can be quantitatively explained by the theory, without an appeal to the active forces or the heterogeneous membrane composition. They stated that two generic geometric features control the mechanics of membranes:

- the sign of the Gaussian curvature, which qualitatively affects cell deformation;
- the existence of singular lines where the Gaussian and normal curvatures simultaneously vanish—the former has been shown to determine localized and extended static deformations and to guide the propagation of undulatory waves on curved surfaces.

A. A. Evans et al. considered the small-amplitude undulations on a surface that is only gently curved (i.e., the local radius of curvature is much greater than the wavelength of deformation) where the full nonlinear problem of elastostatics reduces to the linearized Donnell—Mushtari—Vlasov (DMV) equations for a "shallow shell"

$$B\nabla^4\zeta - \mathcal{L}[\chi] = f(\mathbf{r}) \tag{4.84}$$

$$\frac{1}{S}\nabla^4\chi + \mathcal{L}[\zeta] = 0 \tag{4.85}$$

where ∇^4 is the biLaplacian, and $\zeta(r)$ defines the normal deflection of the shell from its reference state, given by the 3D vector $\mathbf{r}$. The in-plane stress is given in terms of an Airy stress function χ, f is the applied normal load.

The authors demonstrated that thermalized shells containing regions of negative Gaussian curvature naturally develop anomalously large fluctuations. Moreover, the existence of special curves, "singular lines," leads to a breakdown of linear membrane theory. As a result, these geometric curves effectively partition the cell into regions whose fluctuations are only weakly coupled.

In-plane stress is identically satisfied due to the function χ, the linear differential operator

$$\mathcal{L} = \varepsilon^{\alpha\xi}\varepsilon^{\beta\xi}d_{\alpha\beta}\partial\eta\partial\xi \tag{4.86}$$

contains all of the curvature information from the surface, $d_{\alpha\beta}$ is the curvature tensor of the undeformed shape, $\varepsilon^{\alpha\beta}$ is the antisymmetric tensor.

In the limit of a flat membrane where vanishes, this equation reduces to the biharmonic equation, which the Airy stress function must satisfy for the stress balance [1774].

The operator $\mathcal{L}$ is the source for coupling between curvature and elasticity, because it mixes in-plane stress and outof-plane deformation through the curvature of the surface. It takes the form

$$\mathcal{L} = \left(\frac{1}{R_y}\partial_x^2 + \frac{1}{R_x}\partial_y^2\right) \tag{4.87}$$

in a local Cartesian coordinate system x, y aligned with the principal curvature directions on the undeformed membrane with radii of curvature R_x; R_y.

The coupling of the normal to in-plane stress vanishes in the limit of a flat reference state $\mathcal{L} \to 0$ when R_x; $R_y \to \infty$; there is no in-plane stretching of a flat membrane in response to normal loading. The curvature of the stress-free membrane couples stretching and bending, and thus curvature influences the mechanics of normal deformation. The primary cause of this behaviour arises from the Gaussian curvature $K = 1/(R_x R_y)$. When K changes sign, the operator $\mathcal{L}$ changes type, which leads to the different mechanical deformation characteristics [880, 3166].

G. Marcelli et al. [2042] used the model of the mechanical properties of the cell plasma membrane using a finite-temperature particle-dynamics simulation of the whole cell, in which a two-dimensional network of the virtual particles embedded in a three dimensional closed surface that represents the membrane.

The RBC membrane consists of a mixed lipid-protein bilayer, an attached intracellular network called the membrane cytoskeleton, and an extracellular glycocalyx (a layer of glycoproteins covering the outer surface of the cell). The main component of the cytoskeletal network is the protein spectrin. The spectrin cross-links with very short actin filaments are the junctional nodes that are approximately five or sixfold coordinated in spectrin.

A problem in the study of RBC structure is that the observation of thermal fluctuations of nanometre-size seems to be consistent only with a vanishingly small shear modulus [2963], whereas static deformation experiments such as micropipette aspiration [3269] report a large shear modulus.

The authors assumed that the membrane preserves a constant global area under thermal fluctuations and its mechanical behaviour can be described by two material parameters, an out-of-plane bending modulus which describes the resistance to bending, and an in-plane shear modulus which describes the resistance to shape changes of the membrane surface.

The RBC membrane presented as a two-dimensional network consisting of N particles arranged in a regular triangulation. The virtual particles represent discrete areas of the membrane. Each particle is connected to its six nearest-neighbours by the linear springs, giving a potential energy of

$$V_{ij} = \frac{1}{2}k[r_i - r_j - r_0]^2 \tag{4.88}$$

where k is the spring constant, r_i is the position of particle i, r_0 is the equilibrium distance between particles. The value of r_0 fixes the minimum wavelength of the membrane fluctuations. The authors chose r_0 to be close to the average length of the spectrin filaments (100 nm).

A dihedral angle potential is applied between adjacent triangles

$$V_{ijml} = D[1 + \cos(\phi_{ijml})] \tag{4.89}$$

where ϕ_{ijml} is the dihedral angle between the triangles Δ_{ijm} and Δ_{jml}. This potential has a minimum when two adjacent triangles are co-planar.

In general the free energy of a system its given by two contributions, $F = E - TS$. In the case of a spring-network, e.g., the energetic contribution E is given by the sum of V_{ij} over all the particles.

Since the lipid bilayer is relatively incompressible, the authors required that the global surface area of the membrane, defined as the sum of the areas of all triangles by which the surface is tessellated, remains constant. To do this the they applied Lagrange's method of undetermined multipliers.

The evolution of the positions of the particles yields the equilibrium state of the membrane and allows determination of the membrane thermal fluctuations and the elastic moduli.

J. P. Hale et al. [1213] studied the thermal shape fluctuations of the red blood cells using the comparison between the experimentally recorded thermal fluctuation spectra that depend on the spectrin mesh size and their theoretical counterparts determined from the finite temperature coarse-grained molecular dynamics simulations.

The authors used the fast phase contrast video microscopy to quantify the RBC fluctuations. The slight difference in the density between the RBCs and the surrounding buffer stabilised the cells on the bottom of the observation chamber, which allowed for convenient observation of the fluctuations. 2D equatorial contours were extracted using a subpixel resolution algorithm based on the detection of the minimum intensity along the equator of the membrane. A series of contours was then analysed to quantify the membrane fluctuations. Each contour was represented by a Fourier series

$$r(\vartheta) = R\left\{1 + \sum_{n} \left[a_n \cos(n\vartheta) + b_n \sin(n\vartheta)\right]\right\}. \tag{4.90}$$

The contour shape fluctuations around the mean shape was quantified by the mean square values of the Fourier amplitudes for each mode n

$$\left\langle \delta_n^2 \right\rangle = \left\lfloor \left\langle a_n^2 \right\rangle - \langle a_n \rangle^2 \right\rfloor + \left\lfloor \left\langle b_n^2 \right\rangle - \langle b_n \rangle^2 \right\rfloor. \tag{4.91}$$

The authors performed the coarse-grained particle dynamics simulation keeping a constant volume-to-area ratio, $\nu = 3V/(4\pi R_A^3)$, where V is the cell volume and R_A is the equivalent radius, calculated from the cell surface area $R_A = (A/4\pi)^{1/2}$.

Each pair of the neighbouring particles was allowed to interact with the harmonic potential (4.88), in addition, the dihedral angle potential (4.89) is applied between each two adjacent triangles. The volume is kept constant by applying an external pressure to the cell surface. The pressure is applied outwards when the volume exceeds the desired value and inwards in the opposite case. In order to incorporate effects from the surrounding environment, a Nosé-Hoover thermostat was employed to keep a constant temperature. The thermostat mimics the exchange of energy between the simulated cell and the environment. Viscous effects were not taken into account, as they cannot affect the equilibrium fluctuation spectrum [1213].

Comparison between experimentally and computationally obtained fluctuation spectra showed that they are in agreement for values of the spring constant $k_s \approx 5 \times 10^{-6}\,\mathrm{Nm}^{-1}$ and the dihedral constant $D \approx 10^{-18}\,\mathrm{J}$. This allowed Hale et al. made an order of magnitude estimation of the shear and bending moduli using the approximate relationships $\mu = \sqrt{3}/4k_s$ and $k = D/\sqrt{3}$.

The authors stressed that the external pressure acting as a volume constraint in the model does not affect the shear modulus. The external pressure could give rise to membrane tension only via changes in the density of the membrane, which is prevented by the area constraint.

4.6 Haemorheological Properties of Heated Red Blood Cells

The heat denatured erythrocytes (including additionally labelled by Technetium-99 sulfur colloid [845]) due to their compromised deformability are used as the labelled probes to visualize, e.g., the spleen tissue or as the models of the pathological red blood cells in numerous diseases such as the hereditary blood disorders such as spherocytosis, elliptocytosis, ovalocytosis, sickle cell anemia, cardiovascular diseases, liver diseases or the acquired hemotologic diseases such as type II diabetic mellitus and malaria [1001]. The hereditary blood disorders exhibit certain loss of RBC deformability [481, 542, 1814, 2171], leading to severe vascular complications such as vaso-occlusion and spleen sequestration [1814, 1815].

Z. L. Liu et al. [1941] numerically investigated the biomechanics and biorheology of the heated red blood cells. The authors perform a systematic computational study of the haemorheological properties of the heated RBCs with several physiologically relevant static and haemodynamic settings, including optical-tweezers test, relaxation of the prestretched RBCs, the RBC traversal through a capillary-like channel and a spleen-like slit, and a viscometric rheology test.

Compared with the normal healthy RBCs, the heated RBCs show the enhanced membrane stiffness [2273] and up to 10% increase of effective volume apparent haematocrit), depending on the time duration of heating [1941].

At the molecular level, the heat stresses were found to induce the dissociation of the membrane spectrin dimer into the monomers, explaining the stiffening of the RBC membrane [3434].

The RBC membrane could be also "stiffened" chemically. Two chemical agents that are commonly used to study cell deformation are diamide and glutaraldehyde. A. M. Forsyth et al. [987] performed pressure-driven flow experiments with the RBCs in a microfluidic constriction and quantified the cell dynamics using high-speed imaging. The authors found that diamide affects the RBCs by cross-linking spectrin skeletal membrane proteins but does not reduce deformation and show an unchanged effective strain rate when compared to healthy cells. Glutaraldehyde acts on all components of the cell, reduces deformation and shows increased instances of tumbling, both of which are characteristic features of stiffened, or rigidified, cells. Since glutaraldehyde increases the effective viscosity of the cytoplasm and lipid membrane while diamide does not, a possible explanation for obtained results is that viscous effects in the cytoplasm and/or lipid membrane are a dominant factor in dictating dynamic responses of the RBCs in pressure-driven flows.

Z. L. Liu et al. used the dissipative particle dynamics (DPD) [1171] to model the blood plasma . The DPD particles interact with each other through soft pairwise forces, including conservative, dissipative, and random forces. The viscosity of the system is an outcome of the combination of these three forces [1349]. The authors assumed the same viscosity for both plasma and cytoplasm which substantially reduces the computational cost and allows to simulate a physiological haematocrit (45%) of RBC suspensions over a wide range of shear rates.

The RBC membrane visco-elasticity is determined by the membrane cytoskeleton—the spectrin network, whereas bending is controlled by the lipid bilayer. In this model, the RBC membrane is approximated as a one-component collection of membrane vertices connected by hyperelastic springs. The rigorous coarse-graining procedure allows for the flexibility of using a different number of vertices to represent the RBC membrane while preserving the accurate cell membrane mechanical and rheological properties [2466]. This model incorporates the bending energy, in-plane shear energy, membrane viscosity, and area and volume constraints. To be coupled with the fluid phase, the vertices on the RBC membrane are also treated as the DPD particles—each vertex also experiences the DPD forces from the internal and external fluid particles of the cell in addition to the membrane model forces.

The authors considered two types the heated RBC models:

- The short-duration heated RBC (SH-RBC) model considers only the heat-altered membrane properties
- The long-duration heated RBC (LH-RBC) model additionally further assumes a 10% increase in cell volume.

The authors analysed the single cell static responses and the traversal dynamics of the normal and the heated RBCs flowing through the capillary-like microchannels as well as the macroscopic viscosity of RBC suspensions over a wide range of shear rate.

Chapter 5
Regional Models: Skin and Deep Tissues

The difficulties of the mathematical description of the heat transfer in the living tissues originate from

- The complexity of the blood vessels' network [371]—the statistically self-similar fractal tree branching structure [642, 1025, 3459]: the diameter distribution of the blood vessels ranging from the capillaries to the arteries obeys a power-law distribution which is one of the major characteristics of the fractals; the branching patterns of the arterial kidney vessels, for example, are characterized by the fractal geometry with dimension between 2.0 and 2.5 [1265, 2931]. The largest vessels are the aorta and the vena cava having a diameter of a few centimetres, the supply arteries and veins (SAVs) have the diameter up to 1 cm. A diameter of the large vessels (there are about 150 arteries and 40 veins) is 4–10 mm (with up to 25 mm for aorta [647]) while that of small ones (a few hundreds arterioles and a few thousands venules) 0.02–0.05 mm and of the capillaries (about 4500) varies from 5 to 15 μ [106]. The microcirculation is responsible for regulating blood flow in individual organs and for exchange between blood and tissue. Approximately 80% of the total pressure drop between the aorta and the vena cava occurs in small vessels [2487].
- The complex temperature response of the vasculature to external and internal effects. In a living system, the blood flow rate and the vessel size may change as a response to the change of the local temperature, the local pH value, and the concentration of the local O_2 and CO_2 levels [3525].
- The size of the blood vessel network—in an average 10 cm^3 cube of tissue there is more than a thousand kilometres of the blood vessels present, thus, there is a lot of the surface area for the blood to exchange heat with the tissue [3149].
- The small thermal length scale involved in the microvasculature that limit the temperature measuring with sufficient resolution.

A. I. Zhmakin, *Heat Transfer in Vivo*, Springer Series in Biophysics 27,
https://doi.org/10.1007/978-3-031-95459-7_5

5.1 Classical (Fourier) Continuum Models

The continuum models are simplified form of the bioheat equations. In these models, the effect of the blood flow is averaged over the control volume. The blood flow through each individual vessel is ignored. On the other hand, the vascular models are the complex forms of the bioheat equations which accounts for the blood flow in the individual vessels.

The heat conduction model suggested in the beginning of the nineteenth century by Jean Baptiste Joseph Fourier [2269] in his famous work "Analytical Theory of Heat" is based on the constitutive relation

$$\mathbf{q}(\mathbf{r}, t) = -\lambda \nabla T(\mathbf{r}, t). \tag{5.1}$$

leading, after substitution it into the energy conservation equation for the solid body in rest

$$\frac{\partial(\varrho c T)}{\partial t} = \nabla \cdot \mathbf{q} + Q \tag{5.2}$$

to the classical parabolic heat conduction equation (called sometimes the "Fourier heat conduction equation" (FHCE) [3467])

$$\frac{\partial(\varrho c T)}{\partial t} = \nabla \cdot (\lambda \nabla T) + Q \tag{5.3}$$

that for the case of constant thermophysical properties is written as

$$\frac{\partial T}{\partial t} = \varkappa \Delta T \tag{5.4}$$

where $\varkappa = \lambda/\varrho c$ is the thermal diffusivity. Specific heat capacity shows non-negligible variations only in very wide temperature range [2779].

The Fourier law can be derived from the classical irreversible thermodynamics based on the local equilibrium hypothesis [173, 1803] via the entropy production that is written as

$$\Sigma = \mathbf{q} \cdot \nabla \frac{1}{T} \geq 0. \tag{5.5}$$

The Fourier law describes a linear relationship between the generalized force (the temperature gradient) and the generalized flux (the heat flux) [436]. For the isotropic materials the heat flux is

$$\mathbf{q} = \lambda_T \nabla \frac{1}{T} = -\frac{\lambda_T}{T^2} \nabla T. \tag{5.6}$$

Thus $\lambda = \lambda_T/t^2$ is the Fourier heat conduction coefficient with λ_T being a thermodynamic conduction coefficient [2844]. The distinctive property of the heat conduction is that it does not have a reversible part.

The continuum models of the microvascular heat transfer are derived with intention to average the effects of a large number of the blood vessels present in the region of interest. Most continuum models describe the perfused tissue by means of a single tissue temperature.

As the blood moves through the microcapillary system, it has a stabilizing effect that returns the tissue to some natural equilibrium reference state (the state at which the tissue is found in the absence of the externally applied fluxes). This is the effect of the perfusion. In an exchange of the energy between the tissue and the microcapillary network, the perfusion acts to bring the local tissue temperature closer to the body's core temperature. The continuum models cannot predict the effects of the individual vessels on the local tissue temperature.

Almost all (Fourier) bioheat models could be written as

$$\varrho c \frac{\partial T}{\partial t} = \nabla \cdot \lambda \nabla T + + q_{\mathrm{c}} + q_{\mathrm{p}} + \dot{q}_{\mathrm{met}} + Q^{\mathrm{ext}}, \tag{5.7}$$

and differ in the definition of the sources due to the blood convection q_{c} and the blood perfusion q_{p}. Naturally, there are relations between the models, e.g., the local mean blood velocity in the Wulff model is proportional to the square root of the blood perfusion rate due to Pennes [675].

The blood velocities in the arteries are higher during systole than during diastole. The parameter to quantify this difference is the pulsatility index (PI), which is equal to the difference between the peak systolic velocity and the minimum diastolic velocity divided by the mean velocity during the cardiac cycle; this value decreases with the distance from the heart [3073].

5.1.1 *The Pennes Equation*

The perfusion could be viewed as a volumetric phenomenon when the distribution of the microcapillaries is uniform and the relative size scale of the capillary diameter is small compared to the length scale characterizing the transport [3381]: the network's blood flow, when observed from a length scale much larger than the diameter of an individual capillary, has no distinguishable direction and permeates the tissue in a homogenous manner [228].

The best known and certainly the most important continuum model was suggested by Harry H. Pennes in 1948 and referred to as the Pennes equation [2422] (sometimes this model is also called the "heat sink model" [648])

$$\varrho c \frac{\partial T}{\partial t} = \nabla \cdot \lambda \nabla T + \varrho_{\mathrm{b}} c_{\mathrm{b}} \omega_b (T_{\mathrm{a}} - T) + \dot{q}_{\mathrm{met}} + Q^{\mathrm{ext}}, \tag{5.8}$$

where T, ϱ, c and λ are the temperature, the density, the specific heat and the thermal conductivity of the tissue as the homogeneous media, ω_b is the blood perfusion rate, c_b is the blood specific heat, T_a is the temperature of the arterial blood, $\dot{q}_{met}$ and Q^{ext} are the heat sources due to the metabolic reactions (this one usually could be neglected in the cryobiology problems) and the external source of energy. Pennes assumed that in the analogy with the gaseous exchange in the living tissue the heat transfer occurred in the capillaries because of their large area for heat exchange and suggested that the effect of the blood flow in the tissue on the heat transfer can be modelled as a heat source sink term added to the traditional heat conduction equation, the local arterial temperature is assumed to be equal to the body core temperature and the temperature of the local venous blood equals to the local tissue temperature. The Pennes equation is a modified form of the transient heat conduction equation including the effects of the blood flow and the metabolic heat generation rate.

The form of the source term used by Pennes could be derived [3525] under the assumption that the tissue is perfused via a capillary network bifurcating from the transverse arterioles and the blood is drained by the transverse venules, both the artery and vein keep a constant temperature when they pass through this tissue region and thus the total heat released is equal to the total amount of blood perfusing this tissue volume per second multiplied by the blood density, specific heat and the temperature difference between the artery and vein.

However, the assumption that the temperature of the arterial blood does not change when it travels from the heart to the capillary bed is valid only for the blood vessels with a diameter larger than 300 μm and the assumption is that the venous blood temperature equals the local tissue temperature only for the blood vessels with a diameter smaller than 50 μm [3525]. Thus, the Pennes perfusion source term overestimates the effect of the blood perfusion.

J. Okajima et al. [2332] introduced the characteristic length of the bioheat transfer

$$\delta_B = \sqrt{\frac{\lambda}{\varrho c \omega_b}} \tag{5.9}$$

and rewrote the Pennes equation in the dimensionless variables

$$X = \frac{x}{\delta_B}, \quad Fo = \frac{\varkappa t}{\delta_B^2} = \frac{\lambda}{\varrho c} \frac{t}{\delta_B^2}. \tag{5.10}$$

Pennes performed a series of experiments using the thermocouples to measure the temperature distribution as a function of the radial position in the forearms of nine human subjects that were unanaesthetized so as to avoid the effects of anaesthesia on blood perfusion. The thermocouple was scanned transversely across the mediolateral axis to measure the temperature as a function of radial position within the interior of the arm. He also measured the skin temperature distributions along the axis of the upper limb and around the circumference of the forearm. Pennes then modelled the arm as a long cylinder and calculated the steady-state radial temperature profile. The blood perfusion rate could not be directly measured and was adjusted to fit the solution to the experimental data for the fixed ambient temperature and metabolic heating rate.

D. Trucu et al. [3084] determined the space- and time-dependent perfusion coefficient in the one-dimensional transient bio-heat conduction equation using the given boundary and initial conditions and the additional temperature measurements inside the solution domain; both the global (investigation of the inverse problem on the basis of the Tikhonov regularization) and the local (computation of the time first-order and space second-order derivatives by means of the first kind integral equations) approaches were exploited.

G. K. Grabski et al. [1143] also used the inverse problem to determine the time-dependent blood perfusion coefficient exploiting the method of the fundamental solutions together with the Tikhonov regularization.

Pennes' data showed a temperature difference of 3–4° C between the skin and the interior of the arm, which the author attributed to the effects of metabolic heat generation and heat transfer with arterial blood perfused through the microvasculature.

It was later established that in contrast to the Pennes assumption no significant heat transfer occurs in the capillary bed; the exchange of heat occurs in the larger components of the vascular tree—in the terminal arterioles (vessels 60–300 mm in diameter) [489, 491, 503, 3277, 3279, 3280].

Several techniques have been developed that directly measure the heat flux to predict the blood perfusion by exploiting the coupling between the vascular perfusion and the local tissue temperature using the inverse mathematical solutions, e.g., the blood perfusion rate can be determined experimentally by the sinusoidal heating of the skin surface [1902].

An obvious extension leading to the nonlinear (the "modified") Pennes equation is to account for the temperature dependence of the blood perfusion rate $\omega_b = \omega_b^0 + \omega_b^1(T)$ [323, 682, 1762]. The simplest form of $\omega_b^1(T)$ is the linear dependence [1763] leading to the equation

$$\varrho c \frac{\partial T}{\partial t} = \nabla \cdot \lambda \nabla T + c_{\mathrm{b}} \left[\omega_0 + \omega_1 \frac{T - T_{\mathrm{a}}}{T_{\mathrm{a}}} \right] (T_{\mathrm{a}} - T) + \dot{q}_{\mathrm{met}} + Q^{\mathrm{ext}}, \tag{5.11}$$

where ω_0 is the temperature-independent (basal) perfusion component, ω_1 is the temperature-dependent (vasodilation and angiogenesis) perfusion component.

When the heat is deposited locally in tissue, the tissue is altered gradually and damaged eventually, given enough the exposure time. During this process, the microvascular blood vessels become impaired and then destroyed. Consequently, the local blood perfusion decreases till complete cessation. In the early stage when the tissue temperature is only slightly elevated, the hyperthermia occurs to increase the blood flow locally via the vasodilation, as part of the physiological control of the thermo-regulation to maintain the normal temperature. At higher temperatures, the hyperthermia is surpassed by the vascular stasis that accompanies the tissue damage. Thus, the blood perfusion is not constant but the damage dependent. The simplest description of this damage-dependent perfusion phenomenon is a threshold temperature or damage index, above which blood perfusion is absent due to tissue or vascular coagulation [3518].

More complex expressions for the temperature dependent perfusion were also suggested, e.g., [683, 2906, 3121]:

$$\omega_b = \begin{cases} \omega_0 + \omega_1(T - T_a), & T \leq T_{th} \\ \omega_{max}, & T_{th} < T < T_{cg} \\ 0 & T \geq T_{cg} \end{cases} \tag{5.12}$$

where ω_0 is the blood perfusion rate under the normal condition, ω_1 is the increase rate of the blood perfusion with temperature, ω_{max} is the maximum blood perfusion rate, $T_{th} = 45°$ C is the temperature when the blood perfusion achieves its maximum value, and $T_{cg} = 60°$ C is the temperature when the coagulation occurs in the tissue or [1947, 1997]:

$$\omega_b = \omega_{b0} f(T) \exp(-\Omega) \tag{5.13}$$

where $f(T)$ is a dimensionless function that accounts for the vessel dilation at the slightly elevated temperatures and Ω is the damage integral.

D. A. Hodson et al. [1337] simplified the skin heat transfer model introducing the thermal effect of the blood flow through the dermis as a heat-generating interface between the epidermis and the subdermal tissue; however, the heat flux is discontinuous in this model.

H. H. Pennes formulated his equation studying the heat transfer in a human forearm and, most probably, being inspired by the studies on the oxygen transport by the blood to the cells. The blood perfusion term was adjusted using the experimentally measured temperatures in the forearm. The effect of the blood flow in the Pennes equation is reduced to the scalar local heat source; the directionality of the heat transfer by the blood flow is ignored in this model.

It is assumed the heat transfer occurs in the capillaries only where the blood with the temperature T_a flow. In other words, the length of establishing of the thermal equilibrium is assumed to be infinitely large for all the vessels except the capillaries and zero for them. However, S. Weinbaum et al. [3279] and D. E. Lemons et al. [1820] experimentally found that all the arteries of the diameter less than 100 μm and all the veins of the diameter less than 400 μm can be considered fully equilibrated with the surrounding tissue in the normothermic conditions.

The temperature used in the Pennes model is defined as an average over the volume δV:

$$T(\mathbf{r}) = \frac{1}{\delta V} \int_{\delta V} T(\mathbf{r}) dV,$$

where the linear scale $\sqrt[3]{\delta V}$ is assumed to be small compared with the size of the domain considered while large compared with the diameter of the "thermally important" vessels. As was noted by J. W. Baish [156], no such scale exists for the developed blood vascular network.

If the evaporation of the tissue internal water occurs [2569, 2889], the source term $(-Q_E)$ should be added, where $Q_E = -L_W \mathrm{d}W/\mathrm{d}t$, W is the tissue water density and L_W is the water latent heat constant [3400].

W. Wulff [3357] in 1974 listed three flaws of the Pennes model:

- globally estimated convection effect in a local description of energy balance;
- missing additional descriptions for blood flow;
- poor representation of the heat transfer between the tissue and the flowing blood.

The main drawbacks of the Pennes model are neglecting of the vascular architecture and assumption of the isotropic perfusion heat source. Numerous researches modified the "bioheat" equation, trying first of all to account for the directionality of the heat transfer by the blood flow (the lymph flow is never considered): Mitchell and Myers, Keller and Seiler, Wulff, Klinger, Chen and Holmes, Bazett, Scholander, Chato, Weinbaum, Jiji and Lemons, Wissler, Baish. These models were analysed in a number of reviews, e.g., by C. K. Charny [488], H. Arkin et al. [106], M. Stancyk and J. J. Telega [2928], T. C. Shih et al. [2827], A. Zolfhaghari and M. Maerefat [3541], A. Bhowmik et al. [284], L. Zhu [3525], J. Hristov [1369].

Despite the limitations of the Pennes bioheat equation, reasonable agreement between the theory and the experiment has been obtained for the measured temperature profiles in the perfused tissue subject to the various heating protocols. The accuracy of the Pennes equation is higher in the highly perfused tissue such as the brain since a relatively large part of the heat transfer does take in the smallest vessels [3149].

The validity of the Pennes bioheat equation has been largely based on the macroscopic thermal clearance measurements in which the adjustable free parameter in the theory, the blood perfusion rate was chosen to provide a reasonable agreement with the experiments for the temperature decay in the vicinity of a thermistor bead probe: if the limitation of the Pennes bioheat equation is an inaccurate estimation of the strength of the perfusion source term, the adjustment of this term will overcome its limitations and provide a reasonable agreement between the experiment and the theory [3525].

J. C. Chato [491] was the first to theoretically investigate the heat transfer from the individual blood vessels in three configurations: a single vessel, two vessels in counterflow, and a single vessel near the skin surface. It was shown that the Graetz number, proportional to the blood flow velocity and the radius, is the controlling parameter determining the thermal equilibration between the blood and the tissue

$$Gr = RePr\frac{D}{L} \tag{5.14}$$

where Re is the Reynolds number, Pr is the Prandtl number, D is the vessel diameter.

5.1.2 The Weinbaum-Jiji Model

For the blood vessels with the very low Graetz number the blood quickly reaches the tissue temperature. The heat transfer between the countercurrent artery and the vein is affected by the vessel centre-to-centre spacing and the mass transport between them.

S. Weinbaum and L. M. Jiji developed a model for microvascular blood tissue heat transfer, based on the anatomic analysis to illustrate that the predominant mode of the heat transfer in the tissue is the countercurrent heat exchange between a thermally significant artery and vein pair. The near-perfect countercurrent heat exchange mechanism implies that most of the heat leaving the artery is transferred to its countercurrent vein rather than released to the surrounding tissue.

The authors assumed that the vascularized tissue behaves as an anisotropic heat transfer medium and the isotropic blood perfusion term is negligible compared to energy exchange due to countercurrent blood flow and that the complete heat exchange is only possible if the heat leaving from the thermally significant arteries is transferred to its counterflow blood through thermally significant veins and vice versa without releasing to surrounding tissues. Presence of any tissue temperature gradient along the vessel axis due to normal and diseased conditions leads to different amounts of heat transfer from arteries and veins to tissues in a direction perpendicular to the axis. Thus most of the heat conducted away through the wall of arteries is conducted in through the wall of its paired veins and vice-versa.

The countercurrent heat exchange between the paired significant vessels was studied experimentally by D. E. Lemons et al. [1820] who measured the transverse tissue temperature profiles in the rabbit thigh. If the countercurrent heat exchange is dominant in blood tissue heat transfer, the vein must recapture most of the heat leaving the artery; thus, the temperature difference between the countercurrent artery and vein should not vary significantly in the axial direction [3525].

Lemons et al. found the small temperature fluctuations of up to 0.5° C in the deep tissue. The irregularities in the tissue temperature profiles were closely associated with the existence of the blood vessels in the vicinity of the thermocouple wire. It was shown that temperature fluctuation was observed in all the blood vessels larger than $500\,\mu$m, in some vessels between 100 and $500\,\mu$m and no temperature fluctuation was observed in the blood vessels less than $100\,\mu$m in diameter. It was also observed that the temperature field variation rarely exceeded 0.2° C in any countercurrent pair. These results show that the assumption in the Pennes model that the arterial blood reaches the capillary circulation without significant prior thermal equilibration is inaccurate for such vascular architecture, and thus most of the significant blood-tissue heat transfer occurs in the larger vessels upstream.

The similar experiments were performed by Q. He et al. [1279, 1280] who directly measured the temperature decays along the femoral arteries and the veins and their subsequent branches in the rats. The experiments showed that the venous blood in the mid-size blood veins recaptured up to 41% of the total heat released from their countercurrent arteries under the normal conditions. The purpose of their model is to formulate the equations based on the physiological parameters such as the vessel number density, size, the flow velocity and the direction of the arterial-venous arcade in the peripheral tissues for the local energy exchange in the vascularized tissues.

The observations revealed that the heat exchange mechanism between the paired vessels in deep tissue layer is mainly due to the counter-current flow. I. e., the heat is exchanged both through the normal conduction and the capillary bleed-off within the tissue space between the paired vessels in the counterflow. The experimental observation showed that the local tissue temperature fluctuates around a mean temperature whose lengthscale is small compared to the distance between the centres of the artery–vein pair.

K. H. Keller and L. Seiler [1586] suggested a model for the peripheral heat transfer in the subcutaneous fat region for human body. They divided the region of interest, i.e., the subcutaneous region into the constant temperature core and the peripheral region with the assumption that heat transfer is significant in the direction normal to the skin. Their model included the effect of the heat conduction in the tissue and the blood perfusion in the capillaries apart from the countercurrent heat exchange between the blood vessels,viz., the artery and vein pair. Their model coupled an energy conservation equation for the vascularized tissue with the arterial and venous flow energy equations:

Vascularized tissue

$$\lambda_T \left(\frac{\mathrm{d}^2 T}{\mathrm{d}x^2} \right) + (hA''' + c_b\omega_b)(T_a - T) + hA'''(T_b - T) + q_m''' = 0, \tag{5.15}$$

Arterial flow

$$\left[(\dot{m}_a)_0 - \int_0^x \omega_b \mathrm{d}x \right] c_b \frac{\mathrm{d}T_a}{\mathrm{d}x} + hA'''(T_a - T) = 0, \tag{5.16}$$

Venous flow

$$\left[(\dot{m}_a)_0 - \int_0^x \omega_b \mathrm{d}x \right] c_b \frac{\mathrm{d}T_v}{\mathrm{d}x} + (hA''' + c_b\omega_b)(T - T_v) = 0, \tag{5.17}$$

where x is the distance along the direction normal to skin surface, h is the average heat transfer coefficient from vessels to surrounding tissues, A''' is the average area for heat transfer per unit volume, ω_b is the blood perfusion rate, $\dot{m}$ is the blood flow rate, subscripts 0, c and t represent inlet condition, constant core temperature and tissue.

S. Weinbaum et al. [3279] identified the three vascular layers (deep, intermediate, and cutaneous). The authors derived three vascular structures from the anatomic observation that serve as the basic heat transfer units in the thermal equilibration analysis:

1. a large artery and its countercurrent vein in the tapered tissue cylinder oriented obliquely to the skin surface in the deep tissue layer;
2. an isolated counter-current transverse terminal vessel pair embedded in the intermediate tissue layer;
3. a vessel (or vessel pair) in the plexus running parallel to the skin surface in the cutaneous layer.

Weinbaum et al. suggested that the large vessels in the cutaneous layer can be far from the thermal equilibration and, thus, capable of delivering the warm blood from the deep tissue to the skin layer.

Numerous studies (see, for example, [489, 491, 503, 664]) show that heat transfer is mainly occurs in the greater vessels and the blood flowing into the capillaries is already in the thermal equilibrium with the surrounding tissues. Still, in the practical problems a simple Pennes frequently gives a more reasonable predictions than some of the more sophisticated models [1674, 2928, 3328]; its combinations with the different models based on the effective thermal conductivity are also considered [648].

For example, since the blood flow in the vessels gives rise to the convective heat transport, E. H. Wissler [3327] suggested to treat the living tissue as a continuum with the effective convective flux $\mathbf{v}_{eff}$ and to write the governing bioheat transfer equation as

$$\rho_t c_t \frac{\partial T}{\partial t} = \nabla \cdot \lambda \nabla T - \rho_b c_b \mathbf{v}_{eff} \nabla T + q_{met}. \tag{5.18}$$

The model accounting for the effect of the blood vessel's size on the heat transfer in the tissues introduces the "thermal equilibration length", or simply the "thermalization length" L_{e} as the distance over which the difference between the temperature "e" times [503]:

$$\varrho c \frac{\partial T}{\partial t} = \nabla \cdot \lambda \nabla T + (\varrho c)_{\mathrm{b}} \omega_b^* (T_{\mathrm{a}}^* - T) - (\varrho c)_{\mathrm{b}} \bar{u} \cdot \nabla T + \nabla \cdot \lambda_{\mathrm{p}} \nabla T + \dot{q}_{\mathrm{met}} + Q^{\mathrm{ext}}. \tag{5.19}$$

Here the term $(\varrho c)_{\mathrm{b}} \bar{u} \cdot \nabla T$ describes the contribution to the heat transfer of the blood flow that is in equilibrium with the tissue. The temperature (as well as the density and specific heat) in this equation should be understood as the volume-weighted parameters for the composite continuum media comprising both the tissue and the blood.

In the Weinbaum-Jiji equation, the thermal effect of the blood perfusion is described by an enhancement in the thermal conductivity appearing in the traditional heat conduction equation. The term $\nabla \cdot \lambda_p \nabla T$ accounts for the additional "perfusional" thermal conductivity:

$$\lambda_p = n(\varrho c)_b \pi r_b^2 \bar{V} \cos^2 \Gamma \sum_{i=1}^{\infty} \frac{L_e}{L_e^2 \beta_i^2 + 1},$$

where Γ is the angle between the direction of the blood vessels and the tissue temperature, n is the number of vessels and r_b is their radius; the parameter β is assumed to be proportional to the vessel length.

The structure of the Eq. (5.19) resembles that of the Pennes equation (5.8); to strengthen the similarity one can unite the terms with the temperature gradient.

The thermal equilibration length was discussed by A. N. T. J. Kotte et al. in the context of their discrete vessels model [1700, 1812] and an approximate solution for the cylindrical vessels and not too small average blood velocity (allowing to neglect the heat diffusion along the vessel) was derived. The notion of the thermalization length—the length at which the difference between the blood and tissue temperature decreases to 1/e of the initial value—suggested by many researches [491, 648, 1820, 3279] allows one to classify all blood vessels into three groups: the large vessels that should be, as a rule, treated explicitly, the thermally significant and the thermally insignificant vessels.

The idea was to categorize the blood vessels into thermally significant and insignificant blood vessels based on the thermal equilibration length analysis using the geometrical parameter(e.g.,vessel diameter) and flow data.

D. Shrivastava and R.B. Roemer [2849], however, suggested that(a) the concept of vessel diameter and equilibration length, (b) local thermal nonequilibrium/equilibrium conditions of vessels and (c) the criterion of total heat transfer rate are inappropriate to categorize significance of different diameter vessels in heat transfer models. They pointed out important factors which have been overlooked by many investigators while considering the concept of thermally significant/insignificant vessels in the vascularized tissue:

- the thermally significant blood vessel is a strong function of the inlet blood temperature, surrounding tissue temperature distribution and blood mass flow;
- the absolute diameters and heat transfer coefficients do not affect the thermal significance of the vessels much;
- both arteries and veins play important role in the exchange of energy, with the arteries playing major role.

These results lead to the conclusion that, depending on the problem under consideration the thermal significance of vessels can only be established on case-to-case basis, and the site and problem specific vascular network model is necessary to have a realistic prediction.

The elements of the structure of the blood vessels net (in particularity, the presence of the pairs “artery-vein”) are incorporated into the three-temperature model of S. Weinbaum et al. [3280]; the temperature of the vein blood is considered explicitly in this model. Later the model was simplified down to the single equation [3277]:

$$\varrho c \frac{\partial T}{\partial t} = \nabla \cdot \lambda_{\text{eff}} \nabla T + \dot{q}_{\text{met}}.$$

The tensor of the effective thermal conductivity is defined as

$$\lambda_{\text{eff}}^{ij} = k_{\text{t}} \left(\delta_{\text{ij}} + \frac{n\pi^2 r_{\text{b}}^2}{4\sigma} \left(\frac{\lambda_{\text{bl}}}{\lambda_{\text{t}}} \right)^2 Pe^2 l_{\text{i}} l_{\text{j}} \right),$$

where n is the geometric parameter defining the density of the blood vessels, r_{b} is assumed to be the same for arteries and the veins, l_{i} are the direction cosines of the vessels, the Pecle number is Pe $= 2\varrho_{\text{bl}} c_{\text{bl}} r_{\text{b}} u / \lambda_{\text{bl}}$, σ is the parameter that characterizes the heat transfer between the artery and the vein. According to the estimates by the authors of [3277], the anisotropic corrections to the thermal conductivity coefficient are essential for the blood vessels with the diameter greater than 100 μm.

I. A. Lubashevsky et al. [1976] studied the anomalous the heat diffusion in the living tissue caused by the branching artery network. The blood vessels make up a complex network being practically a fractal. The larger is a vessel, the faster is the blood motion in it. The blood flow in the capillaries practically does not affect the heat propagation whereas blood inside the large vessels moves so fast that its heat interaction with the surrounding cellular tissue is ignorable. Thus, there should be vessels of a certain length l_v wherein the blood flow affects heat transfer remarkably. The value of can be estimated as [1971]

$$l_v \approx \sqrt{\left(\frac{D}{\omega f L_n} \right)} \tag{5.20}$$

where $D = \lambda/(c\rho)$ is the temperature diffusivity of the cellular tissue determined by its thermal conductivity λ, specific heat c and density ρ. The value ω is the blood perfusion rate (the volume of the blood going through the tissue region of unit volume per unit time), the factor $L_n \sim \ln(l/a)$ is the logarithm of the mean ratio of the individual length to radius of the blood vessels forming the peripheral circulation. The quantity l_v classifies vessels by their influence on the heat transfer: the value l_v divides all vessels of a microcirculatory bed into two classes according to length. The first class involves the vessels whose length is larger than l_v , and the second class consists of the vessels whose length is smaller than l_v (capillaries, venules, arterioles) [1030].

The blood flowing through the first class vessels has no time to attain the thermodynamic equilibrium with the surrounding cellular tissue and may be considered to take no part in the heat exchange with the cellular tissue [503]. This allows one to suppose that at branching points of the first class vessels the conservation law of the blood current and also the conservation law of the heat current are valid.

Thus in the first class arteries the blood temperature is practically equal to the temperature of blood in the large arteries. In the first class veins blood must be characterized by its own temperature $T^{\star}$. So, when, for example, the tissue temperature $T(r,t)$ is nonuniform on scales of order $l > l_v$ the temperature of blood in a vein, whose length is larger than l, will not coincide with the mean tissue temperature in the vein neighbourhood of the radius l [1030]. In the second class vessels blood is in the thermodynamic equilibrium with the cellular tissue and has no significant direct effect on the heat transfer in the living tissue.

For the vascular networks made up of the paired artery and vein trees where all the vessels are grouped into the pairs of the closely-spaced arteries and veins with the opposite blood currents the coefficient $f \sim L^{-1/2}$ accounts for the counter-current effect [1974].

In the mean-field approximation the effect of the blood flow on the heat transfer is described by the renormalization of the temperature diffusivity [3277] and introduction of the effective heat sink [3281]

$$\frac{\partial T}{\partial t} = \nabla \cdot (D_{eff} \nabla T) - fj(T - T_a) + q_T. \tag{5.21}$$

The renormalization of the temperature diffusivity is mainly determined by the blood vessels of lengths about l_v and due to the fractal structure of the vascular networks the renormalization coefficient $\mathrm{F} = D_{eff}/D$ is a constant of unity order [1976].

Lubashevsky et al. considered the temperature evolution in terms of the random walks whose concentration is $T - T_a$. The part of the vein tree made up of the vessels whose lengths exceed or are about the scale l_v forms the system of traps. The blood streams going through the vein tree merge into the greater streams at the nodes. The mean time during which a walker wanders inside the cellular tissue before being trapped is [1971]

$$\tau \sim \frac{l_v^2}{D} L_n \tag{5.22}$$

The characteristic spatial scale of the walker diffusion in the cellular tissue before being trapped is $l_T \sim \sqrt{D\tau} \sim l_v \sqrt{L_n}$. The scale l_T gives the mean penetration depth of the heat penetration into the cellular tissue from a point source.

The fast heat transport over scales exceeding substantially the length l_T is caused by the effect of the blood flow through the artery tree.

A certain number of random walkers generated inside a cellular tissue neighbourhood of a point $\mathbf{r}$ can be found during the time τ inside a cellular tissue neighbourhood of a point $\mathbf{r}'$ at a distance much larger than l_v, i.e. $|\mathbf{r} - \mathbf{r}'| \gg l_v$.

This effect is the anomalously fast heat diffusion in living tissue, i.e. the heat superdiffusion.

The authors estimated the probability for a walker to make a large jump over the distance l averaged over all the possible realizations of the vascular network for the heat source to be localized inside a domain Q_L of the size $\mathcal{L}$ and found the density of the walker transition rate to a point $\mathbf{r}$ spaced at a distance about l from the domain Q_L to be

$$g_l(\mathbf{r}) \sim \frac{D}{L_n}\frac{\mathcal{L}}{l^3}(T - T_a). \tag{5.23}$$

Summing the transition rate over all the levels of the large artery tree the authors re-wrote the mean-filed bioheat Eq. (5.21) as the equation governing the anomalous heat transfer in living tissue

$$\frac{\partial T}{\partial t} = D\nabla^2 T + \frac{D}{L_n}\int\limits_{\mathcal{M}} \mathrm{d}\mathbf{r}' \frac{T(\mathbf{r}') - T(\mathbf{r})}{|\mathbf{r} - \mathbf{r}'|^5 + l_v^5} - fj(T - T_a) + q_T \tag{5.24}$$

where $\mathcal{M}$ is the region containing the peripheral vascular network as a whole.

The second term on the right-hand side of (5.24) depends weakly on the blood perfusion rate.

The limitations of the Weinbaum-Jiji model are associated with the importance of the countercurrent heat exchange: the assumption that most of the heat leaving the artery is recaptured by its countercurrent vein could be violated in the tissue area containing a big blood vessel; this model was primarily developed for the closely paired micro vessels in the muscle tissue, which may not always be the main vascular structure in other tissues [3525]. In other words, the Weinbaum-Jiji model is applicable to the heat transfer in the peripheral tissues [1369].

Moreover, in contrast to the Pennes bioheat equation, which requires only the value of the local blood perfusion rate, the estimation of the effective thermal conductivity requires the detailed anatomical information on the vessel number density, size, and artery-vein spacing for each vessel generation, as well as the blood perfusion rate [3529]). These anatomic data are usually not available for most blood vessels in the thermally significant range [3525].

The Pennes and Weinbaum-Jiji models represent the two extreme situations of the blood-vessel thermal interaction:

- In the Pennes model the arterial blood releases all of its heat to the surrounding tissue in the capillaries and there is no venous rewarming; in fact, the thermal equilibration is achieved in the vessels at least an order of magnitude larger than the capillaries;
- in the Weinbaum-Jiji model the countercurrent rewarming is assumed to be the main mechanism for the blood-tissue heat transfer—most of the heat lost by the artery is recaptured by its countercurrent vein rather than lost to the surrounding tissue.

The subsequent studies have shown that the assumption that the heat transfer between the artery and the vein does not depart significantly from a perfect countercurrent heat exchanger valid only for the vessels less than 200 μm diameter [489, 3527].

J. C. Chato [491], J. W. Baish [156], H. Brinck and J. Werner [374], S. Weinbaum et al. [3281] suggested to introduce the "correction coefficient" $0 \leq \varepsilon \leq 1$ (the weighting function to correct the overestimation of the Pennes perfusion) in the perfusion term

$$\varrho c \frac{\partial T}{\partial t} = \nabla \cdot \lambda \nabla T + \varepsilon \varrho_{\mathrm{b}} c_{\mathrm{b}} \omega_b (T_{\mathrm{a}} - T) + \dot{q}_{\mathrm{met}} + Q^{\mathrm{ext}}. \tag{5.25}$$

From the anatomic studies of the vascular arrangements of the various skeletal muscles, the correction coefficient was found to vary from 0.6 to 0.8 under normal physiological conditions [3525].

A comparison of the Pennes with a number of the more advanced classic models shows that the differences are smaller in the hyperthermia than at the normal conditions [1257]. Still, the errors of the Pennes could be significant: for example, P. Liu et al. [1936] found that in the endocardiac radiofrequency ablation the lesion size difference ratio can reach 20%.

5.1.3 The Wulff Model

W. Wulff [3357] modified the perfusion term, using the local mean blood velocity and tissue temperature gradient rather than the blood perfusion volumetric rate and the blood temperature difference as in the Pennes model.

Wulff stressed that the Pennes' perfusion term is a global (integral term) that does not corresponds to the other terms (as derivatives) which are the local derivatives. Since the blood is moving through the tissue it may remove or derive the heat by the convection in any directions but not only in the direction defined by of the local temperature gradient. Thus, the convective energy transport through the tissue is

$$\varrho_b h_b \mathbf{u}_b = \frac{1}{4\pi} \int\limits_{\Omega} \varrho_b h_b u \mathrm{d}\omega \tag{5.26}$$

where $\mathbf{u}_b$ is the Darcy velocity or local mean apparent blood velocity, u is the actual blood velocity in the capillary, h_b is the specific enthalpy of blood consisting of both sensible enthalpy and enthalpy of metabolic reaction:

$$\bar{h}_b = \int\limits_{T_0}^{T_b} c_p(T_b^{\star}) + \frac{P}{\varrho_b} + \Delta H_f(1 - \varepsilon) \tag{5.27}$$

Here ω is the solid angle across the control surface of the blood vessel, while Ω is the entire solid angle, ΔH_f is the specific enthalpy of the metabolic reaction and the energy flux is

$$q = -\lambda_t \nabla T_t + \varrho_h h_h u_h, \tag{5.28}$$

thus, the energy balance equation is written as

$$\varrho_b c_b \frac{\partial T_t}{\partial t} = -\nabla \cdot \mathbf{q}. \tag{5.29}$$

Wulff assumed that thermal interaction between the blood and the tissue is efficient which maintains equilibrium with surrounding tissues. Thus, he considered that the blood temperature gradient in this equation equals the gradient of the temperature gradient of the environment surrounding the tissue and thus it is possible to set $T_b = T_t$ and to write the energy equation as

$$\varrho_b c_b \frac{\partial T_t}{\partial t} = \frac{\partial^2 T_t}{\partial x^2} - \varrho_b u_h c_b \frac{\partial T_t}{\partial x} + q_{met} \tag{5.30}$$

where $\varkappa_t$ is the thermal diffusivity of the tissue.

The limitation of this equation is that the tissue and the blood volume are needed to be in thermal equilibrium, which is not the case everywhere.

5.1.4 The Klinger Model

The model of H. G. Klinger [1654–1656] is similar to the Wulff model and focuses on the correct description of the experimental data on the deep tissue temperature changed during the point source heating [2432].

Klinger noted that the measuring devices have finite dimensions that in most of the cases larger than the length scale of the tissue of interest and thus the measured temperature is an average value and cannot be accepted as a local one.

To correct the unidirectional blood flow in the Pennes' model Klinger introduced the notion of the convection multipoles related to the *in vivo* vascular anatomy, thus accounting not only the magnitude of the blood flow but also its direction. The effects of the multipole moments depends on the ratio λ/L where L is the macroscopic length scale, while λ is the length of the side of cube over which the average tissue temperature is determined. Since evidently $\lambda/L < 1$, the countercurrent flow effect on the tissue heat transfer could be considered as a second order effect.

Klinger assumed that blood flow is incompressible ($\nabla \cdot \mathbf{u}_b = 0$) and the thermophysical properties of the tissue are constant

$$\varrho_t c_t \frac{\partial T_t}{\partial t} + \varrho_b c_b \mathbf{u}_b \cdot \nabla T_t = -\lambda_t \nabla^2 T_t + Q. \tag{5.31}$$

5.1.5 The Chen-Holmes Model

M. M. Chen and K. R. Holmes [503] have considered the heat transfer in the living tissue with hierarchical system of vessels assuming that the number of the blood vessels in the tissue domain (volume) over which the energy balance is taken is large and the characteristic length scale of the tissue-blood control volume is much larger than the characteristic dimensions of the individual blood vessels, thus the continuum approach can be applied.

The authors suggested that the perfused blood–tissue continuum must be modelled as two separate sub-continuum based on the length scale analysis. First one is the tissue sub-continuum which includes the scales not larger than one millimetre (such as tissues and smaller blood vessels) and the second one is the vessel sub-continuum which includes the larger vessels. Based on the analysis of the thermal equilibration length,their study led to a concept of the thermally significant blood vessels. They defined the thermal equilibration length as the length of the blood vessel over which the temperature difference between the blood and the solid tissue surrounding it decreases by an exponential factor of its initial value.

Chen and Holmes analysed the thermal equilibration between the blood and surrounding tissue, and found that most of the heat exchange between blood and tissue occurs in the vessels with diameter 0.2–0.5 mm, which link the terminal branch arteries and arterioles in the vascular system. As a consequence, in the large vessels blood temperature remains essentially constant as little thermal equilibration takes place, while the blood in the microvascular vessels (arterioles, capillaries and venules) is in the nearly complete equilibrium with the tissue.

The energy balance equation for both the tissue and the vascular spaces are written as

- the solid tissue space:

$$dV_s \left[\varrho_s c_s \frac{\partial T_s}{\partial t} \right] = dQ_{cond} + dQ_{bs} + dQ_m \tag{5.32}$$

- the vascular space:

$$dV_b \left[\varrho_b c_b \frac{\partial T_b}{\partial t} \right] = dQ_{cond} - dQ_{bs} + \int\limits_S (\varrho_b c_b T_b) \mathbf{u} dS \tag{5.33}$$

where the last integral term is the contribution of the convective energy gain due to the blood flow at the velocity $\mathbf{u}$ across a surface area S, dQ_{bs} is the energy exchange between the solid tissue and the vascular spaces, dQ_m is the metabolic heating.

In the following equation

$$\varrho c \frac{\partial T_t}{\partial t} = \nabla \cdot \lambda_{eff} \nabla T_t + \omega_j^\star (T_a^\star - T_t) - \varrho_b c_b \mathbf{u}(\nabla \cdot \nabla T_t) + \dot{q}_{met} \tag{5.34}$$

the perfusion heat source $\omega_j^\star (T_a^\star - T_t)$ is only formally similar to the perfusion term in the Pennes equation since it describes the total perfusion bleed-off to the tissue only from the microvessels past the $j^\star$th generation of branching. In contrast, the Pennes term considers the bleedoff from all generations of vasculature [1369].

The authors defined the effective heat conductivity as in a two-phase (solid-liquid) medium $\lambda_{eff} = \varepsilon_b \lambda_b + (1 - \varepsilon_b)\lambda_t$ and since the blood volumetric fraction ε_b is small, the effective conductivity is practically independent of the blood flow, and thus it is reasonable to accept it equal to the heat conductivity of the solid tissue.

5.1.6 The Baish Model

J. W. Baish [156] proposed a statistical model of heat transfer in perfused tissue. He suggested that vascular model which explicitly considers a few vessels, is not appropriate since because in reality, near the vessels considered there are other vessels of similar dimensions and the continuum model that provides averaged information of the collective blood does not predict accurately the convective effect. Baish pointed that the problem is that the blood vessels are arbitrarily defined in the vascularized tissue.

Baish suggested a steady-state heat transport model that combined the convection model of blood flow with the conduction model of extravascular tissues and includes both deterministic and statistical approaches,and has three key elements:

- A numerical algorithm is used to model the diffusion of angiogenic factors to generate a complete vascular tree that is geometrically similar to a real vascular tree and accounts for the size, orientation, number density,and branching characteristics of blood vessels. It models growth of different kinds of tissue cells in the vicinity of vessels.
- The rate of heat flow into each vessel, temperature at every point along the vessel and temperature at any point in the tissue are determined by both analytical and numerical approaches. The model combines the heat transfer in tissue sub-volume with that of the blood sub-volume through their common boundary condition at the vessel walls. The steady-state energy equation for the tissue sub-volume and the vessel sub-volume are written as

$$\nabla \cdot \lambda_t \nabla T_t + S_t = 0, \tag{5.35}$$

$$\nabla \cdot \lambda_b \nabla T_b - \rho_b c_b v \cdot \nabla T_b + S_b = 0, \tag{5.36}$$

S_t and S_b are the source terms for tissue and blood representing the rate of volumetric heat generation.

The energy equation for vessel sub-volume is modified as

$$\dot{m}_i c_b \frac{\mathrm{d}T_b}{\mathrm{d}x} = q_b', \tag{5.37}$$

$$q_b' = \pi N u \lambda_b (T - T_b) \tag{5.38}$$

where $\dot{m}_i$ is the mass flow rate of blood, q_b' is the rate of heat flow per unit length in a vessel, Nu is the Nusselt number. The conservation of mass and energy at vessel junction are given as

$$\dot{m}_i = \sum_j \dot{m}_j, \qquad \dot{m}_i T_{bi} = \sum_j \dot{m}_j T_{bj} \tag{5.39}$$

where j represents daughter of vessel i, T_{bi} is the temperature of blood leaving the junction and T_{bj} is the temperature of blood moving towards the junction from daughter vessels. The mass transport across the vessel wall is considered negligible, and a heat sink that convects heat away from the tissue sub-volume lies along the centreline of the vessel. At each location along the vessel, the heat flow per unit length into the vessel is matched with the strength of the heat sink.

The analytical solution of (5.35) obtained using Green's function for an infinite domain is given as

$$T(r) - T_{ref} = \int\limits_{r' \in V_{organ}} S(r')G(r, r')\mathrm{d}V - \int\limits_{r' \in V_{tree}} q_b'(r')G(r, r')\mathrm{d}x \tag{5.40}$$

where $G(r, r') = 1/4\pi |r, r'|$ is the Green's function, the rate of heat generation S is considered the same in both sub-volumes.

Baish provided details on the numerical approach to simulate heat transfer along and normal to the vessel axis.

- By sampling the calculated temperatures at many points in the tissue, statistical measurement is performed to predict the probability density function of temperature at a particular point. This is also done to predict the uncertainty in the temperature due to the random proximity of large number of blood vessels. These predictions are based on the statistical information of the vascular geometry and blood flow.

In spite of being computationally extensive technique, Baish model has advantages over the other models—it permits modelling of realistic vascularized tissue by statistical information of the vascular geometry, without prior knowledge of the vessel architecture.

J. Crezee and J. J. Lagendijk [647] suggested the bio-heat transfer model in a generalized form

$$\varrho c \frac{\partial T}{\partial t} = \nabla \lambda_{eff} \nabla T - f_c \varrho_b c_b (T - T_a) + \dot{q}_{met} \tag{5.41}$$

where λ_{eff} and $0 < f_c \leq 1$ are the phenomenological parameters.

5.1.7 The Shrivastava-Vaughan Model

D. Shrivastava and J.T. Vaughan [2850] used for the tissue and blood sub-domains the energy conservation equation with heated/unheated vascularized tissue as

$$\rho_i c_i \left(\frac{\partial T}{\partial t} + \mathbf{v}_i \cdot \nabla T_i \right) = \nabla \cdot (\lambda_i T_i) + Q_i''', \tag{5.42}$$

where the subscript i is for tissue (t) and blood (b) sub-domains, the velocity of tissue $\mathbf{v}_t$ is zero, Q_i''' is the source term indicating absorbed power deposition.

Integration of the Eq. (5.42) over the tissue and blood sub-volumes separately and applying the Green's divergence theorem gives a modified form

$$\int\limits_{V_i} \rho_i c_i \frac{\partial T}{\partial t} \mathrm{d}V - \int\limits_{V_i} T_i \nabla \cdot (\rho c \mathbf{v})_i \mathrm{d}V = \int\limits_{A_i} [-\lambda_i \nabla T_i - (\rho c)_i \mathbf{v}_i T_i] \cdot \mathbf{n}_i \mathrm{d}A \tag{5.43}$$

$$+ \int\limits_{A_{j-i}} (\lambda_i T_i) \cdot \mathbf{n}_{j-i} \mathrm{d}A + \int\limits_{V_i} Q_i''' \mathrm{d}V, \qquad i, j = t, b, i \neq j \tag{5.44}$$

where $\mathbf{n}_{j-i}$ is the normal vector from j to i, A_{j-i} is the surface area between control volumes i and j.

For the incompressible and constant properties the second term on the left hand side of the Eq. (5.44) becomes zero due to the conservation of mass.

The volume averaged energy equation is obtained by normalizing the Eq. (5.44) by volume V_i

$$\frac{1}{V_i} \int\limits_{V_i} \rho_i c_i \frac{\partial T}{\partial t} \mathrm{d}V = \frac{1}{V_i} \int\limits_{A_i} [-\lambda_i \nabla T_i - (\rho c)_i \mathbf{v}_i T_i] \cdot \mathbf{n}_i \mathrm{d}A \tag{5.45}$$

$$+ \frac{1}{V_i} \int\limits_{A_{j-i}} (\lambda_i T_i) \cdot \mathbf{n}_{j-i} \mathrm{d}A + \frac{1}{V_i} \int\limits_{V_i} Q_i''' \mathrm{d}V. \tag{5.46}$$

The first term on the right hand side of this equation represents the energy gained by the ith tissue (or blood) control volume from the adjacent tissue (or blood) control volume. The second term represents the energy gained from the tissue-blood thermal interaction in the vascularized tissue volume.

The Eq. (5.46) is simplified by applying four assumptions:

- Similar to Chen and Holmes and Weinbaum and Jiji the conductive heat transfer is written as

$$\frac{1}{V_i}\int\limits_{A_i}(-\lambda_i \nabla T_i)\cdot \mathbf{n}_i \mathrm{d}A = C_{i1}\nabla\cdot(\lambda_i \nabla \bar{T}_i) \tag{5.47}$$

  where C_{i1} is the new constant, $\bar{T}$ is the volume averaged tissue (or blood) temperature.
- The heat transfer due to the tissue–vessel interaction is given as

$$\frac{1}{V_i}\int\limits_{A_{j-i}}(-\lambda_i \nabla T_i)\cdot \mathbf{n}_{i-i}\mathrm{d}A = \frac{(UA)_{j-i}}{V_i}(\bar{T}_j - \bar{T}_i) \tag{5.48}$$

  where U is the heat transfer coefficient based on volume averaged temperatures and is not similar to conventional heat transfer coefficient.
- The relative importance of convective term over conductive term for blood volume is given as

$$\frac{1}{V_b}\int\limits_{A_b}((\rho c)_b \mathbf{v}_b T_b)\cdot \mathbf{n}_b \mathrm{d}A \gg \frac{1}{V_b}\int\limits_{A_b}(-\lambda_b \nabla T_b)\cdot \mathbf{n}_b \mathrm{d}A \tag{5.49}$$

- The perfusion term is introduced as

$$\frac{1}{V_b}\int\limits_{A_b}((\rho c)_b \mathbf{v}_b T_b)\cdot \mathbf{n}_b \mathrm{d}A = C_{i2}\nabla\cdot((Pc)_i \bar{T}_i) \tag{5.50}$$

  where C_{i2} is the constant, P is the perfusion vector mass flux.

The final form of the bioheat equations for tissue sub-volume and blood sub-volume are given as

$$\rho_t c_t \frac{\partial \bar{T}_t}{\partial t} = C_{t1}\nabla\cdot(\lambda_t \nabla \bar{T}_t) + \frac{(UA)_{b-t}}{V_t}(\bar{T}_b - \bar{T}_t) + \bar{Q}_t''', \tag{5.51}$$

$$\rho_b c_b \frac{\partial \bar{T}_b}{\partial t} = C_{b2}\nabla\cdot(Pc)_b \bar{T}_b) + \frac{(UA)_{b-t}}{V_b}(\bar{T}_t - \bar{T}_b) + \bar{Q}_b'''. \tag{5.52}$$

5.1.8 Porous Medium Models

The porous medium bioheat models (see, for example, the review by A. R. A. Khaled and K. Vafai [1602], the references therein, [1608, 1609] and the papers by S. Mahjoob and K. Vafai [2009, 2010]) are based on the notion that the biological tissues are composed of the dispersed cells separated by the voids that allow flow of the nutrients, minerals etc. to reach every cell and can be considered as a solid matrix with the interconnected voids—the blood-saturated matrix that includes the cells and the interstitial space [2009, 2010]—and the governing equations are derived by averaging over a representative elementary volume (REV) [1258].

This model is difficult to justify physiologically because the blood perfusion is usually not in any direction; such a model could be successful if applied to the relatively well-defined capillary beds, but it becomes rather meaningless when a model is sought for larger parts such as a limb or the whole human body that contain large blood vessels [2839].

Starting with kinematics, conservation laws and the second law of thermodynamics, a general bio-transport model could be developed in the framework of the mixture theory. The mixture theory provides a thermodynamically consistent way to derive both the governing equations and the constitutive relations.

At the tissue level, e. g., there are the connective tissues, muscles, vessels, and various body fluids, which are composed of the endothelial and epithelial cells at the cellular level; and proteins, lipids, carbohydrates, glucose, metabolites, etc., at the sub-cellular level. There are various types of cells such as the healthy or cancerous cells, from the pathological point of view. Therefore, the mixture concept provides a vehicle for the mathematical representation of a complex collection of the interacting constituents.

At an abstract level, the mixture model contains all constituents of interest, which are present at the same point in space with the different volume fractions or the mass concentrations. The sum of the volume fractions or the concentrations is 100%.

This comprehensive model permits the mass exchange between phases and the deformation of solid tissue, and thus can be used to describe growth phenomena such as angiogenesis and tumour growth.

The derivation of the equations is proceeds by scaling up—both the conservation and the constitutive equations are introduced at the microscale [920], thus the more precise description of the blood-tissue interaction is possible since fewer assumptions are needed for the problem closure [1602, 2259].

The two-temperature (the blood temperature T_b and the tissue temperature T_t) model is formulated as follows

$$\varepsilon(\varrho c)_b \left(\frac{\partial T_b}{\partial t} + \mathbf{v_b} \cdot \nabla T_b \right) = \nabla \cdot \lambda_b \nabla T_b + h_{bt}(T_t - T_b), \quad (5.53)$$

$$(1-\varepsilon)\varrho c \frac{\partial T_t}{\partial t} = \nabla \cdot \lambda \nabla T_t - h_{bt}(T_t - T_b) + (1-\varepsilon)(\dot{q}_{\text{met}} + Q^{\text{ext}}). \quad (5.54)$$

Hear ε is the porosity (the volume fraction of the vascular space), h_{bt} is the heat transfer coefficient.

The one-temperature tissue model is obtained by the summing of the blood and tissue energy equations. In this case the need for the coefficient h_{bt} is avoided.

On the other hand, the three temperature model could be formulated if the arterial and the venous temperatures difference is important [2259, 3218]. The equation for the tissue temperature is somewhat simplified if in the peripheral tissue capillaries provide the connection between the artery and the vein (the arterial-venous anastomoses) [2259].

A. Nakayama and F. Kuwahara [2259] analysed the porous media concept in the bio-heat transfer using the volume averaging theory (VAT) [2259, 2262, 2626, 3390], an approach frequently used for the fluid-saturated porous media. In the porous medium models the blood perfusion term is replaced by the interfacial convective heat transfer term.

In the living tissues, three principle compartments can be identified: the blood vessels, the cells and the interstitium [2009], the interstitial space can be further divided into the extracellular matrix and the interstitial fluid. Usually only two regions: the vascular region and extravascular region (cells and interstitium) are considered. The extravascular region is regarded as a solid matrix (despite the fact that extravascular liquid exists) [1369].

In order for the volume averaging to be meaningful, one should consider a control volume V in a fluid-saturated porous medium, whose length scale $V^{1/3}$ is much smaller than the macroscopic characteristic length L but much greater than the microscopic (anatomical structure) characteristic length $V_f^{1/3}$. The volume average and the *intrinsic* average of a variable φ are defined as

$$\langle\varphi\rangle \equiv \frac{1}{V}\int_{V_f} \varphi \mathrm{d}V, \qquad \langle\varphi\rangle^f \equiv \frac{1}{V_f}\int_{V_f} \varphi \mathrm{d}V. \tag{5.55}$$

Obviously, two averages are related as $\langle\varphi\rangle = \varepsilon\langle\varphi\rangle^f$ where $\varepsilon \equiv V_f/V$ is the local porosity, i.e. the volume fraction of the vascular space.

The authors used the following relationships with the averages

$$\langle\frac{\partial\varphi}{\partial t}\rangle = \frac{\partial\langle\varphi\rangle}{\partial t}, \langle\frac{\partial\varphi}{\partial x_i}\rangle = \frac{\partial\langle\varphi\rangle}{\partial x_i} + \frac{1}{V}\int_{A_{int}} \varphi n_i \mathrm{d}A, \tag{5.56}$$

$$\langle\frac{\partial\varphi}{\partial x_i}\rangle^f = \frac{1}{\varepsilon}\frac{\partial\varepsilon\langle\varphi\rangle^f}{\partial x_i} + \frac{1}{V_f}\int_{A_{int}} \varphi n_i \mathrm{d}A, \tag{5.57}$$

where A_{int} is the local interface between the blood and the solid matrix, n_i is the unit vector pointing outward from the fluid side to the solid side. As the authors noted, the volume averaging is similar to the Reynolds averaging used in the study of the turbulence, but the volume averaging procedure is more complex than the Reynolds averaging procedure, since it involves surface integrals.

A. Nakayama and F. Kuwahara formulated the microscopic governing equations, viz., the continuity equation

$$\frac{\partial u_i}{\partial x_i} = 0, \tag{5.58}$$

the Navier–Stokes equation [1945]

$$\frac{\partial u_i}{\partial t} + \frac{\partial}{\partial x_j} u_j u_i = -\frac{1}{\rho}\frac{\partial p}{\partial x_i} + \frac{\partial}{\partial x_j} \nu_f \left(\frac{\partial u_i}{\partial x_j} + \frac{\partial u_j}{\partial x_i} \right) \tag{5.59}$$

and energy equation

$$\rho_f c_f \left(\frac{\partial T}{\partial t} + \frac{\partial}{\partial x_j} u_j T \right) = \frac{\partial}{\partial x_j} \left(\lambda_f \frac{\partial T}{\partial x_j} \right) \tag{5.60}$$

for the blood phase and the heat conduction equation for the solid matrix phase

$$\rho_s c_s \frac{\partial T}{\partial t} = \frac{\partial}{\partial x_j} \left(\lambda_s \frac{\partial T}{\partial x_j} \right) + \dot{q}_{met} \tag{5.61}$$

where the subscripts f and s refer to the fluid and the solid phases, respectively.

The authors integrated the Navier–Stokes equation over a local control volume to get finally

$$\frac{\partial \langle u_i \rangle^f}{\partial t} + \frac{\partial}{\partial x_j} \langle u_j \rangle^f \langle u_i \rangle^f = -\frac{1}{\rho}\frac{\partial \langle p \rangle^f}{\partial x_i} + \frac{\partial}{\partial x_j} \nu_f \left(\frac{\partial \langle u_i \rangle^f}{\partial x_j} + \frac{\partial \langle u_j \rangle^f}{\partial x_i} \right)$$
$$- \frac{\nu_f}{K_{ij}} \varepsilon \langle u_j \rangle^f - b_{ij} \varepsilon^2 \sqrt{\langle u_k \rangle^f \langle u_k \rangle^f} \langle u_j \rangle^f \tag{5.62}$$

where K_{ij} is the permeability tensor, b_{ij} is the Forchheimer tensor. These tensors that depend on the tissue structure can be determined following the procedure established for the anisotropic porous structure [2260] if sufficient information on the anatomical structure and properties is available.

For the vessels of the sufficiently small diameter the Eq. (5.62) reduces to the Darcy's law

$$-\frac{1}{\rho}\frac{\partial \langle p \rangle^f}{\partial x_i} - \frac{\nu_f}{K_{ij}} \langle u_j \rangle = 0 \tag{5.63}$$

where $\langle u_j \rangle = \varepsilon \langle u_j \rangle^f$ is the Darcian velocity (i.e. apparent velocity). The Darcy law could be used for most tissue regions except the regions where the large arteries or veins are located.

The volume average of the convection term in the energy equation is written as

$$\varepsilon\langle \frac{\partial}{\partial x_j} \rho_f c_f u_j T \rangle^f = \frac{\partial}{\partial x_j} \rho_f c_f \langle u_j \rangle \langle T \rangle^f + \frac{\partial}{\partial x_j} \varepsilon \rho_f c_f \langle \tilde{u}_j \tilde{T} \rangle^f + \frac{1}{V} \int_{A_{int}} \rho_f c_f u_j T n_j \mathrm{d}A. \tag{5.64}$$

where the first term on the right hand-side describes the macroscopic convection, the second term takes account of the thermal dispersion [2261], the last term corresponds to the blood perfusion heat source. Thus, the blood perfusion appears as the extra surface integral term resulting from changing the sequence of integration and derivation.

Thus the energy Eq. (5.60) for the and the conduction Eq. (5.61) for the solid matrix into the volume averaged equations as

$$\varepsilon \rho_f c_f \frac{\partial \langle T \rangle^f}{\partial t} + \rho_f c_f \frac{\partial}{\partial x_j} \langle u_j \rangle \langle T \rangle^f = \frac{\partial}{\partial x_j} \left(\varepsilon \lambda_f \frac{\partial \langle T \rangle^f}{\partial x_j} + \frac{\lambda_f}{V} \int_{A_{int}} T n_j \mathrm{d}A - \varepsilon \rho_f c_f \langle \tilde{u}_j \tilde{T} \rangle^f \right) + + \frac{1}{V} \int_{A_{int}} \lambda_f \frac{\partial T}{\partial x_j} n_j \mathrm{d}A - \frac{1}{V} \int_{A_{int}} \rho_f c_f u_j T n_j \mathrm{d}A, \tag{5.65}$$

$$(1-\varepsilon) \rho_s c_s \frac{\partial \langle T \rangle^s}{\partial t} = \frac{\partial}{\partial x_j} \left((1-\varepsilon) \lambda_s \frac{\partial \langle T \rangle^f}{\partial x_j} - \frac{\lambda_s}{V} \int_{A_{int}} T n_j \mathrm{d}A \right) - - \frac{1}{V} \int_{A_{int}} \lambda_s \frac{\partial T}{\partial x_j} n_j \mathrm{d}A + \frac{1}{V} \int_{A_{int}} \rho_f c_f u_j T n_j \mathrm{d}A + (1-\varepsilon) \dot{q}_{met} \tag{5.66}$$

where $\langle T \rangle^s$ is the intrinsic average of the solid matrix temperature.

The dispersion heat flux $\varepsilon\rho_f c_f \langle \tilde{u}_j \tilde{T} \rangle^f$ appears in the volume averaged energy Eq. (5.65) for the blood phase and may well be modelled under the gradient diffusion hypothesis [2259]:

$$\varepsilon\rho_f c_f \langle \tilde{u}_j \tilde{T} \rangle^f = -\varepsilon \tilde{\lambda}_{kj} \frac{\partial \langle T \rangle^f}{\partial x_k}. \tag{5.67}$$

The different expressions have been proposed for the dispersion thermal conductivity $\tilde{\lambda}$. For example, A. Nakayama et al. [2261] obtained the following expression for the predominant axial component of $\tilde{\lambda}$ for a bundle of vessels of radius R

$$\tilde{\lambda}_{xx} = \begin{cases} \frac{1}{48}\left(\frac{\rho_f c_f \langle u \rangle^f R}{\lambda_f}\right)^2 \lambda_f, & \frac{\rho_f c_f \langle u \rangle^f R}{\lambda_f} < 1 \ (capillary\ blood\ vessels) \\ 2.55\left(\frac{\rho_f c_f \langle u \rangle^f R}{\lambda_f}\right)^{7/8} Pr^{1/8} \lambda_f, & \frac{\rho_f c_f \langle u \rangle^f R}{\lambda_f} > 1 \ (large\ arteries\ and\ veins) \end{cases} \tag{5.68}$$

where Pr is the Prandtl number.

In order to close the macroscopic Eqs. (5.65) and (5.66), the terms associated with the surface integral, describing the interfacial heat transfer and perfusion between the fluid and the solid, must be approximated. For the interfacial heat transfer the Newton's cooling law can be used as

$$\frac{1}{V} \int\limits_{A_{int}} \lambda_f \frac{\partial T}{\partial x_j} n_j \mathrm{d}A = a_f h_f (\langle T \rangle^f - \langle T \rangle^s) \tag{5.69}$$

where a_f is the specific surface area, h_f is the interfacial heat transfer coefficient. For the bundle of the vascular tubes of the radius R $a_f = 2\varepsilon/R$, $h_f = Nu(\lambda_f/2R)$, where Nu is the Nusselt number based on the diameter of the vascular tube (W. Roetzel and Y. Xuan [2626] indicated $Nu = 4.93$ for both the arterial and venous blood vessels). If the local porosity and the specific surface area are known for the complex tissue-vascular structure, the interfacial heat transfer coefficient could be estimated [2259]. The heat transfer coefficients used to calculate the heat transfer to the large blood vessels depend on the entrance effects, the tissue thermal resistance and the vessel shape, and may vary circumferentially around the vessel during the transient increases in the temperature [1669]. The cooling effect of the large vessels is determined by the perfusion in the surrounding tissue [1700, 1701]. Simulation by H. Brinck and J. Werner [374, 375] showed that the temperature gradient around the artery is larger than the temperature gradient around the vein that is explained by the fact that the heat flow between arteries and tissue is larger than the heat flow between veins and tissue. The mean tissue temperature perpendicular to the vessels was thus closer to the temperature of the venous blood, while the mean tissue temperature in between the artery and the vein is represented by the mean arterial and venous blood temperatures.

The fluid lost from the vascular space will be compensated by the flow of extravascular fluids and lymph from the tissue to the vascular space. A. Nakayama and F. Kuwahara assumed that the extravascular fluids and all lymph in the tissue space have the same temperature as the tissue itself and the transcapillary fluid exchange takes place at the rate of ω, thus the blood perfusion term is

$$\frac{1}{V}\int\limits_{A_{int}} \rho_f c_f u_j T n_j \mathrm{d}A = \rho_f c_f \omega(\langle T\rangle^f - \langle T\rangle^s). \tag{5.70}$$

The surface integral terms

$$\frac{\lambda_f}{V}\int\limits_{A_{int}} T n_j \mathrm{d}A, \qquad \frac{\lambda_s}{V}\int\limits_{A_{int}} T n_j \mathrm{d}A \tag{5.71}$$

that present the tortuosity heat fluxes are usually small and their effects may be absorbed in the effective thermal conductivities [2259, 3390].

Finally, the macroscopic energy equations are written for the blood and tissue phases as

$$\varepsilon\rho_f c_f \frac{\partial\langle T\rangle^f}{\partial t} + \rho_f c_f \frac{\partial}{\partial x_j}\langle u_j\rangle\langle T\rangle^f = \tag{5.72}$$

$$\frac{\partial}{\partial x_j}\left(\varepsilon\lambda_f \frac{\partial\langle T\rangle^f}{\partial x_j} + \varepsilon\tilde{\lambda}_{jk}\frac{\partial\langle T\rangle^f}{\partial x_k}\right) - a_f h_f(\langle T\rangle^f - \langle T\rangle^s) - \rho_f c_f \omega(\langle T\rangle^f - \langle T\rangle^s) \tag{5.73}$$

where the four terms on the right hand-side correspond to the macroscopic conduction, the thermal dispersion, the interfacial convective heat transfer and the blood perfusion,

$$(1-\varepsilon)\rho_s c_s \frac{\partial\langle T\rangle^s}{\partial t} = \frac{\partial}{\partial x_j}\left((1-\varepsilon)\lambda_s \frac{\partial\langle T\rangle^s}{\partial x_j}\right) +$$

$$+ a_f h_f(\langle T\rangle^f - \langle T\rangle^s) + \rho_f c_f \omega(\langle T\rangle^f - \langle T\rangle^s) + (1-\varepsilon)\dot{q}_{met} \tag{5.74}$$

where the four terms on the right hand-side correspond to the macroscopic conduction, the interfacial convective heat transfer, the blood perfusion heat source and the metabolic heat source.

The continuity equation, the Darcy's law and the two energy Eqs. (5.73) and (5.74) form a closed set of the macroscopic governing equations.

Several existing bioheat transfer models could be considered as the particular cases of the model based on the theory of the porous media:

- *Pennes model*

  The Pennes model in the porous media terms is written as

$$(1-\varepsilon)\rho_s c_s \frac{\partial \langle T\rangle^s}{\partial t} = \frac{\partial}{\partial x_j}\left((1-\varepsilon)\lambda_s \frac{\partial \langle T\rangle^s}{\partial x_j}\right) + \rho_f c_f \omega_{Pen}(T_a - \langle T\rangle^s) + (1-\varepsilon)\dot{q}_{met} \tag{5.75}$$

  Comparison shows that

$$\rho_f c_f \omega_{Pen}(T_a - \langle T\rangle^s) = a_f h_f(\langle T\rangle^f - \langle T\rangle^s) - \rho_f c_f \omega(\langle T\rangle^f - \langle T\rangle^s) \tag{5.76}$$

  and assuming that $T_a \approx \langle T\rangle^f$ for small vessels one gets

$$\omega_{Pen} \approx \omega + \frac{a_f h_f}{\rho_f c_f} \tag{5.77}$$

  i.e., the Pennes perfusion rate is an effective value including the surface heat transfer.
- *Klinger model* The combination of the Eqs. (5.73) and (5.74) for the case $\langle T\rangle^f = \langle T\rangle^s$ reduces to the Klinger model when the porosity ε is small.
- *Chen-Holmes model* Chen and Holmes take account of the "eddy" conduction due to the random flow of blood, by introducing the thermal conductivity λ_p which corresponds to the dispersion thermal conductivity $\tilde{\lambda}$.

Nakayama and Kuwahara developed a generalized three-equation bioheat models for the vascular and the extravascular space under the local thermal non-equilibrium condition. The authors incorporated the blood perfusion term within the two sub-volume equations. They considered the effect of the heat transfer in the closely spaced countercurrent artery-vein pair. H. C. Bazett et al. [222] found that the axial temperature gradient in the limb artery of the human under the low ambient temperature is an order of magnitude higher than that of in the normal ambient condition due to the effect of countercurrent heat exchange in the bioheat transfer. The countercurrent heat exchange also reduces the heat loss from the extremity to the surroundings [2156]. The model developed by J. W. Mitchell and G. E. Myers [2156] was one-dimensional and was not able to take account of either the metabolic heat source or the perfusion bleed-off from the artery to vein.

The three equations are derived by Nakayama and Kuwahara for the arterial blood phase, the venous blood phase and the tissue phase with three different temperatures

- the arterial blood phase

$$\varepsilon_a \rho_f c_f \frac{\partial \langle T \rangle^a}{\partial t} + \rho_f c_f \frac{\partial}{\partial x_j} \varepsilon_a \langle u_j \rangle \langle T \rangle^a =$$

$$\frac{\partial}{\partial x_j} \left(\varepsilon_a \lambda_a \frac{\partial \langle T \rangle^a}{\partial x_j} + \varepsilon_a \tilde{\lambda}_{a\,jk} \frac{\partial \langle T \rangle^a}{\partial x_k} \right) - a_a h_a (\langle T \rangle^a - \langle T \rangle^s) - \rho_f c_f \omega'_a \langle T \rangle^a \tag{5.78}$$

- the venous blood phase

$$\varepsilon_v \rho_f c_f \frac{\partial \langle T \rangle^v}{\partial t} + \rho_f c_f \frac{\partial}{\partial x_j} \varepsilon_v \langle u_j \rangle \langle T \rangle^v =$$

$$\frac{\partial}{\partial x_j} \left(\varepsilon_v \lambda_v \frac{\partial \langle T \rangle^v}{\partial x_j} + \varepsilon_v \tilde{\lambda}_{v\,jk} \frac{\partial \langle T \rangle^v}{\partial x_k} \right) - a_v h_v (\langle T \rangle^v - \langle T \rangle^s) - \rho_f c_f \omega'_v \langle T \rangle^v \tag{5.79}$$

- the solid tissue phase

$$(1-\varepsilon)\rho_s c_s \frac{\partial \langle T \rangle^s}{\partial t} = \frac{\partial}{\partial x_j} \left((1-\varepsilon)\lambda_s \frac{\partial \langle T \rangle^s}{\partial x_j} \right) + a_a h_a (\langle T \rangle^a - \langle T \rangle^s) + \rho_f c_f \omega'_a \langle T \rangle^a$$

$$+ a_v h_v (\langle T \rangle^v - \langle T \rangle^s) + \rho_f c_f \omega'_v \langle T \rangle^v + (1-\varepsilon)\dot{q}_{met} \tag{5.80}$$

where $\varepsilon = \varepsilon_a + \varepsilon_v$.

If the two temperatures are eliminated, the equation for the single (tissue) temperature could be derived that contains the derivatives of the temperature up to the sixth order in space (eliminating of the tissue or the blood temperature from the two-temperature model leads to the bioheat transfer equation where the blood temperature or tissue temperature is a sole unknown variable that contains temperature derivatives up to the fourth order) [1864, 3109].

A. Ainouz [38] studied the bio-heat transfer model for biological tissues in a micro-scale and periodical settings. The two-component system consisting of solid particles representing the tissue cells and the interconnected pores containing either arterial or venous blood described by two energy equations, one equation for the solid tissues and the other for the surrounding blood. On the interface between these regions the heat transfer is governed by Newton's cooling law.

R. Hochmuth and P. Deuflhard [1334] applied the homogenization technique to a microvascular model consisting of the tissues (solid) regions with the temperature T^ε surrounded by the blood (fluid) with the temperature T_b^ε:

$$-\Delta T^\varepsilon = S^\varepsilon \quad in \qquad \Omega^\varepsilon, \tag{5.81}$$

$$T^\varepsilon = 0 \quad on \qquad \partial\Omega, \tag{5.82}$$

$$\frac{\partial T^\varepsilon}{\partial n^\varepsilon} = \varepsilon_\alpha (T_b^\varepsilon - T^\varepsilon) \quad on \qquad \partial\Omega^\varepsilon, \tag{5.83}$$

where S^ε the source term in the tissues represented by the domain Ω^ε obtained by removing from Ω—a bounded and regular domain—a set of holes where s, n^ε is the unit normal of $\partial\Omega^\varepsilon$ outward to Ω^ε and $\alpha > 0$ is a physiological parameter.

The authors obtained the following homogenized model:

$$-div(\mathcal{A}\nabla T^\star) + \alpha^\star(T^\star - T_b) = S \tag{5.84}$$

where $\mathcal{A}$ is the homogenized tensor, $T^\star$ is the weak limit in $H^1(\Omega)$ of some extended temperature $P^\varepsilon(T^\varepsilon)$, S is the weak limit in $L^2(\Omega)$ of the source term S^ε and $\alpha^\star$ is the effective Helmholtz term.

The biological tissues can be seen as the porous media where the cells (matrix) are separated by voids or pores which are filled with blood. This system of cells-blood can be interpreted as a two-constituent medium. The authors considered the case where the tissue is made of two interacting systems (cells tissues and blood regions) and the conductivities are assumed to be of different order of magnitude. The transition between these two regions on the interface is governed by the Newton's cooling law, i.e. the heat flow through the interface is proportional to the jump of the temperature field. Using the homogenization techniques, it is shown that the obtained macro-model presents some extra terms.

5.2 Non-Fourier Continuum Models

The derivation of the Fourier equation is based on the principle of the local thermodynamic equilibrium and the continuum hypothesis, according to which each small element of the medium has a local equilibrium state and can be described by the local thermodynamic potentials that are dependent on the spatial variable and the time only through the thermodynamic parameters.

The Fourier law is local with respect to time and space [2896]. Accepting the local equilibrium principle is possible only if the rate of change in the system macroparameters due to the external influences is much less than the rate of system relaxation to the local equilibrium [2896]. Based on the local equilibrium principle, it is assumed that the transport laws are valid not only for the entire system, but also for any arbitrarily small part of it. Thus it is possible to perform the limit transition (by tending the volume of the integration to zero) in the integral conservation laws and obtain these laws in the form of the differential equations.

If the characteristic microscopic scale of the system and the characteristic time of its relaxation to equilibrium are significantly less than the characteristic macroscopic scale and the total time of the process, then the differential equations that are based on the local equilibrium principle and the continuum hypothesis will be local both in space and in time. Thus we obtain the transport equations that do not contain the relaxation time τ and the characteristic scale of the microstructure l.

The Fourier law is a prototype of the constitutive equations leading to a parabolic partial differential equation. The major drawback of the parabolic FHCE is the unphysical instantaneous propagation of the disturbances ("the heat transfer paradox" [2844], "the paradox of propagation of the thermal signals" [2897], "the paradox of heat conduction" (PHC) [564, 1174, 2844, 3135], "the heat transfer paradox" (HTP) [1067] or "the paradox of diffusion" [563]) called sometimes *acausality* [44]—it does not forecasts the propagation of the disturbances restricted by the characteristic causal light-cones [1307].

Lars Onsager in 1931 pointed out that "... Fourier law is only approximate description of the process of conduction neglecting the time needed for acceleration of the heat flow" [2341].

The solution of (5.4) in the one dimensional case for the initial conditions $T(x \neq 0, 0) = 0, \quad T(0, 0) = 1$ is

$$T(x, t) = \frac{1}{\sqrt{\pi \varkappa t}} \exp\left(-\frac{x^2}{4\varkappa t}\right). \tag{5.85}$$

Thus, for x as large as one wishes and for any $t > 0$ one gets $T(x, t) > 0$. Hence the heat has propagated from $x = 0$ to x in an interval of time no matter how small. The speed of the propagation of the heat is infinite, which is not acceptable.

However, as noted by G. Fichera [959], the Fourier law is a result of the experimental observations hence it is conditioned by the physical nature of the material under consideration and by the degree of refinement of the measurements. It results that the infinitely small heat signal that propagates at an infinite speed cannot be experimentally seen—the sensitivity of the experimental devices is finite, therefore they can measure only finite speeds [1698, 3135, 3136]. W. A. Day [689] demonstrated that, if an infinitesimal small fraction of the heat signal does propagate at an infinite speed, the propagation of the bulk of the signal—which is physically significant—is at finite velocity. As stressed by F. Márkus [2050], one should distinguish the *signal* and *action* propagation.

J.-L. Auriault [140] claimed that the paradox is inconsistent using the underlying homogenization process to investigate the domains of validity of both the Fourier and the Maxwell–Cattaneo–Vernotte (Sect. 5.2.1.1) equations.

The author used three different characteristic lengths l, L and L_1 associated to the atomic scale, the macroscopic scale where the Fourier law and the Maxwell–Cattaneo–Vernotte law are defined and an upper macroscopic scale: $l/L = \varepsilon \ll$

1, $L/L_1 = \eta \ll 1$. The smallness of the parameter of separation of scales η is the condition for the existence of the equivalent macroscopic models [140]. In the framework of the homogenization theory $T = T^{(0)} + \varepsilon T^{(1)} + \varepsilon^2 T^{(2)} + \cdots$ and the Fourier law is

$$\frac{\partial T^{(0)}}{\partial t} = \varkappa \Delta T^{(0)} + O(\varepsilon). \tag{5.86}$$

Thus an infinite speed of a heat signal obtained from the classical Fourier equation cannot be demonstrated from the last equation since such a signal is very small—it is included in the term $O(\varepsilon)$. Thus, Auriault concluded that since the Fourier equation is an approximation and cannot forecast a heat signal with an infinite speed because such a signal would be infinitely small, it should not be considered paradoxical [140, 3307].

The Fourier law is valid if [1497, 2892, 3515]

- $\frac{L}{\Lambda} \gg O(1)$,
- $\frac{t}{\tau} \gg O(1)$,
- $T \gg 0^\circ K$,

where L is the characteristic size of the system, Λ is the mean free path of the heat carries, τ is the relaxation time.

The ratio $Kn = \Lambda/L$ is called the Knudsen number similar to the rarefied gas dynamics theory. The Fourier law is valid in the limit of the small Knudsen number, i.e., when $\Lambda/L \ll 1$. The Knudsen number may increase either because of an increase of the mean free path of the heat carries as in the rarefied gases or due to the reduction of L as in miniaturized systems. Sometimes the Fourier number is used $Fo = \lambda T_0/\varrho CL$ [778, 3207]. Thus the Fourier law is valid for most macroscopic problems.

However, the Fourier law is invalid if the "microscopic nature of heat" [1170] could not be ignored—if the wave nature of the heat propagation or the time lags become dominant and the memory, nonlinear or the spatial nonlocal effects significant [1169, 3515] as

- in the ultrafast heating:
  - the pulsed laser heating and melting [479, 500, 1766, 2040, 2702, 2704, 2838, 3017, 3108],
  - the welding and drilling of metals,
  - the surface annealing,
  - the sintering of ceramics [3451],
  - the micro-machining [1293],
  - the rapid solidification of liquid metals [1304, 2227, 2228, 3237],
  - the glass transition of supercooled liquids that could involve the multiple steps of the energy relaxation [1314, 2895],
  - the glassy polymers near the glass transition temperature [1501],

  - the heat pulse experiments at the room temperature, including "the book experiment" [334, 800, 941–943, 1017, 1704, 3135, 3138, 3140],
  - the sudden contact of the two liquids such as the uranium oxide and sodium [1575],

- in the heat transfer at temperatures near the absolute zero [133, 2272, 2435, 3261],
- in the heat transfer at nanoscale [224, 423, 424, 499, 1618, 3471]

  - in the microelectronic and the optoelectronic devices [234, 1004, 1348, 1917, 2567, 2579, 2732, 3072, 3182], for example, the hot spots in the nanotransistors [114, 516, 996, 2485, 2486, 2989, 2990, 3403],
  - the nanostructured devices for the solid-sate energy conversion, e.g., the thermoelectric and thermoionic refrigeration [338, 497, 508, 1301, 1528, 2146, 2887, 3173, 3191],
  - in the nano-electromechanical systems (NEMS) [1986],
  - in the two-dimensional materials [1175],
  - in the heterostructures [1175, 3168],
  - in the layered strongly correlated materials (LCM) [2089] such as, e.g., the transition metal oxides (TMOs) [1044] in which the strong short-ranged electronic interactions lead to the failure of the independent-electron approximation and which are characterized by the intertwining of the charge, spin and lattice degrees of freedom and exhibit the wave-like regime for the temperature propagation at the nanoscale [1044, 3497],
  - in the laser plasma in irradiating small targets [3205]),
  - in the nano-engineered suspensions for the radiative cooling [1376] and the volumetric solar thermal energy absorption [2357],
  - in the heat transfer in the macromolecule of the deoxyribonucleic acid (DNA) during its denaturation ("melting")—unravelling of the double-stranded structure into two single strands [3172],[1]
  - in the ultra-high thermal isolation across the heterogeneously layered two-dimensional materials [3167],

- in the heat transfer in the granular and porous materials [79, 209, 1151, 1449, 1474, 1853, 1982, 2345, 2857], in the fractured geological media [187, 857, 1058, 1410, 1980, 2987],
- at the extremely high values of the heat flux [2074] ($>\approx 10^7 W/\text{cm}^2$) and
- **in the heat transfer in the biological tissues** [32, 143, 671, 1898, 1922, 2518, 2865, 2888, 2960, 3510, 3512, 3513, 3519].

The origin of the time lag could be the existence of several energy carriers [3111] or the heterogeneous inner structure. The biological tissues contain the cells, membranes, organelles, superstructures, liquids, solid/soft elements (sometimes the

[1] S. Buyukdagli and M. Joyeux investigated the effect of the finiteness of the length of the sequence on the phase transition that takes place at DNA melting temperature [413].

tissues are referred to as to the *mesoporous* structures [2008]). The heating or cooling of the living tissues induces a series of the chemical, the electrical and the mechanical processes, for example, the diffusion, the electrical potential change and the osmosis across the cell membrane; the cell membranes could store energy [1898]. Thus, the heat propagation involves the multi-mode energy conversion at the different cellular levels [1879, 1898, 2256]. From a purely thermodynamic point of view, the rocks do not differ significantly from the biological materials: they both are porous, having the various heterogeneities and irregularities, thus the parallel time scales are present [2970].

S. L. Sobolev [2894] suggested to distinguish the *strong* nonlocality (the nonlocal effects in the transport processes are described in the integral form of the balance equations) and the *weak* nonlocality (the parameters related to the internal characteristic time and space scales are introduced into the differential balance equations). K. Takahashi [3004] noted that the space nonlocality is related to the introduction of the scale intermediate between the microscale and the nanoscale—the mesoscale.

Some of the non-Fourier heat conduction models are universal, i.e. could be used for the description of the heat transfer in the arbitrary medium (the phase-lag models, the thermodynamic and the fractional derivative models) while others are limited to some class of media (for example, the phonon models based on the Peierls–Boltzmann equation) [3515].

5.2.1 Phase-Lag Models

A number of the non-Fourier models have been developed that are based on the modification of the constitutive relation (sometimes called the flux-gradient relation (FGR) [3184]) between the heat flux and the temperature gradient for the media with the thermal memory (the memory effects could be represented in three different forms: a strong time nonlocality with memory functionals, weak time nonlocality with time derivative expansions, and a directly delayed mathematical description [2844]) or/and with the discrete structure [2890]. Most of these models incorporate the time non-locality for materials with memory [1676, 2317] (the hereditary materials), some are also consider the space non-locality, i.e. materials with the nonhomogeneous inner structure.

The non-Fourier models must be compatible with the kinetic theory of gases—the rarefied gases have the best understood microscopic composition among continua (the macroscopic equations from the moment series expansion of the kinetic theory are instructive in this respect)—and with the second law of the thermodynamics and the spacetime requirements of the nonequilibrium thermodynamics that are universal, independent of the material composition and properties, therefore their consequences are universal as well [1718, 1719, 2844]. The thermophysical fields and the constitutive models should be frame-indifferent (this property is called *objectivity*) [44, 2234, 2239, 3085].

One should remember that strictly speaking, a temperature has a well-established meaning only in the global equilibrium when the heat flux is zero [462]—a temperature at a point can be defined only under the local thermodynamic equilibrium that is valid only for a relatively weak deviation from equilibrium, thus a meaningful temperature can be defined only at points separated on an average by the mean free path of the energy carries [2019, 2159].

The nonlinearity in the heat conduction equation could be expressed in two ways—as the temperature dependence of the material thermophysical properties [2562] and as the presence of the nonlinear products of the temperature gradient or the heat flux [586, 587].

The constitutive relation (5.1) could be formulated in the integral form [603, 605] ("the integrated history of the temperature gradient" [1844]) similar to the theory of the non-linear materials (for example, the viscoelastic materials [602]) in which the stress at the point depends on the deformation gradients at the previous time instants [1157–1159].

A general expression for the heat flux in the nonlinear materials with memory as the convolution integral was written by M. E. Gurtin and A. C. Pipkin half a century ago in analogy with the stress-strain relationship of the viscoelastic materials as [1195] (see also the paper by B. D. Coleman and M. E. Gurtin [604])

$$\mathbf{q} = -\int_{-\infty}^{t} Q(t - t')\nabla T \mathrm{d}t' \tag{5.87}$$

where $Q(s)$ is a positive decreasing function (called the relaxation kernel of the Jeffreys's type [1366, 1497, 2667, 2668] after H. Jeffrey who used similar relations for the rheological description of the Earth core [1450], also called the *memory* kernel) tending to zero as $s \to \infty$.

The Eq. (5.87) is a particular case of the relation between the flux $\mathbf{J}(t)$ at a time t and the conjugated thermodynamic force $\mathbf{X}(t')$ at the previous times t' [1504]

$$\mathbf{J} = \int_{-\infty}^{t} \mathbf{K}(t - t') \cdot \mathbf{X}(t')\mathrm{d}t'. \tag{5.88}$$

In the classical transport theory the fluxes $\mathbf{J}$ and the thermodynamics forces $\mathbf{X}$ are related by the linear law $\mathbf{J}(t) = \hat{L} \cdot \mathbf{X}(t)$ where $\hat{L}$ denotes the phenomenological coefficients.

The different choices of the memory kernel leads to the different constitutive relations and thus the different heat conduction equations. For example, if the kernel is chosen as the Mittag–Leffler function, the fractional order constitutive equation will occur [613].

The condition that the relaxation kernel $Q(t-t')$ is the decreasing function means the decrease of the relevance of the older temperature gradients in comparison with the newer ones, thus the term "fading memory" [607] is used. Note that selection of the integral domain $(-\infty, t]$ instead of $[0, t]$ means that the initial effects are implicitly neglected [1306].

For $Q(s) = \lambda\delta(s)$ where $\delta(s)$ is the one-sided Dirac delta function [1497]

$$\int_0^{0+} \delta(s)ds = 1 \tag{5.89}$$

we get the Fourier law (5.1), i.e. this law corresponds to the heat conduction in the "zero-memory" material.

The Eq. (5.87) could be generalised to include the spatial nonlocality explicitly [3184]

$$\mathbf{q}(\mathbf{r}, t) = -\int_{-\infty}^{t} \int_{-\infty}^{\infty} Q(\mathbf{r} - \mathbf{r}', t - t')\nabla T \mathrm{d}t' \mathrm{d}\mathbf{r}'. \tag{5.90}$$

The different choices of the constitutive relation leads to the numerous "lagging" models (see, e.g., [40, 1898, 1916, 3111] and references therein).

5.2.1.1 Maxwell-Cattaneo-Vernotte Equation

A well-known example of the constitutive heat-flux equation of the Jeffreys type is [1497, 1498]

$$\tau \frac{\partial \mathbf{q}}{\partial t} + \mathbf{q} = -\lambda \nabla T - \tau \lambda_1 \frac{\partial}{\partial t} \nabla T. \tag{5.91}$$

For the case $\lambda_1 = 0$ (5.91) reduces to the well known the Cattaneo (also called the Maxwell[2]-Cattaneo-Vernotte (MCV) [585, 2775, 2844] or the Maxwell-Chester-Cattaneo-Vernotte [52]) constitutive relation that was independently formulated by Philip M. Morse and Hermann Feshbach in their famous monograph 'Methods of Theoretical Physics" [2195], by Harold Grad [1144], by Carlo Cattaneo [467] and by Pierre Vernotte [3185, 3186] (see also papers by O. G. Bakunin [164] and S. L. Sobolev [2898])

$$\tau \frac{\partial \mathbf{q}}{\partial t} + \mathbf{q} = -\lambda \nabla T \tag{5.92}$$

[2] As I. Müller noted [2223], Maxwell indeed wrote the equation for the heat flux with a rate term derived in the kinetic theory of gases in his papers of 1867 and 1879, but he finally dismissed it as being small and uninteresting.

If additionally the relaxation time $\tau = 0$, the Eq. (5.91) reduces to the Fourier law. The term $\tau\partial\mathbf{q}/\partial t$ introduced in a reasonable (but arbitrary) manner is sometimes referred to as the "thermal inertia" [3542]. In the systems with the inertia—a particular form of the memory—the motion does not stop when the force diminishes; from a mathematical point of view, the representation of the inertia requires the additional state variables, for example, the momentum the in mechanics and the heat flux in the heat conduction [2844].

The relaxation time is the intrinsic property [3105]. At present, no direct experimental method exists for the determination of τ [3381].

B. D. Coleman et al. [608, 609] extended the Eq. (5.92) to the anisotropic case

$$\hat{\tau}\frac{\partial\mathbf{q}}{\partial t} + \mathbf{q} = -\hat{\lambda}\nabla T \tag{5.93}$$

where $\hat{\tau}$ and $\hat{\lambda}$ are the non-singular second-order tensors that, as function of temperature, depend on the material under consideration.

Note that similar relation is used to describe the viscoelastic effects [461]

$$\tau\dot{\hat{P}}^{v} + \hat{P}^{v} = -2\eta\hat{V}^{s} \tag{5.94}$$

where $\hat{P}^{v}$ is the viscous pressure tensor, $\hat{V}^{s}$ is the symmetric part of the velocity gradient tensor $\nabla\mathbf{V}$, η is the shear viscosity.

The Cattaneo constitutive relation (sometimes called "modification of the Fourier law" (MFL) [2898]) is obtained if in the general Gurtin-Pipkin relation (5.87) the relaxation kernel is chosen as

$$Q(s) = \frac{\varkappa}{\tau}e^{-\frac{t-t'}{\tau}} \tag{5.95}$$

where $\varkappa = \lambda/(\varrho C)$ is the thermal diffusivity.

Thus the Cattaneo equation for the temperature (also "the hyperbolic heat conduction equation" (HHCE) [2898]) could be written (for the constant material properties) as [556, 577]

$$\tau\frac{\partial^2 T}{\partial t^2} + \frac{\partial T}{\partial t} = \varkappa\nabla^2 T. \tag{5.96}$$

Earlier (in 1934) the similar equation for the diffusion

$$\frac{\partial n}{\partial t} + \tau\frac{\partial^2 n}{\partial t^2} = D\nabla^2 n \tag{5.97}$$

was derived by Boris Davydov [164] who suggested the phenomenological system for the particle density $n(x,t)$ to account for the finite particle velocity during the molecular diffusion introducing an explicit time interval of the mean free path [1397, 1398]

$$\frac{\partial n}{\partial t} + \frac{\partial q}{\partial x} = 0, \quad \frac{\partial q}{\partial t} = \frac{q_0 - q}{\tau} \tag{5.98}$$

where $q_0 = -D(\partial n/\partial x)$.

The Eq. (5.96) in contrast to the parabolic Fourier equation is hyperbolic and can be considered as the particular case of the telegrapher's equation [3537] (sometimes called the *causal* diffusion equation [722]). This equation describes the crossover between the ballistic motion [64, 547] and the onset of diffusion—the transition from the reversible to the irreversible behaviour—at the characteristic time τ:

- for small times $t \ll \tau$ the first term in (5.96) is dominant and it reduces to the wave equation

$$\tau \frac{\partial^2 T}{\partial t^2} = \varkappa \nabla^2 T \tag{5.99}$$

that describes the reversible process since it is invariant with respect to the time inversion. The solution of (5.96) for the initial condition $T(x, 0) = \delta(x)$ is written as

$$T(x, t) = \delta\left(\sqrt{\frac{\varkappa}{\tau}} t - x\right) \tag{5.100}$$

i.e. the initial pulse propagates with the velocity $\sqrt{\varkappa/\tau}$ without changing the shape of the distribution [1398].

- for large time $t \gg \tau$ the first term in (5.96) could be neglected and one get the parabolic equation of the irreversible process—the heat diffusion.

Thus, the hyperbolic nature of (5.96) is the most significant at the short times $t \approx \tau$. From the physical point of view, this corresponds to an initial condition when all particles move in the same direction; after the characteristic time τ due to the effect of the randomization of the particle motion the Eq. (5.96) reduces to the Fourier heat conduction equation [2898]. The value of the relaxation time τ can be considered as the characteristic time for the crossover from the ballistic motion to diffusion [1504] or as the Lyapunov time beyond which the predictability is lost [2297].

D. Jou et al. [1504] suggested to describe the transition to the ballistic regime introducing the generalized conductivity $\lambda(T, l/L)$ in such a way that for all regimes

$$q = \lambda\left(T, \frac{l}{L}\right)\frac{\Delta T}{L} \tag{5.101}$$

with the limiting values

$$\lambda\left(T, \frac{l}{L}\right) \to \lambda(T) \quad for \quad \frac{l}{L} \to 0 \tag{5.102}$$

and

$$\lambda\left(T, \frac{l}{L}\right) \to \frac{\lambda(T)}{a}\frac{l}{L} \equiv \Lambda(T) \quad for \quad \frac{l}{L} \to \infty \tag{5.103}$$

where a is the constant depending on the system.

The dynamic properties of (5.96) are analysed under the assumption that the solution has the form $T(x,t) = t_0 \exp[i(kx - \omega t)]$ where k is a complex wavenumber and ω is a real frequency. The phase speed v_{ph} and the attenuation factor α are given as

$$v_{ph} = \frac{\sqrt{2\varkappa\omega}}{\sqrt{\tau\omega + \sqrt{1+\tau^2\omega^2}}}, \qquad \alpha = \frac{2\varkappa}{v_{ph}}. \tag{5.104}$$

In the high-frequency limit the phase velocity remains finite for the non-zero values of the relaxation time τ

$$v_{ph\infty} = U = \sqrt{\frac{\varkappa}{\tau}}. \tag{5.105}$$

The value of $v_{ph\infty}$ diverges when $\tau \to 0$ leading to the infinite speed of propagation.

The Cattaneo equation (5.96) can be derived in different ways, for example,

- from the Boltzmann transport equation under the relaxation time approximation;
- in the framework of the extended irreversible thermodynamics (EIT).

The basic assumption of the relaxation time approximation written as (in one dimensional case)

$$\frac{\partial f}{\partial t} + v_x \frac{\partial f}{\partial x} = \frac{f_0 - f}{\tau} \tag{5.106}$$

is that the distribution function f is close to the equilibrium distribution function f_0, i.e. one can assume that

$$\frac{\partial f}{\partial x} \approx \frac{\partial f_0}{\partial x} = \frac{\partial f_0}{\partial T}\frac{\partial T}{\partial x}. \tag{5.107}$$

Multiplying (5.106) by $\tau\varepsilon v_x$ (here ε is the particle energy) and integrating over the momentum space gives (5.96).

In the Extended Irreversible Thermodynamics the dissipative fluxes such as the heat flux are considered as the basic independent variables [1803, 2788]. The Classical Irreversible Thermodynamics, the Rational Extended Thermodynamics (RET) [2225, 2662], the Extended Irreversible Thermodynamics, the Thermodynamics with Internal Variables (NET-IV) [2072, 2073] are the special theories of the non-equilibrium thermodynamics [584, 3135].

The most important difference between the Rational Extended Thermodynamics and the Extended Irreversible Thermodynamics is that the former assumes that the state space is strictly local, while the latter allows a nonlocal state space [582].

Similarly to the CIT, the NET-IV uses the balances of mass, linear momentum, angular momentum and internal energy as the constraints to evaluate the entropy production rate density from the balance of entropy, whose positive semi-definiteness is ensured via the Onsagerian force-flux relations. The CIT is based on the hypothesis of local equilibrium.

The NET-IV assumes that the state space is extended by one or more internal–variables. In contrast to the EIT, they are not given any physical interpretation. Usually [3178] it is assumed that entropy is shifted by a concave expression of the internal variables; the simplest choice is a quadratic function.

If ξ_k, where $k = 1, \ldots, M$ are M internal variables with the arbitrary tensorial orders and characters, the extended specific entropy is given as [2997]

$$\tilde{s} = s^{eq} - \sum_{k}^{M} \left(\frac{m_k}{2} \cdot \xi_k \right) \cdot \xi_k \tag{5.108}$$

where m_k are the coefficient functions of the appropriate tensorial orders and dot denotes the full contraction of tensors of an arbitrary order.

In the framework of The NET-IV it is also possible to generalise the relationship between the heat current density $\mathbf{q}$ and the entropy current density $\mathbf{J}$ and instead of the usual expression $\mathbf{J} = 1/T\mathbf{q}$ assume that

$$\mathbf{J} = \left(\frac{1}{T} \hat{I} + \hat{C} \right) \cdot \mathbf{q} + \sum_{k}^{M} \mathbb{C}_k \cdot \xi_k \tag{5.109}$$

where $\hat{C}$ and $\mathbb{C}_k$ are the constitutive tensors of the appropriate orders called the Nyíri multipliers.

The heat current density is frequently identified with the vectorial internal variable, however, M. Szücs et al. [2997] had shown that the identification of the heat current density with the conjugate of the vectorial internal variable is more advantageous.

Physically the new variables differ significantly from the classical ones that obey the conservation laws and change relatively slowly in the system evolution—the fluxes generally do not obey the conservation laws and are the relatively "fast" variables [2896].

The local or the nonlocal models of the heat transport could be derived, depending on whether the gradients of the fluxes are included in the state space, or not [2776].

Thus, an entropy in the EIT depends on the internal energy and the heat flux $s = s(u, q)$ and obeys the evolution equation

$$\frac{\partial s}{\partial t} + \nabla \cdot \mathbf{J}^s = \sigma^s \geq 0, \tag{5.110}$$

where $\mathbf{J}^s$ is the entropy flux, σ^s is the entropy production rate.

The definition of the non-equilibrium temperature as $T^{-1} = \partial s/\partial u$ and an assumption that $\partial s/\partial q = -\alpha T$, where α is a material coefficient, lead (with the energy balance for the rigid conductor at rest—the substantial time derivative is equal to the partial time derivative in case of the rigid heat conductor) to the equation

$$\frac{\partial s}{\partial t} = -\nabla \cdot \frac{\mathbf{q}}{T} + \mathbf{q} \cdot \left(\nabla T^{-1} - \alpha \frac{\partial \mathbf{q}}{\partial t} \right), \tag{5.111}$$

and thus

$$\sigma^s = \mathbf{q} \cdot \left(\nabla T^{-1} - \alpha \frac{\partial \mathbf{q}}{\partial t} \right). \tag{5.112}$$

The simplest way to provide the positiveness of the entropy production rate σ^s is to assume a linear relation between the heat flux and the thermodynamic force in the parentheses

$$\nabla T^{-1} - \alpha \frac{\partial \mathbf{q}}{\partial t} = \mu \mathbf{q}, \tag{5.113}$$

where μ is a positive coefficient.

By introduction of the notation

$$\frac{\alpha}{\mu} = \tau, \qquad \frac{1}{\mu T^2} = \lambda, \tag{5.114}$$

one recovers the Cattaneo constitutive relation (5.92).

Sometimes the Cattaneo number is used [1612] that is defined as

$$Ca = \frac{\varkappa \tau}{L^2} \tag{5.115}$$

and the Vernotte number [2675, 3010]

$$Ve^2 = \frac{\varkappa \tau}{L^2}. \tag{5.116}$$

The Cattaneo constitutive relation (5.92) could be considered as the Taylor's series expansion of the relation with a time lag

$$\mathbf{q}(\mathbf{r}, t + \tau) = -\lambda \nabla T(\mathbf{r}, t) \tag{5.117}$$

which is called sometimes the "improved" Cattaneo model or the single-phase lag (SPL) model [2041]. Evidently, the SPL model (5.117) is Galilean invariant [3369].

L. Chen et al. [527] have derived the SPL model equation from the Boltzmann equation using for the time derivative approximation [3369]

$$\frac{\partial f}{\partial t} \approx \frac{f(t + \Delta t)}{\Delta t} = \frac{f(t + \tau)}{\tau}. \tag{5.118}$$

Recently S. N. Li and B. Y. Cao [1842] noted that the Cattaneo model should not be considered as the consequence of the SPL model since the accuracy of this approximation is uncertain—the remaining higher order terms could be very large even if the relaxation time is very small and the predictions of these models could be different.

The Cattaneo constitutive relation (5.92) could be written as an integral over the history of the temperature gradient

$$\mathbf{q} = -\frac{\lambda}{\tau} \int_{-\infty}^{t} \exp\left(-\frac{t - t'}{\tau}\right) \nabla T t'. \tag{5.119}$$

J. I. Frankel et al. [1002] noted that the alternative formulation—in terms of the heat flux (the scalar equations for the three components in the general case) could be useful for the problems involving the heat flux in the boundary conditions. The temperature distribution is recovered by integration of the energy balance equation over time

$$T(t) = T(0) + \int_{t'=0}^{t} \frac{1}{\varrho c}[-\nabla \cdot q(t') + Q]dt' \tag{5.120}$$

Sometimes the Cattaneo model is called a damped version of the Fourier law (Damped Wave Equation, DWE) [257, 1728, 2043].

The heat conduction is a continuum phenomenon, thus it requires a space-time representation. The substantial time derivatives, the conductive current densities in the internal energy, and the entropy balances are tools to attach the reference frame to the flow of the material. The substantial time derivatives are equal to the partial time derivatives in the case of the rigid heat conductors when the heat-conducting substance does not move. The Fourier's law is objective, but the generalizations with time derivatives such as the Cattaneo equation, the material frame indifference is an open question.

The Cattaneo law removes the paradox of the infinite speed of the propagation of the disturbances but causes another one for the case of the heat transfer in the moving body: the Cattaneo law is not Galilean invariant [2222]—while the Fourier law is frame-indifferent, but does not respect the causality principle, the Cattaneo's model removes the causality problem, but its form depends on the choice of the observer [44].

The Cattaneo law is also not Lorentz-covariant [1841], i. e. the form of the equation changes under the transformations

$$\frac{\partial}{\partial x'} = \gamma\left(\frac{\partial}{\partial x} + \frac{u}{c^2}\frac{\partial}{\partial t}\right), \quad \frac{\partial}{\partial t'} = \gamma\left(\frac{\partial}{\partial t} + u\frac{\partial}{\partial t}\right) \tag{5.121}$$

where $\gamma = 1/\sqrt{1 - u^2/c^2}$.

The heat conduction in the moving medium is governed by the equation (in the dimensionless variables) [564]

$$\frac{\partial^2 T}{\partial t^2} + \frac{\partial(\mathbf{u}\cdot\nabla T)}{\partial t} + \frac{\partial T}{\partial t} + \mathbf{u}\cdot\nabla T = \Delta T. \tag{5.122}$$

In the simplest case of the one dimensional problem and the constant velocity $u(x, t) = U$ (5.122) reduces to

$$\frac{\partial^2 T}{\partial t^2} + U\frac{\partial^2 T}{\partial x \partial t} + \frac{\partial T}{\partial t} + U\frac{\partial T}{\partial x} = \Delta T. \tag{5.123}$$

The speeds of disturbances in the coordinate system moving with a velocity U are the nonlinear functions of U [564]:

$$c_{1,2} = \frac{1}{2}\left[U \pm \sqrt{U^2 + 4}\right]. \tag{5.124}$$

Evidently, $c_{1,2} \neq U \pm 1$, i.e. the speed of disturbances does not equals the sum or the difference of the dimensionless frame velocity and the thermal wave speed.

This paradox is removed when instead of the partial time derivative the material derivative is used [564]. According to the Galilean principle of the relativity the equation should be the same in the inertial moving frame, i.e. do not change under the Galilean transformation

$$\xi = x - Ut, \quad \tau = t. \tag{5.125}$$

In the general three dimensional case the Cattaneo relation in the material framework is written as

$$\mathbf{q} + \tau\left(\frac{\partial}{\partial t} + \mathbf{u}\cdot\nabla\right)\mathbf{q} = \varkappa\nabla T. \tag{5.126}$$

The Eq. (5.126) cannot be resolved with respect to the heat flux and thus a single equation for the temperature cannot be derived [564].

Later I. C. Christov [562] suggested to use the frame-indifferent upper-convected [1611] Oldroyd derivative and to write the Cattaneo equation (5.92) as

$$\tau\left[\frac{\partial \mathbf{q}}{\partial t}+\mathbf{v}\cdot\nabla\mathbf{q}+\mathbf{q}\cdot\nabla\mathbf{v}+(\nabla\cdot\mathbf{v})\mathbf{q}\right]+\mathbf{q}=-\varkappa\nabla T \tag{5.127}$$

that allows elimination of $\mathbf{q}$ to get for the temperature the equation [562]

$$\tau\left[\frac{\partial^2 T}{\partial t^2}+2\mathbf{u}\cdot\nabla\frac{\partial T}{\partial t}+\frac{\partial \mathbf{u}}{\partial t}\cdot\nabla T+\left(\frac{\partial T}{\partial t}+\mathbf{u}\cdot\nabla T\right)(\nabla\cdot\mathbf{u})+\mathbf{u}\cdot\nabla(\mathbf{u}\cdot\nabla T)\right]$$
$$+\frac{\partial T}{\partial t}+\mathbf{u}\cdot\nabla T=\nabla\cdot(\varkappa\nabla T). \tag{5.128}$$

Christov showed that Eq. (5.128) retains its form under the change of variables corresponding to the frame that moves with a constant velocity.

R. Al Nahas et al. [44] proposed to replace the notion of the frame-indifference with the covariance principle introduced by Einstein [2241, 2375] and derived a covariant spacetime heat conduction model using the spacetime formalism to ensure the indifference to changes of the observer.

X. Zhang et al. [3496] studied the heat transfer of the blood vessels subject to the transient laser irradiation, where the irradiation is extremely brief and has high power. J. S. Nelson et al. [2284] investigated the relationship between photothermal damage to blood vessels and laser pulse duration. The authors found that the critical core intravascular temperature, necessary to destroy vessels irreversibly is achieved when laser pulse durations are approximately equal to the estimated thermal relaxation times. The shorter pulse durations are more effective for damaging smaller blood vessels; conversely, longer pulse durations are more effective for damaging larger diameter vessels.

The Cattaneo (Maxwell-Cattaneo-Vernotte) model may give the unphysical predictions such as the negative temperature when the two cooling waves meet [3248] or the jumps at the wave front, indicating the discontinuity of the temperature distributions [2596, 3452].

C. Körner and H. W. Bergmann [1694], M. B. Rubin [2654], M. Criado-Sancho and J. Lebot [649], C. Bai and A. Lavine [153], A. Barletta and E. Zancini [193, 194, 3460] analysed the compatibility of the Cattaneo law with the second law of thermodynamics.

Barletta and Zancini also considered the Taitel's paradox (a temperature exceeding the difference of the boundary values in the slab whose sides are kept at the different temperatures). In the frame of the classical irreversible thermodynamics the entropy production could be written as [1800]

$$\sigma^s=\lambda\frac{(\nabla T)^2}{T^2}+\frac{\tau}{T^2}\frac{\partial \mathbf{q}}{\partial t}+\cdot\nabla T. \tag{5.129}$$

Barletta and Zancini found that the production of the entropy could be negative in the regions where the heat flux decreases so steeply that $|\partial \mathbf{q}/\partial t| > |\mathbf{q}|/\tau$. Note also that the second law does not say anything about fluctuations and for a short time an instantaneous negative production could be observed [580].

However, these results does not contradicts the second law since the classical thermodynamics is based on the local equilibrium hypothesis [1801] that is not fulfilled in this case [614] and the concept of the "temperature" in the hyperbolic heat transfer equation cannot be interpreted in the conventional sense [1800].

The Cattaneo law is compatible with the second law in the Extended Irreversible Thermodynamics that introduces the generalized entropy and the entropy flux depending not only on the classical variables but also on the fluxes [1502–1504].

S. N. Li and B. Y. Cao [1843] studied the thermodynamics problems of the SPL model. Using the expression of the entropy production rate

$$S = -\frac{\mathbf{q} \cdot \nabla T}{T^2} \tag{5.130}$$

from the Classical Irreversible Thermodynamics (CIT) the authors got for the Fourier law

$$S_F = -\frac{q^2}{\lambda T^2} \geq 0 \tag{5.131}$$

and for SPL

$$S_{SPL}^{CIT} = -\frac{\mathbf{q}(t) \cdot \mathbf{q}(t + \tau|)}{\lambda T^2}. \tag{5.132}$$

Evidently, the entropy production rate for the SPL model is not necessary positive or zero. The second law of thermodynamics is satisfied in the EIT [1502] where the entropy production rate is expresses as

$$S_{SPL}^{EIT} = -\frac{\mathbf{q}(t + \tau) \cdot \mathbf{q}(t + \tau)}{\lambda T^2}. \tag{5.133}$$

Still, the SPL model could violate the second law of thermodynamics by the breaking the equilibrium spontaneously in the special circumstances. S. N. Li and B. Y. Cao [1843] have constructed an example of such behaviour. They have considered the case

$$\frac{2\tau n^2 \pi}{\varrho C_V \ell^2} = 1 \tag{5.134}$$

and obtained the solution

$$T(x, t) = C_1 \sin \frac{n\pi x}{l} \sin \frac{\pi t}{2\tau} + C_0 \tag{5.135}$$

of the SPL model equations with the boundary conditions $T(0, t) = C_0$, $T(l, 0) = C_0$ and the initial condition $T(x, 0) = C_0$.

The coefficient C_1 in the solution (5.135) is arbitrary and the initial equilibrium $T = C_0$ could be broken spontaneously. Note that both the Fourier and Cattaneo models preserve the equilibrium [1843]:

$$\frac{dE^F}{dt} = -\int \frac{\lambda}{\varrho C_v}(\nabla T)^2 dV \leq 0, \tag{5.136}$$

D. E. Glass et al. [1094] studied the Cattaneo heat conduction in a semi-infinite slab with the temperature-dependent thermal conductivity $\lambda = \lambda_0(1 + \beta T)$ and found that the nonlinear thermal conductivity alters the speed of the thermal front.

In the biological problems the Cattaneo equation is called the Thermal Wave model or the Thermal Wave Model Bioheat Transfer (TWMBT) [928, 1905, 1938, 2023], sometimes the terms the "heat wave" or the "temperature wave" are also used [2684, 3471]. It should be noted that the thermal wave model captures the microscale response in time [2366, 3110], the wave concept does not capture the microscale response in space [3113] and the thermal wave model can lead to the unusual solutions [1107, 1694].

The Cattaneo equation can be derived using the nonequilibrium statistical mechanics and the assumption of the near equilibrium state (thus accepting the Maxwellian as the nonequilibrium distribution) [51, 1839].

P. Podio-Guidugli [2470] (see also [188]) derived the Maxwell-Cattaneo-Vernotte type equations using the splitting the entropy into the standard and the dissipative parts.

A. Compte and R. Metzler [613] suggested three possible fractional generalization of the Cattaneo equations (GCE) using the fractional generalization of the constitutive equation

$$\mathbf{q}(\mathbf{r}, t) + \tau^{\gamma} \frac{\partial^{\gamma} \mathbf{q}}{\partial t^{\gamma}} = -\lambda \nabla T \tag{5.137}$$

and continuity equation

$$\frac{\partial^{\gamma} T}{\partial t^{\gamma}} = -\mathbf{q}(\mathbf{r}, t). \tag{5.138}$$

The generalization are written as follows

- GCE I

$$\frac{\partial^{\gamma} T}{\partial t^{\gamma}} + \tau^{\gamma} \frac{\partial^{2\gamma} \mathbf{q}}{\partial t^{2\gamma}} = -\lambda \Delta T \tag{5.139}$$

- GCE II

$$\frac{\partial^{2-\gamma}T}{\partial t^{2-\gamma}} + \tau^{\gamma}\frac{\partial^2 \mathbf{q}}{\partial t^2} = -\lambda\Delta T \tag{5.140}$$

- GCE III

$$\frac{\partial T}{\partial t} + \tau\frac{\partial^2 \mathbf{q}}{\partial t^2} = -\lambda\frac{\partial^{1-\gamma}}{\partial t^{1-\gamma}}\Delta T. \tag{5.141}$$

The equation GCE III is equivalent to

$$\frac{\partial^{\gamma}T}{\partial t^{\gamma}} + \tau\frac{\partial^{1+\gamma}\mathbf{q}}{\partial t^{1+\gamma}} = \Delta T. \tag{5.142}$$

Compte and Metzler analysed the behaviour of the solutions of the GCE I–GCE III equations in the short-time and the long-time limits and found that in the former GCE II describes the pure ballistic propagation while the signal propagation velocity is infinite in the case of GCE I and GCE III. In the long-time limit the GCE I and the GCE III models show the subdiffusion and GCE II corresponds to the superdiffusive transport.

5.2.1.2 Dual-Phase-Lag Model

To include the effects of both the relaxation and the microstructure, D. Y. Tzou [3105, 3107, 3110] introduced the dual phase lag (DPL) model

$$\mathbf{q}(\mathbf{r}, t + \tau_q) = -\lambda\nabla T(\mathbf{r}, t + \tau_T) \tag{5.143}$$

where τ_q and τ_T are the phase lags for the heat flux vector and for the temperature gradient, respectively, arising from the "thermal inertia" and the "microstructural interaction" [3380]; these parameters are the intrinsic properties of the material.

In contrast to the SPL model that could lead to the unphysical infinite value of the local heat flux [30] the DPL model smooths the wavefront and eliminates the infinite values [124].

The dual time lag approach is not space-time compatible, for example, it violates the homogeneity of time, one of the basic requirements of any space-time. Also, the separation of time from space, as appears in both the memory integrals and in the time lag equations, cannot survive the relativistic generalizations. A covariant, objective representation is inevitable since the Minkowski form is absolute, but space and time are the reference frame dependent [2844].

The DPL model reduces to the SPL model when $\tau_T = 0$, $\tau_q > 0$ [3104] and to the Fourier law if $\tau_T = \tau_q$. Both relaxation times are very small for the common materials. For example, the phase lags τ_q and τ_T for gold are 8.5 and 90 ps, respectively [665].

There is a lot of the controversies in the literature concerning the values of the relaxation times for the biological tissues [1525, 2157, 3054, 3169]. For the processed meat the values of τ_q and τ_T are estimated as 14–16 s and 0.043–0.056 s[91], the experiments with the muscle tissue from cow give the values 7.36–8.43 s and 14.54–21.03 s [1920]. A. M. Brazhnikov et al. [361] found τ_q =20–30 s for the meat products.

A. J. Stratigos and J. S. Dover [2962] reported the thermal relaxation times of the important cutaneous structures: melanosome (size 0.5–1 μm)—1 μs, cell (size 10 μm)—300 μs, blood vessel (size 50 μm)—1 ms, blood vessel (size 200 μm)—20 ms.

Y. Zhang [3472] estimated the relaxation times in term of the blood and tissue properties and the interphase convective heat exchange coefficient and found they to be close to each other.

K. C. Liu and H. T. Chen [1921] claimed that both phase lag times depend on the various parameters of the problem, such as porosity, the heat capacities of blood and tissues, the coupling factors, and the ratio of thermal conductivity of the tissue and the blood.

A. Vedavarz, S. Kumar and M. K. Moallemi reported that the value of τ_q for biological tissues lies between 10–1000 s and 1–100 s for the cryogenic and the room temperature, respectively [3169].

Equation (5.143) could be re-written as

$$\mathbf{q}(\mathbf{r}, t) = -\lambda \nabla T(\mathbf{r}, t + (\tau_T - \tau_q)) \tag{5.144}$$

Thus, the solution of the DPL model does not depend on the relaxation times τ_T and τ_q separately but only on their difference [1723] and the SPL and the DPL models are mathematically equivalent [899, 1707, 2346].

The DPL model for $\tau = \tau_q - \tau_T > 0$ is the retarded partial differential equation also referred to as the delayed heat conduction model (DH) [1723, 3371]

$$\frac{\partial T}{\partial t'}(\mathbf{r}, t') = \varkappa \Delta T(\mathbf{r}, t' - \tau). \tag{5.145}$$

M. Fabrizio et al. [899, 902] showed that there are mathematical conditions in addition to the physical ones to obtain an exponentially stable equilibrium solution for the DPL equation. Such condition requires the negative time delay τ_d (called the retarded effect) between the heat flux and the temperature gradient $\tau_d = \tau_q - \tau_T \leq 0$. The opposite case, i.e. $\tau_d > 0$ is mathematically ill-posed which enlightens the validity of the MCV equation but excludes the equations based on the arbitrary Taylor expansion.

The DPL model is closely related [3383] to the hyperbolic models that describe the energy exchange in the systems composed of two subsystems, each with its own temperatures [1504], e.g., between the electrons and the phonons in the solids by a pair of the coupled nonlinear equations governing the effective temperatures of the electrons and the phonons [89, 517, 1519, 1856, 2041, 2534, 2890, 2891, 2893] (the two temperature models, TTM).

The two temperature models are valid if the time needed to establish the equilibrium within each subsystem is much less than the time needed to establish the equilibrium between them [2890]. In this case it is possible to assign an individual temperature to each subsystem and reduce the thermal problem in the system to the determination of the space-time evolution of these temperatures with account for the energy exchange between the subsystems [2896].

For example, the energy of the irradiating laser is mainly absorbed by the electrons of the solid and the laser therefore initiates a transient nonequilibrium of the electron gas with the lattice (sometimes referred to as "laser-induced nonequilibrium" [2595]). It is assumed that the electron and phonon subsystems in the solids are at their local equilibrium, the heat conduction by the phonons is neglected and the lattice temperature is near or above the Debye temperature (thus the electron-electron and electron-defect scatterings are insignificant compared with the electron-phonon scattering [3471]). The electrons play a major role in the heat transport in metals; often it is assumed that electrons move with the velocity of the order of the Fermi velocity [2890].

The equations of the two temperature model for the one-dimensional case could be written as [2595]

$$c_e \frac{\partial T_e}{\partial t} = -\frac{\partial q}{\partial x} - G(T_e - T_l) + Q, \tag{5.146}$$

$$c_l \frac{\partial T_l}{\partial t} = G(T_e - T_l), \tag{5.147}$$

$$\tau \frac{\partial q}{\partial t} + q = -\lambda \frac{\partial T_e}{\partial x}. \tag{5.148}$$

where T_e is the temperature of the electron gas, T_l is the temperature of the lattice, c_e and c_l are the heat capacity of the electron gas and the lattice, respectively, G is the electron-phonon coupling factor that is estimated as [1472]

$$G = \frac{\pi^2 m_e n_e c_s^2}{\tau_e(T_e) T_e} \tag{5.149}$$

where m_e is the mass of electron, n_e is the density of the free electrons, τ_e is the electron relaxation time, c_s is the speed of sound in bulk material $c_s = \sqrt{B/\varrho_m}$, B is the bulk modulus, ϱ_m is the density.

There are other physical two-temperature systems, e.g., the protons and the electrons in the astrophysical models of the accretion flow or the ions and the electrons in the partly ionized gas [988].

Y. Zhang [3472] proposed the explicit estimates of the relaxation times for the heat transfer in the tissues

$$\tau_q = \frac{\varepsilon(1-\varepsilon)}{\dfrac{\varepsilon}{C_{tb}} + (1-\varepsilon)} \frac{\varrho_b C_b}{G}, \qquad \tau_T = \frac{\varepsilon(1-\varepsilon)}{\dfrac{\varepsilon}{K_{tb}} + (1-\varepsilon)} \frac{\varrho_b C_b}{G} \tag{5.150}$$

where $C_{tb} = \varrho_t C_t/(\varrho_b C_b)$ is the ratio of the heat capacities of the tissue and the blood, $K_{tb} = \lambda_t/\lambda_b$ is the ratio of the thermal conductivities, ε is the porosity of the tissue, G is "the lumped convection-perfusion parameter".

There are three sources of dissipation in the compound systems:

1. the processes on the macroscopic scale,
2. the processes at the microscopic scale and
3. the interaction processes—the energy exchange—between the two subsystems.

The heat fluxes could be written (in the one dimensional case) as

$$-q_M = L_{11}\frac{\frac{\partial T_M}{\partial x}}{T_M^2} + L_{12}\frac{\frac{\partial T_m}{\partial x}}{T_m^2}, \qquad -q_m = L_{21}\frac{\frac{\partial T_M}{\partial x}}{T_M^2} + L_{22}\frac{\frac{\partial T_m}{\partial x}}{T_m^2} \tag{5.151}$$

where L_{ij} are phenomenological coefficients that obey the reciprocity relations $L_{12} = L_{21}$ for the processes near the equilibrium [1122].

D. Y. Tzou and W. Dai [3111] considered lagging in the multi-carrier systems. The equations for the N-carrier system are written as

$$C_1\frac{\partial T_1}{\partial t} = \lambda_1 \nabla^2 T_1 - \sum_{i=2}^{N} G_{1i}(T_1 - T_i), \tag{5.152}$$

$$C_m\frac{\partial T_m}{\partial t} = \lambda_m \nabla^2 T_m + \sum_{j=1}^{m-1} G_{jm}(T_j - T_m) - \sum_{i=m+1}^{N} G_{mi}(T_m - T_i),$$

$$m = 2, 3, \ldots (N-1), \tag{5.153}$$

$$C_N\frac{\partial T_N}{\partial t} = \lambda_N \nabla^2 T_N + \sum_{i=1}^{N-1} G_{iN}(T_i - T_N). \tag{5.154}$$

Depending on the value of the phase-lags, the alternative orders of the Taylor series expansion can be made to provide a series of the DPL heat conduction models [1864, 3377].

Using the first order expansion for both τ_q and τ_T in (5.143) gives the constitutive relation

$$\mathbf{q} + \tau_q \frac{\partial \mathbf{q}}{\partial t} = -\lambda \left[\nabla T + \tau_T \frac{\partial \nabla T}{\partial t}\right]. \tag{5.155}$$

Inserting this relation in the energy conservation law results in the so called type I [1707, 1915, 3377] (also linear [1932] or first-order [1864]) DPL model.

The first order DPL bioheat transfer equation is written as

$$\frac{\varrho c}{\lambda}\left(\frac{\partial T}{\partial t}+\tau_q\frac{\partial^2 T}{\partial t^2}\right)+\frac{\varrho_b c_b w_b}{\lambda}\left[(T-T_b)+\tau_q\frac{\partial T}{\partial t}\right]=\nabla^2 T+\tau_T\frac{\partial}{\partial t}(\nabla^2 T), \tag{5.156}$$

if the source term the metabolic reactions and the external source of energy are omitted.

The equations of this model could be re-written in terms of the heat flux instead of the temperature [1067]

$$\nabla(\nabla\mathbf{q})+\tau_T\nabla(\nabla\mathbf{q})=\frac{1}{\varkappa}\frac{\partial\mathbf{q}}{\partial t}+\frac{\tau_q}{\varkappa}\frac{\partial^2\mathbf{q}}{\partial t^2} \tag{5.157}$$

and even in terms of the heat flux potential φ defined by the equation $\mathbf{q}=\nabla\varphi$ [3241].

L. Wang et al. [3241] proved the well-posedness of the DPL model on a finite one-dimensional region under the uniform Dirichlet, Neumann or Robin conditions. Later L. Wang and M. Xu [3239] extended this result to the n-dimensional case.

Type II DPL model [1707] is obtained when the first-order and the second-order Taylor series expansions are used for the heat flux q and the temperature T, respectively

$$q+\tau_q\frac{\partial q}{\partial t}=-\lambda\left[\nabla T+\tau_T\frac{\partial\nabla T}{\partial t}+\frac{\tau_T}{2}\frac{\partial^2\nabla T}{\partial t^2}\right]. \tag{5.158}$$

and Type III (second-order [1864]) DPL model [1707, 1933])—when the second-order Taylor series expansions are used for both q and T

$$q+\tau_q\frac{\partial q}{\partial t}+\frac{\tau_q}{2}\frac{\partial^2 q}{\partial t^2}=-\lambda\left[\nabla T+\tau_T\frac{\partial\nabla T}{\partial t}+\frac{\tau_T}{2}\frac{\partial^2\nabla T}{\partial t^2}\right]. \tag{5.159}$$

The second-order DPL bioheat transfer equation is written as follows

$$\frac{\varrho c}{\lambda}\left(\frac{\partial T}{\partial t}+\tau_q\frac{\partial^2 T}{\partial t^2}+\frac{\tau_q^2}{2}\frac{\partial^3 T}{\partial t^3}\right)+\frac{\varrho_b c_b w_b}{\lambda}\left[(T-T_b)+\tau_q\frac{\partial T}{\partial t}+\frac{\tau_q^2}{2}\frac{\partial^2 T}{\partial t^2}\right]$$
$$=\nabla^2 T+\tau_T\frac{\partial}{\partial t}(\nabla^2 T)+\frac{\tau_T^2}{2}\frac{\partial^2}{\partial t^2}(\nabla^2 T). \tag{5.160}$$

Sometimes the other notation is used to distinguish the DPL models, indicating the orders of the Taylor expansions, e.g., the DPL (2,1) [465, 871] or just "1st–1st" order, "1st–2nd" order etc. [125].

S. A. Rukolaine [2664] found the solutions of the DPL model to be unstable. He studied the initial value problem for the three-dimensional equation with the positive localized source of the short duration. The constitutive equation of the Jeffrey-type was used. Rukolaine obtained an analytical solution using the Fourier transform and the Laplace transform for the DPL equation assuming the Gaussian source of finite duration

$$p_\sigma(\mathbf{x}) = \frac{1}{\left(\sqrt{2\pi}\sigma\right)^3} \exp\left(-\frac{|\mathbf{x}|^2}{2\sigma^2}\right) \tag{5.161}$$

The solution for $\tau_q > \tau_T$ gives an unphysical behaviour of the temperature history, it goes into the negative domain in the vicinity of origin at least for the small values of the parameter σ.

Later Rukolaine confirmed his conclusion for the type III DPL model [2665].

R. Quintanilla and R. Racke [2537] (see also [2536]) analysed the stability of the solution of the different DPL versions. Quintanilla introduced the ratio of the two relaxation times of the DPL model $\xi = \tau_T/\tau_q$ as a parameter that controls the stability of the DPL model. The author considered the characteristic polynomial associated to the Laplace operator for the case of the Dirichlet boundary conditions in a bounded domain. The stability or the instability of the solution is determined by the real part of the eigenvalues.

The results of the study could be summarized as

- When approximation up to first order in τ_q and up to first or second order in τ_T is used, the system is always stable.
- When approximation up to second order in τ_q and up to first order in τ_T is used, the system is stable if $\xi > 1/2$ and unstable if $\xi < 1/2$.
- When approximation up to second order in both τ_q and τ_T is used, the system is stable if $\xi > 2 - \sqrt{3}$ and unstable if $\xi < 2 - \sqrt{3}$.
- Whenever $\xi > 1/2$, the several models predict the same behaviour

Restrictions on the ratio $\xi = \tau_T/\tau_q$ were derived from the second law of thermodynamics by M. Fabrizio and B. Laazzari [900].

The compatibility of the DPL with the second law of the Extend Irreversible Thermodynamics was proved by M. Xu [3367]. P. Kishore and S. Kumar [1645] wrote out the nonequilibrium EPR as

$$S_{DPL}^{EIT} = -\left(\frac{\tau_T}{T^2}\frac{\mathrm{d}(\nabla T)}{\mathrm{d}t} + \frac{\tau_q}{kT^2}\frac{\mathrm{d}\mathbf{q}}{\mathrm{d}t} + \frac{1}{T^2}(\nabla T)\right)\cdot\mathbf{q}. \tag{5.162}$$

S. Chiriță et al. [545] investigated the time differential dual-phase-lag model of the heat conduction incorporating the microstructural interaction effect in the fast-transient process of heat transport. The uniqueness results were established under the assumption that the conductivity tensor is positive definite and the delay times τ_q and τ_T vary in the set $0 \le \tau_q \le 2\tau_T$ and $0 \le \tau_T \le 2\tau_q$. The authors have found conditions where the solution is stable and where the solution has an exponential growth in time.

M. A. Castro et al. [466] proved unconditional stability of the finite difference method for the DPL model proposed by J. M. McDonough et al. [2096] by the analysis of its amplification matrix [2601].

Both the TWMBT and the DPL models are used widely to study the heat transfer in the biological objects. For example, interaction of the laser radiation with the tissue, including the laser interstitial thermal therapy (LITT) [2844], have been considered by the numerous researches: J. Zhou et al. [3520, 3521], M. Jaunich et al. [1446], K. C. Liu and P. J. Chen [1922], N. Afrin et al. [22], H. Ahmadikia et al. [32], N. Sahoo et al. [2676, 2677], K. C. Liu and J. C. Wang [1927], H. Z. Poor et al. [2484], P. Hooshmand et al. [1350], S. Kumar and A. Srivastava [1726], M. Mohajer et al. [2168], M. Jasinsky et al. [1444], K. C. Liu and Y. S. Chen [1924], Y. Zhang et al. [3491].

J. Zhou et al. [3521] considered the 2D axisymmetric dual-phase-lag bioheat transfer model

$$\frac{\partial q_r}{\partial t} + \tau_q \frac{\partial^2 q_r}{\partial t^2} = \varkappa \frac{\partial}{\partial r}\left[\frac{1}{r}\left(\frac{\partial r q_r}{\partial r} + \frac{\partial r q_z}{\partial z}\right)\right] + \varkappa \tau_T \frac{\partial^2}{\partial r \partial t}\left[\frac{1}{r}\left(\frac{\partial r q_r}{\partial r} + \frac{\partial r q_z}{\partial z}\right)\right]$$

$$+ \varkappa \omega_b \rho_b c_b \frac{\partial T}{\partial r} + \varkappa \omega_b \rho_b c_b \tau_T \frac{\partial^2 T}{\partial r \partial t} - \varkappa \frac{\partial Q}{\partial r} - \varkappa \tau_T \frac{\partial^2 Q}{\partial r \partial t}, \tag{5.163}$$

$$\frac{\partial q_z}{\partial t} + \tau_q \frac{\partial^2 q_z}{\partial t^2} = \varkappa \frac{\partial}{\partial z}\left[\frac{1}{r}\left(\frac{\partial r q_r}{\partial r} + \frac{\partial r q_z}{\partial z}\right)\right] + \varkappa \tau_T \frac{\partial^2}{\partial z \partial t}\left[\frac{1}{r}\left(\frac{\partial r q_r}{\partial r} + \frac{\partial r q_z}{\partial z}\right)\right]$$

$$+ \varkappa \omega_b \rho_b c_b \frac{\partial T}{\partial z} + \varkappa \omega_b \rho_b c_b \tau_T \frac{\partial^2 T}{\partial z \partial t} - \varkappa \frac{\partial Q}{\partial z} - \varkappa \tau_T \frac{\partial^2 Q}{\partial z \partial t}. \tag{5.164}$$

for the two heating schemes (surface heating and body heating) and two different beam profiles (the flat beam and the Gaussian beam). The authors investigated the effects of the two phase lag times τ_T and τ_q on the temperature transients and noted that in contrast to the heat transfer DPL that reduces to the Fourier equation if $\tau_T = \tau_q$, the *bioheat* transfer DPL reduces to the Pennes equation—due to the presence of the blood perfusion terms—only if $\tau_T = \tau_q = 0$.

A. Banerjee et al. [180] proved the validity of the use of the hyperbolic heat conduction formulation for the case of the bologna meat samples when irradiated with a collimated laser beam from a short pulse or the CW laser source.

Y. Zhang [3472] demonstrated that the phase lag times depended on the intrinsic properties of blood and tissue, blood perfusion rate, and convection heat transfer and their values might vary from place to place in the human body.

The high intensity focused ultrasound ablation was investigated by X. Z. Liu et al. [1938], P. Namakshenas and A. Mojra [2264, 2265] and G. Singh et al. [2868]. The ultrasound irradiation with different powers had been applied to the multilayer model of the neck consisting of the internal organs, extending from the skin toward the thyroid gland, and constructed from the CT-scan images. It was stated that the considered number of layers has a significant effect on the computed parameters. As the power of the transducer increases, the deviation from the Fourier results also increases.

M. J. Noroozi, S. Saedodin and D. D. Gangi used the thermal wave model to study the heat transfer in the slab heated by a laser radiation [2315] and the DPL model for the case of a slab subjected to an internal heat source [2314].

X. Z. Liu et al. [1938] computed the temperature elevation generated by the ultrasonic irradiation using the TWMBT model. S. Özen et al. [2365] applied the TWMBT model to the analysis of the heat transfer in the biological tissue exposed to microwave radiation. The authors considered the skin with three layers with various thermal physical properties were; the bio-heat transfer equations were solved by using finite difference method. Comparative studies on the Pennes equations and thermal wave model of bio-heat transfer were performed. Özen et al. found that the TWMBT model of bioheat transfer gives lower heat rise predictions than that of Pennes equation, initially; when it approaches to the steady state, it overlaps the solution of the Pennes equation.

P. Kumar et al. used the finite element wavelet Galerkin method to study the hyperthermia assuming the Gaussian external heat source [1733, 1734]. J. R. Ho et al. exploited the lattice Boltzmann method (LBM) for solution of the DPL model studying the heat transfer in the two-layer system [1333].

D. Kumar and K. N. Rai [1724] used the DPL model to investigate the heat transfer in the bilayer living tissues during the magnetic fluid hyperthermia treatment exploiting the finite element Legendre wavelet Galerkin method (FEWGM) This produces a higher degree of accuracy. With selecting the Gaussian distribution as the heat source, Kumar and Rai found that the effect of the τ_q on the heat transfer process is negligible while τ_T significantly affects the procedure. The authors also found that FePt magnetic nanoparticles are the most effective MNP used for the thermal treatment; the effect of various magnetic heat source parameters such as the magnetic induction, the frequency, the diameter of the magnetic nanoparticles, the volume fraction of the magnetic nanoparticles, and the ligand layer thickness had been investigated.

R. Kumar et al. [1737] used the nonlocal DPL model to compute the heat conduction approach in a bi-layer tissue (the tumour and the normal tissue) during the magnetic fluid hyperthermia. The effect of variability of the magnetic heat source parameters had been investigated. The effect of the phase lagging times only appear in the tumour region: the temperature increases with the increment of the value of τ_q and decreases with the augmentation of the value of the τ_T in the tumour zone [2844]. The nonlocal parameters are found to affect the first layer consisting of the cancerous cells which are exposed to the injected magnetic nanoparticles. The NL-DPL model predicts that the treatment time, an essential issue for thermal therapy, is less than that of the DPL model. D. Kumar et al. [1736] numerically solved the nonlinear DPL model for the heat transfer within the skin tissue.

K. C. Liu and H. T. Chen [1840], K. C. Liu and T. M. Chen [1923], K. C. Liu and P. J. Cheng [1925], K. C. Liu and Y. C. Yang [1928], I. Raouf et al. [2576]

and Kumar et al. [1734, 1737] also used the DPL model to study the magnetic fluid hyperthermia.

The DPL of the bio-heat transfer is used for the assessment of the protective clothing. L. N. Zhai and Y. Li [3470] have reviewed the different burn prediction techniques used in the clothing evaluation. These methods are founded on the experiments performed on the living bodies which are getting in touch with the hot water or exposed to the radiant source. H. Ye and S. De [3411] published the review on the experimental and numerical studies of the thermal injury of the skin and the subdermal tissues.

A. Moradi and H. Ahmadikia [2179] used the DPL model to study the heat transfer during the extremely fast freezing of the biological tissues (the cooling rate about 1000°/s [29]) when the frozen region tends to form the amorphous ice [3511].

N. A. Alghamdi and H. M. Yuossef [50] applied the DPL model to study the effect of the porosity, the evaporation rate, and the temperature on the heat transfer in the human eye subjected to the laser radiation. N. Afrin, Y. Zhan and J. Chen [20] used the DPL model to investigate the heating of a gas-saturated porous-medium with short laser pulses.

Y. Chou and R.J. Yang [559] studied the two-dimensional problem of the heat transfer in the multi-layered structures using the space-time conservative element and solution element (CESE) method [1237]. This method was developed for solving the Euler and the Navier–Stokes equations of the fluid dynamics [1944] and provides the local and global flux conservation due to using both the flow variables and their derivatives as unknowns. Chou and Yang found four modes of the heat propagation: hyperbolic, wavelike, diffusive and over-diffusive. Earlier these authors [558] used the CESE scheme for computations of the SPL and the DPL thermal waves and found that this method is characterized by the low numerical dissipation and the dispersion errors.

S. Chiriţă et al. [544] investigated the propagation of the plane time harmonic waves in the infinite space filled by a dual-phase-lag thermoelastic material. The authors found that there are six possible basic waves travelling with distinct speeds: two shear waves and four dilatational waves. The shear waves are uncoupled, undamped in time and travels independently with the speed that is unaffected by the thermal effects. The four dilatational waves are coupled, damped in time and dispersive due to the presence of thermal properties of the material.

X. Y. Li et al. [1863] applied the DPL model to investigation of the thermo-viscoelastic behaviour of the biological tissue subjected to the hyperthermia treatment.

J. Fan and L. Wang [920] developed a general bioheat transport model at macroscale for biological tissues based on porous media model. The model shows that both blood and tissue temperatures satisfy the dual-phase-lag energy equations.

K. C. Liu et al. [1937] applied the second-order Taylor expansion of the dual-phase lag model for analysis of the thermal behaviour in the laser-irradiated layered tissue, which was stratified into the skin, the fat, and the muscle.

S.-M. Lin and C.-Y. Li [1883] studied the heat transfer of skin subjected to the pulse laser heating and the fluid cooling and they use an analytical method for solving the general problem of heat conduction.

M. Ciesielski et al. [578] investigated using the DPL model the thermal and electrical processes during the electrosurgical resection of the colorectal polyps—the processes in the polyp-colon system during the electrosurgical polyrectomy to remove the abnormal growths from the colon (to detach the polyp stalk from the colon wall) were studied. The simulation results help the endoscopists by letting them know the optimal time and parameters of the electric current flow depending on the geometry and size of the polyp; the predicted rate of the thermal damage of the tissue obtained by the Arrhenius damage integral, the electric field distribution profile, and the thermal processes occurring in the polyp-colon domain could be used for the optimization of the endoscopic instrument parameters for the clinical applications.

E. Majchrzak and B. Mochnacki [2016] used the second order dual-phase-lag model to investigate the heat transfer in the homogeneous, cylindrical skin tissue.

V. V. Kulish and V. B. Novozhilov [1723] derived using the Laplace transform the integral equation that relates the temperature and its gradient that holds everywhere inside the domain. The authors considered the one dimensional version of the DPL model (5.144)

$$\frac{\partial \vartheta(x,t)}{\partial x} = \alpha \frac{\partial^2 \vartheta(x,t-\Delta\tau)}{\partial x^2}, \quad -\infty < x < \infty, \quad -\infty < t < \infty \tag{5.165}$$

with the following initial and boundary conditions

$$\vartheta(x,0) = 0, \quad \frac{\partial \vartheta(0,t-\Delta\tau)}{\partial x} = -\frac{1}{k} q^n(0,t), \quad \vartheta(\varkappa,t) = 0. \tag{5.166}$$

Taking Laplace transform of the Eq. (5.165) produces the ODE

$$\frac{d^2\vartheta}{dx^2} - \frac{s}{\alpha}\exp(\Delta\tau s)\vartheta = 0. \tag{5.167}$$

with the general solution

$$\vartheta(x,s) = C_1(s)\exp\left(-\sqrt{\frac{s}{\alpha}\exp(\Delta\tau s)x}\right) + C_2(s)\exp\left(\sqrt{\frac{s}{\alpha}\exp(\Delta\tau s)x}\right). \tag{5.168}$$

The solution must be bounded at $t \to \infty$ thus $C_2 = 0$.

Taking the time derivative of the solution and combining with the solution gives

$$\vartheta(x,s) = -\sqrt{\frac{\alpha}{s}}\exp\left(-\frac{\Delta\tau}{2}s\right)\frac{d\vartheta(x,s)}{dx}. \tag{5.169}$$

The inverse Laplace transform involving the table transform produces the relation between the temperature and its gradient [1723] ($t - t^{\star} = \zeta$)

$$T(x,t) = T_0 - \sqrt{\frac{\alpha}{\pi}} \int_0^{t-\frac{\Delta\tau}{2}} \frac{\frac{\partial T(x,\zeta)}{\partial x}}{\sqrt{t^{\star} - \frac{\Delta\tau}{2} - \zeta}} d\zeta. \tag{5.170}$$

D. Kumar et al. [1735] used the analytical solution obtained using Laplace transform technique to compare the Fourier, the single-phase-lag (SPL) and the dual-phase-lag (DPL) models for the heat transfer in biological tissues such as muscle, tumour, fat, dermis and subcutaneous with the experimental data of Stolwijk and Hardy [2949] and found that the DPL transfer model provides better results in comparison to other two models.

5.2.1.3 Nonlocal Dual-Phase-Lag Model

To accommodate the effect of thermomass, the distinctive mass of heat, of the dielectric lattices in the heat conduction and the size-dependency of the thermophysical properties, D. Tzou and Z. Y. Guo [3112] and D. Tzou [3103] have introduced the nonlocal (NL) behaviour of the heat transport in space in addition to the thermal lagging of the temperature gradient and heat flux, in time. The nonlocal dual-phase-lag (NL DPL) heat conduction can be written as:

$$\mathbf{q}(\mathbf{r} + \boldsymbol{\lambda}_q, t + \tau_q) = -\lambda \nabla T(\mathbf{r}, t + \tau_T) \tag{5.171}$$

where $\boldsymbol{\lambda}_q$ is the correlating length of the nonlocal heat flux that has the same form of the length parameter in the thermomass model [42, 3234].

The natural extension of the NL DPL (5.171) is the introduction of another local space scale $\boldsymbol{\lambda}_T$

$$\mathbf{q}(\mathbf{r} + \boldsymbol{\lambda}_q, t + \tau_q) = -\lambda \nabla T(\mathbf{r} + \boldsymbol{\lambda}_T, t + \tau_T). \tag{5.172}$$

Using the first-order Taylor series expansion of Eqs. (5.171) and (5.172), D. Tzou has developed the nonlocal DPL heat conduction models.

The first-order expansion in the lagging time and the second-order effect in the nonlocal parameters of the Eqs. (5.171) and (5.172) gives in the one dimensional case [1864]

$$q(x,t) - \frac{\lambda_q^2}{2} \frac{\partial^2 q(x,t)}{\partial x^2} + \tau_q \frac{\partial q(x,t)}{\partial t} = -\lambda \frac{\partial T(x,t)}{\partial x} \tag{5.173}$$

and

$$q(x,t) - \frac{\lambda_q^2}{2}\frac{\partial^2 q(x,t)}{\partial x^2} + \tau_q \frac{\partial q(x,t)}{\partial t} = -\left[\lambda \frac{\partial T(x,t)}{\partial x} - \frac{\lambda_T^2}{2}\frac{\partial^3 q(x,t)}{\partial x^3}\right]. \tag{5.174}$$

These expressions are used to derive the nonlocal DPL bioheat equations

$$\left(1 + \lambda_q^2 \frac{\varrho_b c_b w_b}{2\lambda}\right)\frac{\partial^2 T}{\partial x^2} + \lambda_q^2 \frac{\varrho c}{2\lambda}\frac{\partial^3 T}{\partial t \partial x^2} = \tau_q \frac{\varrho_b c_b w_b}{\lambda}\frac{\partial T}{\partial t} + \frac{\varrho c}{\lambda}\left[\frac{\partial T}{\partial t} + \tau_q \frac{\partial^2 T}{\partial t^2}\right] + \frac{\varrho_b c_b w_b}{\lambda}(T - T_b) \tag{5.175}$$

and

$$\left(1 + \lambda_q^2 \frac{\varrho_b c_b w_b}{2\lambda}\right)\frac{\partial^2 T}{\partial x^2} + \lambda_q^2 \frac{\varrho c}{2\lambda}\frac{\partial^3 T}{\partial t \partial x^2} - \frac{\lambda_T^2}{2}\frac{\partial^4 T}{\partial x^4} = \frac{\varrho c}{\lambda}\left[\frac{\partial T}{\partial t} + \tau_q \frac{\partial^2 T}{\partial t^2}\right] + \frac{\varrho_b c_b w_b}{\lambda}\left[(T - T_b) + \tau_q \frac{\partial T}{\partial t}\right]. \tag{5.176}$$

X. Li et al. [1864] noted that the coefficient $\lambda_q^2 \varrho c / 2\lambda$ at the term $\partial^3 T / \partial t \partial x^2$ in the Eq. (5.175) is similar to the coefficient τ_t in the first order DPL bioheat transfer Eq. (5.156) indicating that the lagging time terms with τ_T in Eq. (5.156) and the nonlocal items with λ_q in Eq. (5.175) have similar effect on the bioheat transfer. Thus it is possible to model the microstructural interaction effect in space using the lagging time τ_T in the temperature gradient instead of the nonlocal parameter λ_q in the heat flux [1864].

Li et al. [1864] considered the introduction of the nonlocal parameter λ_T into the one dimensional nonlocal DPL model

$$q(x, t + \tau_q) = -\lambda \frac{\partial T(x + \lambda_T, t + \tau_T)}{\partial x} \tag{5.177}$$

whose the first-order expansion in the lagging time and second-order in the nonlocal parameter yields

$$q(x,t) + \tau_q \frac{\partial q(x,t)}{\partial t} = -\lambda \left[\frac{\partial T(x,t)}{\partial t} - \frac{\lambda_T^2}{2}\frac{\partial^3 T(x,t)}{\partial t^3} + \tau_T \frac{\partial^2 T(x,t)}{\partial t \partial x}\right] \tag{5.178}$$

and the corresponding bioheat transfer equation

$$\frac{\varrho c}{\lambda}\left(\frac{\partial T}{\partial t}+\tau_q\frac{\partial^2 T}{\partial t^2}\right)+\frac{\varrho_b c_b w_b}{\lambda}\left[(T-T_b)+\tau_q\frac{\partial T}{\partial t}\right]$$
$$=\frac{\partial^2 T}{\partial x^2}+\tau_T\frac{\partial^3 T}{\partial t\partial x^2}-\frac{\lambda_T^2}{2}\frac{\partial^4 T}{\partial x^4}. \tag{5.179}$$

Li et al. [1864] also investigated the fourth-order expansion λ_T and the second-order expansion in τ_q and τ_T in Eq. (5.177)

$$q+\tau_q\frac{\partial q}{\partial t}+\frac{\tau_q^2}{2}\frac{\partial^2 q}{\partial t^2}=$$
$$-\lambda\left[\frac{\partial T}{\partial t}+\tau_T\frac{\partial^2 T}{\partial t\partial x}+\frac{\tau_T^2}{2}\frac{\partial^3 T}{\partial t^2\partial x}+\frac{\lambda_T^4}{4!}\frac{\partial^5 T}{\partial x^5}-\frac{\lambda_T^2}{2}\frac{\partial^3 T}{\partial x^3}-\frac{\lambda_T^2\tau_T}{2}\frac{\partial^4 T}{\partial t\partial x^3}\right]. \tag{5.180}$$

The corresponding bioheat transfer equation could be found in [1864].

5.2.1.4 Triple-Phase-Lag Model

The triple-phase-lag model obtained by S. Choudhuri [560] extends the thermoelastic model suggested by A. Green and P. Naghdi [1156] by introduction to the DPL additionally to the relaxation times for the heat flux and the temperature gradient the relaxation time for the thermal displacement[3] gradient [40, 41, 560, 1738], sometimes simply the "phase lag of the thermal gradient" [3058] or the "phase lag of the thermal displacement" [893].

$$\mathbf{q}(\mathbf{r},t+\tau_q)=-[\lambda\nabla T(\mathbf{r},t+\tau_T)+\lambda^{\star}\nabla v(\mathbf{r},t+\tau_v)], \tag{5.181}$$

[3] Thermal displacement was introduced by H. von Helmholtz [42, 2469]. It satisfies the definition $\dot{v}=T$. This quantity was used in models by Green and Naghdi [1155, 1156] as " a scalar history variable"

$$v=\int_{t_0}^{t}T(\tau)d\tau+v_0.$$

S. Bargmann and A. Favata [188] used the thermal displacement in their continuum thermomechanical analysis of the laser-pulsed heating in the polycrystals including the time derivative of the thermal displacement $\dot{v}$ (i.e. the temperature), the gradients of the thermal displacement ∇v and its time derivative $\nabla\dot{v}$ as the state space variables for all quantities of the coupled equations—free energy, entropy, first Piola-Kirchhoff stress tensor, chemical potential, defect flux, heat flux.

$\lambda^\star$ is a positive material constant characteristics of the TPL theory ("the rate of thermal conductivity".

S. Falahatkar et al. [914] used the TPL model to study heat conduction in a laser irradiated tooth. The human tooth is composed of enamel, dentine, and pulp with the unstructured shape and the uneven boundaries. Earlier S. Falahatkar et al. [913] studied this problem using the DPL model and found that the heat flux phase lag has the significant effect on the temperature profile at the early stages while the temperature gradient phase lag is more important at the later stages. The authors, using the simulation of the laser irradiated tooth, found the phase lags to be $\tau_q = 16ms$, $\tau_T = 6ms$, $\tau_v = 2ms$.

A. H. Akbarzadeh et al. [41] used the TPL model to investigate the heat transfer in the functionally graded hollow cylinder (earlier these authors exploited the DPL model to study the heat conduction in the one dimensional functionally graded media [39]).

5.2.1.5 Nonlocal Triple-Phase-Lag Model

A. H. Akbarzadeh et al. [42] derived the non-local three-phase-lag (NL TPL) constitutive equation in the following form to include the effects of the thermal displacement and its associated nonlocal length in the NL DPL thermal analysis:

$$\mathbf{q}(\mathbf{r}+\boldsymbol{\lambda}_q, t+\tau_q) = --[\lambda \nabla T(\mathbf{r}+\boldsymbol{\lambda}_T, t+\tau_T) + \lambda^\star \nabla v(\mathbf{r}+\boldsymbol{\lambda}_v, t+\tau_v)] \quad (5.182)$$

where λ_q, λ_T, and λ_v represent the correlating nonlocal lengths of the heat flux, the temperature gradient, and the thermal displacement gradient. The Taylor series expansion of the constitutive Eq. (5.182) with respect to either the nonlocal lengths and/or phase-lags leads to the alternative nonlocal phase-lag heat conduction models.

5.2.1.6 Two Temperature Triple-Phase-Lag Model

Another triple-phase-lag model is based on the two-temperature theory, proposed by P. J. Chen and M. E. Gurtin [501], M. E. Gurtin and W. O. Williams [1196, 1197] and P. J. Chen et al. [509, 510] where two distinct temperatures:—the *conductive* and the *thermodynamic* temperatures—are considered. The difference between these two temperatures is proportional to the applied heat source; in the absence of the heat source, the two temperatures are equal [3264].

The equation connecting the *conductive* φ and the *thermodynamic* ϑ temperatures is given as

$$\vartheta = (1 - a\nabla^2)\varphi$$

where $a > 0$ is the temperature distinction parameter [11].

Three phase lags are introduced: the phase lags of the heat flux τ_q and the conductive temperature gradient τ_φ

$$\mathbf{q}(\mathbf{r}, t + \tau_q) = -\lambda \nabla \varphi(\mathbf{r}, t + \tau_\varphi) \tag{5.183}$$

and the phase lag of the thermodynamic temperature τ_ϑ

$$\vartheta(\mathbf{r}, t + \tau_\vartheta) = (1 - a\nabla^2)\varphi(\mathbf{r}, t + \tau_\varphi). \tag{5.184}$$

The Taylor-series-expansion is applied to both sides of the Eqs. (5.183) and (5.184); the terms up to a suitable higher-order of time-deferential m, n, p in $\tau_q, \tau_\varphi, \tau_\vartheta$ are maintained to produce

$$\left(1 + \sum_{r=1}^{m} \frac{\tau_q^r}{r!} \frac{\partial}{\partial t^r}\right) \mathbf{q} = -\lambda \left(1 + \sum_{r=1}^{n} \frac{\tau_\varphi^r}{r!} \frac{\partial}{\partial t^r}\right) \nabla \varphi, \tag{5.185}$$

$$\left(1 + \sum_{r=1}^{p} \frac{\tau_\vartheta^r}{r!} \frac{\partial}{\partial t^r}\right) \vartheta = (1 - a\nabla^2) \left(1 + \sum_{r=1}^{n} \frac{\tau_\varphi^r}{r!} \frac{\partial}{\partial t^r}\right) \varphi. \tag{5.186}$$

The energy equation for the thermoelastic isotropic material at uniform environmental temperature T_0 is written as [11]

$$\lambda \left(1 + \sum_{r=1}^{n} \frac{\tau_\varphi^r}{r!} \frac{\partial}{\partial t^r}\right) \nabla^2 \varphi = \left(1 + \sum_{r=1}^{m} \frac{\tau_q^r}{r!} \frac{\partial}{\partial t^r}\right) \left(\varrho c \frac{\partial \vartheta}{\partial t} + \gamma T_0 \frac{\partial e}{\partial t} - \varrho Q\right) \tag{5.187}$$

where $\gamma = (3\nu + 2\mu)\alpha_t$ is the thermal coupling parameter, ν, μ are the Lamé constants, α_t is the thermal expansion coefficient, $e = \nabla \cdot \mathbf{u}$ is the cubical dilatation, $\mathbf{u}$ is the displacement vector, Q is the heat source.

This model contains as a special cases a number of the known models, for example,

- the classical coupled thermoelasticity [294] for the case $\tau_q = \tau_\varphi = \tau_\vartheta = 0, a = 0, \varphi = \vartheta$;
- the Lord–Shulman model [1955] for the case $\tau_q = \tau_0 > 0, \tau_\varphi = \tau_\vartheta = 0, a = 0, \varphi = \vartheta, m = 1$;
- the DPL model for $\vartheta = \varphi, \tau_\vartheta = \tau_\varphi, a = 0, n = 1, m = 2$;
- the two-temperature thermoelasticity model with phase-lags by S. Mukhopadhyay et al. [2209] for $\tau_\vartheta = \tau_\varphi, a > 0, m = n = p = 1$.

S. Chiriță et al. [544] showed that there are some restrictions to choosing the higher orders m, n, p, for example, the choice $m \geq 5$ leads to an unstable system.

A. E. Abouelregal et al. [11] applied the described TPL model using the Laplace transform to investigate the behaviour of the isotropic thermoelastic body that is exposed to the thermal shock and is affected by an external force.

5.2.1.7 High-Order Phase-Lag Models

S. Chiriţă [543] suggested a two-parameter (n, m) family of the high-order phase-lag models where the polynomial approximations of the order n for the heat flux vector and of the order m for the temperature gradient are used to get the following class of the constitutive equations relating the heat flux vector q_i to the temperature gradient $T_{,i}$

$$a_0 q_i(\mathbf{x}, t) + a_1 \frac{\partial q_i}{\partial t}(\mathbf{x}, t) + a_2 \frac{\partial^2 q_i}{\partial t^2}(\mathbf{x}, t) + \cdots + a_n \frac{\partial^n q_i}{\partial t^n}(\mathbf{x}, t) =$$

$$= -\lambda_{ij} \left[b_0 T_{,j}(\mathbf{x}, t) + b_1 \frac{\partial T_{,j}}{\partial t}(\mathbf{x}, t) + b_2 \frac{\partial^2 T_{,j}}{\partial t^2}(\mathbf{x}, t) + \cdots + a_m \frac{\partial^m T_{,j}}{\partial t^m}(\mathbf{x}, t) \right]. \tag{5.188}$$

The author noted that the well-posedness question for the time-differential dual-phase lag heat conduction models characterized by (arbitrarily) high expansion orders acquires an enormous significance, in view of the fact that the interactions between the different energy carriers involve the high-order terms: such types of the interaction become non-negligible as the observation scales progressively reduce; a specific reference is made to the transfer mechanisms between the tissues and the blood during the nonequilibrium processes and to the drug delivery in the tumours involving the species diffusion, i.e. to the heat and mass transfer in the biological systems.

By comparison of the suggested models with the time-differential constitutive law for the heat flux vector proposed by D. Y. Tzou

$$q_i(\mathbf{x}, t) + \frac{1}{1!}\tau_q \frac{\partial q_i}{\partial t}(\mathbf{x}, t) + \frac{1}{2!}\tau_q^2 \frac{\partial^2 q_i}{\partial t^2}(\mathbf{x}, t) + \cdots + \frac{1}{n!}\tau_q^n \frac{\partial^n q_i}{\partial t^n}(\mathbf{x}, t) =$$

$$= -\lambda_{ij} \left[T_{,j}t) + \frac{1}{1!}\tau_T \frac{\partial T_{,j} T_{,j}}{\partial t}(\mathbf{x}, t) + \frac{1}{2!}\tau_T^2 \frac{\partial^2 T_{,j}}{\partial t^2}(\mathbf{x}, t) \right.$$

$$\left. + \cdots + \frac{1}{m!}\tau_T^m \frac{\partial^m T_{,j}}{\partial t^m}(\mathbf{x}, t) \right] \tag{5.189}$$

where $\tau_q \geq 0$ and $\tau_T \geq 0$ are the delay times, λ_{ij} is the positive definite conductivity tensor, n and m are natural numbers, Chiriţă obtained for the coefficients a_i, b_i the following relations

$$a_0 = 1, \quad a_1 = \frac{1}{1!}\tau_q, \quad \ldots, \quad a_n = \frac{1}{n!}\tau_q,$$

$$b_0 = 1, \quad b_1 = \frac{1}{1!}\tau_T, \quad \ldots, \quad b_m = \frac{1}{m!}\tau_T. \tag{5.190}$$

The governing equation for the variation of temperature field T for the constitutive law for the heat flux vector by Tzou is written as

$$\left(1 + \tau_q \frac{\partial}{\partial t} + \frac{1}{2!}\tau_q^2 \frac{\partial^2}{\partial t^2} + \cdots + \frac{1}{n!}\tau_q^n \frac{\partial^n}{\partial t^n}\right)\left(\varkappa \frac{\partial T}{\partial t} - \varrho\right) =$$

$$= \left[\lambda_{ij}\left(1 + \tau_T \frac{\partial}{\partial t} + \frac{1}{2!}\tau_T^2 \frac{\partial^2}{\partial t^2} + \cdots + \frac{1}{n!}\tau_T^m \frac{\partial^m}{\partial t^m}\right) T_{,j}\right]_{,i}. \tag{5.191}$$

As noted by D. Y. Tzou [3106], the Eq. (5.191) shows an interchange between diffusive and wave behaviour: when $n = m$, it describes the diffusive behaviour, while when $n = m + 1$ it describes the wave behaviour.

Particular versions (the most often considered case refers to $n = 2$ and $m = 1$) of such equation have been studied intensively in the literature, for example, [10, 899, 900, 902, 1351, 1919, 2536–2538, 3239, 3241, 3370].

The temperature representation is not convenient to use for problems involving a flux specified boundary condition. In this situation, solving the energy equation and the constitutive equation simultaneously instead of the Eq. (5.191) is most convenient in avoiding the complicated conversions between the heat flux vector and the temperature variation gradient [543].

5.2.2 Thermodynamic Models

Thermodynamics models are deduced from the thermodynamic constraints following from the second law of thermodynamics [582, 583, 1500, 1705, 1800, 2628, 2774, 2778–2780, 2787, 2896, 3134, 3137, 3139].

The maximum entropy method of the determination of the field variable is based on the maximization of the entropy density under the constraints of the state variables with the Lagrange multiplier method (*Liu procedure*) [1897]: the constraints may be removed by the use of the Lagrange multipliers [2223]. An alternative approach—*Coleman-Noll procedure* [606] and its generalization [588, 589] according to idea by V. A. Cimmelli [581] that one should consider as constraints for the entropy inequality both the governing equations of the wanted fields and their gradient extensions up to the order which appears in the state space.

5.2.2.1 Jou and Cimmelli Model

D. Jou and V. A. Cimmelli [1500] introduced an extra internal variable represented by a second order tensor $\hat{Q}$ and wrote the balance of the heat flux as follows

$$\tau_1 \dot{\mathbf{q}} + \mathbf{q} = -\lambda \nabla T + \nabla \cdot \hat{Q}, \tag{5.192}$$

where τ_1 is the relaxation time.

The tensor $\hat{Q}$ is assumed to be symmetric and may be split $\hat{Q} = Q\hat{I} + \hat{Q}_s$, where the scalar Q is one-third of the trace of $\hat{Q}$, $\hat{Q}_s$ is the symmetric traceless (deviatoric) part of $\hat{Q}$. In the relaxation time approximation the evolution equations for the tensor parts are written as

$$\tau_0 \dot{\hat{Q}} + \hat{Q} = \gamma_0 \nabla \cdot \mathbf{q} \tag{5.193}$$

$$\tau_2 \dot{\hat{Q}}_s + \hat{Q}_s = 2\gamma_2 \left[(\nabla \mathbf{q})^0 \right]_s \tag{5.194}$$

where $\left[(\nabla \mathbf{q})^0 \right]_s$ is the symmetric traceless part of $\nabla \mathbf{q}$.

Under the assumptions that both τ_0 and τ_2 are small and only the regular processes are considered, for which the time derivatives at the left hand side of the Eqs. (5.193) and (5.194) do not diverge, the equation for the heat flux (5.192) could be re-written as [1500]

$$\gamma_2 \tau_1 \dot{\mathbf{q}} + \mathbf{q} = -\lambda \nabla T + \gamma_2 \nabla^2 \mathbf{q} = \left(\gamma_0 + \frac{1}{3}\gamma_2 \right) \nabla \nabla \cdot \mathbf{q}. \tag{5.195}$$

The thermodynamics underlying the Eq. (5.195) is derived by introduction the constitutive equations [1500] for the specific entropy

$$s = s_{eq} - \frac{\tau_1}{2\lambda T^2} \mathbf{q} \cdot \mathbf{q} - \frac{\tau_2}{4\lambda T^2 \gamma_2} \hat{Q} : \hat{Q} - \frac{\tau_0}{2\lambda T^2 \gamma_0} Q^2 \tag{5.196}$$

and for the entropy flux

$$\mathbf{J}_s = \frac{\mathbf{q}}{T} \left(1 + \frac{Q}{\lambda T} \right) + \frac{\hat{Q}_s \cdot \mathbf{q}}{\lambda T^2} \tag{5.197}$$

where s_{eq} is the local equilibrium entropy. Here the operator " :" means the complete contraction of the tensors producing a scalar, $\hat{Q}_s \cdot \mathbf{q}$ means the contraction over the last index of $\hat{Q}_s$ producing a vector.

The absolute temperature is given by the reciprocal of the derivative of the entropy with respect to the internal energy at the constant values of the other variables. When the extended entropy (5.196) is used instead of the local-equilibrium entropy, the resulting absolute temperature is [462]

$$\frac{1}{T} = \frac{\partial s}{\partial e} = \frac{\partial s_{eq}}{\partial e} - \frac{\partial}{\partial e}\left(\frac{\tau_1}{2\lambda T^2}\right)\mathbf{q}\cdot\mathbf{q} - \frac{\partial}{\partial e}\left(\frac{\tau_2}{4\lambda T^2\gamma_2}\right)\hat{Q}_s : \hat{Q}_s$$
$$- \frac{\partial}{\partial e}\left(\frac{\tau_0}{2\lambda T^2\gamma_0}\right)Q^2 \tag{5.198}$$

that differs from the local-equilibrium temperature

$$\frac{1}{T_{eq}} = \frac{\partial s_{eq}}{\partial e}. \tag{5.199}$$

The local-equilibrium temperature loses its validity in the situations where the deviation from the equilibrium ensemble is significant [650, 1992].

The dispersion relation of the heat waves along the nanowires which are not isolated from the environment allows to compare the different definitions of the nonequilibrium temperature, since the thermal waves are predicted to propagate with the different phase speed [1500, 2777].

The first-order flux as the sole independent variable is not sufficient for the correct description of the high-frequency processes when the frequency becomes comparable to the inverse of the relaxation time of the first-order flux, all the higher-order fluxes will also behave like the independent variables.

The EIT allows one to introduce the higher order fluxes $\hat{J}_2, \ldots, \hat{J}_n$ where $\hat{J}_k$ is the tensor of order k that serves as the flux of the preceding flux $\hat{J}_{k-1}$. Thus the specific entropy is written as [590, 1500]

$$s = s(e, \hat{J}_1, \ldots \hat{J}_n) = s_{eq} + \frac{\alpha_1}{2}\hat{J}_1 : \hat{J}_1 + \cdots + \frac{\alpha_n}{2}\hat{J}_n : \hat{J}_n \tag{5.200}$$

and entropy flux as

$$\hat{J}_s = \frac{\hat{J}_1}{T} + \beta_1\hat{J}_2 \cdot \hat{J}_1 + \cdots + \beta_{n-1}\hat{J}_n \cdot \hat{J}_{n-1} \tag{5.201}$$

where $\alpha_1, \ldots \alpha_n$ and $\beta_1, \ldots \beta_n$ are the functions that can depend on the internal energy, $\hat{J}_k \cdot \hat{J}_{k-1}$ means the contraction over the last $(k-1)$ indices of the tensor $\hat{J}_k$ producing a vector.

The evolution of fluxes is governed by the following equations that are compatible with the positive entropy production

$$\tau_1 \dot{\hat{J}}_1 + \hat{J}_1 = -\lambda \nabla T + \frac{\beta_1 \tau_1}{\alpha_1} \nabla \cdot \hat{J}_2$$

$$\dots$$

$$\tau_n \dot{\hat{J}}_n + \hat{J}_n = \frac{\beta_n \tau_n}{\alpha_n} \nabla \cdot \hat{J}_{n+1} + \frac{\beta_{n-1} \tau_n}{\alpha_n} \nabla \cdot \hat{J}_{n-1}.$$

If the relaxation time of the nth flux $\tau_n \gg \tau_{n+1}$, the hierarchy can be stopped at the nth flux [2896].

Consideration of this hierarchy of the equations for the higher order fluxes in the Fourier space (ω, k) allows one to introduce a generalized thermal conductivity $\lambda(\omega, k)$ as the continued-fraction expansion [954, 1504]

$$\lambda(\omega, k) = \frac{\lambda_0}{1 + i\omega\tau_1 + \dfrac{k^2 l_1^2}{i\omega\tau_2 + \dots}}. \tag{5.202}$$

The authors extended the Guyer-Krumhansl equation to the nonlinear case as

$$\tau_R \dot{\mathbf{q}} + \mathbf{q} = -\lambda \nabla T + \left(\frac{2\tau}{c_V T} \right) + l_p^2 (\nabla^2 \mathbf{q} + 2\nabla\nabla \cdot \mathbf{q}), \tag{5.203}$$

and finally derived a general constitutive equation (that includes the Cattaneo and the Guyer-Krumhansl equations as a special cases) as

$$\tau_R \dot{\mathbf{q}} + \mathbf{q} + (\mu \nabla\nabla \mathbf{q} + \mu' \nabla \mathbf{q}) = -\lambda(1 + \xi \mathbf{q} \cdot \mathbf{q}) \nabla T + l_p^2 (\nabla^2 \mathbf{q} + 2\nabla\nabla \cdot \mathbf{q}), \tag{5.204}$$

where μ, μ', ξ are material coefficients. This equation accounts for the nonlocal, nonlinear and memory effects.

5.2.2.2 Sellitto and Cimmelli Model

A. Sellitto and V. A. Cimmelli [2774] developed a continuum approach to the thermomass model. The authors introduced the absolute temperature ϑ and the vectorial variable $\mathbf{c}$ as the sate variables and postulate similar to the approach by G. Lebon et al. [1802] that $\mathbf{c}$ is proportional to the heat flux.

The authors assumed that the evolution of the variable $\mathbf{c}$ is governed by the balance equation

$$\dot{\mathbf{c}} = \nabla \cdot \hat{\varphi}^{(c)} + \boldsymbol{\sigma}^{(c)} \tag{5.205}$$

where the flux and the production of $\mathbf{c}$ are given as

$$\hat{\varphi}^{(c)} = \Gamma_1(\vartheta)\hat{I} + \Gamma_2(\vartheta)\left[\frac{1}{2}c^2 + \mathbf{cc}\right], \quad \sigma^{(c)} = \Gamma_0(\vartheta)\mathbf{c}. \tag{5.206}$$

Here $\Gamma_{0(1,2)}$ are the scalar functions, $c = |\mathbf{c}|$ and the Eq. (5.205) is re-written as

$$\dot{\mathbf{c}} = \left(\frac{\partial \Gamma_1}{\partial \vartheta} + \frac{1}{2}\frac{\partial \Gamma_2}{\partial \vartheta}c^2\right)\nabla\vartheta + \left(\frac{\partial \Gamma_2}{\partial \vartheta}\cdot\mathbf{c} + \Gamma_2\nabla\cdot\mathbf{c} + \Gamma_0\right)\mathbf{c}$$
$$+ \Gamma_2(\nabla\mathbf{c}\cdot\mathbf{c} + \nabla\mathbf{c}^T\cdot\mathbf{c}). \tag{5.207}$$

In the absence of the heat source the energy conservation is (u is the volumetric internal energy)

$$\dot{u} = -\nabla\cdot\mathbf{q}, \tag{5.208}$$

and the constitutive equation is written as

$$\mathbf{q} = g(\vartheta, c^2)\mathbf{c}. \tag{5.209}$$

Sellitto and Cimmelli listed several reasons to use the variable $\mathbf{c}$ instead of the heat flux $\mathbf{q}$ (see also [1505]):

- This choice allows to get the constitutive equation in the form similar to that in the phonon hydrodynamics.
- $\mathbf{c}$ is proportional to τ , which is a small parameter, it may be very helpful in the determination of the closure of the nonlinear series expansions.
- The compliance with earlier models of the generalized Fourier law [3231].
- The form of the function g could be chosen in accordance of the principle of the frame indifference.

D. Jou et al. [1505] considered $\mathbf{c}$ as the renormalized heat flux. Since $\mathbf{c}$ is collinear to $\mathbf{q}$ but not equal to $\mathbf{q}$ and $\mathbf{c}$ is not identified, $\mathbf{c}$ may be considered as an internal variable, instead of the heat flux. The internal variables constitute an efficient tool when dealing with nonequilibrium processes involving the complex thermodynamical systems. These additional nonequilibrium parameters, whose nature depends on the phenomenon at hand, usually are introduced through the ordinary differential equations, called the kinetic equations.

The Eq. (5.208) could be re-written as

$$\dot{u} = \left(\frac{\partial g}{\partial \vartheta}\nabla\vartheta + 2\frac{\partial g}{\partial (c^2)}\nabla\mathbf{c}\cdot\mathbf{c}\right)\cdot\mathbf{c} - g\nabla\cdot\mathbf{c}. \tag{5.210}$$

The Eqs. (5.207) and (5.210) form a closed set of equations. Solving these equations under the restrictions imposed by the second law of thermodynamics, the authors obtained the general equation for the heat flux that in the one dimensional case is written as [2774]

$$\frac{\dot{q}_x - \frac{\partial \ln g}{\partial \vartheta} q_x \dot{\vartheta}}{g\left[1 - \frac{2}{g}\frac{\partial(1/g)}{\partial(c^2)} q_x^2\right]} = \left[\frac{\partial \Gamma_1}{\partial \vartheta} + \frac{3}{g^2}\left[\frac{1}{2}\frac{\partial \Gamma_2}{\partial \vartheta} - \frac{\Gamma_2 \frac{\partial \ln g}{\partial \vartheta}}{1 - \frac{2}{g}\frac{\partial(1/g)}{\partial(c^2)} q_x^2}\right] q_x^2\right] \frac{\partial \vartheta}{\partial x}$$

$$+ \frac{\Gamma_0}{g} q_x + \frac{3\Gamma_2}{g^2}\left[\frac{1}{1 - \frac{2}{g}\frac{\partial(1/g)}{\partial(c^2)} q_x^2}\right] \frac{\partial q_x}{\partial x} q_x. \tag{5.211}$$

This equation is greatly simplified in the case $g(\vartheta, c^2) = g_0(\vartheta)$:

$$\dot{q}_x - \Gamma_0 q_x = \left[\frac{\partial \Gamma_1}{\partial \vartheta} g_0 + \frac{3}{g_0}\left[\frac{1}{2}\frac{\partial \Gamma_2}{\partial \vartheta} - \Gamma_2 \frac{\partial \ln g}{\partial \vartheta}\right] q_x^2\right] \frac{\partial \vartheta}{\partial x}$$

$$+ q_x \frac{\partial \ln g_0}{\partial \vartheta} \dot{\vartheta} + \frac{3\Gamma_2}{g_0} \frac{\partial q_x}{\partial x} q_x. \tag{5.212}$$

This equation is compatible with the second law of thermodynamics. The entropy flux is written as

$$\boldsymbol{\varphi}^{(s)} = \frac{g_0}{\vartheta}\left(1 - \frac{2\gamma C_V}{3\varrho\vartheta} c^2\right) \mathbf{c} \tag{5.213}$$

and the net entropy production as

$$\sigma^{(s)} = \frac{1}{\lambda \vartheta^2} \mathbf{q} \cdot \mathbf{q} \tag{5.214}$$

where the following relations are used

$$\Gamma_1 = -\vartheta, \quad \Gamma_2 = -\frac{2\gamma C_v}{3\varrho}, \quad \frac{\partial g_0}{\partial \vartheta} = \frac{2\gamma C_v}{\varrho}, \quad \frac{\partial g_2}{\partial \vartheta} = 0. \tag{5.215}$$

5.2.2.3 Kovács and Ván Model

R. Kovács and P. Ván [1705] introduced the heat flux and a second order tensorial variable as the internal field variables and assumed the following form of the entropy flux

$$\mathbf{J} = \hat{b} \cdot \mathbf{q} + \hat{B} : \hat{Q}, \tag{5.216}$$

where $\hat{b}$ is second order and $\hat{B}$ is third order tensorial functions called the current multipliers (the Nyiri multipliers [2321]).

The authors also assumed a quadratic dependence of the entropy density on the additional field variables

$$s = s_{eq}(e) - \frac{m_1}{2}\mathbf{q} \cdot \mathbf{q} - \frac{m_2}{2}\hat{Q} : \hat{Q}, \tag{5.217}$$

where m_1 and m_2 are the constant positive material coefficients.

The entropy inequality could be written as [1705]

$$\begin{aligned} \frac{\partial s}{\partial t} + \nabla \cdot \mathbf{q} &= -\frac{1}{T}\nabla \cdot \mathbf{q} - m_1\mathbf{q} \cdot \frac{\partial \mathbf{q}}{\partial t} - m_2\hat{Q} : \frac{\partial \hat{Q}}{\partial t} + \hat{b} \\ &\quad : \nabla\mathbf{q} + \mathbf{q} \cdot (\nabla \cdot \hat{b}) + \hat{B} : \nabla\hat{Q} + \hat{Q} : (\nabla \cdot \hat{B}) \\ &= \left(\hat{b} - \frac{1}{T}\hat{I}\right) : \nabla\mathbf{q} + \left(\nabla \cdot \hat{b} - m_1\frac{\partial \mathbf{q}}{\partial t}\right) \cdot \mathbf{q} + \left(\nabla \cdot \hat{B} - m_2\frac{\partial \hat{Q}}{\partial t}\right) : \hat{Q} + \hat{B} : \nabla\hat{Q} \geq 0, \end{aligned} \tag{5.218}$$

where $\hat{I}$ is the unit tensor.

The authors identified four generalized forces and four fluxes in this equations and assumed the linear relationships between them that are written as (in the one dimensional case)

$$m_1\frac{\partial q}{\partial t} - \frac{\partial b}{\partial x} = -l_1 q, \tag{5.219}$$

$$m_2\frac{\partial Qq}{\partial t} - \frac{\partial Q}{\partial x} = -k_1 Q + k_{12}\frac{\partial q}{\partial x}, \tag{5.220}$$

$$b - \frac{1}{T} = -k_1 Q + k_2\frac{\partial q}{\partial x}, \tag{5.221}$$

$$B = n\frac{\partial Q}{\partial x}. \tag{5.222}$$

The material coefficients $(l_1, k_1, k_2, k_{12}, k_{21})$ are subjected to the following restrictions from the second law of thermodynamics:

$$l_1 \geq 0, \qquad k_1 \geq 0, \qquad k_2 \geq 0, \qquad k_1k_2 - k_{12}k_{21} \geq 0. \tag{5.223}$$

Using the constraints from the second law of thermodynamics (the nonnegative production of the entropy) and eliminating the internal variables, the authors get the general constitutive equation for the heat flux as

$$m_1 m_2 \partial_{tt} q + (m_) l_1 + m_1 k_1) \partial_t q - (m_1 n + m_2 k_2) \partial_{xxt} q + n k_2 \partial_x^4 q - (l_1 + K) \partial_{xx} q$$

$$+ k_1 l_1 q = m_2 \partial_{xt} \frac{1}{T} + k_1 \partial_x \frac{1}{T} - \partial_x^3 \frac{1}{T}. \tag{5.224}$$

Choosing some of the materials coefficients to be equal to zero, one can get a number of the known models such as

- Fourier: $n = m_1 = m_2 = k_2 = 0$ and ($k_{12} = 0$ or $k_{21} = 0$)
- Cattaneo: $n = m_2 = k_2 = 0$ and ($k_{12} = 0$ or $k_{21} = 0$)
- Ballistic-diffusion: $n = k_2 = 0$
- Jeffrey's type: $n = m_1 = k_2 = 0$ and ($k_{12} = 0$ or $k_{21} = 0$)
- Guyer-Krumhansl: $n = m_2 = 0$
- Cahn-Hillard: $n = m_1 = m_2 = 0$

S. A. Rukolaine [2666] analysed the hyperbolicity of the ballistic-conductive (BC) model of heat conduction [1703, 1706, 1708]—studied the behaviour of solutions to an initial value problem in 1D in the framework of the linearized BC model. The linearized BC model it can be considered a hyperbolic approximation to the Boltzmann equation, which is also hyperbolic. From this point of view, the Cattaneo model (hyperbolic heat equation) is the first hyperbolic approximation to the Boltzmann equation, while the linearized BC model is the second one.

The dimensionless equations of the linearized form of the ballistic conductive model of heat conduction are written as

$$\frac{\partial T}{\partial t} + \frac{\partial q}{\partial x} + \gamma T = f(x, t) \tag{5.225}$$

$$\tau_q \frac{\partial q}{\partial t} + q + \frac{\partial T}{\partial x} + \kappa \frac{\partial Q}{\partial x} = 0 \tag{5.226}$$

$$\tau_Q \frac{\partial Q}{\partial t} + Q + \kappa \frac{\partial q}{\partial x} = 0 \tag{5.227}$$

where f is a heat source, Q is an internal variable, γ is a heat exchange parameter or volumetric heat transfer coefficient, τ_q and τ_Q are relaxation times, κ is a dissipation parameter. The internal variable Q is similar to a second-order dissipative flux in the framework of EIT.

If $\kappa = 0$, then Eq. (5.226) takes the form of Cattaneo's equation. If $\tau_Q = 0$, then the system (5.225)–(5.227) leads to the Guyer–Krumhansl equation. If $\tau_q = 0$ and $\kappa = 0$, then Eq. (5.226) takes the form of the Fourier's law.

The system (5.225)–(5.227) could be written as

$$\frac{\partial \mathbf{w}}{\partial t}+\mathcal{A}\frac{\partial \mathbf{w}}{\partial x}+\mathcal{B}\mathbf{w}=(f,0,0)^T, \tag{5.228}$$

$$\mathbf{w}=\begin{bmatrix}T\\ q\\ Q\end{bmatrix},\quad \mathcal{A}=\begin{bmatrix}0 & 1 & 0\\ 1/\tau_q & 0 & \kappa/\tau_q\\ 0 & \kappa/\tau_Q & 0\end{bmatrix},\quad \mathcal{B}=\begin{bmatrix}\gamma & 0 & 0\\ 0 & 1/\tau_q & 0\\ 0 & 0 & 1/\tau_Q\end{bmatrix}. \tag{5.229}$$

The system (5.225)–(5.227) is symmetric hyperbolic, the eigenvalues of the matrix $\mathcal{A}$ are

$$\lambda_{1,2}=\pm v,\ v=\sqrt{\frac{1}{\tau_q}\left(1+\frac{\kappa^2}{\tau_Q}\right)},\quad \lambda_3=0. \tag{5.230}$$

The finite speed of heat propagation equal to v, the corresponding characteristics are given by the equations $x\pm vt=const$. In addition to the non-zero eigenvalues of the matrix $\mathcal{A}$, there is the zero eigenvalue, and the corresponding characteristic is given by the equation $x=const$.

Excluding in the system (5.225)–(5.227) the variables q and Q, one finds that the temperature satisfies the equation

$$\tau_q\tau_Q\frac{\partial^3 T}{\partial t^3}+(\tau_q+\tau_Q+\gamma\tau_q\tau_Q)\frac{\partial^2 T}{\partial t^2}+[1+\gamma(\tau_q+\tau_Q)]\frac{\partial T}{\partial t}+\gamma T$$
$$-(1+\gamma\kappa^2)\frac{\partial^2 T}{\partial x^2}-(\tau_Q+\kappa^2)\frac{\partial^3 T}{\partial x^2 t}=\tau_q\tau_Q\frac{\partial^2 T}{\partial t^2}+(\tau_q+\tau_Q)\frac{\partial f}{\partial t}+f-\kappa^2\frac{\partial^2 f}{\partial x^2}. \tag{5.231}$$

The author considered the initial value problem for the Eq. (5.231) assuming that at the moment $t=0$ a finite amount of thermal energy was released at the origin $f(x,t)=\delta(x)\delta(t)$ where $\delta(\cdot)$ is the Dirac delta-function.

The Laplace–Fourier transform of the fundamental solution is given

$$\mathcal{LF}T(\xi,s)=\frac{[\tau_q\tau_Q s^2+(\tau_q\tau_Q)s+1+\kappa^2\xi^2]\mathcal{LF}f(\xi,s)}{\tau_q\tau_Q s^3+(\tau_q\tau_Q)s^2+[1+(\tau_Q+\kappa^2)\xi^2]s+\xi^2} \tag{5.232}$$

The Fourier transform is defined as

$$\mathcal{F}\Phi(\xi)=\int\limits_{R^3}\Phi(\mathbf{r})e^{i\xi\mathbf{r}}\mathbf{dr}, \tag{5.233}$$

the Laplace transform is defined as

$$\mathcal{L}\Phi(s) = \int_0^\infty \Phi(t) e^{-st} \mathrm{d}t. \tag{5.234}$$

The author obtained that the fundamental solution can be represented in the form

$$\Phi(x,t) = \begin{cases} \Phi_{sing,1}(x,t) + \Phi_{sing,2}(x,t) + \Phi_{reg}(x,t) & |x| \le vt \\ 0 & |x| > vt \end{cases} \tag{5.235}$$

where

$$\Phi_{sing,1}(x,t) = e^{-\mu_1 t} E_\infty \delta(x), \quad \Phi_{sing,2}(x,t) = e^{-\mu_2 t} F_\infty \frac{1}{2}[\delta(x - vt) + \delta(x + vt)], \tag{5.236}$$

$$E_\infty = \frac{\kappa^2}{\tau_Q + \kappa^2}, \quad F_\infty = 1 - E_\infty, \tag{5.237}$$

expressions for μ_1, μ_2 can be found in the paper by Rukolaine [2666].

One can conclude that

$$\lim_{\varepsilon \to 0} \int_{-\varepsilon}^{\varepsilon} \Phi(x,t) \mathrm{d}x = e^{-\mu_{1,\infty} t} E_\infty. \tag{5.238}$$

This means that part of the initial thermal energy does not spread anywhere, though this part decreases exponentially with time. The reason for this effect is the zero eigenvalue of the matrix $\mathcal{A}$. This effect of the BC model is unphysical.

5.2.2.4 Famá et al. Model

A. Famá et al. [918] introduced the three-dimensional form of the equations of the heat transfer in the isotropic materials, with the second order tensorial internal variable $\hat{Q}$, including the possible reciprocity relations and the second law requirements for the transport coefficients. Their model could be considered as the extension of the Kovács and Ván model (Sect. 5.2.2.3).

The authors consider the balances of the internal energy e

$$\varrho \frac{\partial e}{\partial t} + \nabla \cdot \mathbf{q} = 0 \tag{5.239}$$

and of the entropy s in the rigid heat conductor

$$\varrho\frac{\partial s}{\partial t} + \nabla \cdot \mathbf{J} = \sigma^{(s)} \tag{5.240}$$

where $\sigma^{(s)}$ is the entropy production rate.

The authors introduced an additional internal variable $\hat{Q}$ (a second-order tensor) which incorporates the higher-order effects in the heat transport. This variable $\hat{Q}$ may be interpreted as the flux of the heat flux in the solids [1502, 2778] or as the pressure tensor in the fluids [1502, 1708]. Similar to the Kovács and Ván model it is assumed that the entropy flux is zero if $\mathbf{q}$ and $\hat{Q}$ are zero, the Eq. (5.216) is valid.

Expansion of the entropy function $s(e, \mathbf{q}, \hat{Q})$ up to the second approximation around the equilibrium state gives

$$s(e, \mathbf{q}, \hat{Q}) = s_{eq}(e) - \frac{1}{2\varrho}m_{ij}q_i q_j - \frac{1}{2\varrho}M_{ijkl}Q_{ij}q_{kl} \tag{5.241}$$

where there are symmetry relations

$$m_{ij} = m_{ji}, \quad M_{ijkl} = M_{klij}. \tag{5.242}$$

Note that relation (5.242) is valid for the anisotropic systems, too. For isotropic systems m_{ij} and M_{ijkl} would reduce to a scalar and the three scalar components conjugate to the three scalar invariants of the tensor $\hat{Q}$, respectively [918]. The thermodynamic stability requires that the inductivity tensors $\hat{m}$ and $\hat{M}$ [1201] are positive definite and the authors assumed that they are constant.

The entropy production rate must be nonnegative [918]

$$\varrho\frac{\partial s}{\partial t} + J_{i,j} = \varrho\frac{\mathrm{d}s_{eq}}{\mathrm{d}e}\varrho\frac{\partial e}{\partial t} - \frac{1}{2}m_{ij}\frac{\partial q_i}{\partial t}q_j - \frac{1}{2}m_{ij}q_i\frac{\partial q_j}{\partial t}$$

$$-\frac{1}{2}M_{ijkl}\frac{\partial Q_{ij}}{\partial t}Q_{kl} - \frac{1}{2}M_{ijkl}Q_{ij}\frac{\partial Q_{kl}}{\partial t} + b_{ij,i}q_j + b_{ij}q_{j,i} + B_{ijk,i}Q_{jk} + B_{ijk}Q_{jk,i}$$

$$= \left(b_{ij} - \frac{\delta_{ij}}{T}\right)\left(b_{ji,j} - m_{ij}\frac{\partial q_j}{\partial t}\right)q_i + \left(B_{kji,k} - M_{ijkl}\frac{\partial Q_{kl}}{\partial t}Q_{ij}\right)Q_{ij}$$

$$+ B_{ijk}Qjk, i \geq 0. \tag{5.243}$$

The authors obtained the general three dimensional anisotropic linear relations between the thermodynamic fluxes $b_i j - \delta_i j/T$, $b_{ji,j} - m_{ij}\partial q_i/\partial t$, B_{ijk}, $B_{kij,k} - M_{ijkl}\partial Q_{kl}$ and the forces q_i, $q_{i,j}$, Q_{ij}, $Q_{jk,i}$.

Famá et al. obtained the coefficients taking into account that the body properties are invariant with respect to the rotations and to the inversion of the frame of axes and the rate equations for $\mathbf{q}$ and $\hat{Q}$ accounting for the reciprocity relations. The authors investigated the one dimensional case and showed that the list of the special case includes the ballistic-conductive, the Guyer-Krumhansl, the Cahn-Hillard, the Jeffrey type, the Maxwell-Cattaneo-Vernotte and the Fourier equations.

The authors stressed that the entropy density depends on the internal variables quadratically, in order to preserve the concavity, that is the thermodynamic stability. The entropy flux depends on the internal variables linearly therefore it disappears when they are zero. As long as these two physical conditions and the entropy inequality are valid, the derived consequences are also valid.

5.2.2.5 Rogolino et al. Models

P. Rogolino et al. [2628] based their analysis on the work by P. Ván and T. Fülöp [3137] that used two assumptions:

1. the deviation from the equilibrium state is described by the heat flux and the second-order tensorial internal variable;
2. the deviation from the classical form of the entropy current is described by two tensorial functions, called the current multipliers.

Rogolino et al. developed two versions of the generalized heat conduction equations. The first one ignores the nonlocal effects, the second equation is able to describe the heat conduction in the presence of the nonlocal effects.

The authors used in the first case as the basic field variables the specific (per unitary volume) internal energy e, the heat flux $\mathbf{q}$ and the flux of the heat flux $\boldsymbol{\varphi}_{\mathbf{q}}$. In the second case Rogolino et al. considered the differential consequences in the first case by the spatial and the time derivatives of the equations and obtained the higher-order equation for the heat flux. The authors proved that the entropy flux is nonlocal in both cases while the entropy is local in the first case and nonlocal in the second case.

The authors assumed the following form of the corresponding balance laws (in the one dimensional case)

$$\frac{\partial e}{\partial t} + \frac{\partial \mathbf{q}}{\partial x} = 0 \tag{5.244}$$

$$\frac{\partial \mathbf{q}}{\partial t} + \frac{\partial \boldsymbol{\varphi}_{\mathbf{q}}}{\partial x} = r_q \tag{5.245}$$

$$\frac{\partial \boldsymbol{\varphi}_{\mathbf{q}}}{\partial t} + \frac{\partial \boldsymbol{\Psi}}{\partial x} = r_{\varphi} \tag{5.246}$$

where r_q and r_{φ} are the production rates of the heat flux $\mathbf{q}$ and flux of heat flux $\boldsymbol{\varphi}_q$, $\boldsymbol{\Psi}$ is the flux of $\boldsymbol{\varphi}_q$.

For the closure of the system of Eqs. (5.244)–(5.246) it is necessary to use the constitutive equations for the flux $\boldsymbol{\Psi}$ and the source terms r_q and r_{φ}.

The system of Eqs. (5.244)–(5.246) is the one-dimensional version of the 13-moments system of the extended irreversible thermodynamics [1502, 1503] that is related to the Grad's 13-moments method.

The entropy s and the entropy flux $\mathbf{J}$ depend on the system state

$$Z = \left[e, \frac{\partial e}{\partial x}, \mathbf{q}, \frac{\partial \mathbf{q}}{\partial x}, \boldsymbol{\varphi}_q, \frac{\partial \boldsymbol{\varphi}_q}{\partial x} \right]. \tag{5.247}$$

The authors using the entropy constraints and the Lagrange-Farkas multipliers[4] λ, α and β calculate the entropy inequality and found that the Lagrange-Farkas multipliers are determined by the partial derivatives of the entropy with respect to the basic field variables

$$\lambda = \frac{\partial s}{\partial e}, \qquad \alpha = \frac{\partial s}{\partial q}, \qquad \beta = \frac{\partial s}{\partial \varphi_q}. \tag{5.248}$$

The general expression for the entropy flux is written as

$$\mathbf{J} = \mathbf{J}_0(e, \mathbf{q}, \boldsymbol{\varphi}_q) + \frac{\partial s}{\partial \boldsymbol{\varphi}_q} \tag{5.249}$$

and the entropy inequality is given by

$$\frac{\partial}{\partial x}\left(\frac{\partial s}{\partial \varphi_q}\right)\Psi - \frac{\partial s}{\partial e}\frac{\partial q}{\partial x} - \frac{\partial s}{\partial q}\frac{\partial \varphi_q}{\partial x} + \frac{\partial J_0}{\partial x} + \frac{\partial s}{\partial q} r_q + \frac{\partial s}{\partial \varphi_q} r_\varphi \geq 0. \tag{5.250}$$

The solution of the inequality (5.250) requires the additional assumptions. Rogolino et al. choose as the entropy a quadratic function of the nonequilibrium variables q and φ_q

$$s = \bar{s}(e) - m\frac{q^2}{2} - M\frac{\varphi_q^2}{2}. \tag{5.251}$$

where m and M are the constant positive coefficients and assumed the following compatibility condition to be valid

$$\frac{\partial J_0}{\partial \varphi_q} = \frac{\partial s}{\partial q}. \tag{5.252}$$

[4] The Lagrange multipliers are used to incorporate the constrains related to the entropy [461, 1897]. Some of these multipliers have the classical counterparts such as, e.g., the temperature, the chemical potential, while others related to the nonequilibrium constraints do not have analogous in the equilibrium theory.

The chosen function (5.251) ensures the fulfilment of the principle of maximum entropy at the equilibrium. The authors also required that expression for the entropy flux reduces to the classical value $J = \partial s/\partial e$ at the equilibrium and finally get the simplified form of entropy inequality (5.250) as

$$-M\frac{\partial \varphi_q}{\partial x}\Psi + q\left(\frac{\partial}{\partial x}\left(\frac{\partial s}{\partial e}\right) - mr_q\right) + \varphi_q\left(\frac{\partial}{\partial x}\left(\frac{\partial s}{\partial q}\right) - Mr_\varphi\right) \geq 0. \quad (5.253)$$

The authors introduce the phenomenological coefficients l_1, l_2, l_3 and after the elimination of the production rates r_q, r_φ and the highest order flux Ψ get the system of the transport equations

$$\frac{\partial e}{\partial t} + \frac{\partial q}{\partial x} = 0, \quad (5.254)$$

$$\tau_q\frac{\partial q}{\partial t} + q + \tau_q\frac{\partial \varphi_q}{\partial x} = l_2\frac{\partial}{\partial x}\left(\frac{1}{T}\right), \quad (5.255)$$

$$\tau_\varphi\frac{\partial \varphi_q}{\partial t} + \varphi_q - \tau_\varphi l_1 M\frac{\partial^2 \varphi_q}{\partial x^2} = -l_3 m\frac{\partial q}{\partial x}. \quad (5.256)$$

where $\tau_q = ml_2$ and $\tau_\varphi = Ml_3$.

System of Eqs. (5.254)–(5.256) includes as special cases a number of the known models:

- If $l_3 = 0$ then the Eq. (5.256) yields $\varphi_q = 0$ and the Eq. (5.255) reduces to the Maxwell-Cattaneo-Vernotte equation (5.92)

$$\tau\frac{\partial \mathbf{q}}{\partial t} + \mathbf{q} = -\lambda\nabla T. \quad (5.257)$$

- If τ_φ is negligible, than from the Eq. (5.256) follows that $\varphi_q = 0$ and inserting into the Eq. (5.255) leads to the one-dimensional Guyer-Krumhansl equation (provided $\tau l_3 = 3\ell^2$)

$$\tau\frac{\partial q}{\partial t} + q + \lambda\frac{\partial T}{\partial t} = 3\ell^2\frac{\partial^2 q}{\partial x^2}. \quad (5.258)$$

- If in the last equation q is also negligible, one gets the equation of the Green-Naghdi type

$$\tau\frac{\partial q}{\partial t} + \lambda\frac{\partial T}{\partial t} = 3\ell^2\frac{\partial^2 q}{\partial x^2}. \quad (5.259)$$

Rogolino et al. derived two versions of the generalized heat equation:

1. The second order in space and the first order in time equation neglecting the nonlocal effects.
2. The fourth order in space and second order in time equation incorporating the nonlocal effects.

5.2.3 Random Walk Description of Heat Transfer

V. V. Gafiychuk and I. A. Lubashevsky [1030] used the random walk description of the heat transfer writing the Fokker–Planck equation with sources

$$\frac{\partial C}{\partial t} = \nabla^2(DC) - \nabla(\mathbf{v}(\mathbf{r}) + q \tag{5.260}$$

where

$$C = \frac{T - T_a}{T_a V_n} \quad and \quad C = \frac{T^\star - T_a}{T_a V_n}, \tag{5.261}$$

the cellular temperature tissue and the temperature inside the vessels, respectively,

$$q = \frac{q_h}{\rho_t c_t T_a V_n}, \tag{5.262}$$

$D = \kappa/(\rho_t c_t)$ is the thermal diffusivity of the tissue, and V_N is the volume of fundamental domain of the last level. Evidently that in the cellular tissue $\mathbf{v}(\mathbf{r}) = 0$.

If one ignores effect of the domain Q_0 boundaries on heat transfer, then the solution of (5.260) is written as

$$C(\mathbf{r}, t) = \int_{-\infty}^{t} \mathrm{d}t' \int_{Q_0} \mathrm{d}\mathbf{r}' G(\mathbf{r}, t|\mathbf{r}', t') q(\mathbf{r}', t') \tag{5.263}$$

where the Green function admits the path integral representation

$$G(r, t|r', t') = \int D\left\{\mathbf{r}|[t'']\right\} \exp\left\{-\frac{1}{4D} \int_{t'}^{t} \mathrm{d}t''[\mathbf{r}'' - \mathbf{v}(\mathbf{r}'', t'')]^2\right\} \tag{5.264}$$

where $D\left\{\mathbf{r}|[t'']\right\}$ is the Wiener integral measure of the paths and $\mathbf{r}''$ is a given realization of paths connecting the initial and terminal points.

The known relationship between the Fokker–Planck equation and the random motion of the Brownian particles enables the authors to mimic the heat transfer in the living tissue described by (5.260) or the path integral (5.264) as the random motion of the certain walkers governed by the stochastic equation

$$\mathbf{r} = \mathbf{v}(r, t) + \mathbf{f}(t) \tag{5.265}$$

where $\mathbf{f}(t) = (f_x(t), f_y(t), f_z(t))$ is a random force such that

$$\langle \mathbf{f}(t) \rangle = 0, \qquad \left\langle \mathbf{f}_\alpha(t)\mathbf{f}_{\alpha'}(t') \right\rangle = 2D\delta(t - t')\delta_{\alpha\alpha'} \tag{5.266}$$

where $\alpha, \alpha' = x, y, z$, $\delta_{\alpha,\alpha'}$ is the Kronecker delta and $\delta(t)$ is the Dirac delta function.

Equation (5.265) determines the path of a walker in the living tissue. The walker created in the cellular tissue randomly moves until it reaches the vessel boundary which is permeable for it due to the boundary conditions at the vessel interface σ

$$T|_\sigma = T^\star, \qquad \nabla_n T|_\sigma = \nabla_n T^\star \tag{5.267}$$

where for the sake of simplicity the density, the heat capacity and the thermal conductivity of the blood are assumed to be the same as for the cellular tissue.

Then, the walker is transported with until it either goes out of the vessels into the cellular tissue again or leaves the domain Q_0 containing the microcirculatory bed, through the host vein with blood. Thus, each path of a walker involves a sequence of portions associated with its migration either inside the cellular tissue or in the vessels.

The value of q can be negative, thus, there are two types of walkers:"positive" and "negative" ones. Since the equations and the boundary conditions are linear, the motion of a walker can be considered without regard to the arrangement of other walkers.

V. Gafiychuk and I. A. Lubashevsky [1030] considered two examples of the heat transfer in the living tissue:

- the random walks in the tissue phantom that contains the vessel system involving straight identical pipes parallel to the z-axis. The authors assumed that the pipes make up a hexagonal array of spacing d_p, which is well above the pipe radius a ($d_p \gg a$), and in the xy-plane normal to the pipes this array forms a hexagonal lattice of discs. The velocity field of the in the given vessel system $\mathbf{v}(\mathbf{r}) = (0, 0, v)$ is described as

$$v(\rho)) = \pi a^2 v \sum_i \chi_i U(\rho - \rho_i) \tag{5.268}$$

  where v is the mean velocity of blood in the pipes taken individually, the sum runs over all the pipes, the quantity χ_i specifying the direction of in the pipe i takes the values +1 and −1 for the positive and negative z-directions, respectively, $\boldsymbol{\rho}$ is the projection of the vector $\mathbf{r}$ into the xy-plane (S_{xy}).

The function $U(\rho)$ normalized to unity describes the blood velocity distribution over the pipe cross section satisfies the equality

$$\int\limits_{S_{xy}} \mathrm{d}\rho U(\rho - \rho_i) \equiv \int\limits_{S_{xy}} \mathrm{d}\rho_i U(\rho - \rho_i) = 1. \tag{5.269}$$

The pipe centres line coordinates ρ_i^+, ρ_i^- are considered to be pairwise independent random variables characterized by the distribution functions

$$\Phi_i(\boldsymbol{\rho}) = \frac{1}{\pi b^2} \exp\left\{-\frac{(\boldsymbol{\rho} - \boldsymbol{\rho}_i)^2}{b^2}\right\}. \tag{5.270}$$

Here $b = \mu a$ is the mean distance between pipes of a single pair and $\mu \geq 2$ is a system parameter.

The quantities χ_i are regarded as pairwise independent random variables obeying the conditions

$$\langle \chi_i \rangle = \chi \tag{5.271}$$

$$\langle \chi_i \chi_i' \rangle = \delta_{ii'} + \chi^2(1 - \delta_{ii'}) \tag{5.272}$$

where χ is a constant such that $|\chi| < 1$, $\delta_{ii'}$ is the Kronecker delta.

Thus the random walks in the given medium can be described in terms of the two-dimensional random walks $\boldsymbol{\rho}[t]$ in the xy-plane and the one-dimensional random walks $z[t]$ along the z-axis, with the former being independent of the latter. The authors analysed the characteristic properties of the distance $l_{\|}(t, \boldsymbol{\rho})$ that walkers starting from the point $r_0 = \boldsymbol{\rho}_0$ travel along the z-axis during the time t

$$l_{\|}(t, \boldsymbol{\rho}) = \int\limits_0^t \mathrm{d}t' \left\{ f_z + (t') \int\limits_{S_{xy}} D\boldsymbol{\rho}')v(\boldsymbol{\rho}')\delta(\boldsymbol{\rho}' - \boldsymbol{\rho}[t']) \right\}. \tag{5.273}$$

- the cooperative effect of the pipe system on walker motion.

The authors pointed out that in the case under consideration the walker random motion along the pipes does not obey the regular diffusion law because the dispersion of the random variable depends on the time t as $t \ln(t/\tau_a\ )$ rather than t. Thus from the standpoint of random walks the tissue phantom containing straight parallel pipes cannot be described in terms of a medium with an effective diffusion coefficient . However, in the general case, i.e. for the tissue that contains vessels variously oriented in space, this conclusion does not true.

V. Gafiychuk and I. A. Lubashevsky [1030] considered the effective medium description of the tissue phantom with pipes differently oriented in space—the case of a tissue phantom that contains, in addition, pipes parallel to the xy-plane. In particular, they showed that the influence of pipes parallel to the xy-plane allows one to introduce the effective diffusion coefficient and estimate its value. with the mean distance between these pipes being equal to d.

When the tissue phantom contains pipes variously oriented in space random walks in the xy-plane depend on the walker motion along the z-axis because during motion along the z-axis the walker can meet a pipe parallel to the xy-plane, which gives rise to its fast transport along the xy-plane with blood in these pipes. Since, the characteristics of walker motion in the xy-plane determine properties of their motion along the z-axis walker motion in the tissue phantom with the undirected pipes and the pipes variously oriented in space differ in properties.

There two limiting cases of the walker motion in the on the two-dimensional random walks in the xy-plane.

- When the time t of the walker motion satisfies the condition $t \ll d_2(2D)$, one may ignore the influence of the in the vessels on the two-dimensional random walks—because the walkers at the initial time near a vessel parallel to z-axis, practically cannot reach other vessels during this time.
- When $t \gg d_2/(2D)$, each walker in its motion visits a large number of the vessels including the vessels parallel to the xy-plane. In this case one can roughly describe the walker motion along the xy-plane in terms of the two- dimensional random walks in a medium with an effective diffusion coefficient D_{eff}.

V. Gafiychuk and I. A. Lubashevsky [1030] in considering walker motion in the z-direction on a time scale $t \gg d - 2$ chosen a time scale τ that $d/(2D) \ll \tau \ll t$ and represented the probability $P(z, z'|t, t')$ for the walker to reach the point z at time t provided it has been at the point z' at time t' in the form

$$P(z, z'|t, t') = D_z[t''] \exp\left\{ -\frac{1}{4} \int_{t'}^{t} dt'' \frac{(\dot{z} - v_{eff})^2}{D_{eff}} \right\} \tag{5.274}$$

and noted that the distribution of walkers in such a tissue phantom evolves according to the equation

$$\frac{\partial C}{\partial t} = \nabla[D_{eff} \nabla C - v_{eff} C]. \tag{5.275}$$

5.2.3.1 Continuous Time Random Walks

The random walk description of the heat transfer by V. V. Gafiychuk and I. A. Lubashevsky [1030] is a particular case of the more general approach—Continuous Time Random Walks.

Diffusion is essentially the macroscopic manifestation of the random microscopic motion of the particles. The word *particle* is used in the generic sense and, depending on the problem, may refer to the real particles (e.g., electrons) or to the spatially localized entity that can be transported, such as the fluid mass or the energy packet. This idea has been generalized in the continuous time random walk (CTRW) formalism, which leads to the generalized master equation (GME) that provides the useful models of the transport. Under some conditions the fluid limit of the GME corresponds to the Fokker–Planck (FP) diffusion equation in the inhomogeneous systems, which reduces to Fick's law when the system is homogeneous [3155]. The CTRW formalism does not enter into the details of the physical processes underlying the particle motion.

The continuous time random walks model introduced by Montrol and Weiss [2177, 2729] is based on the description of the energy transport in the system by the random motion of the particles each carrying a single quantum of energy. Thus, the local energy density and the energy current at any point are directly proportional to the particle density and current, respectively [744, 2122, 2123].

In the general random walk models [1114, 1650, 3284] a single particle located at the space (x', t') has a certain probability $p(x - x', t - t')$ to move to a new point (x, t). Each step of the walk consists in

- choosing a time of walk τ from a given distribution
- and then moving it at speed v over a distance $x = v\tau$.

CTRW arises [1132] by a sequence of the independently identically distributed (IID) positive random waiting times $T_1, T_2, \ldots,$ with probability distribution function (PDF) $\psi(x - x', t - t'))$, $t > 0$ and sequence of the IID random jumps $X_1, X_2, \ldots, \in \mathcal{R}$ with the PDF $\psi(x - x', t - t')$. Frequently these probability distribution functions are temporal and spatial invariant—$\psi(t)$ and $w(x)$—that enables the use of the Laplace-Fourier transform [2700].

An essential assumption is that the waiting time distribution and the jump length distribution are independent of each other. The stochastic process is *Markovian* if and only if the wating time PDF is of the form $\psi(t) = m \exp(-mt)$ with positive m [1132].

The CTRW model is a generalization of the compound Poisson model in which the underlying Poisson process is replaced by the renewal process [854]. In the Poisson process the waiting times are the exponentially distributed IID random variables. In the renewal process the waiting times have the general distribution. The only "memoryless" distribution is the exponential one and thus the only Markov renewal process is the Poisson one.

The transition from the microscopic level to the macroscopic level by scaling produces four classes of the motion: the Brownian motion, the Lévy motion, the subordinated Brownian motion, the subordinated Lévy motion [854]. These classes are similar to the random motions derived by Eliazar and Schlesinger [854] from the Langevin motion (two Markov motions—the Brownian motion and the Lévy motion, two non-Markov motions—the fractional Brownian motion, the fractional Lévy motion).

Eliazar and Schlesinger noted that these classes of motion are the self-similar processes, but have the different statistical behaviour: the increments of the subordinated Brownian/Lévy motions are not Gauss/Lévy distributed and are non-stationary while the increments of the fractional Brownian/Lévy motions are Gauss/Lévy distributed and stationary.

The evolution equation for the probability $p(x, t)$ of finding the random walker at the position x at the time instant t (the Montrol-Weiss equation [703]) can be written (for the initial condition $p(x, 0)$, i.e. the walker is initially at the origin) as

$$p(x,t) = \delta(x)\Psi(t) + \int_0^t \psi(t-t')\left[\int_{-\infty}^{\infty} w(x-x')p(x',t')\mathrm{d}x'\right]\mathrm{d}t', \qquad (5.276)$$

where

$$\Psi(t) = \int_t^{\infty} \psi(t')\mathrm{d}t' = 1 - \int_0^t \psi(t')\mathrm{d}t'.$$

The function $\Psi(t)$ is the *survival* function—the probability that at the instant t the walker is still sitting in the starting position $x = 0$.

The special choice $w(x) = \delta(x - 1)$ gives the "pure renewal process" with the position $x(t) = N(t)$ denoting the *counting function* with all jumps of the length 1 in the positive direction at the renewal instants [1133].

The resulting equation can be interpreted as the evolution equation of the generalized Fokker–Planck–Kolmogorov type [2726]

$$\int_0^t \Phi(t-t')\frac{\partial p(x,t')}{\partial t'}\mathrm{d}t' = -p(x,t) + \int_{-\infty}^{\infty} w(x-x')p(x',t')\mathrm{d}x'$$

where the auxillary function $\Phi(t)$ is such as

$$\Psi(t) = \int_0^t \Phi(t-t')\psi(t')\mathrm{d}t'.$$

If the CTRW is Markovian, by the choice of unit of time $\Phi(t) = \delta(t)$ [2726] we get the exponential waiting time $\phi(t) = e^{-t}$ and the most general master equation for the Markovian CTRW (also called the Kolmogorov-Feller equation [1133])

$$\frac{\partial p(x,t)}{\partial t} = -p(x,t) + \int_{-\infty}^{\infty} w(x-x')p(x',t')\mathrm{d}x'.$$

The choice

$$\Phi(t) = \frac{t^{-\beta}}{\Gamma(1-\beta)}, \quad 0 < \beta < 1$$

gives the Mittag–Leffler waiting time

$$\phi(t) = -\frac{\mathrm{d}}{\mathrm{d}t} E_\beta(t^{-\beta})$$

and the time-fractional Kolmogorov-Feller equation

$${}_tD_\star^\beta p(x,t) = -p(x,t) \int\limits_{-\infty}^{\infty} w(x-x')p(x',t)\mathrm{d}x'$$

where ${}_tD_\star^\beta$ is the fractional derivative of order β in the Caputo sense [1133].

The Montrol-Weiss equation (5.276) can be written as the generalized master equation [703]

$$\frac{\partial p}{\partial x} = \int\limits_0^t \psi(t-t')\mathrm{d}t' \int_{-\infty}^{\infty} [w(x-x')p(x',t') - w(x-x')p(x,t')]\mathrm{d}x' \tag{5.277}$$

and also in the Fourier–Laplace domain

$$\mathcal{F}(f(x);k) = \hat{f}(k) = \int\limits_{-\infty}^{\infty} e^{ikx} f(x)\mathrm{d}x, \quad \mathcal{L}(g(t);s) = \int\limits_0^{\infty} e^{-st} g(t)\mathrm{d}t$$

as [1132, 1135]

$$\hat{\tilde{p}}(k,s) = \tilde{\Psi}(s) + \tilde{\psi}(s)\hat{w}(k)\hat{\tilde{p}}(k,s), \tag{5.278}$$

thus the solution is

$$\hat{\tilde{p}}(k,s) = \frac{\tilde{\Psi}(s)}{1-\hat{w}(k)\tilde{\psi}(s)} = \frac{1-\tilde{\psi}(s)}{s}\frac{1}{1-\hat{w}(k)\tilde{\psi}(s)}. \tag{5.279}$$

Barkai [189, 190] considered the long time behaviour of the Eq. (5.279) using the long wavelength approximation

$$\psi(s) = 1 - s^\alpha, \quad \psi(s)f(k) = 1 - s^\alpha - \frac{k^2}{2}, \quad 0 < \alpha < 1.$$

Sánchez et a. [2700] called the considered above CTRW model *integrable* due to the spatial and temporal invariant dependencies of the waiting time and the jump probability distribution functions that simplifies the use of the Laplace-Fourier transform.

The authors proposed to extend the model to the *nonintegrable* CTRW case using for the step size PDF the function $w(x - x', h(x', t))$ where h is an arbitrary nonlinear function of the form

$$h(x', t) = \left(x', t; n(x', t), \frac{\mathrm{d}n}{\mathrm{d}x}(x', t), \frac{\mathrm{d}^2 n}{\mathrm{d}x^2}(x', t), \ldots\right),$$

where $n(x', t)$ is the *walker density*—the probability of the walker being at the point x at time t.

This extended CTRW model could describe the nonlinear interaction between the multiple transport mechanisms. As an example, the authors considered the case of the two transport channels. The step size PDF is written as

$$w(x - x', x', t) = \lambda_1(x', t) P_{[\alpha,\beta,\sigma]}(x - x') + \lambda_2(x', t) P_{[2,0,\sigma']}(x - x'), \tag{5.280}$$

where it is assumed that the first process is characterized by the Lévy PDF[5] with the index $\alpha < 2$

$$P_{[\alpha,\beta,\sigma]}(k) = \exp\left[-\sigma^\alpha |k|^\alpha \left[1 - i\beta sgn(k) \tan\left(\frac{\pi\alpha}{2}\right)\right]\right] \tag{5.281}$$

while the second process is diffusive characterized by the Gaussian.

---

[5] The Lévy PDF is characterized by three parameters [2700]:

1. α gives the asymptotic behaviour of the function at large k and for $\alpha \neq 1$

$$P_{[\alpha,\beta,\sigma]}(k) \propto \begin{cases} C_\alpha \left(\dfrac{1-\beta}{2}\right) \sigma^\alpha |k|^{-(1+\alpha)} & k \to -\infty \\ C_\alpha \left(\dfrac{1+\beta}{2}\right) \sigma^\alpha |k|^{-(1+\alpha)} & k \to \infty, \end{cases}$$

where the constant

$$C_\alpha = \frac{(\alpha - 1)\alpha}{\Gamma(2-\alpha)\cos\left(\dfrac{\pi\alpha}{2}\right)}.$$

If $\alpha = 1$

$$P_{[1,\beta,\sigma]}(k) \propto \frac{\sigma}{\pi} |k|^{-2}.$$

2. β measures the asymmetry of the function.
3. σ is the scale parameter.

The "projectors" $\lambda_1(x,t)$ and $\lambda_2(x,t)$ are arbitrary as long as they are positive and satisfy

$$\lambda_1(x,t) + \lambda_2(x,t) = 1.$$

As Sánchez et al. [2700] noted, the CTRW model defined by the Eq. (5.280) even it is Markovian in time by choosing the waiting time exponential could keep some sort of the system memory—the past events could affect the system later evolution—due to the nonlinearity hidden in the functions $\lambda_1(x,t)$ and $\lambda_2(x,t)$. This model was used to study the turbulent particle transport in the plasma confined in the tokamak [3155].

The authors considered the *fluid* limit of the CTRW model that is achieved in the limit of the large distances ($k \to 0$ in the Fourier space) and of the long times ($s \to 0$ in the Laplace space). The Eq. (5.281) in the fluid limit simplifies to

$$p(k) \propto 1 - \sigma^{\alpha}|k|^{\alpha}\left[1 - i\beta sgn(k)\tan\left(\frac{\pi\alpha}{2}\right)\right].$$

The nonintegrable CTRW model (5.280) is based on the assumption that all transport channels share the same temporal dynamics, i.e. are characterized by the same waiting-time probability distribution function. Sánchez et al. showed that this restriction could be removed and that the finally obtained fractional differential equation with the Riesz derivative for the walker density could be considered as the simplest nonlinear generalization of the distributed order fractional equation (Sect. 5.277).

The diffusion equation is obtained from the CTRW model by the rescaling of the waiting times and jumps [1135]. It is achieved by making smaller the waiting times by a positive factor τ and jumps by a positive factor h.

The reduced waiting times τT_n have pdf $\phi_\tau(t) = \phi(t/\tau)/\tau$ and the reduced jumps hX_n have pdf $w_h(x) = w(x/h)/h$, thus the generalized master equation in the Fourier–Laplace domain is written as [1135]

$$\hat{\tilde{p}}_{h,\tau}(k,s) = \frac{1 - \tilde{\phi}_\tau(s)}{s} + \tilde{\phi}_\tau(s)\hat{w}_h(k)\hat{\tilde{p}}_{h,\tau}(k,s),$$

where $\tilde{\phi}_\tau(s) = \tilde{\phi}(s\tau)$, $\hat{w}_h(k) = \hat{w}(kh)$ and the solution is

$$\hat{\tilde{p}}_{h,\tau}(k,s) = \frac{1 - \tilde{\phi}_\tau(s)}{s}\frac{1}{1 - \tilde{\phi}_\tau(s)\hat{w}_h(k)}.$$

Metzler and Klafter [2119, 2120, 2122] presented the derivation of the Fokker–Planck equation for the normal diffusion in an external force field $F = -\partial V/\partial x$

$$\frac{\partial P(x,t)}{\partial t} = \left[\frac{\partial}{\partial x}\frac{V'}{m\eta} + K\frac{\partial^2}{\partial x^2}\right]P(x,t) \tag{5.282}$$

where m s the mass of the diffusing test particle, η is the friction constant characterising the interaction between the test particle and its embedding, K is the generalized diffusion constant [2119], from the CTRW model.

The Eq. (5.282) is often written as

$$\frac{\partial P(x,t)}{\partial t} = K\left[\frac{\partial^2}{\partial x^2} - \frac{\partial}{\partial x}\frac{F(x)}{k_B T}\right] P(x,t) \tag{5.283}$$

and called the linear Fokker–Planck equation or the Smoluchowski equation [191]. In the absence of the external force $F(x) = 0$ the Eq. (5.283) describes the Gaussian evolution. This equation describes the overdamped motion and thus ha no explicit dependence on the velocity of the test particle. The fractional Fokker–Plank equation was suggested to model the quantum particle interacting with the chaotic bath: Kuznetsov et al. [1746] considered the dynamics of a system coupled to a chaotic subsystem described through random matrix theory and found that the reduced density matrix can display dynamics given by Lévy stable laws.

Denoting the local probabilities to jump right or left, $A(j)$ and $B(j)$, the discrete master equation is written as [3284]

$$P_j(t + \Delta t) = A_{j-1} P_{j-1}(t) + B_{j+1} P_{j+1}(t). \tag{5.284}$$

The Taylor expansions in time and in space

$$A_{j-1} P_{j-1}(t) = A(x)P(x,t) - \Delta x \frac{\partial A(x)P(x,t)}{\partial x} + \frac{(\Delta x)}{2}\frac{\partial^2 A(x)P(x,t)}{\partial x^2} + O(\Delta x^3)$$

leads to the Fokker–Planck equation with the definition of the limits

$$\frac{V'}{m\eta} = \lim_{\Delta x \to 0,\, \Delta t \to 0} \frac{\Delta x}{\Delta t}[B(x) - A(x)], \quad K = \lim_{\Delta x \to 0,\, \Delta t \to 0} \frac{\Delta x^2}{2\Delta t}.$$

The discrete master Eq. (5.284) that describes the Markovian systems with local jumps could be extended to the case of the non-local jumps from any site $j \pm n$ to the site j [2120]

$$P(t + \Delta t) = \sum_{n=1}^{\infty} A_{j,n} P_{j-n}(t) + \sum_{n=1}^{\infty} A_{j,n} P_{j-n}(t)$$

with the normalization

$$\sum_{n=1}^{\infty} (A_{j,n} + B_{j,n}) = 1$$

to get finally the fractional Fokker–Planck equation [2122]

$$\frac{\partial P(x,t)}{\partial t} = {}_0 D_t^{1-\alpha} \left[\frac{\partial}{\partial x} \frac{V'}{m\eta_\alpha} + K_\alpha \frac{\partial^2}{\partial x^2} \right] P(x,t). \tag{5.285}$$

The generalised diffusion coefficient K_α and the generalised friction coefficient η_α with the dimension $[\eta_\alpha] = sec^{\alpha-2}$ are connected by the generalised Einstein–Stokes–Smoluchowski relation [190, 2119]

$$K_\alpha = \frac{k_B T}{m\eta_\alpha}.$$

The normal (Gaussian) diffusion—that can be considered as the unbiased random walk model for the Brownian motion [1134]—is the Markovian stochastic process whose PDF satisfies the Cauchy problem

$$\frac{\partial p(x,t)}{\partial t} = \mathcal{D} \frac{\partial^2 p(x,t)}{\partial t^2}, \quad p(x,0) = p_0(x) \tag{5.286}$$

where $\mathcal{D}$ is the diffusion coefficient.

The fundamental solution of the Eq. (5.286) (the Green function)

$$f(x,t) = \frac{1}{\sqrt{4\pi \mathcal{D} t}} \exp\left(-\frac{x^2}{4\mathcal{D}t} \right)$$

represents the propagator that allows to express a general solution via the convolution integral involving the initial condition

$$p(x,t) = \int_{-\infty}^{\infty} f(\xi,t) p_0(x-\xi) \mathrm{d}\xi.$$

The Eq. (5.286) is a special case of the Fokker–Planck equation (also known as the Kolmogorov forward equation) [2370]

$$\frac{\partial p(x,t)}{\partial t} = \left[\mathcal{D}_1 \frac{\partial}{\partial t} + \mathcal{D}_2 \frac{\partial^2}{\partial t^2} \right] p(x,t).$$

A non-Markovian generalization of this equation is obtained by introducing the memory effects [2370]

$$\frac{\partial p(x,t)}{\partial t} = \int_0^t \left[\mathcal{D}_1 \frac{\partial}{\partial t} + \mathcal{D}_2 \frac{\partial^2}{\partial t^2} \right] p(x,t) D\tau.$$

Gorenflo et al. [1134, 1135] proposed the generalized space-time fractional diffusion equation to describe the evolution in time of the spatial probability density related to the self-similar stochastic process as

$$D_\star^\beta P(x,t) = D_\theta^\alpha P(x,t), \quad -\infty < x < \infty, \quad t \geq 0, \tag{5.287}$$

where the real parameters α, β, θ are restricted as $0 < \alpha \leq 2, \quad 0 < \beta \leq 1, \quad |\theta| \leq min[\alpha, 2-\alpha]$.

In the Eq. (5.287) D_θ^α is the Riesz-Feller space-fractional derivative of order α and skewness θ, $D_\star^\beta$ is the Caputo time-fractional derivative of order β. The particular cases of this equation are the space-fractional diffusion equation ($0 < \alpha < 2, \quad \beta = 1$), the time-fractional diffusion equation ($\alpha = 2, \quad 0 < \beta < 1$) and the standard diffusion equation ($\alpha = 2, \beta = 1$) [1135].

The fundamental solution (the Green function) of the Eq. (5.287) $G_{\alpha,\beta}^\theta(x,t)$ is the solution with the initial condition $u(x, 0^+) = \delta(x)$—the probability density evolving in time with the scaling property

$$G_{\alpha,\beta}^\theta(x,t) = t^{-\frac{\beta}{\alpha}} K_{\alpha,\beta}^\theta\left(\frac{x}{t^{-\beta/\alpha}}\right).$$

The *reduced Green function* could be expressed through the Mellin-Barnes integral [2014]

$$K_{\alpha,\beta}^\theta(x) = \frac{1}{\alpha x}\frac{1}{2\pi i}\int\limits_{\gamma-i\infty}^{\gamma+i\infty} \frac{\Gamma(s/\alpha)\Gamma(1-s/\alpha)\Gamma(1-s)}{\Gamma(1-(\beta/\alpha)s)\Gamma(\rho s)\Gamma(1-\rho s)} x^s \mathrm{d}s,$$

where $\rho = (\alpha - \theta)/2\alpha$, γ is a real constant.

5.2.3.2 Lévy (Khintchine-Lévy) Walks

The Lévy (also Khintchine–Lévy or Lévy–Gnedenko [2700]) walks [1614, 3114, 3464] or the Lévy flights[6] are the random walks the step length of which follow the probability with the heavy power-law tails such that the clusters of the short steps are connected by the rare long steps [3455] and are able to describe the anomalous heat transport [746, 1651, 2121]. Lévy walks model combines the ability to generate the fast diffusion and a finite velocity of the random walker [3455].

It is claimed that the Lévy walks occur in nature: the wandering albatrosses (*Diomedea exulans*) perform Lévy flights which are more efficient than the Brow-

[6] Shlesinger [2842] noted that while the term *Lévy flight* only addresses the end points of the jump and does not account for the time spent in the flight or points crossed, the term *Lévy walk* implies that the time of the flight and all points on the flight are accounted for.

nian search when searching for the prey on the ocean surface; the similar search strategies are used by the deer and the bumblebees[495, 3455]. However, the closer examination of the foraging trips of the albatrosses does not confirm this statement [839]—the flight durations are consistent with the shifted gamma distribution

$$f(t) = \frac{r^s}{\Gamma(s)} \gamma^{s-1} e^{ry}$$

where s is the shape parameter, r is the rate parameter.

The Lévy walks occur in other cases, for example, in the energy space for the single molecules interacting with the two-level systems via the long-range interactions [495].

There are two approaches to incorporate the finite velocity of the random walkers [3455]:

- after spending the waiting time, the walker does not jump instantaneously, but moves with a constant speed to its destination;
- waiting time periods are eliminated, the walker is always on the move.

Thus the second approach requires the speed of the particle v and only one distribution function—the PDF of the duration of movement $\psi(\tau)$. The frequency of the velocity changes at a given point is [3455]

$$\nu(x, t) = \int_{-\infty}^{\infty} \mathrm{d}y \int_{0}^{t} \phi(y, \tau)\nu(x - y, t - \tau)\mathrm{d}\tau + \delta(t) P_0(x)$$

where the coupled transition probability $\phi(y, t)$ accounts for the fact that only particles flying from $x - v\tau$ and $x + v\tau$ can reach the point x in time τ and change the direction of their velocities

$$\phi(y, \tau) = \frac{1}{2}\delta(|y| - v\tau)\psi(\tau).$$

The actual amount of particles at a given point is

$$P(x, t) = \int_{-\infty}^{\infty} \mathrm{d}y \int_{0}^{t} \Phi(y, \tau)\nu(x - y, t - \tau)\mathrm{d}\tau$$

where

$$\Phi(y, \tau) = \frac{1}{2}\delta(|y| - v\tau)\Psi(\tau)$$

is the probability density to travel a distance y and remain in the state of flight.

Zaburdaev et al. [3455] solved this equation using the combined Fourier–Laplace transform

$$P(k,s) = \frac{[\Psi(s+ikv) + \Psi(s-ikv)]P_0(k)}{2 - [\psi(s+ikv) + \psi(s-ikv)]}.$$

This solution can be used for the analysis of the particular cases. For example, for the exponentially distributed flight time $\psi(\tau)$ after the substitution of its Laplace transform and the subsequence inversion the telegraph equation results

$$\frac{\partial P}{\partial t} + \tau_0 \frac{\partial^2 P}{\partial t^2} = D\Delta P(x,t)$$

where $D = v^2\tau$. This equation describes the almost ballistic spreading of the particles on the very short times. As time goes, the diffusive evolution start to dominate for $t \gg \tau_0$.

The Lévy flights are the Markovian stochastic processes with the individual jumps having PDF $w(x)$ decaying at large x as $W(x) \propto |x|^{-1-\alpha}$ with $0 < \alpha < 2$, the extremely long jumps may occur, the typical trajectories are self-similar, on all scales showing clusters of the short jumps interspaced by the long excursions.

In contrast to the "area-filling" nature of the Gaussian random walk, Lévy flights have a fractal dimension with the exponent α and consists of self-similar clusters connected by the long-distance jumps; this happens on all length scales creating slowly decaying spatial correlations [2122].

The Lévy stable PDF can be defined through its characteristic function of the probability density, i.e. its Fourier transform [495]

$$p_{\alpha,\beta}(k;\mu,\sigma) = \mathcal{F}\{p_{\alpha,\beta}(x;\mu,\sigma)\} \equiv \int_{-\infty}^{\infty} e^{ikx} p_{\alpha,\beta}(x;\mu,\sigma)\mathrm{d}x =$$

$$\exp\left[i\mu k - \sigma^\alpha |k|^\alpha \left(1 - i\beta \frac{k}{|k|} \varpi(k,\alpha)\right)\right]$$

where

$$\varpi(k,\alpha) = \begin{cases} \tan\left(\dfrac{\pi\alpha}{2}\right) & \alpha \neq 1, \quad 0 < \alpha \leq 2 \\ -\dfrac{2}{\pi}\ln|k| & \alpha = 1. \end{cases}$$

Thus the Lévy stable PDF are determined by four real parameters:

- $\alpha \in [0, 2]$ is the index of stability (the Lévy index) defines the asymptotic decay of the PDF;
- $\beta \in [-1, 1]$ is the skewness parameter defines the asymmetry;

- μ is the shift parameter;
- $\sigma > 0$ is the scale parameter.

The shift and the scale parameters can be eliminated by the shift and scale transformations

$$p_{\alpha,\beta}(x;\mu,\sigma) = \frac{1}{\sigma} p_{\alpha,\beta}\left(\frac{x-\mu}{\sigma};0,1\right).$$

There are three cases when the Lévy stable PDF can be expressed in terms of the elementary functions:

- Gaussian distribution ($\alpha = 2$)

$$p_2(x) = \frac{1}{\sqrt{4\pi}} \exp\left(-\frac{x^2}{4}\right),$$

  the parameter β is irrelevant since $\tan(\pi) = 0$;
- Cauchy distribution ($\alpha = 1, \beta = 0$)

$$p_{1,0}(x) = \frac{1}{\pi(1+x^2);}$$

- Lévy–Smirnov distribution [3114] ($\alpha = 1/2, \beta = 1$)

$$p_{1/2,1}(x) = \begin{cases} \frac{1}{\sqrt{2\pi}} x^{-3/2} \exp\left(-\frac{1}{4x}\right) & x \geq 0 \\ 0 & x < 0. \end{cases}$$

The Lévy stable PDF for the arbitrary α can be expressed via the Fox H-functions [495].

There are two generalizations of the Lévy walks model [3455]:

- Lévy walks with random velocities;
- Lévy walks with rests where the waiting periods alternate with the periods of the ballistic motion.

The Lévy fractional diffusion equation and the Lévy fractional FokkerPlanck equation are written as [2123]

$$\frac{\partial P(x,t)}{\partial t} = K_\alpha \frac{\partial^\alpha P(x,t)}{\partial |x|^\alpha}$$

and

$$\frac{\partial P(x,t)}{\partial t} = \left[\frac{\partial}{\partial x}\frac{V'}{m\eta_\alpha} + K_\alpha \frac{\partial^\alpha}{\partial |x|^\alpha}\right] P(x,t).$$

Here the fractional Riesz derivative is defined as

$$\frac{\mathrm{d}^\alpha f(x)}{\mathrm{d}|x|^\alpha} = \begin{cases} -\dfrac{D^\alpha_{a+} f(x) + D^\alpha_{b-} f(x)}{2\cos\left(\dfrac{\pi\alpha}{2}\right)} & \alpha \neq 1 \\ -\dfrac{\mathrm{d}}{\mathrm{d}x}\mathcal{H} f(x) & \alpha = 1 \end{cases}$$

where D^α_{a+} and D^α_{b-} are the left and right Riemann–Liouville derivatives (5.300) and (5.301) (see Appendix) and the Hilbert transform operator is [2123]

$$(\mathcal{H}f)(x) = \frac{1}{\pi}\int\limits_{-\infty}^{\infty} \frac{f(\xi)}{x-\xi}\mathrm{d}\xi.$$

Mantegna and Stanley [2037] proposed a truncated Lévy flight to capture the natural cutoff in real physical systems.

5.2.4 *Fractional Derivative Models*

5.2.4.1 A Brief Introduction to Fractional Calculus

The fractional calculus (FC) is now an efficient tool for the problems in science and engineering [452, 1643, 2137, 2263, 2336, 2471, 2695, 3019]. The term "fractional" is kept for the historical reasons—it is a misnomer since the order can be real [247, 1132].

The historical survey of the development of FC starting from the letter by Gottfried Leibniz to Guillaume l'Hôpital (1695) including contributions by Joseph Liouville, Bernhard Riemann, Niels Abel, Grünwald, Aleksey Letnikov, Gerasimov, Marcel Riesz, Magnus Mittag–Leffler, Paul Lévy, Raoul Nigmatullin, Yuri Rabotnov, Arthur Erdélyi and others during the nineteenth and twentieth centuries could be found in refs. [1315, 2641, 3034, 3035, 3130].

Lorenzo and Hartley [1956] analysed the minimal set of criteria for the generalized calculus formulated by B. Ross:

- *analyticity*: if $f(z)$ is an analytic function of the complex variable z, the derivative $D^\nu_z f(z)$ is an analytic function of z and ν;
- *backward compatibility*: the operation $D^\nu_z f(z)$ must produce the same result as ordinary differentiation when $\nu = n$ is a positive integer; the operation $D^\nu_z f(z)$ must produce the same result as ordinary n-fold integration when n is a negative integer; $D^\nu_z f(z)$ must vanish along with its $n-1$ derivatives at $x = c$;
- *zero property*: the operation of order zero leaves the function unchanged $D^0_z f(z) = f(x)$;

- *linearity*: the fractional operators must be linear $D_z^\nu[af(x) + bg(x)] = aD_z^\nu f(x) + D_z^\nu g(x)$;
- *composition (index law)*: the law of exponents for integration of arbitrary order $D_z^\nu D_z^\mu f(x) = D_z^{\nu+\mu} f(x)$.

The fractional derivatives are based on the extension of the repeated integration and are defined either by the continuation of the fractional integral to the negative order or by the integer order derivatives of the fractional integrals [1315].

There is no unique definition [2471, 2695, 3019] (and notation is not standardized [701, 1315]). There are several kinds of definitions of the fractional derivatives (Riemann–Liouville, Caputo, Grünfeld-Letnikov, Riesz, Weyl, Marchaud, Caputo–Fabrizio, Yang, Chen, He and others) that are not equivalent [1846, 1852]. The initial conditions for the Caputo derivative are expressed in terms of the initial values of the integer order derivatives; for the zero initial conditions Riemann–Liouville, Caputo and Grünwald–Letnikov derivatives coincide [2564]. Most of the fractional derivatives are defined through the fractional integral thus derivatives inherent some non-local behaviour [1419].

The following relation is valid for all types of the fractional derivatives [3464]

$$\frac{d^{\alpha+\beta}}{dt^{\alpha+\beta}} = \frac{d^\alpha}{dt^\alpha}\frac{d^\beta}{dt^\beta} = \frac{d^\beta}{dt^\beta}\frac{d^\alpha}{dt^\alpha}. \tag{5.288}$$

The most frequently used are the Riemann–Liouville (e.g., in the calculus), the Caputo (e.g., in the physics, the numerical computations) and Grünwald–Letnikov (e.g., in the signal processing, the engineering) fractional derivatives [27, 701].

E. C. Grigoletto and E. C. de Oliveira [1167] considered the generalization of the fundamental theorem of calculus—Fundamental Theorem of Fractional Calculus (FTFC) for the cases of the Riemann–Liouville, Liouville, Caputo, Weyl and Riesz derivatives.

D. Baleanu and A. Fernandez [171] considered the possible classification of the fractional operators into the broad classes under some restrictions and criteria. In particularity, many operators could be considered as the special cases of the following operator [949]

$$^{A}_{c}I^{\alpha,\beta}_{x} \int\limits_c^x (x-t)^{\alpha-1} A\left((x-t)^\beta\right) f(t)dt, \qquad A(z) = \sum_{k=0}^{\infty} a_k z^k \tag{5.289}$$

where c is a constant often taken as zero or ∞, α and β are the complex parameters with the positive real parts, and $A(z)$ is a general analytic function whose coefficients $a_k \in C$ are permitted to depend on α and β, x as a real variable larger than c.

Riemann–Liouville Fractional Integral

Both the Riemann–Liouville and the Caputo fractional derivatives are based on the Riemann–Liouville fractional integral that for any $\alpha > 0$ is defined as [708, 2013]

$$J_{a+}^{\alpha} f(x) = \frac{1}{\Gamma(\alpha)} \int_{a}^{x} (x-t)^{\alpha-1} f(t) dt, \quad (5.290)$$

If $\alpha = 0$, $J_{a^+}^0 = I$, I is the identity operator. Here

$$\Gamma(\alpha) = \int_{0}^{\infty} \exp(-\alpha) u^{\alpha-1} du \quad (5.291)$$

is the Euler Gamma function. This integral exists if $f(t)$ is the locally integrable function and for $t \to 0$ behaves like $O(t^{-\nu})$ with $\nu < \alpha$. To get the strict mathematical rigor it is possible to use the framework of the Lebesgue spaces of the summable functions or the Sobolev spaces of the generalized functions [1131].

The Riemann–Liouville integral is a generalization of the Cauchy's formula for an n-fold integral

$$\int_{a}^{x} dx_1 \int_{a}^{x_1} dx_2 \ldots \int_{a}^{x_{n-1}} dx_n = \frac{1}{(n-1)!} \int_{a}^{x} (x-t)^{n-1} dt \quad (5.292)$$

using the relation

$$(n-1)! = \prod_{k=1}^{n-1} k = \Gamma(n). \quad (5.293)$$

The Eq. (5.290) is *left-sided* Riemann–Liouville integral. The *right-sided* Riemann–Liouville integral is written as

$$J_{b-}^{\alpha} f(x) = \frac{1}{\Gamma(\alpha)} \int_{x}^{b} (t-x)^{\alpha-1} f(t) dt. \quad (5.294)$$

The Riemann–Liouville integral is a case of the convolution integral of the Volterra type [2012]

$$K * f(x) = \int_{a}^{b} k(x-t) f(t) dt. \quad (5.295)$$

The Riemann–Liouville integral has the *semi-group* property (also called *additivity law* [1315]):

$$J^{\alpha}_{a+} J^{\beta}_{a+} f(x) = J^{\alpha+\beta}_{a+} f(x), \quad \alpha > 0, \quad \beta > 0 \tag{5.296}$$

which implies the *commutative* property [2012]:

$$J^{\beta}_{a+} J^{\alpha}_{a+} = J^{\alpha}_{a+} J^{\beta}_{a+} \tag{5.297}$$

The Riemann–Liouville fractional integral coincides with the classical definition in the case $\alpha \in \mathcal{N}$. The fractional integration improves the smoothness of functions [753].

Sometimes the Riemann–Liouville integral could be expressed via the elementary functions, e.g.,

$$J^{\alpha}_{a+} (x-a)^{\mu} = \frac{\Gamma(\mu+1)}{\Gamma(\alpha+\mu+1)} (x-a)^{\alpha+\mu}. \tag{5.298}$$

A particular case of the Riemann–Liouville fractional integrals is the Liouville fractional integrals [1167] that is obtained by transitions $a \to -\infty$ and $b \to \infty$ in Eqs. (5.290) and (5.294) as

$$J^{\alpha}_{+} f(x) = \frac{1}{\Gamma(\alpha)} \int\limits_{-\infty}^{x} (x-t)^{\alpha-1} f(t) dt, \tag{5.299}$$

Riemann–Liouville Fractional Derivative

The left and the right Riemann–Liouville fractional derivatives are defined as [2471]

$$D^{\alpha}_{a+}[f(x)] = \begin{cases} \dfrac{1}{\Gamma(1-\alpha)} \dfrac{d}{dx} \displaystyle\int\limits_{a}^{x} (x-t)^{-\alpha} f(t) dt, & \alpha \in (0,1) \\ \dfrac{df(x)}{dt}, & \alpha = 1 \end{cases} \tag{5.300}$$

and

$$D^{\alpha}_{b-}[f(x)] = \begin{cases} \dfrac{1}{\Gamma(1-\alpha)} \dfrac{d}{dx} \displaystyle\int\limits_{x}^{b} (t-x)^{-\alpha} f(t) dt, & \alpha \in (0,1) \\ \dfrac{df(x)}{dt}, & \alpha = 1 \end{cases} \tag{5.301}$$

The operator D^{α}_{a+} is left-inverse meaning that $D^{\alpha}_{a+} J^{\alpha}_{a+} = I$, I is the identity operator. Thus $D^{\alpha}_{a+} J^{\alpha}_{a+} f = f$ but the unconditional semigroup property of the fractional differentiation in the Riemann–Liouville sense does not hold: K. Diethelm [753] gives examples where $D^{\alpha_1}_{a+} D^{\alpha_2}_{a+} f = D^{\alpha_2}_{a+} D^{\alpha_1}_{a+} f \neq D^{\alpha 1+\alpha_2}_{a+} f$ and $D^{\alpha_1}_{a+} D^{\alpha_2}_{a+} f \neq D^{\alpha_2}_{a+} D^{\alpha_1}_{a+} f = D^{\alpha 1+\alpha_2}_{a+} f$.

A. Atangana and A. Secer [131] presented the tables of the Riemann–Liouville derivatives of the trigonometric and some special functions. The fractional Riemann–Liouville derivative of the power function is

$$D^{\alpha}_{a+} t^{\nu} = \frac{\Gamma(1+\nu)}{\Gamma(1+\nu-\alpha)} t^{\nu-\alpha} \tag{5.302}$$

and, particular, the derivative of a constant

$$D^{\alpha}_{a+} 1 = \frac{t^{-\alpha}}{\Gamma(1-\alpha)} \tag{5.303}$$

Since the fractional Riemann–Liouville derivative of a constant is not zero, thus the magnitude of the fractional derivative changes with adding of the constant.

G. Jumarie [1508] suggested a modification to remove this drawback. He started with a fractional derivative (*F-derivative*)

$$f^{\alpha}(x) = \lim_{h \to 0} \frac{\Delta^{\alpha} f(x)}{h^{\alpha}} \tag{5.304}$$

based on the fractional difference $\Delta^{\alpha} f(x)$ of order α, $\quad \alpha \in \mathfrak{R}, \quad 0 < \alpha \leq 1$.

Jumarie proposed the modification of the fractional Riemann–Liouville derivative

$$\frac{1}{\Gamma(1-\alpha)} \frac{d}{dx} \int\limits_{a}^{x} (x-t)^{-\alpha} (f(t) - f(0)) dt. \tag{5.305}$$

Leibniz' Formula

The classical Leibnitz' formula for the first-order derivative (i.e. when $n \in \mathcal{N}$) is

$$D^{n}[f(x)g(x)] = \sum_{k=0}^{n} \binom{n}{k} [D^{k} g(x) D^{n-k} f(x)] \tag{5.306}$$

where $f(x)$ and $g(x)$ are the n-time differentiable functions.

The fractional derivatives violate the classical Leibnitz' rule [3020, 3023]. Generalization of the Lebnitz' formula was developed by T. J. Osler [2355, 2356].

The Leibniz' formula for the differentiation of the product of the functions for the Riemann–Liouville operators for the functions that are analytic on $(a-h, a+h)$ is written as [753]

$$g](x) = \sum_{k=0}^{\lfloor n \rfloor} \binom{n}{k} (D_{a+}^k f)(x)(D_{a+}^{n-k} g)(x) + \sum_{k=\lfloor n \rfloor+1}^{\infty} \binom{n}{k} (D_{a+}^k f)(x)(J_{a+}^{n-k} g)(x). \tag{5.307}$$

where $\lfloor \quad \rfloor$ denotes the floor function. Jumarie studied the Leibniz' formula for the differentiation of the product of the non-differentiable functions [1511].

Faá di Bruno Formula (the Chain Rule)

For the functions f and g with a sufficient number of the derivatives and $n \in \mathcal{N}$ [708, 753, 3024]

$$D^n[g(f(\cdot))](x) = \sum (D^k g) \prod_{m=1}^{n} (D^m f(x)^{b_m} \tag{5.308}$$

where the sum is over all partitions of $\{1, 2, \ldots, n\}$ and for each partition k is its number of the blocks and b_j is the number of the blocks with exactly j elements.

V. E. Tarasov [3022] analysed the simplified chain rules suggested by G. Jumarie [1509, 1510, 1512] and found that these simplifications are not universally valid.

Fractional Taylor Expansion

The fractional Taylor expansion is written as [753, 2327, 2531]

$$f(x) = \frac{(x-a)^{n-m}}{\Gamma(n-m+1)} \lim_{z \to a+} J_a^{m-n} f(z) + \tag{5.309}$$

$$\sum_{k=1}^{m-1} \frac{(x-a)^{k+n-m}}{\Gamma(k+n-m+1)} \lim_{z \to a+} D^k J_a^{m-n} f(z) + J_a^n D_a^n f(x). \tag{5.310}$$

Symmetrised Space Derivative

B. Vermeersch and A. Shakouri [3183] formulated the symmetrised space derivatives of the fractional order between 1 and 2 and between 0 and 1:

- $1 < \alpha < 2$. The symmetrised space derivative of the function $g(x)$ that is integrable over the entire real axis is

$$\frac{\partial^\alpha g}{\partial |x|^\alpha} = \frac{\partial}{\partial x} \left[w_\alpha \star \frac{\partial g}{\partial x} \right] \tag{5.311}$$

where $\star$ denotes the convolution and w_α is an unknown kernel function with the Fourier image found to be $W_\alpha = 1/|\xi|^{2-\alpha}$. The Fourier inversion yields

$$w_\alpha = \frac{1}{2\pi} \int_{-\infty}^{\infty} \frac{exp(j\xi x)d\xi}{|\xi|^{2-\alpha}} = \frac{|x|^{-(\alpha-1)}}{2\Gamma(2-\alpha)\cos[(2-\alpha)\frac{\pi}{2}]}. \tag{5.312}$$

Thus

$$\frac{\partial^\alpha g}{\partial |x|^\alpha} = \frac{1}{2\Gamma(2-\alpha)\cos[(2-\alpha)\frac{\pi}{2}]} \frac{\partial}{\partial x} \int_{-\infty}^{\infty} \frac{\frac{\partial g}{\partial x}(x')dx'}{|x-x'|^{\alpha-1}}. \tag{5.313}$$

- $0 < \alpha < 1$. The symmetrised space derivative of the function $g(x)$ is

$$\frac{\partial^\alpha g}{\partial |x|^\alpha} = w_\alpha \star \frac{\partial g}{\partial x}. \tag{5.314}$$

The Fourier image of the kernel function w_α is $W_\alpha = j \cdot sgn(\xi)/|\xi|^{1-\alpha}$. Performing the Fourier inversion, the authors get finally

$$\frac{\partial^\alpha g}{\partial |x|^\alpha} = \frac{-1}{2\Gamma(1-\alpha)\cos[(1-\alpha)\frac{\pi}{2}]} \frac{\partial}{\partial x} \int_{-\infty}^{\infty} \frac{-sgn(x) \cdot \frac{\partial g}{\partial x}(x')dx'}{|x-x'|^{\alpha}}. \tag{5.315}$$

In the case $\alpha = 1/2$ the fractional integrals and derivatives are also called *semi-integrals* and *semi-derivatives* [2899].

Caputo Fractional Derivative

The fractional derivatives in the Caputo sense on the left $({}_C D^\alpha_{a+})$ and on the right $({}_C D^\alpha_{b-})$ are defined via the Riemann–Liouville fractional integral [1167] $({}_C D^\alpha_{a+} f) = (J^{n-\alpha}_{a+} f^{(n)})(x)$ and $(-1)^n ({}_C D^\alpha_{b-} f) = (J^{n-\alpha}_{b-} f^{(n)})(x)$. It was introduced independently in 1948 by M. Caputo and by A.N. Gerasimov [1063]; later by M. M. Dzherbashyan and A. D. Nersesian [829].

The major difference of the Caputo fractional derivative is that the derivative act first on the function and after the integral is evaluated while in the Riemann–Liouville approach the derivative act on the integral.

The Caputo fractional derivative is defined as [2471]

$$D_{\star}^{\alpha} f(t) = \begin{cases} \dfrac{1}{\Gamma(1-\alpha)} \displaystyle\int_{0}^{t} (t-x)^{-\alpha} \frac{df(x)}{dt} dt, & \alpha \in (0,1) \\ \dfrac{df(x)}{dt}, & \alpha = 1 \end{cases} \tag{5.316}$$

The definition of the Caputo derivative (5.316) is more restrictive than of the Riemann–Liouville one (5.300) and (5.301) since it requires the absolute integrability of the derivative $df(x)/dt$ [1131].

The Caputo fractional derivative can be considered as the regularization in the time origin for the Riemann–Liouville derivative [1132]

$$D_{\star}^{\alpha} f(t) = D^{\alpha} f(t) - f(0^{+}) \frac{t^{-\alpha}}{\Gamma(1-\alpha)} \tag{5.317}$$

and satisfies the property of being zero when applied to a constant.

Yuan and Agrawal and Singh and Chatterjee suggested the alternative definitions of the Caputo fractional derivative [753]. The first approach is based on the introduction of the auxiliary bivariate function $\phi : (0, \infty) \times [a, b] \to \mathcal{R}$ as

$$\phi(w, x) = (-1)^{\lfloor n \rfloor} \frac{2 \sin \pi n}{\pi} w^{2n - 2\lceil n \rceil + 1} \int_{a}^{x} f^{(\lceil n \rceil)}(\tau) e^{-(x-\tau) w^{2}} d\tau \tag{5.318}$$

where $\lceil \quad \rceil$ denote the ceiling function, and, finally

$$D_{\star a}^{n} f(x) = \int_{0}^{\infty} \phi(w, x) dw. \tag{5.319}$$

The second approach is based on expressing the fractional derivative of the given function in the form of the integral over $(0, \infty)$ with the integrand that can be obtained as the solution of the first-order initial value problem

$$\frac{\partial \phi^{\star}(w, x)}{\partial x} = -w^{\frac{1}{n - \lceil n \rceil - 1}} \phi^{\star}(w, x) + \frac{(-1)^{\lfloor n \rfloor} \sin \pi n}{\pi (n - \lceil n \rceil - 1)} f^{(\lceil n \rceil)}(x) \tag{5.320}$$

with the initial condition $\phi^{\star}(w, a) = 0$. Thus

$$\phi^{\star}(w, x) = \frac{(-1)^{\lfloor n \rfloor} \sin \pi n}{\pi (n - \lceil n \rceil - 1)} \int_{0}^{x} f^{(\lceil n \rceil)}(\tau) \exp(-(x-\tau) w^{\frac{1}{n - \lceil n \rceil - 1}}) d\tau \tag{5.321}$$

$$D_{\star a}^{n} f(x) = \int_{0}^{\infty} \phi^{\star}(w, x) dw. \tag{5.322}$$

Matrix Approach

Operations of the fractional calculus can be expressed by matrix [2472, 2475]. E.g., the left-sided Riemann–Liouville or Caputo derivative can be approximated in the nodes in the equidistant discretization net with the help of the upper triangular strip matrix $B_n^{(\alpha)}$ as [2472]

$$\left[v_n^{(\alpha)}\ v_{n-1}^{(\alpha)}\ \dots\ v_1^{(\alpha)}\ v_0^{(\alpha)}\right]^T = B_n^{(\alpha)}\left[v_n\ v_{n-1}\ \dots\ v_1\ v_0\right]^T \tag{5.323}$$

where

$$B_n^{(\alpha)} = \frac{1}{\tau^\alpha}\begin{bmatrix} \omega_0^{(\alpha)} & \omega_0^{(\alpha)} & \dots & \dots & \omega_{n-1}^{(\alpha)} & \omega_n^{(\alpha)} \\ 0 & \omega_0^{(\alpha)} & \omega_0^{(\alpha)} & \dots & \dots & \omega_{n-1}^{(\alpha)} \\ \dots & & & & & \\ \dots & & & & & \\ 0 & 0 & 0 & \dots & \dots & \omega_0^{(\alpha)} \end{bmatrix} \tag{5.324}$$

Similarly, the right-hand Riemann–Liouville or Caputo fractional derivative can be approximated with the help of the corresponding lower triangular strip matrix.

Caputo and Fabrizio Fractional Derivatives

M. Caputo and M. Fabrizio [438, 439] proposed the fractional derivatives without the singular kernel [1958] by replacing the kernel $(t-\tau)^{-\alpha}$ with the function $\exp(-\alpha/(1-\alpha))$ that does not have singularity for $t=\tau$ in the definition of the Caputo derivative and replacing the factor $1/\Gamma(1-\alpha)$ with $M(\alpha)/(1-\alpha)$.

For example, the fractional time derivative for $\alpha \in [0, 1]$ and the function $f \in L^1(-\infty, b)$ is

$$\mathcal{D}_t^\alpha f(t) = \frac{\alpha M(\alpha)}{1-\alpha}\int_{-\infty}^{t}(f(t)-f(\tau))\exp\left[-\frac{\alpha(t-\tau)}{1-\alpha}\right]d\tau \tag{5.325}$$

where $M(\alpha)$ is a normalization function such as $M(0) = M(1) = 1$.

GC and GRL Derivatives

D. Zhao and M. Luo [3502] suggested to divide the fractional derivative with different—singular and non-singular—kernels (e.g., Riemann–Liouville, Caputo,

Caputo–Fabrizio, Atangana–Baleanu [129][7] with the kernel

$$k(x,\alpha) = E_\alpha\left(-\frac{\alpha}{1-\alpha}x\right), \tag{5.326}$$

Atangana–Gomez [130] with the kernel

$$k(x,\alpha) = \exp\left(-\frac{\alpha}{1-\alpha}x^2\right) \tag{5.327}$$

derivative with the stretched exponential kernel [2979] (that is useful in the study of the water diffusion in the human brain using the magnetic resonance imaging [245])

$$k(x,\alpha) = \exp\left(-\frac{\alpha}{1-\alpha}x^\beta\right), \quad \beta > 0, \quad \beta \neq 1) \tag{5.328}$$

into two classes—GC (general, Caputo sense) and GRL (general, Riemann–Liouville) derivatives that obeys the principles formulated by V. Volterra in his "general laws of heredity" [3206]: the linearity principle, the invariance principle, the fading memory principle, the compatibility principle. The compatibility principle requires the validity of two limits: $D_\alpha f(x) \to f(x)$ when $\alpha \to 0$ and $D_\alpha f(x) \to f'(x)$ when $\alpha \to 1$.

The principle of *nonlocality* was suggested by Tarasov [3026].

GC Derivatives

D. Zhao and M. Luo [3502] defined the GC derivative by

$$D^{GC}_{a,\alpha} f(x) = N(\alpha)\int_a^x k(x-t,\alpha)\frac{df(t)}{dt}dt \tag{5.329}$$

The fading memory principle requires that the remote time and position has less effect: $k(x-t,\alpha)$ decreases when x increases and $k(x-t,\alpha) \to 0$ when $x \to \infty$.

The compatibility principle requires that $N(\alpha)/k(x,\alpha) \to 1$ when $\alpha \to 0$ and $N(\alpha)/k(x,\alpha) \to \delta(x)$ when $\alpha \to 1$.

GRL Derivatives

$$D^{RL}_{a,\alpha} f(x) = \frac{d}{dx}N(\alpha)\int_a^x k(x-t,\alpha)f(t)dt. \tag{5.330}$$

The restrictions on $k(x-t,\alpha)$ and $N(\alpha)$ are the same.

[7] The equation with the Atangana–Baleanu operator is related to the derivatives of distributed order [3027].

Marchaud–Hadamard Fractional Derivatives

Marchaud's approach is based on the analytic continuation of the fractional integrals to the negative orders using the Hadamard's finite parts of the divergent integrals (Hadamard's idea is to ignore the unbounded contribution to the integral and to assign the value of the remaining—finite—expression [753]).

The Marchaud fractional derivative with the lower limit a is

$$(M_{a+}^{\alpha} f)(x) = \frac{f(x)}{\Gamma(1-\alpha)(x-a)^{\alpha}} + \frac{\alpha}{\Gamma(1-\alpha)} \int_a^x \frac{f(x)-f(y)}{(x-y)^{\alpha+1}} dy \tag{5.331}$$

and with the upper limit b is

$$(M_{b-}^{\alpha} f)(x) = \frac{f(x)}{\Gamma(1-\alpha)(b-x)^{\alpha}} + \frac{\alpha}{\Gamma(1-\alpha)} \int_x^b \frac{f(x)-f(y)}{(x-y)^{\alpha+1}} dy. \tag{5.332}$$

Marchard's method is to extend the Riemann–Liouville integral to $\alpha < 0$

$$(J_+^{-\alpha} f)(x) = \frac{1}{\Gamma(-\alpha)} \int_0^{\infty} y^{-\alpha-1} f(x-y) dy \tag{5.333}$$

and to subtract the divergent part of the integral in (5.333)

$$\int_{\epsilon}^{\infty} y^{-\alpha-1} f(x-y) dy = \frac{f(x)}{\alpha \epsilon^{\epsilon}} \tag{5.334}$$

to get finally

$$(M_+^{\alpha} f)(x) = \lim_{\epsilon \to 0+} \frac{1)}{\Gamma(-\alpha)} \int_{\epsilon}^{\infty} \frac{f(x)-f(y)}{y^{\alpha+1}} dy. \tag{5.335}$$

There are two approaches to extend the definition (5.335) to the case $\alpha > 1$ [1315]:

1. To apply (5.335) to the nth derivative $d^n f/dx^n$ for $n < \alpha < n+1$.
2. To consider $f(x-y) - f(x)$ as the first-order difference and to generalize to nth order difference (difference quotient)

$$(\Delta_h^n f)(x) = (\mathcal{I} - Th)^n f(x) = \sum_{k=0}^{n} (-1)^k \binom{n}{k} f(x-kh) \tag{5.336}$$

where $\mathcal{I}$ is the identity operator and $T_h = f(x-h)$ is the translation operator.

Thus Marchard fractional derivative for $0 < \alpha < n$ is written as

$$(M_+^\alpha f)(x) = \lim_{\epsilon \to 0+} \frac{1}{C_{\alpha,n}} \int_\epsilon^\infty \frac{\Delta_y^n f(x)}{y^{\alpha+1}} dy \tag{5.337}$$

where

$$C_{\alpha,n} = \int_0^\infty \frac{(1 - e^{-y})^n}{y^{\alpha+1}}. \tag{5.338}$$

Grünwald–Letnikov Derivative

The approach suggested independently by Grünwald in 1867 and A. V. Letnikov [1828] in 1868 is based on the use the limits of the difference quotients (5.336) similar to the classical definition of the derivatives for $n \in \mathcal{N}, \quad f \in C^n[a, b], \quad a < x \leq b$

$$\tilde{D}^n f(x) = \lim_{h \to 0} \frac{\Delta_h^n f)(x)}{h^n} \tag{5.339}$$

and extension to the case of the arbitrary n.

Since $\binom{n}{k} = 0$ if $n \in \mathcal{N}$ and $n < k$ the expression (5.336) is equivalent to

$$(\Delta_h^n f)(x) = \sum_{k=0}^\infty (-1)^k \binom{n}{k} f(x - kh). \tag{5.340}$$

The series (5.340) is uniformly convergent for any bounded function if $n > 0$ [1114].

The use of (5.340) introduce two problems [753]:

- the function f needs to be defined on $(\infty, b]$;
- the function f should be such that the series converges.

These problems are solved by the introduction a new function $f^\star$

$$f^\star = \begin{cases} f(x) & x \in [a, b] \\ 0 & x \in (-\infty, a) \end{cases} \tag{5.341}$$

that is used instead of the original f.

It is also assumed that in the tending to zero h takes only the Grünwald–Letnikov fractional derivative of order n defined as

$$\tilde{D}_a^n = \lim_{N\to\infty} \frac{\Delta_{h_N}^n f(x)}{h_N^n} = \lim_{N\to\infty} \sum_{k=0}^{N} (-1)^k \binom{n}{k} f(x - kh_N). \tag{5.342}$$

The Grünwald–Letnikov derivative is called *pointwise* or *strong* derivative depending on whether the limit is taken pointwise or in the norm of a suitable Banach space [1315].

The binomial coefficient can be generalized to the non-integer arguments

$$(-1)^j \binom{q}{j} = (-1)^j \frac{\Gamma(q+1)}{\Gamma(j+1)\Gamma(q-j+1)} = \frac{\Gamma(j-q)}{\Gamma(-q)\Gamma(j+1)}. \tag{5.343}$$

The (left-sided) Grünwald–Letnikov derivative could be written as $(nh = x - a)$

$$\tilde{D}_a^\alpha f(x) = \lim_{h\to 0} \frac{1}{h^\alpha} \sum_{k=0}^{\lfloor n \rfloor} (-1)^k \frac{\Gamma(\alpha+1) f(x - kh)}{\Gamma(k+1)\Gamma(\alpha - k + 1)} \tag{5.344}$$

and right-sided $(nh = b - x)$ as

$$\tilde{D}_b^\alpha f(x) = \lim_{h\to 0} \frac{1}{h^\alpha} \sum_{k=0}^{\lfloor n \rfloor} (-1)^k \frac{\Gamma(\alpha+1) f(x + kh)}{\Gamma(k+1)\Gamma(\alpha - k + 1)}. \tag{5.345}$$

The Grünwald–Letnikov integral of the order n of the function f is written as

$$\tilde{J}_a^n f(x) = \frac{1}{\Gamma(n)} \lim_{N\to\infty} h_N^n \sum_{k=0}^{N} \frac{\Gamma(n+k)}{\Gamma(k+1)} f(x - kh_N). \tag{5.346}$$

Riesz Fractional Operators

The fractional integral of the order α in the Riesz sense (also known as the Riesz potential) is defined by the Fourier convolution product

$$(\mathcal{I}^\alpha f)(\mathbf{x}) = \int_{\mathcal{R}^n} \mathbf{K}_\alpha(\mathbf{x} - \boldsymbol{\xi}) f(\boldsymbol{\xi}) d\boldsymbol{\xi}, \tag{5.347}$$

where $Re(\alpha) > 0$. The Riesz kernel

$$\mathbf{K}_\alpha = \frac{1}{\gamma_n(\alpha)} \begin{cases} \|\mathbf{x}\|^{\alpha-n}, & \alpha - n \neq 0, 2, \ldots, \\ \|\mathbf{x}\|^{\alpha-n} \ln\left(\frac{1}{\|\mathbf{x}\|}\right), & \alpha - n = 0, 2, \ldots \end{cases} \tag{5.348}$$

where $\gamma_n(\alpha)$ is defined by

$$\frac{\gamma_n(\alpha)}{2^\alpha \pi^{\frac{n}{2}} \Gamma(\alpha/2)} = \begin{cases} \left[\Gamma\left(\frac{n-\alpha}{2}\right)\right]^{-1}, & \alpha - n \neq 0, 2, \ldots, \\ (-1)^{\frac{n-\alpha}{2}} 2^{-1} \Gamma\left(\frac{\alpha-n}{2}\right), & \alpha - n = 0, 2, \ldots. \end{cases} \tag{5.349}$$

The Riesz fractional integral is [1167]

$$(\mathcal{I}^\alpha f)(\mathbf{x}) = \frac{\Gamma\left(\frac{1-\alpha}{2}\right)}{2^\alpha \pi^{\frac{n}{2}} \Gamma(\alpha/2)} \int_{-\infty}^{\infty} f(\xi)|x-\xi|^{\alpha-1} d\xi. \tag{5.350}$$

The Riesz fractional derivative is [701]

$$D^\alpha[f(x)] = -\frac{1}{2\cos(\alpha\pi/2)} \frac{1}{\Gamma(\alpha)} \tag{5.351}$$

$$\frac{d^n}{dx^n}\left[\int_{-\infty}^{x} (x-\xi)^{n-\alpha_n-1} f(\xi) d\xi + \int_{x}^{\infty} (x-\xi)^{n-\alpha_n-1} f(\xi) d\xi\right]. \tag{5.352}$$

The Riesz derivative is the generalization of the Laplace operator [3464]

$$(-\Delta)^{\frac{\alpha}{2}} = -\frac{1}{2\cos(\alpha\pi/2)}\left[\frac{d^\alpha}{dx^\alpha} + \frac{d^\alpha}{d(-x)^\alpha}\right], \quad \alpha \neq 1. \tag{5.353}$$

The Riesz derivative could be expressed in terms of the Marchaud derivative

$$D^\alpha[f(x)] = -\frac{1}{2\cos(\alpha\pi/2)}[(M_+^\alpha f)(x) + (M_-^\alpha f)]. \tag{5.354}$$

The related Riesz-Feller derivative [1134] has an additional parameter—"skewness" θ

$$D_\theta^\alpha f(x) = \frac{\Gamma(1+\alpha)}{\pi} \times$$

$$\left[\sin\left[(\alpha+\theta)\frac{\pi}{2}\right] \int_0^\infty \frac{f(x+\xi) f(x)}{\xi^{1+\alpha}} d\xi + \sin\left[(\alpha-\theta)\frac{\pi}{2}\right] \int_0^\infty \frac{f(x-\xi) f(x)}{\xi^{1+\alpha}} d\xi\right]. \tag{5.355}$$

The allowed region of the parameters α and θ turn out to be a diamond in the plane $\{\alpha, \theta\}$ with the vertices in the points (0,0), (1,1), (2,0), (1, -1) called the "Feller–Takayasu diamond" [1132, 2123].

Weyl Fractional Derivative

The Weyl derivative is based on the generalization of the differentiation in the Fourier space [1114]—the integer derivative of the nth order $(ik)^n$ of the absolutely integrable function on $[-\pi, \pi]$ presented as the Fourier series is extended to the noninteger n.

The Weyl fractional derivative is defined as [2011]

$$D_{\pm}^{\alpha} = \begin{cases} \pm \dfrac{d}{dx}[I_{\pm}^{1-\alpha} f(x)] & 0 < \alpha < 1, \\ \dfrac{d^2}{dx^2}[I_{\pm}^{2-\alpha} f(x)] & 1 < \alpha < 2, \end{cases} \tag{5.356}$$

where the Weyl fractional integrals are ($\mu > 0$)

$$I_{+}^{\mu} = \frac{1}{\Gamma(\mu)} \int_{-\infty}^{x} (x - \chi)^{\mu-1} f(\chi) d\chi. \tag{5.357}$$

Erdélye–Kober Fractional Operators

The Erdélye–Kober integral for a well-behaved function $\phi(t)$ is defined as [1979, 2369]

$$I_{\eta}^{\gamma,\mu}\phi(t) = \frac{\eta}{\Gamma(\mu)} t^{-\eta(\mu+\gamma)} \int_{0}^{t} \tau^{\eta(\gamma+1)-1} (t^{\eta} - \tau^{\eta})^{\mu-1} \phi(\tau) d\tau, \tag{5.358}$$

where $\mu > 0, \quad \eta > 0, \quad \gamma \in \mathcal{R}$.

In the special case $\gamma = 0, \eta = 1$ the Erdélye–Kober fractional integral is related to the Riemann–Liouville fractional integral of the order μ as

$$I_{1}^{0,\mu}\phi(t) = \frac{t^{-\mu}}{\Gamma(\mu)} \int_{0}^{t} (t - \tau)^{\mu-1} \phi(\tau) d\tau = t^{-\mu} J^{\mu} \phi(t). \tag{5.359}$$

The Erdélye–Kober fractional derivative for $n - 1 < \mu < n, \quad n \in \mathcal{N}$ is defined as

$$D_{\eta}^{\gamma,\mu}\phi(t) = \prod_{j=1}^{n} \left(\gamma + j + \frac{1}{\eta} t \frac{d}{dt} \right) (I_{\eta}^{\gamma+\mu, n-\mu} \phi(t)). \tag{5.360}$$

The Erdélye–Kober fractional derivative reduces to the identity operator when $\mu = 0$

$$D_{\eta}^{\gamma,0}\phi(t) = \phi(t) \tag{5.361}$$

and for $\eta = 1$ and $\gamma = -\mu$ is related to the RL fractional derivative as

$$D_{\eta}^{\gamma,\mu}\phi(t) = t^{\mu} D_{RL}^{\mu}\phi(t). \tag{5.362}$$

5.2.4.2 Interpretation of Fractional Integral and Derivatives

The integer-order and integrals have a clear physical and geometrical interpretation that simplify their use in practice. The numerous different interpretations of the fractional derivatives and integrals have been proposed [1316]: the probabilistic [2930, 3032, 3033], geometric [242, 243, 591, 2197], physical interpretations [591, 1312, 2197, 2299, 2671].

However, as I. Podlubny [2473] noted, "since the appearance of the idea of the differentiation of arbitrary (not necessary integer) order there was not any acceptable geometric and physical interpretation of these operations for more than 300 years".

J. A. Teneiro Machado [3032] wrote the Günwald–Letnikov derivative of $x(t)$ as

$$D^{\alpha}[x(t)] = \lim_{h\to 0}\left[\frac{1}{h^{\alpha}}\sum_{k=0}^{\infty}\gamma(\alpha,k)x(t-kh)\right], \tag{5.363}$$

where h is the time increment. The author noted that

$$\gamma(\alpha,0) = 1, \quad -\sum_{k=0}^{\infty}\gamma(\alpha,k) = 1 \tag{5.364}$$

thus the "present" (P) is constituted by $x(t)$ the probability 1 while the totality of the "past/future" (PF) is constituted by the samples $x(t), x(t-h), x(t-2h), \ldots$; each sample is weighted with a probability $-\gamma(\alpha,k)$.

R. R. Nigmatullin [2299, 2300] interpreted the fractional integral in terms of the fractal Cantor set. The author considered the evolution of the state of the physical system

$$J(t) = \int_0^t K(t,\tau)f(t)d\tau \tag{5.365}$$

where the memory function $K(t,\tau)f(t)$ accounts for the loss of some states of th system; the fractional index of integration equals the fractal dimension of the Cantor set.

Podlubly [2473] and Podlubny et al. [2474] suggested the geometrical interpretation of the left-sided (Eq. (5.290)) and right-sided (Eq. (5.294)) Riemann–Liouville integrals and of the Riemann–Liouville (Eqs. (5.300)–(5.301)) and the Caputo (Eq. (5.316)) derivatives, as well as of the Riesz potential that is the sum of the left-sided and right-sided Riemann–Liouville fractional integrals

$$R_b^\alpha f(x) = \frac{1}{\Gamma(\alpha)} \left[\int_a^x (x-t)^{\alpha-1} f(t) dt + \int_x^b (t-x)^{\alpha-1} f(t) dt \right] \tag{5.366}$$

and of the Feller potential

$$\Phi^\alpha f(x) = c J_{a+}^\alpha f(x) + d J_{b-}^\alpha f(x). \tag{5.367}$$

The geometric interpretation by Podlubny is based on adding the third dimension

$$g_x(t) = \frac{1}{\Gamma(\alpha+1)} [x^\alpha + (x-t)^\alpha] \tag{5.368}$$

to the pair (t, f (x)) and considering the three-dimensional line $(t, g_x(t), f(t))$ as the top edge of the "fence" that gives shadow on the wall in the (g,f) plane.

V. E. Tarasov [3025] proposed the "informatic" ("computer science") interpretation of the Riemann–Liouville and the Caputo derivatives of the non-integer orders using the reconstructions from the infinite sequence of the derivatives of the integer orders. Such reconstructions are based on the Kotel'nikov theorem (also known as the sampling theorem) proved by Vladimir Kotel'nikov in 1933 and also by Claude Shannon 1949: under the certain restrictive conditions, function $f(t)$ can be restored from its samples $f[n] = f(nT)$ according to the Whittaker-Shannon interpolation formula. The author stressed that infinity of the sequences of the integer derivatives plays a fundamental role in representation of the fractional derivatives that describe nonlocality and memory.

J. F. Gómez–Aguilar et al. [1117] analysed the Caputo differentiation using the RC circuit for which the fractional version of the Ohm's law and Kirchhoff's law are written as

$$v(t) = \frac{1}{\sigma^{1-\gamma}} \frac{d^\gamma q}{dt^\gamma}, \qquad R\frac{dq}{dt} + \frac{1}{C} q(t) = v(t) \tag{5.369}$$

where q is the electric charge, v is the voltage, R is the resistance of the conductor, C is the capacitance. The parameter σ is introduced in order to be consistent with the dimensionality; it characterizes the fractional structures (the components that show the intermediate behaviour between conservative (capacitor) and dissipative (resistor) [1117]. The authors derived the fractional differential equation for the RC circuit

$$\frac{d^\gamma}{dt^\gamma} + \frac{1}{\tau_\gamma} q(t) = \frac{C}{\tau_\gamma} v(t), \qquad \tau_\gamma = \frac{RC}{\sigma^{1-\gamma}} \tag{5.370}$$

where τ_γ is the time constant. Gómez–Aguilar et al. claimed that the differentiation is related to the memory effects that reflect the intrinsic dissipation in the system.

D. Sierociuk et al. [2854] used the RC network to model the fractional order diffusion based on the analogy between the heat and electrical conduction. The authors showed that the equations for the capacitor and for the resistor in the transmission line could be used to get the diffusion equation; the loosing of heat was represented by the additional resistors connected parallel to capacitors.

A. Carpinteri et al. [453] considered the mechanical interpretation of the Marchaud fractional derivative using the body springs connecting the non-adjacent points of the body with the stiffness decaying with the distance between the material points.

5.2.4.3 Local Fractional Derivatives

The fractional derivatives are nonlocal. Several researches introduced the local variants [3476] that are useful for study of the pointwise behaviour of the fractal and multifractal functions that describe, e.g., the stress and deformation patterns in materials exhibiting the fractal-like microstructure [453] or the velocity field of turbulent fluid [1679].

K. M. Kolwankar and A. D. Gangal [1677–1680] defined the derivative via the Riemann–Liouville derivative as

$$\mathfrak{D}^q f(y) = \lim_{x \to y} \frac{D^q(f(x) - f(y))}{(x - y)^q} \tag{5.371}$$

if the limit exists and finite.

The local fractional Taylor formula is written as [3397]

$$f(x) = \sum_{i=0}^{n} \frac{f^{(n)(y)}}{\Gamma(1+n)}(x-y)^n \frac{\mathfrak{D}^\alpha}{\Gamma(n!+\alpha)}(x-y)^\alpha + R_\alpha(x, y). \tag{5.372}$$

X.-J. Yang et al. [3407, 3507] used similar definition

$$\mathfrak{D}^{(k)} f(\tau) = \lim_{\tau \to \tau_0} \frac{f(\tau) - f(\tau_0)}{\tau^k - \tau_0^k}. \tag{5.373}$$

Y. Chen et al. [519] proposed the local derivatives based on the integrals of the difference-quotient (IDQ) or the singular integrals of the difference-quotient (SIDQ). For example, the right SIDQ local derivative is

$$\mathfrak{D}^\alpha f(y) = \frac{1}{\Gamma(1-\alpha)} \lim_{h \to 0_+} \int_0^1 (1-t)^{-\alpha} \frac{f(th+y) - f(y)}{h^\alpha} dt. \tag{5.374}$$

The local fractional derivative is essentially the *fractal* derivative [498, 1274]. In contrast to the purely analytical approach of the fractional calculus, the fractal calculus follows the physical-geometric approach; to avoid the confusion it is suggested to call the latter the *scaled* calculus [2342].

The fractal ("Hausdorff") derivative on the time fractal is defined as [1275]

$$\frac{\partial f}{\partial t^{\sigma}} = \lim_{t_B \to t_A} \frac{f(t_B) - f(t_A)}{(t_B)^{\sigma} - (t_A)^{\sigma}} \tag{5.375}$$

where σ is the fractal dimension of time.

A more general definition is formulated as [128, 506]

$$\frac{\partial^{\tau} f}{\partial t^{\sigma}} = \lim_{t_B \to t_A} \frac{f^{\tau}(t_B) - f^{\tau}(t_A)}{(t_B)^{\sigma} - (t_A)^{\sigma}} \tag{5.376}$$

Since the fractal derivative is the local operator, the numerical solution of the fractal derivative equations can be performed by the standard numerical techniques for the integer-order differential equations [518]. The similar properties have the so called "conformable" fractional derivatives.

"Conformable" Fractional Derivative

Most fractional derivatives do not have the desirable properties [6, 1570, 1604]: the derivative of a constant is not zero; they do not obey the product rule $D^{\alpha}(fg) = fD^{\alpha}(g) + gD^{\alpha}f$; they do not obey the quotient rule $D^{\alpha}(f/g) = (gD^{\alpha}(f) - fD^{\alpha}(g))/g^2$; they do not obey the chain rule $D^{\alpha}(fg) = f^{\alpha}(g(t)g^{\alpha}(t)$; they do not obey in general $D^{\alpha}D^{\beta}f = D^{\alpha+\beta}f$.

R. Khalil et al. [1604] and U. N. Katugampola [1569, 1570] suggested the so called "conformable" limit based [76] derivatives

$$D^{\alpha}f(t) = \lim_{\epsilon \to 0} \frac{f(t + \epsilon t^{1-\alpha}) - f(t)}{\epsilon}, \qquad 0 < \alpha < 1, \tag{5.377}$$

and

$$D^{\alpha}f(t) = \lim_{\epsilon \to 0} \frac{f(te^{\epsilon t^{-\alpha}}) - f(t)}{\epsilon}, \qquad 0 < \alpha < 1. \tag{5.378}$$

Since the conformable derivative is the extension of the classical derivative definition, this derivative obeys the product rule, the quotient rule, the linearity property, the zero derivative for the constant and are valid for some extensions of the classical calculus such as the Rolle's Theorem or Mean Value Theorem [1419].

5.2.4.4 Fractional Fourier Model

S. X. Deng and X. X. Ge [711] studied the steady heat transfer using the two dimensional fractional Helmholtz equation in the fractal media (note that the fractal media cannot be described as the fractals since the main property of the fractal is the non-integer Hausdorff dimension that should be observed on all scales—the fractal structure of the real media cannot be observed on all scales [3021])

$$\frac{\partial^{2\alpha} T}{\partial x^{2\alpha}} + \frac{\partial^{2\alpha} T}{\partial y^{2\alpha}} + k^2 = f, \tag{5.379}$$

where $0 < \alpha \leq 1, 0 < \beta \leq 1$.

J. H. He and F. Liu [1276] and F.-J. Liu et al. [1904] used the He's fractional derivative

$$\frac{\mathrm{d}^\alpha}{\mathrm{d}x^\alpha} f = \frac{1}{\Gamma(n-\alpha)} \frac{\mathrm{d}^n}{\mathrm{d}x^n} \int_{t_0}^{t} [f_0(s) - f(s)]\mathrm{d}s \tag{5.380}$$

where $f_0(s)$ is a known function and the fractional form of the Fourier law

$$\lambda^{2\alpha} \frac{d^\alpha T}{dx^\alpha} = q \tag{5.381}$$

to study the heat transfer in a fractal medium arising in the silk cocoon hierarchy; the structure of the cocoon could be probably duplicated in the biomimetic fabric design.

The similar approach was used by V. D. Beybalaev [277] to investigated the heat conduction in the fractal medium and V. D. Beybalaev et al. [278] to study of the ground freezing.

J. H. He et al. [1282, 1283] used the stationary space-fractional equation

$$\frac{\partial^\alpha}{\partial x^\alpha} \left(\lambda \frac{\partial^\alpha T}{\partial x^\alpha} \right) = 0 \tag{5.382}$$

and Q. L. Wang et al. [3249] the nonstationary equation

$$\frac{\partial T}{\partial t} + \frac{\partial^\alpha}{\partial x^\alpha} \left(\lambda \frac{\partial^\alpha T}{\partial x^\alpha} \right) = 0 \tag{5.383}$$

to study the heat transfer in the fractal medium of the polar bear hair.

R. P. Meilanov and M. R. Shabanova [2110] solved the one-dimensional problems for the space-time fractional equation

$$\frac{\partial^\alpha T}{\partial t^\alpha} = \lambda \frac{\partial^\beta T}{\partial x^\beta} \tag{5.384}$$

where the time nonlocal effects are accounted by the Caputo fractional derivative and the spatial nonlocal effects—by the Riesz derivative

$$\frac{\partial^\beta T}{\partial x^\beta} = \frac{1}{2\Gamma(2-\beta)\cos\left(\frac{\pi}{2}(2-\beta)\right)} \frac{\partial^2}{\partial x^2} \int\limits_0^\infty \frac{T(x',t)}{|x-x'|^{\beta-1}} dx'. \tag{5.385}$$

Earlier these authors considered the Eq. (5.384) with the Riemann–Liouville fractional time derivative $\partial^\alpha T/\partial t^\alpha$ and used the Fourier transform [3057] in the spatial coordinate and the Laplace transform in the time to get the solution [2109].

V. R. Voller et al. [3204] used the time-fractional and R. P. Meilanov et al. [2108] the time-space fractional models to solve the Stefan problem.

Voller et al. considered the cases of both the sharp and diffuse interfaces between the liquid and the solid phases. The authors neglected the surface tension and the kinetic effects and assumed that the heat flux, temperature and the latent heat are zero. Thus the heat balance is

$$-\int\limits_{\Sigma_l} \int\limits_0^t \mathbf{q} \cdot \mathbf{n} dt' dA = \int\limits_{\Omega_l} (\varrho c T + \varrho L) dV, \tag{5.386}$$

where L is the latent heat of fusion, Σ_l is the part of the liquid domain surface that coincides with the total domain surface, $\mathbf{n}$ is the outwards pointing normal.

To account for the memory effects the heat flux is expressed as

$$\mathbf{q} = \frac{1}{\Gamma(\alpha)} \frac{\partial}{\partial t} \int\limits_0^t (t-t')^{\alpha-1} \mathbf{q}^l dt', \qquad 0 < \alpha \le 1, \tag{5.387}$$

where $\Gamma()$ is the gamma function, α is the measure of the memory retention (the value $\alpha = 0$ corresponds to the instantaneous flux).

Inserting (5.387) into (5.386) and taking the Caputo derivative the authors got

$$-\int\limits_{\Sigma_l} \mathbf{q}^l \cdot \mathbf{n} dA = \varrho c \frac{d^\alpha}{dt^\alpha} \int\limits_{\Omega_l} T dV + \int\limits_{\Sigma_l} \varrho L v_f^\alpha dA, \tag{5.388}$$

here v_f^α is the normal "fractional speed" at a point at the phase interface pointing into the solid defined by the relationship

$$\int\limits_{\Sigma_f} v_f^\alpha dA = \frac{1}{\Gamma(1-\alpha)} \int\limits_0^t (t-t')^{-\alpha} \int\limits_{\Sigma_f} \mathbf{v} \cdot \mathbf{n}_f dA dt' \tag{5.389}$$

and after some transformations obtained

$$-\int_{\Sigma_l+\Sigma_f} \mathbf{q}^l \cdot \mathbf{n} dA = \int_{\Omega_l} \varrho c \frac{\partial^\alpha T}{\partial t^\alpha} dV + \int_{\Sigma_f} (\varrho L v_f^\alpha - \mathbf{q}^l \cdot \mathbf{n}) dA. \tag{5.390}$$

The above equations are derived for the case of the sharp phase boundary. In case of the diffuse interface Voller et al. use the enthalpy $H = \varrho c T + \varrho L f$ where the liquid fraction f changes smoothly from $f = 1$ to $f = 0$ across the temperature interval and get

$$-\int_{\Sigma} \mathbf{q}^l \cdot \mathbf{n} dA = \int_{\Omega} \frac{\partial^\alpha H}{\partial t^\alpha} dV. \tag{5.391}$$

To compare the sharp and the diffusive formulations the authors considered the one dimensional problem where the melting is driven by imposing the fixed temperature at $x = 0$:

- The sharp phase boundary: the governing equation is

$$\varrho c \frac{\partial^\alpha T}{\partial t^\alpha} = \lambda \frac{\partial^2 T}{\partial t^2} \tag{5.392}$$

  with the boundary condition at the melt front $x = s(t)$

$$\varrho L \frac{d^\alpha s}{dt^\alpha} = -\lambda \frac{\partial T}{\partial x}. \tag{5.393}$$

- The diffuse phase boundary: the governing equation is

$$\frac{\partial^\alpha H}{\partial t^\alpha} = \lambda \frac{\partial^2 T}{\partial t^2}. \tag{5.394}$$

The memory of the latent heat accumulation is lumped on the interface through the fractional interphase speed $d^\alpha s/dt^\alpha$ in the sharp interface model while in the diffusive model each point in the melt retains the memory of both the sensible and the latent heat—the memory of the latent heat accumulation is distributed.

R. P. Meilanov et al. [2108] considered the nonlocal Stefan problem and solved the Eq. (5.384) written for the solid and the liquid phases coupled by the condition at the front $\zeta(\tau)$ (the dimensionless variables $\xi = x/L$ and $\tau = t/\tau_0$ are used)

$$T_s(\zeta, \tau) = T_l(\zeta, \tau), \tag{5.395}$$

where Q is the rate of heat release during the phase , $\gamma = \beta/2$ [2108].

D. Sierociuk et al. [2853] exploited the time fractional Fourier equation to investigate the heat transfer in the nonuniform semi-infinite beam.

W. Cheng and G. Pang [507] reviewed the existing definitions of the fractional Laplacian $(-\Delta)^s$ [502] (when $s = 2$ the standard Laplacian is recovered) that is the spatial integro-differential operator describing the spatial nonlocality and the power law behaviour in the heat conduction:

- Definition A. The three-dimensional operator ($x \in R^3$) with an approximate finite difference

$$(-\Delta)^s u(x) = c_1(s) \int\limits_{R^3} \frac{\Delta_y^l u(x)}{||y||^{3+s}} dy, \qquad l > s > 0, \tag{5.396}$$

where $\Delta_y^l u(x) = \sum_{k=0}^{l} (-1)^k \binom{l}{k} u(x - ky)$ and $|| \cdot ||$ is the Euclidean norm.

- Definition B. Operator of the Riesz-Marchaud type

$$(-\Delta)^s u(x) = c_2(s) \int\limits_{R^3} \frac{u(y) - u(x)}{||x - y||^{3+s}} dy, \qquad 0 < s < 2 \tag{5.397}$$

- Definition C. Operator that combines the Laplacian

$$(-\Delta)^s u(x) = \begin{cases} -\Delta(I^{2-s} u(x)) & (Riemann - Liuville), \\ I^{2-s}(-u(x)) & (Caputo) \end{cases} \tag{5.398}$$

with the Riesz potential (the inverse operator of the fractional Laplacian [507])

$$I^t u(x) = c_3(s) \int\limits_{R_3} \frac{u(x)}{||x - y||^{3-t}} dy, \qquad 0 < t < 1 \tag{5.399}$$

- Definition D. Operator with directional derivatives ($1 < s \leq 2$)

$$(-\Delta)^s u(x) = \begin{cases} c_4(s) \int\limits_{||\vartheta||=1} D_\vartheta^s u(x) d\vartheta & (Riemann - Liuville), \\ c_4(s) \int\limits_{||\vartheta||=1} D_\vartheta^{\star s} u(x) d\vartheta & (Caputo). \end{cases} \tag{5.400}$$

Cheng and Pang [507] suggested a new implicit definition of the fractional Laplacian using the Riesz fundamental solution, The implicit fractional Laplacian is defined by the integro-differential operator that satisfies the Fourier transform

$$F[(-\Delta)^s u(x), k] = ||k||^s F[u(x), k], \qquad x, k \in R^3 \tag{5.401}$$

where the Fourier transform and the inverse transform are defined as

$$\hat{u}(k) = F(u(x), k) = \int_{R^3} u(x)e^{-i\mathbf{k}\cdot\mathbf{x}}d\mathbf{x} \tag{5.402}$$

and

$$u(x) = F^{-1}(\hat{u}(x), x) = \frac{1}{(2\pi)^3}\int_{R^3} \hat{u}(k)e^{i\mathbf{k}\cdot\mathbf{x}}dk. \tag{5.403}$$

and that has a fundamental solution of the form

$$u^\star(x) = c_6(s)\frac{1}{||x||^{3-s}}, \quad c_6(s) = \frac{\Gamma(\frac{3-s}{2})}{2^s\pi^{3/2}\Gamma(\frac{s}{2})} \qquad x \in R^3 \tag{5.404}$$

where $c_6(s)$ is the normalizing function, Γ is the Euler gamma function.

The Laplacian can be also defined as the pseudo-differential operator given by the Fourier transform

$$\widehat{(-\Delta)^s u}(\xi) = |\xi|^{2s}\widehat{u}(\xi) \tag{5.405}$$

and considered a symmetric differentiation operator of the fractional order $2s$.

5.2.4.5 Nonlinear Diffusivity

K. S. Fa [897] used the Fokker–Planck equation in the Stratonovich approach that follows from the Langevin equation with a multiplicative replacing the probability distribution function by the temperature

$$\frac{\partial T(x,t)}{\partial t} = \varkappa a(t^\star)\frac{\partial}{\partial x}\left(D(x)\frac{\partial}{\partial x}(D(x)T(x,t))\right) \tag{5.406}$$

where $a(t) = 1 - \exp(-t^\star)$.

Fa used the dimensionless variables

$$\xi = \frac{x}{L}, \quad t^\star = \frac{t}{\tau}, \quad \vartheta = \frac{T - T_0}{T_1 - T_0}, \quad \varphi = \frac{q}{C_V(T_1 - T_0)} \tag{5.407}$$

where $\tau = l/v$, l is the mean free path, v is the velocity of sound.

Using the transformations $du/d\xi = 1/D(\xi)$ and $ds/dt^\star = a(t^\star)$, $u(0) = 0$, $s(0) = 0$ the author obtained the solution of the problem of the heat conduction in a slab with the different sides temperatures

$$\vartheta(\xi, t^\star) = \frac{D_0}{D(\xi)} \left[1 - \frac{u(\xi)}{u_1} - \frac{2}{\pi} \sum_{m=1}^{\infty} \frac{\sin\left(\frac{m\pi u(\xi)}{u_1}\right) \exp\left(-\frac{Kn^2 (m\pi)^2 s(t^\star)}{3u_1^2}\right)}{m} \right] \tag{5.408}$$

where D_0 is the value of $D(\xi)$ at $\xi = 0$ and u_1 is the value of $u(\xi)$ at $\xi = 1$, $Kn = l/L$.

The author stated that solution is close to the results of the simulation of the Radiative Phonon Transport model. When $D = 1$ the model by Fa reduces to the model suggested by K. R. Naqvi and S. Waldenstream [2268]

$$\vartheta(\xi, t^\star) = 1 - \xi - \frac{2}{\pi} \sum_{m=1}^{\infty} \frac{\sin(m\pi\xi) \exp\left(-\frac{Kn^2 (m\pi)^2 s(t^\star)}{3}\right)}{m}. \tag{5.409}$$

F. Falcini et al. [915] used the limiting case of the Stefan problem—the moisture infiltration into the porous medium—to study three effects that can extend the applicability of the Fourier equation to account for the anomalous heat diffusion:

- time nonlocality (memory)
- space nonlocality
- non-linear spatial dependent thermal diffusivity

and action of their combinations.

The authors considered the heat conduction equation

$$\frac{\partial^\gamma T}{\partial t^\gamma} = \frac{\partial}{\partial x} \left(\varkappa(x) \frac{\partial^\alpha T}{\partial x^\alpha} \right) \tag{5.410}$$

where $\alpha \quad (1 < \alpha \leq 1)$ is the degree of the space nonlocality, $\gamma \quad (1 < \gamma \leq 1)$ is the measure of the memory effect, and the nonlinear diffusivity is described as

$$\varkappa(x) = \varkappa_0 x^{\frac{2\beta - 1}{\beta}}, \tag{5.411}$$

thus $\beta \quad (1 < \beta \leq 1)$ could be considered as a measure of nonlinearity.

The relation between the characteristic time and length scales in the normal heat conduction (normal diffusion) $l \sim \sqrt{\tau}$. In the general anomalous case this relation is written as $l \propto \tau^n$ where $n \neq 1/2$.

F. Falcini et al. [915] showed that three non-Fourier effects combine and compete to determine the time exponent

$$n = \frac{\gamma}{1 + \alpha + \left(\frac{1}{\beta} - 2\right)}. \tag{5.412}$$

Thus the deviations from the Fourier law behaviour could be described by the different non-Fourier effects, for example, the roles of the fractional derivatives and the suitable nonlinearity of the thermal diffusivity are interchangeable [915] and the anomalous diffusion effects could be obtained by the different combinations:

- subdiffusion ($n < 1/2$):
  - memory only
  - nonlinearity only, $\beta \in (0, 0.5]$
  - memory + nonlocality, $1 + \alpha > 2\gamma$
  - memory + nonlinearity, $\beta\gamma < 1/2$
  - nonlocality + nonlinearity, $\alpha + 1/\beta > 3$
- superdiffusion ($n > 1/2$):
  - nonlinearity only, $\beta \in (0.5, 1]$
  - memory + nonlocality, $1 + \alpha < 2\gamma$
  - memory + nonlinearity, $\beta\gamma > 1/2$
  - nonlocality + nonlinearity, $3 > \alpha + 1/\beta > 2$
- superdiffusion+ ($n > 1$):
  - nonlocality + nonlinearity, $\alpha + 1/\beta < 2$

5.2.4.6 Fractional Pennes Equation

The time-fractional generalization ($\partial^\alpha/\partial t^\alpha$ is the Caputo fractional derivative of order $\alpha \in (0, 1]$ T, α is called the *order of fractionality*) of the Pennes bioheat Eq. (5.8)

$$\varrho c \frac{\partial^\alpha T}{\partial t^\alpha} = \nabla \cdot \lambda \nabla T + c_\text{b}\omega_b(T_\text{a} - T) + \dot{q}_\text{met} + Q^\text{ext}, \tag{5.413}$$

was used by R. S. Damor et al. [669] to investigate the hyperthermia and the anomalous diffusion in the skin tissue with the constant and the sinusoidal heat flux at the boundary [668, 670] and by M. Ezzat et al. [887] to study the temperature transient in the skin exposed to the instantaneous surface heating. M. Ezzat et al. [896] investigated the tissue responses to the fractional transient heating with the sinusoidal heat flux condition on the skin surface.

As noted by L. L. Ferras et al. [952], the Eq. (5.413) is not dimensionally consistent and one has to either redefine the coefficients of these equations or to introduce the factor $\tau^{1-\alpha}$ to get a "new" thermal conductivity.

J. Singh et al. [2869] used the space-time fractional bioheat equation

$$\varrho C \frac{\partial^{\beta}}{\partial^{\beta}} = \lambda \frac{\partial^{\alpha}}{\partial^{\alpha}} + Q_p, \qquad 0 < \beta \leq 1 < \alpha \leq 2 \tag{5.414}$$

to study the heat transfer in the tissues during the thermal therapy.

5.2.4.7 Fractional Cattaneo and SPL Models

The hyperbolic models of the non-Fourier heat conduction suffer from the unrealistic singularity of the temperature gradient across the wavefront; the fractional calculus for differentiation and integration can remove the thermal wave singularity [42]. The fractional version of the Cattaneo equation is also called the "nonlocal" Cattaneo-Vernotte equation, reflecting the basic properties of the fractional derivatives [406].

L. Liu et al. [1939] used the modification of the Cattaneo model by Christov to develop the space fractional equation with the Riesz derivative.

A time-fractional SPL (FSPL) model for the bioheat transfer is formulated as [3445]

$$\varrho c \left(\frac{\partial T}{\partial t} + \tau \frac{\partial^{1+\alpha} T}{\partial t^{1+\alpha}} \right) = \nabla \cdot \lambda \nabla T + c_{\mathrm{b}} \omega_b (T_{\mathrm{a}} - T) + \dot{q}_{\mathrm{met}} + Q^{\mathrm{ext}}. \tag{5.415}$$

Computations show that the fractional SPL equation give the same temperature distribution as the DPL model [3445].

The space-fractional SPL model was formulated for the one dimensional case by P. Kumar et al. [1732]

$$\varrho c \frac{\partial T}{\partial t} = - \frac{\partial^{\alpha} q}{\partial x^{\alpha}} + c_{\mathrm{b}} \omega_b (T_{\mathrm{a}} - T) + \dot{q}_{\mathrm{met}} + Q^{\mathrm{ext}}, \tag{5.416}$$

where

$$q(x,t) + \tau \frac{\partial q(x,t)}{\partial x} = -\lambda \frac{\partial q(x,t)}{\partial x}. \tag{5.417}$$

M. Fabrizio [898] formulated the fractional Cattaneo equation as

$$\frac{\gamma \alpha}{\Gamma(1-\alpha)} \int\limits_{-\infty}^{t} \frac{\mathbf{q}(x,t) - \mathbf{q}(x,\tau)}{(t-\tau)^{1+\alpha}} \mathrm{d}\tau = \mathbf{q}(x,t) + \lambda \nabla T(x,t). \tag{5.418}$$

X. Jiang and H. Qi [1471] derive the fractional thermal wave model of the bioheat transfer changing the Cattaneo relation to

$$\frac{\tau^{\alpha}}{\alpha!} D_t^{\alpha} \mathbf{q} + \mathbf{q} = -\lambda \nabla T, \tag{5.419}$$

where D_t^{α} is a modified Riemann–Liouville derivative of the order α.

H. T. Qi et al. [2531] studied the laser heating generalizing the Cattaneo relation as

$$\tau^p D_t^p \mathbf{q} + \mathbf{q} = -\lambda \nabla T; \qquad 0 < p < 1, \tag{5.420}$$

where D_t^p is the Caputo derivative of order p; the factor τ^p is introduced to keep the dimensionality in order.

Y. Z. Povstenko [2494, 2495] and X. Y. Jiang and M. Y. Xu [1473] proposed the time fractional Fourier law as the constitutive relationship

$$\mathbf{q} = -\varkappa D^{1-\alpha} \nabla T \tag{5.421}$$

where $D^{1-\alpha}$ the Caputo fractional derivative. Y. Z. Povstenko [2496] showed that many fractional generalizations of the Cattaneo relation could be obtained from the time-nonlocal generalization where the kernels are functions of the Mittag–Leffler type.

H. Xu et al. [3386] formulated the fractional Cattaneo equation using two fractional derivatives (Caputo) of the different order. The authors started with the generalized constitutive equation

$$\frac{\partial^{\beta-1} \mathbf{q}}{\partial t^{\beta-1}} + \tau \frac{\partial^{\alpha-1} \mathbf{q}}{\partial t^{\alpha-1}} = -\lambda \nabla T, \qquad 0 < \beta \leq \alpha \leq 2. \tag{5.422}$$

and using the Laplace transform obtained the time-nonlocal constitutive relation

$$\mathbf{q}(t) = -\frac{\lambda}{\tau} \int_0^t (t-t')^{\alpha-2} E_{\alpha-\beta,\alpha-1} \left(-\frac{(t-t')^{\alpha-\beta}}{\tau} \right) \nabla T(t') dt' \tag{5.423}$$

where the generalized Mittag–Leffler function is defined as

$$E_{\mu,\nu}(z) = \sum_{k=0}^{\infty} \frac{z^k}{\Gamma(\mu k + \nu)}. \tag{5.424}$$

The Mittag–Leffler function E_{μ} reduces to the exponential function for $\mu = 1$. It provides the interpolation between the stretched exponential pattern and the inverse power behaviour [2123]

$$E_\mu(-\lambda_n t^\mu) \propto \begin{cases} \exp\left(-\dfrac{\lambda_n t^\mu}{\Gamma(1-\mu)}\right) & t \ll \lambda^{\frac{1}{\mu}} \\ (\lambda_n t^\mu \Gamma(1-\mu))^{-1} & t \gg \lambda^{\frac{1}{\mu}} \end{cases} \tag{5.425}$$

H. Xu et al. [3386] wrote the generalized Cattaneo heat equation as

$$\frac{\partial^\beta T}{\partial t^\beta} + \frac{\partial^\alpha T}{\partial t^\alpha} = \mathcal{D}\Delta T, \tag{5.426}$$

where the generalized thermal diffusivity $\mathcal{D}$ has dimension $[\mathcal{D}] = m^2/s^\beta$.

J. Hristov [1367, 1368] developed the space-fractional equation for the transient heat conduction with the damping term expressed via the Caputo–Fabrizio fractional derivative [438] that is the extension of the Caputo fractional derivative with the singular kernel

$$D_t^\alpha f(t) = \frac{M(\alpha)}{1-\alpha}\int_0^t exp\left[-\frac{\alpha(t-s)}{1-\alpha}\right]\frac{df(t)}{dt}ds, \tag{5.427}$$

where $M(\alpha)$ is a normalization function such as $M(0) = M(1) = 1$. Numerical solution of this problem was considered by B. S. Alkahtani and A. Atangana [56].

X. J. Yang et al. [3404] (see also the paper [3406]) for the investigation of the fractional heat transfer introduced a new fractional derivative without the singular kernel that is the extension of the Riemann–Liouville fractional derivative with the singular kernel

$$D_{\alpha+}^{(\nu)} T = \frac{\Re(\nu)}{1-\nu}\frac{d}{dx}\int_a^x \exp\left(-\frac{\nu}{1-\nu}(x-x')\right)Tdx', \tag{5.428}$$

where $\Re(\nu)$ is the normalization function.

H. R. Ghazizadeh and M. Maerafat [1069] used the generalized Taylor's formula developed by Z. M. Odibat and N. T. Shawagfeh [2327]

$$q(\mathbf{r}, t+\tau) = q(\mathbf{r}, t) + \frac{\tau^\alpha}{\Gamma(1+\alpha)}, \quad 0 < \alpha \leq 1. \tag{5.429}$$

Inserting this equation into the energy balance equations and eliminating the heat flux the authors get the "nonlocal" Cattaneo equation

$$\varrho C\frac{\partial T}{\partial t} + \frac{\varrho C\tau^\alpha}{\Gamma(1+\alpha)}\frac{\partial^{\alpha+1} T}{\partial t^{\alpha+1}} = \lambda\nabla^2 T. \tag{5.430}$$

H. R. Ghazizadeh et al. [1068] employed the FSPL model for the simulation of the non-Fourier conduction in processed meat.

Y. Povstenko and Y. Povstenko [2497] studied the fractional heat conduction in the composite medium consisting of the spherical inclusion. The heat conduction was described by the time-fractional heat conduction equation with the Caputo derivative; the Laplace transform was used to solve equations.

W. Zhang et al. [3489] investigated the one-dimensional heat conduction problem where two cold waves collide in a layer and compare the solutions of the Fourier equation, the hyperbolic heat equation and the fractional heat equation.

B.Yu et al. [3445] formulated a fractional thermal wave model for a bi-layered spherical composite medium tissue. The authors the estimated relaxation parameters and the Caputo fractional derivative using the inverse analysis with the Levenberg–Marquardt method.

P. Goudarzi and A. Azimi [1137] considered two problems: the three-layer skin tissue in contact with a hot water source and the single-layer skin tissue exposed suddenly to a heat source generated by a laser. In both cases the super-diffusion fractional non-Fourier model is used to simulate the heat transfer diffused through the skin tissue.

The authors used the generalised Cattaneo relation (5.137) to write the fractional single-phase-lag equation in the form

$$\rho c \frac{\partial T}{\partial t} + \rho c \tau_q^\alpha \frac{\partial^{1+\alpha} T}{\partial t^{1+\alpha}} = \lambda \nabla^2 T + \tau^\alpha \frac{\partial^\alpha q_g}{\partial t^\alpha} + q_g, \tag{5.431}$$

$$q_g = w_b \rho_b c_b (T_b T) + q_{met} + q_{ext}. \tag{5.432}$$

The authors that the FSPL model's results indicate that this model encompasses all the results of the dual-phase-lag model.

5.2.4.8 Fractional DPL Model

C. C. Ji et al. [1467] derived the fractional DPL (FDPL) model starting with the one dimensional version of the original general definition (5.143)

$$q(x, t + \tau_q) = -\lambda T_x(x, t + \tau_T) \tag{5.433}$$

and using the fractional Taylor series expansion [2327, 2531] up to the first two terms as

$$q(x, t + \tau_q) = q(x, t) + \frac{\tau_q^\alpha}{\Gamma(1 + \alpha)} D_\star^\alpha q(x, t) + \dots \tag{5.434}$$

and

$$T_x(x, t + \tau_T) = T_x(x, t) + \frac{\tau_T^\alpha}{\Gamma(1+\alpha)} D_\star^\alpha T_x(x, t) + \ldots . \tag{5.435}$$

where $0 < \alpha < 1$ and $D_\star^\alpha$ denotes the Caputo fractional derivative of the order α to get

$$q(x, t) + \frac{\tau_q^\alpha}{\Gamma(1+\alpha)} D_\star^\alpha q(x, t) = -\lambda \left(T(x, t) + \frac{\tau_T^\alpha}{\Gamma(1+\alpha)} D_\star^\alpha T(x, t) \right)_x . \tag{5.436}$$

and finally

$$\varrho C \left(\frac{\partial T}{\partial t} + \frac{\tau_q^\alpha}{\Gamma(1-\alpha)} D_\star^\alpha T \right) = \lambda \left(T + \frac{\tau_T^\alpha}{\Gamma(1-\alpha)} D_\star^\alpha T \right) . \tag{5.437}$$

H. Y. Xu and X. Y. Jiang [3373] studied the bioheat transfer using the following equation in the Caputo derivatives of the orders α and β

$$\mathbf{q} + \tau_q^\alpha \frac{\partial^\alpha \mathbf{q}}{\partial t^\alpha} = -\lambda \left(\nabla T + \tau_T^\beta \frac{\partial^\beta}{\partial t^\beta} \nabla T \right) . \tag{5.438}$$

The authors have changed the relaxation times of the DPL model τ_q and τ_T to τ_q^α and τ_T^β to maintain the dimensionality in order.

Xu et al. used the non-linear least-square method to estimate simultaneously two relaxation times and two or up to degrees of fractionality.

D. Kumar and K. N. Rai [1725] solved the time fractional DPL model to study the heat transfer within the skin tissue during the thermal therapy. To reduce the time fractional equation into the system of the ordinary differential equations, the spatial discretization in space is applied. Then the time fractional ODEs are converted into the Sylvester matrix equation by using the finite element Legendre wavelet Galerkin method with utilization of the block pulse function in terms of the Caputo fractional order derivative.

A. M. Famy [904] proposed a new hybrid algorithm based on integrating the local radial basis function collocation method and the general boundary element method for studying the time fractional DPL problems in the functionally graded tissues. The author claims that the general solution, T, of the time fractional DPL equation is the sum of the T_f , the solution of the fractional-order governing equation without the dual phase lags, and T_d, the solution of the DPL governing equation without the fractional derivative $T = T_f + T_d$.

L. Liu et al. [1940] used the convected derivative introduced by I. Christov [562] and the Caputo fractional derivative of the order α to formulate the model

$$\tau_q \left[\frac{\partial^\alpha \mathbf{q}}{\partial t^\alpha} + \mathbf{v} \cdot \nabla \mathbf{q} + \mathbf{q} \cdot \nabla \mathbf{v} + (\nabla \cdot \mathbf{v}) \mathbf{q} \right] + \mathbf{q} = -\lambda \nabla \left(1 + \tau_T \frac{\partial^\alpha}{\partial t^\alpha} \right) T. \tag{5.439}$$

5.2.4.9 Fractional TPL Model

M. A. Ezzat et al. [893, 894] obtained the fractional TPL model (FTPL) taking the Taylor series expansion on the both sides of the constitutive relation of the TPL model (5.181) and retained the terms up to $2\alpha_F$-order for the relaxation time τ_q and up to α_F-order for τ_T and τ_v to get

$$\left(1+\frac{\tau_q^{\alpha_F}}{\alpha_F!}\frac{\partial^{\alpha_F}}{\partial t^{\alpha_F}}+\frac{\tau_q^{2\alpha_F}}{(2\alpha_F)!}\frac{\partial^{2\alpha_F}}{\partial t^{2\alpha_F}}\right)\mathbf{q}=-\left[\left(\tau_v^\star+\lambda\frac{\tau_T^{\alpha_F}}{\alpha_F!}\frac{\partial^{\alpha_F}}{\partial t^{\alpha_F}}\right)\nabla T+\lambda^\star\nabla v\right] \tag{5.440}$$

where

$$0<\alpha\leq 1, \tag{5.441}$$

$$\tau_v^\star=\lambda+\lambda^\star\frac{\tau_v^{\alpha_v}}{\alpha_F!}\frac{\partial^{\alpha_F-1}}{\partial t^{\alpha_F-1}}. \tag{5.442}$$

The fractional TPL model as well as the ordinary TPL one was used to study the problems of the thermoelasticity [893, 894] and the piezoelectric thermoelasticity problems. M. A. Ezzat et al. [893] use the energy conservation equation for the homogeneous isotropic thermoelastic solid

$$-\nabla\cdot\mathbf{q}+\varrho Q=\varrho C_E\dot{T}+\gamma T_0\dot{e}_{ii} \tag{5.443}$$

where Q is the heat source, C_E is the specific heat at constant strain, e_{ii} are the components of the strain tensor, $\gamma=(3\nu+2\mu)\alpha_T$, ν, μ are the Lame's constants, α_T is the coefficient of the linear thermal expansion.

Taking the divergence and time derivative of the Eq. (5.440) Ezzat et al. got

$$\left(1+\frac{\tau_q^{\alpha_F}}{\alpha_F!}\frac{\partial^{\alpha_F}}{\partial t^{\alpha_F}}+\frac{\tau_q^{2\alpha_F}}{(2\alpha_F)!}\frac{\partial^{2\alpha_F}}{\partial t^{2\alpha_F}}\right)\nabla\cdot\dot{\mathbf{q}}$$
$$=-\left[\left(\tau_v^\star+\lambda\frac{\tau_T^{\alpha_F}}{\alpha_F!}\frac{\partial^{\alpha_F+1}}{\partial t^{\alpha_F+1}}\right)\nabla^2\dot{T}+\lambda^\star\nabla^2\dot{v}\right]. \tag{5.444}$$

and finally obtain the modified fractional heat transport equation

$$\left(1+\frac{\tau_q^{\alpha_F}}{\alpha_F!}\frac{\partial^{\alpha_F}}{\partial t^{\alpha_F}}+\frac{\tau_q^{2\alpha_F}}{(2\alpha_F)!}\frac{\partial^{2\alpha_F}}{\partial t^{2\alpha_F}}\right)(\varrho C_E\ddot{T}+\gamma T_0\ddot{e}-\varrho\dot{Q})=$$
$$\tau_v^\star\nabla^2\dot{T}+\lambda\frac{\tau_T^{\alpha_F}}{\alpha_F!}\frac{\partial^{\alpha_F+1}}{\partial t^{\alpha_F+1}}\nabla^2T+\lambda^\star\nabla^2T. \tag{5.445}$$

The authors consider several limiting cases:

- When $\alpha_F = 1$ and the thermal conductivity λ is much smaller than $\lambda^\star$, $q = -\lambda^\star \nabla v$, $\tau_v^\star \lambda^\star \tau_v$ and (neglecting τ_q^2) the Eq. (5.445) simplifies to

$$\left(1 + \tau_q \frac{\partial}{\partial t}\right) (\varrho C_E \ddot{T} + \gamma T_0 \ddot{e} - \varrho \dot{Q}) = \lambda^\star \left(1 + \tau_v \frac{\partial}{\partial t}\right) \nabla^2 T. \tag{5.446}$$

This equation is an *extension* of the Green-Naghdi theory of type II. The heat conduction equation of the Green-Naghdi theory of type II [1156]

$$\varrho C_E \ddot{T} + \gamma T_0 \ddot{e} - \varrho \dot{Q} = \lambda^\star \nabla^2 T \tag{5.447}$$

is obtained if $\tau_T = \tau_v = \tau_q = 0$.

- When $\alpha_F = 1$, $\tau_T = \tau_v = \tau_q = 0$ (hence $\tau_v^\star = \lambda$) the Eq. (5.445) reduces to the heat conduction equation of the Green-Naghdi theory of type III [560, 892]

$$\varrho C_E \ddot{T} + \gamma T_0 \ddot{e} - \varrho \dot{Q} = \lambda^\star \nabla^2 T + \lambda \nabla^2 T. \tag{5.448}$$

- When $\alpha_F = 1$ and $\lambda^\star = 0$ the Eq. (5.445) reduces to the heat conduction equation of the DPL model [890, 3110].
- When $\alpha_F = 1$ and $\lambda^\star = 0$, $\tau_T = \tau_v = 0$, $\tau_q > 0$, the Eq. (5.445) reduces (neglecting τ_q^2) to the Lord-Shulman model [1955].
- When $\alpha_F = 1$ the Eq. (5.445) reduces to the heat conduction equation of the TPL model.
- When $0 < \alpha_F \leq 1$, $\tau_T = \tau_v = 0$ and neglecting $\tau_q^{2\alpha_F}$ the fractional model due to H. H. Sherief et al. [2816] is obtained.

5.2.4.10 Nonlocal Fractional TPL Model

A. H. Akbarzadeh et al. [42] suggested the nonlocal fractional three-phase-lag (NL FTPL) heat conduction model. The constitutive equation of the NL FTPL heat conduction is written as

$$\begin{aligned} \mathbf{q}(\mathbf{r} + \boldsymbol{\lambda}_q, t + \tau_q) = -[&\lambda J^{\alpha_F - 1} \nabla T(\mathbf{r} + \boldsymbol{\lambda}_T, t + \tau_T) \\ &+ \lambda^\star J^{\alpha_F - 1} \nabla v(\mathbf{r} + \boldsymbol{\lambda}_v, t + \tau_v)] \end{aligned} \tag{5.449}$$

where $0 \leq \alpha_F \leq 2$, J^α is the Riemann–Liouville fractional integral.

Alternative forms of the constitutive Eq. (5.449) can be derived by the Taylor series expansion of the correlation lengths $\lambda_q, \lambda_T, \lambda_v$ in the space domain and of the relaxation times τ_q, τ_T, τ_v in the time domain.

The heat conduction equation is obtained by the elimination of the heat flux from the constitutive equation and the energy conservation equation. In the case of the second order Taylor expansion for both the correlation lengths and the relaxation times it is written (in the absence of the volumetric heat source) as [42]

$$\left(1+\boldsymbol{\lambda}_q\cdot\nabla+\left(\frac{\boldsymbol{\lambda}_q\cdot\boldsymbol{\lambda}_q}{2}\right)\nabla^2+\tau_q\frac{\partial}{\partial t}+\frac{1}{2}\tau_q^2\frac{\partial^2}{\partial t^2}\right)\left(\frac{1}{\varkappa}\tau_q^2\frac{\partial^2 T}{\partial t^2}\right)=\nabla\cdot\left(F_T+\frac{\lambda^\star}{\lambda}F_v\right)J^{\alpha-1}\nabla T, \quad (5.450)$$

where

$$F_T=1+\boldsymbol{\lambda}_T\cdot\nabla+\left(\frac{\boldsymbol{\lambda}_T\cdot\boldsymbol{\lambda}_T}{2}\right)\nabla^2+\tau_T\frac{\partial}{\partial t}+\frac{1}{2}\tau_T^2\frac{\partial^2}{\partial t^2} \quad (5.451)$$

and

$$F_v=1+\boldsymbol{\lambda}_v\cdot\nabla+\left(\frac{\boldsymbol{\lambda}_v\cdot\boldsymbol{\lambda}_T}{2}\right)\nabla^2+\tau_v\frac{\partial}{\partial t}+\frac{1}{2}\tau_v^2\frac{\partial^2}{\partial t^2}. \quad (5.452)$$

A. H. Akbarzadeh et al. [42] also derived the wave-like NL FTPL for the homogenous medium by simplifying the NL FTPL heat conduction Eq. (5.450) using the first-order Taylor series expansion of λ_q in space and the second-order Taylor series expansion of τ_q and the first-order Taylor series expansion of τ_T and τ_v in time:

$$\left(1+\boldsymbol{\lambda}_q\cdot\nabla+\tau_q\frac{\partial}{\partial t}+\frac{1}{2}\tau_q^2\frac{\partial^2}{\partial t^2}\right)\left(\frac{1}{\varkappa}\tau_q^2\frac{\partial^2 T}{\partial t^2}\right)=\left[1+\tau_T\frac{\partial}{\partial t}+\frac{\lambda^\star}{\lambda}\left(1+\tau_v\frac{\partial}{\partial t}\right)\right]J^{\alpha-1}\nabla^2 T.$$

The authors noted that the relaxation times τ_q, τ_T, τ_v could present the delayed thermal responses due to the collisions of electrons and phonons, the temporary momentum loss, the normal relaxation in the phonon scattering, and the internal energy relaxation [40, 3112] and the time-fractional derivative removes the thermal wave singularity across the thermal wavefront and is used to present the subdiffusion ($0\le\alpha<1$), normal diffusion ($\alpha=1$), and superdiffusion phenomena ($1<\alpha<2$) [3439].

The NL FTPL heat conduction shows that the nonlocal length of heat flux in the absence of heat generation only contributes in the mixed derivative term implying the negligible effect of heat flux nonlocality on the thermal responses for long-term and steady-state heat transport analysis and the reduction of thermal behaviour of NL FTPL to transient Fourier heat conduction [39, 40].

5.3 Vascular Models

The concept of discrete vascular model was developed by J. Lagendijk [1754] with the goal to study the temperature distribution around large blood vessels in the heated tissues during the hyperthermia. His observation showed that the in larger vessels within the heated tissue causes local nonuniformity in the temperature distribution.

The vascular models describe the heat transfer between the blood in individual vessels and tissue. No assumption is made *a priori* about the possible relation between these two temperatures—the full range of the thermal equilibration regimes can be reproduced. The modelling of physiological phenomena such as vasodilation and vasoconstriction, blood viscosity changes etc. can be done in a more straightforward manner than in the case of the continuum models [2929].

V. V. Gafiychuk and I. A. Lubashevsky [1030] stressed that in the living tissue vessels make up a highly branching network and the in the vessels of one level are related to in all other vessels. Thus, in order to describe the influence of the blood flow on the heat propagation the vascular network should be taken into account as a whole rather than in terms of the individual vessels—one can speak about the cooperative response.

V. V. Gafiychuk and I. A. Lubashevsky introduce the parameter n_{cc} which divides all vessels into two groups:

- the arteries and veins of the first group whose level number $n \leq n_{cc}$ are located in the immediate vicinity of each other and form the countercurrent pairs;
- the arteries and veins of the second group for which $n > n_{cc}$ are formally not considered to correlate in the orientation.

The authors assumed that all vessels of the level n and the blood flows in them are equivalent and are described by the same set of parameters, such as their length $l_n = l_0 2^{-n}$, the radius a_n and the mean velocity v_n of the blood in these vessels.

Due to the blood current conservation law at the branching points for the given vascular network these parameters satisfy the relation

$$\pi a_n^2 v_n M_n = J_0 \tag{5.453}$$

where $M_n = 2^{3n}$ is the total number of the arteries or veins belonging to the level n, and J_0 is the total blood current in the host artery or vein. Each countercurrent pair of level n is also characterized by the distance b_n between its vessels; V. V. Gafiychuk and I. A. Lubashevsky assumed that the ratio b_n/a_n is a constant $b_n/a_n = \sigma, \sigma \geq 2$.

The last level vessels of the arterial and venous beds (called arterioles and venules) are interconnected by the capillary system: each arteriole branches out into $m(m \gg 1$ capillaries forming a "chimney brush" structure [1030]. Each capillary is described by the set of the parameters l_c, a_c, v_c obeying a relation similar to (5.453).

In order to estimate the number of capillaries in the of region of size of order $\mathcal{R}$ the authors assumed that

- each arteriole is joined to inside its neighbourhood of order $\mathcal{R}$ by an approximately equal number of capillaries;
- each capillary keeps its spatial direction within the scale $\lambda \geq l_N$, is the mean curvature radius of capillary regarded as a line;
- on scales above λ different parts of a capillary are oriented independently from one another.

Thus, every arteriole and venule spaced at distance r are joined to each other, on the average, by $m(r)$ capillaries

$$m(r) = \frac{V_N m}{(2\pi \mathcal{R})^{3/2}} \exp\left\{-\frac{r^2}{\mathcal{R}}\right\} \tag{5.454}$$

where $V_N = (2l_N/\sqrt{3})^3$ is the volume of the fundamental domain of the last level.

Gafiychuk and Lubashevsky estimated the distance between the arteries (veins) of the level n as

$$d_n = \left(\frac{2}{\sqrt{3}}\right)^{3/2} l_n \tag{5.455}$$

and between the capillaries as

$$d_c = \left(\frac{2}{\sqrt{3}}\right)^{3/2} \left(\frac{l_N^3}{ml_c}\right). \tag{5.456}$$

In the vascular models the blood vessels are represented as the rigid tubes buried in the tissue; the complicate vascular geometry allows one only consider several blood vessels and neglect the others. The blood flow in the large, thermally unequilibrated vessels is the main cause for temperature nonhomogeneity during the hyperthermia treatment [647, 791]. The large blood vessels may significantly cool the tissue volumes around them, making it very difficult to cover the whole tumour volume with the therapeutic thermal exposure [3525].

The vascular models differ greatly by their complexity [2928]. There are models based on the concrete anatomical information. It is possible to explicitly consider the contribution to the heat transfer of the large blood vessels either by the specification of the vessel's wall temperature [2509, 2848] or by the solution of the conjugated problem for the blood flow in the vessel. The latter could be accomplished using the Navier–Stokes equations of the viscous incompressible flow [2509, 2848] or assuming that the analytic model of the Poiseuille flow [2984] is accurate enough for the problem in question. The high level of the complexity results in the high computational power needed to perform the calculations on such models. As a rule, the specific features of the blood rheology are disregarded [640, 717, 720, 2169, 2828, 3480].

H. P. Kok et al. [1669] suggested to divide the discrete vessel models into three categories:

- Basic discrete vessel models
  - Basic heat transfer studies;
  - Predicts the most significant cooling effects by large vasculature.
- Networks of straight vessels
  - More realistic heat transfer studies;
  - Predicts more realistic significant cooling effects by large vasculature.
- Realistic vessel networks
  - Fully realistic hyperthermia planning;
  - Evaluation of hpothermia effects.

A simple description of the heat flux q from the blood with the temperature T_b flowing through the vessel of diameter d to the cylindrical region (diameter D) of the tissue having the temperature T is possible via the overall heat transfer coefficient h as $q(s) = h\pi d(T(s) - T_b)$, where s is the coordinate along the vessel centreline. This coefficient is expressed via the so called conduction shape factor $h\pi d = \lambda_t\sigma$ where [157] $\sigma = 2\pi/\ln(D/d)$. Similarly, the heat exchanged between the adjacent arteries and veins with the countercurrent flow is written with the corresponding conduction shape factor [157]

$$\sigma_\Delta = \frac{2\pi}{\cosh^{-1}\left(\frac{w^2 - r_a^2 - r_v^2}{2r_a r_v}\right)} .$$

Similar relation is used for the description of heat transfer inside the blood vessels $q(s) = h\pi d(T_w - T_b)$, where T_w is the temperature of the vessel's wall.

The value of this coefficient is frequently determined from the empirical relations $\mathrm{Nu} = \mathrm{Nu}(\mathrm{Gz})$ between the Nusselt number

$$\mathrm{Nu} = \frac{hd}{\lambda_b}$$

and the Graetz number [157] (this similarity criterium rarely used in the "non-biological" fluid dynamics is expressed through the more common ones—the Reynolds and Prandtl numbers—as $\mathrm{Gz} = d\mathrm{RePr}/L_0$)

$$\mathrm{Gz} = \frac{\varrho_b c_b u_0 d^2}{\lambda_b L_0} .$$

The relations Nu = Nu(Gz) are usually obtained by the approximation of the experimental data for the flow of the Newtonian incompressible fluids in the straight pipes of the circular cross section. The blood vessels in the living organisms are more complex objects, to say nothing about the blood rheology. Evidently, the heat transfer description based on the shape conduction factors (determined also for other geometrical arrangements such as the vessel near the skin surface or near a junction of vessels [157]) is rather crude and more sophisticated simulation is necessary.

The large blood vessels embedded into the tumour region through the contribution to the heat transfer by the blood flow could lead to the significant temperature non-uniformity around the vessel and the possible thermal under-dosage. Moreover, the heat transfer along the vessel may result in the damage of the surrounding healthy tissues. This problem is common for the cryosurgery and the hyperthermia [1702, 2828, 3099] with an obvious change between the "cooling" and "heating". J. Crezee and J.J.W. Lagendijk [648] and M.C. Kolios et al. [1673] were the first to study this issue.

M. C. Kolios et al. used the finite difference method to compute the temperature distribution in the cylindrical blood vessels and in the surrounding tissue shape as a co-axial cylinder of a larger diameter. The authors showed, inter alia, that the contribution of the blood flow in the central vessel to the global heat transfer in the considered region does depend on the blood perfusion level in the tissue (modelled as a source term in the Pennes equation (5.8)).

M. C. Kolios et al. in the subsequent paper [1675] reported the results of the experimental study of the temperature gradients in the biological tissues (the porcine kidney) near the large vessels heated either locally via the needle perfused with the hot water or using the distributed ultrasound source. The authors have measured the temperature gradients up to 6° C/mm and found that the effect of the vessel on the heat transfer is primarily depends on its orientation with respect to the heat source. Some of the authors conclusions are hardly unexpected—fo example, the observation that the temperature gradients caused by the local heating are larger than those caused by the distributed heat source. X. Chen and G. M. Saidel [514] studied the thermal processes that occur during the radio frequency ablation of a tissue near a large blood vessel, which should not be damaged. The authors noted that the optimal estimation model parameters, especially those associated with changes in the perfusion distribution, require support by the MRI image data. The limiting factor now is not the speed of the model simulation, but the time required for the image processing.

A. N. T. J. Kotte et al. [1700, 1812] suggested to consider the discrete vessels (more exactly, the segments of the vessels) as the geometrical objects that could be characterized by a three dimensional parametric curve with the associated diameters and consider a simplified analytical model of the heat transfer along the vessel neglecting the longitudinal diffusion of heat and assuming the circular cross section. It is also suggested to characterize the blood by the so called “mixing cup temperature” that is defined as [1700]

$$T_{\text{mix}}(u)\pi R^2 = \int_0^R u(r)T(r)2\pi r\mathrm{d}r,$$

where $T(r)$ is the real blood temperature, and to define the heat exchange between the tissue and the blood using Nusselt number.

The experimental imitation of the heat transfer around the blood vessel embedded into tissue was performed by Y. T. Zhang et al. [3480] who used the pre-frozen in liquid nitrogen massive copper block to freeze the gel volume that contained two tubes with the countercurrent warm water flow. The author have also derived an analytic solution for the heat transfer in the cylindrical tissue with a single embedded vessel (see Sect. 7.10).

Z.-S. Deng and J. Liu [717] have considered two models—with the single artery transiting the tumour (SATT) and with the counter-current flow vessels transiting the tumour (CVTT). The authors have neglected the conduction along the blood flow inside the vessel and assumed that heat transfer between the blood and the surrounding tissues could be described by the heat transfer coefficient $h = Nuk_b/D$, where D is the diameter of the vessel and the Nusselt number Nu is considered to be constant. Then the temperature of the blood T is governed by the equation:

$$\varrho_b c_b \frac{\partial T}{\partial t} = \frac{hP}{S}(T\mathrm{w} - T) - \varrho_b c_b v \frac{\partial T}{\partial z} + Q_r,$$

where $T\mathrm{w}$ is the wall temperature of the vessel, $h = Nuk_b/D$ is the convective heat transfer coefficient, P and S are the perimeter of the vessel and its cross-sectional area, v is the mean blood velocity along the vessel (in the direction z). V. Shahidi and P. Savard [2797] and S. Labonte [1751] estimated the values of the convective heat transfer coefficient from mathematical calculations assuming the blood flow is laminar, the swine endocardial convective heat exchange coefficient was measured *in vivo* by C. Tangwongsan et al. [3014].

Y. Mohammed and J. F. Verkey [2169] in their study of the effects of heat transfer by the large blood vessels such as carotid artery in the neck region during the laser-induced interstitial thermotherapy use the incompressible Navier–Stokes [1945] for the flow of blood that was assumed to behave as the Newtonian fluid:

$$\varrho \frac{\partial u(r,t)}{\partial t} - \mu\nabla^2 u(r,t) + \varrho u(r,t)\nabla u(r,t) + \nabla p = F \tag{5.457}$$

$$\nabla \cdot u(r,t) = 0\,. \tag{5.458}$$

The authors additionally have assumed (not stating it explicitly) that the dynamic viscosity μ is constant.

The non-stationary form of the Navier–Stokes equations allows to simulate the time-periodic blood flow, accounting for the beat cycle effect. For the vessels far away from the heart, the pumping cycle could be disregarded and the stationary Navier–Stokes equations used:

$$-\mu\nabla^2 u(r) + \varrho u(r)\nabla u(r) + \nabla p = F \tag{5.459}$$

$$\nabla \cdot u(r) = 0\,. \tag{5.460}$$

When the blood flow in the vessel is computed directly, it is usually assumed that the flow entering the region of interest is equilibrated and could be described by the well-known Poiseuille velocity profile [1945], that for the flow in the cylindrical vessel is written as [640]

$$u(x, y, 0) = 2u\left(1 - \frac{x^2 + y^2}{R^2}\right).$$

P. Balogh and P. Bagchi [176, 177] performed the direct numerical simulation of the 3D cellular-scale blood flow in physiologically realistic microvascular networks. The vascular networks were generated following in vivo images and data, and are comprised of bifurcating, merging,and winding vessels. The model resolved the large deformation and dynamics of each individual red blood cell flowing through the networks with high fidelity, while simultaneously retaining the highly complex geometric details of the vascular system. The architecture of the network is different depending on which organ or tissue is considered. Evidently, computational models for blood flow in straight tubes are not sufficient to address cell motion in such geometrically complex microvascular networks. The model must account for large deformation of individual blood cells, and the geometric complexity of the network architecture.

The ability to resolve the cellular-scale details that are important in many pathophysiological events, such as, fo example, sequestration and localized adhesion of platelets, leukocytes, is very attractive in the blood flow simulation but such simulations have high computational requirements.

P. Balogh and P. Bagchi exploited the immersed boundary method (IBM), which allow modelling flows in arbitrary geometry while resolving the large deformation and dynamics of every blood cell with high fidelity. This methodology seamlessly integrates different modelling components dealing with stationary rigid boundaries of complex shape, moving rigid bodies, and highly deformable interfaces governed by nonlinear elasticity.

The complex vascular walls are modelled using a sharp-interface ghost nodes, while the deformable interfaces of the RBCs—using the continuous forcing IBM. This approach permits the simulation of the actual geometry without simplifications, while resolving the cellular-scale details.

The original IBM [2436] is based on three groups of equations [1001]:

- the incompressible Navier–Stokes equations describing the motion of the fluid in an Eulerian coordinate system;
- the elasticity equations describing the deformation of the immersed bodies in terms of the change of the total elastic energy in a Lagrangian coordinate system;
- the equations that transform Eulerian into Lagrangian quantities and vice versa.

5.3.1 *Large Scale Simulations of Red Blood Cell Flows Through Vascular Networks*

The high-resolution blood flow simulations have potential for developing better understanding biophysical phenomena at the microscale, such as vasodilation, vasoconstriction and overall vascular resistance. The direct simulation of the blood flow is an extremely challenging task. Even simulating the blood flow in the smaller vessels requires modelling millions of cells along with a complex blood vessel. The RBCs are highly deformable and cannot be well-approximated by rigid particles; the volume fraction of cells in human blood flow is about 45% (36–48% in healthy women and 40–54% in healthy men [1970]), which means that a very large fraction of cells are in close contact with other cells or the vessel walls [1007]. L. Lu et al. [1970] performed one of the largest simulation enforces a contact-free state between 4 billion surface elements and solves for 3 billion degrees of freedom on 1 million RBCs and a blood vessel composed from 2 million patches.

The large-scale simulation of the RBC flows typically fall into several categories [1970]:

- Immersed Boundary (IB) and immersed interface methods [3415];
- particle-based methods such as Dissipative Particle Dynamics (DPD) [940] and Smoothed Particle Hydrodynamics (SPH) [3414] or their combination Smoothed Dissipative Particle Dynamics (SDPD);
- multiscale network-based approaches [2428];
- Moving Particle Semi-implicit (MPS) method [127];
- Boundary Integral Equation (BIE) approaches [3429].

L. Lu et al. [1970] developed the simulation system that includes handling the RBC-blood vessel interaction with a fully parallel, high order boundary integral equation solver and explicit handling of collisions [1969] with a parallel constraint-based resolution and detection algorithm. The analysis is restricted to the low Reynolds number regime, i.e., small arteries and capillaries; a simplified model for the RBCs is used that assumes the cell membranes to be inextensible and with no inplane shear rigidity.

The authors simulated the flow of N cells with deformable boundary surfaces in a viscous Newtonian fluid in a domain $\Omega \in \mathfrak{R}^3$ with a fixed boundary Γ. The governing partial differential equations describing the conservation of momentum and mass are the incompressible Stokes equations for the velocity u and pressure p, combined with velocity boundary conditions on Γ

$$-\mu\Delta\mathbf{u}(\mathbf{x}) + \nabla p(\mathbf{x}) = \mathbf{F}(\mathbf{x}), \qquad \nabla\cdot\mathbf{u}(\mathbf{x}) = 0, \quad \mathbf{x}\in\Omega, \tag{5.461}$$

$$\mathbf{u}(\mathbf{x}) = \mathbf{g}(\mathbf{x}) \quad \mathbf{x}\in\Gamma, \tag{5.462}$$

where μ is the viscosity of the ambient fluid.

The right-hand side force in the momentum equation is due to the sum of tension and bending forces $\mathbf{f} = \mathbf{f}_\sigma + \mathbf{f}_b$; it is concentrated on the cell surfaces.

The cell membranes are assumed to be massless, thus the velocity of the points on the cell surface coincides with the flow velocity. The cell surfaces are considered to be inextensible, with bending forces determined by the Canham-Helfrich model [428, 1290, 3517] based on the surface curvature, and surface tension determined by the surface incompressibility condition $\nabla_{\gamma_i}\cdot\mathbf{u} = 0$ resulting in

$$\mathbf{F}(\mathbf{x}) = \sum_i \int_{\gamma_i} \mathbf{f}(\mathbf{x})\delta(\mathbf{x}-\mathbf{y})d\mathbf{y} \tag{5.463}$$

(see, for example, [2563] for the expressions for **f**).

In the largest test the authors simulated 1,048,576 RBCs in a blood vessel composed of 2,097,152 patches. In each time step, this test uses over 3 billion degrees of freedom and over 4 billion surface elements (triangles) for collision. They were able to study the realistic human blood flows with RBC volume fractions over 47%.

5.3.2 *Vascular Network Structure*

V. V. Gafiychuk and I. A. Lubashevsky [1030] considered a hierarchical system of the interacting vessels of the different level (size); it was shown, *inter alia*, that the blood velocity could non-locally depend on the temperature distribution in the tissue. The number of the vessel levels in the vascular network is usually large $N \approx$10–20 [2099, 2908].

Since the vessel system is hierarchically organized the blood flow distribution has to be characterized by correlations between the different hierarchy levels, thus in order to describe the blood flow effect on the heat transfer one should take into account the vascular network as a whole rather than the vessels of the different levels individually [1030]—it is not acceptable to consider the vessels of the different levels practically independently of each other as is done in some models [158].

Usually a vascular network involves arterial and venous parts in the tree form as well as the system of the artery-artery, vein-vein and artery-vein anastomoses [1030, 2099]. However, some organs contain a few number of anastomoses or practically no one at all.

The fractal structure is an approximation to the real vascular network that includes both symmetrical and asymmetrical vessels but is sufficient for most heat transfer problems (see [3254] and the references therein). The generation of the vascular net could be done via a constrained optimization [1543]. This methods yields a binary branching network of the segments of the rigid cylindrical tubes that is contained in a three-dimensional convex volume (the perfusion volume) and subjected to a number of the physiological conditions and constraints. In 1926 Murray questioned whether vascular networks grow to optimize a certain quantity. He showed that at a symmetric vascular bifurcation, minimizing a "cost function" equal to the sum of the viscous dissipation and the energy required to maintain the vascular volume yields the ratio of the diameters of parent to daughter branches $D_{parent}/D_{daughter} = 2^{1/3}$ [2487]. This relationship is referred to as the Murray's law. The general empirical Murray's law—the relation connecting the radii of the parent and the child segments—has the form $r_{\text{parent}}^{\gamma} = r_{\text{left}}^{\gamma} + r_{\text{right}}^{\gamma}$. In reality the value of γ ranges from 2 to 3; for the large arteries it is close to 2.33 [1024].

A consequence of Murray's law is that wall shear stress should be constant throughout the vascular network. However, a combination of experiment and theory shows that wall shear stress tightly correlates with intravascular pressure along the network and therefore decreases gradually from the arterial to the venous ends of the network [2487, 2514].

Another empirical relation due to Murray gives the angles between the child and the parent vessels

$$\cos\varphi_{\text{right}} = \frac{r_{\text{parent}}^4 - r_{\text{left}}^4 + r_{\text{right}}^4}{(2r_{\text{parent}}r_{\text{right}})^2}, \qquad \cos\varphi_{\text{left}} = \frac{r_{\text{parent}}^4 - r_{\text{right}}^4 + r_{\text{left}}^4}{(2r_{\text{parent}}r_{\text{left}})^2}.$$

The total branch angle $(\varphi_{\text{right}}+\varphi_{\text{left}}) \in [75^\circ, 90^\circ]$ and is about 75° for the symmetric case [1024].

J. Shi et al. [2824] constructed a vascular tree in the cryosurgery study using a set of rules: the each level of the tree is obtained by the symmetric dichotomous division (a bifurcation angle as well as the ratios of the lengths $\gamma = L_{k+1}/L_k$ and diameters $\beta = D_{k+1}/D_k$ at two successive branching levels are fixed).

I. Li et al. [1855] derived an expression for the effective thermal conductivity of tissue containing the randomly distributed fractal vascular trees—the symmetrical two-branching networks—assuming that the parent channel length, the branching angle, length and diameter ratios are the same for every tree. The effective thermal conductivity of tissue depends on the thermal conductivities of the solid tissue matrix and blood, porosity and the structural parameters of the vascular trees.

Similar tree-like branched fractal vascular network was considered by K.J. Chua [566] who derived an expression for an effective blood perfusion rate

$$\omega_e = \omega_b \left[\frac{1-\gamma^{m+1}}{1-\gamma}\right]\left[\frac{1-N\beta^2\gamma}{1-(N\beta^2\gamma)^{m+1}}\right]$$

where m is the number of N-branchings, ω_b is the blood perfusion rate at 0th branching level.

5.3.3 Discrete Vasculature Model

DIscrete VAsculature model (DIaVA) [2542] was developed for the heat transfer in the system of the blood vessels.

The DIVA model can calculate the three dimensional temperature distributions in the tissue while accounting for the individual vessels in an extensive discrete unidirectional vascular tree [1700, 1701]. It embeds vessels as one dimensional line segments into a three dimensional domain. The flow of blood is modelled along these segments and then transferred to the tissue domain via the source and sink terms that correspond to a local perfusion This is combined with the convective heat transfer from the direct interfacial contact of these segments with the adjacent tissue.

The model avoids simulating the small vessels as these have been determined to be "thermally insignificant" [491]. Due to their small diameters, these vessels rapidly reach the equilibrium with their surroundings and are then modelled using a perfusion term with the arterial temperature based on the local surrounding arteries. This model was then validated with the direct measurements from an artificially perfused bovine tongue [2540].

The correctness of the DIVA model has been validated by comparison with analytical solutions [1700]. Simulations of a single vessel embedded in a coaxial tissue cylinder showed deviations less than 1% between the simulated and theoretical thermal equilibration lengths and between the simulated and theoretical axial blood temperature profiles.

The DIVA model was applied by G. M. Van Leeuwen et al. [3151] to a neonatal head, by V. Flyckt et al. [976] to the computation of the temperature distributions in the human eye and by L. A. van Lier et al. [3153] to the radio-frequency heating of the head.

The DIVA model estimates the three-dimensional temperature distribution by combining the blood perfusion continuum heat model for tissue volume and energy flux equation for the flow of blood in large vessel through common boundary conditions in the same global space. The elements in global space are either tissue sinks (known as voxels) or vessel segments. Voxels are discretized tissues having regular cuboid shape and the vessel segments are three-dimensional curved tubes embedded into certain voxels. The vessel segments are discretized into sub-elements (buckets) having similar nodal distanceas in the tissue voxel.The buckets can be visualized as cylindrical elements inside the vessel segments. All buckets form a temperature profile inside the vessel segment. The heat flow from the surrounding tissues to each bucket in vessel segment embedded inside the tissue voxel can be estimated by dividing that particular tissue voxel into boundary elements surrounding the vessel segment. The boundary elements are essentially the intermediate sets that bring heat flow balance between the outside sink set,i.e., tissue voxel with buckets inside vessel segment.

To handle the tissue–vessel interaction two sets of voxels are defined for each bucket: the estimation set and the exchange set. The estimation set voxels are used for calculation of the tissue–vessel interaction. The exchange set voxels are then used to effectuate the heat exchange in for tissue. The heat exchange with for tissue is distributed evenly over the exchange set voxels. Thermal equilibration of the blood from the final generation of for discrete vessels is completed by attaching for user defined sub-volumes to for vessel end points and applying a local heat sink.

The energy transfer between smaller blood vessels and tissue is described by the simplified perfusion model as

$$\rho_t c_t \frac{\partial T_t}{\partial t} = \nabla \cdot (\lambda_t \nabla T_t) + q_m''' + S - q_p \tag{5.464}$$

where T_t is the volume averaged tissue temperature, q_m''' is the rate of heat generation per unit volume due to metabolic heat, S describes the source due to power deposition per unit volume caused by heating, q_p is the amount of heat withdrawn due to the local blood flow.

The convective heat transfer between flow of blood within large vessel segments and vessel walls is modelled separately. The vessel segment embedded in the voxel is represented as the cylindrical tube with the radius r_{ves} surrounded by the coaxial tissue cylinder (vessel wall) having the radius R. The energy equation in the cylindrical coordinates for vessel wall is given as

$$\rho_t c_t \frac{\partial T_t}{\partial t} = \frac{1}{r} \frac{\partial \left(\lambda_t \frac{\partial T_t}{\partial r} \right)}{\partial r} + S = 0. \tag{5.465}$$

The radial outward heat flux $q_{r_{ves}}$ from the blood to the inner vessel walls is expressed as

$$q_{r_{ves}} = \frac{Nu \lambda_t}{2 r_{ves}} (T - T_{mix}), \tag{5.466}$$

The mixing cup blood temperature T_{mix} for each bucket in the vessel segments is written as

$$T_{mix} v_m \pi r_{ves}^2 = \int_{r=0}^{r_{ves}} 2\pi r v(r) T(r) \mathrm{d}r \tag{5.467}$$

where $T(r)$ is the radial temperature of blood, $v(r)$ is the radial velocity of blood, v_m is the mean velocity of blood

$$v_m = \int_{r=0}^{r_{ves}} 2\pi r v(r) \mathrm{d}r. \tag{5.468}$$

The heat flow in the vessel related to the mass flow of the blood within the vessel segment is simulated by shifting and re-sampling the bucket temperature profile. The shifting is achieved by assuming fully developed condition with negligible thermal instability along the flow direction by a distance $\delta t = v_m \Delta t$ in the time interval Δt.

A drawback of the initial formulation of the DIVA model is using many small time steps to calculate the evolving temperature distribution [1669]. For the treatment planning purposes, the steady-state temperature distribution is most relevant since it reflects the thermal dose in the tumour and the location and magnitude of the hot spots in the normal tissue. However, calculation of a steady-state temperature distribution can take up to many hours that makes it impractical for the simulations in the routine clinical applications.

H. P. Kok et al. [1668] developed a method for the fast simulation of the steady-state temperature. The bioheat transfer and the interaction between the vasculature and the tissue were represented by a linear system $Ax =$. The coefficients in the matrix A describe the thermal processes, the right-hand side vector b holds the boundary conditions and the power deposited by the heating system. The linear system was solved using an iterative method implemented on a graphical processing unit (GPU).

B. W. Raaymakers et al. [2541] discussed how a thermal model could incorporate the individual impact of discrete vessels when not all vessels can be reconstructed individually. The authors investigated the possible strategies to model the thermal impact of the missing vessels that all use a local or global heat sink model in addition to the discrete vasculature.

The intracellular transport the is predominantly heterogeneous in both time and space, exhibiting varying the non-Brownian behaviour. Characterization of this movement through averaging methods over an ensemble of trajectories or over the course of a single trajectory often fails to capture this heterogezneity.

The HAT (heterogeneously anomalous transport) models Developed, e.g., by , a deep learning feedforward neural network trained on the fractional Brownian motion, providing a novel, accurate and efficient method for resolving heterogeneous behaviour of intracellular transport in space and time. The neural network requires significantly fewer data points compared to established methods. This enables robust estimation of Hurst exponents for very short time series data, making direct, dynamic segmentation and analysis of experimental tracks of rapidly moving cellular structures such as endosomes and lysosomes. By using this analysis, fractional Brownian motion with a stochastic Hurst exponent was used to interpret, for the first time, anomalous intracellular dynamics, revealing unexpected differences in behaviour between closely related organelles. Such model are more relevant to the mass transport than to the thermal transport.

5.3.4 Vascular Porous (VaPor) Model

In the Vascular Porous (VaPor) model the full cerebral circulatory system is modelled using the vascular channels embedded in a porous tissue simulating the blood vessels and capillaries—the tortuous nature of the vasculature could not be captured by the porous media alone. It was used by S. Blowers to study an effect cooling the head has on brain temperature [319]. Initially, the blood vessels were modelled in the 3D to fully capture the effects of flow, however, this is computationally inefficient so was subsequently replaced with a system of the one dimensional line segments.

C. D'Angelo and A. Quarteroni [672] showed how theory for modelling the water flow through a cracked porous rock bed could be applied to the vascular perfused capillary bed. This approach combines the concepts of the discrete vasculature and porous media. A similar approach has been performed T. Secomb et al. [2766, 2767] for simulating the oxygen perfusion in a capillary bed with the embedded vascular structure. J. Reichold et al. [2590] presented a "vascular graph" model which included both one dimensional line segments in a porous two dimensional area and described transport of an arbitrary scalar throughout the system. Z. Z. He and J. Liu [1278] proposed a coupled continuum-discrete bioheat model for implementing the vasculature in the porous tissue.

This model is similar to the DIVA in the heat transfer to the capillary region being condensed into a source term, and no advection from blood flow present at this level. The main difference between the two models is the resolution of interfacial heat transfer between the vessels and the adjacent tissue.

The DIVA model has its own method to generate the vasculature, weighted by the perfusion [3150]. Most methods for generating the vasculature implement a form of the conditional growth algorithm with the diameters of the generated segments relying on the established rules to dictate the spread and size of the vessels such as Murray's Law [2236] and other scaling laws [997].

S. Blowers considered four inlets and one outlet which represent the major vasculature within the brain: the anterior cerebral arteries, the left and right middle cerebral arteries, the posterior cerebral arteries and the transverse sinuses at the rear of the brain.

The brain tissue was assumed to be a porous material. The momentum loss associated with travelling through this medium was assumed to adhere to the Darcy's law

$$\frac{\partial P}{\partial l} = \frac{\mu_b}{K} u \tag{5.469}$$

where μ_b is the viscosity of blood, the permeability K is given by

$$K = \frac{\varepsilon_b d_c^2}{32\tau} \tag{5.470}$$

where ε_b is the volume fraction of domain in porous media, d_c is the capillary diameter, the parameter τ represents the tortuosity of the capillaries, which is defined as the ratio between the length of the capillary vessel and the distance between the two end points.

The author used the three dimensional hemispherical model of the brain and found that assigning the inlets and outlets in the arbitrary locations creates regions with the large variances in the flow rates. This is due to all the blood entering the system through a small number of inlet points, and the low permeability of the porous media limiting the distribution to distant areas; the regions of lower velocities towards the surface are more susceptible to temperature changes. The inclusion of directional flow for the scalp cooling has a larger impact on cerebral temperatures than seen with previous bioheat models.

To create a more realistic flow pattern, the model requires vasculature to properly distribute the blood around the brain without resorting to the arbitrary source points. This can be done by creating the three dimensional domains of the blood vessels taken from scans (it is possible to extract geometries from scans such as angiograms and venograms) and embedding them within the tissue domain. The denser vasculature was obtained by Blowers from an online database of the cerebral arterial vessels [3344]. This database contains the various trees stored as the line segments with connectivity.

The VaPor model replaces the three dimensional vessels with the one dimensional segments but maintains the porous media domain separating the arteries and veins. This facilitates control over the distribution of the blood throughout the system. The individual segments within the vessel trees can be isolated to control how much blood is distributed to the porous domain in the different regions. Additionally, the trees themselves can be adjusted by adding and removing the segments to determine the effects of the vasculature on the cerebral temperatures.

Modelling the dense vasculature captures the counter-current heat exchange incurred by the tortuous vessels, which is not present when only the porous media is used, leading to the increased core cooling observed by the VaPor model [319].

M. Peyrounette et al. [2448] used similar hybrid—vascular-porous—model to study the blood flow in the cerebral microcirculation. The capillary bed, which is a space-filling network, is treated as a porous medium using a homogenized continuum approach. The larger arteriolar and venular trees, which cannot be homogenized because of their fractal-like nature, are treated as a network of interconnected tubes. These descriptions overlay in space and need matching at a large number of coupling sites. Thus, there are the main model components

- *The capillary bed (microscale).*

  The capillary bed, which is a space-filling network, is treated as the porous medium where the capillary vessels represent an interconnected network of pores embedded in the cerebral tissue and modelled using a homogenized continuum approach. The capillary bed is dense and space-filling over a cut-off length of 50 μ [1957].

In the homogenized descriptions velocity and pressure fields are averaged in space over a large number of vessels and the detail of the architecture at microscopic scale is not fully described. These microscopic details are implicitly represented via effective parameters (for example, the permeability) that characterize the flow at a larger scale—the mesoscopic scale [2448].

The momentum transport is governed by the Darcy law, which can be obtained from homogenization via the volume averaging [687, 3310] of the creeping flow at the microscopic scale

$$\mathbf{U} = -\frac{1}{\mu^{eff}} \hat{K}^{eff} \cdot \nabla P \tag{5.471}$$

where $\mathbf{U}$ is the Darcy velocity (a spatially averaged velocity), P is the pressure, μ^{eff} is the effective viscosity of the blood, $\hat{K}^{eff}$ is the permeability tensor. When the capillary network is isotropic and homogeneous $\hat{K}^{eff} = K^{eff} \hat{I}$. The effective permeability represents the permeability of the network with uniform viscosity, while the effective viscosity represents the impact of the Fahrnus-Lindqvist effect [2448].

The mass balance in the porous medium could be written as

$$\nabla \cdot \mathbf{U} = -\sum_{s \in S} \delta(\mathbf{x} - \mathbf{x}_s) q_s \tag{5.472}$$

where q_s the source term resulting from the coupling condition applied at the coupling point s located at the position x_s at any endpoint of feeding arteriolar or draining venular trees. S is the set of all sources and δ is the Dirac delta function.

A system of the equations involving only pressures is obtained by eliminating the Darcy velocity

$$\frac{\hat{K}^{eff} h}{\mu^{eff}} = -\sum_{s \in S} \delta(\mathbf{x} - \mathbf{x}_s) q_s. \tag{5.473}$$

- *The arteriolar and venular networks.*

  The arteriolar and venular vessels are treated as a network of interconnected tubes where blood flow is described by a succession of linear relationships between the flow rate and the pressure drop. The geometry and topology of the vascular network can be either extracted from available anatomical databases or generated as was proposed by A. A. Linninger et al. [1893].

  The flow $q_{\alpha\beta}$ in the vessel connecting vertices α and β is written as

$$q_{\alpha\beta} = G_{\alpha\beta}(\pi_\alpha - \pi_\beta) \tag{5.474}$$

where π_α, π_β are the pressures at the vertices and the vessel conductance

$$G_{\alpha\beta} = \frac{\pi d_{\alpha\beta}^4}{128\mu_{\alpha\beta}^{app} l_{\alpha\beta}} \tag{5.475}$$

where $d_{\alpha\beta}$ is the vessel mean diameter and $l_{\alpha\beta}$ is the vessel length. The apparent viscosity $\mu_{\alpha\beta}^{app}$ accounts for the emergence of the complex microvascular flow structures (for example, the cell free layer) and their impact on the average viscous dissipation at the vessel scale [2448].

The mass balance at each "inner" vertex α—the vertex where no boundary condition is imposed—is given as

$$\sum_{\beta \in \mathcal{N}_{\alpha,in}} G_{\alpha\beta}(\pi_\alpha - \pi_\beta) = q_s \tag{5.476}$$

where $\mathcal{N}_{\alpha,in}$ represents the set of the inner neighbouring vertices that are connected to α and q_s is a flow rate source term that describes the boundary conditions for the arteriolar inlets ($q_s > 0$) and the venular outlets ($q_s < 0$) or the coupling conditions for the vertices connected to the homogenized capillary bed.

- *Coupling of continuum and vascular descriptions*

  A coupling condition must be developed to relate the pressure at the coupled endpoint of the arteriolar or venular vessel and the pressures of the local finite volume cells within the capillary bed.

  M. Peyrounette et al. developed an approach to couple the two frameworks which is based on a "well model" for petroleum engineering by D. W. Peaceman [2404]. The wellbore model is coupled with a continuum reservoir description, in which variations at the scale of the underlying structure are not described. The idea is to represent the strong gradients in the vicinity of the wellbore, i.e., within a length scale much smaller than the size of a finite volume cell. An analytical expressions of the pressure field in the vicinity of the coupling points are used to derive relationships between variables of both frameworks.

  The authors decomposed the pressure field in the neighbourhood of a coupling point into two parts

  - In Ω_{lin} region (the closest region to the coupling point)—a linear variation captures the network structure at the scale of the capillary vessels connected to this point.
  - In Ω_{sph} region up to two finite volume cells away from the coupling point—a spherical solution of the continuum equations describes the variation at the scale of h, the side of a finite volume cell.

  The advantage of the coupling condition is that it expresses a linear relationship between the pressure at the coupled endpoint of the arteriolar or venular vessel and the pressures of the local finite volume cells and can be applied to each coupling point independently.

For example, for a single coupling point s at any arteriolar or venular coupled endpoint the local flow rate q_s corresponds to the coupling term that appears in both (5.476) for the vasular network approach and (5.473) for the continuum representation. The coupling condition aims at describing the pressure drop between the pressure π_s at the endpoint of the arteriolar or venular vessel, and the cell pressure P_j.

5.4 Temperature Fluctuations in Living Tissues

In all the models discussed in the preceding sections it was implicitly assumed the blood perfusion is stationary. A model accounting for the pulsative character of the perfusion was formulated by Z.-S. Deng and J. Liu [712]. The authors use the Pennes' equation (5.8). They proposed that the blood perfusion should be treated as a transient term due to external perturbation. As a result of pulsating blood flow, the temperature in living tissues also appears fluctuating. They introduced a perturbation model that is capable of giving quantitative relation between temperature fluctuation, pulsating blood perfusion,and arterial blood temperature and heat generation in living tissues. Their model incorporates small perturbations to arterial blood temperature, tissue temperature, blood perfusion term and metabolic heat generation term by adding a transient fluctuating term with the mean terms in traditional Pennes model.

The model in the case of the constant thermal diffusivity of the tissues and neglecting the external source is written as:

$$\varrho c \frac{\partial T}{\partial t} = \lambda \nabla^2 T + c_{\mathrm{b}} \omega_b (T_{\mathrm{a}} - T) + \dot{q}_{\mathrm{met}},$$

and represents all variables as a sum of a mean and a fluctuation:

$$T = \overline{T} + T', \qquad T_{\mathrm{a}} = \overline{T_{\mathrm{a}}} + T_{\mathrm{a}}', \qquad \omega_b = \overline{\omega_b} + \omega_b', \qquad \dot{q}_{\mathrm{met}} = \overline{\dot{q}_{\mathrm{met}}} + \dot{q}_{\mathrm{met}}',$$

where the mean values denoted by overline are averaged over the time interval Γ:

$$\overline{A(\mathbf{r}, t)} = \frac{1}{\Gamma} \int_{t-\Gamma/2}^{t-\Gamma/2} A(\mathbf{r}, \tau \mathrm{d}\tau.$$

Evidently, the averaged values of fluctuation should be zero:

$$\overline{T'} = 0, \qquad \overline{T_{\mathrm{a}}'} = 0, \qquad \overline{\omega_b'} = 0, \qquad \overline{\dot{q}_{\mathrm{met}}'} = 0.$$

The authors follow the well known in the fluid dynamics procedure (the derivation of the Reynolds equation for the turbulent flow) to write out the equation for the mean temperature:

$$\varrho c \frac{\partial \overline{T}}{\partial t} = \lambda \nabla^2 \overline{T} + c_{\mathrm{b}} \overline{\omega_b} (\overline{T_{\mathrm{a}}} - \overline{T}) + c_{\mathrm{b}} \overline{\omega_b' T_{\mathrm{a}}'} - c_{\mathrm{b}} \overline{\omega_b' T'} + \overline{\dot{q}_{\mathrm{met}}} \tag{5.477}$$

and for the temperature fluctuations:

$$\varrho c \frac{\partial T'}{\partial t} = \lambda \nabla^2 T' + \omega_b' c_{\mathrm{b}} (\overline{T_{\mathrm{a}}} - \overline{T})$$

$$- c_{\mathrm{b}} \overline{\omega_b' T_{\mathrm{a}}'} + c_{\mathrm{b}} \overline{\omega_b' T'} - c_{\mathrm{b}} \omega_b' T' - c_{\mathrm{b}} \overline{\omega_b} T' + c_{\mathrm{b}} \overline{\omega_b} T_{\mathrm{a}}' + c_{\mathrm{b}} \omega_b' T_{\mathrm{a}}' + \dot{q}_{\mathrm{met}}'. \tag{5.478}$$

Two additional terms in (5.477) $c_{\mathrm{b}} \overline{\omega_b' T_{\mathrm{a}}'}$ and $c_{\mathrm{b}} \overline{\omega_b' T'}$ represent the mean transferred energy due to the perfusion perturbations and temperature fluctuations in the arterial blood and in the tissue, respectively. Assuming the mean quantities to be $O(1)$ while the fluctuations to be $O(\delta)$, it is possible to neglect in (5.477) all terms smaller than $O(1)$ and in the (5.4)—all terms smaller than $O(\delta)$—to get the final equations:

$$\varrho c \frac{\partial \overline{T}}{\partial t} = \lambda \nabla^2 \overline{T} + c_{\mathrm{b}} \overline{\omega_b} (\overline{T_{\mathrm{a}}} - \overline{T}) + \overline{\dot{q}_{\mathrm{met}}} \tag{5.479}$$

and for the temperature fluctuations

$$\varrho c \frac{\partial T'}{\partial t} = \lambda \nabla^2 T' + \omega_b' c_{\mathrm{b}} (\overline{T_{\mathrm{a}}} - \overline{T}) - c_{\mathrm{b}} \overline{\omega_b} T' + c_{\mathrm{b}} \overline{\omega_b} T_{\mathrm{a}}' + \dot{q}_{\mathrm{met}}'. \tag{5.480}$$

The study by Deng and Liu indicates that the small perturbation in the blood perfusion causes significant temperature fluctuation in the living tissues.

M. I. A. Othman et al. [2358] performed a normal mode analysis for characterizing the temperature fluctuation in tissues due to the propagation of the ultrasonic waves. The authors used the single-phase lag version of the Pennes equation

$$\varrho C_t \left(1 + \tau \frac{\partial}{\partial t}\right) \frac{\partial T}{\partial t} = \lambda \nabla^2 T - \omega_b C_b \varrho_b (T - T_b) + \left(1 + \tau \frac{\partial}{\partial t}\right) Q \tag{5.481}$$

where ϱ is the density of tissue, C_t is the heat capacity of tissue, C_b is the heat capacity of blood, ω_b is the perfusion due to blood flow, ϱ_b is the density of blood, T_b is the arterial blood, τ is the relaxation time.

Othman et al. introduced the dimensionless parameters (L is the tissue length)

$$\vartheta = \frac{T - T_b}{L^2/\lambda}, \ x' = \frac{x}{L}, \ t' = \frac{\varkappa t}{L^2}, \ \tau' = \frac{\varkappa \tau}{L^2}, \ \varkappa = \frac{\lambda}{\varrho C_t}, \ p_f = \frac{\omega_b C_b L^2}{\lambda}, \ G = \frac{Q}{Q_0} \tag{5.482}$$

and got the dimensionless Pennes equation

$$\left(1 + \tau \frac{\partial}{\partial t}\right) \frac{\partial \vartheta}{\partial t} = \nabla^2 \vartheta - p_f \vartheta + \left(1 + \tau \frac{\partial}{\partial t}\right) G. \tag{5.483}$$

The solution is decomposed in terms of the normal modes as

$$\vartheta(x, y, t) = \vartheta^\star(x) e^{(\omega t + iky)}, \ G(y, t) = G^\star e^{(\omega t + iky)} \tag{5.484}$$

where k is the wave number in the y-direction, ω is a complex time constant, $\vartheta^\star$ and $G^\star$ are the amplitude of the functions ϑ and G.

The solution for the temperature is written as

$$\vartheta(x, y, t) = \left[S_1 e^{-mx} - \frac{1 + \omega \tau}{m^2} G^\star \right] e^{(\omega t + iky)} \tag{5.485}$$

where $m^2 = k^2 + p_f + 1 + \omega \tau$ and from the boundary condition $\vartheta(x, y, t) = \vartheta_0$

$$S_1 = \vartheta_0^\star + \frac{1 + \omega \tau}{m^2} G^\star. \tag{5.486}$$

The authors noted that the soft tissue can be treated as a viscous liquid for propagation of the ultrasound. The amount of heat generated in the tissue is dependent on the absorption coefficient of the tissue. The acoustic absorption due to the shear relaxation invariably can not be accounted for by the single relaxation times instead a continuous distribution of relaxation times must be used. This is because the relaxation process arises from the finite time taken for molecules to diffuse between the adjacent shearing layers in the medium [2400]. The time taken for a particular molecule to diffuse depends upon the local environment because the random nature of medium is different for each molecule. Generally, the more dense materials such as bone and teeth have higher acoustic absorption coefficients than less dense tissues such as, e.g., liver or muscle and are, therefore, heated to a greater extent than soft tissue (for example, the absorption coefficient of bone may be as much as 50 times greater than most soft tissue [572]).

Every person has a varied blood perfusion at various health statuses. Othman et al. noted that large blood perfusion tends to prevent the biological body from burn injury.

The tissue temperature increases with the increase of the thermal conductivity of tissue, the temperature of the surface tissue will be closer to the body core temperature when the thermal conductivity becomes high enough. The thermal conductivity of the tissue depends on its components, it will decrease with the loss of water. Thus the thermal injury will be more serious if taking into concern the decrease of the thermal conductivity due to water loss.

V. V. Gafiychuk and I. A. Lubashevsky [1030] studied the vessel temperature variations which are described as variations in the vessel resistances R_i to the blood flow. When the tissue temperature is constant over the domain and coincides with the arterial blood temperature T_a. In this case all vessels of the same level are assumed to be equivalent and are characterized by the resistance

$$R_n^0 = R_0 2^{3n} \rho(n) \tag{5.487}$$

where n is the number of the given level, $\rho(n)$ is a smooth function of n such that $\rho(0) = 1$ and $\rho(0) = 1$ as $n \to \infty$. This dependence follows from the requirement of the blood flow redistribution over the vascular network controlled by an artery group comprising vessels of the different lengths; this property is typical for the real microcirculatory beds [1030].

This region can be treated as a basic fundamental domain. In fact, the main aim of systemic circulation is to maintain the arterial-venous pressure drop at a given constant. A regional circulation, i.e. the vascular network of a single organ, varying its resistance to blood flow supplies different points of the organ with such amount of blood that is needed. For a relatively simple organ its whole regional circulation is a microcirculatory bed. In other organs a microcirculatory bed is a certain part of the organ regional circulation, for example, brain arterial network forms a single microcirculatory bed. In this case blood flow in different microcirculatory beds can be analyzed independently of each other.

When the resistances of vessels belonging to one level are equal at each branching point the blood current splits into eight equal parts. Thus, for the arbitrary path on the vascular network leading from the arterial bed stem to the capillaries and then to the venous bed stem one can write

$$2P = 2\left\{J_0 R_0 + \frac{1}{8} J_0 R_1 + \frac{1}{8^2} J_0 R_2 + \dots\right\} = 2J_0 \sum_{n=0}^{N} 2^{-3n} R_n^0 \tag{5.488}$$

where the capillary bed resistance is neglected.

Thus the capillary bed resistance is

$$R^\star = R_0 \sum_{n=0}^{N} \rho(n). \tag{5.489}$$

It follows from (5.490) that the total resistance of the microcirculatory bed is determined by an artery group comprising vessels of different levels if $\rho(n)$ is a smooth function of n and not by the vessels of one level.

If the effective coefficient μ_{eff} of the blood viscosity the vessel resistance to the blood flow is given as [1773]

$$R = \frac{8}{\pi}\mu_{eff}\frac{l}{a^4}. \tag{5.490}$$

If, for example, $\mu_{eff} \propto a^{\alpha}$, where α is a constant, (5.490) leads to the formula (5.487).

When the tissue temperature is nonuniform, the blood temperature is also nonuniform and the resistances of all the vessels can become different. In this case, the response of any artery to temperature variations is assumed to coincide with that of the corresponding vein and to be governed by the temperature of blood in this vein. The response of such two vessels, e.g., of an artery or vein i, is described by the following model.

The response of a vessel to temperature variations can be described in terms of the dependence of the vessel radius a on the temperature $T^{\star}$ of blood in this vessel by the phenomenological equation

$$\tau^{\star}(a)\frac{\mathrm{d}a}{\mathrm{d}t} + a_0 f^{\star}\left(\frac{a}{a_0}\right) = a_0\frac{|T^{\star} - T|}{\Delta} \tag{5.491}$$

The first term in (5.491) describes delay of the vessel response to the dimensionless "signal"

$$\frac{|T^{\star} - T|}{\Delta} \tag{5.492}$$

generated by the "temperature" receptors, $\tau^{\star}(a)$ is the characteristic time delay of this process, a_0 is the vessel radius for $T^{\star} = T_a$, the second term represents the vessel counteraction to its expansion.

V. V. Gafiychuk and I. A. Lubashevsky assumed that the resistance R of the given vessel depends on its radius a as:

$$R = \frac{Const}{a^{\beta}} \tag{5.493}$$

where the constant $\beta > 0$.

Accounting for (5.491), one gets

$$\tau\frac{\mathrm{d}R}{\mathrm{d}t} + (R - R^0)f^{\star}\left(\frac{R}{R_0}\right) = R_0\frac{|T^{\star} - T|}{\Delta}. \tag{5.494}$$

The authors introduced in (5.494) the quantities

$$R^0 = R|a = a_0, \quad \tau = \tau^{\star}(a)\frac{1}{\beta}\left(\frac{a}{a_0}\right)^{\beta+1} \tag{5.495}$$

and the function

$$f(x) = \frac{1}{1-x} f^{\star} \left[\left(\frac{1}{x} \right)^{1/\beta} \right]. \tag{5.496}$$

The value of τ in (5.494) is determined by the time delay in the response of both the vessel muscles and the nervous system. There are also other mechanisms of the time delay, related, e.g., to the time required for a heated portion of the blood to reach a given vein or the response of the vessels to the temperature variations occurs through their response to the variations in the concentrations of K+, H+, etc.

Gafiychuk and Lubashevsky suggested combine all the mechanisms of the time delay in one term $\tau \frac{\partial R_i}{\partial t}$ and describe dynamics of the vessel response assuming $\tau = const$

$$\tau_n \frac{\mathrm{d}R_i}{\mathrm{d}t} + (R_i - R_n^0) f \left(\frac{R_i}{R_n^0} \right) = -R_n^0 \frac{|T_i^{\star} - T|}{\Delta} \tag{5.497}$$

where n is the level number of the given vessel pair, τ_n is the characteristic time of the nth level vessel response; Δ is the halfwidth of the temperature vital interval, $f(x)$ describes the vessel capacity for responding.

Gafiychuk and Lubashevsky stated that usually the blood flow in a vessel of length l directly controls the mean blood flow rate in a tissue domain whose size is about l, whereas the smaller vessels are responsible for the blood flow redistribution over different parts of this domain. Thus, fluctuations in the vessel resistance to the blood flow in it caused, e.g., by the time variations in its radius are bound to give rise to the spatiotemporal fluctuations in the blood flow rate j(r, t) in the tissue domain which are correlated on the spatial scales of order l and on temporal scales determined by the vessel characteristics. These fluctuations in j(r, t), in their turn, cause the spatiotemporal fluctuations in the tissue temperature. Since, the vascular network involves vessels of different lengths, both the tissue temperature and distribution of these components can exhibit the fluctuations characterized by a wide range of the spatial and temporal scales.

To analyse the characteristics of the temperature fluctuations the authors used the correlation function

$$G_{\mathbf{r},t} = \langle\langle \delta T_{\mathbf{r}'+\mathbf{r},t'+t} \delta T_{\mathbf{r}',t'} \rangle\rangle \tag{5.498}$$

where symbol $\langle\langle \ldots \rangle\rangle$ denotes averaging over both the time t' and the tissue points $\mathbf{r}'$ under the conditions $t = constant$ and $|\mathbf{r}| = constant$ and the correlation function of the blood flow rate fluctuations

$$\Omega_{\mathbf{r},t} = \langle\langle \delta j_{\mathbf{r}'+\mathbf{r},t'+t} \delta j_{\mathbf{r}',t'} \rangle\rangle. \tag{5.499}$$

Using the linearized Pennes equation in the form

$$\frac{\partial}{\partial t}\delta T = D_{eff}\nabla^2\delta T - \left[j_0 + \frac{\partial J}{\partial T}|_{T=T_0}(T_0 - T_a) \right] \delta T - \delta j(T_0 - T_a) \tag{5.500}$$

where δj is the blood flow rate fluctuations inherent to the living tissue, and neglecting the term

$$\frac{\partial J}{\partial T}|_{T=T_0}(T_0 - T_a)$$

the authors got relationship between the Fourier transforms of the above two correlation functions with respect to both the time t and the coordinates $\mathbf{r}$:

$$G(\mathbf{k}, w) = (T_0 - T_a)^2 \frac{\Omega(\mathbf{k}, w)}{w^2 + (D_{eff}k^2 + j_0)^2} \tag{5.501}$$

where w and $\mathbf{k}$ are the variables conjugate to t and $\mathbf{r}$.

The authors pointed out that when averaging the product $\delta j \delta j$ over the time t' only one gets the function

$$\Omega_{\mathbf{r}',\mathbf{r}} = \delta j_{\mathbf{r}',t'+t}\delta j_{\mathbf{r}',t'} \tag{5.502}$$

which depends on both the variables $\mathbf{r}$, $\mathbf{r}'$.

Chapter 6
Thermal Properties of the Biological Tissues

The model-based optimization algorithms for the planning of the thermal treatment, both the hyperthermia and the cryosurgery, will give the reliable predictions if the thermal properties of the different tissues—liver, fat, cartilage, brain, myocardium, muscle, bone, skin, pancreas tissues (for example, thermal conductivity, specific heat, volumetric heat capacity, density, thermal diffusivity) are known with a sufficient accuracy [285, 1951]. Unfortunately, these properties are poorly known, especially for the temperatures below $-40°$ C, and are often approximated by using the properties of the water/ice.

6.1 Composition of the Biological Tissue

The two main constituents of the biological tissue are the protein and the water. The biological tissue generally contains the blood vessels, the cells and the interstitial space [2009] which can be classified as the vascular and extra-vascular regions. The whole anatomical structure can be modelled as a porous medium—the blood perfused tissues resemble biphasic porous media, i.e., solid tissue matrices in the infiltrated with flowing blood.

Generally, the pressure is uniformly higher throughout the tumour as compared to the peripheral values, which leads to an extremely slow interstitial flow inside the tumour [3350]. Hence, the blood flow with in the tumour region can be represented by the Darcy flow model [623, 2009, 2010].

All biological tissue is made of the cells that vary in the size and shape depending on the function they fulfil. The cells are usually made up of a central nucleus, surrounded by a mass of protoplasm and bounded by a membrane. The interior of a cell is approximately 75% water and the lipids, the proteins, the salts, the sugars and the nucleic acids (RNA, DNA). The membrane made of the lipids is semi-permeable

A. I. Zhmakin, *Heat Transfer in Vivo*, Springer Series in Biophysics 27,
https://doi.org/10.1007/978-3-031-95459-7_6

and regulates the exchange of the ions and the molecules with the extracellular water [385].

The water constitutes 50–70% of the total body weight in the human male and 45–65% of the female body weight [1224]. The fat tissue has a low water content and hence people with the large amounts of fat will have proportionally less water compared to the total body weight.

Usually the body water is divided into the intracellular that exists within the tissue cells and the extracellular water outside the cell. The intracellular water makes up two thirds of the total mass of the body water occurs in two forms, free and bound. The free water is able to move as a liquid inside the cell, but the bound water is loosely attached to the protein molecules by the hydrogen bonding caused by the dipole attraction. The free water is available for the metabolic processes in contrast to the bound water.

The proteins are the major part of the solid matter in the human body tissues. The proteins are the polymer molecules with the molecular weight from several thousand to several million a. m. u. made from the amino acids (there are 20 types in the human body) the peptide links. The angles of these bonds and the hydrogen bonding within the protein molecule are responsible for the helical shape of proteins.

The lipids are the fats and the fat like substance mainly in the cells of the fatty tissues and in the membranes of the cells. Most lipids are the esters of fatty acids and have an average molecular weight of 700 a. m. u. The lipids are insoluble in the water, but are a vital part of the cell membrane [1224].

Both the intracellular and extracellular waters contain many different types of the dissolved salts, which are account for the approximately 1% of the total body weight; the ionic profile of the water present in the body is of importance in relation to the tissue dielectric properties [68, 1021, 2479, 2682, 2798]. In the aqueous solution the dielectric behaviour of both the proteins and the water are altered by the hydrogen bonding. The huge size of the protein macromolecules compared to the water molecules means that the protein molecules are not affected by the hydrogen bonding. Thus the biological tissue can be thought of as containing three distinct materials, which are the free water, the bound water and the protein [1136]. The dielectric constant of the bound water is closer to that of the water than of the ice indicating that the bonding between water and protein in the biological tissues is far weaker than the strong bonding found in the ice [1224]. The thermal properties of the bound water are also close to that of the free water [991].

The thermal conductivity of the proteins can not be measured directly due to the porous structure of the protein—the protein sample is filled with the air or the water. Thus it is measured indirectly, usually by measuring the thermal conductivity of the protein-water gels at the various concentrations of the protein and then use a mixture equation to find the "true" thermal conductivity. This process introduces an extra source of uncertainty to the thermal conductivity of the protein, since the different mixture equations will give the different estimates of the thermal conductivity [1224].

One approach to estimate the thermal properties of the biological tissues is the volume averaging that is frequently use for the heterogeneous media, such as, e.g., the soil [2103]:

$$\lambda = \varepsilon S_w \lambda_w + \varepsilon S_{ice} \lambda_{ice} + (1 - \varepsilon)\lambda_s, \tag{6.1}$$

where ε is the porosity, λ_w, λ_{ice} and λ_s are the thermal conductivities of the liquid water, the ice and the solid matrix, respectively, S_w and S_{ice} are the saturation of the liquid water and the ice.

The averaging based on the mass fractions of the water, the proteins and the fats in the biological objects with the different weights is common in both the medicine and the food industry; frequently the coefficients suggested many year ago by T. E. Cooper and G. J. Trezek [628] are used for the thermal conductivity, the specific heat and the density.

Similar weighting approach is used to determine the properties of the tissues with the injected nanoparticles [2394]. For example, the thermal conductivity λ_{mix} is given by

$$\frac{1}{\lambda_{mix}} = \frac{\eta}{\lambda_{np}} + \frac{1 - \eta}{\lambda_{tissue}} \tag{6.2}$$

where η is the volumetric concentration of the infused nanoparticles.

C. Rossmann and D. Haemmerich [2644] reviewed the thermal and dielectric properties while S. Gabriel et al. [1021–1023] the dielectric properties of the biological tissues at the hyperthermic and the ablation temperatures.

The properties of the healthy tissue and the tumour can greatly differ. For example, there is more than 10-fold difference in the conductivity and the relative permittivity of the breast and the tumour tissue at 2.45GHz [1181]. Then studies have also shown that various conditions or the diseases, such as the hydrocephalus and the Alzheimer's disease, cause a variation in the mechanical properties of the brain [1217, 3223].

A. Bhattacharya and R. L. Mahajan [280] studied the thermal conductivity of the biological tissues using a transient technique based on the principles of the cylindrical hot-wire method. The authors investigated the sheep collagen in the range of 25° C–55° C and found that the thermal conductivity is a linear function of the temperature. The similar experiments were also conducted for determining the thermal conductivity of the cow liver in the range of 25° C–55° C, the thermal conductivity, as for the sheep collagen, varied linearly with temperature. The authors also made in vivo the measurements of the different organs of a pig.

6.2 Experimental Techniques

One the first measurement of the thermal conductivity of the biological tissues appears to be that carried out by Bordier in 1898, who found that the beef muscle conducted heat almost twice as fast as fat [3381].

The methods of the measurement of the thermal conductivity in vivo are divided into three groups: the invasive, the semi-invasive and the non-invasive techniques.

It should be noted that the effective in vivo thermal conductivity is measured since it depends not only on the type of the tissue being sampled but other factors, such as the rate of the blood flow in the volume [1224]—it includes the contribution to the heat transfer due to the intrinsic conduction added to the contribution caused by the transport of the blood through the tissue. Moreover, the in vivo thermal conductivity can be affected by the act of measuring: the thermoregulatory system will respond to the changes in the skin temperature caused by measuring the in vivo thermal properties of the tissue. The standard method to measure the thermal diffusivity that is suitable due to the wide range of the time scales: the sample is thermally excited on its front side by a short heat pulse, and the temperature is recorded on the other side [2381].

In the semi-invasive techniques, the thermocouples are inserted into the tissue and the tissue surface is heated, either with the contact heater or the electromagnetic radiation. In the non-invasive techniques, the electromagnetic radiation is used to heat the tissue, with the subsequent variation in the tissue surface temperature with time being measured using the infra-red detector or the thermocouples.

Evidently, the non-invasive methods can only be used close to the skins surface. The semi-invasive techniques may give the information about the variation of the temperature with the depth, if many thermocouples are used, but there is the added difficulty of accurately determining the position of the thermocouples. The non-invasive techniques can give no information about the variation of the temperature with the depth and will just give an average thermal conductivity—the thermal properties of the tissues close to the surface of the skin can vary widely [655].

The microwave thermography is the technique of measuring the tissue temperatures of the human body by the detection of the microwave radiation which is naturally emitted by the body tissues. The microwave radiation can penetrate the distances of several centimetres through the biological tissues and hence the part of the radiation generated from within the body will reach the surface where it can be detected [1224].

The typical equipment for measuring the microwave thermal signals, and hence the body's temperature, consists of an antenna—a dielectrically loaded waveguide section placed on the surface of the skin—linked by a coaxial cable to a radiometer receiver. The radiation from the body couples with the antenna and a signal is transmitted to the radiometer which measures its intensity. The radiometer converts the signal to an equivalent Celsius temperature which can be displayed digitally, or recorded by a computer connected to the radiometer. A. Samsonov and S. V. Popov [2698] measured the temperature in the immediate vicinity of the individual cells during exposure to the electromagnetic field by coating their samples with molecular layers with the well-defined melting temperatures.

The skin temperature can be also measured using the liquid crystal thermography that uses the liquid crystal compounds exhibiting the colour-temperature sensitivity. The thin, blackened sheet, containing the liquid crystals encapsulated in a polymer matrix, is applied to the surface being examined. This system has many of the drawbacks: the sheet may fail to make the proper contact with the skin or the pressure applied by the sheet may affect the blood circulation just below the skin, either of which would give an incorrect temperature pattern [1224]. The range of temperatures observed by each plate is limited, so more than one plate may be needed to observe the full temperature pattern. Despite these drawbacks, this is the cheap and reasonably effective system [1490, 2608].

The non-contact method of determining the skin temperature is the infra-red thermography. The detector is placed in front of the area of interest and the skin surface is imaged by the detector. This technique is more accurate than the liquid crystal plate thermography, but it is also more expensive. There are two groups of detectors used: the thermal (usually pyroelectric) and the photon detectors.

R. C. Webb et al. [3274] used the mechanically soft arrays of the thermal actuators and sensors that laminate onto the skin to provide the rapid, quantitative in vivo determination of both the thermal conductivity and the thermal diffusivity, in a completely non-invasive manner. The sensor with the thickness, modulus and thermal mass matched to the epidermis, for the spatiotemporal mapping of temperature on the surface of the skin with the precision equal to or better than that of the state-of-the-art infrared thermography enables mapping of not only temperature but also the thermal transport properties, including the thermal conductivity and the thermal diffusivity (and, therefore, the heat capacity per unit volume via the ratio of these two quantities) their in-plane directional anisotropies [3273]. The authors noted that obtaining similar data with the alternative measurement techniques would require either the expensive and complex optical thermography and the laser heating systems, or the bulky single-point probes that can have the undesired effects on the skin properties of interest due the pressure that must be applied to the skin when measured in vivo.

The most reliable and accurate method to determine the temperature at a point inside the human body is inserting a thin temperature probe, such as a thermistor or a thermocouple, into the tissue—a probe equipped with a single thermistor, which performs a dual function: measuring the tissue temperature and providing the thermal power to the tissue [3133].

The dual-needle technique is an experimental measurement method that allows for the quantification of thermal conductivity, thermal diffusivity, and volumetric heat capacity of biological tissues as temperature changes [285]. A probe with two parallel needles is used both to provide heat and measure the temperature in the tissue, once it is inserted in the prepared tissue sample. The basic principle of the dual-needle technique involves the imposition of a current to the first needle for some time, hence providing a certain heat to the specimen, and the measurement of the tissue temperature variations, exerted by the second needle with temperature sensing capabilities. The dual-needle technique could be employed also to study the thermal properties of interest during cooling processes—following the heating process, it is sufficient to let the thermal bath cool naturally and perform subsequent measurements as temperature subsides [285].

Differential Scanning Calorimetry (DSC) is a widely utilized method for the investigation of the thermal properties of materials, including the measure of the specific heat capacity of tissues. The temperature-dependent variation of the physical and energetic material properties is monitored versus temperature and time. The DSC instruments typically consist of a single or a double furnace, two crucibles, one containing the specimen of the material, whereas the other one is left empty and serves as a reference, and the thermocouple sensors for the temperature monitoring [596, 1077].

Evidently, the invasive techniques are limited by the discomfort the probes produce in a patient and the danger of the damage to the tissue. A number of the non-invasive technique exist:

- The applied potential (electrical impedance) tomography is based on the measurement of the electrical resistivity variations. An array of the electrodes are placed on the skin surface. The resolution is determined by the number of the electrodes [621, 2198].
- The microwave tomography reconstructs the dielectric properties of a body illuminated with the microwaves from measurement of the scattered fields: the radiation is applied to one side of the region under investigation, where it is scattered and attenuated as it propagates through the tissue [1224]. The change in the microwave dielectric properties of the biological tissues are due to the change in the water permittivity in the tissues (the tissues can have the water contents which are less than 10% to greater than 80% of the total mass) with temperature.
- The ultrasonic methods are based on the temperature dependence of the acoustic properties of the biological tissues.

The increase of the temperature alters the characteristics of the tissue. The tissue appearance changes above 50° C–60° C due to the coagulation processes, and above 100° C the tissue carbonises [2858] due to the tissue desiccation induced by the vaporisation of the water content [2570, 3401]. The shrinkage of the tissue was reported to occur at the ablative temperatures [1953, 1954].

N. P. Silva et al. [2858] used the electromagnetic-based hyperthermia to study the thermal properties of the ex vivo ovine liver in the wide temperature range—from 25° C to 97° C. A significant increase in the thermal properties was observed only above 90° C. Silva et al. suggested to approximate the temperature dependence of the thermal conductivity as $\varkappa(T) = a + be^{cT}$ where a, b and c are the regression coefficients that best fit the data.

The authors observed the exponential growth of thermal conductivity in the ovine liver beyond 90° C. N. P. Silva et al. also performed the measurement of thermal properties during cooling cycles from 95° C to the body temperature 37° C and found no hysteresis.

The studies of the temperature-dependent thermal properties at hyperthermic temperatures (above 40° C) using the different measurement techniques leading to the different results [280, 281, 555, 1186, 1208, 1954, 3133]. Most of these studies have investigated the thermal properties in ex vivo biological tissues in the steady state heat transfer conditions neglecting the influence of the blood perfusion. An "effective" thermal conductivity and an "effective" thermal diffusivity should be considered, instead, to account for the contribution of the perfusion to the steady state heat transfer in vivo [3132].

A self-heated thermistor inserted into the tissue was used by J. W. Valvano et al. [3133] to study the thermal conductivity of the different ex vivo tissues in a temperature range between 3° C and 45° C. A similar method was used to determine the thermal conductivity of the porcine myocardium, at temperatures between 25° C and 76° C by N. C. Bhavaraju and J. W. Valvano [281] to measure the thermal conductivity of the ex vivo human and porcine liver for a range of temperatures between 20° C and 85° C.

L. Bianchi et al. [285] presented an extensive review of the thermophysical and mechanical properties of the biological tissues as a function of temperature. The temperature range was different for the different tissues up to 20° C–112 ° C. The human compact bone (the axial direction, the radial direction, the tangential direction), the bovine trabecular bone (both the defatted and non defatted state), the bovine liver, the porcine liver ex vivo, bovine liver ex vivo and in vivo, the human liver ex vivo, the ovine collagen ex vivo, the pig forelimb (across fibres and along fibres), the porcine myocardium, the human brain (white and gray matter), the calf brain, the ovine kidney were considered, among others.

The experiments on the hog liver tissue showed that the heat capacities and the thermal conductivities increase insignificantly ($\approx$ 3%) for temperatures up to 75 ° C, after which the significant changes occur due to the appreciable water loss [3266, 3267].

The experimental methods to determine the thermal properties of the biological tissues are rather advanced for the case of the homogeneous samples. The measurements of the spatially varying properties are more involved, especially when simultaneously both the thermal conductivity and the heat capacity are to be determined. The additional difficulty of the measurement of the thermal properties of the biological tissues in vivo is the need to estimate and separate the blood perfusion contribution to the heat transfer. Finally, the measurement techniques should be the non-invasive and be able to operate with the small temperature differences acceptable for the tissues. As a rule, one should solve the inverse problem and frequently exploit some model of the heat transfer in the biological media.

Sometimes it is assumed that the thermal conductivity depends exponentially on temperature as [2861]:

$$\lambda = \lambda_0 e^{\frac{E_0}{k_B T}} \tag{6.3}$$

where E_0 is the activation energy, i.e. the amount of energy which must be provided to the biological tissue to undergo the heating process and the denaturation process.

J. M. J. Huttunen, T. Huttunen, M. Malinen and J. P. Kaipio reviewed approaches to the determination of the heterogeneous thermal properties of the biological tissue [1389]. The authors themselves used the US induced heating to provide the necessary temperature gradients in the tissues and the MR imaging the temperature evolution. This technique was used earlier to measure the properties of homogeneous medium [525] and [3162]. Huttunen and co-workers considered the target domain that contains several sub-domains in which the tissue parameters, both thermal and acoustic, can be assumed constant. Note that this approach could provide the temperature-independent properties only. To extract the thermal properties from the temperature measurements, the authors used the classical Pennes' equation neglecting the metabolic heat source. The semi-discrete Galerkine FEM with the piecewise linear basis functions was used.

The thermophysical properties of some specific tissue are known to depend on the number of the factors such as the age, the gender, the ethnicity, the circadian rhythm, the state of thermoregulatory sweating. For example, V. Lopresto et al. [1951] implemented a simulation model of the MWA, considering the temperature-dependent dielectric and thermal properties of tissue; the results showed the especially relevant influence of the specific heat of tissue on the temperature increase and the extension of the thermally attained coagulation zone.

The state of the blood vessels greatly effect the measured values of the tissue parameters. For example, the thermal conductivity of the human skin could vary from 0.2–0.3 W/mK in the vasoconstricted state to 0.4–0.9 W/mK in the vasodilated state [2498].

The high-fidelity tissue thermodynamic models are needed to predict the tissue thermal response and could be used in the robot-assisted surgery and for the real-time monitoring and estimation of the heat spread underneath the tissue surface when performing the minimally invasive energy-based surgery. Such thermodynamics models must be fit to the actual physical process before they can be employed for the simulation synthesis. This process of the model or the parameter estimation was considered by H. El-Kebir et al. [850] who used the physics-based learning of the parametrized thermodynamics from the real-time infrared thermography.

6.3 Human

The first measurements of the heat conduction in the human skin were performed by F. Klug in 1874 and by Lefevre in 1901, who reported data on the thermal conductivity of the peripheral tissues in man [600, 3381]. The thermal conductivity and the specific heat of some human tissues are presented in Table 6.1.

Yu. I. Luchakov et al. [1978] noted that the differences in the thermal conductivity of the soft tissues (except fat) are relatively small due to the large fraction of water.

Table 6.1 Properties of human tissues

Tissue	Density, 10^3 kg/m^3	Thermal conductivity, W/mK	Specific heat 10^3 J/kgm^3	Reference
Adipose tissue	0.95	0.27	3.1	[378]
Blood	1.05	0.52	3.84	[3008]
Blood	1.058	0.45	3.96	[3253]
Bone, cancellous	1.7	0.582	1.59	[2498]
Bone, cortical	1.92	0.79	1.30	[378]
Bone	1.36	0.75	1.70	[3297]
Bone	1.038	0.436		[3253]
Bone	1.908	0.32	1.31	[3008]
Brain	1.08	0.49	3.85	[3297]
Brain	1.04	0.54	3.64	[378]
Brain		0.527		[628]
Brain		0.501–0.573		[350]
Brain, white matter		0.443–0.55		[1224]
Brain, grey matter		0.529–0.574		[1224]
Brain, grey matter		0.7		[2913]
Brain, cortex		0.512		[3133]
Cartilage		0.21 ± 0.06	3.2 ± 0.13	[2167]
Cartilage		0.518 ± 0.026		[2458]
Cornea	1.05	0.58	4.178	[2271, 2293]
Dentine	1.96–2.18	0.45–0.96	1.17–1.59	[1889]
Dentine		5.69	1.59	[387]
Enamel	2.80–2.95	0.93	0.75	[1889]
Enamel		9.34	0.71	[387]
Fat	0.85	0.45	2.3	[1998]
Fat		0.171–0.321		[1224]
Fat	0.85	0.16	2.3	[3321]
Fat, surface		0.268		[350]
Fat, intermediate		0.248		[350]
Fat, deep		0.219		[350]
Gland, mammary	1.05	0.48	3.77	[70]
Iris		1.004		[576]
Heart	1.08	0.47	3.55	[3297]
Kidney	1.05	0.54	3.9	[2968]
Kidney		0.54	3.7	[378]
Kidney	1.05	0.546	3.74	[167]
Kidney		0.417–0.59		[1224]
Kidney		0.53 ± 0.04	3.6	[2616]
Lens		0.40		[1757]
Liver	1.06	0.57	3.6	[2968]
Liver	1.06	0.52	1.06	[378]
Liver		0.5122		[157]

(continued)

Table 6.1 (continued)

Tissue	Density, 10^3 kg/m^3	Thermal conductivity, W/mK	Specific heat 10^3 J/kgm^3	Reference
Liver	1.05	0.567		[167]
Liver		0.5	4.18	[1192]
Liver		0.49	3.6	[2420]
Liver		0.438–0.592		[1224]
Lung		0.4506		[157]
Lung	0.56	0.28	3.52	[3297]
Muscle	1.05	0.642	3.75	[2498]
Muscle	1.27	0.53	3.8	[3321]
Muscle	1.05	0.50	3.46	[378]
Muscle, heart		0.431–0.574		[1224]
Muscle, heart	1.2	0.55		[1936]
Pancreas		0.325–0.523		[1224]
Pulp	1.00	0.63	4.2	[1889]
Sclera	1.0	0.603	4.178	[2271]
Sclera	1.05	1.04	3.18	[50]
Sclera	1.10	1.0004	3.18	[2293]
Skin, epidermis		0.21	4.32	[655]
Skin, epidermis		0.24	3.59	[1475]
Skin, epidermis, corium		0.335		[2623]
Skin, subcutaneous		0.19	2.50	[1475]
Skin, subcutaneous	1.21	0.194	2.24	[799]
Skin, subcutaneous	1.0	0.16–0.21	2.29–3.06	[3074]
Skin, dermis	1.2	0.53	3.8	[3321]
Skin, dermis		0.53	4.56	[655]
Skin, dermis	1.09	0.42	3.35	[799]
Skin, dermis		0.45	3.30	[1475]
Skin, dermis		0.294		[2623]
Skin, dermis	1.2	0.37–0.52	3.20–3.40	[3074]
Spleen		0.5394		[157]
Spleen		0.493–0.559		[1224]
Spleen	1.05	0.546		[167]
Tarsal	1.25	0.47	3.6	[1328]
Viscera[a]	1.08	0.49	3.51	[2752]
Vitreous	1.0	0.603	4.178	[2271]

[a]Viscera is a model mixture of liver, kidney, stomach, gut, spleen, bladder, connective tissue [953]

The data on the temperature dependence of the thermal conductivity is even more scarce. One of the few results is the thermal conductivity of the human liver suggested by Trujillo and Berjano [3087] as $\lambda(T) = \lambda_{T_{ref}} + 0.0013(T - T_{ref})$ where $\lambda_{T_{ref}} = 0.502$ and $T_{ref} = 25°$ C.

6.4 Animals

The animal models that are currently in use include the nonhuman primates, the rodents (the rats, the mice, the guinea pigs), the large animal models (the pig, the dog, the sheep), the chick, and the simple animals, including the fish, the insects, and the round worms. Each model system has the strengths and the weaknesses, depending on the question being addressed. Physiologically the pig is the closest to the human and is the preferred model compared to other organisms [2214, 3100].

The thermal conductivities of some animal tissues adapted from [600, 686, 856, 1224, 1260, 1261, 1748, 1783, 1865, 2188, 2481, 2858] are presented in the Table 6.2.

Table 6.2 Properties of animal tissues

Tissue	Animal	Thermal conductivity W/mK
Cortea	Rabbi	0.58
Fat, blubber	Whale	0.21–0.31
Fat, subcutaneous	Lamb	0.212–0.261
Fat, kidney	Lamb	0.156–0.201
Fat, subcutaneous	Cow	0.166–0.314
Fat, kidney	Cow	0.097–0.144
Fat, subcutaneous	Pig	0.152–0.338
Kidney, whole	Rabbit	0.502
Kidney, cortex	Rabbit	0.465
Kidney, cortex	Dog	0.491
Liver	Rabbit	0.493
Liver	Rat	0.498
Liver	Sheep	0.495
Liver	Sheep, 25° C	0.50
Liver	Sheep, 70° C	0.55
Liver	Sheep, 92° C	0.61
Liver	Sheep, 97° C	1.08
Liver	Dog	0.55
Lung	Dog	0.57–0.63
Muscle	Rat	0.505
Muscle	Lamb	0.440–0.557
Muscle	Pig	0.518
Muscle	Pig	0.451–565
Muscle	Cow	0.410
Muscle	Cow	0.411–0.575
Skin	Giraffe	0.442
Skin	Pig	0.414
Skin	Chicken	0.356
Skin	Crocodile, back	0.432
Skin	Crocodile, tail	0.334

6.5 Latent Heat

One more parameter characterizing the biological solutions should be known to provide the reliability of the numerical simulations—the latent heat of the phase change that is frequently taken equal to that of the pure water. B. Han and J. C. Bishof using the DSC thermograms, measured the latent heat of the different aqueous mixtures of the biological relevance [1229], including the sodium chloride and the phosphate buffered saline solution with the different chemical additives. They found that the latent heats do not correlate directly (that is, in a linear fashion) with the water content, but could be correlated with amount of the water that participates in the phase change. The measurements gave the value 303.7 J/g and 233.0 J/g for the water-ice and the eutectic phase change, respectfully. The reason for the former being smaller than that for the pure water (335 J/g) warrants further study. The latent heat of the crystallization is found to sharply drop for the large concentration of the glycerol [552]. The authors also found the disappearance of the eutectic phase transition when even a small amount of glycerol was added that again was observed after addition of the AFP. The eutectic crystallization occurs only under a significant supercooling in contrast to the melting that is in good agreement with the phase diagram temperature.

The similar studies have been performed by R.V. Devereddy et al. [735] who based the analysis of the measurements, as the authors themself state, "on the full set of heat and mass transport equations". The authors wrote out the equations for the case of the spherical crystals growing in the solution that are assumed all be identical and grow independently each in its "own pool of liquid". In fact, only the mass transfer equation is solved, since estimates of the similarity parameters showed that the heat transfer occurs so rapidly that the temperature could be assumed to be uniform throughout the system. The authors suggested several explanations for the observed reduction of the latent heat in the considered aqueous solutions in comparison with pure water including the transformation of some amount of water into the "unfreezable" form due to binding to the solutes and the possible entropic effects due to the ordering of the water in the presence of the solutes prior to the phase transition.

6.6 Optical Properties

To study the thermal response of laser-irradiated tissue, one needs to know two parameters, the absorption and scattering coefficients, which describe the optical absorption of light in the tissue [2298]—they allows one to determine the optical penetration of light into the tissue and the fraction of light will be absorbed [1427]. These two coefficients are specific to each tissue and depend on the radiation wavelength. The inverse of the absorption coefficient is defined as the optical absorption length.

Knowledge of these coefficients is necessary to correctly plan and carry out a laser treatment, i.e. to specify the number of parameters that could be varied in the laser action on the tissues: the laser wavelength, the beam waist, the number and the duration of pulses, the energy per pulse.

The main optical parameters of the tissues are the absorption coefficient, the scattering coefficient, the real refractive index (the complex refractive index includes the real refractive that describes energy storage and hence affects the speed of light in the medium and the imaginary refractive describing the energy dissipation and thus the absorption coefficient). The terms *Rayleigh scattering* and *Mie scattering* are commonly used in the field of the biomedical optics, with Rayleigh scattering referring to scattering by the small particles or the mass density fluctuations much smaller than the wavelength of light, and the Mie scattering referring to scattering by the particles comparable to or larger than the wavelength of light [1427].

There are several reviews that tabulate the optical properties (absorption, scattering, anisotropy, reduced scattering, refractive index) of the various tissues measured at some wavelengths: [202, 528, 1624, 2005, 2701]. M. Jaunich et al. [1446] presented the absorption and scattering coefficients for the human and mouse skin tissue (the epidermis, the dermis, the fatty tissue) at two values of the radiation wavelength.

There are direct and indirect for determining the optical parameters of the tissues [3094]. Direct methods include those based on some fundamental concepts such as the Bouguer-Beer-Lambert law, the single-scattering phase function for thin samples, or the effective light penetration depth for slabs. The parameters measured are the collimated light transmission the angular dependence of the scattered light intensity for the thin samples. These methods are advantageous in that they use the simple analytic expressions for the reconstruction of the optical parameters of the tissue. Their disadvantages are related to the necessity to strictly fulfil the experimental conditions dictated by the selected model (e.g., the single scattering in the thin samples, exclusion of the effects of the light polarization and the refraction at the cuvette edges).

The indirect methods obtain the solution of the inverse scattering problem using a theoretical model of the light propagation in a medium. They are, in turn, divided into the iterative and non-iterative models. The former uses the equations in which the optical properties are defined through the parameters directly related to the quantities being evaluated. The latter are based on the two-flux Kubelka-Munk model and the multi-flux models. These methods are cumbersome, but the optical models currently in use may be even more complicated than those underlying the non-iterative methods.

S. L. Jacques [1427] suggested an expression for the absorption coefficient of the *generic* tissue that is based on the content of the absorbing chromophores (blood, water, melanin, fat, yellow pigments) in the tissue and by the balance of the Rayleigh and the Mie scattering

$$\mu_\alpha = BS\mu_{\alpha,oxy} + B(1-S)\mu_{\alpha,deoxy} + W\mu_{\alpha,water} + F\mu_{\alpha,fat} + M\mu_{\alpha,melanosome} + 2.3C_{bili}\varepsilon_{bili} + 2.3C_{\beta C}\varepsilon_{\beta C} \tag{6.4}$$

where S is the haemoglobin (HGb) oxygen saturation of the mixed arterio-venous vasculature, B is the average blood volume fraction, W is the water content, $bili$ is the bilirubin concentration, βC is the β-carotene concentration, F is the fat content, M is the melanosome volume fraction, or alternatively the molar concentration of the melanin monomers.

Since the optical properties of the living tissue can vary considerably from individual to individual, site to site, and even time to time [2740, 3201]—for example, the optical properties of the lung are a function of the respiration [230]—it is desirable to measure the tissue optical properties directly during a procedure, so that the physician could regulate the laser settings accordingly.

A. Arnold and L. Fichera [111] suggested the implementation of the Ensemble Kalman Filter [885, 929, 1571] (a type of the Bayesian filtering algorithm for data assimilation) for the estimation of the tissue optical properties based on observations of the tissue thermal response to the laser irradiation—the method that has the potential for the straightforward implementation in a clinical setup, since it only requires a simple thermal sensor and the tissue temperature is routinely monitored during many medical laser procedures [2674].

The authors showed that it is possible to estimate the additional unknown tissue physical properties (the volumetric heat capacity and the thermal conductivity) along with the time-varying absorption and scattering coefficients.

Chapter 7
Exact Solutions

The exact analytical solutions are usually obtained for the one dimensional problems under the assumption that the thermal tissue properties are constant. The latter, however, is as a rule acceptable in the case of local hyperthermia. A few examples are presented in this section; numerous solutions of both Pennes equation and DPL bioheat transfer model are considered in the review [919].

Among the analytical approaches used to investigated the heat conduction in the biological tissues are the finite Fourier transform (FFT), Green's functions (GFs) method, Laplace transform method (LTM), and separation of variables method (SOV).

For example, A. Shitzer and J. C. Chato [2839] used the FFT to study the transient heat conduction in a domain with the combination of skin and skeletal muscle (the transient times for reaching the "locally fully developed" temperature profile were found to depend on the geometry—rectangular and cylindrical coordinates were considered—and on the parameter representing the ratio of heat supplied by blood flow to heat conducted in the tissue), while E. Kengne et al. [1594] analysed the influence of the blood flow on the temperature distribution in the skin tissue for the different spatial heating—heating both on the skin surface and inside the biological bodies. E. Kengne and A. Lakhssassi [1591] solved the bioheat transfer equation with the constant blood perfusion for a simplified one-dimensional model of the spherical living biological tissues.

The GFs method is used for determining the thermal behaviour in the living tissues under the non-homogeneous boundary conditions. R. Vyas and M. L. Rustgi [3217] found the Green's function solution to the tissue bioheat equation including the blood flow in the cylindrical geometry in the bovine muscle and the liver tissue that is exposed to the neodymium–yttrium–aluminum garnett (Nd:YAG) laser with the Gaussian radiation profile capable of being operated in both the pulsed and the continuum wave modes. The authors considered the cylindrical symmetric model and used the Hankel transform with respect to the radial coordinate.

A. I. Zhmakin, *Heat Transfer in Vivo*, Springer Series in Biophysics 27,
https://doi.org/10.1007/978-3-031-95459-7_7

B. Gao et al. [1047] used a time-dependent Green's function and the Fourier transform techniques to describe both the steady-state and time-dependent effects of the inhomogeneous blood perfusion and to model the thermal conduction hyperthermia system and the RF interstitial system assuming the constant thermal conductivity; a possibility for measuring the blood perfusion parameter and the thermal conductivity from the steady state temperature distribution of a point source was also proposed.

W. H. Newman et al. [2292] applied a Green's function solution of the tissue bioheat equation to the analysis of the temperature decay curves following the local energy deposition representative of the ultrasound and microwave hyperthermia therapy devices. For the measurements of the effective blood flow following heating, the errors are dependent on the measurement position within the heated volume, the heating geometry, and duration of the heating prior to the decay. Thus, as the authors stressed, the effective tissue property calculated without correctly modelling the heating geometry, the boundary conditions and the initial conditions will be of a qualitative rather than the quantitative nature, and may lead to the erroneous and misleading conclusions concerning the tissue and the tumour response.

Z. S. Deng and J. Liu [713] used the analytical solutions to the bioheat transfer problems with the space or transient heating on the skin surface or inside the biological bodies obtained using the Green's function method to study several selected typical bioheat transfer processes, which are often encountered in the cancer hyperthermia, the laser surgery, the thermal comfort analysis, and the tissue thermal parameter estimation.

H. G. Klinger [1654, 1655] calculated the Green's function for the heat and matter transport in an infinite medium in which a convection field v (r, t) makes a contribution to the total heat and matter current. It is given by the uniformly and absolutely convergent series in which every term is calculated from the preceding one merely by the integration.

H. Ahmadikia et al. [31, 32] and B. Kundu and D. Dewanjee [1742] used the TWMBT model (as well as the Pennes equation) to simulate the action of the constant and the pulse train heat flux on the surface of the skin tissue.

The DPL model was used by H. Askarizadeh and H. Ahmadikia [122, 123] and by S. M. Lin [1880] to investigate the transient heating of the skin tissue. Askarizadeh and Ahmadikia considered the cases of the constant, the pulse train and the periodic heat flux. The biomechanics of the skin under the thermal treatment was studied also by F. Xu et al. [3380, 3384].

V. V. Kulish and J. L. Lage [1722] developed the relation between the temperature and the heat flux at the same point thus avoiding the need to solve the three dimensional time-dependent heat conduction equation

$$\frac{\partial T(x_1, x_2, x_3, t)}{\partial t} - \varkappa \nabla^2 T(x_1, x_2, x_3, t) = 0. \tag{7.1}$$

The authors considered the unidirectional problems within a semi-infinite domain assuming the constant and uniform properties. Using the change of variables $\vartheta = T - T_0$ and $(\xi_1, \xi_2, \xi_3) = \varkappa^{-1/2}(x_1, x_2, x_3)$ the authors get

$$\frac{\partial T(\xi_1, \xi_2, \xi_3, t)}{\partial t} - \nabla^2 T(\xi_1, \xi_2, \xi_3, t) = 0 \tag{7.2}$$

and after performing the Laplace transform [774] accounting for the initial condition $\vartheta(\xi_1, \xi_2, \xi_3, 0) = 0$ obtained

$$s\vartheta^\star(\xi_1, \xi_2, \xi_3, s) - \nabla^2 \vartheta^\star(\xi_1, \xi_2, \xi_3, s) = 0. \tag{7.3}$$

The separation of variables $\vartheta^\star(\xi_1, \xi_2, \xi_3, s) = X_1(\xi_1)X_2(\xi_2)X_3(\xi_3)$ reduces (7.3) to the system of equations

$$\frac{1}{X_j}\left[\frac{1}{\prod_j h_j}\frac{\partial}{\partial \xi_j}\left(\frac{h_p h_q}{h_j}\frac{\partial X_j}{\partial \xi_j}\right)\right] = \lambda_j^2 s, \qquad p \neq q \neq j = 1, 2, 3. \tag{7.4}$$

where h_1, h_2, h_3 are multipliers of the orthogonal coordinate system. The solution of (7.4) in the Cartesian coordinates is

$$X_j(\xi_j, s) = \sum_{m_j=-\infty}^{\infty} [A_{jm_j}(s)e^{s^{1/2}\lambda_{jm_j}\xi_j} + B_{jm_j}(s)e^{-s^{1/2}\lambda_{jm_j}\xi_j}]. \tag{7.5}$$

Thus the Laplace image of the temperature is

$$\vartheta^\star = \sum_{m_1=-\infty}^{\infty}\sum_{m_3=0}^{m_1}\sum_{m_2=0}^{m_1}\left[\prod_{j=1,2,3}(A_{jm_j}(s)e^{s^{1/2}\lambda_{jm_j}\xi_j} + B_{jm_j}(s)e^{-s^{1/2}\lambda_{jm_j}\xi_j})\right]. \tag{7.6}$$

It is follows from (7.6) that the equality

$$\frac{\partial \vartheta^\star(\xi, s)}{\partial \xi} = -s^{1/2}\vartheta^\star(\xi, s) \tag{7.7}$$

holds when, depending on the sign of λ_{jm_j}, $A_{jm_j} = 0$ or $B_{jm_j} = 0$ and when λ_{jm_j} is unity.

J. Dutta and B. Kundu [825] investigated the one dimensional bioheat transfer in a single layer of the living tissue using the Fourier, the single-phase-lag, and the dual-phase-lag models. The authors applied the separation of variables and the finite Fourier transform method to get the closed-form solution. The skin surface is heated by the constant or unique oscillating heat sources to simulate the non-invasive therapeutic treatment. The authors concluded that a sinusoidal heat source is a better option for the longer duration of the therapeutic heating to reach the thermal equilibrium, while a cosine heat source is more applicable for the short time span of the treatment with a range of the time-lags in the dual-phase-lag bioheat model.

7.1 Fundamental Solutions of the Pennes Equation in the Principal Orthogonal Coordinate Systems

M. A. Giordano et al. [1082] obtained the fundamental solutions of the Pennes equation in the rectangular, cylindrical, and spherical coordinates. The isotropic medium with the effective thermal properties is considered. The authors wrote the Pennes equation in the form

$$\left(\frac{\partial}{\partial t} - \alpha\nabla^2 + \gamma^2\right)\vartheta(\mathbf{x},t) = L(\vartheta(\mathbf{x},t)) = \frac{q(\mathbf{x},t)}{\varrho c} \tag{7.8}$$

where $\gamma^2 = \varrho_b c_b w_b/\varrho c$ is a constant which characterises the blood perfusion rate, $\alpha = \lambda\varrho c$ is the thermal diffusivity of the tissue, $q(\mathbf{x},t) = \dot{q}^{met}(\mathbf{x},t) + Q(\mathbf{x},t)$.

The equation for the Green's function $G(\mathbf{x},t)|\mathbf{x}',t'))$ is written as

$$L(G(\mathbf{x},t)|\mathbf{x}',t'))) = \delta(\mathbf{x}-\mathbf{x}')\delta(t-t'). \tag{7.9}$$

7.1.1 Cartesian Coordinates

The fundamental solution for the one dimensional problem in the Cartesian coordinates is the solution of the equation

$$L(G) = \left(\frac{\partial}{\partial t} - \alpha\frac{\partial^2}{\partial x^2} + \gamma^2\right)G = \delta(x-x')\delta(t-t'). \tag{7.10}$$

The Fourier transform with respect to the spatial coordinate is used to eliminate it

$$\hat{G}(\omega,t) = \mathcal{F}[G(x,t)] = \frac{1}{\sqrt{2\pi}}\int_{-\infty}^{\infty} G(x,t)e^{-i\omega x}\mathrm{d}x, \tag{7.11}$$

$$G(x,t) = \mathcal{F}^{-1}[\hat{G}(\omega,t)] = \frac{1}{\sqrt{2\pi}}\int_{-\infty}^{\infty} \hat{G}(\omega,t)e^{i\omega x}\mathrm{d}\omega. \tag{7.12}$$

and get an equation in time

$$\frac{\partial \hat{G}}{\partial t} + \left(\alpha\omega^2 + \frac{\gamma^2}{\alpha}\right)\hat{G} = \frac{\delta(t-t')}{\sqrt{2\pi}} e^{-i\omega x}. \quad (7.13)$$

Finally the authors got

$$G(x-x', t-t') = \frac{H(t-t')}{\sqrt{4\pi\alpha(t-t')}} \exp\left[\frac{-(x-x')^2}{4\alpha(t-t')} - \gamma^2(t-t')\right]. \quad (7.14)$$

7.1.2 Cylindrical Coordinates

The authors used the Hankel transform to get

$$G(r-r', t-t') = \frac{H(t-t')}{2\alpha(t-t')} \exp\left[\frac{-(r-r')^2}{4\alpha(t-t')} - \gamma^2(t-t')\right] I_0\left(\frac{rr'}{2\alpha(t-t')}\right) \quad (7.15)$$

where I_0 is the modified Bessel function of the first kind.

7.1.3 Spherical Coordinates

The fundamental solution for the spherical coordinates is the solution of the equation

$$L(G) = \left(\frac{\partial}{\partial t} - \alpha\frac{\partial^2}{\partial r^2} - \frac{2\alpha}{r}\frac{\partial}{\partial r} + \gamma^2\right) G = \delta(r-r')\delta(t-t'). \quad (7.16)$$

The final expression for the fundamental solution is

$$G(r-r', t-t') = \frac{H(t-t')e^{-\gamma^2(t-t')}}{2rr'\sqrt{\pi\alpha(t-t')}} \left[e^{\frac{-(r-r')^2}{4\alpha(t-t'}} - e^{\frac{-(r+r')^2}{4\alpha(t-t'}}\right]. \quad (7.17)$$

7.2 One Dimensional Tissue Under Surface Heating

F. Xu et al. [3381] used the Green's function method to investigate the one dimensional heat transfer in the skin by the Pennes equation that had been transformed by the separation of the time-independent part in the temperature

$$T(z,t) = T_0(z) + W(z,t)e^{-(\omega_b \varrho_b c_b/\varrho c)t} \quad (7.18)$$

into the following form

$$\frac{\partial W}{\partial t} = \varkappa \frac{\partial^2 W}{\partial z^2} + \frac{Q(z,t)}{\varrho c} e^{-(\omega_b \varrho_b c_b/\varrho c)t} \tag{7.19}$$

where $\varkappa$ is the thermal diffusivity of the skin tissue.

The authors considered the case of no external heating and neglected the metabolic heat generation for six boundary conditions. For example, when the skin is heated at the surface by a heat source with a constant temperature $T(-h/2, t) = T_\infty$, e. g., in contact with a hot plate, while the bottom of the skin is kept at body temperature $T(h/2, t) = T_c$, the Green's function is written as ($\beta_m = m\pi/h$)

$$G(z, t|z', t') = \frac{2}{h} \sum_{m=1}^{\infty} e^{-\varkappa \beta_m^2 (t-t')} \sin(\beta_m z) \sin(\beta_m z') \tag{7.20}$$

and the temperature as

$$T(z,t) = T_0(z) + \frac{2\varkappa}{h} \left[T_\infty - \frac{\mathrm{d}T_0(z)}{\mathrm{d}z}|_{z=0} \right] \times \tag{7.21}$$

$$\times \sum_{m=1}^{\infty} \beta_m \sin[\beta_m (z + h/2)] \frac{1}{\varkappa \beta_m^2 + \frac{\omega_b \varrho_b c_b}{\varrho c}} \left(1 - e^{-\varkappa \beta_m^2 t - \omega_b \varrho_b c_b/\varrho c} \right). \tag{7.22}$$

D. W. Tang and N. Araki [3011] used the Cattaneo equation to study the one dimensional heat transfer in the finite medium under the laser-pulse heating and in the semi-infinite medium under the periodic heating. The authors exploited the dimensionless variables

$$\eta = \frac{x}{2\sqrt{\varkappa \tau}}, \qquad \xi = \frac{t}{2\tau}, \qquad \vartheta(\eta, \xi) = \frac{\lambda \sqrt{\tau}}{q_0 \sqrt{\varkappa}}. \tag{7.23}$$

The following boundary and initial conditions were used

$$\frac{\partial T(0,t)}{\partial x} = \frac{\partial T(L,t)}{\partial x} = 0, \quad T(x,0) = \frac{\partial T(x,0)}{\partial t} = 0. \tag{7.24}$$

The solution was obtained in the form of the infinite series.

The boundary and initial conditions for the periodic heating case are

$$-\lambda \frac{\partial T(0,t)}{\partial x} = q_0 e^{i\omega t}, \quad T(\infty, t) = \frac{\partial T(\infty, t)}{\partial x} = 0, \quad T(x,0) = \frac{\partial T(x,0)}{\partial x} = 0. \tag{7.25}$$

The solution is written with the use of the Heaviside unit step function and the modified Bessel function of zero order.

J. Ordonez-Miranda and J. J. Alvarado-Gil [2344] solved the Cattaneo-Vernotte equation for the system of the finite and semi-infinite layers in the perfect thermal contact

$$\frac{\partial^2 T(x,t)}{\partial x^2} - \frac{1}{\varkappa}\frac{\partial T(x,t)}{\partial x} - \frac{\tau}{\varkappa}\frac{\partial^2 T(x,t)}{\partial t^2} = -\frac{1}{\lambda}\left(S(x,t) + \tau\frac{\partial S(x,t)}{\partial t}\right), \tag{7.26}$$

where

$$S(x,t) = F(x)(1+\cos(\omega t)) = Re[F(x)(1+\exp(i\omega t))] \tag{7.27}$$

is the rate per unit volume of heat generation. The temperature in the sample is written as

$$T(x,t) = T_0 + T_{dc}(x) + T_a c(x,t), \tag{7.28}$$

T_0 is the ambient temperature, $T_{dc}(x)$ and

$$T_a c(x,t) = Re[\vartheta(x)e^{i\omega t}] \tag{7.29}$$

are the stationary and periodic components of the temperature.

Thus the general solution of (7.26) for the spatial part of the oscillatory temperature ϑ is given by $\vartheta(x) = Be^{qx} + Ce^{-qx}$, where the constants B and C depend on the boundary conditions and q is [2344]

$$q = \sqrt{\frac{i\omega}{\varkappa}}\sqrt{i\omega\tau} = \frac{\chi + i\chi^{-1}}{\mu}, \quad \chi = \sqrt{\sqrt{1+(\omega\tau)^2} + \omega\tau}, \quad \mu = \sqrt{\frac{2\varkappa}{\omega}}. \tag{7.30}$$

The perfect contact means the continuity of both the temperature and the heat flux at the interface $x = L$ between the layers

$$\vartheta(x = L_-) = \vartheta(x = L_+), \quad \frac{\lambda}{1+i\omega\tau}\frac{d\vartheta(x=L_-)}{dx} = \frac{\lambda}{1+i\omega\tau}\frac{d\vartheta(x=L_+)}{dx}, \tag{7.31}$$

where L_- and L_+ denotes the limit $x \to L$ from the left and from the right side.

B. B. Mikic [2127], K. J. Baumeister and T. D. Hamil [212] and D. E. Amos and P. J. Chen [69] derived the temperature distribution in a semi-infinite medium due to a step change in the temperature at its surface.

D. E. Glass et al. [1095] and M. J. Maurer and H. A. Thompson [2074] considered the heat conduction in a semi-infinite slab with the periodic temperature at the surface.

Y. Taifel [3002] and M. Lewandowska and L. Malinowski [1837] obtained the solution for the heat transport in a thin layer due to a step change of the temperature on the surfaces.

M. N. Özisik and B. Vick [2367] derived an analytical solution for the temperature distribution in a finite layer with the insulated boundaries due to a volumetric energy source. B. Abdal-Hamid [5], D. W. Tang and N. Araki [3010] investigated the heat conduction in a finite slab under periodic surface thermal disturbance and J. Gembarovich and V. Majernik [1061] under the heat pulse flux boundary conditions.

J. Gembarovich and J. Gembarovich, Jr. [1060] considered the temperature distribution in an isotropic homogeneous medium ($0 \leq x \leq L$) with zero initial temperature and the adiabatically insulated boundaries when one surface is heated by the stepwise heat pulse of duration t_1 using the Cattaneo model. The solution is written as

$$T(x,t) = \frac{1}{t_1}\int\limits_0^{t_1}\left[F(x,t-t') - \tau\frac{\partial F(x,t-t')}{\partial t'}\right]\mathrm{d}t' \tag{7.32}$$

where

$$F = \frac{L}{\sqrt{\varkappa\tau}}e^{\left(-\frac{t}{2\tau}\right)}\sum_{k=0}^{\infty}\left[I_0\left(\frac{1}{2\tau}\sqrt{t^2-\gamma^2}\right)H(t-\gamma) + I_0\left(\frac{1}{2\tau}\sqrt{t^2-\gamma_1^2}\right)H(t-\gamma_1)\right] \tag{7.33}$$

where $\gamma = (2kL+x)\sqrt{\tau/\varkappa}$, $\gamma_1 = (2kL+2L+x)\sqrt{\tau/\varkappa}$, $H(t)$ is the Heaviside unit step function, $I_0(z)$ is the modified Bessel function of the first kind of zero order.

F. Jiang [1469] considered the radial heat conduction in the hollow sphere with the constant properties, r_i is the inner radius, r_o is the outer radius. The initial temperature is uniform and equals to T_0, The heat transfer is caused by the sudden change of the inner surface temperature to T_{wi} and the outer surface temperature—to T_{wo}.

The steady-state equation using dimensionless variables

$$\eta = \frac{r}{r_o}, \quad \xi = \frac{\varkappa t}{r_o^2}, \quad \vartheta = \frac{T - T_o}{T_{wo} - T_o} \tag{7.34}$$

is written as [1469]

$$\frac{\partial^2\vartheta(\eta,\xi)}{\partial\eta^2} + \frac{2}{\eta}\frac{\partial\vartheta(\eta,\xi)}{\partial\eta} = \varepsilon\frac{\partial^2\vartheta(\eta,\xi)}{\partial\xi^2} + \frac{2}{\eta}\frac{\partial\vartheta(\eta,\xi)}{\partial\xi} \tag{7.35}$$

where $\varepsilon = \varkappa\tau/e_o^2$, the relaxation time $\tau = 3\varkappa/v^2$, v is the phonon velocity.

The initial and boundary conditions are

$$\vartheta(\eta, 0) = 0, \quad \frac{\partial \vartheta}{\partial \xi}|_{\xi=0} = 0, \quad (r_\gamma \leq \eta \leq 1), \quad \vartheta(r_\gamma, \xi) = T_\gamma, \vartheta(1, \xi) = 1 \ (\xi > 0) \tag{7.36}$$

where $r_\gamma = r_i/r_o, \quad T_\gamma = (T_{wi} - T_0)/(T_{wo} - T_0)$.

Applying the Laplace transform to (7.35) with respect to the variable ξ and accounting for the initial conditions, Jiang got

$$\frac{\partial^2 \bar{\vartheta}(\eta, s)}{\partial \eta^2} + \frac{2}{\eta}\frac{\partial \bar{\vartheta}(\eta, s)}{\partial \eta} - (s + \varepsilon s^2)\bar{\vartheta}(\eta, s) = 0 \tag{7.37}$$

with the boundary conditions $\bar{\vartheta}(r_\gamma, s) = T_\gamma/s, \quad \bar{\vartheta}(1, s) = 1/s$.

The solution of (7.37) was transformed with the use of the expansion

$$\frac{1}{1 - \exp[-2\sqrt{s + \varepsilon s^2}(1 - r_\gamma)]} = \sum_0^\infty \exp[-2\sqrt{s + \varepsilon s^2}(1 - r_\gamma)]. \tag{7.38}$$

and a table of inverse Laplace transform to get a cumbersome expression for $\eta\vartheta(\eta, \xi)$ that reduces to the solution for the solid sphere when $r_\gamma \to 0$ (i.e. when $r_i \to 0$ or $r_o \to \infty$) and to the solution of the corresponding parabolic problem for $\varepsilon \to 0$.

7.3 One Dimensional Model of RF Thermal Ablation

J. A. L. Lopéz Molina et al. [2174] investigated one of the energy-based ablative techniques—the radiofrequency thermal ablation—using the model based on the electrode of the radius r_0 embedded into the biological tissue. The authors considered the Fourier, the hyperbolic and the "relativistic" heat conduction equations. The latter is written as

$$-\varkappa\left(\frac{\partial^2 T(r, t)}{\partial r^2} + \frac{2}{r}\frac{\partial T(r, t)}{\partial r}\right) + \frac{\partial T(r, t)}{\partial t} + \frac{\varkappa}{C^2}\frac{\partial^2 T(r, t)}{\partial t^2} = Q(r) \tag{7.39}$$

where $Q(r) = Pr_0/4\pi r^4, \quad r \geq r_0$ is the Joule heat produced per unit volume of the tissue with the initial and boundary conditions

$$T(r, 0) = t_0, \quad \frac{\partial T}{\partial t}(x, t) = 0, \quad \lim_{r \to \infty} T(r, t) = 0. \tag{7.40}$$

The boundary condition at the interface between the electrode and the tissue ($r = r_0$) was obtained under assumption that it mainly governed by the thermal inertia of the electrode and the heat increment in the mass of the electrode equals to

$$\varrho_0 c_0 \frac{4\pi r_0^3}{3} \frac{\partial T}{\partial t}(r_0, t)$$

where ϱ_0 and c_0 are the density and the specific heat of the electrode, thus

$$4\pi \varkappa r_0^2 \sqrt{-\frac{1}{C^2}\left(\frac{\partial T}{\partial t}(r_0, t)\right)^2 + \left(\frac{\partial T}{\partial r}(r_0, t)\right)^2} = \varrho_0 c_0 \frac{4\pi r_0^3}{3} \frac{\partial T}{\partial t}(r_0, t). \tag{7.41}$$

The authors used the dimensionless variables

$$\varrho = \frac{r}{r_0}, \ \xi = \frac{\varkappa t}{r_0^2}, \ \mu = \left(\frac{\varkappa}{r_0 C}\right)^2, \ V = \frac{4\pi \varkappa r_0}{P}(T - T_0) \tag{7.42}$$

to solve the equation

$$-\left(\frac{\partial^2 V}{\partial \varrho^2} + \frac{2}{\varrho}\frac{\partial V}{\partial \varrho}\right) + \frac{\partial V}{\partial \xi} + \mu \frac{\partial^2 V}{\partial \xi^2} = \frac{1}{\varrho}. \tag{7.43}$$

Applying the Laplace transform $\mathcal{L}(\varrho, s)$ the authors obtained

$$-\left(\frac{\partial^2 \mathcal{L}}{\partial \varrho^2} + \frac{2}{\varrho}\frac{\partial \mathcal{L}}{\partial \varrho}\right) + (s + \mu s^2) = \frac{1}{\varrho^4 s} \tag{7.44}$$

with the general solution

$$\mathcal{L}(\varrho, s) = \frac{e^{\nu\varrho}}{2\varrho\nu}\left(\frac{1}{s}\int_1^{\varrho} \frac{e^{-\nu u}}{u^3} du + M_1(s)\right) + \frac{e^{-\nu\varrho}}{2\varrho\nu}\left(\frac{1}{s}\int_1^{\varrho} \frac{e^{\nu u}}{u^3} du + M_2(s)\right) \tag{7.45}$$

where $\nu = \sqrt{s + \mu s^2}$, $M_1(s)$ and $M_2(s)$ determined by the boundary conditions [2174].

J. A. L. Lopéz Molina et al. [1950] considered the cylindrical geometry to study the RF ablation. The device consisted of a metallic electrode with an internal liquid cooling system that cools the electrode surface. The authors obtained (the rather cumbersome) analytical solution in terms of the Bessel functions and the modified Bessel functions of the complex argument.

7.3.1 Solutions of the DPL Models in a Semi-Infinite Domain

M. A. Castro et al. [465] considered the DPL heat transfer model for a plate of infinite thickness, $x \in [0, \infty]$, that can be heated either at its surface $x = 0$, or up to a certain depth, $x \in [0, l]$. The authors treated the cases of the Dirichlet, $T(0, t) = 0$, and the Neumann, $T_x = 0$, boundary conditions and also assumed that

$$\lim_{x \to \infty} T(x, t) = 0, \qquad t \geq 0. \tag{7.46}$$

Appropriate initial conditions must be provided for the different models [465]. Thus, for the DPL (1, 1) the initial values for temperature and its time derivative have to be specified, $T(x, 0) = \varphi(x, 0)$, $T_t(x, 0) = \varphi_1(x, 0)$, $0 \leq x < \infty$, while for the DPL (2, 1) and the DPL (2, 2) models the second derivative also has to be given, $T_{tt}(x, 0) = \varphi_2(x, 0)$, $0 \leq x < \infty$. The authors used the Fourier transform to eliminate the derivatives in the spatial domain and to obtain the expressions for the exact solutions in the form of an infinite integral, either using the Fourier sine transforms for the Dirichlet conditions,

$$T(x, t) = \frac{2}{\pi} \int_0^\infty \mathcal{F}(\omega, t) \sin(\omega x) \mathrm{d}\omega \tag{7.47}$$

or the cosine transforms for the Neumann conditions,

$$T(x, t) = \frac{2}{\pi} \int_0^\infty \mathcal{F}(\omega, t) \cos(\omega x) \mathrm{d}\omega \tag{7.48}$$

where the functions $\mathcal{F}(\omega, t)$ are the corresponding the Fourier sine or cosine transforms of $T(x, t)$ obtained as the solutions of the transformed temporal problems, depending on the continuous set of eigenvalues ω.

The transformed problems are the initial-value problems for the linear differential equations with constant coefficients. For example, for the DPL (1, 1) model one gets

$$\tau_q \mathcal{F}''(\omega, t) + (1 + \omega^2 \varkappa \tau_T) \mathcal{F}'(\omega, t) + \omega^2 \varkappa \mathcal{F}(\omega, t), \tag{7.49}$$

with the initial conditions

$$\mathcal{F}(\omega, 0) = F(\omega), \quad \mathcal{F}'(\omega, 0) = (\omega), \tag{7.50}$$

for the DPL (2, 1) model

$$\frac{\tau_q^2}{2} \mathcal{F}'''(\omega, t) + \tau_q \mathcal{F}''(\omega, t) + (1 + \omega^2 \varkappa \tau_T) \mathcal{F}'(\omega, t) + \omega^2 \varkappa \mathcal{F}(\omega, t), \tag{7.51}$$

with the initial conditions

$$\mathcal{F}(\omega, 0) = F(\omega), \quad \mathcal{F}'(\omega, 0) = G(\omega), \quad \mathcal{F}''(\omega, 0) = H(\omega), \tag{7.52}$$

for the DPL (2, 2) model

$$\frac{\tau_q^2}{2}\mathcal{F}'''(\omega, t) + \left(\frac{\tau_q + \omega^2 \varkappa \tau_T}{2}\right)\mathcal{F}''(\omega, t) + (1 + \omega^2 \varkappa \tau_T)\mathcal{F}'(\omega, t) + \omega^2 \varkappa \mathcal{F}(\omega, t), \tag{7.53}$$

with the initial conditions as for the DPL (2, 1) model. Here $F(\omega)$, $G(\omega)$, $H(\omega)$, are the Fourier sine or cosine transforms, according to the type of boundary conditions, of the initial functions $\varphi(x, t)$, $\varphi_1(x, t)$, $\varphi_2(x, t)$.

7.3.2 *Analytical Solution for Radiofrequency Heating Using Hyperbolic Pennes Bioheat Transfer Equation in Perfused Tissue*

M. R. Talaee and A. Kabiri [3007] considered the hyperbolic bioheat equation [2829]— Pennes Bioheat Transfer Equation (PBHTE) [2837, 2844]

$$\tau_q \frac{\partial^2 T}{\partial t^2} + \left(1 + \tau_q \frac{\varrho_b \omega_b c_b}{\varrho_t c_t}\right)\frac{\partial T}{\partial t} = \varkappa \nabla^2 T \frac{\varrho_b \omega_b c_b}{\varrho_t c_t}(T_b - T) + \frac{1}{\varrho_t c_t}\left(\tau_q \frac{\partial Q}{\partial t} + Q\right) \tag{7.54}$$

in the spherical coordinates with the nonuniformly distributed heat source term (H is the Heaviside function)

$$Q(r, t) = P\frac{r_i}{4\pi r^4}H(t) \tag{7.55}$$

that models the radiofrequency heating technique where a spherical electrode with the radius r_i is inserted into the diseased tissue leading to the heat generation there due to the imposed electromagnetic field, P is the applied radiofrequency heating power. The boundary conditions are

$$\frac{\partial T}{\partial r}(r_i, t) = 0, \qquad T(r_0, t) = 37^\circ\,\mathrm{C} \tag{7.56}$$

where r_0 is the (arbitrary) radius of the unaffected tissue.

The authors used the following dimensionless quantities ($w = \sqrt{\varkappa/\tau_q}$ is propagation speed of the temperature wave, T_m is the reference temperature)

$$R = \frac{wr}{2\varkappa}, \quad \tau = \frac{t}{2\tau_q}, \quad \vartheta = \frac{T - T_b}{T_m - T_b}, \quad \psi = \frac{Q\tau_q}{\varrho_t c_t (T_m - T_b)},$$

$$P_0 = \frac{P}{8P_r\sqrt{(\tau_q \varkappa)^3}}, \quad h = \frac{\tau_q \varrho_b \omega_b c_b}{\varrho_t c_t} \tag{7.57}$$

to get the governing equation in the dimensionless form as

$$\frac{\partial^2 \vartheta}{\partial \tau^2} + 2(1+h)\frac{\partial \vartheta}{\partial \tau} = \frac{\partial^2 \vartheta}{\partial R^2} + \frac{2}{R}\frac{\partial \vartheta}{\partial R} - 4h\vartheta + f(R, \tau), \tag{7.58}$$

$$f(R, \tau) = 2\frac{\partial \psi}{\partial \tau} + 4\psi. \tag{7.59}$$

The authors obtained the solution using during the derivation the spherical Bessel equation of order zero as

$$\vartheta(R, \tau) = \sum_{n=1}^{\infty} A_n(\tau) R^{-\frac{1}{2}} \left[K_n J_{\frac{1}{2}}(\mu_n R) + J_{-\frac{1}{2}}(\mu_n R) \right] \tag{7.60}$$

where

$$K_n = \frac{J_{-\frac{3}{2}}(\mu_n R_i)}{J_{\frac{3}{2}}(\mu_n R_i)} \tag{7.61}$$

is a constant and eigenvalues μ_n are the positive zeroes of the characteristic equation

$$(J_{\frac{3}{2}}(\mu_n R_i) J_{-\frac{1}{2}}(\mu_n R_0)) + (J_{-\frac{3}{2}}(\mu_n R_i) J_{\frac{1}{2}}(\mu_n R_0)) = 0. \tag{7.62}$$

7.3.3 Thermal Response to RF Exposure During Magnetic Resonance Procedures

G. Brix et al. [378] determined the thermal response to the RF exposure during MR procedures at the tissue level using the Green's function method to solve the non-stationary Pennes equation. The authors considered two cases:

- the temperature field at the periphery of the idealized volume RF resonator;
- the temperature of the homogeneous tissue variation in time during the MR examination with time-varying specific absorption rates (SARs).

The governing equation in 1D is

$$\frac{\partial T(x,t)}{\partial t} - \alpha \frac{\partial^2 T(x,t)}{\partial x^2} + \beta(T(x,t) - T_a) = q(x,t) \tag{7.63}$$

where

$$\alpha = \frac{\lambda}{\rho_t c_t}, \quad \beta = \frac{w \rho_b c_b}{\rho_t c_t}, \quad q(x,t) = \frac{SAR(x,t)}{c_t}, \tag{7.64}$$

T_a is the temperature of the arterial blood, w is the volumetric perfusion rate.

The general solution for the case of constant coefficients is

$$T(x,t) = T_a + \int_{-\infty}^{\infty} \mathrm{d}x' G(x - x', t - t_0)(T_0(x') - T_a)$$

$$+ \int_{-\infty}^{\infty} \mathrm{d}x' \int_{-\infty}^{\infty} \mathrm{d}t' G(x - x', t - t') q(x', t') \tag{7.65}$$

where the unbounded-space Green's function is

$$G(x - x', t - t') = \frac{1}{2\sqrt{\pi}} \Theta(t - t') \exp(-\beta(t - t')) \frac{\exp(-(x - x')^2 / 4\alpha(t - t'))}{\sqrt{\alpha(t - t')}}. \tag{7.66}$$

Here $\Theta(t - t')$ is the unit step function.

7.4 Steady-State Solution of the Modified Pennes Equation

A. Lakhssassi et al. [1763] considered the steady-state modified Pennes equation (5.11) assuming the temperature of the arterial blood T_{a} to be constant

$$\frac{\mathrm{d}^2 T}{\mathrm{d}x^2} = \frac{\omega_1}{\alpha T_{\mathrm{a}}} T^2 + \frac{\omega_0 - 2\omega_1}{\alpha} T + \frac{T_{\mathrm{a}}(\omega_1 - \omega_0)}{\alpha}. \tag{7.67}$$

After multiplication of (7.67) and integration the authors got

$$\left(\frac{\mathrm{d}T}{\mathrm{d}x}\right)^2 = 4\beta t^3 + 6\gamma t^2 + 4\delta T + \varepsilon = R(T) \tag{7.68}$$

where

$$\beta = \frac{\omega_1}{6\alpha T_a}, \quad \gamma = \frac{(\omega_0 - 2\omega_1}{6\alpha}, \quad \delta = \frac{T_a(\omega_1 - \omega_0)}{2\alpha}, \tag{7.69}$$

ε is the constant of integration.

The general solution of (7.68) is [3314] (the prime denotes the differentiation)

$$T(x) = T_0 + \frac{\sqrt{R(T_0)}\frac{d\mathcal{P}(x,g_2,g_3)}{dx} + \frac{R'(T_0)}{2}\left[\mathcal{P}(x,g_2,g_3) - \frac{1}{24}R''(T_0)\right] + \frac{1}{24}R(T_0)R'''(T_0)}{2\left[\mathcal{P}(x,g_2,g_3) - \frac{1}{24}R''(T_0)\right]^2 - \frac{1}{48}R(T_0)R''''(T_0)} \tag{7.70}$$

where T_0 is a real constant that is not necessary a zero of $R(T)$. The invariants of the Weierstrass' elliptic function $\mathcal{P}$ g_2 and g_3 are related to the coefficients of $R(T)$ as [480]

$$g_2 = -4\beta\delta + 3\gamma^2. \quad g_3 = 2\beta\gamma\delta - \gamma^3 - \varepsilon\beta^2. \tag{7.71}$$

The discriminant $\Delta = g_2^3 - 27g_3^2$ is used to classify the behaviour of $T(x)$ [13, 1763, 2747]:

- $\Delta \neq 0$ or $\Delta = 0, g_2 > 0, g_3 > 0$: the periodic solution;
- $\Delta = 0, g_2 \geq 0, g_3 \leq 0$: solitary wave-like solution.

Lakhssassi et al. [1763] listed some particular solutions of (7.68)

- $\Delta > 0$. In this case the Weierstrass' elliptic function is expressed as [13]

$$\mathcal{P}(x, g_2, g_3) = e_3 + \frac{e_1 - e_3}{sn^2(\sqrt{e_1 - e_3}x, m)} \tag{7.72}$$

where $sn(x, m)$ is the Jacobi elliptic sine function with modulus m [417, 1592]. The solution is

$$T(x) = \frac{(4\beta T_0^2 + 2e_3T_0 + 5\gamma T_0 + 2\delta)sn^2(\sqrt{e_1 - e_2}x, m) + 2(e_1 - e_3)T_0}{(2e_3 - 2\beta T_0 - \gamma)sn^2(\sqrt{e_1 - e_2}x, m) + 2(e_1 - e_3)}. \tag{7.73}$$

- $\Delta < 0$. In this case the Weierstrass' elliptic function is expressed as [13]

$$\mathcal{P}(x, g_2, g_3) = e_2H\frac{1 + cn(2\sqrt{H}x, m)}{1 - cn(2\sqrt{H}x, m)} \tag{7.74}$$

where where $cn(x, m)$ is the Jacobi elliptic sine function with modulus m, $H^2 = 3e_2^2 + g_2/4$.

The solution is

$$T(x) =$$

$$\frac{[(2H - 5\gamma - 2e_2T_0 - 4\beta T_0^2 - 2\delta]\xi + 2\delta + 4\beta T_0^2 + (2H + 5\gamma + 2e_2)T_0}{(2\beta T_0 + 2H + \gamma - 2e_2)\xi + 2H + 2e_2 - 2\beta T_0 - \gamma}. \tag{7.75}$$

where $\xi = cn(2\sqrt{H}x, m)$

- $\Delta = 0$. In this case the Weierstrass' elliptic function is expressed by the trigonometric functions and the solution is

$$T(x) = T_0 + \frac{R'(T_0)\sin^2\sqrt{\frac{3}{2}e_1}x}{12e_1 - \frac{24e_1 + R''(T_0)}{6}\sin^2\sqrt{\frac{3}{2}e_1}x}. \tag{7.76}$$

7.5 An Integral-transform Solutions of the Pennes' Equation

K. Murase [2231] applied an integral-transform approach to the analytical solutions of the Pennes' bioheat transfer equation for the hyperthermia with the magnetic nanoparticles (the magnetic hyperthermia).

The author introduced the "core body" temperature that is maintained by the balance between the metabolic heat generation and the heat transfer by the blood perfusion

$$T_c = T_a + \frac{\dot{q}_{met}}{\varrho_b c_b . \omega_b}. \tag{7.77}$$

The core temperature can be assumed equal to the arterial blood temperature if $\dot{q}_{met}/(\varrho_b c_b . \omega_b) \ll T_a$ [1047, 1082].

The Pennes equation (10.101) in the spherical coordinates is written as

$$\varrho c \frac{\partial T}{\partial t} = \frac{\lambda}{r^2}\frac{\partial}{\partial r}\left(r^2 \frac{\partial T}{\partial r}\right) + \varrho c_b \omega_b (T_c - T) + Q^{ext}. \tag{7.78}$$

Introducing $\vartheta = T - T_c$ and $\vartheta = r\vartheta$, the author got

$$K\frac{\partial \vartheta}{\partial t} = \frac{\partial \vartheta}{\partial r^2} - \alpha^2 \vartheta + \frac{rQ^{ext}}{\lambda} \tag{7.79}$$

where

$$K = \frac{\varrho c}{\lambda}, \quad \alpha^2 = \frac{\varrho c_b \omega_b}{\lambda}. \tag{7.80}$$

Applying the Fourier sine transform [458] to (7.79) yields

$$K\frac{\partial \bar{\vartheta}}{\partial t} = -(\alpha^2 + \beta^2)\bar{\vartheta} + \frac{F(\beta)}{\lambda} \tag{7.81}$$

where

$$\bar{\vartheta} = \sqrt{\frac{2}{\pi}}\int_0^\infty \vartheta \sin(\beta r)\mathrm{d}r, \quad F(\beta) = \sqrt{\frac{2}{\pi}}\int_0^\infty r Q^{\mathrm{ext}} \sin(\beta r)\mathrm{d}r. \tag{7.82}$$

The solution of (7.81) is

$$\bar{\vartheta} = e^{-\frac{\alpha^2+\beta^2}{\lambda}t}\bar{\vartheta}(0) + \frac{F(\beta)}{\lambda(\alpha^2+\beta^2)}\left[1 - e^{-\frac{\alpha^2+\beta^2}{\lambda}t}\right] \tag{7.83}$$

and ϑ is given by the inverse Fourier transformation

$$\vartheta = \sqrt{\frac{2}{\pi}}\int_0^\infty \bar{\vartheta}(\beta)\sin(\beta r)\mathrm{d}\beta. \tag{7.84}$$

Murase considered the several heat sources:

- Point source

$$Q^{\mathrm{ext}} = \frac{Q_0}{4\pi r^2}\delta(r), \tag{7.85}$$

$$T = T_c + \frac{Q_0}{2\pi^2\lambda r}\left[\frac{\pi}{2}e^{-\alpha r} - \int_0^\infty \frac{\beta}{\alpha^2+\beta^2}e^{-\frac{\alpha^2+\beta^2}{K}t}\sin(\beta r)\mathrm{d}\beta\right] \tag{7.86}$$

and in the steady state ($t = \infty$)

$$T = T_c + \frac{Q_0}{4\pi\lambda r}e^{-\alpha r}; \tag{7.87}$$

- Shell source

$$Q^{\mathrm{ext}} = \frac{Q_0}{4\pi r^2}\delta(r - r_0), \tag{7.88}$$

$$T = T_c + \frac{Q_0}{2\pi^2 \lambda r r_0} \left\{ \frac{\pi}{4\alpha} \left[e^{-\alpha|r-r_0|} - e^{-\alpha|r+r_0|} \right] - \int_0^\infty \frac{\beta}{\alpha^2+\beta^2} e^{-\frac{\alpha^2+\beta^2}{K}t} \sin(\beta r) \mathrm{d}\beta \right\} \tag{7.89}$$

and in the steady state ($t = \infty$)

$$T = T_c + \frac{Q_0}{8\pi \lambda r r_0 \alpha} \left[e^{-\alpha|r-r_0|} - e^{-\alpha|r+r_0|} \right]; \tag{7.90}$$

- Gaussian-distributed source

$$Q^{\mathrm{ext}} = Q_0 e^{-\frac{r^2}{r_0^2}}, \tag{7.91}$$

$$T = T_c + \frac{Q_0 r_0^3}{2\sqrt{\pi} \lambda r} \int_0^\infty \frac{\beta e^{-\frac{\beta^2 r_0^2}{4}}}{\alpha^2+\beta^2} \left[1 - e^{-\frac{\alpha^2+\beta^2}{K}t} \right] \sin(\beta r) \mathrm{d}\beta \tag{7.92}$$

and in the steady state ($t = \infty$)

$$T = T_c + \frac{Q_0 r_0^3}{2\sqrt{\pi} \lambda r} \int_0^\infty \frac{\beta e^{-\frac{\beta^2 r_0^2}{4}}}{\alpha^2+\beta^2} \sin(\beta r) \mathrm{d}\beta. \tag{7.93}$$

The energy dissipation of the MNPs subjected to the alternating magnetic field is expressed as [2638]

$$Q^{\mathrm{ext}} = \pi \mu_0 \chi_0 H_0^2 f \frac{2\pi f \tau}{1+(2\pi f \tau)^2} \tag{7.94}$$

where χ_0 the equilibrium susceptibility, H_0 and f are the amplitude and frequency of the AMF, τ is the effective relaxation time given by

$$\frac{1}{\tau} = \frac{1}{\tau_N} + \frac{1}{\tau_B} \tag{7.95}$$

where τ_N and τ_B are the Neel relaxation and Brownian relaxation times [2638].

7.5.1 *Hybrid Integral Transforms Analysis of the Pennes Equation with Variable Properties*

R. M. Cotta et al. [636] developed the Generalized Integral Transform Technique (GITT) to produce the hybrid numerical-analytical solution of the bioheat model in heterogeneous media that reduces to the exact solution obtained via the Classical Integral Transform Technique (CITT) [635, 2276] for a linear equation with uniform coefficients.

The authors considered the heterogeneous tissue consisting of the epidermis, dermis, subcutaneous and inner tissue. The heat conduction equation is written as

$$w(x)\frac{\partial T(x,t)}{\partial t} = \frac{\partial}{\partial x}\left(k(x)\frac{\partial T(x,t)}{\partial x}\right) - d(x,T)T(x,t) + P(x,t,T) \tag{7.96}$$

where $w(x) = \rho(x)c_t(x)$, $d(x,T) = \omega(x,T)\rho_b c_b$, $P(x,t,T) = q_m(x,t) + \omega(x,T)\rho_b c_b T_a$.

Before providing the solution of (7.96), the authors introduce the filtering solution for the improved convergence behaviour of the eigenfunction expansions in the form:

$$T(x,t) = T_f(x,t) + T^\star(x,t) \tag{7.97}$$

where

$$T_f(x,t) = T_\infty(t) + \frac{q_0}{h}. \tag{7.98}$$

The formal solution is obtained with the Generalized Integral Transform Technique and is written as the inverse formula

$$T^\star(x,t) = \sum_{i=1}^{\infty} \tilde{\psi}_i(x)\bar{T}_i(t) \tag{7.99}$$

where the transformed potentials are defined with the integral transformation operation as

$$\bar{T}_i(t) = \int_0^L w^\star(x)\tilde{\psi}_i(x)T^\star(x,t)\mathrm{d}x. \tag{7.100}$$

Here the normalized eigenfunctions $\tilde{\psi}_i(x) = \psi_i(x)/\sqrt{N_i}$, the normalization integrals are

$$N_i = \int_0^L w^\star \psi_i^2(x)\mathrm{d}x. \tag{7.101}$$

The eigenvalues μ_i and eigenfunctions $\psi_i(x)$ are obtained from the eigenvalue problem

$$\frac{\mathrm{d}}{\mathrm{d}x}\left[k^\star(x)\frac{\mathrm{d}\psi_i(x)}{\mathrm{d}x}\right] + \left[\mu_i^2 w^\star(x) - d^\star(x)\right]\psi_i(x) = 0 \tag{7.102}$$

with boundary conditions

$$\begin{cases} -k^\star(0)\frac{\mathrm{d}\psi_i(x)}{\mathrm{d}x} + h^\star\psi_i(x) = 0 & x = 0 \\ \frac{\mathrm{d}\psi_i(x)}{\mathrm{d}x} = 0 & x = L. \end{cases} \tag{7.103}$$

Finally the authors obtained the following transformed ordinary differential system

$$\frac{\mathrm{d}\bar{T}_i(t)}{\mathrm{d}t} = \bar{g}_i(t, \bar{T}_j) \tag{7.104}$$

where the transformed source terms are

$$\bar{g}_i(t, \bar{T}_j) = \int_0^x G(x, t, T)\tilde{\psi}_i(x), \tag{7.105}$$

$$G(x, t, T) = \frac{w^\star(x)}{w(x)}\left[\frac{\partial}{\partial x}\left(k(x)\frac{\partial T^\star(x, t)}{\partial x}\right) - d(x, T)T^\star(x, t) + P^\star(x, t, T)\right]. \tag{7.106}$$

The solution of the ODE system (7.104) numerically to a sufficiently large finite order N gives the transformed temperatures, which could be combined with the inverse formula to provide the original temperature field.

7.6 Multilayer Composite Model of Skin Tissue

X. Jiang and S. Chen [1470] studied the heat transfer in the composite medium consisting of M parallel layers in the perfect thermal contact that could be considered as the skin tissue model. The heat conduction problem is governed by the equation

$$\varrho c_0 D_t^{\alpha} T_i = \lambda_i \nabla(\nabla T_i) + g_i(x,t), \quad 0 < \alpha \leq 1, \quad i = 1, 2, \ldots M, \tag{7.107}$$

where ${}_0D_t^{\alpha}$ is the Caputo derivative, $g_i(x,t)$ is the heat generation in the i-th layer.

The boundary conditions are formulated as follows:

- inner surface of the first layer $T_i = T_b$
- interface of the i-th layer (i = 1, 2, ... M - 1)

$$T_i = T_{i+1}, \quad \lambda_i \frac{\partial T_i}{\partial x} = \lambda_{i+1} \frac{\partial T_{i+1}}{\partial x} \tag{7.108}$$

- outer surface of the M-th layer

$$\lambda_M \frac{\partial T_M}{\partial x} + h_{M-1} T_M = h_M f_M(t). \tag{7.109}$$

and the initial conditions $T(x,0) = F_i(x), \quad x_{i-1} < x < x_i$.

The authors introduced dimensionless variables

$$z = \frac{x}{x_M}, \quad \tau = \left(\frac{\lambda_M}{x_M^2 \varrho c}\right)^{\frac{1}{\alpha}}, \quad \vartheta_i = \frac{T_i - T_b}{T_0}, \quad \lambda_i^{\star} = \frac{\lambda_i}{\lambda_M} \tag{7.110}$$

and performed the integral transform for the function $\vartheta_i(z,\tau)$

$$\bar{\vartheta}_i(\xi_n, \tau) = \sum_{i=1}^{M} \int_{z_{i-1}}^{z_i} \psi_{in}(\xi_n, z') \vartheta_i(z', \tau) dz', \tag{7.111}$$

where ξ_n are the eigenvalues and $\psi_{in}(\xi_n, z)$ are the eigenfunctions

$$\psi_{in} = A_{in} \sin\left(\frac{\xi_n z}{\sqrt{\lambda_i^{\star}}}\right) + B_{in} \cos\left(\frac{\xi_n z}{\sqrt{\lambda_i^{\star}}}\right). \tag{7.112}$$

The inversion formula is

$$\vartheta_i(z,\tau) = \sum_{n=1}^{\infty} \frac{\psi_{in}(\xi_n, z) \bar{\vartheta}_i(\xi_n, \tau)}{N(\xi_n)}, \tag{7.113}$$

where $N(\xi_n)$ is the normalization integral.

Thus the transformed equation could be written as

$$ {}_0D_\tau^\alpha \bar{\vartheta} + \lambda_n^2 \bar{\vartheta}(\lambda_n, \tau) = A(\lambda_n, \tau) \tag{7.114} $$

with the initial condition

$$ \bar{\vartheta}(\lambda_n, 0) = \bar{F}_0(\lambda_n, 0) \tag{7.115} $$

where

$$ \begin{aligned} A(\lambda_n, \tau) &= \sum_{i=1}^{M} \int\limits_{z_i-1}^{z_i} \psi_{in}(\lambda_n, z') g_i^\star(z', \tau) dz' + \lambda_M^\star \left[\psi_{Mn}(z')_{z'}\right]_{z'} \\ &= z_M \left[h^\star \frac{f_M - T_b}{T_0}\right], \end{aligned} \tag{7.116} $$

$$ F_0 = \sum_{i=1}^{M} \int\limits_{z_i-1}^{z_i} \psi_{in}(\lambda_n, z') F_{0i} dz'. \tag{7.117} $$

Applying the Laplace transform to (7.114) and the initial condition (7.115) one gets

$$ s^\alpha \bar{\bar{\vartheta}} - s^{\alpha-1} \bar{F}_0 + \lambda_n \bar{\bar{\vartheta}} = \bar{A}(\lambda_n, s) \tag{7.118} $$

and thus

$$ \bar{\bar{\vartheta}} = \frac{s^{\alpha-1} \bar{F}_0 + \bar{A}(\lambda_n, s)}{s^\alpha + \lambda^2}, \qquad \bar{A}(\lambda_n, s) = \int\limits_0^\infty e^{-s\tau} A(\lambda_n, \tau) d\tau. \tag{7.119} $$

X. Jiang and S. Chen [1470] used the formula for the inverse Laplace transform [2471]

$$ \mathcal{L}^{-1}\left[\frac{k! s^{\lambda-\mu}}{(s^\lambda \mp c)^{k+1}}; t\right] = t^{\lambda+\mu-1} E_{\lambda,\mu}^{(k)}(\pm c t^\lambda) \tag{7.120} $$

where

$$ E_{\lambda,\mu}^{(k)}(z) = \frac{d^k}{dz^k} E_{\lambda,\mu}(z) \tag{7.121} $$

and $E_{\lambda,\mu}(z)$ is the generalized Mittag-Leffler function.

Finally the authors obtained the temperature distribution in each region

$$\vartheta_i = \sum_{i=1}^{\infty} \frac{\psi_{in}(\lambda_n, z)}{N(\lambda_n)} \left[E_{\alpha,1}(-\lambda_n^2 \tau^\alpha) \bar{F}_0 + \int_0^\tau A(\lambda_n, \tau') G(\tau - \tau') d\tau' \right] \tag{7.122}$$

where

$$G(\tau) = \tau^{\alpha-1} E_{\alpha,\alpha}(-\lambda_n^2 \tau^\alpha). \tag{7.123}$$

C.-Y. Li et al. [1861] investigated the temperature field and the thermal damage of the multilayer skin tissues subjected to the time-varying laser heating and fluid cooling by a semianalytical method. For each layer $L_{i-1} < x \le L_i$ the authors used the Pennes equation

$$\lambda_i \frac{\partial^2 T}{\partial x^2} - \rho_i c_i \frac{\partial T}{\partial t} - \omega_b \rho_b c_b (T - T_a) + q_{met,i} + q_{ext,i} = 0. \tag{7.124}$$

The authors used a semianalytical method based on the separation of variables and decomposed the dimensionless temperature $\vartheta(\xi, \tau)$ as

$$\vartheta_i(\xi, \tau) = \nu_i(\xi, \tau) + \sum_{m=1}^{2} g_{i,m}(\xi) f_{i,m}(\tau) \tag{7.125}$$

where $g_{i,m}(\xi)$ are the shifting functions to be specified and $\nu_i(\xi, \tau)$ is the transformed function.

The temperature field and the thermal damage of the skin tissue with the different laser heating cycles and the different value of the convective heat transfer coefficient were determined. The authors found, in particularity, that the smaller the convective heat transfer coefficient is, the faster the temperature of the skin tissue rises and the larger the convective heat transfer coefficient is, the smaller the heat flux of the skin tissue.

7.7 Solution of One-Dimensional Multiregion Bioheat Equation

J. W. Durkee, Jr., P. P. Antich and C. E. Lee rewrite the classical Pennes equation (5.8) for the case of the one dimensional composite tissue having the regionwise constant thermomechanical properties. The equation governs the temperature evolution $T_m(r, t)$ in each region m that occupies the interval $R_{m-1} < r < R_m$ for $m = 1, 2, \ldots, M$ and are written as

$$\varrho_{tm} C_{pm} \frac{\partial T_m}{\partial t} = k_m \frac{1}{r^G} \frac{\partial}{\partial r} \left(r^G \frac{\partial T_m}{\partial r} \right) + m_{bm} C_{bm} (T_{am} - T_m) + Q_m + P_m, \quad (7.126)$$

where $G = 0, 1, 2$ correspond to the Cartesian, cylindrical and spherical geometry.

The authors suggested to refer to (7.126) as to the multiregion bioheat equation (MBE).

The equations in the different regions are coupled by the continuity of the temperature

$$T_m(R_m, t) = T_{m+1}(R_m, t) \quad (7.127)$$

and of the heat flux

$$-k_m \frac{\partial T_m}{\partial r}|_{r=R_m} = -k_m \frac{\partial T_{m+1}}{\partial r}|_{r=R_m} \quad (7.128)$$

at the interfaces between the regions.

The different boundary conditions (Dirichlet, Neumann or other) could be specified at the external boundaries of the multiregion domain. For example, in the case of the spherical symmetry the boundness of the temperature at the centre could be required

$$\lim_{r \to 0} T_1(r, t) < \infty \quad (7.129)$$

At the outer surface the authors used the so called conductive-convective heat transfer condition

$$-\hat{k}_M \frac{\partial T_M}{\partial r}|_{r=R_m} = h(T_M(R_M, t) - T_\infty). \quad (7.130)$$

Here h and T_∞ are the convective heat transfer coefficient at the outer surface and the ambient temperature; $\hat{k}_M$ is not a true thermal conductivity but just the boundary condition coefficient.

The initial condition is a trivial one $T_m(r, 0) = T_m^0(r)$

Durkee et al. introduce dimensionless variables for the temperature $\psi_m(\varrho, \tau)$, position (and, hence, the region boundaries) $\varrho = r/R_M, \quad \varrho_m = r_m/R_M$, and time

$$\tau = \frac{k_D}{\varrho_{tD} c_{pD} R_M^2} t \quad (7.131)$$

to obtain the dimensionless form of the MBE

$$\frac{\partial \psi_m}{\partial \tau} = K_{1,m} \frac{1}{\varrho^G} \frac{\partial}{\partial \varrho} \left(\varrho^G \frac{\partial \psi_m}{\partial \varrho} \right) - K_{2,m} \psi_m + \mathcal{L}_m \tag{7.132}$$

and of the interface conditions for the continuity of temperature and heat fluxes

$$\psi_m(\varrho_m, \tau) = \psi_{m+1}(\varrho_m, \tau), \qquad k_m \frac{\partial \psi_m}{\partial \varrho}|_{\varrho=\varrho_m} = k_{m+1} \frac{\partial \psi_{m+1}}{\partial \varrho}|_{\varrho=\varrho_m} \tag{7.133}$$

The boundary conditions at the outer boundaries and the initial condition are transformed similarly.

The two family of the constants $K_{1,m}$ and $K_{2,m}$ named by the authors the *dispersion* and the *removal* constants, respectively, are defined as

$$K_{1,m} = \frac{k_m \varrho_{tD} c_{pD}}{k_D \varrho_{tm} c_{pm}}, \qquad K_{2,m} = \frac{m_{bm} c_{pbm} R_M^2 \varrho_{tD} c_{pD}}{k_D \varrho_{tm} c_{pm}}. \tag{7.134}$$

Heat source is then expressed as

$$\mathcal{L}_m = \frac{\varrho_{tD} c_{pD}}{\varrho_{tm} c_{pm}} \frac{R_M^2}{k_D T_r^0} (m_{bm} c_{pbm} T_{arm} + Q_m + P_m). \tag{7.135}$$

The subscript "D" in the above expression indicate the reference region.

The authors additionally introduced the dimensional homogeneous H_1 and inhomogeneous H_2 heat transfer coefficients

$$H_1 = h, \qquad H_2 = -h \frac{T_\infty}{T_r^0} \tag{7.136}$$

and the conduction coefficients $\varkappa_m = k_m / R_m$.

To solve the governing equations that are linear and separable Durkee et al. decompose the solution into the transient ψ_m^{tr} and the stationary parts ψ_m^{s}

$$\psi_m(\varrho, \tau) = \psi_m^{\mathrm{tr}}(\varrho, \tau) + \psi_m^{\mathrm{s}}(\varrho). \tag{7.137}$$

The former satisfies the following equation

$$\frac{\psi_m^{\mathrm{tr}}}{\partial \tau} = K_{1,m} \frac{1}{\varrho^G} \frac{\partial}{\partial \varrho} \left(\varrho^G \frac{\psi_m^{\mathrm{tr}}}{\partial \varrho} \right) - K_{2,m} \psi_m^{\mathrm{tr}} \tag{7.138}$$

while the steady state solution—the equation

$$0 = K_{1,m} \frac{1}{\varrho^G} \frac{\partial}{\partial \varrho} \left(\varrho^G \frac{\psi_m^{\mathrm{s}}}{\partial \varrho} \right) - K_{2,m} \psi_m^{\mathrm{s}} + \mathcal{L}_m. \tag{7.139}$$

The solution is sought as a series expansion

$$\psi_m^{\mathrm{tr}}(\varrho, \tau) = \sum_{I=1}^{N_\alpha} \mathcal{R}_{m,i}(\varrho) \exp(-\alpha_I^2 \tau) \tag{7.140}$$

where the eigenfunctions are determined from the equation

$$\frac{1}{\varrho^G} \frac{\mathrm{d}}{\mathrm{d}\varrho} \left(\varrho^G \frac{\mathrm{d}\mathcal{R}_{m,i}}{\mathrm{d}\varrho} \right) + \left(\frac{\alpha_i^2 - K_{2,m}}{K_{1,m}} \right) \mathcal{R}_{m,i} = 0. \tag{7.141}$$

The eigenvalues α_i are region independent (hence the subscript "m" is omitted).

The details could be found in the cited papers. Authors [823] presented the eigenfunctions for the cases of the Cartesian and the spherical geometry that are expressed via sine and cosine and hyperbolic sine and cosine. In the cylindrical geometry the solution is written out using the Bessel and the modified Bessel functions [821].

The transient heat sources and the transient boundary conditions have been considered by the first two of these authors in [822]. The two forms of the heating term $P_m(t)$ and of the ambient temperature T_∞ were analysed. The first one is the stepwise source and the temperature variation that can be described as

$$P_m(t) = \sum_{l=1}^{L} H_P(t - t_{m,l}) \Delta P_{m,l}, \tag{7.142}$$

for the sources localized in the specific region and

$$T_\infty(t) = \sum_{k=1}^{K} H_T(t - t_k) \Delta T_{\infty,k}, \tag{7.143}$$

for the ambient temperature. Here $H_P(t - tm, l)$ and $H_T(t - t_k)$ are Heaviside functions. The second form considered is the sinusoidal perturbation of the ambient temperature $T_\infty(t) = T_\infty^{\mathrm{av}} + T_\infty^a \sin(\omega_b t)$ where T_∞^{av} is the average ambient temperature, T_∞^a is the amplitude of its perturbation and ω_b is the frequency.

The authors proceed similarly to the previously considered case of the stationary source and boundary conditions with some modifications. The separation of the solution into the non-steady ψ_m^{tr} and steady ψ_m^{s} parts is written as

$$\psi_m(\varrho, \tau) = \psi_m^{\mathrm{tr}}(\varrho, \tau) + \psi_m^{\mathrm{s}}(\varrho) H_2(\tau), \tag{7.144}$$

where in contrast to the steady state case an explicit time dependence due to the variation of the ambient temperature introduced via the nonhomogeneous surface heat transfer coefficient $H_2(\tau)$ that is defined as

$$H_2(\tau) = -h\frac{T_\infty(\tau)}{T_r^0}. \tag{7.145}$$

For the transient part one than have instead of Eq. (7.138) the following relation

$$\frac{\psi_m^{\mathrm{tr}}}{\partial\tau} = K_{1,m}\frac{1}{\varrho^G}\frac{\partial}{\partial\varrho}\left(\varrho^G\frac{\psi_m^{\mathrm{tr}}}{\partial\varrho}\right) - K_{2,m}\psi_m^{\mathrm{tr}} + \hat{S}_m(\tau) \tag{7.146}$$

where the source term is written as

$$\hat{S}_m(\tau) = S_m(\tau) - \left[\psi_m^{\mathrm{s}}\frac{\mathrm{d}H_2(\tau)}{\mathrm{d}\tau}\right]. \tag{7.147}$$

Performing the rather lengthy calculations, the authors wrote out the solution in terms of the multi-regional Green's function $G_{m,n}$. The details could be found in the cited paper. As in the steady state case, the authors note that the intended use of their results is as the benchmarks for the numerical methods verification only and does not try to compare the computations with the experimental data or draw the general conclusions on the peculiarities of the heat transfer in tissues relevant to the physiological problems.

J. W. Durkee et al. [824] computed the solutions to the time-dependent multiregion bioheat equation are for cartesian and spherical models. Principle and alternate branch contributions to the solutions were examined. The calculations include a parametric investigation and analyses of temperature behaviour in idealised upper thigh and tumour models.

However, the solutions by Durkee et al. do not include the case in which the value of the perfusion term in one layer is exactly zero. Instead, within layers with the nonperfuse tissues, the perfusion rates are set to the small nonzero values. Moreover, the solution presented in these studies does not vary the value of the arterial blood temperature between layers. To treat the discontinuity in the perfusion at the interfaces, D. A. Hodson et al. [1338] replaced the inner perfuse layer (the perfuse layer adjacent to nonperfuse layer) by an interface condition (effectively reducing the number of layers by one). Because the model represents the perfuse dermis by an interface condition, all spatial information within that layer is lost. An analytic solution to one-dimensional transient transport through a two-layer composite region in which one layer lacks perfusion and experiences pure diffusion transfer, while the second layer consists of perfuse tissue and experiences perfusion aided diffusion was developed by S. M. Becker [227].

However, that study only considered a two-layer system. Later Becker [228] developed the analytic solution to the bioheat equation in a composite system of N layers that

- allows for variations in all parameter values (including the arterial blood temperature);
- allows adjacent layers to have discontinuities in the perfusion rate.

The parameter values of the layers can experience a discontinuity at the interface. The systems' two outer faces (layer 1 and layer N) exchange the thermal energy by the convection with the two different flow fields that are at the two different ambient temperatures.

The one-dimensional transient heat conduction in each layer i is governed by an equation of the form

$$AT_i \frac{\partial T}{\partial t} = \frac{\partial}{\partial x_i}\left(B_i \frac{\partial T}{\partial x_i}\right) - C_i(T - E_i) + D_i \tag{7.148}$$

where $i = 1, \ldots, N$, the constants $A_i, B_i, C_i.D_i, E_i$ are the thermophysical parameter values that characterize the ith layer.

The dimensionless representation of the Pennes bioheat equation is [228]

$$\frac{1}{\delta_i}\frac{\partial \vartheta_i}{\partial \tau} + m_i^2 \vartheta_i - \varphi_i = \frac{\partial^2 \vartheta_i}{\partial x_i} \tag{7.149}$$

where the perfusion term $m_i^2 = (\varrho_b C_b \omega L^2/\lambda)_i$, the dimensionless constant Fourier number is δ_i, the dimensionless volumetric energy source term is $\varphi_i = g_i L^2/(T_0 - T_L)/\lambda_i$, g is the volumetric heating that can result from the metabolic heat generation or from some externally induced heating source, ω is the volumetric blood perfusion rate per unit volume of tissue.

To use the separation of variables, the author described the solution as a sum of a nonhomogenous steady spatially dependant component and a homogenous component that depends on the position and time

$$\vartheta_i(x_i, \tau) = \Psi_i(x_i) - \vartheta_i(x_i, \tau). \tag{7.150}$$

The separation of variables method assumes that each layer's homogenous transient solution may be represented as the product of a time dependent component and a spatially dependent component:

$$\vartheta_i(x_i, \tau) = X_i(x_i)\Gamma(\tau). \tag{7.151}$$

The author based the solution presentation on the characteristic of periodic and hyperbolic functions which allows their integrals to be represented in terms of their first derivatives. These approaches provide two simplifications

1. the eigenfunctions $X_i(x_i)$ of layers $i = 2, \dots, N$ are described in terms of those of the previous layer, thus can be "built" sequentially;
2. the constants of integration are described in terms of the eigenfunctions and their first derivatives, which are evaluated at the layer boundaries.

The similar problem was considered by D. B. Rodrigues et al. [2622]. The authors allowed the metabolic heat generation to depend linearly on the temperature. The boundary conditions of the first, second and third kinds to the temperature at the inner and outer surfaces were assumed. The equation in the dimensionless variables was written in the following form

$$\frac{\partial T_i}{\partial t}(r,t) = \alpha_i \nabla^2 T_i(r,t) - \beta_i T_i(r,t) + \sigma_i(r) \tag{7.152}$$

To solve Eq. (7.152) by the method of separation of variables, the solution of this nonhomogeneous equation is decomposed in two components as

$$T_i(r,t) = \bar{T}_i(r,t) + \hat{T}_i(r) \tag{7.153}$$

where $\bar{T}_i$ and $\hat{T}_i$ are the transient (homogeneous) and steady state (nonhomogeneous) components. The corresponding equations are

$$\frac{\partial \bar{T}_i}{\partial t}(r,t) = \alpha_i \nabla^2 \bar{T}_i(r,t) - \beta_i \bar{T}_i(r,t) \tag{7.154}$$

and

$$\alpha_i \nabla^2 \hat{T}_i(r,t) - \beta_i \hat{T}_i(r,t) = -\sigma_i(r). \tag{7.155}$$

The solution of the linear and nonhomogeneous ordinary differential Eq. (7.155) is given by:

$$\hat{T}_i = H_i(r) + S_i(r) \tag{7.156}$$

where H_i is the solution of the homogeneous equation and S_i is the particular integral of (7.155). Rodrigues et al. wrote out the solution in terms of the modified Bessel functions of the first and second kinds, the order of which is given by $-1/2$, 0 and 1/2 for the Cartesian, cylindrical and spherical coordinates, respectively.

To solve the Eq. (7.154) by the method of separation of variables, the authors assumed the following relationship

$$\bar{T}_i(r,t) = R_i(r)\Psi_i(t). \tag{7.157}$$

The solution was written via the Bessel functions of the first and second kinds.

The authors considered two examples of the clinical applications:

1. the heating in a tumour and its surrounding tissue, induced during a magnetic fluid hyperthermia;
2. the effect of the different environmental conditions in a multilayered human head model (brain , bone and scalp).

7.8 Heat Transfer with a Sinusoidal Heat Flux on Skin Surface

The problem of the temperature response of the semi-infinite sample of the biological tissue to the periodic (sinusoidal) heat flux applied to its surface was considered by T.-C. Shih et al. [2830]. This problem was earlier considered by J. Liu and L.X. Xu [1902] who were interested, however, not in the transient response, but in the long run steady periodic temperature oscillations and in the phase shift between the heat flux and the surface temperature that could be utilized for the diagnostics. Phase shifts in the surface temperature response were revealed to be related to local blood perfusion rate and heating frequency. The influence of the thermal contact resistance on the perfusion estimation was investigated. The mapping of the temperature of the skin surface as the diagnostic tool was also considered by Z.S. Deng and J. Liu [715].

The one-dimensional Pennes equation (5.8) is considered

$$\varrho_t c_t \frac{\partial T}{\partial t} + w_b c_b (T - T_a) = k \frac{\partial^2 T}{\partial x^2}. \tag{7.158}$$

The oscillatory heat flux with the amplitude q_0 and the frequency ω_b is given at the boundary $x = 0$

$$-k \frac{\partial T}{\partial x} = q_0 \mathrm{e}^{\mathrm{i}\omega_b t}. \tag{7.159}$$

Introducing the dimensionless variables

$$z \equiv \omega_b t, \quad \alpha \equiv \frac{k}{\varrho_t c_t}, \quad X \equiv \sqrt{\frac{\omega_b}{\alpha}}, \quad c_1 \equiv \frac{w_b c_b}{\varrho_t c_t \omega_b}, \quad \vartheta \equiv \frac{k(T - T_a)}{q_0} \sqrt{\frac{\omega_b}{\alpha}}, \tag{7.160}$$

Shih et al. get the dimensionless form of the equation

$$\frac{\partial \vartheta}{\partial z} + c_1 \vartheta = \frac{\partial^2 \vartheta}{\partial X^2} \tag{7.161}$$

and of the boundary condition

$$\frac{\partial \vartheta}{\partial X} = -\mathrm{e}^{\mathrm{i}z}. \tag{7.162}$$

Using the Laplace transform method, the authors finally obtained the solution for the temperature evolution as [2830]

$$\vartheta = \frac{\mathrm{e}^{\mathrm{i}z}}{2\sqrt{c_1+\mathrm{i}}}\left[\mathrm{e}^{-\sqrt{c_1+\mathrm{i}}X}\operatorname{erfc}\left(\frac{X}{2\sqrt{z}} - \sqrt{(c_1+\mathrm{i})z}\right)\right.$$
$$\left. -\mathrm{e}^{\sqrt{c_1+\mathrm{i}}X}\operatorname{erfc}\left(\frac{X}{2\sqrt{z}} + \sqrt{(c_1+\mathrm{i})z}\right)\right] \tag{7.163}$$

and for the temperature response on the skin surface

$$\vartheta(X=0) = \frac{\mathrm{e}^{\mathrm{i}z}}{\sqrt{c_1+\mathrm{i}}}\operatorname{erf}\left[\sqrt{(c_1+\mathrm{i})z}\right] \tag{7.164}$$

Using the obtained solutions, the authors studied the effect of the frequency of the heating on the affected depth of the tissues and discussed implications for the optimal choice of the boundary heat flux for the measurements of the blood perfusion rate.

The results showed that the temperature oscillation due to the sinusoidal heating on the skin surface is unstable in the initial period. The lower frequency of sinusoidal heat flux on the skin surface induces a more sensitive phase shift response to the blood perfusion rate change, but extends the beginning time of sampling because of the avoidance of the unavailable first cyclic oscillation.

7.9 Thermal Response in Biological Skin Tissue Subjected to Multiple Laser Beams

Q. Zhang et al. [3499] investigated the temperature distribution in a three-dimensional biological tissue when irradiated by multiple lasers using the dual-phase-lag bio-heat conduction model. The laser-induced interstitial thermotherapy (LITT) is based on the ability to locally coagulate biological tissue. S. Bown [351] noted that the possible precision for local treatment of the solid tumours with lasers is greater than for almost any other techniques, but the quantitative studies are necessary to establish the appropriate treatment parameters.

Zhang et al. used the dual-phase-lag model to simulate the thermal behaviour with application of the multiple laser beams to the skin tissue considered as a three-dimensional cuboid (the length, width and depth are L, L, h). Four identical laser beams irradiate on the top surface of the skin and they are placed at the corners of a square, the laser beams were simplified as square columns.

The effect of laser irradiance was presented as the volumetric heat source [1350]

$$Q^l(x, y, z, t) = \mu_a \psi(z, t)\varphi(x, y) \tag{7.165}$$

where μ_a, $\psi(z, t)$, $\varphi(x, y)$ are the absorption coefficient of material, local light distribution along depth direction and transverse direction. According to the Beer law [3270]

$$\psi(z, t) = (1 - R)I(t)e^{-\mu_t z} \tag{7.166}$$

where $I(t)$, R, μ_t are the laser intensity, the light reflectance of surface and the attenuation coefficient.

The governing DPL equation for the temperature increment $\vartheta = T - T_a$, T_a is the temperature of arterial blood, is [1927]

$$\varrho_t c_t \tau_q \frac{\partial^2 \vartheta}{\partial x^2} = (\varrho_t c_t + w_b \varrho_b c_b \tau_q)\frac{\partial \vartheta}{\partial x} + w_b \varrho_b c_b \vartheta = \lambda \left(1 + \tau_T \frac{\partial}{\partial t}\right) \nabla^2 \vartheta + \left(1 + \tau_q \frac{\partial}{\partial t}\right) Q \tag{7.167}$$

where the subscripts t and b denote tissue and blood, w_b is the perfusion rate of arterial blood which means the volume of blood flowing into unit volume tissue in unit time, $Q(x, y, z, t) = \dot{q}^{met}(x, y, z, t) + Q^l$ (x, y, z, t).

The authors used the method of separation of variables and series solution that, with account for the boundary conditions, yields the eigenfunctions along various direction as

$$X_m = \sin(\lambda_m x), \quad \lambda_m = \frac{m\pi}{L}(m = 1, 2, 3, \dots), \tag{7.168}$$

$$Y_n = \sin(\lambda_n y), \quad \lambda_n = \frac{n\pi}{L}(n = 1, 2, 3, \dots), \tag{7.169}$$

$$Z_s = \cos(\lambda_s z). \quad \lambda_s = \frac{(2s - 1)\pi}{2h}(s = 1, 2, 3, \dots), \tag{7.170}$$

thus the temperature increment is

$$\vartheta(x, y, z, t) = \sum_{m=1}^{\infty} \sum_{n=1}^{\infty} \sum_{s=1}^{\infty} \vartheta_{mns}(t). \tag{7.171}$$

The governing equation for the time term $\vartheta_{mns}(t)$ is

$$\varrho_t c_t \tau_q \frac{d^2\vartheta_{mns}}{dt^2} + (\varrho_t c_t + w_b \varrho_b c_b \tau_q + \lambda \tau_T \lambda_{mns}^2)\frac{d\vartheta_{mns}}{dt} + (w_b \varrho_b c_b + \lambda\lambda_{mns}^2)\vartheta_{mns} = \tag{7.172}$$

$$= \frac{1}{M_m N_n S_s} \int\limits_{z=0}^{h} \int\limits_{y=0}^{L} \int\limits_{x=0}^{L} q_L X_m(x) Y_n(y) Z_s(z) \mathrm{d}x\mathrm{d}y\mathrm{d}z \tag{7.173}$$

where $\lambda_{mns}^2 = \lambda_m^2 + \lambda_n^2 + \lambda_s^2$, $q_L = \left(1 + \tau_q \frac{\partial}{\partial t}\right) Q(x, y, z, t)$,

$$M_m = \int\limits_0^L X_m^2 \mathrm{d}x = \frac{L}{2}, \quad N_n = \int\limits_0^L Y_n^2 \mathrm{d}y = \frac{L}{2}, \quad S_s = \int\limits_0^h Z_s^2 \mathrm{d}z = \frac{h}{2}. \tag{7.174}$$

Finally, the temperature response is

$$T(x, y, z, t) = \sum_{m=1}^{\infty} \sum_{n=1}^{\infty} \sum_{s=1}^{\infty} \vartheta_{mns}(t) + T_a. \tag{7.175}$$

The authors found that the effect of multiple laser beams on the temperature response differs from those of the single laser beam distinctly. Due to the finite propagation speed of heat energy, the energy of the multiple laser beams arrive at a specified point at different times, which leads to a second heating phenomenon that occurs and the contour of a square with curved corners are obtained. Different irradiated zone can be obtained and accomplished by changing the spot size, the arrangement layout and the interval distance of the laser beams.

7.10 Freezing of the Cylindrical Tissue Region with a Single Embedded Coaxial Blood Vessel

This problem considered by Y. T. Zhang et al. [3480] is stated as follows. The long (so the end effects could be neglected) cylindrical sample of the tissue contains the single cylindrical blood vessel that is coaxial to the sample. The steady state case is studied with the thermal properties of the tissue being constant throughout the frozen and unfrozen tissue subregions. The exterior cylindrical boundary of the tissue ($r = b$) is maintained at the constant freezing temperature T_0, the boundaries $z = 0$ and $z = L_t$ are adiabatic. The authors suppose that this problem could serve as a crude model of the blood vessel deep inside the frozen tumour region. The similar problem relevant to the hyperthermia under a few additional assumptions was considered by A. N. T. J. Kotte et al. [1700, 1812].

The equation for the dimensionless tissue temperature $T^\star = (T_t - T_0)/(T_a - T_0)$, where T_a is the fixed body core temperature is written as

$$\frac{1}{r}\frac{\partial}{\partial r}\left(r\frac{\partial T^\star(r,z)}{\partial r}\right) + \frac{\partial^2 T^\star(r,z)}{\partial z^2} = 0, \tag{7.176}$$

with the Dirichle boundary conditions at $r = a$ ($f(z)$ is the temperature along the vessel wall), $r = b$ and Neumann conditions at $z = 0, L_t$.

Under assumption that the blood rheology could be approximated by the viscous Newtonian fluid, the Poiseuille velocity profile

$$u = u_0\left(1 - \left(\frac{r}{a}\right)^2\right) \tag{7.177}$$

was used for the flow in the vessel, where u_0 is the average blood velocity over the vessel cross-section. The equation for the dimensionless blood temperature $T_\mathrm{b}^\star$ is formulated as

$$\frac{\lambda_\mathrm{b}}{\varrho_\mathrm{b} C_\mathrm{b}}\frac{1}{r}\left(\frac{\partial}{\partial r}\left(r\frac{\partial T_\mathrm{b}^\star}{\partial r}\right)\right) = u_0\left(1 - \left(\frac{r}{a}\right)^2\right)\frac{\mathrm{d}T_\mathrm{c}}{\mathrm{d}z}, \tag{7.178}$$

where λ_b is the blood thermal conductivity, ϱ_b is the blood density, C_b is the blood specific heat, and T_c is the blood bulk temperature defines as

$$T_\mathrm{c} = \frac{4}{a^2}\int\limits_0^a T_\mathrm{b}\left(1 - \left(\frac{r}{a}\right)^2\right) r\mathrm{d}r. \tag{7.179}$$

The thermal resistance of the wall of the blood vessel is neglected; at the boundary between the blood and tissue regions the continuity of both the temperature and the heat flux is required.

Using the separation of variables, the authors sought the tissue temperature distribution in the form

$$T^\star(r,z) = R_0(r)Z(z) \tag{7.180}$$

with $Z(z)$ satisfying

$$\frac{\mathrm{d}^2 Z(z)}{\mathrm{d}z^2} + \beta_m^2 Z(z) = 0 \tag{7.181}$$

and $R_0(r)$ satisfying

$$\frac{\mathrm{d}^2 R_0(r)}{\mathrm{d}r^2} + \frac{1}{r}\frac{\mathrm{d}R_0(r)}{\mathrm{d}r} - \beta_m^2 R_0(r) = 0. \tag{7.182}$$

Here β_m are the positive roots of the equation $\sin(\beta L_t) = 0$, i.e. $\beta_m = m\pi/L_t$.

The final solution is expressed as

$$T^\star(r,z) = A_0 \ln\left(\frac{r}{b}\right)\cos(\beta_0 z)$$

$$+\sum_{m=1}^{\infty} A_m \left[-\frac{K_0(\beta_m b)}{I_0(\beta_m b)} I_0(\beta_m r) + K_0(\beta_m r)\right]\cos(\beta_m z) \tag{7.183}$$

The coefficients A_m are determined as follows:

$$A_0 = \frac{1}{\ln\left(\frac{a}{b}\right) L_t} \int_0^{L_t} \cos(\beta_0 z) f(z) \mathrm{d}z, \tag{7.184}$$

for $m = 0$ and

$$A_m = \frac{2\int_0^{L_t} \cos(\beta_m z) f(z)\mathrm{d}z}{\left[-\frac{K_0(\beta_m b)}{I_0(\beta_m b)} I_0(\beta_m a) + K_0(\beta_m a)\right] L_t}, \tag{7.185}$$

for $m = 1, \ldots$.

7.11 Heat Transfer in a Finite Tissue Region with Two Embedded Blood Vessels

D. Shrivastava and R. Roemer [2846–2848] considered analytically heat transfer in the uniform tissue region with two imbedded blood vessels and found the so called Poisson conduction shape factors for both the vessel-vessel heat transfer (VVPCSF) and the vessel-tissue heat transfer (VTPCSF) that could be used for the relatively low cost computations of heat transfer in the tissue. Some of the energy leaving the blood vessel goes to the other vessel; the rest goes to the tissue boundary that can be expressed as $Q_{\text{total,v1}} = Q_{\text{v1}\to\text{v2}} + Q_{\text{v1}\to\text{t}}$, $Q_{\text{total,v2}} = Q_{\text{v2}\to\text{v1}} + Q_{\text{v2}\to\text{t}}$.

Mathematically the problem reduces to the solution of the two dimensional Poisson equation in the domain that contains the circular holes, if the conduction is the only heat transfer mechanism.

Such problem has been considered earlier by L. Zhu and S. Weinbaum [3526] for the case of a single hole imbedded into a finite medium who derived fundamental solution for a Green's function for this flow geometry. This approximate solution is exact when the ratio K of the blood to tissue conductivity is unity and a highly accurate approximation when $K \neq 1$. This basic solution was applied to develop a model for the heat transfer from a countercurrent artery-vein pair in an exteriorized rat cremaster muscle preparation.

Y. L. Wu et al. [3348] considered problem for the two holes in the infinite medium. The vessels of diameter ≈ 100–$400\,\mu$m make the majority of the countercurrent vessels pairs in organism that significantly influence the heat transfer in tissues.

D. Shrivastava and R. Roemer considered the circular domain with the two circular holes of the smaller (not necessarily equal) diameter. The heat source P is assumed to be uniform in the tissue part of the domain. Starting with the Poisson equation for the temperature distribution

$$\frac{1}{R}\frac{\partial}{\partial R}\left(R\frac{\partial T}{\partial R}\right)+\frac{1}{R^2}\frac{\partial^2 T}{\partial \psi^2}=P \tag{7.186}$$

with Dirichlet conditions specified on all boundaries, the authors by substitution

$$T=T_1-\frac{PR^2}{4} \tag{7.187}$$

convert it into the Laplace equation

$$\frac{1}{R}\frac{\partial}{\partial R}\left(R\frac{\partial T_1}{\partial R}\right)+\frac{1}{R^2}\frac{\partial^2 T_1}{\partial \psi^2}=0 \tag{7.188}$$

where P is the power deposition.

Using the linearity of the governing equations, the authors subdivide the original problem of the disk with two eccentric holes into two subproblems for the disk with a single hole. These latter problems are further simplified by the conformal transformation that makes the disk and the hole to be co-centric. The final equation for the first subproblem reads as

$$\frac{1}{R_1^\star}\frac{\partial}{\partial R_1^\star}\left(R_1^\star\frac{\partial T_{11}}{\partial R_1^\star}\right)+\frac{1}{R_1^{\star 2}}\frac{\partial^2 T_{11}}{\partial \alpha_1^2}=0 \tag{7.189}$$

where $R_1^\star$ is the dimensionless radius in the transformed coordinate system and α_1 is the corresponding angular coordinate. The general solution to this equation is expressed as

$$T_{11}=A_{01}+A'_{01}\ln(R_1^\star)+\sum_{n=1}^{\infty}\left[A_{n1}(R_1^\star)^n+A'_{n1}(R_1^\star)^{-n}\right]\sin(n\alpha_1) \tag{7.190}$$

$$+\sum_{n=1}^{\infty}\left[B_{n1}(R_1^\star)^n+B'_{n1}(R_1^\star)^{-n}\right]\cos(n\alpha_1) \tag{7.191}$$

Similar calculations are performed to obtain the analytic solution to the second subproblem for T_{12}.

Once the temperature field is determined, the heat transfer rates Q are found by integration, e.g., $Q_{\mathrm{v1}\to\mathrm{t}}$ is expressed as

$$Q_{\mathrm{v1}\to\mathrm{t}} = -\int_0^{2\pi} \frac{\partial T}{\partial R}\Big|_{R=R_{\mathrm{v1}}} R_{\mathrm{v1}}\,\mathrm{d}\vartheta, \tag{7.192}$$

where ϑ is the angular position relative to the coordinate system with origin at the vessel's centre.

The analytical solution procedure developed by the authors could be extended to the general case of N vessels imbedded into the tissue with their arbitrary location [2848] by the decomposition into N subproblems for the single embedded vessel.

Part III
Heat Transfer in Medicine

Chapter 8
Hyperthermia

The bian stone (*bianshi, stone needle*) is the earliest (about 4000 years ago) instrument of the thermal therapy [3352], also used for the acupuncture (later replaced by the metal needles—silver, copper, brass, iron, stainless steel; the difference in the thermal conductivity could be essential in the medical efficiency of the acupuncture). The bian stone consists of the nanometre-sized calcite crystals and emits the infrared radiation in the 8–14 μ spectral range that is easily absorbed by the human body and improved the energy metabolism and promotes the microcirculation.

The cauterization using the red-hot iron is a very old procedure to stop the bleeding or to destroy the tumour—it is described in The Hippocratic Corpus, the collection of the early Ancient Greek medical works. The effect of the blood flow on the heat transfer in the living tissues have been studied by °C. Bernard in the nineteenth Century ("Lecons sur la chauleur animale", Bailliere, Paris, 1876) [3525]; the data on the characterization of the blood vessels were published in 1888 by Prof. Mall of Saxony [494]. W. Coley, who noted the tumour regression in the patients with the fever, mixed the bacterial toxins to intentionally induce the fever in the late nineteenth century [743].

The general definition of the hyperthermia is a therapy in which the tissue temperature is elevated above normal by the external means. In the systemic, or the whole body, hyperthermia the temperature of the body is raised using the RF heating, a water bath or the extracorporeal induction, or the extracorporeal hyperthermia (the blood is pumped out of the patient's body, heated to 42° C or more, then put back in the body while still hot) [2140, 2617, 2956].

The local heat sources are introduced in the targeted tissue regions by several energy deposition techniques using the laser, the RF current, the microwave or the focused ultrasound. Despite the different mechanisms involved in each case, the net thermal effects can be described by a single quantity: the heat-generation rate per unit volume. The blood flow accounts for up to 90% of the heat removal [1756] and can increase up to a factor of 10 in response to the hyperthermia, especially in

A. I. Zhmakin, *Heat Transfer in Vivo*, Springer Series in Biophysics 27,
https://doi.org/10.1007/978-3-031-95459-7_8

the normal tissue [2906]. The non-uniform heat removal by the blood flow further aggravates the temperature heterogeneity [1669].

The maximum temperature is usually between 41.8° C and 42° C [1224]. If the volume being heated does not include a sensitive organ, such as the heart or the liver, the temperatures higher than those associated with the systemic hyperthermia, can be used. Research show that the cell killing effect doubles every centigrade [961, 1666].

J. Crezee and J. J. W. Lagendijk [648] studied the impact of the single and counter-current paired large vessels on the temperature uniformity during hyperthermia. The vessel wall temperature determines the minimum tumour temperature, and it was shown that even when the blood temperature is low, the vessel wall temperature may reach the therapeutic levels in specific cases. A high tissue blood flow reduces the temperature gradient near the large vessels, which implies that it is more difficult to achieve a uniform dose in the poorly perfused than in the well perfused tumours. They also showed that including a relatively large margin of the normal tissue in the heated volume will improve the temperature homogeneity in the tumour region.

The time needed to attain an irreversible cell injury at hyperthermic temperatures is typically from 30 min up to one hour [565]. Most normal tissues are undamaged by treatment for 1 hr at a temperature of up to 44° C [912]. At higher thermal intervals (temperatures of 48–60° C) the required exposure time to attain the irrecoverable thermal damage is a few seconds [285].

Several techniques are applied in clinical hyperthermia depending on the tumour volume and location. Local (also locoregional [1666]) and whole body hyperthermia can be used. The applied frequency depends on tumour location and heating technique.

Usually the local hyperthermia involves heating a volume of the tissue to between 40° C and 45° C [315]. Local hyperthermia includes intraluminal/intracavitary, interstitial and superficial heating techniques. Energy deposition is localised to the tumour, thereby minimising normal tissue heating. A drawback of this technique is the limited penetration depth (only a few millimetres) which may result in an underdosage of some parts of large tumours. The penetration depth can be slightly improved by applying active cooling of the applicator surface [1666]. With interstitial hyperthermia, heating sources are placed invasively in the tumour.

For the safe thermal ablation, two constraints should be valid [514]:

- the peak tissue temperature must be less than 92° C to prevent charring and vaporization,
- the maximal temperature of the large vessel wall must be less than 43° C to avoid protein denaturation and cell death [408].

The three variants of the hyperthermia are distinguished according to the goal temperature level [1202, 1203, 3210]:

- the diathermia (usually in the physiotherapy): heating up to 41° C;
- the hyperthermia: heating into the interval of 41° C–45° C (according to W. C. Dewey [741], the ideal therapeutic temperature is higher than 42° C);
- the thermodestruction/thermoablation: heating over 45° C.

The duration of the diathermia is usually 6–72 hr, the hypethermia—15 to 60 min, the thermoablation—4 to 6 min [742, 1202]. K. Overgaard and J. Overgaard [2362] established that for the hyperthermia the longer heat exposure is required where a lower temperature of the treatment is used, and vice versa. The combination of these two parameters (the duration and the temperature) can be used to define the minimum heating time required to achieve the damage of the tumour at the specific temperature [751, 2982]. This relation is formulated in the form of *the thermal isoeffect dose* [1064] that could be estimated using an Arrhenius plot [2982]. The Arrhenius plot depicts the time–temperature relationship of the cells survival phenomenon due to the applied heat and is affected, among other factors, by the thermotolerance, the pH, the cell phase in the cell cycle [741, 2982]. For example, the skin damage can be represented as a chemical rate process, which is calculated by using a first-order Arrhenius rate equation, whereby damage is related to the rate of protein denaturation and exposure time at a given absolute temperature [1295, 2186].

Another approach is to use "iso-effect relationship" [1254] between two different thermal exposure conditions to quantify the thermal isoeffect dose via the cumulative equivalent minutes at 43° C (CEM43) which is a method for the conversion of the heating duration at the specific temperature to an equivalent duration that generates the same heating outcome [2712].

The "critical thermal load" approach assumes that the total damage is a function only of the total cumulative dosage, so that equal doses produce equal injury. However, A. M. Stoll [2947] demonstrated that a large amount of energy delivered over an extended period of time may produce no damage whatever, while the same dose delivered nearly instantaneously may destroy the skin.

The toxicity of heat generated during the thermal therapy in general is low; the burns represent typical the thermal-therapy-associated toxicity with low incidence that can be avoided via the correct heating techniques [1202]. The hazards of the thermal therapy are due to either the increased body core temperature or the increased temperature in specific organs.

The diathermia is used in the physiotherapy for the treatment of the rheumatic diseases [1202]; it could cause the different processes in the human organism, e.g., stimulates the antigen-specific T cell responses [1663].

The local hyperthermia is safer than the systemic hyperthermia. since the risks associated with maintaining the whole body temperature at such high levels for a long period of time are removed [1233, 2955].

The hyperthermia is used for the treatment of the cancer tumours, the benign prostate hyperplasia (BPH) treatment in the urology, in the cardiology for the heart stimulations [1202], the retina corrections and for some diseases [3210] by the process of heating the regions of the body containing the carcinoma tissue—

the cancer cells are more sensitive to the heat than the normal healthy cells [388, 1224, 2361] due to the high acidity of the cancer cells as a consequence of the high glycolytic activities inside cancer cells [1065]. The thermal based therapies are widely used to cure the disease/injury involving the skin tissue, such as the photothermal epilation, the port wine stains, the pigmented and cutaneous lesions, the tattoos, the hemangiomas, the rhytides, the nonablative skin resurfacing, the acne treatment, the scars, the photorejuvenation of the photodamaged skin, etc. [3381].

The hyperthermia treatment is considered as the fourth adjunct cancer therapy technique following the surgery, the chemotherapy, and the radiation techniques. The hyperthermia is particularly effective in hypoxic tumours, i.e. tumours with a low level of oxygen that are more resistant to ionising radiation than well oxygenated tumours; the complementary effect of hyperthermia and radiotherapy is also due to the fact that cells in the S-phase of the cell cycle are more sensitive to hyperthermia, whereas cells are more resistant to radiotherapy in the S-phase [1666].

Hyperthermia is used for many cancer types such as breast cancer, sarcomas, melanomas, bone metastases, carcinomas of the lung, stomach, pancreas, gallbladder, kidneys, neck, brain tumours, prostate tumours, and cervical cancer [2009]. The hyperthermia can be used as a stand alone cancer technique. However, the hyperthermia has demonstrated most success when it used in conjunction with other cancer treatment techniques such as the radiotherapy or/and chemotherapy [1224, 2955]—the so called *multimodality treatment* [677, 2982]. For example, the heating increases the effectiveness of chemotherapy by increasing the local blood supply and the oxygenation of tissues in response to the body thermoregulation in removing the excessive heat which enhanced the uptake of the chemotherapeutic drugs in the tumour [2833]. The heating increases the cells sensitivity to the radiation [1527, 3144] due to the increase of the blood flow that drives the tissue oxygenation enrichment and the enhancement of the radio sensitivity by facilitating formations of the oxygen radicals [288, 3216].

In contrast to the chemotherapy and the radiotherapy, the hyperthermia is not limited by the number of the doses of the radiation [3210].

The clinical hyperthermia is divided into the three almost separate domains: the whole body hyperthermia (WBH), the regional hyperthermia (RHT), and the local hyperthermia (including the superficial local and the interstitial local hyperthermia) (LHT) [917]. The WBH causes the release of the tumour necrosis factor (TNF) [1657]. The increased levels of the TNF have the direct cytotoxic effects and induce the tumour heat-induced damage [1407]. The WBI has tremendous impact on the tissue metabolism, the blood flow, the organ function, and the tissue repair [1203].

Usually the LHT is performed using the single microwave or ultrasound applicators, whereas the deep RHT is performed using the arrays of the multiple applicators for deep heating. The historical background of the hyperthermia treatment can be found in [1049].

There are different ways to heat the target tissue [34, 699, 2624, 3213]:

- the microwaves (MW) and the radiofrequency (RF) radiation [353, 657, 1952, 2637, 2964, 3244, 3253]—at the molecular level, two types of the heating mechanisms are known: the resistive heating by the ionic conduction and the dielectric heating by the inter-molecular friction of the dipole molecules—any polar solution, for example, the living tissue, is heated by the rotations the polar molecules (for example, water and the proteins [1485]) induced by the electromagnetic field.

  Usually the microwave radiation is considered to cover the wavelength range from 1 mm to 1 m (from 300 MHz to 300 GHz) being a part of the RF radiation (from 0.5 MHz to 300 GHz) [145, 2559] (still, however, MW and RF sometimes are considered as the separate ablation techniques [1423, 2598]).

  MW penetrates only a few millimetres into the body surface, thus the energy of the MW is absorbed by the skin surface, resulting in a temperature increase mainly on the body surface [1396, 2717] and causes subsequent changes in the skin gene expression [2134].

  For example, E. Ijima et al. [1396] used the MW radiation a 26.5 Hz. The penetration depth of the absorbed power generally increases with decreasing frequency, but for intraluminal applicators the upper limit of the penetration depth is determined by the radius of the cavity. The depth where the absorbed power is reduced to half the value at the antenna surface is at most equal to the radius of the cavity [3212]. To achieve minimally required therapeutic temperature levels over the complete tumour volume, a high surface temperature will be necessary, which may induce thermotoxicity [1666].

  The relatively low frequencies (from 10 to 100 MHz) are needed for deep penetration into the biological tissue (i.e. up to 10 centimetres under the body surface)—such treatments are based either on the principle of the dielectric filled waveguide or on the principle of the evanescent modes (i.e. the waveguides excited below its cut-off frequency) [999, 3210, 3211].
- the ultrasound (US) [154, 531, 752, 1012] and the high intensity focused US (HIFU) [307, 1451, 1595, 3038]; the longitudinal wave is of primary importance for the medical applications of ultrasound—though the shear acoustic waves can propagate in solids, and may therefore travel in the calcified tissues such as bone or tooth, they are of little relevance in the soft tissue, which can barely support them at the ultrasonic frequencies [809];
- the infusion of the fluids containing the magnetic nanoparticles (MNPs) with the subsequent thermal activation—the magnetic fluid hyperthermia (MFH) [1539, 2982] or the nanomagnetic hyperthermia (NMH) [457, 826, 1082, 1493, 2532, 2690, 3454]—the magnetic particles function as a heat source, which generates heat due to the hysteresis loss, the Néel relaxation, the brownian motion, or the eddy currents [3525];

- the various laser techniques [785, 2298, 2900, 3288] including the ultrafast lasers (with the pulse duration range from sub-nanoseconds down to femtoseconds [43, 569, 985, 1598] or even attoseconds [2508])— the photodynamic therapy [180, 2322], the hyperthermia [1456, 1740, 1858, 1914, 1943, 2376, 2837, 3482], the laser thermokeratoplasty (LTK), the interstitial laser photocoagulation therapy (ILP) [143].

V. Milani and E. Noessner [2128] emphasized that the response to the thermal stress is not a one-time point event, but rather a time period starting with the heat exposure and extending over several days of the recovery; the response of the tumour cells and their susceptibility to the immune effector cells is strongly dependent on the magnitude and the duration of the thermal stress and on the time of recovery after the heat application.

In response to the temperature rise, the blood flow rate in the healthy tissues can increase up to 20 times higher than its preheating rate, while the heating would only cause an increase in the tumour blood flow rate up to double its flow rate [810, 2907].

Hyperthermia allows to treat the embedded tumours—to damage or even exterminate the malignant cells—in the vital regions where the classical resection surgery is not feasible or risky. However, in order to reach the underlying tumours, the activating energy source must sufficiently penetrate through healthy tissues, while minimizing damage to the later.

The RF and microwave technologies frequently require insertion of the probes directly into the target tumour and the laser therapy needs the optical fibres for delivery of the laser light [81] while the HIFU is completely noninvasive—it does not require puncturing the skin and typically has minimal or no side-effects. In the HIFU therapy the focused ultrasound beams are used to create a rapid temperature rise at the focal point, which results in the irreversible tissue damage due to the coagulative thermal necrosis [3229, 3343].

There is a variant of the hyperthermia that enhances the efficiency of conventional hyperthermia by additional, nonequilibrium thermal effects [1071, 1203]. In the *extracellular hyperthermia (electrohyperthermia, oncothermia, electrosurgery* [2057, 2935]) the targeted tissue is heated by the electric field with the energy absorption localized within the extracellular liquid [2996]. The extracellular hyperthermia is based on a capacitively coupled energy transfer applied at a frequency that is primarily absorbed in the extracellular matrix due to its inability to penetrate the cell membrane [1699]. Sometimes only non-thermal effects of the ultrasound are exploited, for example, the high-pressure-amplitude shockwaves are used for mechanically resolving kidney stones [2138].

In the medical applications two general bio heat transfer characteristics are important:

- the thermal penetration depth which is the deepest depth where the heat effect reaches;
- the time to reach the steady state.

J. Okajima et al. [2332] presented the dimensionless expression for these parameters obtained from the solution of the one dimensional Pennes equation

$$\varrho c \frac{\partial T}{\partial t} = \frac{\varkappa}{x^n} \frac{\partial}{\partial x} \left(x^n \frac{\partial T}{\partial x} \right) + \varrho_b c_b \omega_b (T_b - T) \tag{8.1}$$

where $n = 0, 1, 2$ indicates the Cartesian, axisymmetric and spherical symmetric coordinates, respectively, T_b is the steady-state temperature of the biological tissue defined as

$$T_b = T_a + \frac{q_{\text{met}}}{\varrho_b c_b \omega_b} \tag{8.2}$$

where T_a the arterial blood temperature.

The authors assumed that

1. the biological tissue is isotropic and homogeneous
2. the tissue properties are independent of the tissue temperature;
3. the metabolic heat generation rate is constant per unit volume and unit time [568];
4. the blood perfusion rate is uniform spatially and temporally and independent of tissue temperature [3483];
5. the arterial blood temperature is constant.

The three kinds of the boundary conditions on the surface of the biological tissue (the boundary between biological tissue and the fluid at a constant temperature T_f) were considered: 1st kind: $T_{surface} = T_w$, 2nd kind: $-\lambda(\partial T/\partial x)_{surface} = q_w$, 3rd kind: $-\lambda(\partial T/\partial x)_{surface} = h(T_f - T_{surface})$.

The thermal effects are present in other medical procedures. For example, the Joule heating by the electric current accompanies the both modes of the electroporation [1057, 2402] that is based on the alteration or the destruction of the cell membranes by the electric impulses [1639, 3271]—the irreversible electroporation (the increase in the membrane permeability is associated with the formation of the nanoscale defects—the pores—in the membrane when the strength of the electric field exceeds the threshold value that do not reseal [679, 835, 2659]) and the reversible electroporation (in which the pores reseal) used for the transdermal drug delivery [710] or for the delivery of the normally impermeable drugs, for example, in the chemotherapy [2149].

8.1 RF Hyperthermia

The hyperthermia treatment could be performed using either the invasive or noninvasive systems [3060]—it can be performed via the open surgery, laparoscopy or percutaneous approaches [1346, 2097, 2349].

The interstitial (invasive) hyperthermia involves inserting the heat sources—the coaxial microwave antenna arrays or the RF needle electrodes of the diameter 1–2 mm—into the targeted tumour tissue [2488] with the guidance of ultrasound, computed tomography (CT) or magnetic resonance (MR) imaging device. The RF ablation technique usually uses a pair of electrodes: an active electrode with a small surface area that is placed on the target zone, and a larger dispersive electrode needed to close the electrical circuit [262]. Recently R. Ortega–Palacios et al. [2351] suggested a novel 2.45 GHz double short distance slot coaxial antenna for the minimally invasive microwave ablation therapy.

While these invasive methods allow for more the accurate heating of the tumour than the non-invasive systems, there is always the discomfort associated with placing the foreign objects in the human body [1224]. Rates of the complications are low (less than 5%) with both the radiofrequency and the microwave ablation [272].

In the RF heating, the currents pass through the insulated flexible wire arranged in the coils that generate the oscillating magnetic field which produces the currents in the tissue, heating it.

The electromagnetic field also directly affects the deoxyribonucleic acid conformations [2117]; the molecular dynamics simulations show that the electric field causing a re-orientation of the protein's dipole moment results in the changes of the secondary structure of the protein and aids its denaturation [864].

Usually the frequencies in the range 20–30 MHz are used. When the deep seated tumours are treated, the surface cooling is required to prevent the damage to the skin [1224, 1235, 1431]. The different cooling methods, such as the contact cooling [63] and cryogen spray cooling (a liquid cryogen is sprayed onto the skin surface) [94, 97], are used to reduce the epidermal thermal damage [3225, 3381]. Similar skin protection measures are used in the laser hyperthermia providing reduction in the epidermal damage [94–97, 107, 546, 971, 2285, 2452, 2969, 3181], increased light dose at the safe level [26, 1589], and reduction in pain caused by the laser treatments [409, 1161, 3227, 3285].

The RF ablation has been used for the treatment of cardiac diseases, such as arrhythmia (abnormal heart rate), and several types of tumours. It employed for treating tumours in kidney, lung, rectum, breast, prostate and musculoskeletal system, it is most commonly used for liver tumours [3518].

In some cases, for example, in the cardiac and the hepatic ablation, the irrigated electrodes are used in which the heated saline solution is simultaneously infused into the tissue [1125]. In this case, the tissue can be considered as a porous medium into which saline is infused, and hence the additional equations should be employed. Sometimes the internally cooled RF electrodes (the dual channel probes with an electrode at the tip, in which the cooling fluids are continuously delivered through one channel and removed through another [262]) are used that can be approximated by means of a single temperature boundary condition at the electrode surface (value fixed to the same value as the cooled fluid temperature) [1436].

At the frequencies employed in the RF ablation and within the area of interest—the region of a small radius around the active electrode, the tissues are purely resistive, because the displacement currents are negligible. Thus, the quasi-static approach is usually employed to solve the electrical problem equations [793, 2468] and the heat source q is given by $q = \mathbf{j} \cdot \mathbf{E}$ where $\mathbf{j}$ is the current density, $\mathbf{E} = \nabla\varphi$ is the electric field governed by the Laplace's equation: $\nabla \cdot \sigma\nabla\varphi = 0$ where σ is the electrical conductivity.

The absorption process of the electromagnetic radiation is divided into the categories that correspond to modes of the molecular energy storage—the thermal, the vibrational, the rotational, and the electronic modes [1202]. The low-level EM radiation could affect the nervous system (the effects on the blood-brain barrier (BBB), the morphology, the electrophysiology, the neurotransmitter activity, and the metabolism) [765, 1760]. The EM radiation at such levels might also affect the immune system, the gene and the chromosomal morphology, the enzyme activity, the neurological function, the cell morphology, the membrane ion permeability, the intracellular ion concentration, the mutation rates, the tumour promotion, the endocrine secretion rates [1202].

The heat absorbed by a human body depends on the tissue properties such as the dielectric properties and conductivity, the shape of the different tissues and their interfaces. E.g., the materials with high water content (muscle or the brain tissue) have higher energy absorption than materials with low water content (fat or the and bone tissue). Moreover, the complex shape of biological tissues and their interfaces may cause standing waves which result in hot spots [1200] and lead to a deep burn of the tissue under the skin.

The radiofrequency hyperthermia is broadly employed in the clinical areas such as the elimination of the cardiac arrhythmias, the destruction of the tumours in the different locations—liver [1822], breast [25, 2360, 3200], kidney [2090], lung [1043], bone [435], prostate [2806], the treatment of gastroesophageal reflux disease [1105], the thermal remodelling of the oral cavity tissue to cure the sleep obstructive apnea [2939], the therapeutic heating of the articular cartilage [3422] and the heating of the cornea for refractive surgery [266].

In addition to the temperature increase, the RF power causes other effects, for example, the change of the permeability of the blood-brain barrier (BBB) [1884, 2187] or an increase in the leukocyte adhesion to the endothelium [2061, 2062, 2077, 3119].

The major problem of the RF hyperthermia is the inability to produce a large ablation zone that envelopes the large tumours because the employment of the larger RF power induces the undesirable tissue burns [2802]. Another problem of the application of the RF hyperthermia as well as cryoablation to the large-size tumours is the heat sink effect due to the blood flow in the highly complex vascular system [3500]—the adjacent large blood vessels cause tissue cooling/heating, reducing the effectiveness of ablation [1967].

There are several approaches to this problem:

- the development of the internally cooled probes [182];
- the sequentially overlapping of the single electrode [779];
- the use of the multiple electrodes [1375, 2211, 2944].

In the microwave heating, the radiation is applied using the contact microstrip applicator, the frequencies between 300–1000 MHz are usually used [1666]. It is applicable for the heating of the larger volumes of the tissues than the ultrasound (usually in the frequency range 1–20 MHz [809]) and can permit heating of the tissue which are inaccessible to the ultrasound [1224]. The microwave hyperthermia is most effective for the tissues with the high water content [2492].

To improve the uniformity of the temperature distribution and to provide the precise tumour ablation, the multiple antenna configurations with the simultaneous power delivery are used. Such devices are considered to be less invasive [5, 6, 22] than the single antenna configuration with the sequential power delivery [1244, 1739, 1753].

M. Cifra et al. [579] reviewed the effects of the low and high (above 1 MHz) frequency electric fields (EFs) on the living cells. The very low frequency EFs (< 2 kHz) are used to stimulate the excitable cells while the high frequency EFs are used to heat the cells and tissues.

The electromagnetic fields in both the extremely low frequency (ELF) and the RF ranges activate the cellular response [18], a protective mechanism that induces the expression of the stress response genes, for example, HSP70, and increased levels of the heat stress proteins, for example, hsp70—induction of the stress response involves activation of the DNA, and despite the large difference in the energy between the ELF and the RF, the same cellular pathways respond in the both frequency ranges [311, 1124, 1313, 1826, 2649].

While the low energy field interacts with the DNA to induce the stress response, increasing the EMF energy in the RF range can lead to breaks in the DNA strands. The possibility of the small energy in the ELF frequency range to activate the DNA is related to the large hydration energy tied up in the protein and the DNA structures that could be released by the small changes in charge [310].

E. D. Kirson et al. [1641, 1642] reported that EFs at frequencies of 100–300 kHz have an antimitotic effect, inhibiting the proliferation rate of the human and the rodent tumour cell lines while having no effect on the non-cancerous cells. The inhibitory effect was reported to be frequency-dependent with a peak effect at 200 kHz for the glioblastoma multiforme (GBM) cells labelled these EFs as the "Tumour Treating Fields (TT Fields)" (the GBM is the most common and lethal adult brain tumour [206]).

M. C. Ziskin et al. [992, 3538] and A. Hirata et al. [1329] showed that at frequencies above 6 GHz the heating effect occurs predominantly in the superficial tissue, particularly in the skin.

R. Morimoto et al. [2185] investigated the relation between the specific absorption rate (SAR) [1666] and the temperature of the human head models with the frequency up to 30 GHz. The strong correlations between the peak averaged SAR and the temperature rise were observed at the frequencies up to 3–4 GHz, and the weaker correlations were observed at the higher frequencies.

Despite the numerous studies the mechanism(s) of the action of the "TT Fields" is not well understood. The abnormal cell mitotic morphology including the blebbing followed by the apoptosis [206, 1062, 1642]) was observed, suggesting the biophysical anti-mitotic mechanism of the action. At the ultra-low frequency, the EFs are largely screened and do not penetrate the cell membrane, whereas in the low to intermediate (100–500 kHz) frequency range, the EFs penetrate cells and could have the biophysical effect [3101, 3294, 3295]: an interaction of these EFs with the electric dipoles of the tubulin dimers occurs which results in changes in the tubulin dimer rotational dynamics, disrupting the formation and function of the mitotic spindle [685, 1641].

Another mechanism proposed by E. D. Kirson et al. [1641, 1642] is based on the dielectrophoretic (DEP) forces that result in the electrokinetic effects on the polarizable bio-macromolecules. During the cytokinesis, a cleavage furrow appears between the dividing daughter cells. The DEP forces could potentially cause the molecules to migrate toward or away from the furrow, impeding the cell cleavage [267, 3101, 3294].

Recent studies support the additional potential mechanisms [1074, 2582]: the mitotic checkpoint inhibition, the tubulin mediated conductance effects, the permeability changes, the DNA damage, the migratory inhibition.

The contacting delivery of the EF field suffers from the poorly characterized EF profile and the conductive heating that limits the duration and the amplitude of the applied EFs. R. Ravin et al. [2582] suggested the device—the air-core solenoid coil—that delivers the EFs with a well-characterized radial profile in a non-contacting manner, eliminating the conductive heating and enabling the thermally regulated EF delivery.

In the RF hyperthermia problems the transient effects in the Maxwell equations can be neglected because the EM fields in the tissues achieve the steady-state in micro-seconds, thus, the Maxwell equations can be approximated by the time-harmonic Maxwell equations corresponding to the forcing frequency [1621]. For example, the unknown electric field is written as

$$\mathbf{E}(\mathbf{x}, t) = \Re(\mathbf{E}(\mathbf{x}) e^{i\omega t}) \tag{8.3}$$

where ω is the angular frequency, $\mathbf{E}(\mathbf{x})$ is the complex-valued phasor, being only a function of the position $\mathbf{x}$ and the frequency. The phasor can be also identified as the Fourier transform (in time) of the real-valued electric field $\mathbf{E}(\mathbf{x}, t)$ [1621].

Thus the time-harmonic Maxwell equations are written as

$$\nabla \mathbf{E} = -i\omega \mathbf{B}, \quad \nabla \mathbf{H} = \mathbf{J} + i\omega \mathbf{D}, \quad \nabla \cdot \mathbf{D} = 0, \quad \nabla \cdot \mathbf{B} = 0. \tag{8.4}$$

$\mathbf{E}$ and $\mathbf{H}$ fields are related to the corresponding fluxes $\mathbf{D}$ and $\mathbf{B}$ through the constitutive laws $\mathbf{D} = \varepsilon \mathbf{E}$ and $\mathbf{B} = \mu \mathbf{H}$, the impressed current $\mathbf{J}^{imp}$ and free current $\mathbf{J}$ are related as

$$\mathbf{J} = \mathbf{J}^{imp} + \sigma \mathbf{E}. \tag{8.5}$$

The permittivity, permeability and conductivity are tensors for the anisotropic media; they can be functions of both the position and frequency, and may also depend on the field intensities. For the case of the lossy media, the material parameters become complex-valued.

When the wavelength of the RF current in the tissues is much greater then the depth of a human body, the quasistatic electric field approximation can be applied [2017] and the solution of the Maxwell's equations be avoided. The quasistatic electric field is irrotational, thus the electric potential φ can be introduced that is described by the Laplace equation

$$\nabla \cdot (\varepsilon \nabla \varphi) = 0 \tag{8.6}$$

where ε is the dielectric permittivity of the tissue.

The magnetic field within an air-core current-carrying coil is readily approximated by the Ampere's $B_z(t) = \mu_0 I(t) N/L$ where $B_z(t)$ is the axial component of the applied magnetic field, μ_0 is the permeability of a vacuum, I is the current amplitude, N is the number of the coil turns, L is the length of the coil. The Faraday's law in the integral form

$$\varepsilon = \oint_l \mathbf{E} \cdot \mathbf{l} = -\frac{\mathrm{d}}{\mathrm{d}t} \int_S \mathbf{B} \cdot \mathrm{d}\mathbf{S}. \tag{8.7}$$

If the surface is chosen to be a circular cross section of the coil with radius r from the centre of the coil then [2582]

$$|\mathbf{E}(r,t)| = E_\vartheta(r,t) = \begin{cases} \dfrac{r}{2}\dfrac{\mathrm{d}B_z(t)}{\mathrm{d}t} = \mu_0 \dfrac{N}{L}\dfrac{r}{2}|\dfrac{\mathrm{d}I}{\mathrm{d}t}| & r < R, \\ \dfrac{R^2}{2r}\dfrac{\mathrm{d}B_z(t)}{\mathrm{d}t} = \mu_0 \dfrac{N}{L}\dfrac{R^2}{2r}|\dfrac{\mathrm{d}I}{\mathrm{d}t}| & r \geq R \end{cases} \tag{8.8}$$

where R is the radius of the coil.

The rate of the heat generation, Q, delivered by the current-carrying coil to the material of the height L_z is

$$Q = \sigma \int_V |\mathbf{E}(r)|^2_{rms} \mathrm{d}V = \int_0^{L_z} \mathrm{d}z \int_0^{2\pi} \mathrm{d}\vartheta \int_0^R r|\mathbf{E}(r)|^2_{rms} \mathrm{d}r. \tag{8.9}$$

The energy-deposition rate deceases rapidly with increasing the distance from the RF electrode. Thus, the solution of the whole electric field is not necessary sometimes: one can consider only the near-field pattern around the RF applicator. A simple model of this type approximates the applicator as the infinitely-long line source, resulting in the approximation [3518]

$$|\mathbf{E}|, |\mathbf{J}| \propto \frac{1}{d} \tag{8.10}$$

and the inverse-square law for the energy-deposition rate [1486, 2965]

$$r = \mathbf{J} \cdot \mathbf{E} = \frac{C}{d^2} \tag{8.11}$$

where C is a constant, and d is the radial distance from the centreline of the RF applicator (therefore, d is always larger than zero). This model does not account for the finite length or the attenuation due to the dissipation. An improved model that addressed these issues was given by L. Zhu et al. [3528]

$$r = C\frac{\exp(-2\gamma d)}{d^2} \exp\left(-\frac{z^2}{z_0^2}\right) \tag{8.12}$$

where the first term on the right-hand side includes both the inverse-square geometric spreading and the attenuation over the path with γ being the attenuation constant. The second term describes the end effects of energy applicator, where z is the axial distance from the middle of applicator and z_0 is an adjustable parameter.

In the microwave hyperthermia heating of the target is achieved by exploiting the constructive wave interference from a number of the antennas [802, 1029, 2396]. The system consists of an applicator holding multiple antenna elements to transfer the electromagnetic energy into the tissue allowing excellent spatial and temporal control and a temperature measuring system that provides the real-time, three dimensional information on the temperature distribution [3157]. The heating of the skin is inevitable; to avoid this, a bag filled with the circulating water is introduced between the applicator and the body [2606].

The effectiveness of a hyperthermia treatment is strongly related to the possibility to plan the heating [1670], especially for the challenging anatomical sites such as the head and neck region [2397, 2399].

During the thermal therapy, the non-uniform temperature field in the tissue can induce the internal forces leading to the thermomechanical deformation governed by the thermoelastic wave equation [2393, 3389]. A minute alteration in the mechanical stress can cause the abnormality in the hormone release and reduced the immune response [2394]. The induced thermal stress can decrease the energy barrier [2393] of acquiring the protein denaturation boasting the necrosis of the tissue. The tissue deformation during the thermal treatment can activate the pain sensing receptors to stimulate the thermal pain [2093, 2393, 3389].

The anti-tumour immunity is induced by the HT—the natural killer cells and phagocytes like granulocytes are directly activated against the tumour by the HT, the heat also activates dendritic cells [1009].

8.2 High Intensity Focused Ultrasound (HIFU)

The ultrasound radiation can be focused to heat the very small volumes of the tissue. It appears well suited for producing the sustainable, uniform hyperthermia in the tumours up to 5 cm below the surface in the tissue volumes which do not contain bone or air [1224]. The ultrasound at the high power can lead to the tissue damage due to the cavitation [134]. Tissues from a particular organ have a range of properties including the wave speed of ultrasound that may depend on age, sex, disease state, perfusion; an increase in either water or fat content leads to a decrease in wave speed [809].

The penetration of the acoustic pulse into the tissue depends on the effectiveness of the coupling to the tissue. For skin-coupling the attenuation coefficient of the dermal and sub-dermal layers may also have a strong effect, since it may be high depending strongly on hydration, and fat and collagen content [809]. The acoustic pulse contains a broad spectrum of frequencies centred approximately at the resonant frequency of the piezoelectric source, the amplitude and intensity of the wave reduces with distance. The remaining energy is scattered, effectively spreading the beam, and this scattered energy may undergo further scattering interactions. An extremely small fraction of the energy returns to the transducer.

As the wave propagates farther into the tissue it may reach an acoustic interface between media of the differing acoustic properties. If the second medium is bone, about half the energy in the wave is reflected and half enters the bone. The pattern of reflected energy will depend on the scattering properties of the tissue-to-bone boundary, and the subsequent propagation of the wave through soft tissue is difficult to predict. The remaining energy that enters the cortical bone may propagate as longitudinal, shear or surface waves, all of which are rapidly absorbed, resulting in a local temperature rise [809].

Almost all of the incident wave energy is reflected from the boundary between soft tissue and gas. The mechanical stress experienced by soft tissue at a tissue-to-gas interface can cause the permanent damage to membranes (e.g., causing lysis of erythrocytes in the presence of bubbles) or to weak connective tissue structures, especially tissues with low shear strength (causing, for example, lung capillary bleeding). Apart from mechanical effects, the interaction between the acoustic wave and bubbles can also generate heat locally, because of an increase in absorption coefficient.

The problem of the HIFU therapy is the real-time temperature monitoring and evaluation of the treatment efficacy [1634, 2981, 3524] in order to avoid the overheating of the surrounding normal tissue and the skin damage [731]. Tissues with the higher absorption coefficients can get warmer than those with less absorption. Thus, the surfaces of calcified bone absorb energy strongly, and heat more than soft tissues—transmission into the bone, and hence its increase in temperature, may be reduced for angles of incidence other than those near normal; the adjacent soft tissues experience the secondary heating from the thermal conduction into the tissue from the bone [809].

To avoid such problems, the intensity of the acoustic energy for the complete tissue ablation can be reduced by introducing the energy absorbing materials. Hence, the different micron to nano level bubbles [321] and the different type of the nanoparticles [731] are infused into the target site. The nanoparticles, due to their nano size can easily accumulate at the targeted cells since they can easily perfuse through the blood capillaries. The nanoparticles can be delivered to the targeted site either via direct intratumoural infusion [1330] or intravenously injected through the cardiovascular network applying the effect of the enhanced permeability.

The non-Fourier effects are significant in case of the HIFU treatment since the very high temperature increment occurs at a very short period [1191, 1905, 2264].

The pressure field generated by the acoustic source in the attenuation media is governed by the Helmholtz equation [707]

$$\nabla \cdot \left(\frac{1}{\varrho} \nabla p \right) + \frac{\omega^2}{\varrho c_s} p = 0 \tag{8.13}$$

where c_s is the speed of sound, $\omega = 2\pi f$ is the angular frequency, f (Hz) is the frequency.

The generated heat is

$$Q = \frac{2\alpha |p|^2}{\varrho c_s}. \tag{8.14}$$

During the HIFU treatment the tumour damage mechanism mainly consists of the cavitation, the thermal and mechanical effects [3258].

The various non-thermal effects are observed in the intense sonic field such as the secondary flows formation in the large and medium blood vessels—the finite amplitude nonlinear acoustic wave causes the time-average blood element's oscillatory velocity that has the time-independent (DC) component velocity in addition to the sinusoidal oscillatory velocity; this DC velocity is called the "acoustic streaming" (the acoustic streaming only occurs in the viscous fluid) [1249, 3346, 3347]—and the cavitation due to the presence of the gas bubbles in the blood [2065, 3347] (these bubbles greatly increase the contrast in the ultrasonic images [1818]). The strong absorption of the ultrasound energy by the bubbles leads to their oscillation and finally to the instability—the inertial cavitation [3159] (the probability of occurrence of the inertial cavitation is indicated by the so called the mechanical index that is proportional to the maximum value of the peak rarefaction pressure and inversely proportional to the square root of the frequency [2808]; for large values of the mechanical index the ultrasound devices have been shown to produce capillary hemorrhage in the lungs [2136, 2139] and intestines [573] of some mammals)—for low values of peak acoustic pressure, oscillations in bubble radius largely follow variations in pressure; as the peak acoustic pressure increases, the bubble becomes unstable as it contracts, collapsing catastrophically under the inertia of the surrounding liquid. The term *acoustic cavitation* is also used to refer to the creation of bubbles in a liquid by an acoustic field at the nucleation sites, such as microscopic impurities, surface roughness on the container or even small-scale local density variations; the adiabatic conditions associated with the extremely rapid bubble compression during the inertial cavitation result in the very high instantaneous temperatures within the bubble [809].

Both the mechanical stress on the epithelial cells that cover the vessels' inner surface caused by the shear stress in the blood flow (that is forming in the viscous fluid flow near the wall or due to the microstreaming surrounding bubbles [3345]) and the activation of the processes in the cells at the gene level occur [3159].

The HIFU can also be used for the acoustic haemostasis—the procedure to seal a bleeding site [1393, 2904, 3124, 3125]—and for the occlusion of the small blood vessels [1292, 2791]. The haemostasis is essential to maintain the closed blood-circulatory system and involves three individual components [2627]:

- blood cells (platelets),
- cells that line the blood vessels (endothelial cells),
- blood proteins (blood-clotting proteins).

The haemostatic mechanism includes three reactions: (1) the formation of a blood clot, (2) the formation of a platelet plug, and (3) changes associated with the wall of the blood vessel after injury of its cells. The defects in any of these processes may result in the persistent bleeding from the slight injuries.

The heating by the ultrasound in the thermoviscous medium in which the absorption increases as the frequency squared was investigated numerically by J. Huang et al. [1377] and by M. A. Solovchuk et al. [2903].

The ultrasound has been reported to have other beneficial bioeffects in the dissolving the blood clots (the thrombolysis), activating the drug delivery, opening the blood–brain barrier, and increasing the cell membrane and the skin permeability to the molecules [637]. Use all the focused ultrasound systems are guided by the magnetic resonance imaging (MRI) [1391].

The ultrasound is also widely used for the medical imaging, including, inter alia, the photoacoustic tomography [3233, 3235]. The encapsulated microbubbles are used as the contrast agent, which can significantly enhance the contrast of the ultrasound image.

The diameter of the HIFU transducer has to be relatively large, typically above 10 cm in diameter, which allows the pre-focal ultrasound beam energy to be spread over a wider area—this reduces the pre-focal heating and the possible effect of shielding, allows for greater focal lengths (up to 15 cm), thus the treatment of the deep-lying organs such as the kidney [2981].

The power of the acoustic wave is related to the operating parameters of the transducer as [2264]

$$P = \frac{p_0^2 S}{2 \varrho_w c_w} \tag{8.15}$$

where S is the transducer aperture area, ϱ_w and c_w are the density and sound speed of the water, p_0 is the source sound pressure originated from the vibration of the transducer.

The US frequency needs to be low enough that the attenuation from the intervening tissue layers does not remove much energy—usually in the frequency range of 0.5–1.5 MHz to maximise the ultrasound penetration depth with the high enough intensity [3037].

The experimental studies [3123, 3125, 3126] showed that the thermal effect of the focused ultrasound is responsible for the haemostasis. The absorbed ultrasound energy in the tissues is transformed into the thermal energy and this energy deposition can quickly cause the tissue temperature to increase. The temperature elevation in excess of 70° C in about 1 second [3125] allows sealing of the bleeding site.

The acoustic field generated by the HIFU source is governed by the coupled system of the partial differential equations that take into account the effects of the diffraction, the absorption, the nonlinear propagation and the relaxation effects [2902, 2904]

$$\nabla^2 p - \frac{1}{c_s^2}\frac{\partial^2 p}{\partial t^2} + \frac{\delta}{c_s^4}\frac{\partial^3 p}{\partial t^3} + \frac{\beta}{\varrho_0 c_s^4}\frac{\partial^2 p^2}{\partial t^2} + \sum_i P_i = 0, \tag{8.16}$$

$$\left(1 + \tau_i \frac{\partial}{\partial t}\right) P_i = \frac{2}{c_s^3} c_i \tau_i \frac{\partial^3 p}{\partial t^3} \tag{8.17}$$

where p is the sound pressure, β is the coefficient of the nonlinearity, δ is the diffusivity of sound originating from the fluid viscosity and the heat conduction, τ_i and c_i are the relaxation time and the small signal sound speed increment for the ith relaxation process.

The first two terms in the Eq. (8.16) describe the linear lossless wave propagating at the sound speed, the third term denotes the loss resulting from the thermal conduction and the fluid viscosity, the fourth term—the acoustic nonlinearity, the last term is due to the relaxation processes.

The intensity of the linear process is $I = p^2/\varrho_0 c_s$, in the nonlinear case the total intensity is

$$I = \sum_{n=1}^{\infty} I_n \tag{8.18}$$

where I_n are the intensities for the harmonics nf_0.

The ultrasound power deposition per unit volume is [2904]

$$Q = \sum_{n=1}^{\infty} 2\alpha(nf_0) I_n \tag{8.19}$$

where the absorption coefficient in the tissue depends on the frequency as $\alpha = \alpha_0(f/f_0)^\eta$. According to F. A. Duck [808], $\alpha_0 = 8.1Np/m, \eta = 1.0, f_0 = 1MHz$.

With the spatially-dependent ultrasonic deposited power, the thermal dose will be increased on the medium surface leading to the elevation of the medium temperature [2819]. Frequently the Pennes bioheat transfer equation is used that involves the heat source produced by the incident acoustic wave and the heat sink due to the perfusion in the capillaries averaged over all the tissues. Evidently, the homogenization assumption is no longer valid in modelling the temperature rise in the regions containing the large vessels, inside which the blood is flowing. The heat convection in large vessels is significant for thermal ablative therapies [619, 1374].

Both the convective cooling in the large blood vessels and the perfusion cooling in microvasculatures need to be taken into account together. F. P. Curra et al. [658] developed a model based on the nonlinear acoustic equation and the bioheat equation; the blood vessel was on the acoustic axis. The HIFU in the pulsed mode with the lower duty cycle ($\leq$ 10%) can provide better control over the rapid temporal tissue temperature rise [3092].

P. Hariharan et al. [1249] considered two cases: the ultrasound beam propagating in a direction parallel to the blood vessel and perpendicular to the blood vessel. For example, the non-zero component of the force along the blood axis is given as $f_v = (2\alpha_b/c_b)I_s$ where α_b is the acoustic attenuation coefficient, c_b is the sound speed of blood, I_s is the acoustic intensity.

It is well known that the acoustic properties of the soft tissue depend on the tissue temperature, thus, the localized heating induces a change in the acoustic properties of the targeted tissue and tissue surrounding it. This phenomenon distorts the incoming acoustic wavefront, and has been termed the thermal lens effect (also the *thermo-acoustic lensing* [1218, 3318]) for this reason—the heating of the tissue influences the speed of propagation of the sound waves and, in turn, the position of the focal region [2301]. Furthermore, the nonlinear effects in the acoustic propagation become non-negligible at the ultrasound intensities required for the therapeutic action. For the diagnostics, the exposure that produces a maximum temperature rise of 1.5° C above normal physiological levels does not appear to present a risk from thermal effects in humans for an imaging session of less than 30 min [572].

The significance of the thermal lens effect is characterized by comparing the simulation of the coupled acoustic and thermal propagation with the uncoupled treatment; neglecting the thermal lensing typically produces a movement of 1–2 mm in the predicted position of the focus towards the transducer. The results also show that the classical methods of the acoustic propagation can produce the grossly erroneous results under the certain clinically relevant transducer configurations [616]. V. NiIkolić and B. Said-Houari [2301, 2302] proved that the quasi-linear model—the Westervelt wave equation coupled to the Pennes bioheat equation that captures the thermal lensing effect—is well-posed locally in time and does not degenerate under a smallness assumption on the pressure data.

8.3 Magnetic Fluid Hyperthermia (MFH)

The magnetic nanoparticles [1412, 2189, 2377] could be used design for the medical diagnosis and therapy. The control of the temperature in the treated tissues could be enhanced by the injection of the nanostructures [24, 526, 1358, 1903, 2815], for example, of the gold nanoparticles (GNPs) [784, 2392, 2453, 3428], the gold-coated (gold-silica core-shell) nanoparticle [818] or the nanoparticles of other materials (for example, two phases of the iron oxide with a spinel structure, i.e., the magnetite Fe_3O_4 and maghemite γFe_2O_3, the alumina Al_2O_3 [279], the titanium oxide TiO_2 [1739], magnetic hydroxyapatite nanoparticles [1357], manganese perovskite nanoparticles [397]), the nanoshells [1106]—the fluid containing the nanoparticles (*nanofluid*) can have much greater thermal conductivity [1376, 2357]. The unwanted hot spots formation in the non-targeted area can be avoided in the MFH because the use of the nanoparticles offers a better control of the energy deposition in the tumours by heating the vicinity of particles [1318].

The idea of the MFH is to use the heating of the magnetic nanoparticles (MNPs) under an alternating magnetic field (AMF) [1492, 2638, 3046] and to localize the heating to the tumour regions by accumulating the MNPs within the tumours.

The term "magnetic" refers to temporarily magnetic material with the ability to comprise ferromagnetic, paramagnetic or superparamagnetic materials:

- When an external magnetic field is applied, ferromagnetic materials become magnetised and remain so even when the magnet is removed.
- Paramagnetic materials have magnetic moments even in the absence of a magnetic field.
- The superparamagnetic materials only are magnetic in the presence of an external magnetic field and revert to a non-magnetic state as soon as the external magnet is removed.

The origin of the power absorption in the single-domain MNPs under an AMF is essentially the dephasing between the magnetic moment of the MNPs and the magnetic component of the AMF [1878, 3120]. The net magnetic moment of the single-domain MNPs couples to the magnetic field component of the applied AMF, absorbing energy and transforming it into heat by a thermally assisted relaxation process [2710].

The typical magnetic field frequency used in the MFH is between 0.05 and 1.2 MHz [2982]. The strength of the magnetic field must be sufficient to induce hyperthermia but it is also limited by the human ability to safely withstand it [1539]. The amount of heat generated in the MFH and the toxicity effect on the cancer cells strongly depend on the concentration of the nanoparticles [2982]. As the concentration of the nanoparticles increases, the temperature reached by a tumour increases [2401, 3361], as showed the experiments using the collagen, linearly [1521]. The MNPs frequently are coated with a biocompatible polymer to prevent their aggregation and biodegradation for in vivo applications [1539], for example, the human-like collagen protein-coated MNPs (HLC-MNPs) [483].

L. Chang e al. [483] found that the HLC-MNPs possess rapid heating capacity upon alternating magnetic field exposure compared to that of the MNPs without HLC coating, irrespective of the size of the MNPs. The significant enhancement of specific absorption rate is favourable for larger sized nanoparticles. The authors attributed such behaviour to the reduced aggregation and increased stability of the HLC-MNPs.

Depending on the amplitude of the magnetic field and the particle dosage, the rate of the temperature increase could be as high as several degrees Celsius per minute [3525]; temperature up to 71° C was recorded at the tumour centre [1317]. Usually the MFH consists of heating the target ill tissue to a temperature of 43–48° C [1483, 2710].

The nanoparticles can be delivered to the tumour through four different approaches [2982]:

- the arterial injection;
- the direct injection into the tumour region;
- the in situ implant formation using entrapped nanoparticle gel;
- the active targeting by coating with the targeting ligands [1789].

After the delivery of the magnetic nanoparticles, an alternating magnetic field is activated to generate heat in the nanoparticles through an energy conversion based on the principle of the hysteresis loss or the Néelian and Brownian relaxation [2982].

The heating in the magnetic particles caused by changes of the magnetic field can occur via several mechanisms. The most well-known one is by generating the heat loss due to the induced eddy currents which are the flows of electrons against the electrical resistance of the material. For the nano-sized materials, the eddy current heating is extremely weak since the magnetic nanoparticles have a poor electrical conductivity to produce a noticeable induction heating and the nano-sized dimension of particles cannot generate a substantial electrical voltage [827]. For the particles having a size of the nano-scale, the heating due to the hysteresis loss is dominant. Heat is generated by the thermal energy dissipated through the magnetic domain reversal process that occurs in the ferromagnetic material.

The power dissipation from a nanoparticle depends upon its magnetic susceptibility, which in turn is related to the ferrofluid relaxation time. There are three magnetic loss mechanisms which result in heat generation within mNP's: Néel relaxation, Brownian relaxation, and magnetic hysteresis—the first two are single domain mechanisms, while hysteresis is considered the dominant loss mechanism for particles with multiple magnetic domains within a single magnetic core [2941]. Within the bulk ferromagnetic material, the magnetic domains are defined as sub-regions within the material for which the magnetic dipoles are aligned in a single direction. The region of the transition between two adjacent domains is referred to as a domain wall. Hysteresis heating occurs when, under an applied magnetic field, the domain walls shift and the domains themselves grow and shrink. This is a lossy process and thus energy is transferred to the environment as heat. There are many approaches to modelling magnetic hysteresis but there is no one universal model [2941].

The Fe_3O_4 NPs smaller than 20 nm generally exhibit a single magnetic domain with the superparamagnetic behaviour, whereas NPs larger than 20 nm have multiple domains with the ferromagnetic behaviour [826, 874]. Thus, heating due to the relaxational losses through both the Néelian mechanisms (rotation of the magnetic moments within the NP) and the Brownian mechanisms (due to the viscous effects as the entire NP aligns with the external magnetic field) are present. These losses depend on the NP size, concentration, magnetic properties, shape, and crystalline structure [298]. The research has demonstrated that the aggregation state of the superparamagnetic particles in the high concentration protein or the salt solutions can reduce the SAR by up to 50% [1454, 1455].

The Brownian relaxation occurs when the particle spins to align itself with the magnetic moment, while the Néel relaxation occurs when the magnetic moment spins within the crystal structure of the nanoparticle [1539]. The Néelian relaxation refers to the heating due to the energy loss produced by the rotation of individual magnetic moments within the particles. On the other hand, the Brownian relaxation refers to the heat generated by the physical rotation of particles due to the alignment process of magnetic moments with the external applied magnetic field.

Both the Brownian τ_B and Néel τ_N relaxation times influence the total relaxation time of the particle $1/\tau = 1/\tau_B + 1/\tau_N$ where [2941]

$$\tau_B = \frac{3\eta V_H}{kT}, \qquad \tau_N = \frac{\sqrt{\pi}}{2}\tau_0 \frac{e^{\Gamma}}{\Gamma^{3/2}}, \tag{8.20}$$

η is the viscosity coefficient of the matrix fluid, $\Gamma = \gamma V_M/kT$, γ is the anisotropy constant, accounting for both shape and magnetocrystalline anisotropy, V_M is the volume of the magnetic core, $\tau_0 = 10^{-9}$ s.

More detailed information about the different types of the heating mechanisms can be found in refs [457, 1302, 1303, 1494].

The injected nanoparticles must have the high values of the imaginary part of the dielectric constant (to convert the microwave irradiation to the thermal energy via the polarization hysteresis) [1739] and the biological compatibilities [2802]. Both the spherical and the irregularly shaped (e.g., $\gamma' - Fe_4N$ [3355]) nanoparticles are used. The superparamagnetic nanoparticles are the nano-sized particles with single domain configuration due to the energetically instability of domain wall [1142]. The use of such particles for the hyperthermia offers benefits of allowing better dispersion of particles [2982].

In order to qualify as a superparamagnetic nanoparticle, the particle' diameter has to be smaller than its superparamagnetic critical diameter. Suriyanto et al. [2982] listed the values of the superparamagnetic critical diameter for a number of materials (including barium-ferrite, iron, iron-cobalt, magnetite).

R. E. Rosensweig developed the formulation for the heat generation by superparamagnetic nanoparticles [2638]. R. Kappiyoor et al. [1539] performed the thermodynamic analysis of ferrofluid magnetic heating and compare the performance of the six different types of ferrofluids—colloidal suspensions of the magnetic nanoparticles in a nonpolar medium, namely those containing magnetite, iron-platinum, iron-cobalt, barium-ferrite, and cobalt-ferrite and found that the iron-cobalt MNPs induce temperature changes that are too large, whereas barium-ferrite and cobalt-ferrite MNPs do not provide enough heat to treat a tumour.

The local magnetic hyperthermia for the cancer treatment using the magnetic particles has been first reported in 1957 by R. Gilchrist et al. [1076] who injected the small particles (around 1 mm) of a magnetic iron oxide in the lymph nodes of animals, obtaining the differential heating upon application of an oscillating magnetic field [1082].

The main advantage of the MFH is the ability to increase the temperature of the human body locally. The low-frequency AMF is desirable because it avoids the non-specific thermal generation due to excitation of the electrolytes dissolved in the interstitial fluid [473, 2244].

Z. Iszály et al. [1411] showed that the localization can be further increased ("superlocalization") by using a combination of the inhomogeneous static and the alternating external magnetic fields. The gold (Au) nanoparticles and carbon nanotubes (CNTs) are good choice since they not only promote the RF hyperthermia (RF ablation, also referred to as to the external electromagnetic method [3157]) but also enable the effective absorption of the excessive RF radiation [1096, 1994].

J. Ortega-Julia et al. [2352] descried the magnetization dynamics by the Landau-Lifshiz-Gilbert equation

$$\frac{\mathrm{d}\mathbf{m}}{\mathrm{d}t} = -\frac{\gamma}{1+\alpha^2}[\mathbf{m} \times \mathbf{B}_{eff} + \alpha \mathbf{m} \times (\mathbf{m} \times \mathbf{B}_{eff})] \tag{8.21}$$

where γ is the gyromagnetic ratio, α is the Gilbert damping constant, $\mathbf{B}_{eff}$ is the effective magnetic field.

The magnetic nanoparticles tend to form aggregates when injected into the living tissues, which alters their response to the applied alternating magnetic field and complicates predicting the released heat accurately. In the biological tissues these aggregates can be accurately described as the fractals characterized by the fractal equation [2352]

$$n_p = k_f \left(\frac{R_g}{r_p}\right)^{D_f} \tag{8.22}$$

where n_p and r_p are the number and the radius of the nanoparticles, R_g is the radius of gyration, D_f is the fractal dimension, k_f is fractal prefactor. Most fractals found in the biological tissue have the fractal dimension ranging from 1.6 to 2.5. The physical meaning of the fractal prefactor and its value are still a topic under discussion [2352].

J. Ortega-Julia et al. found that the average heat per particle is described well by a plane, defined by the linear dependencies on both fractal parameters

$$Q(D_f, k_f) = AD_f + Bk_f + C. \tag{8.23}$$

The heat generated from the plasmonic—the near infrared-absorbing [748, 1330, 2338]—nanoparticles can be utilized in the plasmonic photothermal therapy [745, 1615]. A combination of the near-infrared laser and the metallic nanoparticles are compelling for the treatment of the brain cancer, due to their efficient light-to-heat conversion and the bio-compatibility.

One of the challenges of the plasmonic photothermal therapy [745] is to minimize the damage of the surrounding brain tissue. The adjacent tissue can be damaged as the results of the absorption of the laser light, the thermal conductivity or the nanoparticles diffusion from the tumour. Y. He et al. [1284] measured the temperature change in the three different porcine cerebral tissues (the stem, the cerebrum, and the cerebellum) under the laser treatment and found that the different tissues have different optical and thermal properties and confirmed the enhancement of the heating from adding the plasmonic nanoparticles.

The electromagnetic field of a laser beam resonates with the plasmonic field of nanoparticles, which absorb the incident light; the absorbed energy is dissipated as heat within a distance comparable to the particles diameter. If the nanoparticles are embedded in the tumours, the laser irradiation will cause the hyperthermia and, thus, the localized irreversible damage of the cancerous tissue. This effect is enhanced by the fact that the cancer cells are more susceptible to the thermal treatment than the healthy cells [3143].

M. Nabil and P. Zunino [2244] numerically analysed the hyperthermic treatment of the cancer tumour based on the systemically injected the vascular magnetic nanoconstructs carrying the super-paramagnetic iron oxide nanoparticles (IONPs). An alternating magnetic field is used to excite the nanoconstructs and generate localized heat within the tissue [2638]. The advantage of using the IONPs is that the magnetic resonance imaging (MRI) scanners can readily identify the particles within the body organs [402, 1788, 2378].

The model used by Nabil and Zunino incorporated the microvasculature configurations based on the physiological data combined with the coupled capillary flow, the interstitial filtration and the heat transfer. The authors considered two different types of the nanoparticles

- the small, 20 nm individual IONPs, which accumulate within the tumour tissue via the enhanced permeability and retention effect;
- the large, sub-micrometre vascular magnetic nanoconstructs (VMNs) which accumulate within the tumour tissue by adhering firmly to the diseased vasculature.

They used the model previously developed by M. Nabil et al. [2245] for the IONPs case and adopted the general approach to model flow, mass and heat transfer using the theory of the porous media combined with a multiscale approach to modelling the microcirculation [1579, 2763] to investigate the hyperthermia with the vascular magnetic nanoconstructs. The authors exploited the embedded multiscale method that involves modelling capillaries as one-dimensional channels carrying flow and special mathematical operators to model the interaction of the capillaries with the surrounding tissue [468, 672]. This method facilitate the solution of the unit cell problems defined in the asymptotic homogenization approaches [2424, 2425, 2836]. The numerical solutions obtained in this way may be processed by means of the suitable averaging operators in order to evalbbbbbbuate the impact of the microscale flow and the transport parameters, as well as the microvascular configuration such as the tortuosity, on the perfusion properties of the whole tumour.

Nabil and Zunino considered the two components of the flow, i.e. the microcirculation and the flow in the interstitial tissue. The microvascular wall is considered as a semipermeable membrane; the malignant interstitial volume is treated as an isotropic porous structure. The flow through the tumour interstitium is governed by the Darcy's law.

The microcirculation is characterized by the low Womersley[1] and Reynolds numbers. Thus, the general Navier–Stokes equations reduce to steady Stokes flow. Under the additional assumptions of (1) the straight channels, (2) the rigid walls, (3) the constant radius, (4) the no-slip boundary conditions for the velocity and (5) the absence of the body forces the Stokes flow further simplifies to Poiseuille equation.

The pressures p_t, p_v and the velocity fields u_t, u_v for the microcirculation in Ω and the flow in the interstitial volume Λ are governed by the equations [2244]

- in Ω

$$-\nabla \cdot \left(\frac{\lambda}{\mu}\nabla p_t\right) + L_p^{LF}\frac{S}{V}(p_t - p_L) - f_b(\bar{p}_t, p_v)\delta_\Lambda = 0, \tag{8.24}$$

$$\mathbf{u}_t = -\frac{\lambda}{\mu}\nabla p_t, \tag{8.25}$$

- in Λ

$$-\frac{\pi R^4}{8\mu}\frac{\partial^2 p_v}{\partial s^2} + f_b(\bar{p}_t, p_v) = 0, \tag{8.26}$$

$$f_b(\bar{p}_t, p_v) = 2\pi R L_p((p_v - \bar{p}_t) - \sigma^P(\pi_v^P - \pi_t^P)), \tag{8.27}$$

$$\mathbf{u}_v = -\frac{R^2}{8\mu}\frac{\partial p_v}{\partial s}\boldsymbol{\xi}, \tag{8.28}$$

[1] The Womersley number is a dimensionless number in the biofluid mechanics—the ratio of the transient inertia force to the viscous force $Wo^2 = \omega L^2/\nu$ where ω is the angular frequency of the oscillations, ν the kinematic viscosity. It is named after John R. Womersley (1907–1958) for his work with blood flow in the arteries [3337, 3338]. The Womersley number characterizes extent of deviation of the viscous fluid flow under consideration from a stationary mode [496]. It can also be expressed in terms of the Reynolds and Strouhal numbers $Wo = (2\pi ReSt)^{1/2}$, the square root of this number is also referred to as the Stokes number, $Stk = \sqrt{Wo}$.

The Womersley number is about 16.6 for vena cava, 13.8 for aorta, 2–3 for arteries and veins, about 0.01–0.02 for venule and arteriole, approximately $4 \cdot 10^{-3}$ for capillary.

- at the boundary $\partial\Omega$

$$-\frac{\lambda}{\mu}\nabla p_t \cdot \mathbf{n} = \beta_b(p_t - p_0) \tag{8.29}$$

where s is the arc length coordinate along with each capillary segment, λ is the tissue thermal conductivity, μ is the blood viscosity, L_p is the capillary permeability, $L_p^{LF} s/v$ is the lymphatic permeability, β_b is the heat exchange coefficient, σ^P is the sieving coefficient that accounts for the difference of the semipermeable membrane compared to the case of ideal permeability (i.e. the membrane for which there are no resistance force on the molecules passing through the membrane), the oncotic pressure is defined as $\pi = R_g T c$, where c is the concentration of the given osmotic agent, R_g is the universal gas constant, $\bar{p}_t$ is an average of the interstitial pressure acting on the capillary surface, viz.,

$$\bar{p}_t(s) = \frac{1}{2\pi R}\int_0^{2\pi} p_t(s, \vartheta)\mathrm{d}\vartheta \tag{8.30}$$

where ϑ is the angular coordinate on the cylindrical surface representing the capillary wall.

Nabil and Zunino considered the corresponding mass transfer equations to determine the distribution of the nanoconstructs within the microvascular network. The authors assumed that the injecting of the nanoparticles into the circulation does not alter the blood properties, they are not damaged during the circulation and that they do not release any sub-products. The adhesion of the nanoparticles to the capillary walls is described following the earlier research [705, 1355].

The distribution of the temperature field within the tumour tissue is governed by the equation that includes the convection and the diffusion of the heat through the interstitial flow, the heat sink by the blood flow, the lymphatic drainage

$$\varrho C_p\left[\frac{\partial T}{\partial t} + \nabla\cdot\left(T\mathbf{u}_t - \frac{\lambda}{\varrho C_p}\nabla T\right) + L_p LF\frac{s}{v}(p_t - p_L)(T - T_{bl})\right] \tag{8.31}$$

$$+ 2\pi R\beta_T(T - T_{bl})\delta_\Lambda = Q. \tag{8.32}$$

The heat loss from the outer boundaries of the tumour slab is determined by the boundary condition

$$(-\lambda\nabla T + \varrho C_p T\mathbf{u}_t)\cdot\mathbf{n} = \beta_T(T - T_{bl}). \tag{8.33}$$

The authors found that the network topology and the particle distribution along the microvasculature are the key factors in the VMN delivery approach, because they affect the temperature field within the tumour tissue; the density of particles decorating the arterial wall is directly proportional to time.

Evidently, the accurate knowledge of the spatial distribution and the amount of the MNPs accumulated in the targeted region is crucial for designing optimal treatment planning of the MFH to prevent the insufficient heating of the targeted region and overheating of the healthy tissue. Such information could be provided by the magnetic particle imaging (MPI) [1098].

The central quantity in the MNP based hyperthermia is the absorbed power related to the specific absorption rate (SAR) [1236], or the specific power absorption (SPA) [2710], or the specific loss power (SLP) [1160]. The SAR is a measure of the rate at which the energy is absorbed per unit mass by the human body when exposed to the radio frequency electromagnetic field. It can also refer to the absorption of other forms of the energy by the tissue, including the ultrasound.

The classical definition of the SAR is based on the calorimetric measurements and is defined as [298, 3250]

$$SAR = \frac{c_s m_s}{m_{NP}} \frac{\mathrm{d}T}{\mathrm{d}t}|_{t=0}. \tag{8.34}$$

where c_s and m_s are the specific heat and the mass of the sample, m_{NP} is the mass of the MNP.

For the electromagnetic energy the SAR is given by

$$SAR = \frac{1}{V} \int_V \frac{\sigma(\mathbf{r})|E(\mathbf{r})|^2}{\varrho(\mathbf{r})} d\mathbf{r}. \tag{8.35}$$

where σ is the electrical conductivity, E is the electric field, ϱ is the density, V is the volume of the sample.

Sometimes a local value of the SAR is used [1029]

$$SAR(\mathbf{r}) = \frac{\sigma(\mathbf{r})}{\varrho(\mathbf{r})} |\mathbf{E}(\mathbf{r})|^2. \tag{8.36}$$

I. Gresits et al. [1160] suggested a method to determine the power absorbed during the ferrite based hyperthermia (the particle dipole interaction essentially influences the AC magnetically induced heating characteristics of the ferrite nanoparticles [1465]) using the quality factor Q of the irradiation circuit by placing a ferrite sample in the resonator.

8.3.1 Gene Expression by Magnetic Nanoparticle-mediated Heat Induction

Another application of the magnetic field-directed, nanoparticle-mediated heat induction is the production of the potential therapeutic protein from the genetically modified heterologous encapsulated cells inside the patient's body at the place of the therapeutic relevance without rejection by the host's immune system [1559, 2353].

The externally induced therapeutic gene expression is used for the treatment of many diseases using the cell-based therapy approaches [2905]. For example, the genetically modified harbouring a highly inducible artificial heat shock promoter could be co-encapsulated together with magnetite nanoparticles in the biologically inert sodium cellulose sulphate (SoCS) [1559]. These capsules can be instilled at the site of the therapeutic relevance bearing the advantage of not being rejected by the patient's immune system because of immunoisolation by a semipermeable the SoCS capsule membrane—preventing cellular attacks of the immune system, encapsulation of cells within a permeable membrane enables continuous nutrition of the cells.

The alternating magnetic field induces heat within the capsules due to co-encapsulated magnetite nanoparticles: This in turn switches on the gene expression of encapsulated cells. This approach accomplish a spatial and temporal regulation of the gene expression inside a patient's body.

Microcapsules are almost exclusively generated as hydrogels that exhibit [1305]:

- a highly hydrated microenvironment for encapsulated cells that can mimic biochemical, cellular, and physical stimuli that direct cellular processes;
- a soft characteristic that diminishes mechanically irritation of the surrounding tissue;
- virtually no interfacial tension with the surrounding tissues;
- permeability for low molecular weight molecules as nutrients, therapeutics and metabolic waste.

These properties result in the high biocompatibility of hydrogels.

Encapsulation technology has two advantages [1559]:

- the transplantation of potential therapeutic cells and tissue without the need of immune-suppressive agents;
- the potential use of cells and tissue from a variety of different sources such as primary cells and stem cells.

The magnetite nanoparticles are often used for the biologically applications such as thermotherapy because they appear to be minimally toxic and exhibit reduced irritation of healthy tissue. Magnetite (Fe_3O_4) belongs to the iron oxide family. It is a hard black magnetic mineral that is widespread in natural rocks. Maghemite ($\gamma - Fe_2O_3$) is a red-brown magnetic mineral, isostructural with magnetite but with cation vacancies. Their global properties are quite similar. The magnetite particles can be either ferromagnetic or superparamagnetic at room temperature depending on their diameter.

C. Kaspar [1559] used the magnetic field coil that was driven by a current of 1 to 30 A and created an alternating magnetic field with frequencies of 10 to 100 kHz and a magnetic field strength of up to 40,000 A/m. A set of different nanoparticles was characterised with respect to their heat generation capacity in an alternating magnetic field and their ability to be encapsulated.

8.4 Laser-Tissue Interaction

Light interacts in many ways with the biological tissues—from the simple daily phenomena (such as the production of the melanin or the synthesis of the vitamin D, as a result of the skin reacting to the light [77]) to the light-induced therapies currently used in most fields of medicine including the photodynamic therapy (PDT) and the cancer treatment.

The radiation must pass through the stratum corneum before reaching the viable tissues, then it is scattered and absorbed by the structures and the chromophores which vary dynamically and between the individuals. The component chromophores of the epidermis and stratum corneum determine the attenuation of the radiation in these layers, more so than does the optical scattering. The epidermal thickness and the melanization are the important factors for the UV wavelengths less than 300 nm, whereas the attenuation of the UV A (320–400 nm) and the visible radiation is primarily via the melanin [838, 886]. The blood chromophores Hb, HbO_2, and bilirubin determine the dermal absorption of the wave-lengths longer than 320 nm, scattering by the collagen fibres largely determines the depths to which these wavelengths penetrate the dermis [77].

In the direct sunlight, the temperature of human skin rises to about 40° C following the conversion of the absorbed IR into heat. Recent work demonstrates that IR and heat exposure each induces the cutaneous angiogenesis and the inflammatory cellular infiltration, disrupts the dermal extracellular matrix by inducing the matrix metalloproteinases, and alters dermal structural proteins, thereby adding to the premature skin aging [551].

The infrared (IR) radiation consists of wavelengths from 760 nm to 1 mm and is subdivided into three regions of increasing wavelength, IR-A (760–1400 nm), IR-B (1400–3000 nm), and IR-C (3000 nm–1 mm). Since almost half of the solar energy reaching the earth's surface is in the IR range, solar IR is expected to have significant biological effects on human skin. The IR-A radiation can penetrate epidermal and dermal layers and reach subcutaneous tissues without increasing the skin temperature significantly, whereas IR-B and IR-C radiation are absorbed mostly in the epidermal layers and increase the skin temperature significantly [2730] (the temperature of the human skin can increase to more than 40° C under direct IR irradiation [1808]). The IR radiation induces the cutaneous angiogenesis and inflammatory cellular infiltration, disrupts the dermal extracellular matrix by inducing the expression of the collagen and the matrix metalloproteinases, and alters the dermal structural proteins, thereby adding to the premature skin aging [551].

The infrared radiation could be used for the tissue ablation [1506]. For example, the picosecond infrared laser (PIRL) [1310] was investigated for the use in the possible therapeutic application in cutting of the dental roots (in oral surgery, amputation of an inflamed tooth root is a well-established intervention) [1010].

The infrared radiation could also cause the non-thermal immune responses that are difficult to assess due to its close association with heat [1811].

The characteristics of the electromagnetic radiation and the optical properties of the biological tissues determines the treatment outcome for the patient.

Laser (Light Amplification by Stimulated Emission of Radiation) possesses several unique properties:

- monochromaticity (spectral purity of the laser beam);
- coherence
- directivity
- low divergence
- brightness

The coherence could be either spatial (interference of the laser radiation at two different points) or temporal (a superposition of laser beam at certain points in a different time line).

The directivity and the divergence are similar in a sense they both describe the effect of the laser beam. The directivity is defined as the minimum divergence of a laser beam or uniformity of the beam cross section along the small distances. The divergence describes quantitatively and the directivity qualitatively the concentration of the laser beam along the long distances, the divergence of laser beam differs from one mode to other.

It is still not clear, whether the coherence and the monochromatic light have advantages in the therapy in comparison to the conventional or the light emitting diode (LED) light with the same centre intensity and the wavelength [1603]. The experimental data suggested that at the cellular level the coherent and the non-coherent light (for example, emitted by the LEDs) with the same wavelength, intensity and irradiation provides the same biological effect [271, 1556, 1557]. The directivity of the laser beam play the major role in the beam penetration power, its role becomes more prominent for the tissues which have the fibre like structure [1603].

The First Law of the photobiology (the Grotthus-Draper law) states that in order to have an effect by the light beam photons must be absorbed by the electronic bands of the molecular photoreceptors or the chromophores [2985]. The chromosphores are a part of molecule which imparts its parent colour to the compound, e.g., chlorophyll, haemoglobin, cytochrome oxydase, flavins, myoglobin, flavroprotein, porphyrins [1130, 1567, 2388, 2985].

When a photon with energy $h\nu$ reach a molecule A, the energy could be absorbed, bringing A to the excited state $A^{\star}$

$$h\nu + A \rightarrow A^{\star}. \tag{8.37}$$

The excited molecule $A^\star$ suffers an inelastic collision with the neighbour B (an electron, an atom or a molecule), transferring part of its energy and decay to the lower energy state:

$$A^\star + B(\varepsilon) \rightarrow A + B(\varepsilon + \Delta\varepsilon). \tag{8.38}$$

This process changes part of the laser energy (the high frequency electromagnetic waves) into the thermal energy (the low frequency electromagnetic waves) [3469]. As a consequence, there is a temperature increment by the energy absorbed from the photon.

During the continuous illumination, the heat absorbed power density per unit volume per unit time is governed by the Beer–Lambert law

$$Q \propto \exp\left(-d(z)/\delta\right) \tag{8.39}$$

where the penetration depth $\delta = 1/\sqrt{3\mu_a\mu_r}$ [205], μ_a and μ_r are the absorption and scattering coefficients.

When the light is applied to the biological tissues, it can undergo absorption, scattering, penetration (transmission) and fluorescence (reflection). The multiple scattering and absorption of the laser beam should be considered taking into the account the presence of the pigments, the tissue organelles, and the blood vessels [656]. Sum of the absorption and scattering coefficient is defined as the attenuation, the inverse determines the mean free path of light in the medium. The different cells and elements have the different wavelengths for the absorption of light, thus there exists an optical window in the tissues at which the penetration of light is maximized.

The equation that governs the behaviour of the radiative heat transfer within the biological medium in the direction $\hat{s}$ about the elemental solid angle $\Delta\Omega$ is given, e.g., by M. F. Modest [2165], C. S. Mishra et al. [2151] and J. R. Howell et al. [1364]

$$\frac{1}{c}\frac{\partial I}{\partial t} + \frac{\mathrm{d}I}{\mathrm{d}s} = -\beta I + k_a I_b + \frac{\sigma_s}{4\pi}\int_{4\pi} I\Phi(\hat{s}, \hat{s}_i)\mathrm{d}\Omega_i \tag{8.40}$$

where I is the intensity that is a function of the direction $\Omega(\theta, phi)$ and the wavelength λ, θ is the polar angle, ϕ is the azimuthal angle, s is the geometrical distance in the direction $\hat{s}$, c is the speed of light in the medium k_a is the absorption coefficient, σ_s is the scattering coefficient, $\beta = k_a + \sigma_s$ is the extinction coefficient, $\Phi(\hat{s}, \hat{s}_i)$ is the scattering phase function, subscript i is for incident direction.

The first term on the right hand side of the Eq. (8.40) is due to the absorption and the out-scattering, the second and third terms due to the emission and the in-scattering.

In the case of the short-pulse laser irradiation of the biological tissues during the medical diagnosis and the thermal therapeutic techniques, the external collimated intensity of laser undergoes the attenuation while propagating within the medium. The attenuation of the collimated intensity of laser I_c within the medium, gives rise to the diffuse intensity I_d. Thus,in the participating medium, the intensity I is composed of two components $I = I_c + I_d$.

The attenuation of collimated intensity is given as

$$\frac{1}{c}\frac{\partial I_c}{\partial t} + \frac{\mathrm{d}I_c}{\mathrm{d}s} = -\beta I_c, \tag{8.41}$$

thus a modified radiative transfer equation is written as

$$\frac{1}{c}\frac{\partial I_d}{\partial t} + \frac{\mathrm{d}I_d}{\mathrm{d}s} = -\beta I_d + S_c + S_d. \tag{8.42}$$

The source terms resulting from the collimated intensity S_c and the diffuse intensity of radiation S_d are given as

$$S_c = \frac{\sigma_s}{4\pi}\int_{4\pi} I_c \Phi(\hat{s}, \hat{s}_i)\mathrm{d}\Omega_i, \qquad S_d = k_a I_b + \frac{\sigma_s}{4\pi}\int_{4\pi} I_d \Phi(\hat{s}, \hat{s}_i)\mathrm{d}\Omega_i. \tag{8.43}$$

The scattering of the light in the biological medium is mostly in the forward direction [3146] due to which the scattering phase functions $\Phi(\hat{s}, \hat{s}_i)$ are approximated either using the Legendre decomposition phase function [2165] or the Henyey–Greenstein phase function [1299]. The Legendre decomposition scattering phase function in terms of the Legendre polynomial P_n is given as

$$\Phi(\hat{s}, \hat{s}_i) = 1 + \sum_{n=1}^{N} A_n P_n(\hat{s}, \hat{s}_i), \tag{8.44}$$

the Henyey–Greenstein phase function in terms of the Legendre polynomial is given as

$$\Phi(\hat{s}, \hat{s}_i) = 1 + \sum_{n=1}^{N} (2n+1) g^n P_n(\hat{s}, \hat{s}_i) \tag{8.45}$$

where A_n is the anisotropy coefficient, g is the asymmetry factor/average cosine of phase function, N is the order of approximation, $\hat{s}$ and $\hat{s}_i$ are the direction vectors. The coefficients A_n and g are related to the Mie-scattering coefficients a_n and b_n through the complicated formulae given by M. F. Modest [2165].

The consideration of the blood vessels is essential due to the different optical properties, for example, the absorption coefficient of the blood vessels is six times larger then of the surrounding brain grey matter [2913].

The diffuse reflection is a common phenomenon of all tissues, since none of them have the highly polished surfaces such as the optical mirrors. Only in the special cases such as the wet tissue surfaces might the specular reflection occurs.

The different cellular responses to the laser radiation is possible [331], for example, the increase of the mast cell numbers and the degranulation [851], the enhancement of the procollagen production in the human skin fibroblast cultures [1765] and the stimulation of the fibroblast proliferation and the macrophage responsiveness [3436].

The main power source at the cellular level is the mitochondria that plays an important role in the energy generation and the metabolism by converting the food molecules into the ATP during the process of the oxidative phosphorylation. The light increases the production of the ATPs, the modulation of the reactive oxygen species and the induction of the transcription factors by acting on the mitochondria [74, 520, 1300, 1558, 2038, 2387]. These transcription factors induce the protein synthesis, the cell proliferation [3447], the migration [1266], the cytokines modulation and increase the tissue oxygenation [520, 2910, 3447]. The absorption of the near infrared light or the red light absorbed chromophores or the photoreceptor in the mitochondria triggers the oxygenation that produce more ATPs and results in the production of the reactive oxygen species and generation of the nitric oxide [1603].

The visible or near infrared light brought changes in cells due to the physical or the chemical changes in the photo respiratory molecules. After the absorption of the light the electrons excite to the higher states and the following processes are possible [1603]: the acceleration of the electron transfer, the local heating of the chromophores inducing changes in the biochemical activity, O_2 and O_2^1, the one-electron auto oxidation.

The laser interaction with the tissues can be categorized into several mechanisms [1425, 1446, 3199]:

- photo-thermal;
- photo-mechanical (photo-acoustical)—an explosive expansion and the mechanical disruption of the tissue;
- photo-chemical;
- plasma-mediated ablation (photo-ablation) [970].

The photo-thermal mechanism of the tumour irradiation has been used for the laser-induced hyperthermia (HT) [2955], the laser-induced interstitial thermotherapy (LITT) [2033, 2742] and the interstitial laser photocoagulation therapy (ILPT) [66] all of which involve the localized heating.

LITT is based on the local supply of energy via an optical fibre, located in a water-cooled applicator placed directly into the tumorous tissue. The LITT treatment can take place under MRI control because the laser applicator does not include any metal parts and the patient is not exposed to radiation, in contrast to some other treatments that can only be carried out under CT control.

The laser interstitial thermo therapy is aimed at the tissue destruction by he vaporization of the water in the tissue and the thermal coagulation [313], but simultaneously the photochemical processes at the molecular level could occur; if the pulsed lasers [3199] are used, the mechanical (photoacoustic) effects are also take place [3179]. LITT could be used, e.g., for the thermotherapy of the brain lesions [2707].

The beam delivery technique depends on the location of the tumour in the body—if it is located deep inside the body then the fibre-optic delivery system is used to transport the laser light to the tumour site [2409, 3462] in the form of the collimated beam or in the form of the diffused beam [1446]. The depth of the penetration during the laser–tissue interaction depends on the wavelength of the irradiation and can reach 1 cm [3145].

There are three near-infrared (NIR) optical windows have been identified that and allow for an increased penetration depth for the thermo therapy and the optical imaging [304, 2919]. The NIR region with wavelengths from 650 to 950 nm is most commonly used due to the abundance of laser sources and detectors that function well in this range [77, 1000]. The second NIR window has been defined to cover the wavelength range of 1100–1350 nm, while the third window covers 1600–1870 nm—these ranges fall on either side of a strong water absorption band centred near 1450 nm [304].

A. Paul and A. Paul [2394] investigated using the DPL model the effects of the nanoparticles in the tissues under the pulsed and continuous heating during the high intensity focused ultrasound and the laser interstitial thermal therapy. An *in vitro* study on the agar based tissue phantom was performed to validate the present numerical results of the focused ultrasound heating. The authors found that the presence of the nanoparticles and the multilevel artery and vein affects both the thermal and mechanical response under the external heating.

A. Phadnis et al. [2453] investigated numerically the effect of the picosecond laser irradiation on the biological tissue phantom embedded with the gold nanoshells. The authors solved the generalized form of the Radiative Transfer Equation using the discrete ordinate method to obtain the radiation intensity distribution. They noted the significant changes in the optical properties leading to the localized increase in the temperature of the embedded inhomogeneity.

In the case where tumours are not directly accessible to the laser beam from the ambient, the indirect heating via the gold nano particles (GNPs) turns out to be particularly useful for treatment efficiency and reduction of the collateral effects by targeting tumour tissues with a great accuracy. B. Faslaet al. [927] performed computations of the direct and indirect via the GNPs laser hyperthermia for the spherical tumour and found that the heat metabolism contributes to increasing the blood temperature by a couple of degrees while blood perfusion significantly damps the heating process; in the early stages of laser treatment, the temperature rises fast but quickly reaches an upper limit which is systematically lower as the blood flow rate is higher.

N. Sahoo et al. [2677] numerically and experimentally investigated the heat transport in the collagen microstructures under the near-infrared laser irradiation both for cases with and without the gold nanostructures infused. The authors used the DPL model and the Pennes bioheat transfer equation and stressed that the non-

Fourier computation is necessary while thermally treating tissues; otherwise, the finding will be unrealistic from the clinical perspectives and lead to the damage of the healthy tissues.

The therapy with the pulsed laser provides the higher tissue temperature than can be achieved by the continuous wave laser [1046, 1881]. D. M. Simanovski et al. [2860] proposed that the threshold temperature of the cell viability decreases with the laser pulse duration.

Due to the concentration of the energy in the focused laser beam, it can penetrate the skin and reach greater depths without the significant attenuation. As a result of the peak absorption at the focal plane, the desired subsurface temperature higher than the surface temperature can be achieved. This technique has been used, for example, for the non-ablative dermal remodelling [1606, 2036] and the treatment of the striated muscles [1152].

There are numerous parameters affecting the light-tissue interaction in the laser therapy [331]:

- the *laser* variables: the laser power; the light delivery protocol; the power density; the profile of the beam (Gaussian profile, top-hat profile); the irradiation spot size, the energy density of a light source, the local beam angle of incidence with the tissue, the collimated or diffuse irradiation, the wavelength, the pulse repetition rate, the pulse length, the light delivery pulse modulation, the coherence, the polarization [571] and
- the *tissue* variables: the optical properties; the index of refraction; the depth of tissue; the temperature; the thermodynamic properties; the blood flow and the blood content.

For example, in the transcranial low level laser therapy the 810 nm radiation performs much better than the laser radiation with the wavelength 980 and 1064 nm providing the much deeper and wider photon penetration into the cerebral tissue [1193, 3236].

Frequently the neodymium-doped yttrium aluminium garnet (Nd:YAG) and titanium-sapphire (Ti:Sapph) lasers are used. The CO_2 laser radiation is well appropriate for the thermal therapy because the components of the human organism are made up largely of the water and the water absorbs the frequency of this laser very well; an advantage of the CO_2 laser is the shorter surgery time [2861].

The biological tissues are characterized by the strong scattering and the weak absorption in the so-called therapeutic window (wavelengths 650÷1300 nm) [782, 1443, 2298, 3288].

The *in vivo* optical properties variation is tissue-type dependent and the absorption spectra are function of the wavelength [331]. The main factor, which affects the accuracy of the determined fluence rates in the tissue, is the heterogeneous optical properties distribution. The impact of the optical properties on the light fluence rate is also dependent on the tissue geometries, including superficial, interstitial, or whether located within a cavity. The fluence rate is also a function of the beam radius [2701]. The refraction, reflection and back-scattering also affect the laser absorption. The media in which both the absorption and the scattering are present are called the turbid media and an additional parameter is defined—the optical albedo $a = \mu_r/(\mu_a + \mu_r)$.

The interactions between the tissue and the laser beam often lead to the irreversible damage of the tissue that could cause the alteration of the thermophysical and optical properties of tissue, for example, the perfusion coefficient is often treated as the main indicator of the tissue injury (the decrease of the perfusion during the damage process of the vascular network); in addition, an increase of the scattering due to the tissue denaturation has a visual effect of the tissue "whitening" [1443]. Thus, the thermophysical properties of the biological tissue are the tissue damage-dependent [12, 1099]. The thermal damage of the tissue also affects the optical properties of the tissue: during the process of the coagulation, the changes in tissues lead to the higher scattering while the absorption remains the same.

For the application of any "light" (the electromagnetic radiation) therapy, the photons must penetrate the tissue and deposit their energy, via the optical absorption properties of that specific tissue. The varied presence and the number of the absorbing chromophores, such as the blood components (the haemoglobin, the oxyhaemoglobin), the water, the melanin, the fat, the yellow pigments (the bilirubin, the beta-carotene), in the different types of the living tissue, and the tissue scattering properties, determine its wavelength dependence [1427]. The chromophore absorbs the certain wavelengths of the visible light and transmits or reflects others wavelengths—giving the molecule a colour.

For example, the visible and ultraviolet light are absorbed mainly by the haemoglobin that has the absorption peaks around 280, 420, 540, and 580 nm, the absorption coefficient of the melanin monotonically increases across the visible spectrum toward the UV, the infrared light is absorbed primarily by the water—the main constituent of the most tissues. The proteins mainly absorb in the UV and visible range of the spectrum and have an absorption peak at approximately 280 nm. As the wavelength decreases toward the blue-violet and ultraviolet scatter—which limits the depth that light may penetrate into the tissue—becomes more significant. The cellular organelles such as the mitochondria are the main scatterers in the various tissues.

The pigment-specificity can affect the penetration and the absorption: B. C. Wilson and S. L. Jacques found that the visible laser light (between 400 and 700 nm) is highly absorbed by the melanin, the haemoglobin and the myoglobin [3320], whilst the infrared wavelengths have little pigment-specificity and the main absorbing media are the proteins and the water [2329].

The monochromaticity of the laser light is an important factor for the interaction with the biological tissue, since the light needs to be absorbed in order to interact with the tissue. The biological tissues have the light receptors—the chromophores—that are highly selective to the wavelength they absorb [331].

The polarized characteristics of the laser light also influences this interaction, as the different polarizations of the light can be absorbed to the different degrees by the different biological tissue or matter. V. Sankaran et al. [2706] investigated how the light scatters in a variety of the birefringent and nonbirefringent tissues contribute to the depolarization of linearly versus circularly polarized light propagating through that tissue. The experiments were performed on the porcine blood, fat, tendon, artery, and myocardium. The authors found that the two incident polarization states are depolarized differently: for tissues containing the dilute Mie scatterers, the circularly polarized light is maintained preferentially over the linearly polarized light while for the more dense tissues the reverse is true.

J. L. Boulnois [343] suggested to distinguish three manifestation of the photothermal effect of the light-tissue interaction

- the vaporization;
- the coagulation [1973, 1975];
- the protein denaturation.

M. A. Ansari et al. [90] and A. Vogel and V. Venugopalan [3199] noted that the laser induced effects such as coagulation, vaporization (volatilization [3381]), carbonization or melting don't depend only on the thermal properties of the biological tissue, but also on the peak power and the wavelength of the laser radiation.

R. M. Verdaasdonk et al. [3177] investigated the ablation in the aortic and the myocardial bovine tissues under the continuous wave radiation of the Nd-YAG and argon lasers and found three stages of the process:

1. the tissue discolouring and shrinkage due to the denaturation of the proteins and the dehydration during the surface temperature increase to about 150 °C;
2. the explosive subsurface vapour formation ("popcorn") resulting in the enhanced scattering and reflection at the multiple tissue-vapour transitions followed by the tissue water boiling;
3. the tissue vaporisation and carbonisation.

S. Natarajan et al. [2274] also reported the tissue shrinkage after the focal laser ablation.

The "popcorn" phenomenon is possibly related to the layered structure of the tissue and to the tissue structure changes at the surface during irradiation. During the tissue vaporisation, the ablation velocity was constant (the authors defined ablation as the vaporisation of the non-water tissue elements). The large thermal vaporization of water helps to remove the excess heat and prevents overheating of the adjacent tissues.

If the water molecules evaporate and the laser effect continues, a further increase in the tissue temperature will follow. After the water evaporates, the dehydrated tissues are quickly heated to a temperature of 150° C, which initiates the process of the carbonization of the tissues [2224, 3095]. The coagulation and vaporization processes require seconds to accomplish, the carbonization and smoke generation—fractions of a second [1202].

The photodynamic therapy is a form of the phototherapy involving the light and the photosensitizing chemical substance, used in conjunction with the molecular oxygen to cause the cell death [795]. The thermal effect of the photodynamic therapy is not a major one. The PDT involves three components [3242]:

- the light source;
- the photosensitiser;
- the tissue oxygen.

The PDT is a multi-stage process [1373]. First the photosensitiser, ideally with the negligible toxicity other than its phototoxicity, is administered in the absence of the light radiation. The sensitizers accumulate in the various cell compartments (for example, the mitochondria, the lysosomes, the endoplasmic reticulum, the Golgi apparatus) [61].

When a sufficient amount of the photosensitiser appears in the diseased tissue, the photosensitiser is activated by exposure to the light to stimulate the photosensitiser to produce the radicals and/or reactive oxygen species—the free radicals and highly reactive state of oxygen known as singlet oxygen. The reactive oxygen kills the target cells [2878].

One of the ways to generate the singlet oxygen $^1O_2$—the electronically excited state of the triplet oxygen—is the direct optical excitation of the triplet oxygen form; since the molecular oxygen weakly absorbs the light, the high power is required to obtain the sufficient concentrations of the singlet oxygen. The reactivity of the excited state is several orders of magnitude greater than that of the triplet form of the oxygen [1581]. The $^1O_2$ oxygen exhibits the considerable reactivity toward the electron-rich organic compounds—the lipids, the proteins, the nucleic acids and the ribonucleic acids [732, 2055]. The basic triplet state of the oxygen has several absorption bands in the infrared and the visible regions at which the singlet oxygen can be produced [314, 799].

V. Dremin et al. [799] noted that the increase in the radiation power of the laser induces the local temperature increase around the laser spot that may be critical since the temperature governs the biological reaction within the living cells.

F. Asllanaj et al. [126] studied the fluorescence and the elastically scattered light propagation through the biological tissue illuminated by the collimated beam. The fluorescence molecular imaging is an investigative tool that uses the fluorescent sources inside the tissue as the imaging contrast [113, 1562, 2316].

8.5 Thermoelastic Deformation of Tissues

The heating process in the thermal ablation of the soft tissues is coupled with the thermal-induced mechanical deformations and should account for the constitutive elastic material law under thermal loads as well the as non-rigid motion dynamics [1857]. The tissue deformation due to heating and cooling might produce pain sensation; even a small increment of the heat-induced stress can destroy the immune response; the protein cell organelle structures can be changed, resulting in the cell death [3440].

During the thermal heating surgical procedures the soft tissue deformation due to the surgical tool-tissue interaction [3490] (for example, in the percutaneous RFA, the liver experiences deformations due to needle insertion) and the patient movement [2242] can affect the distribution of the thermal energy [3473]. The clinical outcome of these surgical procedures relies on accurate control of the soft tissue temperature for the deformed tissue to provide the precise delivery of the thermal energy. The variation of the thermal conductivity and the specific heat capacity with the temperature should be accounted for [3494]. Even a small change of the thermoelastic stress can suppress the immune response, alter the production of the hormones and the protein denaturation [1787]. The elastic parameters of the tumour tissue have an important effect on the magnetic nanoparticles distribution in the magnetic hyperthermia [1403].

S. Singh and R. Melnik [2866] used the SPL and the DPL models incorporating both the tissue contraction and expansion during the procedure. The authors considered the DPL model along with the Helmholtz harmonic wave equation, the modified stress-strain and the thermoelastic wave equations to simulate the radiofrequency ablation (RFA) and the microwave ablation (MWA) of the cancerous tissues. The authors reported that neglecting the mechanical coupling results in an underestimated temperature distribution (they emphasized the significant contribution of the mechanical terms in the internal energy) and that the effect of the non-Fourier based coupled thermo-electro-mechanical model is less pronounced in the RFA as compared to the MWA.

Y. Hu et al. [1371] used the time-fractional dual-phase-lag model combined with the one-dimensional thermoelastic theory to determine the thermoelastic response of a biological tissue during the hyperthermia treatment by a moving laser heating (increasing the moving laser beam speed will reduce the magnitude of the temperature and the deformation but enlarge the thermo-mechanical response zone [2844]). The authors studied the effects of the blood perfusion rate and the laser velocity on the temperature distribution and the thermal stress response during the percutaneous laser thermal ablation.

F. Xu and T. Lu [3374] and B. Zhou et al. [3522] investigated a mathematical approach to examine the heat transfer process, the thermal damage and the heat-induced mechanical response of the human skin. W. S. Shen et al. [2811, 2812] reported a quasi-steady state numerical model, to predict the thermo-elastic deformation of the skin tissue at the high temperature.

H. M. Youssef and N. A. Alghamdi [3440] investigated the thermoelastic stress in the skin tissue in the context of the dual phase lag heat conduction model. One-dimensional skin tissue have been considered with a small thickness L. The equation of motion of a one-dimensional thermoelastic material is written as [1859]

$$(\nu + 2\mu)\frac{\partial^2 e}{\partial x^2} - \gamma\frac{\partial^2 \vartheta}{\partial x^2} = \varrho\frac{\partial^2 e}{\partial t^2} \tag{8.46}$$

and the stress-strain relation as

$$\sigma = (\nu + 2\mu)e - \gamma\vartheta, \tag{8.47}$$

$e = \partial u/\partial x$.

The case of the traction free inner and outer surfaces $\sigma(0, t) = \sigma(L, t) = 0$ was considered.

Three different types of the thermal loading were analysed: the thermal shock, the ramp type heating, and the harmonic heating.

The authors used the non-dimensional variables introduced earlier by H. M. Youssef [3438] and the Laplace transform to obtain the solution and found that the thermal shock time parameter, the ramp-type heat parameter, and the angular thermal parameter have significant effects on the temperature increment, the strain, the displacement, and the stress distributions, and play the vital roles in the speed propagation of the thermomechanical waves through the skin tissue.

A. McBride et al. [2092] performed the thermoelastic modelling of the finite deformations of the skin.

Recently S. Singh and R. Melnik published the review of the computational models and the future directions of the thermal ablation [2867].

The human tissue is an elastic homogeneous isotropic medium [883, 2393], thus the stress-strain relations under the thermal load in the three dimensional Cartesian coordinates are

$$\sigma_{xx} = \frac{E}{1+\nu}\varepsilon_{xx} + \frac{\nu E}{(1+\nu)(1-2\nu)}(\varepsilon_{xx} + \varepsilon_{yy} + \varepsilon_{zz}) - \frac{\beta E}{1-\nu}\vartheta, \tag{8.48}$$

$$\sigma_{yy} = \frac{E}{1+\nu}\varepsilon_{yy} + \frac{\nu E}{(1+\nu)(1-2\nu)}(\varepsilon_{xx} + \varepsilon_{yy} + \varepsilon_{zz}) - \frac{\beta E}{1-\nu}\vartheta, \tag{8.49}$$

$$\sigma_{zz} = \frac{E}{1+\nu}\varepsilon_{zz} + \frac{\nu E}{(1+\nu)(1-2\nu)}(\varepsilon_{xx} + \varepsilon_{yy} + \varepsilon_{zz}) - \frac{\beta E}{1-\nu}\vartheta, \tag{8.50}$$

$$\sigma_{xy} = \frac{E}{1+\nu}\varepsilon_{xy}, \quad \sigma_{xz} = \frac{E}{1+\nu}\varepsilon_{xz}, \quad \sigma_{yz} = \frac{E}{1+\nu}\varepsilon_{yz} \tag{8.51}$$

where σ_{zz} is the normal strain, ε_{xx} is the normal stress, ϑ is the temperature difference between the two consecutive steps of time, E is the Young's modulus of elasticity, ν is the Poison's ratio, β is the coefficient of linear thermal expansion.

The strain in terms of the displacement vector is written as

$$\varepsilon_{xx} = \frac{\partial u_x}{\partial x}, \quad \varepsilon_{yy} = \frac{\partial u_y}{\partial y}, \quad \varepsilon_{zz} = \frac{\partial u_z}{\partial z}, \tag{8.52}$$

$$\varepsilon_{xy} = \frac{1}{2}\left(\frac{\partial u_x}{\partial y} + \frac{\partial u_y}{\partial x}\right), \quad \varepsilon_{xz} = \frac{1}{2}\left(\frac{\partial u_x}{\partial z} + \frac{\partial u_z}{\partial x}\right), \quad \varepsilon_{yz} = \frac{1}{2}\left(\frac{\partial u_y}{\partial z} + \frac{\partial u_z}{\partial y}\right). \tag{8.53}$$

The displacement vector is determined by the equation that govern the mechanics of the non-rigid motion

$$\varrho\frac{\partial^2 u_x}{\partial t^2} = \frac{\partial \sigma_{xx}}{\partial x} + \frac{\partial \sigma_{xy}}{\partial y} + \frac{\partial \sigma_{xz}}{\partial z}, \qquad \frac{\partial \sigma_{yy}}{\partial y} + \frac{\partial \sigma_{yz}}{\partial z}, \tag{8.54}$$

$$\varrho\frac{\partial^2 u_x}{\partial t^2} = \frac{\partial \sigma_{zx}}{\partial x} + \frac{\partial \sigma_{zy}}{\partial y} + \frac{\partial \sigma_{zz}}{\partial z}. \tag{8.55}$$

Finally, the thermally induced displacement is determined by the equations

$$\varrho\frac{\partial^2 u_x}{\partial t^2} = \frac{E}{2(1+\nu)}\left(\frac{\partial^2 u_x}{\partial x^2} + \frac{\partial^2 u_x}{\partial y^2} + \frac{\partial^2 u_x}{\partial z^2}\right) \tag{8.56}$$

$$+\frac{E}{(1+\nu)(1-2\nu)}\frac{\partial}{\partial x}\left(\frac{\partial u_x}{\partial x} + \frac{\partial u_y}{\partial y} + \frac{\partial u_z}{\partial z}\right) - \frac{\beta E}{1-2\nu}\frac{\partial \vartheta}{\partial x}, \tag{8.57}$$

$$\varrho\frac{\partial^2 u_y}{\partial t^2} = \frac{E}{2(1+\nu)}\left(\frac{\partial^2 u_y}{\partial x^2} + \frac{\partial^2 u_y}{\partial y^2} + \frac{\partial^2 u_y}{\partial z^2}\right) \tag{8.58}$$

$$+\frac{E}{(1+\nu)(1-2\nu)}\frac{\partial}{\partial y}\left(\frac{\partial u_x}{\partial x} + \frac{\partial u_y}{\partial y} + \frac{\partial u_z}{\partial z}\right) - \frac{\beta E}{1-2\nu}\frac{\partial \vartheta}{\partial y}, \tag{8.59}$$

$$\varrho\frac{\partial^2 u_z}{\partial t^2} = \frac{E}{2(1+\nu)}\left(\frac{\partial^2 u_z}{\partial x^2} + \frac{\partial^2 u_z}{\partial y^2} + \frac{\partial^2 u_z}{\partial z^2}\right) \tag{8.60}$$

$$+\frac{E}{(1+\nu)(1-2\nu)}\frac{\partial}{\partial z}\left(\frac{\partial u_x}{\partial x} + \frac{\partial u_y}{\partial y} + \frac{\partial u_z}{\partial z}\right) - \frac{\beta E}{1-2\nu}\frac{\partial \vartheta}{\partial z}, \tag{8.61}$$

A. Paul and A. Paul [2394] investigated the non-linear stress-strain profile for the focused ultrasound and the interstitial laser heating. A higher stress-strain value was predicted in the case of the nanoparticle mixed tumour compared to the tumour in the absence of the nanoparticles. Moreover, for an applied strain a lower stress is induced in case of the pulsed energy activation compared to the continuous mode for all the considered tissue models. Thus, the thermally induced nociceptive pain in the tissue had been found to be more for continuous agitation of energy compared to pulse mode. A hysteretic stress-strain profile with the dissimilar loading and unloading path had been observed. The authors also found that the tissue had deformed more under thermal load for the LITT compared to the HIFU.

A. González-Suárez et al. [1121] investigated the thermoelastic response of the tissues including the subcutaneous tissue to the RF hyperthermia. The subcutaneous tissue consists of a fine, collagenous and fibrous septa network enveloping clusters of the adipocyte cells; the density and orientation of the septa, varies among patients. The authors found that during the first seconds of the RF exposure, the fibrous septa were subjected to the thermal expansion regardless of the orientation. In the long term, the fibrous septa contracted due to the thermal expansion of fat. The skin and muscle were subjected to the significantly higher von Mises stresses (the measure of yield) or the distortion energy than the subcutaneous tissue.

The importance of the inclusion deformation in the dose calculations was shown by K. K. Brock et al. [380] who compared (using the liver tissue as an example) the three models:

- the traditional static model;
- the model that includes the rigid body motion;
- the model that incorporates both the rigid body motion and the deformation.

The usual Pennes equation describes the tissue temperature only in the static non-moving state of tissue. In order to apply it to the deformed soft tissue, Brock et al. proposed to use a transformation of the unknown deformed system configuration to the known initial static non-moving system configuration via a mapping function. Using the deformed system configuration Ω^t at the time t, the Pennes equation can be integrated over the tissue volume to get [380]

$$\int_{\Omega^t} \varrho c \frac{\partial T}{\partial t} \mathrm{d}\,\Omega^t = \int_{\Omega^t} \left[\nabla \cdot \lambda \nabla T + \varrho_{\mathrm{b}} c_{\mathrm{b}} \omega_b (T_{\mathrm{a}} - T) + \dot{q}_{\mathrm{met}} + Q^{\mathrm{ext}}\right] \mathrm{d}\,\Omega^t .() \tag{8.62}$$

The authors defined the transformation of system configuration $\Omega^0 \to \Omega^t$ by the mapping function

$$\chi_t : \Omega^0 \times \mathbb{R}_0^+ \to \Omega^t, \quad \mathbf{x}^t = \chi_t(\mathbf{x}^0) \tag{8.63}$$

where $\mathbf{x}^0$ and $\mathbf{x}^t$ are the spatial coordinates of the material point $\mathbf{x}(x, y, z) \in \Omega$ in the initial Ω^0 and current Ω^t system configurations at time 0 and t. The mapping function χ_t is continuous in time, orientation-preserving, continuously differentiable and globally invertible at all times.

The change of variables transforms the integral in the form in the initial system configuration Ω^0 as

$$\int_{\Omega^t} (\ldots) \mathrm{d}\,\Omega^t = \int_{\Omega^t} (\ldots) \det\left(\frac{\partial \mathbf{x}^t}{\partial \mathbf{x}^0}\right) \mathrm{d}\,\Omega^0 \tag{8.64}$$

thus [380] the heat conduction term in the Pennes equation is expressed in the form of the initial system configuration Ω^0 as

$$\int_{\Omega^t} \nabla^t \cdot \lambda \nabla^t T(\mathbf{x}) \mathrm{d}\,\Omega^t = \int_{\Omega^0} \left(\nabla^0 \left(\frac{\partial \mathbf{x}^0}{\partial \mathbf{x}^t} \right) \cdot \lambda \nabla^0 \left(\frac{\partial \mathbf{x}^0}{\partial \mathbf{x}^t} \right) T(\mathbf{x}) \right) \det \left(\frac{\partial \mathbf{x}^t}{\partial \mathbf{x}^0} \right) \mathrm{d}\,\Omega^0 = \tag{8.65}$$

$$= \int_{\Omega^0} \left(\nabla^0 (\hat{F}_0^t)^{-1} \cdot (\lambda \nabla^0 (\hat{F}_0^t)^{-1}) T(\mathbf{x}) \right) \det(\hat{F}_0^t)\, \mathrm{d}\,\Omega^0 \tag{8.66}$$

where the deformation gradient tensor is the fundamental measure of deformation

$$\hat{F}_0^t = \frac{\partial \mathbf{x}^t}{\partial \mathbf{x}^0} \tag{8.67}$$

where the displacement from $\mathbf{x}^0$ to $\mathbf{x}^t$ from time 0 to t is governed by the non-rigid mechanics of motion for the soft tissue deformation

$$\varrho \frac{\partial^2 \mathbf{x}^t}{\partial t^2} + \gamma^t \frac{\partial \mathbf{x}^t}{\partial t} + \frac{\delta W^t}{\delta \mathbf{x}^t} = \mathbf{f}^t(\mathbf{x}) \tag{8.68}$$

where γ^t is the system damping constant, W^t is the instantaneous strain energy and $\mathbf{f}^t(\mathbf{x})$ is the net external force at $\mathbf{x}$.

In the majority of investigations the biological tissues are assumed to have the linear-elastic properties. Z. Liu and L. Bilston [1918] experimentally investigated the linear viscoelastic properties of bovine liver. S. Ocal et al. [2325] obtained the viscoelastic properties of tissue in the frame of the general Maxwell model for viscoelastic materials. A. Majeed [2018] analysed the heat transfer process in ferromagnetic viscoelastic fluid. M. M. Attar et al. [135] investigated the thermo-viscoelastic behaviour of tumorous and healthy bovine liver tissue neglecting the inertia effect.

M. A. Ezzat [888] studied the thermomechanical behaviour of the cancerous layer using the thermo-viscoelasticity model of Lord-Shulman and considering the varying thermal conductivity and the rheological properties of the volume. The Laplace transform technique was employed to solve the problem in the transformed domain. The author found that the thermal conductivity and the elasticity parameters affect the circulations significantly.

X. Li et al. [1862] developed the dual phase lag thermoviscoelastic model in order to capture the micro-scale responses of the biological tissue and studied the effects of the relaxation times and the viscoelastic properties on the responses of the tumour and the normal tissues.

The authors considered the one dimensional problem of the ablation of the tumour using an ultrashort pulse laser [2679]. The coordinate system was chosen that the x-axis (the laser beam direction) is perpendicular to the tumour surface, the size of spot was assumed to be much larger than the thickness of the biological tissue, the metabolic heat generation was neglected.

The only nonzero component of the displacement vector is $u_{xi} = u_i(x, t)$, $u_{yi} = u_{zi} = 0$, where $i = 1$ and $i = 2$ indicate the variables in the tumour and normal tissues, the strain $\varepsilon_{xxi} = \partial u_i/\partial x$, is the cubical dilatation $e_i = \varepsilon_{xxi}$.

It is assumed that the tumour and normal biological tissues are the linear isotropic viscoelastic media with constant material parameters; no phase change and no chemical reactions occur within the biological tissue. The stress components are determined by the constitutive equations [889, 895]

$$\sigma_{xxi} = \int_0^t R_i(t-\tau)\frac{\partial}{\partial \tau}\left(\varepsilon_{xxi} - \frac{e_i}{3}\right)\mathrm{d}\tau + \int_0^t R_{vi}(t-\tau)\frac{\partial}{\partial \tau}\left(e_i - a_{ti}\vartheta_i\right)\mathrm{d}\tau, \tag{8.69}$$

$$\sigma_{yyi} = \sigma_{zzi} = -\frac{1}{3}\int_0^t R_i(t-\tau)\frac{\partial e_i}{\partial \tau}\mathrm{d}\tau + \int_0^t R_{vi}(t-\tau)\frac{\partial}{\partial \tau}\left(e_i - a_{ti}\vartheta_i\right)\mathrm{d}\tau \tag{8.70}$$

where a_{ti} is the linear thermal expansion coefficient.

The entropy density S_i is given by

$$\varrho_i S_i = a_{ti}\int_0^t R_{vi}(t-\tau)\frac{\partial e_i}{\partial \tau}\mathrm{d}\tau + \frac{\varrho_i c_i \vartheta_i}{T_0}, \tag{8.71}$$

here $\vartheta_i = T_i - T_0$, T_0 is the reference temperature, T_i is the biological tissue temperature.

The relaxation functions $R_i(t)$ and $R_{vi}(t)$ are expressed as [891]:

$$R_i(t) = 2\mu_i\left[1 - A_1\int_0^t e^{-\beta t}t^{\alpha-1}\mathrm{d}t\right],\ R_{vi}(t) = K_i\left[1 - A_2\int_0^t e^{-\beta t}t^{\alpha-1}\mathrm{d}t\right] \tag{8.72}$$

where $K_i = (\nu_i + 2\mu_i/3)$, ν_i and μ_i are the Lame constants, $\alpha(0 < \alpha < 1)$, $\beta(\beta > 0)$, $A_1, A_2(A_1, A_2 > 0)$ are the empirical constants.

When $A_1 = A_2 = 0$ in the Eqs. (8.72), the Eqs. (8.69) and (8.70) reduce to the linear thermoelastic Eq. [1863].

Li, Qin and Tian compared the results of the dual phase lag thermo-elastic and the dual phase lag thermo-viscoelastic models at different instants to study the influence of viscoelastic parameter on the transient responses. The absolute values of displacement and stress in the thermo-viscoelastic model are found to be smaller than in the thermo-elastic model as time goes on, thus the thermal deformation damage of thermoelastic model is more serious than that of the thermo-viscoelastic model during hyperthermia treatment.

The jump of the stress was detected at the tumour-normal tissue interface. The authors found that the thermo-viscoelastic responses in the biological tissue depend not only on the ratio of the relaxation times τ_T/τ_q but also on the absolute values of τ_T and τ_q.

J. Zhang et al. [3498] performed the numerical thermo-visco-elastodynamic analysis of the anisotropic soft tissues undergoing large deformations for the surgical simulation of the thermal ablation. Due to body organ movements (such as respiratory motion) and tool-tissue interactions, soft tissues are constantly in a state of motion consisted of dislocations and deformations; the former can be described by rigid-body displacements including translations and rotations, while the latter involves change in organ shapes that affect the distribution of thermal energy. Furthermore, due to the variation of temperatures, the effects of the thermal expansion [1952] and the shrinkage [922, 923, 2645] induce the thermal stresses, leading to the deformations that also affect the energy distributions. The thermally-induced deformations have also been evidenced in the studies of the thermo-poroelastic modelling for the fluid transport in tumours [82], the thermal strain effects on the low-density lipoprotein deposition in the arterial walls [570], and the thermo-diffusion effects [1394].

The authors developed the visco-hyperelastic biomechanics model based on the finite elasticity theory. The variation of strain at a material point is expressed by the Green-Saint Venant strain tensor ${}_0^t\hat{E} = \frac{1}{2}({}_0^t\hat{C} - \hat{I})$ where ${}_0^t\hat{C}$ is the right Cauchy-Green tensor and $\hat{I}$ is the identity matrix of the second rank. The corresponding strain energy density is expressed by ${}_0^t\Psi$ which is typically a function of the strain invariants of ${}_0^t\hat{C}$, the energy conjugated stress measure is the second Piola-Kirchhoff stress tensor

$$ {}_0^t\hat{S} = \frac{\partial_0^t \Psi({}_0^t\hat{C})}{\partial_0^t\hat{E}}. \tag{8.73} $$

For anisotropic soft tissues exhibiting the directional-dependent behaviours, ${}_0^t\Psi$ can be modified by employing unit vectors to describe the local fibre directions [3286]. For viscoelastic behaviours, it can be modified by employing a time-dependent ${}_0^t\tilde{\Psi}$ such as

$$ {}_0^t\tilde{\Psi} = \int_0^t \varphi(t - t') \frac{\partial_0^t\Psi}{\partial t'} \mathrm{d}t' \tag{8.74} $$

where

$$ \varphi(t - t') = \varphi_\infty + \sum_{i=1}^{N} \varphi_i e^{(t-t')/\tau_i} \tag{8.75} $$

is the relaxation function of a generalised Maxwell model [1345] and φ_∞, φ_i, τ_i are the positive constants.

The equation of motion is written as [3498]

$$\nabla \cdot_0^t \hat{P} +_0^t \hat{B} = \varrho^t \dot{V} \tag{8.76}$$

where ${}_0^t\hat{P}$ is the first Piola-Kirchhoff stress tensor, ${}_0^t\hat{B}$ is the reference body force, and ${}^t\dot{V}$ is the material acceleration field.

J. Zhang and S. Chauhan performed the coupling of the bioheat and biomechanics models using a transformation of the system configurations to achieve bioheat computation on the deformed soft tissues [3475].

The approach used by the authors is based on the finite-strain thermoelasticity and total Lagrangian explicit dynamics and considers the coupled nonlinear analysis of the bioheat transfer under the soft tissue deformations and the soft tissue deformations due to thermal expansion/shrinkage. The presented method accounts for the anisotropic, finite-strain, temperature-dependent thermal (for example, tissue thermal conductivity, specific heat, volumetric heat capacity, density, thermal diffusivity)), and viscoelastic properties of soft tissues, and it is implemented using GPU acceleration for the real-time computation.

The key advantage of the presented method is that it enables full nonlinear modelling of the anisotropic, finite-strain, temperature-dependent, thermal, and viscoelastic behaviours of the soft tissues, instead of the linear elastic, linear viscoelastic, and thermal-only modelling in the most known methods.

The authors noted that although the clinically relevant application was demonstrated by a simulation of the thermal ablation for the local hyperthermia, the principle of the coupled bioheat and biomechanics modelling is generic and can be straightforwardly adapted for the regional and whole-body hyperthermia simulations. They also stress that it is necessary to account for the anisotropic properties (the thermal conductivity, the thermal expansion coefficient, and the hyperelasticity) and the temperature-dependent specific heat capacity and thermal conductivity properties of the soft tissues.

Chapter 9
Cryomedicine

The origins of the cryomedicine [2519] could be traced down to the ancient Egyptians: the application of cold to the wounds and use of the thermal therapy for the breast ulcers and tumours is mentioned in the "chirurgical papyrus" of E. Smith, an ancient Egyptian text on the medicine and the surgery (about 3500 years B.C.) [363].

Probably the first scientific account of this science is the monograph published by Sir Robert Boyle "New Experiments and Observations Touching Cold, Or, An Experimental History of Cold, begun. To which are added An Examen of Antiperistasis, And an Examen of Mr. Hobs's Doctrine about Cold. Whereunto is annexed An Account of Freezing, brought in to the Royal Society, by the learned Dr. C. Merret, a Fellow of it. Together with an Appendix, containing some promiscuous Experiments and Observations relating to the precedent History of Cold." (London, Richard Davis, 1683).

The effect of the water supercooling was discovered by Daniel Gabriel Fahrenheit in the beginning of the eighteenth century [1179]. The experiments on the freezing of the cells started in the end of the eighteenth century. In 1776 Spallanzani studied the motility of the human spermatozoa and noted that the spermatozoa, cooled in the snow, became inactive but were revived on warming [659, 690].

In the middle of the nineteenth century Julius von Sachs [759] developed a low-temperature stage for his microscope that allowed him to detect the dehydration of cells in the presence of the extracellular ice. Hans Molish was the first to study in the end of the nineteenth century the formation of the intracellular ice and to measure the cell volume change [760].

The experiments in the middle of the nineteenth century by Michael Faraday who achieved a temperature of $-163\,\mathrm{K}$ by mixing the solid carbon dioxide and alcohol under vacuum, the development in 1877 by Cailletet of France and Pictet of Switzerland of the adiabatic expansion systems for cooling gases and in 1892 by Dewar of Great Britain the first vacuum flask for storage and handling of the liquefied gases as well as invention in 1895 by Linde of Germany and Hampson

A. I. Zhmakin, *Heat Transfer in Vivo*, Springer Series in Biophysics 27,
https://doi.org/10.1007/978-3-031-95459-7_9

of England the continuously operating air liquefiers based on the Joule-Thompson effect provide the technical support for the cryomedicine development.

In 1833 Openchowski reported using a low-temperature system for freezing the portions of the cerebral cortex of dogs, however, not for therapy but as a tool for inducing the lesions in the brain for research.

The twentieth century witnessed the rapid development of the cryobiology related to the progress of the cryogenic equipment (in 1887 Lui Cailletet and Raul Pectet obtained the liquid oxygen and nitrogen; the closed systems based on the liquid nitrogen evaporation or on the Joule-Tohomson effect—cooling by the rapid gas expansion from a high to a low pressure zone through a small orifice—had been constructed), developments of the monitoring techniques, extension of the list of the diseases that have been successfully treated by the cryomedicine.

In contrast to the high temperatures in the hyperthermia used exclusively to destroy the undesirable tissues the low temperatures in the cryomedicine are used

1. to liquidate the pathological tissues [1034, 2413, 2655] (cryosurgery; also cryodestruction, cryoablation) as well as
2. to suppress the metabolic and other biochemical processes (sometimes an increase of the rate of the biochemical reactions is observed in the course of freezing of the tissues that is attributed to the increase of the solution concentration ("freeze-concentration effect") and/or to the catalytic properties of ice probably determined by the high mobility of protons in the solid phase [3129])

   - in cells (including the stem cells [1115] needed for the stem-cell-mediated clinical therapies and the tissue engineering [2784, 3260, 3431]),
   - in tissues (the natural and composite tissue transplants [2609] and the engineered tissues [286]),
   - in organs with a subsequent revitalization for the transplantation [216, 691, 911, 1432, 1551, 2444, 2656, 2973],
   - during testing of the new pharmaceutical products [695, 696, 698, 2003],
   - in the conservation of the genetic resources—both the domestic livestock [599, 659, 2060] and the rare and endangered species [306, 611, 789, 841, 862, 1554, 2585, 2653] as well as for facilitating the distribution of the genetically superior domestic species,
   - in the cryopreservation of the living bacteria to stabilize the cultures for the long-term storage and exclude or significantly reduce the contamination, the phenotypic and genetic drifts and enables the cultures to be used as the standards [1597].

The biological objects are routinely stored at the liquid-nitrogen temperature, therefore the only source of the damage will be the slow accumulation of the physical damage of the DNA from the background ionizing radiation [119]. D. G. Whittingham et al. [3316] showed that the mouse embryos survive 2 years exposure to the ionizing radiation which intensity is 100-times of the background radiation.

Sometimes higher temperatures are used for the storage of the biological objects [1404, 2368, 2650]. At the temperatures about − 80° C the viabilities decline over weeks to months while at the temperatures below − 140° C no degradation is observed [2083]. The bovine spermatozoa that have been stored for over 50 years show no reduction in the viability [2368].

The most reliable data on the admissible storage duration is probably on the long-term storage of the human red cells [2754]. The study of the samples stored for the period from 11 to 34 years gives the estimate of the shelf life for the conserved human erythrocytes as 113 and 276 years for the storage temperatures − 65° C and − 196° C, respectively [2972].

9.1 Hypothermia

Almost all cerebral processes are sensitive to the temperature fluctuations [1176, 1648, 3252]. There is the destructive relationship between the brain temperature elevation and the cerebral injuries [3252]. The cerebral tissues have a remarkable tolerance to the brain temperature reduction. In the patients undergoing the operative deep hypothermia, the brain well tolerates the temperatures of 15°–18° for the intervals of 2–3 hours [2054, 2427].

The hypothermia is used in the massive stroke and the traumatic brain injury patients [319, 2048]. After the brain injury, the brain temperature is often higher than and can vary independently of the core temperature; the brain is extremely sensitive and vulnerable to the small variations in the temperature [2203]. The prevention of the fever has been proposed as a therapeutic tool to limit the neuronal injury [425]. The reduction in the brain temperature as small as 1° C can substantially reduce the ischemic cell damage [592, 3265]. The hypothermia is used for the protection against the excitotoxicity, including the glutamate- and dopamine-associated cerebral cytotoxicity of the global ischemia [754, 3252], the hypoxic ischemic neonatal encephalopathy [1104, 2801]. The hypothermia reduces the rate of metabolism in cerebral tissue [2124] and suppresses the biochemical responses to cells following the onset of ischemia; the cooler temperatures can also suppress the inflammatory response of tissue following trauma and additionally reduce intracranial pressure which can prevent damage caused by inflamed tissue [319].

An effective method to achieve this is to cool the entire body ("the full body hypothermia") via either the internal methods through injecting the cold saline solution intravenously [270] or externally through surrounding the patient with the cooling blankets or applying the evaporative solutions such as alcohol to the skin [2477].

However, this can lead to the further complications such as the spread of the infection in other organs [1553]. Alternatively, to avoid such developments, the brain temperature could be lowered independently to the rest of the body by cooling the surface of the head. Directly determining the effectiveness of this type of the intervention is difficult due to the invasiveness of measuring temperatures in the brain [319]. Thus, researchers have to rely on the mathematical models to assess the outcomes of the hypothermia.

The moderate low temperatures are used in the therapy, e.g., after the cochlear implantation [872, 3008] or the intracarotid cold saline infusion in the human brain to induce the local hypothermia (the gray matter is cooling twice as fast as the white matter) having advantages over the hypothermia that is induced by the surface cooling in the much shorter time to action and the lesser probability of such adverse effects as the impaired immune function, the decreased cardiac output, the pneumonia, and the cardiac arrhythmias/bradycardias [1690].

The mild to moderate hypothermia, during which the brain temperature is reduced to 30° C–35° C, has been investigated as an adjunct treatment for the protection from the cerebral ischemia during the cardiac bypass injury [2318] and the resection of the aneurysms [3221].

The hypothermia causes a decrease in the nerve conduction. For example, the conduction velocity of the human peripheral nerves decreases from 30 m/s at 35° C to 12 m/s at 21° C [2499]. These decreases explain the observed motor incoordination and decrease in the manual dexterity. There is also some cold induced muscle stiffness and decrease in the blood flow to the limbs, which contribute to the incoordination and the loss of the strength at that core temperature. The significant attenuation and the frequency alterations in the brain's electrical activity are observed at the temperatures below 34° C [261, 1779, 2934]. The prolonged hypothermia of the brain affects the cerebral functioning: the cerebral cortex function is initially impaired, then the subcortical structures. The cessation of the respiration occurs when the medullary cellular activity is suppressed. The complete absence of the electrical activity is usually observed at temperatures below 20° C.

The blood flow is important in controlling the brain temperature. The CBF is the main mechanism to maintain the brain and body temperature coupling [3252]. The CBF changes with the temperature, and these changes are proportional to the changes in the cerebral metabolism induced by the temperature variations [2678]. During mild hypothermia in the humans the CBF is low [303]. The relation of the brain hypothermia induced through whole body cooling to the reduction in the CBF was demonstrated in several investigations [411, 1784, 2124].

An induction mechanism relies on the convective heat exchange through the cerebral circulation to accomplish the brain temperature alterations and is considered intravascular in nature [3252]. For example, the whole body hypo-or hyper-thermia induces the brain temperature changes convectively through the cerebral circulation and, thus, is intravascular as well as manipulation of the perfusing blood in the cervical carotid arteries to accomplish the selective brain cooling or warming.

The extravascular mechanisms do not rely on convective heat exchange through the cerebral circulation include conductive (for example, the heating or cooling devices applied externally), convective (for example, the forced cold or hot air circulation), and other methods (e.g., the exposure to the radiofrequency electromagnetic fields).

The "sub-physiological" temperatures are known to trigger the disassembly of microtubules thus potentially undermining the endothelial cell structure and disrupting the basic cellular functions [3252]. Some studies accept the temperatures below 29° C as the "profound hypothermia" [755]. It should be noted that the CBF and the brain tissue temperature differ significantly between the in vitro and in vivo conditions. For example, the experiments using the brain slices cannot examine the CBF and are typically conducted at the significantly sub-physiological temperatures.

9.2 Cryosurgery

The term "cryosurgery" (*κρύο* + *χειρουργική*) is not exact in its second part since "hand work" is limited to the preliminary stage of the operation—placing of the cryoprobes in the vicinity of the target tissue region.

The cryosurgery has a number of advantages over the alternative techniques such as the relatively painless and bloodless treatment, ease of the performance and the low cost, the high cure rate, the few post surgical complications, the short recovery period. Sometimes the cryoablation is preferable over the RF ablation even when the local recurrences after these treatment techniques are equivalent due to the greater risk of the insufficient ablation by the RF in case of the metastases located near to the major blood vessels [787]. The drawbacks of the cryosurgery are the scarce information on the long-term efficiency, the bulky equipment, the expensive MRI or other monitoring, the difficulty in the precise temperature control.

In dermatology the cryosurgery is used for removal of the diverse benign and malignant lessons such as the actinic keratosis, the warts, the hypertrophic scars, the keloids, the mucoceles solar lentigos [1721]. The cryosurgery treatment of the surface tissues is performed by spraying the liquid nitrogen [2977]; the simplicity of this procedure makes it accessible for a family physician [80]. The low temperatures are used in the cryoplasty—a procedure to open the arteries narrowed due to the atherosclerosis by exploiting the liquid nitrous oxide to simultaneously inflate and cool the dilatation balloon [3174].

The cryo-insults are used as the research tool in the animal models as pioneered by Openchowski with freezing the portions of the cerebral cortex of dogs almost two century ago. Now they are applied to get the information on the early stages of the pathogenesis of the osteonecrosis since the resulting repair responses are similar to those for the naturally occurring lesions [2176]. For example, K. L. Reed, T. D. Brown and M. G. Conzemius followed this approach to study the femoral head osteonecrosis which is an important problem in the orthopaedic surgery of the hip using a novel animal model of the femoral head osteonecrosis—emu [622]—in which the cryosurgically induced necrotic lesions consistently progress to the human-like structural collapse [2586].

The cryoaction as well as the hyperthermia could be combined [678, 1491, 1526, 1625, 2445, 3143, 3157, 3254, 3360, 3360, 3503, 3505] with the traditional therapeutic procedures—the adjuvant cryosurgery [219, 2455] (for example, with the immune therapy [2001], the radiotherapy [917, 1572] or the chemotherapy—the chemo-cryo combination therapy [593, 594, 984, 2148])—as well as with other physical methods of the tissues' treatment. The efficiency of the combined hyperthermia treatment clearly depends on the scheduled sequence of the two types of treatment and the time interval between them; the best outcome for the patient is expected from a simultaneous application [700].

The tissue fate in the peripheral zone of the ice ball is critical to the success of the treatment because of the possibility that some tumour cells may survive, thus the adjunctive therapy is beneficial. The freezing and the chemotherapy differently activate the apoptotic cascades [594] and thus act in parallel. For example, the ^{125}Iodine seed implantation can destroy the residual surviving cancer cells after the cryosurgery [3379].

The simultaneous use of the heater probes [2547] or the laser radiation [2633, 2634] enhances the possibility to control the extent of the frozen domain; for the same purpose the injections of the solutions with the special thermophysical properties, for example, the mixtures of the nanoparticles made of metal (such as gold) or the metal oxide (such as the iron oxide magnetite Fe_3O_4 or the vascular magnetic nanoconstructs carrying the super-paramagnetic iron oxide nanoparticles [1412, 2232, 2244, 2287]) with water that are exploited in the RF-magnetic hyperthermia [1995, 2333][1] could be used [719, 3450]; spraying of the cryogenic fluids is applied for the protection of the epidermis during the laser operations on the surface tissues;[2] the application of the impulse lasers in the cryoconservation in order to increase the cooling rate is proposed by D. Kandra [1530]. An aggressive pre-cooling could prevent the thermal tissue damage associated with the Joule heating in

---

[1] The effect of the nanoparticles is twofold:

- First, the presence of the nanoparticles with the high thermal conductivity greatly increases the thermal conductivity of the base liquid [3375, 3484] thus one could expect the increase of the thermal conductivity of the biological tissues and the greater effect of freezing; the asymmetrical injections of the particles allows the additional control over the size and shape of the ice ball, and, hence, the greater flexibility of the cryosurgery action.
- Secondly, these nanoparticles could serve as the sites for the heterogeneous nucleation. The nanoparticles also improve the image contrast for such cryosurgery monitoring techniques as the magnetomotive optical coherence tomography or the NMR imaging [71, 1632, 1633]. It is proposed to refer to this mode of cryosurgery as to the *nanocryosurgery* [1906, 3394].

[2] The laser surgery is used for treating some dermatological lesions such as the port wine stain (PWS) birthmarks that are the vascular malformation of the dermis. The laser energy at the appropriate wavelength causes the permanent thermal damage to the PWS blood vessels. Some part of the energy is, however, absorbed by the epidermal melanin causing the local heating that leads to the complications such as the hypertrophic scarring or the skin dyspigmentation. The cryogen spray cooling (CSC) provides the protection of the epidermis from the non-specific thermal injury by the cryogen evaporation from the skin surface [998, 2283]. The different cryogens could be used for this purpose—from the tetrafluoroethane-1,1,1,2 (R134a) with a boiling temperature of $-26°$ C at the atmospheric pressure to the liquid nitrogen temperature. The heat extraction from the skin depends on the average droplet diameter and velocity, the mass flow rate, the spray density [1907, 2461] and is rather non-uniform in both time and space [998].

the skin electroporation (especially injury to the highly electrically resistive stratum corneum and epidermis) [229].

The RF heating during the thawing of the cryoconserved biological objects increases the uniformity of the temperature field and prevents the ice recrystallization [2619, 3479]. The RF radiation could also be used for the action at the molecular level. Breaking of the hydrogen bonds inhibits the ice crystal nucleation thus reducing the crystallization temperature [1241]; this mechanism is also considered as a way to reduce the amount of the bound water in the cell and thus to increase the thermal conductivity of the tissue that will enhance the cryodestruction [2796]. The considerable (two-three times) enlargement of the zone of necrosis in the presence of the RF radiation was observed by A. Hines-Peralta et al. [1325]. The cryosurgery is also used to complement the dissections and resections, e.g., in treating of the cardiac arrhythmias [2482]. In this case the cryosurgery is preferable over the alternative RF ablation, being less thrombogenic [1601].

James Arnott of Aberdeen [295] was probably the first to apply the extreme cold locally for the anesthesia as well as for the pallation of the tumours in the breast and the uterine cavity and the destruction of the tissue using a mixture of the salt and the crushed ice [219]. He used a water cushion cooled by the flow of a solution from a reservoir of the brine. The cushion was applied to the tumours, which were frozen to temperatures of about $-24°$ C with no haemorrhage. His work was published by J. Bennet of Edinburg in his book "Cancerous and Canceroid Growth" (1846).

Campell White of New York and William Pusey of Chicago were the first to use for the treatment the cotton swabs dipped in the liquid air and the solid carbon dioxide (the carbon dioxide snow, or the carbonic acid snow), respectively. C. White used the liquid air for the treatment of a large range of diseases. W. Pusey recognized the low scarring potential of the cryosurgery although he attributed this to the regeneration of the residual epidermal cells rather than to the collagen's resistance to cold.

After the liquid oxygen became commercially available, it began to be used in the treatment of the skin diseases in 1929. However, since the liquid oxygen is a fire hazard, it has never become the popular cryogen for the cryosurgery. In the early 1940s P. L. Kapitsa in the USSR and Collins in the USA began developing the commercial techniques for the large-scale liquefaction of the hydrogen and helium with the liquid nitrogen as an abundant and the low-cost by-product. Soon after the liquid nitrogen became readily commercially available, Allington introduced it in the clinical practice in 1950. The liquid nitrogen was applied with a cotton swab, and soon became common in the treatment of the verrucae, the keratoses and the various non-neoplastic lesions.

During the early part of the twentieth century the cryosurgery was primary used to treat the dermatological and the gynecological diseases. Treatment of the deep tissues was started by a neurosurgeon Temple Fay of Philadelphia [2738] using the self-made closed system in 1938 for treatment of the carcinoma, glioblastoma, and Hodgkin's disease. M.J. Gounder and colleagues were the first to use the cryosurgery for the urological disorders.

A number of cryogens was used for medical applications including the dichloodifluoromethane (Freon-12), with the boiling point −29.8° C, the solid carbon dioxide (− 79° C), the liquid nitrous oxide (− 88.5° C), liquid argon (− 187° C), the liquid air, the liquid nitrogen (− 195.8° C) [218].

Sometimes no cryosurgery equipment at all is necessary: for the treatment of the open tissues such as the malignant bone tumour the so called "direct pour" (the open system—pouring the liquid nitrogen through the stainless-still funnel into the tumour cavity) method [2045] could be used. However, even in this case exploitation of the technological advances (the pressurized spraying of liquid nitrogen, the "semi-open" system [662]) is preferable since it allows to greatly reduce the various complications (the skin necrosis, the delayed healing, the nerve injury etc.) due to the rapid evaporation of the liquid nitrogen from the cavity [662].

Probably the first cryomedicine system was designed by J. Arnott in the middle of the nineteenth century. It consisted of a waterproof cushion applied to the skin, two long flexible tubes to convey water to and from the affected part of the patient and a reservoir for the ice/water mixture with a pump [624]. Using the salt/ice mixture, Arnott achieved temperature of − 24° C.

The various techniques were used to administer cold including the pre-cooled metal blocks, the pre-cooled needles, the dry ice applications, the thermoelectric methods and the cryogenic heat pipes [2178]. The solid copper cylindrical disks were cooled by immersion in the liquid nitrogen before application to the skin. These cylinders by far exceed the cotton applicators in the thermal capacity and the heat exchange characteristics. The additional effect follows from the pressure exerted on the lesion that reduces the blood flow and the related heat source. This approach allowed to double the depth of the tissue destruction to 4–5 mm.

A spray for cryosurgery was invented by a dermatologist H. Whitehouse of New York in 1907. He studied the effects of the liquid air on the normal skin and treated the epitheliomata, the lupus erythematosus and the vascular naevi. J. T. Bowen and H. P. Towle reported the successful use of the liquid air for the vascular lesions in 1907. The liquid nitrogen spray developed by Torre in 1965 was later modernized by S. Zacarian who suggested the interchangeble tips allowing variation of the spray diameter and designed a hand-held device.

The close systems based on the liquid nitrogen were introduced by I. S. Cooper and A. Lee in 1961 [625, 626]. The low temperature was achieved by evaporation of the liquid nitrogen in a hollow insulated metal probes close to the tip attached to the circulatory pump. The liquid nitrogen transport to the tip is provided by the use of the so called the *Leidenfrost flow* (the droplet film-boiling flow). The heat transfer from the liquid nitrogen to the tube wall in the Leidenfrost flow is very low because the liquid droplets are separated and thermally insulated from the wall by gas, thus little coolant is boiled off during the transport to the tip.

The totally closed system allows the surgeon apply the cold to any part of the body accessible to the probe.

Originally developed for the treatment of Parkinson and other neural disorders, the probe was found with great value for use in destroying the undesirable tissues deep in the body. R. W. Rand et al. [2574] reported the use of cryosurgery in the neurology. R. Markcover and T. Miller extended the use of cryosurgery in the orthopedics [2045]. D. Torre made contributions in the cutaneous cryosurgery [3069].

Later Cooper modified the original cryoprobe by adding a heating element that facilitates probe removing from the tissue after freezing relieving the surgeon from the need to use special tricks such as flushing with saline to assist the detachment of cryoprobe tip from the frozen tissue [3228].

The liquid nitrogen based cryoprobes provide the lowest temperature of the probe tip ($-$ 196° C that could be lowered down to $-$ 209° C if supercooling of the liquid nitrogen is used [2686]). They have, however, two drawbacks:

- the diameter of the cryoprobe, due to a rather complex construction providing the efficient two-way flow of nitrogen, could not be made smaller than approximately 3 mm that limit the number of the probes that can be used simultaneously (usually 5–6); a small number of the cryoprobes forces operators to reposition the cryoprobes during the operation to create the composite volume of the overlapping spheres (ellipsoids) of the frozen tissue;
- the liquid nitrogen systems react slowly to the changes in user input: an experimentally determined time delay is about 1–2 min [2686]; in addition, the transfer tubes and their associated valves should be pre-cooled before operation; the cryoprobes for the short operations that have a small storage Dewar incorporated onto the probe itself are free of this minor disadvantage [2350].

The modern cryoprobes using the Joule-Thomson effect—a constant enthalpy expansion of gas rapidly cooling it down to the boiling point—have a significantly smaller diameter. The modern systems based on the ultra-thin gauge needles (down to 1.47 mm diameter) allow the direct transperinal insertion and, if necessary, the relatively easy reposition of the probes during the operation reducing the disturbances of the tissues. The large number of the cryoprobes used simultaneously (up to 17 [2178]) provides the more flexible control of the shape and extension of the freezing zone, allowing treatment of the irregularly shaped tumours. The additional flexibility of the treatment results from the ability of the independent modulation in time of each of the multiple simultaneously operated cryoprobes.

Since some gases such as helium, under the Joule-Thompson effect warm up rather than cool when expanded, it is possible to create the systems for both the cooling and the heating with the rapid conversion between the regimes. The gas cryoprobes using argon and helium can be modulated between $-186°$ C and $+40°$ C in about 30 s [2686]. The different mixtures of the hydrocarbons and the synthetic refrigerants with argon and nitrogen are also considered as the potential working fluids [1005, 2281].

J. Liu et al. [1912, 3505] developed the novel minimally invasive probe system capable of performing both the cryosurgery and the hyperthermia. In contrast to the systems using the Joule-Thompson effect, in the new system one can alternatively supply the cryoprobe tip with either the liquid nitrogen or the hot water vapour. The drawback of the system described by the authors is a rather high diameter of the probe (5 mm) needed to provide the transport of the large amount of the working fluid.

Y. Rabin and A. Shitzer [2545] have modified the conventional cryoprobe by adding an electric heater to develop the computer-controlled system that provide such temperature of the probe that a specified constant cooling rate at the crystallization front is maintained. The warming catheter ("cryoheater") also protects the important organs that are placed in the close vicinity to the frozen malignant tissues such as the urethral mucosa.

A special cryoprobe for the study of the early stages of the osteonecrosis by introducing the cryo insult in the animal models that mimicks the naturally occurring osteonecrosis lesions was design by K. L. Reed et al. [2586] that in contrast to all commercial cryoprobes based on the "passive" vacuum tube insulation incorporates the "active" resistive electric heating that emulate the insulation, giving two independent operator-controlled parameters: the mass flow rate of the liquid nitrogen and the heating coil current.

In the early years of the cryosurgery monitoring was essentially reduced to the temperature measurement and a platinum resistance thermometer designed by D.E. Pegg and A.R. Hayes [2415] for the temperature range from 37° C down to −200° C with accuracy of 0.1° C was an essential achievement. Local monitoring of the cryosurgery in addition to the thermometry could be based on the impedancemetry where the electrode needles are placed inside the target tissue to detect the freezing-induced changes in the local impedance.

Now cryosurgeon could use a number of advanced techniques such as the US and the MRI imaging. The freezing interfaces can be conveniently monitored with the conventional ultrasound because there is a large difference in the acoustic impedance between ice and water. Ice essentially reflects all the acoustic energy, therefore entire freezing area looks dark. Ultrasound can only capture the image of the freezing interface in front of transducer. Magnetic resonance imaging produces an image of the proton density, which closely relates to the tissue structures. The MRI produces a precise three-dimensional image of the freezing interface; therefore it can be used to calculate the temperature distribution in the frozen region. MRI has solved the problem encountered by ultrasound as it is able to produce the real-time three-dimensional image of the frozen tissue without the acoustic shadowing. However, the high cost and the special surgical environment may limit MRI usage. Electrical impedance tomography is the technique that may provide an inexpensive and flexible supplement to the existing cryosurgical monitoring techniques. Injecting the small sinusoidal electrical currents into the body and measuring the resulting voltages through the electrode array produces a typical image. An impedance image of the tissue is then produced from the voltage data using a reconstruction algorithm [2655].

A more comprehensive treatment of the history of the cryosurgery could be found in the papers [236, 624, 1032, 1696, 2517, 2555, 2738, 3368, 3385] and of the history of the cryosurgery equipment in [760, 2178, 2555].

Evidently, however, that the progress in the development of the cryosurgery systems and the operation cooling/thawing protocols could neither diminish the crucial role of the surgeon nor liberate she/he from the responsibility for the treatment. New cryosurgical tools along with modern intraoperative monitoring techniques and the mathematical modelling (preferably using the patient-specific anatomical data) just help the surgeon to make the right decisions.

9.2.1 Destruction of Biological Tissues

The response of the living tissues under the thermal action is determined by the thermal history: the extent of the domain of the action changes due to the heat conduction and the heat transfer by the blood flow.

The behaviour of the tissues in the cryosurgery is more complex than during the hyperthermia treatment since

- the ice crystals are formed and grow;
- the ice recrystallization [817] and the formation of the gas bubbles [120] can occur during the thawing;
- the concentration of the solutes in the cytoplasm and in the extracellular space changes in time;
- the mechanical stresses arising due to both the non-uniformity of the temperature distribution and the volumetric expansion at the phase change can induce the fractures in the frozen tissues [2416] causing, inter alia, the postoperative bleeding.

The tissue response depends on the severity of the freezing [217]. There are three stages of the living tissue's destruction in the cryosurgery [1033, 3254, 3511]:

1. The immediate "direct cell" injury—the destruction of the cell due to the ice crystals formation and the mechanical forces acting on the cell membrane [1013] and the intracellular ice blades chopping the cell organelles, especially the mitochondria, the crystals' size enhancement due to the recrystallization [179, 2539]—the growth of the large crystals at the expense of the smaller ones—the during thawing (the free energy of the crystal surface decreases with its curvature) or/and due to the cell solute concentration growth (the "solution effect"—the exposure of cells to the high solute concentrations changes the hydrophobic and hydrophilic interactions and thus the functional properties of the membrane lipids, the proteins and the nucleic acids). Another mechanism of the cell damage during the thawing (the "osmotic shock") is related to the decrease of the osmolarity of the surrounding solution due to the ice melting

leading to the shift of the free water into the cell and swelling of the latter leading to the cell membrane rapture.
2. The blood microcirculation failure that is observed in several hours or days after the operation. The vascular stasis and the resulting cellular anoxia are developed through a number of the successive changes in the blood circulation. Initially the freezing of the tissues causes the vasoconstriction that lead to the decrease of the blood flow. The process could include such phenomena as the damage of the vascular endothelium resulting in the increased permeability of the capillary wall, the aggregation of platelets and formation of the microtrombs. Recovering of the circulation during thawing could be accompanied by the compensatory vasodilatation and the hyperperfusion of the tissues that is thought to stimulate the free radical formation which causes the additional endothelial damage via the peroxidation of the membrane lipids [246].
3. The apoptosis (the gene regulated cell death) [984] caused by the low temperatures that is essential for the partially damaged tissues in the peripheral regions of the ice ball where the tissue's temperature could be not low enough for the direct ablation [219].
The factors causing the cell's death could be of the physical (e.g., the ionizing radiation, the temperature, the mechanical stress), the chemical (drugs, toxins) or of the biological (the cytokines, the viral infection) nature [598]. The apoptosis (from Greek: apo—from, ptosis—falling, earlier the term "programmed cell necrosis" was used)—the "normal" (programmed) cell's death (the cell's suicide) [1070, 2560, 3398]—differs form the necrosis ("accidental", traumatic cell's death as a result of the physical or the chemical action) by the morphological features (such as the membrane deformation in the first case and loss of its integrity in the second) and by the character of the biochemical reactions (e.g., the random digestion of the DNA in the necrosis vs. the mono- and oligonucleosomal length fragmentation of the DNA during apoptosis). The necrosis usually affects groups of the contiguous cells while apoptosis—individual cells. The principles of the apoptosis were described by Carl Voigt in 1842, in most cases apoptosis is accompanied by the degradation of the chromosomal DNA [73].
The apoptosis could be initiated by both the extracellular and intracellular processes; in the last case the mitochondria act as the intermediate agent [219, 2076, 3398], involving the activation of the certain membrane channels and release of the apoptogenic proteins from the mitochondria into the cytosol. The mechanism of action of most common chemotherapeutic agents is to initiate apoptosis in the cancer cells; the temperature variation including the cold shock is also one of the signals (several dozens of the different apoptotic stimuli are described [2076]) that could trigger the apoptosis [2735].
The cells are susceptible to entering the apoptotic state up to 8 hours after re-warming [217, 3398].

The frostbite (the accidental cryosurgery) is characterized by formation of the ice crystals inside the organism under the low temperature environment conditions. It is almost always occurs for the outer extremities, the feet and hands, and face. The frostbite often occurs in conjunction with the hypothermia.

The IIF does not usually occur in the frostbite, except perhaps for the skin cells in contact with the cold metal. Unlike the cryosurgery which frequently performed in close vicinity to the vital organ, the frostbite does not damage the important tissues. The necrosis is observed within several days of the injury. The severe frostbite kills the nerve cells, so no great pain is associated with it (however, the mild frostbite could be painful).

The damage to the cells and tissues from freezing by the different mechanisms is cumulative.

An important issue for both the cryosurgery and cryopreservation (as well as for the hyperthermia) is the rapid and reliable characterization of the damage to the biological objects. Different approaches are used, both general ones such as the morphological and physiological observation, the NMR spectroscopy, the electron spin resonance spectroscopy, the oxygen or glucose consumption analysis and the cryobiology-specific such as the freezing curve-based monitoring or the dynamic low-frequency electrical impedance detection of the damaged tissues [3449].

Evidently, the conditions causing the tissue's destruction should be avoided in the cryopreservation.

9.3 Cryopreservation

The low temperature preservation usually is subdivided into the

- the hypothermic preservation at the temperatures above the water freezing point that is mostly used for the short-term preservation of the whole organs [2656] and
- the cryopreservation with the use of the special substances—the cryoprotective agents (CPAs).

The short safe storage time causes a large number of the healthy organs being discarded after exceeding the preservation time [522, 1838].

At present, the main methods used clinically to extend the organ hypothermic preservation time are

- static cold storage (SCS) [1541]
- machine perfusion (MP)[1011, 1298, 3056]

The preservation of most organs involves immersion in a preservation solution at a temperature range of 0° C–4° C, which reduces the organ metabolism [1482]. E.g., the SCS can typically preserve the heart for 4 h and the liver for 12 h [951]. But the metabolic activity does not completely cease during the SCS, and the accumulation of the toxic metabolites over time can cause the organ damage. Recently, the MP has attracted great interest due to its ability to provide oxygen and nutrients for the cellular metabolism and remove the toxic metabolites, thereby reducing the ischemia-reperfusion injury.

Two main modes of cryopreservation are the slow controlled freezing and the vitrification—the direct transition to the glassy state without crystallization [1548]. P. Kilbride et al. [1619] determined the temperature necessary at the end of the controlled slow cooling before transfer to the cryogenic storage which ensures the optimal recovery of the processed mammalian cell samples.

The technique of the directional freezing is used to regulate the growth of ice crystals, where the sample moves through a temperature gradient at a constant velocity to limit the gradual growth of ice crystals across that gradient. The controlled ice crystals appear as lamellae, with cells trapped between the lamellae to reduce the mechanical damage [2715]. The tight contact between the sample and the highly heat-conductive metal block enables the efficient heat dissipation, avoiding the damage to cells caused by the freeze-thaw cycles. The directional freezing can be used for both the slow and rapid cooling, and has been extended to the directional vitrification [104]. This technology demonstrates the capability to achieve the uniform cooling rates in tissues and organs, and has been applied in the cryopreservation experiments of various types of the tissue and organs, such as ovaries, rat hindlimbs, and pig livers [522].

During the cooling and rewarming processes in cryopreservation, the ice crystal formation and growth is fatal to the tissue and organ preservation [2414]. For the programmable slow freezing, the ice crystals are difficult to avoid even with the best preservation protocols [2081, 2082], which can cause damage to the intercellular connection and the blood vessels in the tissues and organs. It is difficult to achieve the optimal cryopreservation conditions for the multiple cell types in an organ simultaneously [2414]. The phase transition process of the ice crystal formation is accompanied by the heat release, which can cause the serious damage to the tissue cells beyond a certain limit [1900].

The vitrification is based on the notion that the homogeneous nucleation temperature is not a thermodynamic constraint, just a kinetic one and thus is not an absolute but a practical limit depending on the cooling rate [704].

To achieve the vitrification, the cooling rates faster than the critical cooling rate (CCR) are required to prevent the ice crystal formation, and it has been achieved after perfusion of the small organs with the CPAs. Similarly, a very high warming rate is required during the recovery process to avoid the devitrification [875, 909]. The critical warming rate (CWR) is usually an order of magnitude higher than the CCR [1230], and with the traditional convective warming is difficult to achieve such a rate. Moreover, these rates also need to be sufficiently uniform throughout the material to avoid the large thermal gradients.

The recently reported magnetic nanoparticle-induced heating technology [1616, 2039] effectively prevents the devitrification and the thermal stress during the rewarming process. Ideally, the organs would be uniformly loaded with the CPAs and the gold or iron oxide (IO) nanoparticles [1616] through the optimized hypothermic MP, and then rapidly heated uniformly in a radiofrequency-induced magnetic field after the vitrification. The GNPs and IONPs can be used for nanoscale and bulk heating thanks to their careful design for the optimal interaction with the electromagnetic fields such as RF and light. A benefit of the nanowarming [298] is that the RF fields penetrate with the negligible attenuation in comparison to the laser, microwave, or other electromagnetic fields.

The nanowarming has been demonstrated to be feasible in the cryopreservation of the various tissues and organs, such as the porcine arteries, the rat hearts, the rat kidneys, and rat livers; however, the capillary network of the frozen kidneys can be damaged after the nanowarming [522].

The first studies in the low temperature preservation were motivated by the development of the artificial insemination (AI) procedures. Beginning of the AI is attributed to Arabs stealing the stallion semen from a rival tribe [639]. The horses also were the subject of the modern development of the AI started by Ivanov in Russia in the beginning of the twentieth century [2145]. He studied the AI in other domestic animals (dogs, rabbits, poultry) as well. Ishikawa, working with Ivanov, began a similar program in the horses after return to Japan in 1912 [982]. At the same time work on the AI in the horses was done in Denmark, at The Royal Veterinary College in Copenhagen, by Eduard Sorensen who also was familiar with Ivanov's work. Sorensen is also known for his "invention" of the straw for the packaging semen [982].

In 1949 C. Polge, A. U. Smith and A. S. Parkes reported for the first time the use of the glycerol as a cryoprotectant. This discovery has started a fast development of the both cryopreservation techniques and the cryoprotective solutions. The cryopreservation was reported for the fowl spermatozoa suspended in a glycerol—albumen solution surviving freezing to $-79°$ C and later the cryopreservation for the bull spermatozoa [2478].

The establishment of the human sperm banks suggested in 1866 by Mantegazza [1817, 2368] was realized a century later almost simultaneously in France (Centre d'Etude et Conservation du Sperme, 1973) and USA (American Association of Tissue Banks, 1976). The human embryos are now routinely cryopreserved as an adjunct to other techniques of the assisted reproduction technology (ART) but still this approach remains the secondary one: babes born from the cryopreserved embryos represent less than 8% of the total of the ART babies born [1681].

In 1938 Jahnel reported the survival of the human spermatozoa stored at the tempereltature of the solid carbon dioxide. The systematic use of the frozen-thawed semen techniques was pioneered by Christopher Polge in the late 1940s who worked with the chicken and the bull semen [639]. The early cryopreservation procedures were developed in ad hoc manner and consist, for example, in the preliminary keeping the semen in the glycerol for a very long time ("the glycerol equilibration time") or the dilution in the sucrose—the egg yolk mixture and the vapour freezing as the pellets.

The sperm motility was, and essentially remains, the main criterion to assess the success of the freezing procedure. C. Polge was the first to show that too large concentration of the glycerol in the liquid semen decreases the fertility by the damage of the sperm cell membrane [348]. The first mammal—a calf—was produced with the cryopreserved spermatozoon in 1951 [1851]; the first pregnancy obtained with the frozen stallion semen was reported by C. A. V. Barker and J. C. C. Gandlier in 1957 [192].

First experiments on the cryopreservation of the human spermatozoa were reported in 1940 by L. B. Shelters (see [1817]) who used the direct plunging of the thin-walled capillaries with the undiluted semen into alcohol cooled down to $-79°$ C with the solid carbon dioxide, into the liquid nitrogen at $-196°$ C or into the liquid helium at $-269°$ C. The first pregnancy from a cryopreserved human embryo was reported in 1983 A. O. Trounson and L. Mohr [3083]. A significant progress was achieved by J. F. Lovelock and M. W. H. Bishop by discovering the cryoprotective properties of the dimethylsulfoxide in 1959 [1960].

Probably the most important biological object subjected to the low temperature preservation is the human erythrocyte—the red blood cell (RBC). There are different reasons for the RBC transfusions—the large RBC loss (the traumatic or surgical hemorrhage), the decreased bone marrow production, the defective haemoglobin, the reduced RBC survival (the hemolytic anemias).

The low temperature preservation of the RBC ex vivo for the clinical use allows to separate the donor and the patient in space and time. Two obstacles to the successful transfusion are the blood clotting and in vitro loss of the RBC viability and function. To overcome them, the citrate as the anticoagulant was introduced by A. Hustin in 1914 and glucose as a preservative by P. Rous and J. R. Tumer in 1916; the cryopreservation of the RBC was pioneered by A. U. Smith and J. Lovelock in the early 50s [2755]. Later both the vitrification and the lyophillization also have been considered as the preservation approaches for the RBC; the latter having the evident advantages due to the stability of the dried cells at the room temperature for the long time periods.

However, the standard cryopreservation procedures can fail for some biological objects, for example, they could severely damage the hepatocytes, thus, the new strategies for long term storage with adequate protection of the hepatocytes are needed [1942, 2158]. K. Yamada et al. [3392], for example, used the thermo-reversible gelation polymer (TGP) that can be converted into a gel state upon warming and liquid upon cooling for the encapsulation method for cryopreserving the hepatocytes.

The advances in the modern cryopreservation procedures could be divided into two directions [2280]:

- optimization of the cooling/thawing protocols, including the vitrification-based approaches, and development of the special procedures;
- improvement of the cryoprotective solutions.

There is a number of the special carriers or the vessels developed for the cryopreservation such the open pulled straws (OPS), the flexipet-denuding pipette (FDP), the microdrops, the hemistraw systems, the cryoloop and others [2348].

Cryopreservation protocols are increasingly required in the regenerative medicine applications but must deliver the functional products at the clinical scale and comply with the Good Manufacturing Process (GMP). While the GMP cryopreservation is achievable on a small scale, successful large-scale GMP cryopreservation is more challenging due to the heat transfer issues and the control of the ice nucleation, both the complex events that impact success. I. Massie et al. [2059] studied the cryopreservation of the large volumes of cells.

9.3.1 CP Procedures

Several vitrification-based procedures used for the conservation of the genetic resources could be distinguished [861]:

- the encapsulation-dehydration—the cells are encapsulated in the alginate beads kept for several days in the concentrated sucrose solution [149, 2032, 2591]; the encapsulation-dehydration could be used with the two-step cooling method [3485];
- the vitrification—the dehydration of sample with the highly concentrated vitrification solution and the subsequent rapid freezing;
- the encapsulation-vitrification—the alginate beads are preliminary dehydrated at 0° C before plunging in the liquid nitrogen [2031];
- the pregrowth—cultivation of the samples in the presence of the cryoprotectant, then the direct immersion in the liquid nitrogen;
- the pregrowth-desiccation—the dehydration of the samples pregrown in the presence of the CPA under the airflow cabinet or with the silica gel, then the rapid freezing [814];
- the desiccation—probably the simplest procedure—the direct immersion of the dehydrated samples into the liquid nitrogen; sometimes the ultra-rapid drying in a stream of the compressed dry air is performed [861];
- the droplet freezing—the small drops of the solution with samples placed on the aluminium foil and frozen by the direct immersion in the liquid nitrogen;
- in the channel-based methods the cells are mixed with the CPAs inside the quartz channel; the whole channel is immersed into the liquid nitrogen for freezing [3382].

The classical cryopreservation involves the slow cooling down to a given prefreezing temperature, followed by the rapid immersion in the liquid nitrogen. In the vitrification-based procedures, the cell dehydration is achieved before freezing by exposure of the samples to the concentrated CPA solutions and/or by the air desiccation and the subsequent rapid cooling. The critical stage in the classical approach is the freezing step and in the vitrification it is the dehydration step. The vitrification has economic advantages, since it is relatively simple, does not requires the expensive programmable freezing equipment and very fast (this procedure requires several seconds).

A multistep freezing process consists, as a rule, of the initial slow freezing followed by the fast freezing to reach the final low temperature. Such procedure provides the preliminary dehydration of the tissues or the organs that is beneficial in two aspects:

1. it minimize the amount of ice that can grow within the organ capillaries reducing the risk of the injury due to the ice expansion within these vessels—such damage is a known problem in the cryosurgery [2658];
2. it increases the concentration of the CPA within cells.

The multistep freezing procedure is also preferable with respect to the development of the thermal stresses that may cause the fracture in the tissues [2822].

The slow freezing is characterized by a relatively low cooling rate and the low concentrated solutions of CPA. Thus, this method has the low cytotoxicity for the biological cells. On the other hand, the risk of the cell injury due to ice crystal formation is high [3382].

The Fick's law is commonly used to study the CPA diffusion, but this simple modelling framework does not account for the complexity of the mass transfer in tissues, such as [3263].

- mass transfer in the extracellular space;
- coupling between extracellular and cell membrane mass transfer;
- fixed electrical charges on the extracellular matrix;
- tissue volume changes.

E.g., the tissue volume changes after exposure to the CPA have been observed for various tissue types, including cartilage, pancreatic islets, ovarian tissue, heart valves [3, 248, 1232, 3164].

A. Abazari et al. [3] developed the model that consider nearly all of the mentioned phenomena except the coupling between extracellular and cell membrane mass transfer. This decoupling is acceptable for the cell-sparse tissues such as cartilage, but it is not suited for the cell-dense tissues.

R. M. Warner et al. [3263] considered one-dimensional transport in a thin slab of tissue, with a zeroflux condition on one end and the other end exposed to an infinite bath solution. In the case of articular cartilage, this condition reflects that one end of the cartilage is exposed to the bath and the other to effectively impermeable bone. The insulated side of the tissue is fixed in space, and the bath side is free to move as governed by the fluid fluxes and mechanical properties of the tissue. Thus it is a classical one-dimensional moving boundary problem.

The tissue was divided into four compartments: extracellular fluid, intracellular fluid, intracellular solids, extracellular solids. Water, the CPA, and salt can move through the extracellular space in the x-direction, and this can result in movement of solids in the x-direction. The cells were considered to be embedded in the extracellular solids and hence move with the solids. The intracellular fluid can exchange water and the CPA with the extracellular fluid. The authors used the following subscripts to specify the chemical species and the compartment; iw, ic, in, s, n, w, and c represent the intracellular water, the intracellular CPA, the intracellular salt, solids, the extracellular salt, the extracellular water, and the extracellular CPA.

The mass concentration and volume fraction were defined per tissue volume and are used to write the continuity equations. The volume fraction of species k can be calculated from the mass concentration as follows $\phi_k = \rho_k/\bar{\rho}_k$ where $\bar{\rho}_k$ is the density of pure species k. The volume fractions of extracellular and intracellular salt were considered negligible.

The mole concentrations were defined for the extracellular fluid

$$C_w = \frac{\rho_w}{M_w(\phi_w + \phi_c)}, \quad C_c = \frac{\rho_c}{M_c(\phi_w + \phi_c)} \tag{9.1}$$

and for the intracellular fluid

$$C_{iw} = \frac{\rho_{iw}}{M_w(\phi_{iw} + \phi_{ic})}, \quad C_{ic} = \frac{\rho_{ic}}{M_c(\phi_{iw} + \phi_{ic})}. \tag{9.2}$$

Here M_k represents the molecular weight of species k.

The extracellular matrix is negatively charged and contains the fixed positively charged sodium counterions that also contribute to the driving forces for transport. These fixed charges cannot move freely and move with the matrix. The mole concentration of chloride ions can be calculated from the mass concentration:

$$C_{Cl} = \frac{\rho_n}{M_n(\phi_w + \phi_c)}. \tag{9.3}$$

The mole concentration of sodium ions includes the counterions for the chloride ions as well as the fixed charges: $C_{Na} = C_{Cl} + C_{fc}$. Thus, the sodium concentration is higher than the chloride concentration.

The concentration of the fixed charges can change as the tissue shrinks and swells, but the number of the fixed charges per volume of solids does not change. This allows to express the fixed charge concentration, C_{fc} in terms of a reference fixed charge concentration C_{fc}^0

$$C_{fc} = C_{fc}^0 \left(\frac{\phi_w^0 + \phi_c^0}{\phi_s^0}\right)\left(\frac{\phi_s}{\phi_w + \phi_c}\right). \tag{9.4}$$

The mole fraction of species k in the extracellular fluid in terms of the mole concentrations is written as

$$x_k = \frac{C_k}{C_w + C_c + C_n}. \tag{9.5}$$

Warner et al. used the continuity equations for the six mass concentrations (excluding intracellular salt). Movement of each species in the x-direction is expressed in terms of its average velocity, u whereas exchange of water and the CPA across the cell membranes is expressed in terms of the two-parameter membrane transport model

$$\frac{\partial \rho_k}{\partial t} + \frac{\partial \rho_k u_k}{\partial x} + \chi_k \frac{\partial V_{ik}}{\partial t} \eta \phi_s \bar{\rho}_k = 0 \tag{9.6}$$

where $k = iw, ic, s, n, w$, or c, $\chi_k = -1$ for $k = iw$ and ic, $\chi_k = 0$ for $k = s$ and n, and $\chi_k = 1$ for $k = w$ and c. The variables V_{iw} and V_{ic} are the intracellular water and CPA volume per cell, η is a constant term describing the cell density in the tissue.

The cell membrane transport terms are based on a classic two-parameter membrane transport model [1653], which gives the rate of change of intracellular water volume and intracellular CPA volume for a single cell in terms of concentration driving forces:

$$\frac{\partial V_{iw}}{\partial t} = L_p A_{cell} RT(C_{ic} + C_{in} - C_c - C_n), \tag{9.7}$$

$$\frac{\partial V_{ic}}{\partial t} = P_c A_{cell} \bar{v}_c (C_c - C_{ic}) \tag{9.8}$$

where L_p is the hydraulic conductivity, P_c is the CPA permeability, A_{cell} is the cell membrane surface area, v_c is the CPA molar volume, R is the universal gas constant, T is temperature. The intracellular salt concentration C_{in} represents the total osmotic contributions of all intracellular salts. The salt is considered to be impermeable to the cell membrane.

To use these membrane transport equations in the continuity equations, it is necessary to relate the volume changes for a single cell to the rate of transport between the intracellular fluid and extracellular fluid per tissue volume. This was accomplished by defining a constant term η to describe the cell density, which represents the number of cells per volume of solids, η is also used to interconvert between single cell volumes and tissue volume fractions as follows:

$$V_{iw} = \frac{\phi_{iw}}{\eta \phi_s}, \quad V_{ic} = \frac{\phi_{ic}}{\eta \phi_s}. \tag{9.9}$$

The velocities are related to the chemical potential gradients via the multicomponent momentum balance [3263]

$$-\rho_w \frac{\partial \mu_w}{\partial x} = f_{cw}(u_w - u_c) + f_{ws}(u_w - u_s) + f_{nw}(u_w - u_n) \tag{9.10}$$

$$-\rho_c \frac{\partial \mu_c}{\partial x} = f_{cw}(u_c - u_w) + f_{cs}(u_c - u_s) \tag{9.11}$$

$$-\rho_n \frac{\partial \mu_n}{\partial x} = f_{nw}(u_n - u_w) \tag{9.12}$$

where μ_k is the chemical potential of species k, f_{mn} is the frictional coefficient between species m and n. A volume balance can be used to define an additional relation

$$u_s(\phi_s + \phi_{iw} + \phi_{ic}) + u_c \phi_c + u_w \phi_w. \tag{9.13}$$

The frictional coefficients are functions of the diffusivities of the CPA and salt within water D_{cw} and D_{nw} as well as the permeabilities of the CPA and water in the K_{cs} and K_{ws}

$$f_{cw} = \frac{RT(\phi_w + \phi_c)C_c}{D_{cw}}, \quad f_{nw} = \frac{RT(\phi_w + \phi_c)C_{Cl}}{D_{nw}}, \quad f_{cs} = \frac{\phi_c^2}{K_{cs}},$$

$$f_{ws} = \frac{\phi_w^2}{K_{ws}}. \tag{9.14}$$

The authors compared predictions to the experimental data for the articular cartilage and the pancreatic islets, demonstrating that the model can predict the observed changes in tissue size and the CPA concentration for both of these tissue types.

A specific method of the oocyte and the embryo cryoconservation was reported by H. Nagashima et al. [2250, 2251] that consists of the preliminary stage of the polarization and removal of the cytoplasmic lipids from the cells before the vitrification. The authors thus avoided a negative aftereffect caused by the cooled intracellular lipids. While the intracellular lipids are considered as an energy source for the oocytes and as the building material for the membranes of the future embryos, their removal did not adversely affect the development of the oocytes and embryos after the cryopreservation [1868].

The success of the cryopreservation depends on the ability to control, in addition to the cooling rate, the nucleation temperature T_n [3076, 3466]. T_n depends on the sample volume, the properties of the container wall and the composition of the solution. T_n is also found to coincide with the onset of the liquid crystalline to gel the phase transition in the cell membranes [3336]. The addition to the solution (similar to the approach developed in the freeze-tolerant animals) of the ice nucleating substances such as *Pseudomonas Syringae* bacterium [3317, 3458] or the nucleator, known for over a half a century—the silver iodide AgI [3209] shifts T_n to the higher temperature but does not allows its precise control.

In practice the initiation of the crystallization (the *seeding*) is achieved manually by touching the sample with the cold tweezers or the cold rod. Some years ago A. Petersen, I. Heschel, G. Rau and B. Glasmacher described the device for controlling the nucleation temperature with a high accuracy for the cryopreservation system containing up to 8 samples [2440]. The action of the device is based on the phenomenon of the electrofreezing—an ability of the high voltage applied to the metal electrodes to cause the nucleation in the supercooled water, as was demonstrated for the first time by W. Rau in 1951 [2439]. A. Petersen and colleagues studied the influence of common in the cryobiology additives such as the glucose, hydroxyethyl starch, glycerol on this effect and showed that it is also observed in the solutions of the non-ionic additives. However, the presence of the ionic NaCl makes the electrofreezing in the solutions with the physiological salt concentration impossible. This problem was solved by the authors by designing a separate volume of the pure water inside the cap of the platinum electrode that makes the nucleus formation to be independent of the solution composition. It should be noted that the cryopreservation is not the only field to benefit from the precise control of the nucleation temperature—this parameter is known to strongly affect the rate of the lyophillization [2757].

The liquid nitrogen could be exposed to either the atmospheric pressure or to a vacuum. The latter approach decreases the temperature by several degrees of Centigrade (at 300 mm Hg pressure down to $-200°$ C). This improvement in the CP was shown to significantly increase the development rate of the bovine oocytes after the vitrification [702].

For a long time all attempts of the fish embryo conservation have failed. Only in 2007 V. Robles et al. [2620] have reported some success in the cryopreservation of the two-cell embryos of the seabream *Sparus aurata* using the technique of the microinjection of the CPA into the embryo to overcome the permeability barrier [1440]. The results also support the hypothesis that the AFP (the authors used a natural antifreeze protein type I), in addition to inhibition of ice growth and recrystallization, stabilizes the cellular membrane [2620]. The laser heating of the gold nanorods has recently been used to improve the cryopreservation of the fish embryos [1616]. This area is increasing in importance because the laboratories around the world have produced tens of thousands of the mutant, transgenic, and wild-type zebrafish (*Danio rerio*) lines for a wide range of the vertebrate genetics and the biomedical research [298]. The main barrier to the successful zebrafish embryo cryopreservation is the embryo's large volume (800 μm diameter) [1209, 1440]. Still, both the mouse oocytes (80 μm) and the zebrafish embryos can reach a cryogenically stable state in the liquid nitrogen through the use of the ultrafast cooling at roughly 90,000° C/min [1480, 1481].

There are three ways to increase the cooling rates [2348] viz.

1. to minimize the volume of the solution surrounding the cells;
2. to minimize the thermo-insulation;
3. to avoid the liquid nitrogen vapour formation.

The simplest way to minimize the thermo-insulation is to remove it by dropping the sample directly into the liquid nitrogen. However, this approach has a number of disadvantages. To form the drop, a rather large amount of the solution is needed. The drop will float for a long period of time on the surface of the liquid nitrogen. A strong evaporation on the surface will form the "vapour coat" as the thermo-insulated layer [1478]. One of the approaches to solve this problem is to drop the solution containing the sample cells onto the pre-cooled metal plate (the "solid surface vitrification") [770].

It is known for a long time that the ice nucleation temperature can be reduced by an increase of the hydrostatic pressure [1533] while the glass transition temperature rises with the increased pressure [2000]. This approach to the reduction of the needed CPA concentration is, however, limited since the high pressure could cause the damage to the biological objects. For example, while the dogs kidneys survive a 20-min exposure to 1000 atm, the rabbit kidneys were severely damaged after 20 min at the pressure of 500 atm [2348]. It should be noted that the high pressure is necessary for the vitrification stage only—the atmospheric pressure is sufficient for the storage.

A device for the ultra-fast cooling has been suggested by A. Jiao et al. [1478] that is based on the oscillating motion heat pipe (OHP) [1477, 1479] and produces the extremely high heat exchange coefficient using the thin film evaporation effect in the oscillation motion of the liquid plugs and the vapour bubbles. The authors stated that the heat transfer coefficient over 10^4 W/m^2 K could be reached that will provides, at least for some cooling methods such as the ultra-thin straw (100 μm in diameter), the cooling rates exceeding $2 \cdot 10^4$ K/min could be attained that will allows the fivefold reduction of the passage time though the dangerous temperature region (DTR) (typically from 240 to 200 K) where most ice nucleation occurs, in comparison with the conventional cooling techniques and thus significantly reduce the concentration (and, hence, the toxicity) of the CPA solutions.

A special procedure has been developed for the CP of the blastocysts that contains the fluid-filled cavity called the blastocoele. Since the probability of the ice formation is proportional to the sample volume, this cavity is considered a weak point for the CP. In addition, the CPA permeation into the blastocoele is slow: the resulting CPA concentration is insufficient after 3-min exposure of the blastocysts to the EG solution [2348]. P. Vanderzwalmen et al. showed that the survival rate could be improved by the artificial reduction of the blastocoele with the needle or the pipette before the vitrification [3160].

Recently I. I. Katkov et al. [1563, 1565] suggested the fast kinetic vitrification method using the spray cooling that avoids the Leidenfrost effect (LFE)—formation of the vapour thermoinsulative coat around the sample placed in a liquid cryogenic coolant such as the liquid nitrogen—and eliminates the need of the potentially toxic and the osmotically damaging vitrificants such as the dimethyl sulfoxide or the ethylene glycol usually used in the vitrification. However, the method could not be applied in the case of the thick tissues, the large 3-D structures such the cartilages or the whole organs.

I. I. Katkov [1564] listed the basic methods of the long-term biostabilization:

- the high-temperature vitrification of a highly dehydrated sample (the desiccation) and its stabilization by the air or vacuum drying;
- the slow freezing to the moderately low (− 60° C to − 40° C) temperatures that consists of the primary drying and the sublimation of the bulk of ice at the very high vacuum temperatures (the lyophilization);
- the equilibrium (slow) freezing that allows the freeze-out of both the extracellular and intracellular water to vitrify the cells in the "inter-ice channels" that are surrounded by the ice;
- the ice-free equilibrium vitrification of the cell suspensions or the tissues, at the very low temperatures and the moderate to high rates of freezing that requires the use of the high concentrations of the cryoprotectors;
- the intracellular ice-free kinetic vitrification that requires the very fast (abrupt) plunging into a cooling agent such as the liquid nitrogen; the extremely high rate of cooling (10^{4}° C/min–10^{6}° C/min) prevents the ice formation inside cells.

9.3.2 *Vitrification*

The vitrification as the preservation method was suggested by the Jesuit priest B. J. Luyet in 1937 [1990] and demonstrated by him (in collaboration with R. Hodapp) a year later for the frog spermatozoa vitrified in the liquid air after the dehydration in the solution of sucrose [1991]; the fundamental monograph "Life and Death at Low Temperatures" by B. J Luyet and P. M. Gehenio was published in 1940. The authors described the extensive experiments on the freezing of a wide range of organisms from bacteria and nematodes to the tissues isolated from the higher organisms, including the frog muscle and the mammalian erythrocytes. B. J Luyet and P. M. Gehenio concluded that frequently the cellular damage and death were caused by the ice formation.

The experiments show that further increase of the cooling rate again increases the cells survival. F. Dumont et al. [815] varied this parameter in the range 5° C/min—30,000° C/min and observed the monotonous increase of the fraction of the survived cells starting the cooling rate equal to 5000° C/min. The authors suggested two alternative explanations:

1. the ice crystals formed in the cells are too small to cause the cell damage (the characteristic crystal size decreases with the cooling rate [2407]);
2. the vitrification occurs.

K. Shimada and F. Asahina reported that formation of the intracellular ice crystals smaller than a critical size is tolerable in the rapidly frozen cells [2831]. This exact critical diameter was not cited, but it is believed to be less than 0.05 μm. Some protecting effect of the intracellular ice could be related to preventing the cell dehydration during the slow cooling [15].

The vitrification is based on the notion that the homogeneous nucleation temperature (depending on the pressure, for pure water being equal to approximately $-39°$ C at the atmospheric pressure with a minimum $-92°$ C at 2 kbar) is not a thermodynamic constraint, just a kinetic restriction and thus not an absolute but a practical limit depending on the cooling rate [704]. The rate of the nucleation in water is a non-monotonic function of the temperature due to the competition between the driving force for the nucleation that is increases with the supercooling and the viscous slow-down of the water molecules mobility with the temperature decrease. The vitrification occurs if the temperature interval within which the nucleation occurs can be bypassed rapidly.

The vitrification could be achieved by either of the following approaches (or their combination):

- A superfast cooling (for the tiny volumes about one nanolitre using the super-cooled down to $-200°$ C liquid nitrogen the cooling rate 130,000° C/min has been demonstrated [105]); the computations show the possibility to reach the cooling rate over 300,000° C/min using the specially designed cooling systems with the increased heat transfer coefficients [1478];

- Use of the highly concentrated (up to 50% by mass) solutions of the permeating glass-inducing cryoprotective agents [382, 2809]. The primary role of these additives is increase of the solution viscosity. The higher concentrations of the CPA are more needed for the extracellular vitrification than for the intracellular one since cells naturally contain the high concentration of proteins that are beneficial for the vitrification.

In the latter case the co-existence of the domains of the crystalline and the amorphous ice in the same sample is possible [2936]. It could be also observed in the conventional cryopreservation by the slow control freezing of the cells: a part of the system vitrifies due to the increase of the solutes concentration up to the point when the concentration of the residual solution is so high that it could vitrify in the presence of ice. The vitrification inside the cells is assisted by the presence of the intracellular macromolecules.

Usually the traditional slow cooling methods of the cryopreservation are referred to as the equilibrium cooling, and the rapid procedures (the vitrification) as the non-equilibrium one [1681].

Sometimes a state usually referred to as the *doubly unstable glass* [87] is achieved. In this state the vitrous system contains the ice nuclei and the probability of the vitrification is very high if the warming is not rapid enough.

One of the effects of the cryoprotector's (especially the non-permeating macromolecules such as polyethylene glycol or polyvinylpyrolidone [1717]) presence in the solution is the increase of the solution viscosity that assists in the transition to the glassy state [849]. The specimens should be sufficiently dehydrated to avoid the intracellular ice formation during the rapid cooling.

In the vitrification, the cells are dehydrated mainly before the cooling by the exposure to the high concentration of the cryoprotective agents [1717]. The principal benefit of the vitrification is the elimination of the need of the search for the compromise cooling protocol for the different types of the cells present in the tissue sample.

The drawbacks of the cryoprotectors are their high toxicity [2212] that manifests itself after the transplantation [163] and the difficulty in the uniform tissue saturation in the cryosurgery and in the cryopreservation of the large samples (the microinjection techniques [2620] could only meliorate the latter problem). The toxicity could have the chemical or the osmotic nature [695, 2348]; in addition, the intracellular cryoprotectors could directly interact with the intracellular organelles [2444]. The osmotic toxicity is mainly defined by the strength of the hydrogen bond of the polar part of the cryoprotector with water [910].

The reduction of the toxicity by the minimization of the perturbations of the cytoplasm consists in using the cryoprotectors mixtures [484, 910, 1717, 3050] including the tregalose or the glucose addition to the concentrated cryoprotector solutions [484, 634] or the preliminary saturation of the tissues in the ethylene glycol solution [2348]. The combination of the CPAs may result in the synergetic enhancement of the cell survival [381].

The effect of the type of the cryoprotector on the formation of the amorphous ice and its stability was studied by P. Boutron [345] using the erythrocytes as an example. The main parameter that determines the behaviour of the vitrified sample during the storage is the glass transition temperature: above this temperature the sample is the fluid with the extremely high viscosity in the metastable state; below this threshold the sample is the amorphous solid. The comparison of the different models that predict the glass transition temperature for the mixture "water - tregalose" [1566] showed that the general approach to estimation of the mixture property Y_{mix} as the weighted average of the components properties Y_i

$$Y_{mix} = \left(\sum Y_i f_i\right) / \left(\sum f_i\right),$$

where f_i is the fraction in the corresponding units (mass, molar or volume), does not gives the acceptable accuracy. The authors considered the Couchman-Karasz and Gordon-Tayor models for the binary solutions and their extension to the mixtures of cryoprotectors of the similar chemical nature (for example, the low molecular weights carbohydrates).

The water glass transition temperature has been for a long time assumed to be equal to 136 K. However, the recent investigations indicate the higher temperatures—from 145 K [2672] up to 160 [2932]–165 K [1083].

The thawing should be conducted in such a regime that it prevents the devitrification [1547]—the nucleation and growth of the ice crystals. The nucleation of the cubic I_c ice along with the ordinary hexagonal ice was observed in the high concentration solutions [3190]. The nucleation processes in the ethylene glycol solutions were considered [384] using the Johnson-Mel-Avrami-Kolmogorov model[3] that describes the evolution of the fraction x of the crystalline ice as $x = 1 - \exp(-(Kt)^n)$, where the parameter n depends on the crystal morphology. The authors proposed to write the reaction rate constant as $K \propto D(T)N^{2/3}$, where $D(T)$ is the diffusion coefficient of the water molecules at the temperature T, N is the density of the crystallization nuclei. The stability of the amorphous state is characterized by the critical heating rate [382], above which there is insufficient time for a vitreous sample to crystallize before the melting temperature is reached. The vitrified materials, which may contain the significant residual thermal stresses developed during cooling, may require an initial slow warming stage to relieve these stresses [382].

J. O. M. Karlsson [1547] in his analysis of the intracellular devitrification estimated the number of the nuclei in the cell as the integral over the cell volume

[3] This model was formulaed independently by W.A. Johnson and R.F. Mehl [1489] and by M. Avrami [142] to describe the isothermal transformation of the mother phase to the daughter phase by the nucleation, growth and impingement (the restriction of the growth of the transformed region by other already transformed regions) assuming the random nucleation, the isotropic growth and the infinite volume available for transformation. The last restriction was later removed by M. Weinberg and R. Kapral [3278].

and over the time interval of the homogeneous nucleation rate, assuming that the heterogeneous nucleation is suppressed by the presence of the CPAs, and represented the evolution of the radius of the crystal nucleated at the time moment τ using the diffusion-limited crystal growth model by D. R. MacFarlane and M. Fragoulis [1999] as

$$r(t;\tau) = \sqrt{\int_{\tau}^{t} \alpha^2 D_{\text{eff}} \mathrm{d}t'}$$

where α is the dimensionless crystal growth parameter, D_{eff} is the effective water diffusivity in the depletion zone near the growing crystal.

The risk of the devitrification during thawing depends on the nature of the tissue. For example, M. C. Wusterman, J. Simmonds et al. have found, using the same vitrification procedures and the same CPA mixtures, that the devitrification is a serious problem for the rabbit arteries while it is not observed in the case of the rabbit cornea [3363]. The probability of the devitrification could be reduced with the help of the ice blocking agents [3342].

The cryomicroscope allows one to observe the formation of the intracellular ice, the cell deformation due to the dehydration and interaction with the ice crystals and other above mentioned phenomena [761, 1013]. The ice initiation and propagation in plants could be monitored using the infrared video thermography [2940].

Under the normal physiological conditions, cells maintain a balance between oxidation and reduction reactions. However, this balance is disrupted during the cryopreservation by decreased enzyme activity, ATP deficiency, and Ca^{2+} accumulation, leading to the oxidative stress and the production of a large number of reactive oxygen species (ROS) [376]. Excessive ROS can induce a multitude of the deleterious effects—DNA damage, protein oxidation, and lipid peroxidation [1821]. The adverse effects of the ROS are further aggravated by the fact that mitochondria, the source of the ROS, are also attacked by the ROS. Damaged mitochondria produce less ATP and more ROS, which further exacerbates the cellular oxidative damage [302]. The series of damages caused by the oxidative stress can impair the cellular functions and decrease vitality post-cryopreservation, accelerating cell apoptosis and necrosis in conjunction with other types of damage [178]. An additional problem during ex vivo cryopreservation is ischemia-reperfusion injury, which is caused by redox imbalance [522].

9.3.3 Cryoprotective Agents

The optimal cooling rate in the cryopreservation could be modified with the use of the cryoprotective agents [1960, 2413, 2530, 2755, 2809, 2925] that could be divided into the two groups depending on their ability to penetrate the plasma membrane:

- The intracellular (permeating) CPA such as the propylene glycol (propane-1, 2-diol), the trehalose [16, 398, 515], the ethylene glycol (EG), the glycerol, the dimethyl sulfoxide) [870] reduce the concentration of the electrolytes and prevent the cell shrinking in the hypertonic solution. If the intracellular cryoprotectant cannot diffuse out of the cell quickly enough during the thawing to prevent the excessive influx of the free water and swelling of the cell, the osmotic shock may occur resulting in the cell rupture [1717] due to the extremely low intrinsic elasticity of the lipid bilayers [3334]. The rehydration at the high temperatures, at which the membrane are in the fluid state, is less dangerous than at the low temperatures at which the gel state of the membrane is observed [393].
- The extracellular (non-permeating) cryoprotectors (e.g., the sucrose, the d-mannitol, the lactose) activate the water loss by the cell and decrease the probability of the intracellular ice formation in the fast cooling process. Some cryoprotectors (the glycerol, the saccharides) prevent the partial denaturation of the proteins in the freezing solution [2959] and the eutectic crystallization [1228]. Both the low molecular weight compounds (the dimethylsulfoxide [2959], the glycerol, the disaccharides and others) and the high molecular weight polymers could serve as the cryoprotectors [2348]; the metallic nanoparticles also could be used for the modification of the cryoaction [3047].

The cells placed into the CPA solution are losing the intracellular water due to the hypertonic character of the solution [2213], and the cell volume decreases. After the initial shrinkage, the water and the CPA enter the cells to recover the equilibrium and the cells swell. The reverse behaviour is observed during the CPA removing: the initial cell swelling in response to the hypotonic extracellular environment is followed by the shrinkage when water and the CPA are leaving the cell [2687]. If the cell volume excursions are too large, the irreversible membrane injury occurs. For example, the osmotic cell shrinkage could cause an irreversible endocytic vesiculation that will result in the lysis during the osmotic expansion since the new membrane material is not rapidly enough available [2591].

One of the parameters that determine the quality of the cryoprotective agent is the number of the water molecules it could bound [305]: for example, one molecule of the propylene glycole is surrounded by over thirty molecules of the water [1717].

Glycerol is usually used for the mammalian species, ethylene glycol, the DMSO—for the avian, glycerol, the DMSO, methanol—for the fish species, EG (due to its low molecular weight, the high permeating ability and the low toxicity)—for the oocytes and embryos [73].

The efficiency of the CPA to increase the cell osmotic tolerance varies greatly between even the closely related species. The recent comparative study of the avian spermatozoa [309] showed the sixfold difference in the extreme acceptable hypertonicity between the crane and the turkey sperm in accordance with the earlier study [308] comparing the turkey, the chicken and four raptor species. The mammalian species also exhibit a wide range of variability in the osmotic tolerance of the sperm [175, 1080, 2670]. The authors state that the cell survival is determined by two factors viz.

- the final critical volume which is a function of the cell morphology and dimensions;
- the hydraulic conductivity of the membrane.

The species-specific differences were found to be especially significant with respect to the toxicity of the CPA during the long-term storage in the isotonic media. The effect of the CPA is related to its interaction with the cell membrane that markedly alters the transition behaviour of the membrane components and their protein kinetics [2379]. The molecules of the cryoprotective agent start to interact with the cellular membrane at the room temperature changing the conformations of the integral proteins in the lipid bylayer [653, 807]. The cryoprotective properties of the DMSO are attributed to the electrostatic interaction between its polar sulfoxide moiety with the phospholipids of the membrane, at the higher temperatures a hydrophobic association between the DMSO and the lipid bilayer is also significant [2809]. The cryoprotective properties of the DMSO—the highly polar aprotic solvent that mixes with water at the ambient conditions in all proportions—are also attributed to the exposed sulfonyl oxygen that acts as a strong hydrogen bond acceptor [2917].

The comparative analysis of the disaccharides as potential cryoprotectors was done using the MD by A. Lerbret [1827]; it was shown that the tregalose (α-D—glycopyranosyl–α-D—glycopyranosid) provides bonding for the greater number of the water molecules and more homogeneous solution reducing thus the probability of the nucleation. The ability of the tregalose to destroy the tetrahedral structure of the water and prevent the ice crystallization is confirmed by C. Branca et al. in the study of the thermodynamic properties of the disaccharides solution using the DSC method [356].

There is a problem of use of the trehalose since the mammalian cells do not synthesize trehalose or express trehalose transporters in their membranes, and the lack of an approach for the efficient intracellular delivery of trehalose has been a major hurdle for its use in cell cryopreservation. Y. Zhang et al. [3495] suggested to use a cold-responsive polymer to synthesize nanoparticles for the encapsulation and intracellular delivery of trehalose.

The properties of the lipid membranes are determined by the physical and chemical processes with the spatial scales from the atomistic to mesoscale (microns) in the wide range of the characteristic times. The different models focus on the different spatial and temporal scales; ideally, these models should be coupled—the multiscale modelling—that require the specification of the two-way information flow between the levels of description. The review of the MD simulation for the bilipid membranes performed before 2000 could be found in the paper by M. Pasenkiewicz-Gierula et al. [2386]; later models were considered in [2464, 2734, 3189].

The model lipid systems [93] (a monolayer at the air-water interface called the Langmuir layer, the planar bilayers, the assemblies of bilayers in the lamellar phase), in addition to their main purpose—study of the properties of the biological membranes—could serve, due to their relative simplicity, as ideal objects for the development and verification of the simulation methods. The simulation of the membrane protein behaviour is performed in [213]. At the present time the hottest topic is the simulation of the ion channels [2132, 2133, 3003]; i.a., the possibility of using of the carbon nanotubes for the emulation of the biological transport processes and the collective dynamics of the ion channels [1322] are studied.

The MD method is used to study the interaction of the cryoprotectors (the tregalose, the glycose, the methanol, the DMSO) with the model membrane that differ in the lipid saturation in the isothermal conditions [1807, 2464]; it was proved that the small organic molecules reduce the membrane thickness, increase the area per the lipid molecule and increase the membrane permeability. The protective mechanism of the saccharides including the sucrose and the tregalose is acts through the substitution of the water molecules [1836] in the hydrogen bonds with the lipid heads and stabilization of the lipid bilayer structure—a considerable shift of the temperature of the main phase transition. This effect could be described qualitatively by the order parameter

$$S_{CD} = -\left(\frac{2}{3}S_{xx} + \frac{1}{3}S_{yy}\right),$$

where $S_{\alpha\beta} = \langle 3\cos\vartheta_\alpha \cos\vartheta_\beta - \delta_{\alpha\beta}\rangle$ $(\alpha, \beta = x, y, z)$, $\cos\vartheta_\alpha = \mathbf{e}_\alpha \mathbf{e}_\beta$, $\mathbf{e}_z$—a unit vector in the direction z in the laboratory coordinate system, and $\mathbf{e}_\alpha$ is the unit vector in the local system defined by the three successive carbon atoms C_{i-1}, C_i, C_{i+1} in the tail of the lipid molecule, $\mathbf{e} = \mathbf{r}_{i+1} - \mathbf{r}_{i+1}$. This parameter that could be found experimentally using the NMR method characterizes the orientation of the lipid tails with respect to the normal to the lipid bilayer.

At the low temperatures the ordering of the tails increases that assist the transition to the gel phase. Note that the problem of the interaction of the low molecular weight organic compounds with the lipid membrane arises in the quite different environment—the wine production: the interaction of the alcohol molecules with the yeast membrane alters the membrane protein conformations and interferes with their function (arrests the fermentation))[1807].

The above mentioned models treat macromolecules as the flexible chains. Sometime more simple models (the so called spin models) could be used that describe the orientation of the amphiphilic molecule as a whole entity [2734]. Finally, it is possible to use a continuum description of the amphiphiles through the functional of the free energy depending on the small number of parameters, e.g., the local concentrations,—models of the Landau-Ginzburg type.

The presence of the cryoprotector decreases the membrane fluidity [1087] and the temperature of the main phase transition [393, 1823] as well as the temperature of the transition from the lamellar liquid crystal phase to the inverse hexagonal one [3442], changes the hydraulic permeability of the membrane [2449] and its temperature dependence (the activation energy) [733, 734, 1081], the thermophysical properties of the solution [552] (including the depression of the crystallization temperature [2085]), the morphology of the growing crystals; the cryoprotectors reduce the temperature at which the formation of the intracellular ice is being registered [1180].

The small solutes such as the sugar remain in the very narrow spaces (at approximately 20% of water, the average separation between the membranes is about 1 nm [393]) between the membranes at the low hydration resulted from either freezing or, in problems of the anhydrobiology, from the desiccation.

The oligosaccharides and other low molecular weight compounds prevent too close convergence of the membranes when the strong dehydration occurs that delays their transition from the liquid to the gel phase [393].

The effect of the cryoprotector nature (the sorbitol, the DMSO, the tregalose, the saccharose) on the freezing of the lamellar phase [3170] of the model lipid membrane was studied by Y. H. Yong, J. M. Pope and J. Wolfe using the nuclear magnetic resonance (NMR) [3430] that allows to measure a fraction of the unfrozen water.

The severe cell dehydration (the water content reduces to about 10%) results, in addition to the increase of the intracellular solution concentration, to the contact of the non-aqueous intracellular components, including the membranes. When these components are in the close proximity (the gap could be as small as 1 nm or even less), the repulsive hydration forces that are characterized by the exponential decay with the distance will reach the extremely high values that can lead to the membrane deformation. In these conditions the stacks of the membranes often are observed that resemble the lamellar phases [3430]. The role of the CPAs is to disrupt the water structure and thus to reduce the repulsive hydration interaction between the membranes.

It was shown that all the chemicals considered increase the hydration of the lamellar phase. However, it was found that the pure osmotic effect dominates in the tregalose and saccharose solutions of the low concentration. The effect of solutes in the lamellar phases is separated into two components: the purely osmotic effect that is approximately similar for the different solutes and the specific effects that are determined by the peculiarities of the interaction of the given solute with water and with the lipids.

When water is in equilibrium with ice, the lamellar phase contains no solutes. One could write for the chemical potentials of the water μ_{w} and the ice μ_{i} $\mu_{\mathrm{i}} = \mu_{\mathrm{w}} = \mu_{\mathrm{w}}^0 + P\nu_{\mathrm{w}}$, where μ_{w}^0 is the standard chemical potential of the water and ν_{w} is the partial molecular volume of water that is assumed to be equal to its bulk value. The mechanical equilibrium requires the interlamellar force per unit area $F = -P$. The experimental data could be fitted to the exponential decay for the force as [3430] $F = F_0 \mathrm{e}^{-R/R_c}$ where R is the number of the water molecules per the lipid molecule, R_c is the characteristic value and F_0 is the extrapolated force per unit area at contact or using a characteristic length λ as $F = F_0 \mathrm{e}^{-y/\lambda}$. The forces observed in the experiments are rather high: a values of the preexponential factor F_0 up to $4 \cdot 10^9\,\mathrm{Nm}^{-2}$ have been reported with the typical decay length λ about 0.2–0.3 nm [391].

Assuming that at the low hydration the interlamellar force is dominated by the exponentially decreasing hydration force, Y. H. Yoon et al. obtained the relation for the chemical potential of the water in the form

$$\mu_w = \mu_w^0 + kT \ln \gamma X_w - (F_0 e^{-R/R_c}) \nu_w,$$

where the activity of the water is represented as the product of the number fraction of the water X_w and the activity coefficient γ. The presence of any solute will decrease X_w which, in turn, will result in the increase in the interbilayer separation y—this is called the purely osmotic effect. The nature of the solute could manifest itself in its effects on the activity coefficient, the characteristic length and the force at zero separation. These effects are called by the authors the specific effects. The authors also assume that the interlamellar layer is in liquid state and note that if it is vitrified, the model should account for the ability of the interlamellar layer to support the lateral stress. The lateral stress is important since it could influence the temperatures of the phase transitions of the lipid bilayers [392, 393]: the membrane lipids in the gel phase have a lower surface area per molecule than those in the fluid phase, thus the compression favours the fluid-gel transition. The estimate of the transition temperature shift ΔT_m could be obtained from the Clausius-Clapeyron equation as [393]:

$$\Delta T_m = \frac{T_0 \Delta a}{2L} \pi,$$

where T_0 is the transition temperature at full hydration, Δa is the difference of the lipid surface area between the fluid and gel phases, L is the enthalpy of the phase transition, and π is the lateral stress in the bilayer.

If vitrification occurs in the bulk solution, it has little effect on the state of the membranes. However, if a vitrified layer is formed between the two fluid phase membranes during freezing, it greatly increases the stability of the fluid membrane phase and depresses the transition temperature of the main phase transition by more than 20 degrees of Centigrade [393].

The tregalose provides significantly more effective protection of the biomolecules than having the same chemical formulae $C_{12}H_{22}O_{11}$ maltose and sucrose due to the different configuration [356]. Among these compounds, the tregalose has the smallest partial molar volume indicating a more compact conformational arrangement, especially at the low temperatures, and characterized by the highest values of the interaction strength parameter and of the hydration number and thus a greater efficiency in destroying the tetrahedral network of water compatible with the structure of ice. Stabilization action of tregalose on the biological membranes is attributed to the interaction of the tregalose hydrogens with the polar head groups of the lipid macromolecules resulting in the replacement of the water molecules at the membrane-solution interface [356].

The antifreeze proteins [346, 684, 803, 1145, 2295, 3129] manifest their presence in two aspects. On the one hand, these chemicals inhibit the growth of ice crystals and the ice recrystallization during the thawing [505, 1166, 2066, 2620]. The AFP are extremely effective at inhibiting the ice recrystallization even at the very small concentrations (less than 1 $\mu g/ml$ [451]).

The molecules of the AFP are incorporated into the growth steps on the surface of the ice crystal splitting the crystallization front into the numerous segments; the increase of the front curvature results in the increase of the free surface energy and the front propagation slowing; a significant part of the protein surface participate in this interaction [684]. Using the scanning tunnel microscope one can see the system of grooves—the sites of the protein adsorption. The depth of these grooves—about 2–10 nm—allows to estimate the curvature of the crystallization front [523]. The MD computations show that the AFP increases the energetic barrier for the homogeneous nucleation [2556].

On the other hand, these proteins change the morphology of the growing crystal (or, in other words, modify the growth habit [1659]) of the extracellular ice, increasing the cell destruction by the ice crystals that have the specular form [451, 2966]. This property is exploited in the cryosurgery [1339, 2210]: when the AFP concentration is high, the needle-like (bi-pyramidal) ice crystals grow and mechanically destroy the cellular membrane [369, 2655]. The AFPs could be injected into the tumour region prior to freezing to increase the destruction of malignant tissues [2455]. The micronsize highly destructive specular shape of ice crystals is observed independently of the thermal conditions if the AFP concentration is higher than about 5 mg/ml [369].

Crystals formed by two hollow hexagonal pyramids could grow in the pure water if the supercooling is greater than 2.7° C [358]. The temperature of the ice-water phase boundary T_i could be written as

$$T_i = T_m(1 - d_1(\vartheta, \varphi)k_1 - d_2(\vartheta, \varphi)k_2 - \beta(\vartheta, \varphi, V_n)),$$

where T_m is the melting temperature, ϑ, φ are the boundary orientation with respect to the two orthogonal planes, d_1, d_2 and k_1, k_2 are the corresponding values of the capillary length and the interface curvature, V_n is the normal velocity of the crystallization front.

The kinetic supercooling $\beta(\vartheta, \varphi, V_n)$ describes the deviation from the equilibrium for the given orientation and the normal velocity. The variation of the crystallization front morphology is determined by the dependencies of the surface tension and the kinetic supercooling coefficient on the crystallographic surface orientation. The more complex crystal shapes are observed for the higher supercooling exceeding 5.5° C.

The "growth-melt asymmetry" observed for the pure ice under the pressure of 2000 bar was reported for the crystallization-melting under the normal conditions in the presence of the spruce budworm *Choristoneura fumiferana* antifreeze protein [2434] (the growth and melt shapes show the faceted hexagonal morphology, but they are rotated 30° relative to one another) and attributed to the different adsorption of the AFPs to the different crystallographic planes of the ice.

9.3.4 CPA Solutions

As a rule, there is not a distinct difference between the cryoprotective solutions for the hypothermic preservation and the cryopreservation. Usually any hypothermic preservation solution as well as a buffered physiological solution could be used for the cryopreservation [2656]. Still, there are exceptions: for example, the so called St. Thomas Hospital solution for the heart preservation is not good as a long storage medium [2656].

The different additive are aimed at reducing the detrimental effects that accompanies the cooling and freezing. The toxicity of conventional CPAs has been a persistent obstacle to the development of cryopreservation, prompting researchers to explore solutions [522].

Some solutions have been developed specially for the preservation of a particular organ. For example, the UW solution (see below) was developed first for the liver preservation while the original Collins solution—for the kidney preservation. The latter tries to mimic the intracellular composition and is rich in K^+. The cardioplegic solutions differ in the basic ionic composition, which can be either intracellular (rich in K^+) or extracellular (rich in Na^+).

Sometimes different solution is used at the preliminary stages of the CP: for example, the spermatozoa are incubated in the cholesterol-loaded cyclodextrin since loss of the cholesterol from the plasma membrane is one of the reasons of its destabilization [1851].

A number of the CPA mixtures has been suggested aiming primary at the reduction of the solution toxicity: the low ionic-strength vehicle solution (LSV) containing propanediol and trehalose [3363], the Euro-Collins solution [3028], the Hank balanced salt solution (HBSS), the histidine-lactobionate (HL) solution [2974], the histidine-tryptophan-ketoglutanate (HTK) solution, the sodium-lactobionate-sucrose (SLS) solution [3061], to name a few [2580].

Sometimes other components call extenteders such as the citrates or the egg yolk are added to the CPA solution; these compounds are thought to have an additional protective effect during the freezing and the thawing [2348]. The addition of the polymers with high molecular eight such as the polyvinylpyrrolidone (PVP) or the polyethylene glycol (PEG) is aimed at assisting the vitrification, primary the extracellular one since cells contain the large number similar in action macromolecules.

The variety of the cryoconservation procedures and even greater variety of the cryoprotective solutions is rightfully considered as an obstacle to the development of this technique since, in the absence of the evident best approach, it disperse the research efforts instead of focusing on perfecting a single approach [2348].

A milestone in the development of the cryoprotective solutions is the invention in 1986 by F. O. Belzer and J. A. Southard from the University of Wisconsin—

Madison the synthetic solution called the "UW-solution" [3224] that later became the practical standard in the cryopreservation—it is usually a starting solution when a new preservation protocol should be developed [2656]. The solution contains the potassium lactbionate, the adenosine, the glutamine, the allopurinol, the hydroxyethyl starch, the dexamethasone, the insulin [3224]. The comparative study proving its preference over other solutions was reported 2 years later [1437]. It was also found that the solution still could be improved by addition of the natural factors (small proteins called the trophic factors) increases the storage time and decreases the organ damage by simulation of the growth and the DNA repair [1231]. One of the recent natural components used in the storage solution is probably the green tea polyphenolic compounds (GTPC) [172]. The GTPC could act as the biological antioxidant and protect the mammalian cells and tissues from the oxidative stress-induced damage [1231].

The study by M. Wusterman, U. Rauen et al. had shown that a significant reduction of the CPA solution toxicity could be attained by replacing the sodium in the vehicle solution with the choline [3362]. In the experiments the twofold and fourfold decrease of the fraction of the cells losing their functional capacity were found for the porcine endothelial cells and the muscle cells in suspension, respectfully. As the authors state, the molecular mechanism by which the sodium leads to the cell death is still unclear, but it is probably mediated by the hydrogen peroxide or the nitride oxide. This opinion is supported by the earlier studies of the sodium effects in the cardiomyocytes [2465, 3399], the hepatocytes [2442] and the fibroblasts [1516]. The advantage of the choline chloride over the sodium chloride is believed to be its extracellular character with the similar colligative properties that prevents the damaging increase in the intracellular sodium concentration during the cryoconservation using the conventional solutions. The beneficial choline properties were also reported for the cryopreservation of the oocytes [2924] and the embryos [3066].

In spite of the great progress in the investigations of the cryopreservation and it's role in the assisted reproduction (now it is possible, e.g., to separate in the course of the preservation X- and Y-bearing spermatozoa, thus allowing the selective fertilization of the oocytes to produce either the female or male offspring) there are many unanswered question such as variability of the reaction of spermatozoa from different men to the same cryopreservation procedure or why the fraction of the sperm cells being damaged by the freezing is about one half regardless of other conditions [1817].

The cryopreservation is the cornerstone of the rapidly developing branch of medicine—the regenerative medicine. Of particular importance is the ability to preserved the engineered tissues such as the veins, the arteries, the cartilage and, in perspective, even the whole laboratory-produced organs to provide the needed supply for the transplantation and allow the necessary gap in time between the moment a replacement is created and its final transplantation [911] . Another fascinating issue is the cryopreservation of the human embryonic stem cells that could be lately used for the regeneration of the certain tissue or given back to the donor to correct the age-related deficits (sometimes this emerging discipline is referred to as the rejuvenatory medicine) [3260, 3393].

The success of the cryopreservation of the cell suspensions stems from the relative uniformity of cell size, shape and cytoplasmic content of the majority of cell types of interest. The important diffusion distances for solutes, solvents and heat are short for these suspended cells with little cell-to-cell differences and so provide uniform responses to imposed conditions. Responses to applied cryoprotectant whether physiological, osmotic or related to toxicity are also relatively uniform within a single cell type [855, 1620].

The much more difficult is the cryopreservation of the larger subjects ranging from the cell spheroids and organoids to the tissue slices and, eventually, the entire organs [1710, 2243, 2418, 3030]. The benefits of the cryopreservation of the multicellular structures range from facilitating population-wide biopsy studies to supporting large-scale manufacturing and providing economies of scale within organoid preservation [1620]. Realising these benefits would support a sizable fraction of the needs of regenerative medicine.

For even larger tissues and whole organs, success is largely limited to those which can operate as discrete units when dissected, for example, ovarian tissue and thymic slices. These can be removed from the body and cut into smaller functional units, which can each be successfully cryopreserved, thawed and transplanted independently [1620]. Mammalian organs lacking this ability such as the heart and kidneys cannot be cryopreserved successfully.

The difficulties of the whole organ cryopreservation, as was already mentioned, are the primary different nature of the cells that constitute the organ (and, hence, different optimal preservation procedures) and the significant size that prevents the attaining of the spatial uniformity in both the temperature distribution and the tissues saturation with the CPAs [522, 1620].

Compared to the cell suspensions, the processes of heat and mass transfer in tissues and organs appear to be much more complex. In terms of mass transfer, the loading and removal of the CPAs and the distribution of water are the main considerations [1550]. The mass transfer process within the organ becomes complicated due to the different types and arrangements of cells, as well as the interactions between cells. This limits the diffusion of water and the CPAs, resulting in difficulties achieving the uniform distribution of the CPAs in organs.

Following the addition of a permeating CPA an initial, cellular response of exposed cells is to shrink due to the osmotic gradient the CPA exerts. As the CPA then permeates into the cell, the gradient is diminished and cell volume recovers. In larger structures, exposure to the gradient, and the responses to it, will be delayed for the cells embedded deeper in the structure. This generates a risk of the insufficient CPA protection if cooling proceeds before the CPA equilibration is reached in the central regions of the structure. However, an extended incubation time in the CPA to ensure deep equilibration can lead to damaging levels of the toxicity for the peripheral cells. The larger and more complex the structure the more challenging this issue becomes, with the extracellular channels, the cell membrane parameters, viscosity, temperature and physical distance all playing a role; a similar issue, but reversed in direction, is encountered on warming and subsequent the CPA removal [1620].

A similar problem to the mass transfer is faced in the heat transfer, which is limited by the macroscopic volume and the thermal conductivity of the organ, making it difficult to achieve the uniform cooling and rewarming [166, 548]. As a result, organs experience the thermal stress damage during the cryopreservation, which often leads to the rupture of the biological samples. The cracks also promote the ice crystal formation, especially during the rewarming process [846, 908]. The uneven distribution of the CPA tends to lead to the appearance of temperature gradients and aggravates the thermal stress damage [1038]. The complex heat and mass transfer problems seriously hinder the tissue and organ cryopreservation.

The mechanical stresses could arise due to both the nonuniformity of the temperature distribution and the volumetric expansion at the phase change, since they could result in fractures in the frozen tissues [1048, 1713, 2416] causing the irreversible organ damage and the postoperative bleeding [3366, 3506] or the stresses in the cryopreserved samples. The latter could be caused, in addition to the temperature gradients, by the nonuniformity of the tissue saturation by the CPAs which are known to have the different thermal expansion coefficients [2551]. The danger related to the crack appearance in the vitrified samples is also related to the risk of devitrification since defects could serve as the nucleation sites.

Some progress was reported, but up to now no vital organ (particularly the heart or the liver) has been deeply frozen to the low enough temperature needed for the long term storage and later thawed, transplanted and proved to be functionally consistent [911]. One of the major reasons of the failure is the vascular damage due to the mechanical action of ice crystals on the vascular wall. Thus the vitrification seems to be the only promising way for the whole organ cryopreservation. Since the ultrarapid cooling for the large samples is evidently physically impossible, use of the highly concentrated vitrification solutions that provide the extreme increase of viscosity is inevitable. Review of the progress in the vitrification of the organ can be found in the papers by G. M. Fahy et al. [907, 911] and in J. Chen et al. paper [522].

9.4 Forensic Medicine

The osmotic effect and the crystallization of the ice cause the irreversible changes in the tissue: the destroyed cellular bonds are not restored after the thawing that results in the change of the microstructure of the tissue as well as of the mechanical and thermophysical properties [2711, 3349]; the voids appear [1871] originating from the ice cavities that arise either due to the cell membrane fracture or by the pushing cells apart [369, 1605] that in some cases (e.g., in the tissues of the liver) form the regular structures.

It is proposed by A. Th. Schäfer and J. D. Kauffmann to use the tissue destruction during the freezing to inquire whether the tissues were exposed to freezing in the past in the forensic investigations [2727]).

9.5 Cryofixation

Fast freezing is used (sometimes under the high pressure [2226]) to study the biological tissues with the cryomicroscope [1819, 2101] (the "freezing-shearing" or the "freezing-etching" [328, 1988]). The key problem of the electron tomography is to immobilize the object under study for the measurement's duration: placing the object into the amorphous ice provides the absence of the changes in neither the conformations nor the relative location of the macromolecules and the cellular organelles in space. The study of the structure of the proteins and other macromolecules by methods of the neutron crystallography [538, 2786] or the X-ray analysis [1985, 3128] using the synchrotron sources [3039] requires a preliminary crystallization of the object [460, 1079, 2266].

It is implicitly assumed that conformations of the macromolecules in the solution when they are biologically active and in the crystal are identical since "some water molecules are incorporated into the crystal during the growth" and "the crystal is placed into the water during measurements" [3511]. Indeed, a significant number of the water molecules (from a half to two for every amino acid residue, or about 200 for a typical protein [1834], according to other estimates from 34 to 39% by mass [2257]) is incorporated into the growing protein crystal. However, some of these molecules are incorporated into the structures not intrinsic for the protein itself, but formed during the crystal growth [2257]. It should be also noted that

- some complexes (such as lipid-protein) are extremely difficult to crystallize [2733];
- up to 30–40% of the area of the macromolecule surface that is in the contact with the solvent under the normal conditions (SAS—solvent accessible surface) is hidden inside the crystal in the case of small proteins [1220];
- no cold denaturation is observed for the proteins incorporated into the crystal, presumably due to the stabilizing action of the van der Waals forces and the hydrogen bonds between the molecules [2257];
- the crystallization of the proteins from the solution requires a supersaturation to be extremely high, orders of magnitude greater than is usual for the growth of the inorganic crystals [529]; some proteins forms the aggregates since, in contrast to the nucleation, there is no an energetic barrier to this process and these aggregates act as "the growth units" (for example, the white hen egg lysozyme forms the dimers, tetramers, octamers and 16-mers) while even formation of the simplest aggregate—the dimer—leads to modifications of the protein structure—in particularity, the hydrophobic groups are shifted to the protein surface [2446].

Another advantage of the cryoelectron microscopy is its applicability to the study of not macromolecules only but of the heterogeneous structures as well [138], e.g., the subcellular (and even subnuclear [2328]) objects in their mutual disposition corresponding to the living cell.

The rate of the freezing is critical: C. E. Jeffree et al. [1448] using a number of the plant samples showed that otherwise a redistribution of the water in the tissues occurs resulting in an appearance of the ice deposits in the extracellular space. Sometimes the artifacts are more difficult to detect as, e.g., the six- and seven-member ring structures in the hydration shell of the protein that exist at the temperatures below 200 K [1219, 1220].

The rapid freezing preserves the structure of the proteins to the atomic resolution [2102]. The electron cryomicroscopic tomography (Cryo ET), however, encounters two difficulties of its own:

- the image contrast is low reflecting the small differences in the density (more strictly, in the electron scattering) between the clusters of the macromolecules and the cytosol;
- the samples are sensitive to the damage done by the beam electrons.

The frozen tissues are also used for macroscopic studies: e.g., E. M. Brey et al. described the quantitative three-dimensional analysis of the microvascular structure that allows to determine such parameters of the system of the blood vessels as the turtuosity and percentage of the endothelial cell area [371].

Chapter 10
Thermal Ablation in Silico

Numerical simulation is useful for both the study of the fundamental processes such as, e.g., the protein denaturation at high temperatures, intracellular and extracellular ice formation, ice propagation in the tissues etc. and solution of the practical problems—in both the hyperthermia and the cryosurgery the control of the intensity and location of the action is of the paramount importance. Due to the restrictions of the real-time temperature monitoring, the mathematical modelling is extremely important. There are three stages:

1. The simulation of the evolution of the physical field and its absorption by the tissues to determine the heat source term.
2. The heat transfer in the target and surrounding tissues.
3. The assessment of the thermal damage.

The main difficulties of the mathematical description of the heat transfer in the vascular tissues stem from the complexity of the vasculature—the branching blood vessels' network, the presence of the complex network of pairs of arteries and veins with the countercurrent flow and variation of the perfusion intensity due to the change of the organism state [2588]. The heat transfer is affected by the vessel geometry, the local blood flow rates, and the thermal capacity of the blood. Somewhat simpler is the simulation of the heat transfer in the avascular tissues, for example, in the human tooth, in the bone-cement-prothesis system, in the knee cartilage under the cyclic mechanical loading, in the bones during grinding with the high-speed abrasive tools and drilling in the surgery.

The patient-specific discrete vessel input for the thermal simulations can be obtained using the CT or the MR angiography. However, not all thermally significant vasculature might be visible on the angiogram. To overcome this problem, the vessel generation software has been developed to create the vessel networks or add the thermally significant vasculature to the incomplete networks derived from an angiogram. A 3D vessel network construction algorithm, modelling the diffusion of the angiogenic factors in tissue has been developed by J. W. Baish [156]. The

A. I. Zhmakin, *Heat Transfer in Vivo*, Springer Series in Biophysics 27,
https://doi.org/10.1007/978-3-031-95459-7_10

generated vessel trees closely match the real vasculature in number and diameter of the vessels as well as the degree of asymmetry in branching.

"Modelling" and "Simulation" are frequently used as synonyms. In the Computational Heat Transfer (CHT) and the Computational Fluid Dynamics communities, however, the former usually refers to the development or modification of a model while the latter is reserved for model application [1183] (this usage problem is language-dependent: e.g., in Russian both terms are translated by the same word "modelirovanie" [3514]).

The necessary stages of the code development are

- the verification (an assessment of the correctness of the model implementation) and
- the validation (an assessment of the adequacy of the model to the real world) [725, 2612].

The verification is primarily a mathematical issue [2612]. The five major sources of the errors in the numerical solution have been listed by W. L. Oberkampf and T. G. Trucano [2323]:

1. the insufficient spatial discretization convergence;
2. the insufficient temporal discretization convergence;
3. the insufficient convergence of an iterative procedure;
4. the computer round-off errors;
5. the programming errors.

The errors of the last type are the most difficult to detect and fix when the code executes without an obvious crash yielding the "moderately incorrect results" [2323]. The study reported in ref. [1262] revealed a surprisingly large number of such faults in the tested scientific codes (in total over a hundred both the commercial and research codes regularly used by their intended users).

The verification is performed by comparison of the numerical results with the "highly accurate" (benchmark) solutions. There are three main sources of such solutions:

1. the exact analytical solutions;
2. the benchmark solutions of the ordinary differential equations (ODEs);
3. the benchmark solutions of the partial differential equations (PDEs).

The usual use of ODE solution is based on exploiting the symmetry properties: one can solve an essentially one-dimensional problem (e.g., having spherical symmetry) using a general three dimensional grid. Additional tests could be constructed via a so called "method of manufactured solutions".

This method [1162, 2613, 2683] is a simple process, although it involves the cumbersome algebra: a chosen analytical "exact" solution $\varphi_{ex}(t, x, y, z)$ is substituted into the equations to produce the source terms $f_{ex} = L(\varphi_{ex}, \varphi'_{ex}, \varphi''_{ex})$ and the initial/boundary conditions.

This approach is attractive since it could be applied to the complex multidimensional domains while allowing the direct assessment of the solution accuracy.

However, it usefulness depends on the similarity of the guessed "exact" solution to the typical solution of the real problem in question.

An availability of the analytic exact solution greatly reduces the CPU time needed for testing of the nonstationary codes: one can use an exact solution for the specification of the initial and, if necessary, (nonsteady) boundary conditions and avoids the need of using the large time integration intervals.

The validation is the second step in the assessment of the software quality. It should be stressed that the verification should not be skipped: the successful validation alone does prove the reliability of the code due to the limitness of a set of the validation cases, a "graphical" comparison with the experimental data in the most cases and a possible cancellation of the multiple errors that can give can an impression of a success.

Sometimes it is stated that nature of the code (research, pilot, production) is determined by its maturity with respect to the validation level [1101]. This seems to be oversimplification. If one needs a single criteria (ideally, quantifiable) to estimate the practical usefulness of simulation, the best choice is probably the reliability of a computer prediction [2326].

It should be stressed, however, that this parameter characterizes not the code itself, but the simulation, being depending on the adequacy of the model and the accuracy of the computations as well as on the particular aim of the simulations. Evidently, the same results could be considered successful if one is interested in unveiling some general trend - and unsatisfactory if the goal is to find, for example, the exact position of the cryoprobes.

There is a danger related to the numerical simulations that now easily produce huge datasets. A half a century ago Richard W. Hamming in his classical monograph "Numerical Methods for Scientists and Engineers" (McGraw-Hill, New York, 1962) warned that aim of computations is understanding, not numbers. It is not hard to escape drowning in the ocean of numbers: one just should remember that more data does not necessary mean more information, more information—more knowledge, and more knowledge - more wisdom.

The cryoaction cardinally differs from the hyperthermic action (ultrasound [403, 848, 1377], laser [201, 2169, 3179] or RF radiation [727, 728, 848, 1961] heating). The latter is based on the *volume* absorption of the acoustic or electromagnetic signal; the cooling of the tissues is possible through *the surface* only. Thus the spatial non-uniformity of the temperature and the rate of its variation, i.e. different thermal history of the cells in cryosurgery and cryopreservation is inevitable. In the numerical simulation of the hyperthermia a source term arises in the energy equation; description of the cryoaction is contained in the boundary conditions. Another distinction is the role of cell dehydration and the local problems of the extracellular and intracellular ice formation.

10.1 Microscopic Models

Statistical methods are the powerful tool for study of the phenomena characterized by the intrinsic stochasticity that are very difficult if not impossible to describe by the pure analytical methods. These approaches are divided into two groups—Monte Carlo (MC) and Molecular Dynamics (MD) simulations [1441]. With still ongoing advances in computer hardware (even if some deviations from the Moore's law on the constant rate of gaining the computing speed are noticed) it is expected that their importance will increase in future allowing simulation of the more complex systems for the longer physical time.

When the subject of the study is the phase transition, the MC or the MD approach could be used to generate the relation between the order parameters and the free energy of the system via the bottom-up approach. If N order parameters $\varphi_1, \varphi_2, \ldots, \varphi_N$ are specified, the free energy density $\Lambda[\varphi_1, \varphi_2, \ldots, \varphi_N)]$, also called the potential of the mean force or the Landau free energy, is related to the microscopic Hamiltonian H as [2558]:

$$\Lambda[\varphi_1, \varphi_2, \ldots, \varphi_N)] = \int \mathbf{dr}_1\mathbf{dr}_2 \cdots \mathbf{dr}_N \exp(-\beta H(\mathbf{r}1, \mathbf{r}_2, \ldots, \mathbf{r}_N)) \times \delta(\varphi_1 - \varphi_1')\delta(\varphi_2 - \varphi_2') \cdots \delta(\varphi_N - \varphi_N'). \tag{10.1}$$

Here δ is the Dirac delta function and the free energy F is expressed as

$$\exp(-\beta F) = \int \mathrm{d}\varphi_1 \mathrm{d}\varphi_2 \cdots \mathrm{d}\varphi_N \exp(-\beta\Lambda[\varphi_1, \varphi_2, \ldots, \varphi_N)]). \tag{10.2}$$

In the course of the simulation, the free energy is computed by collecting the histograms of the distribution of the order parameters. For the sufficiently long simulation these histograms are proportional to the joint probability of the order parameters $P[\varphi_1, \varphi_2, \ldots, \varphi_N)]$ which is related to the free energy density as

$$\beta\Lambda[\varphi_1, \varphi_2, \ldots, \varphi_N)] = -\ln(P[\varphi_1, \varphi_2, \ldots, \varphi_N)]) + C, \tag{10.3}$$

where C is a constant.

The growth rate of the crystal could be related to the increase of the number of "solid" particles N_s as

$$R = \frac{d}{A/a}\frac{\mathrm{d}N_s}{\mathrm{d}t} \tag{10.4}$$

where d is the interplane distance parallel to the interface, A is the total area of the interface, a is the specific area per particle. The problem is no net growth in the equilibrium. Thus, to compute the growth rate from the equilibrium simulations, an 's hypothesis is invoked that state that equilibrium fluctuations decay according to the macroscopic laws [372].

10.1.1 The Monte Carlo Method

The Monte Carlo simulation methods are especially useful in studying the systems with a large number of the coupled degrees of freedom. The major components of the MC algorithm are the following ones

- the random number generator—a source of the random numbers uniformly distributed on the unit interval;
- the sampling rule—a prescription for choosing from the possible states of the model physical system;
- the scoring—accumulation of the computational results to get estimates for the quantities of interest;
- the error estimation - an estimate of the statistical error (variance) as a function of the number of trials and other quantities.

The algorithm implementation is easily divided into two largely separated stages—the generation of the "raw" data and the data analysis.

The MC method based on the random sampling is very simple, however, its naive application will be extremely inefficient. One of the most important aspects of the MC simulation is the use of an advanced *sampling technique* that could be rather sophisticated. The idea is that one should not pick configurations randomly, but should draw them according to their Boltzmann weight [1441]. The different update rules are usually subdivided into the more conceptually simple local and the more involved nonlocal algorithms. The important ingredient of the MC simulation is the use of the various variance reduction techniques to decrease the computational time for the Monte Carlo simulation.

10.1.1.1 Update Algorithms

The most popular update algorithm is undoubtly the local algorithm that was suggested over a half a century ago by N. Metropolis [2118] and later named after him. The algorithm is formulated as follows. If E and E' are the current energy and the energy after the tentative update, respectively, the probability of the acceptance of this update (the transition probability $W(\sigma_i \to \sigma_i')$ from the state σ_i to the state σ_i') is given by

$$W(\sigma_i \to \sigma_i') = \begin{cases} 1, & E' < E \\ \exp[-\beta(E' - E)], & E' \geq E, \end{cases} \tag{10.5}$$

or, more compactly, as

$$W(\sigma_i \to \sigma_i') = \min\{1, \exp[-\beta(E' - E)]\}. \tag{10.6}$$

The key feature of the Metropolis algorithm is the finite probability of the acceptance of the unfavourable updates that increase energy (this probability tends to zero in the case of zero temperature expressed via the parameter β as $\beta \to \infty$). It is this feature that prevents getting trapped in the local minimum. This algorithm has a wider range of applicability, in particularity, it is a basis of the simulation annealing—the general optimization method that provides under some rather mild conditions finding the global minimum while avoiding the local minima [1640, 1832].

There are other local update algorithms such as the Heat-Bath or the Glauber, for example; the former is usually applied to the lattice models and in its simplest form to the systems with the discrete degrees of freedom with a small number of the allowed states.

The different patterns could be used to update the degrees of freedom including the random one or the random permutation, or just the lexicographical (sequential) order. This choice has small effect on the qualitative algorithm behaviour, but could significantly influence the computational time.

The non-local algorithms such as the multigrid schemes or the cluster-update methods improve the performance of computations at the cost of the generality. Being more specified than the local update algorithms, they could encounter difficulties in the application to the new models.

Among the analysis approaches the different reweighting techniques (the single-histogram reweighting, the multi-histogram technique) should be mentioned.

One of the most important reasons for the frequently large time needed for the simulation to give the results with the acceptable accuracy is the presence of the so called “rare events”. Thus, two strategies for the computations’ acceleration were developed. The first one that could be named “rare event avoidance” tries to connect the important parts of the phase space by paths which go around the rare-event regions that, hence, could not be studied directly. The second strategy—let us call it “rare event tolerance”—is aimed at enhancing (frequently artificially) the probability of the rare event states.

In the early 90s B. A. Berg and T. Neuhaus suggested the multicanonical algorithm (MUCA) for the MC method [255]. The method, in contrast to the conventional MC, is based on the non-Boltzmann weight factors allowing to get the uniform potential energy distribution. The random walks in this energy space require the significantly smaller number of the simulation steps then the conventional canonical MC. The MUCA MC splits simulation into two stages: the estimation of the weight factors and the production run itself. However, while the canonical MC gives estimates of the observable physical variables for the fixed temperature and other parameters (the extrapolation in the vicinity of this temperature value is possible using the re-weighting techniques), the MUCA MC provides the estimates for the range of temperatures or other parameters. The methods proved to be very efficient for a number of problems, including the phase transition of the first order because it allows to enhance the weights of the rare configurations [255, 801].

A. C. Maggs [2007] suggested a generalization of the Metropolis update algorithm called by the author the "multiscale Monte Carlo" that is based on the collective updates that shift the blocks of the particles for the distance that is intermediate between the particle size and the dimensions of the computational cell and avoids the so called hydrodynamic slowing down [2007].

The discussion of the intricate features of the MC methods and algorithmic tricks developed over years to accelerate computations is beyond the present work. The reader is kindly asked to consult numerous books and papers on this subject, for example, the above cited introduction by W. Janke [1441] or the monograph by C.P. Robert and G. Casella [2614].

10.1.2 Molecular Dynamics Methods

Molecular dynamics (MD) *per se* and as an element of the more complex computational methods (see, for example, [866, 1240, 2799, 2843] and references therein) is one of the main numerical approach to the problems of biophysics; its applications to the cryobiology problems are scarce up to day.

10.1.2.1 Classical Molecular Dynamics

In the classical MD method the motion of the atoms (or, sometimes the molecules or other units such as the specific residues of proteins) that interact via the empirical potential are monitored using the Newtonian equations for the point mass motion. Usually the potential energy of the system could be represented as a sum of the several contributions as [2733]

$$E(X) = \sum_{bonds\ i} S_i^b(b_i - \bar{b}_i) + \sum_{bond\ angles\ i} S_i^\vartheta(\vartheta_i - \bar{\vartheta}_i)^2 + \sum_{dihedral\ angles\ i} \sum_n \left(\frac{Vn_i^\tau}{2}[1 \pm \cos(n\tau_i)]\right) + \sum_{atoms\ i<j} \left(-\frac{A_{ij}}{r_{ij}^6} + \frac{B_{ij}}{r_{ij}^{12}}\right) + \sum_{atoms\ i<j} \left(\frac{Q_I Q_j}{D(r_{ij})r_{ij}}\right) + \cdots . \qquad (10.7)$$

Here the symbols b, ϑ, τ and r refer to the bond lengths, bond angles, dihedral angles and interatomic distances, respectively, the bar symbol represent the equilibrium value. Each dihedral angle could be associated with several rotational indices n (the third term in the right hand side). The last term corresponds to the Coulomb interaction where Q are the atomic partial charges and $D(r)$ is the spatially dependent dielectric function. X represents a collection of the Cartesian coordinates of atoms. Other terms also could be added to account, for example, for hydrogen bonds or planar moieties.

The Newtonian equations of motion that had to be solved at every time step of the MD method are written as

$$\mathbf{M}\dot{V}(t) = -\nabla E(X(t)), \tag{10.8}$$

$$\dot{X}(t) = V(t), \tag{10.9}$$

where $\mathbf{M}$ is the diagonal mass matrix, dot denotes differentiation in respect to time.

Sometimes the frictional and/or the random-force terms are added a la the phenomenological Langevin equations to yield

$$\mathbf{M}\dot{V}(t) = -\nabla E(X(t)) - \gamma \mathbf{M} V(t) + R(t), \tag{10.10}$$

$$\dot{X}(t) = V(t). \tag{10.11}$$

Here γ is the damping constant. The random Gaussian force $R(t)$ should have a zero mean value

$$\langle R(t) \rangle = 0 \tag{10.12}$$

and the specified autocovariance matrix

$$\langle R(t) R(t')^T \rangle = 2\gamma k T \mathbf{M} \delta(t - t'). \tag{10.13}$$

Here superscript T denotes the matrix transpose.

The leap-frog integration method introduced by L. Verlet half a century ago for the Newtonian equations of motion is trivially extended to the above modified equations incorporating the damping and the random force. This method is based on the use of the intermediate time level $t^{n+1/2} = t^n + \Delta t/2$ and can be written as

$$V^{n+\frac{1}{2}} = V^n + \mathbf{M}^{-1} \frac{\Delta t}{2} \left[-\nabla E(X^n) - \gamma \mathbf{M} V^n + R^n \right], \tag{10.14}$$

$$X^{n+1} = X^n + \Delta t V^{n+\frac{1}{2}}, \tag{10.15}$$

$$V^{n+1} = V^{n+\frac{1}{2}} + \mathbf{M}^{-1} \frac{\Delta t}{2} \left[-\nabla E(X^{n+1}) - \gamma \mathbf{M} V^{n+1} + R^{n+1} \right]. \tag{10.16}$$

10.1.2.2 Ab initio Molecular Dynamics (AIMD)

The classical MD simulations based on use of the pre-computed empirical interaction potentials are not able to describe the quantum phenomena that are underlying such processes as the proton transfer among the water molecules or between the water molecule and a foreign atom (however, in some cases an approximate description of the arising the intermolecular H-bonding is possible in the frame of the classical approach with the use of the electrostatic interactions between the partially charged atoms [3243]) or other chemical processes involving the bond breaking and forming.

Another deficiency of the current force fields is their failure to correctly include the polarization effects, although the polarizable force fields have been suggested.

The variant of the MD—the ab initio MD (AIMD)—solves these problems by combining the "on the fly" electronic structure calculations with the finite temperature dynamics. Naturally, the AIMD simulations are substantially more expensive than calculations based on the empirical force fields.

The most important element in the AIMD calculation is the representation of the electronic structure. Clearly, the calculation of the exact ground-state electronic wavefunction is intractable, and the approximations must be used. The electronic structure theory employed should be reasonably accurate yet not too computationally demanding. One formulation of the electronic structure problem that satisfies these criteria is the density functional theory (DFT). The DFT formulates the many-electron problem in terms of the electron density, rather than the many-body wavefunction. Thus, in principle, the central quantity is a function of just three rather than of 3N variables, a fact that renders calculations based on the DFT computationally tractable. The basic problem of the DFT is that the energy of a quantum many-body system can be expressed as a unique functional of its density. By minimizing the density functional over all densities that give rise to a particular number of electrons, one obtains the ground state density and the energy for a given system. Unfortunately, the explicit and unique form of this functional is not known. However, in the orbital-based formulation of the DFT by Kohn and Sham, reasonable approximations to the density functional have been developed. In the Kohn-Sham formulation, the energy is expressed in terms of a set of the n occupied single-particle orbitals and the N nuclear positions.

10.1.2.3 Acceleration Methods

One is the most common acceleration methods allowing to increase of the integration time step consists in constraining the fastest degrees of freedom by augmenting the equations of motion via the Lagrange multipliers [2733]. The iterative procedure of this approach is formulated in the following way

$$V^{n+\frac{1}{2}} = V^n - \mathbf{M}^{-1}\frac{\Delta t}{2}\nabla E(X^n) + \mathbf{G}(X^n)^T\lambda^n, \tag{10.17}$$

$$X^{n+1} = X^n + \Delta t V^{n+\frac{1}{2}}, \tag{10.18}$$

$$G(X^{n+1}) = 0, \tag{10.19}$$

$$V^{n+1} = V^{n+\frac{1}{2}} - \mathbf{M}^{-1}\frac{\Delta t}{2}\nabla E(X^{n+1}). \tag{10.20}$$

where the vectors λ^n are determined from the equation $G(X^{n+1})$. The matrix **G** is $m \times m$ Jacobian of the vector g which specifies the m constraints and λ is the vector of the m constraint multiplies.

One of the critical components that determines the overall computational cost is the treatment of the long-range forces. While the number of the bonded forces—presented in the energy expression by the terms involving the bond lengths, the bond angles, the dihedral angles—grow linearly with the number of atoms, the cost of the computing the non-bound—van der Waals and electrostatic—forces depends quadratically on the size of the system. The common cut-off procedures that consist in specifying the distance over which the interactions are neglected could lead to the errors that can not be estimated a priori and are used now for the van der Waals forces only while the electrostatic interactions are treated using such techniques as the multipoles or the Ewald summations [2733].

10.1.2.4 Parallel Computations

The computer-intensive character of the MD forces one to exploit the simplified and hybrid models of the macromolecules and/or to implement the simulation codes on the multiprocessor systems. The efficiency of the parallel computations is characterized by the speed-up $S_n = T_1/T_n$ or an efficiency $E_n = T_1/nT_n$ (T_n is the execution time on the system consisting of n processors) that can be written as a product of the three factors [2431]: the parallel efficiency, the numerical efficiency and the uniformity of the load balance. The first one is the ratio of the execution time to the total time that includes the computations, the data exchange and the waiting time for the messages $E_n^{par} = T_{comp}/(T_{comp} + T_{comm})$ (the time needed to send/receive messaged could be reduced using the asynchronous regime—overlapping the computations and the communications). The numerical efficiency reflects the increase of the volume of the computations (for example, the number of iterations) due to the parallelization of the algorithm and varies broadly depending on the problem and the algorithm type. The load balance could be considered statically or dynamically. Ideally,

- The number of the cells or the particles for a given processing unit should be proportional to its productivity;
- The volume and the frequency of the interprocessor data exchanges should be minimal.

One could additionally require the number of the processor pairs in need of the data exchange to be minimal. Evidently, some trade off should be find: the minimal volume of communications is reached for a single processor. Thus, one should solve a minimization problem for $T = T_{comp} + T_{comm}$ subjected to certain restrictions. For a wide class of problems $T_{comp} = \max_q(N_q T_q)$, where qth processor is responsible for N_q cells or particles, T_q is the time needed to perform one step or iteration. For the most common high performance computational systems—the clusters of workstation connected by the Ethernet or the Myrinet—it is possible to write

$$T_{comm} = \sum_{pq} (C_{pq}/b + \delta(C_{pq})L) \tag{10.21}$$

where C_{pq} is the communication matrix, b is the effective bandgap of the net, L is its latency [1832].

The MD with the near range potentials is well suited for the parallelization, about the methods for the effective implementation of the parallel computations for the long range electrostatic interactions see, for example, [1466, 2733, 3118]. Thus, the efficiency of the MD software code NAMD (Not Another Molecular Dynamics) developed by the Theoretical and Computational Biophysics group in the University of Illinois at Urbana Champaign [2456] for the biophysical problems is as high as 85 and 70% for the computer systems having 1024 [2699] and 2250 processors [3118], respectively. The implementation of the MD approach using the distributed GRID resources is more complex, first of all it is hard to provide the load balance due to the dynamic character and the heterogeneity of the computational environment [2438]. There is the computational system (MDM [2173]) that have the hardware architecture specially designed for the large scale MD computations.

10.1.3 Water Models

The water molecule seems to be very simple and easy to model—there are only two hydrogen atoms and one oxygen with the overall charges of $1e$ for the hydrogen atoms and $-2e$ for the oxygen. However, to succeed in this task one should decide where these charges are located, how far is the hydrogen atoms from oxygen, whether this charges are distributed in the spherical or some other arrangement, whether it is a rigid or a flexible molecule, how it interacts with the neighbours, whether the quadrupole moment of the molecule should be reproduced and so on. The many body effects are significant, but the most important are the pair interaction potentials since as the analysis shows these interactions comprise at least about 3/4 of the total energy for the water trimer, tetramer, pentamer ring-like structures and the heptamer chains [1111].

10.1.3.1 Interaction Potentials

A significant number of the models of the water molecules was suggested (see, for example, [410, 950, 1495, 1835, 2716, 3059, 3171, 3443] and references therein) differing in the structure (a number of the charges, their position, the character of the bond, account for the water molecule polarizability (the dipole moment of the water molecule is about $1.85D$ [3365] and the quadrupole moment is also significant) etc.), in the accuracy of the reproduction of the experimental data and the suitability for the high performance computing; the recent experiments on the X-ray Compton scattering [2320] indicate that it is necessary to include in the model the correlation of the length of the O–H bond with the structure of the hydrogen bonds net. The potential parameters frequently are fitted to reproduce the water phase diagram as well as the raw experimental data such as the spectroscopic data (including the vibration-rotation-tunneling data) obtained for the small water clusters [1111].

The main model parameters are the number and value of the charges, their geometrical arrangement, type of the interaction potentials. The first models were introduced in early 30s by J. D. Bernal and R. H. Fowler [268] such as simple point charge (SPC) and transferrable intermolecular potentials (TIPS). The most common ones are probably SPC and TIP3P, both having three point charges but differing in their locations (for example, the equilibrium HOH angle is 109.47° in SPC and 104.52° in TIP3P).

In the SPC/E model the water molecule is represented as a set of point charges with Lennard-Jones interactions between the oxygen atoms. The intermolecular potential between two water molecules is expressed as

$$U = 4\varepsilon_{\mathrm{OO}}\left[\left(\frac{\sigma_{\mathrm{OO}}}{r_{\mathrm{OO}}}\right)^{12} - \left(\frac{\sigma_{\mathrm{OO}}}{r_{\mathrm{OO}}}\right)^{6}\right] + \frac{1}{4\pi\varepsilon_0}\sum_{i=1}^{3}\sum_{j=1}^{3}\frac{q_i q_j}{r_{ij}}, \tag{10.22}$$

where σ_{OO} and $\varepsilon_{\mathrm{OO}}$ are the parameters of the Lennard-Jones potential, r_{ij} is the distance between the charge site i and j on two different molecules, ε_0 is the permittivity of the free space.

The comparison of the models on the basis of the reproduction of the ice melting temperature and other parameters of the water phase diagram could found in [950, 3171], on the basis of applicability to the dynamics of proteins and amino acids in the water solutions—in [1835].

As was already noted, the classical presentation of the water molecule by a tetrahedral disposition of the two negatively and the two positively charged regions is an oversimplification and, being used as a basis for the interaction potentials, will overemphasize the angular dependence of the intermolecular potential. The more adequate potential functions represent the lone-pair region by a single negative point charge located close to or even at the molecular (oxygen) centre [964].

The TIP5P pair potential is formulated under the assumption that the water molecule is a rigid non-polarizable tetrahedral structure presented by five point charges [1730]. Two positive charges ($q_H = 0.241e$) are located on the hydrogen atoms at the distance 0.09572 nm from the oxygen atom forming the HOH angle equal to 104.52°. The negative charges ($q_e = -q_H$) represent the lone pair of the electrons in the plane perpendicular to the HOH plane; the angle e^-Oe^- is equal to 109.47°. The fifth interaction site introduced to prevent the overlapping of the water molecules is located on the oxygen atom and interacts via the Lennard-Jones potential with the parameters $\sigma_{\mathrm{OO}} = 0.312$ nm and $\varepsilon_{\mathrm{OO}} = 0.6694$ kJ/mol.

The TIP5P is known to accurately reproduce some water anomalies such as the density anomaly at $T = 277\,\text{K}$ and the atmospheric pressure and some other peculiarities [1730].

The water potentials could be very complex, for example, the so called ASP-W potential has 72 parameters corresponding to the electrostatic interactions, the two-body exchange-repulsion, the two-body dispersion, the many-body induction [1111].

10.1.4 Bulk Water

Z. Yan et al. [3395] performed a detailed analysis of the water anomalies of the different nature—structural, thermodynamic, dynamic. The authors used two order parameters:

- translational

$$Q_t \equiv \int_0^{c_s} |g(s) - 1| ds, \tag{10.23}$$

  where $s \equiv r\varrho_n^{1/3}$ is scaled by the average distance between the particles, $g(s)$ is the pair correlation function, s_c is the cut-off parameter (it was set to a half of the computational domain size in the paper) and
- orientational

$$Q_{li} \equiv \left[\frac{4\pi}{2l+1} \sum_{m=-l}^{m=l} |\bar{Y}_{lm}|^2 \right]^{1/2}. \tag{10.24}$$

A definition of the last order parameter is based on the characterization of the particle position relative to the twelve closest neighbours by the angles ϑ, φ that are used to calculate the corresponding spherical harmonics $Y_{lm}(\vartheta, \varphi)$ and the average $\bar{Y}_{lm}$. When $l = 6$ the parameter Q_{6i} could serve as a measure of the orientational order in the system of particles. The necessary condition for the appearance of the anomalous structural properties is proved to be the presence of two linear scales in the interaction potential [3395].

Detailed study of the first- and second-neighbour shells defined, respectively, by four and twelve neighbours, reported in the subsequent paper by Z. Yan et al. [3396] showed that the anomalous decrease of the translational order upon the compression occurs mainly in the second neighbour shell. The authors analysed the oxygen-oxygen pair correlation function $g(r)$ and divide it into three regions to study the average effect of the density on the different shells. The average number of the neighbours of the central molecule at a distance r was computed as

$$N(r) \equiv 4\pi n \int_0^r (r')^2 g(r')\mathrm{d}r' \tag{10.25}$$

The above mentioned regions of water are defined as

1. first shell: $0 < r \leq r_1$
2. second shell: $r_1 < r \leq r_2$
3. rest water: $r > r_2$

where r_1 and r_2 depending on the temperature and the pressure are implicitly given by the relations $N(r_1) = 4$ and $N(r_2) = 12$.

To characterize the tetrahedrality of the first shell, the following orientational order parameter was used

$$q_i \equiv 1 - \frac{3}{8}\sum_{j=1}^{3}\sum_{k=j+1}^{4}\left[\cos\vartheta_{jik} + \frac{1}{3}\right]^2 \tag{10.26}$$

as well as its average value

$$q \equiv \frac{1}{N}\sum_{i=1}^{N} q_i. \tag{10.27}$$

Here ϑ_{jik} is the angle formed by the central molecule i with the neighbours j and k. For the perfect tetrahedral order $q = 1$, for the uncorrelated system $q = 0$.

The introduced previously Q_{li} (10.24) was averaged for $l = 6$

$$Q_6 \equiv \frac{1}{N}\sum_{i=1}^{N} Q_{6i} \tag{10.28}$$

to quantify the orientational order of the system based on the molecules in the second shell (for the uncorrelated system $Q_6 = 1/\sqrt{12}$).

The authors performed MD simulations using 512 water molecules interacting via the five-site transferable interaction potential TIP5P and found that anomalies of water are related to the alteration of the arrangement of the water molecules in the second shell.

The nonequilibrium MD simulations were used by W. Evans et al. [882] to study the thermal conductivity of the pure water and its dependence on the various degree of the orientational and translational order. The degree of the order in the water was altered by the applied electrical field [882]. The authors used the planar heat source and heat sink within a rectangular periodic simulation box. A simple point charge flexible water molecule was used. The steady state temperature profile was established after 100 000 MD steps, after which the averaging was performed over about 300 000 MD steps. The strength of the uniform electric field was varied from 0 to 7 V/nm.

It was found that while the orientational ordering of the water dipole moments has a minor effect on the thermal conductivity, the translational ordering results in approximately threefold increase of the water thermal conductivity making it comparable with that of the crystalline ice.

10.1.4.1 Water Confined Between Surfaces

P. Kumar et al. [1730] have considered the structure and the behaviour of the pure water confined between the two hydrophobic plates for several temperatures from 220 to 300 K and several densities from 0.8 to 1.32 g/cm^3, in total over fifty state points. In the MD simulations $N = 512$ water molecules were used that interact via TIP5P potential in the computational domain $L_x \times L_y \times L_z$, where separation between the plates $L_z = 1.1$ nm corresponds to two or three layers of the water molecules.

The interactions between the water molecules and the smooth surfaces of the plates were modelled to mimic the water interaction with solid paraffin and described as [1730]

$$U(\Delta z) = 4\varepsilon_{\mathrm{Ow}} \left[\left(\frac{\sigma_{\mathrm{Ow}}}{\Delta z} \right)^9 - \left(\frac{\sigma_{\mathrm{Ow}}}{\Delta z} \right)^3 \right], \qquad (10.29)$$

where Δz is the distance from the oxygen atom of the water molecule to the wall and $\varepsilon_{\mathrm{Ow}} = 1.25$ kJ/mole, $\sigma_{\mathrm{Ow}} = 0.25$ nm are the parameters of the interaction potential.

The pressure along the transverse direction $P_\perp$ is determined via the total force $\mathbf{F}_{\mathrm{w}}$ projection on the normal to the wall

$$P_\perp = \frac{F_{\mathrm{w}}}{L_x L_y} = \frac{\left| \sum_{i=1}^{N} \mathbf{F}_{i,\mathrm{w}} \right|}{L_x L_y}. \qquad (10.30)$$

Here $\mathbf{F}_{i,\mathrm{w}}$ is the force of the oxygen atom of the ith molecule acting on the wall; it is assumed that hydrogen atoms do not interact with the wall.

Since the repulsive interactions with the wall prevents molecules from coming to close to the wall, the authors introduced the effective separation between the plates L'_z that is estimated as

$$L'_z = L_z - \frac{\sigma_{\mathrm{Ow}} + \sigma_{\mathrm{OO}}}{2} \approx 0.819\,\mathrm{nm} \qquad (10.31)$$

and is used to calculate the effective density $\varrho \equiv Nm/L_x L_y L'_z$.

For the comparison with the bulk water, Kumar et al. computed the lateral oxygen-oxygen radial distribution function (RDF)

$$g_{\parallel} = \frac{1}{\varrho^2 V} \sum_{i \neq j} \delta(r - r_{ij}) \left[H\left(|z_i - z_j| + \frac{\delta z}{2}\right) - H\left(|z_i - z_j| - \frac{\delta z}{2}\right)\right], \tag{10.32}$$

where V is the volume, r_{ij} is the distance between the ith and jth molecules parallel to the walls, z_i is the z coordinate of the oxygen atom of the ith molecule, $\delta(x)$ is the Dirac δ function. The Heaviside functions $H(x)$ are introduce to provide that the sum is calculated for the pairs of molecules located in the same layer of the thickness $\delta z = 0.1$ nm.

The computations showed that the RDF at the low temperature and the low density exhibits the distinctive maxima and minima and indicates that the liquid is highly structured, with the structure parallel to the walls similar to that of the bulk water. As the density increases, the liquid becomes less structured as is evident from the decreasing of the second and third peaks of $g_{\parallel}$.

The lateral oxygen-oxygen RDF $g_{\parallel}(r)$ (Eq. (10.32)) is used to calculate the lateral static structure factor $S_{\parallel}(q)$ which is defined as the Fourier transform of the former

$$S_{\parallel}(q) = \frac{1}{N} \sum_{i,j} \left\langle e^{i\mathbf{q}\cdot(\mathbf{r}_i - \mathbf{r}_j)} \right\rangle, \tag{10.33}$$

where the **q** vector is the inverse space vector in the xy plane and r is the projection of the position vector onto the xy plane.

The study of the diffusion of the water molecules in the different layers at the high density showed that the molecules in the layer adjacent to the wall move faster than the molecules in the middle layer.

P. Kumar et al. have found the effect of the confinement to be similar to the effect of the temperature elevation with the same temperature shift about 40 K for both the thermodynamic and dynamic properties. Such behaviour is qualitatively explained by the alterations in the hydrogen bonds network. Since the hydrogen bonds are not formed with the walls, the average number of the hydrogen bonds per molecule in the confined system is smaller than in the bulk water that is similar to having the water at the higher temperature.

J. Wang et al. [3243] have found the confined water behaviour to resemble not that of the bulk water under the elevated or reduced temperature, but under the higher pressure for the somewhat different case of two hydrophilic surfaces modelling (001) surface of the mineral brucite $Mg(OH)_2$ [3243]. The authors performed the MD simulation using the SPC interaction potential. The essential distinction of this work is the consideration of the surface which was not assumed rigid, but also was simulated using the MD approach. The relaxation of the atom positions were performed in two steps. First the position of all atoms in the brucite substrate were fixed and only the water subsystem was allowed to relax. Then all atoms—both in the liquid water and the solid brucite—were relaxed. To characterize the local structure of the hydrogen bonds network near the surface, the total number of the HBs per water molecule, the number of the HBs donated to and accepted from other water molecules, the number of the HBs donated to and accepted from the surface OH groups were calculated. The orientational order parameter (Eq. (10.26)) also was computed with an evident modification—the inclusion of the angles ϑ_{jik} that are formed by the HBs with the surface OH groups.

The authors found that both the HB donation to the surface oxygen atoms and the HB acceptance from the surface OH groups determine the structure of the first water layer that shows the two-dimensional hexagonal arrangement in the plane parallel to the surface and reflects the brucite structure. The oxygen and hydrogen atomic densities and the dipole orientations of the water molecules deviate from the corresponding parameters of bulk water up to the distance of about 15 Å from the surface ($\approx$ 5 molecular water layers). The evolution of the structure of the water layers could be described as a gradual decrease of the substrate-induced ordering. The translational ordering decays faster, than the orientational one [3243].

The systematic MD simulation of the effects of the pressure ($-0.15GPa \leq P \leq 0.2GPa$), the plates separation ($0.4\,\text{nm} \leq d \leq 1.6\,\text{nm}$) and the nature of the surface (hydrophobic or hydrophilic) on the structure and the phase state of the confined water was performed by Giovambattista et al. [1084]. The computations were conducted for 3375 water molecules, the temperature was fixed at 300 K using the Berendsen thermostat. The interaction of the water molecules was modelled using the extended simple-point-charge pair potential (SPC/E). The structure of both the hydrophobic and hydrophilic surfaces corresponded to four layers of SiO_2. The unit cell of SiO_2 was represented by a perfect tetrahedron. In the case of the hydrophobic surface silicon and the oxygen atoms interact with the water molecules via the Lennard-Jones potential. The hydrophilic surface corresponding to the fully hydroxylated silica was obtained by attaching a hydrogen atom to each surface oxygen atom. The silicon and oxygen atom positions were fixed (as in the case of the hydrophobic surface) but hydrogen atoms were allowed to move in the plane without violation of the bond length. The local tetrahedral order in the water was characterized by the average coordination number and the orientational order parameter (Eq. (10.26)). Computations showed that in the case of the hydrophobic plates water molecules are pushed towards the plates and the liquid order increases with the pressure while no effect was observed for the water orientation near the surface. No changes in the structure of the liquid with increasing the pressure were registered in the case of the hydrophilic surfaces.

As was already mentioned, the effect of the wall proximity on the crystallization temperature is known from observations on the soil freezing and the water behaviour in the pores of minerals. R. Zangi [3461] simulate using the MD method the water crystallization in the narrow (several diameters of the water molecule) gap for the case of the inert surface that does not form the hydrogen bond with the water molecule. The variation of the gap width at the room temperature results in the drastic (three-four orders of magnitude) variation of the diffusion coefficient that has been considered as a transition to the crystalline state named the "confinement induced freezing". Two phases of the liquid water were observed. The author stressed a synergetic action of the space confinement and the electric field.

10.1.4.2 Aqueous Solutions

In the continuum description of the ions in the solution, the free energy of the solvation of an ion of the radius R_i and the charge q is given by the Born equation

$$\Delta F = -\frac{q^2}{8\pi\varepsilon_0 R_i}\left(1 - \frac{1}{\varepsilon}\right), \tag{10.34}$$

where ε_0 is the permittivity of the free space and ε is the dielectric constant of the medium, while the entropy of the solvation is defined as [2577]

$$\Delta S = \frac{q^2}{8\pi\varepsilon\varepsilon_0 R_i}. \tag{10.35}$$

This theory thus predicts a parabolic dependence of the free energy, the entropy and the energy of solvation $E = \Delta F + T\Delta S$ on the charge for both the anions and the cations that is in contradiction to the experimental data. The continuum theory based on the hydrodynamic friction given by the Stokes law predicts decrease of the ion mobility, irreversible to the sign of the charge, with the ion radius. In the experiments, however, the reduction of the mobility for both the small and large ions is observed, with the maximum mobility occurring at the different radii for the positively and negatively charged ions. One of the proposed explanation of the low mobility of he small ions—the so called "solventberg" model—assumes that such ions (Li^+, for example) diffuse as a single entity with some solvated water molecules and thus its effective radius should be increased by the size of the water molecule (if it is assumed that solventberg includes the water molecules of the first hydration shell). This model fails, however, to explain the mobilities of other small ions such as Na^+, Cs^+ or Br^-—probably, due to the different "times of life" of the hydration shell.

Another suggested model exploit the so called dielectric friction accounting for the inertia of the polarization of the water molecules caused by the ion. However, this model is also symmetrical with respect to the sign of charge and thus unable to cope with differences in the behaviour of the anions and the cations. About other attempts to modify the continuum theory such as correction of the effective hydrodynamics radius and the molecular theories based on the Langevin equation see the paper by J. C. Rasaiah and R. M. Lynden-Bell [2577]. These authors performed the MD simulations of the ions in the aqueous solution at the infinite dilution that reproduce the main experimental trends.

The SPC/E model of the water was used. The potential of the interaction between the ion and the water molecule was written similar to the water model (10.22) (there is a misprint in (3.2) in the paper [2577]—a single, not a double, sum should be present):

$$U_{\mathrm{iw}} = 4\varepsilon_{\mathrm{io}} \left[\left(\frac{\sigma_{\mathrm{io}}}{r_{\mathrm{io}}} \right)^{12} - \left(\frac{\sigma_{\mathrm{io}}}{r_{\mathrm{io}}} \right)^{6} \right] + \frac{1}{4\pi\varepsilon_0} \sum_{i=1}^{3} \frac{q_i q_j}{r_{ij}}. \tag{10.36}$$

The computations were performed for a single ion and 215 water molecules at 298 K and 512 molecules at 683 K. The diffusion coefficients were determined by the asymptotic slope of the mean square displacement

$$D_i = \frac{1}{6} \lim_{t\to\infty} \frac{\mathrm{d}|\mathbf{r}_i(t) - \mathbf{r}_i(0)|^2}{\mathrm{d}t} \tag{10.37}$$

and from the integral of the velocity autocorrelation function

$$D_i = \frac{1}{3} \int_0^{\infty} \langle \mathbf{v}_i(t) \cdot \mathbf{v}_i(0) \rangle \, \mathrm{d}t. \tag{10.38}$$

J. C. Rasaiah and R. M. Lynden-Bell [2577] have found that

- with the possible exception of the smallest solute Li, the Stokes law is valid for the movement of the neutral solutes even thought their size is comparable with that of the solvent;
- the charge asymmetry originates from the differences in the orientation of the water molecules in the first solvation shell: the oxygen atom of the water molecule in the hydration shell of the cation is located closer to the ion than the hydrogen atoms, while the reverse configuration is observed for the anion.

The authors also determined the average hydration number for the different ions as a volume integral of the ion-oxygen pair distribution function g_{io}

$$N_h = \varrho_{\mathrm{w}} \int_0^{R_h} g_{\mathrm{io}} 4\pi r^2 \mathrm{d}r \tag{10.39}$$

and the residence times of the water molecules in the first hydration shell that were found to decrease with size for the anions and increase for the neutral solutes. The smaller diffusion coefficient of the neutral solute compared to that of the charged one is attributed to the formation the solvent cage around the hydrophobic solute.

10.1.5 *Ice Crystallization*

10.1.5.1 Pure Water

The simulation of the homogeneous nucleation of the pure water is rather difficult: the first MD computations of freezing water from scratch were reported as late as 2002 [2064]. These extensive computations required the integration for a period about 300 nsec with the time step 1 fmsec. The problem is the complexity of the potential energy surface (PES) in the 6N-dimensional space (N is the number of particles in MD computations) that has a large number of local minima, i.e. numerous possible configurations of the hydrogen bonds net [2182], and the need to simulate the rare event of the spontaneous formation of an ice nucleus.

One of the problems in crystallization simulation is the choice of the criterium to distinguish water molecules that belong to the liquid and solid phases. Since the bulk water has an average number of hydrogen bonds per molecule close to four, the instantaneous number of hydrogen bonds is not a good parameter to discriminate between ice and water. M. A. Carignano et al. [444] suggested to monitor the time average of the number of bonds on each individual molecule. The period of time over which the average is taken should be short for the solid phase, but large enough for the liquid phase to allow molecules to diffuse. By numerical experiments the authors found the period of 20 ps to be adequate. Hydrogen bonds in computations are defined by two conditions involving the distance cut-off and the angular cut-off related to the angle formed by the hydrogen, the donor oxygen and the acceptor oxygen. The typical values of these cut-off parameters are $r_{HB} = 0.35$ nm and $\vartheta_{HB} = 30°$ [444].

W.J. Briels and H.L. Tepper [372] use another approach. They inquire whether the particle surroundings have octahedral symmetry and construct a "recognition function" (the order parameter)

$$\Psi = \sqrt{\frac{\langle X^2 \rangle - \langle X \rangle^2}{\langle X^2 \rangle}}, \tag{10.40}$$

where

$$X(\mathbf{r}) = \frac{x^4}{r^4} + \frac{y^4}{r^4} + \frac{z^4}{r^4} - \frac{3}{5}, \tag{10.41}$$

with the averages taken over all nearest neighbours residing within a given cutoff radius and coordinates are in respect to the centre of mass of the neighbours, not to the central particle; for "liquid" particles $\Psi > 0.5$. Since the crystalline structure will partly persist in the liquid near the interface, the authors expect that their discriminator will slightly overestimate the number of "solid" particles.

Topological properties of the PES are related to the thermodynamic and dynamic properties of the supercooled water [2750]. The authors of the cited paper studied this problem using SPE/E interaction potential in MD simulations of a system of $N = 216$ water molecules for six different densities and six temperatures in the range from 210 to 300 K. The PES can be partitioned into a numer of local *basins* which are uniquely defined by the set of points connected to a local minimum via a steepest-descent paths; the configurations corresponding to the local minima are called *inherent structures* (IS) and their energies are denoted by E_{IS}. The free energy of the system F at the temperature T can be expressed as [2750]

$$F(T) = -TS_{conf}(e_{IS}(T)) + f_{basin}(e_{IS}(T), T), \tag{10.42}$$

where e_{IS} is the average value of the IS energy, $S_{conf}(e_{IS}(T)) = log(\Omega(e_{IS}(T)))$ is the entropic contribution arising from the number $\Omega(e_{IS})$ of basins of depth e_{IS}. The separation of the free energy into two parts reflects the separation of time scales in supercooled liquids: two-scale relaxation takes place with intrabasins dynamics being much faster than interbasins dynamics [2750]. Slowing-down of the dynamics approaching the glass transition temperature can be attributed both to the reduction of the connectivity between local mimima and to the decrease in the number of available local minima.

The results of MD computations are highly sensitive to the choice of the interaction potential [2150] and could underestimate as well as overestimate the degree of the water ordering [2725]. A set of the order parameters serves as a measure of the latter. The order parameters based on the expansion of the pair correlation function in the spherical coordinates describe the geometrical properties of the bonds of neighbour molecules; an additional three-particle order parameter is related to the tetrahedral structure. The nucleation could be represented by the minimal free energy trajectory in the space of the order parameters [2557]. Two- and three-particle order parameters were used by Radhakrishnan and Trout for calculation of the nucleation energy barrier in the different environments: homogeneous liquid; water in the hydrophobic pore; water in the constant electric field; mixture of water and CO_2; water with antifreeze protein [2556].

Simulations of the melting, of the crystallization in the spatially restricted systems or of the crystallization of the water that is in the contact with ice [3165] are less computer-intensive. M. A. Carignano et al. [444] considered growth of ice from the supercooled water with the explicit presence of ice-vapour and water-vapour interfaces using the six-cite water model of H. Nada and J. P. M. van der Eerden [2246]. These sites are the oxygen and hydrogen atoms, the point charge located at the bisector of the angle $\angle HOH$ and two point charges on the plane perpendicular to HOH introduced to emulate the water lone pairs. The authors' computations show that the growth rate at the prismatic plane of the ice is approximately twice as large as the growth rate at the basal plane. The rough ice-water interface was registered at the prismatic plane, while smooth interface due to the layer-by-layer growth mechanism was observed at the basal plane. Contrary to the experimental data, the computations do not reveal the growth rate dependence on the supercooling.

A free surface of a solid has a quasi-liquid layer. M. A. Carignano et al. [444] proposed a theoretical definition of the thickness of this layer to be equal to four times the standard deviation calculated from the liquid density profile $\varphi_L(z)$:

$$\delta_I = 4\left[\frac{\int \varphi_L(z)(z-\bar{z})^2 dz}{\int \varphi_L(z)dz}\right]^{1/2}, \tag{10.43}$$

where

$$\bar{z} = \frac{\int \varphi_L(z)z dz}{\int \varphi_L(z)dz}. \tag{10.44}$$

The thickness of this layer could be measured experimentally [444]. The authors of the cited paper also studied the dynamics of the quasi-liquid layer. The decrease of the initially wide layer on the ice-vacuum surface is found to be linear with the time and rather fast, reaching the equilibrium value—the same as that of the layer that melted from the solid side—in about 200 ps.

I. M. Swishchev and P. G. Kusalik [2992] performed MD simulations of the crystallization of the liquid water subjected to a homogeneous static electric field. The authors used TIP4P model of water; the field strengths considered were 0.1 V/Å and 0.5 V/Åwhich are smaller than the average internal electric fields within the condensed phases of water. The crystal of cubic ice I_c was grown that in the absence of the electric field melts at 250 K. The growth of the cubic ice in the electric field instead of the hexagonal I_h is explained by the authors by the better parallel arrangement of the molecular dipoles in the diamond-like packing of I_c in comparison with the isomorphous hexagonal packing of I_h. The authors conclude that local electric fields occurring at the surfaces of various materials could play a crucial role in promoting the crystallization of the liquid water.

C. Muguruma et al. [2205] applied the multicanonical MC method to the bulk water system containing 64 molecules as a basic unit cell. The determination of the weight factors required about 10^5 MC sweeps and the production run about $5 \cdot 10^6$ sweeps. The authors obtained the water transition into the amorpous ice at 190 K and noted that for simulation of the formation of the crystalline ice the more accurate weight factors in the low energy region should be determined.

10.1.5.2 Aqueous Solutions

The ice nucleation in the liquid water and the aqueous solution is a common natural process occurring annually in mid and high latitudes. In spite of the development of the sophisticated experimental techniques enabling the high speed visual and infrared imaging of the nucleation and growth of ice in the small droplets such as the free-fall tubes, the emulsified droplets, the jet expansions, the droplets hanging on the wire or the acoustically suspended droplets free of the wall or the substrate contacts [211, 1324], the most detailed information on the process, especially its early stages, is provided by the numerical simulation.

L. Vrbka and P. Jungwirth [3214, 3215] simulate the crystallization of the aqueous solution of NaCl that is in the contact with the cubic ice I_c that occupies one fourth of the computational domain comprising 720 water molecules [3214, 3215]; the non-polarizable potential SPC/E was used. The system geometry was a slab with a bulk region between two air/water interfaces. The size of the system was sufficient for the study of the spontaneous formation of the ice nucleus.

The time needed for the nucleus formation in the salty water turned out to be over an order of magnitude larger than in the case of pure water. This result indicate that the effect of the salt presence is not only the thermodynamic one (the freezing point depression), but also kinetic.

The system freezes as pure ice almost completely, the ions being expelled into a small volume of the unfrozen brine, but sometimes an ion (chloride anion) gets trapped inside the ice crystal. The instability of the crystallization front results in the formation of the voids containing the concentrated solution—the brine. The front propagation is related to the local fluctuations of the salt concentration: the local reduction of the salt density close to the crystallization front is followed by building a new ice layer. Note that the problem considered is of a considerable interest for the geophysics: an increase of the solution density under the surface ice significantly influences the ocean circulation.

10.1.6 Transport in Small Pores

The MD simulations are used to study of the water and the small molecules in the pores of the different nature. The most popular objects for the investigation are the carbon nanotubes and the channel formed in the cellular membranes by the integral proteins such as the aquaporin and the gramicidin. The CTNs are of interest by themselves for the various technological applications and also as a potential models of the biological channels. The discussion of the simulation of the metabolically pumped ion channels and their gating is beyond the scope of the present book.

The integral proteins from the aquaporin family form in the cellular membrane the tetramers with the functionally independent channel in each monomer that is permeable for the water but not for the charged molecules and ions while the fifth channel in the centre of the tetramer is permeable for both the water and the charged particles [2713].

The experimentally osmotically driven fluxes through the biological channels are measured by mounting these channels into the liposomes and monitoring the volume change of the latter in response the concentration of the environment media. A major experimental challenge is the need to determine exactly the protein-to-lipid composition, i.e. the density of the water channels to be able to extract the single-channel properties from the measurements of the membrane permeability [1452]. Evidently, these difficulties are not encountered in the MD simulation study.

The properties of the small pore are characterized by the two parameters: the osmotic permeability p_f and the diffusion permeability p_d. The determination of the latter from the MD simulations is trivial. It could be found by counting the water molecules passing from one side of the channel to the other as [2029]

$$p_d = \frac{V_w}{N_A} q_0 = v_w, \tag{10.45}$$

where V_w is the molar volume of the water, N_A is Avogadro number, q_0 is the unidirectional permeation rate, v_w is the volume of a single water molecule.

In the dilute solutions, the total water flux j_w through a single channel is linear proportional to the solute concentration difference Δc_S: $j_w = p_f \Delta c_S$. The second permeability parameter, p_d, corresponds to the case of $\Delta c_s = 0$.

To determine p_f from the MD computations, one should produce the difference of the osmotic or the hydrostatic pressures on the two sides of the membrane containing the pore. F. Zhu et al. [3531] proposed to use the periodic system in which the unit cell is replicated in all three dimensions and thus the water layers and membranes alternate in the direction z normal to the membrane plane. The authors define three zones (I, II, III) for each water layer with the zone III being in the centre of the layer. A constant force f acting on all molecules in the zone III along z-direction is added to generate a pressure gradient in this region thus providing the pressure difference between the zones I and II: $\Delta P = P_I - P_{II} = nf/A$, where N is the number of water molecules in the zone III and A is the area of the membrane. Therefore, the net water flux across the membrane is induced and p_f could be calculated from j_w and ΔP.

The computations performed by F. Zhu et al. using the aquaporin molecule in the lipid bilayer composed from the palmitoyl-oleyoyl-phosphatidyl-ethanolamine (POPE) molecules (the complete system, including water, contains 81 065 atoms) gave the p_f/p_d ratio about 12.

This approach was used later by M. O. Jensen and O. G. Mouritsen [1452] to study the single-channel water permeability of the *Escherichia coli* aquaporin AqpZ and the aquaglyceroporin GlpF and by A. B. Mamonov et al. [2029] to compute the water and the deuterium oxide permeabilities through the channel formed by the aquaporin [2029]. The authors used the protein structure from the Protein Data Bank (http://www.rcsb.org) and built the tetramer that was preequilibrated in 177 POPE molecules and solvated in 15 079 water molecules; the TIP3P interaction potential was used for the water [2029]. Four different values of the force were used in the method of Zhu et al. to produce four pressure differences. The computations were performed using the above mentioned NAMD package.

F. Zhu, E. Tajkhorst and K. Schulten suggested for the determination of the permeabilities in the channels where the so called water file—the chain of the water molecules with the highly correlated movement along the channel—exists to apply the continuous-time random-walk model developed by A. Berezhkovskii and G. Hummer [254] for the water transport through the carbon nanotube. The model is developed under the assumption that the channel is always filled by N water molecules that move together in hops with the left and right hopping rates k_l and k_r being equal k_0 in the equilibrium.

The complete permeation of the single water molecule from the one side of the channel to the other requires at least $N+1$ hops, thus it is expected that the number of the unidirectional permeation events per unit time will be given by [3531] $q_0 = k_0/(N+1)$, then the diffusion permeability $p_d = v_w k_0/(N+1)$.

In the measurements of p_f the net water flux is induced by the different solute concentrations on the two sides of the membrane and thus instead of $k_l = k_r$ one should have

$$\frac{k_r}{k_l} = \exp\left(\frac{-\Delta\mu}{kT}\right). \tag{10.46}$$

Since under the physiological conditions the difference in the chemical potential is small compared with kT, it is reasonable to use the linear approximation and write

$$k_r = k_0\left(1 + \alpha\frac{-\Delta\mu}{kT}\right), \quad k_l = k_0\left(1 + \beta\frac{-\Delta\mu}{kT}\right), \tag{10.47}$$

where $\alpha = -\beta$ for the symmetrical channel. Then the net water flux is

$$j_w = \frac{1}{N_A}(k_r - k_l) = \frac{k_0(\alpha - \beta}{N_A}\frac{-\Delta\mu}{kT}. \tag{10.48}$$

M. Hashido et al. [1256] argue that a scalar permeability could not describe correctly the water transport through the channels formed by the aquaporin and the similar integral membrane proteins and the osmotic permeability matrix should be introduced to account for the contributions of all local regions of the channel on the water transport. The diagonal elements of this matrix named the p_f- matrix are equivalent to the local permeability at each region of the channel, and the off-diagonal elements represent the correlated motions of the water molecules in the different regions. Averaging both the diagonal and off-diagonal elements recovers the scalar permeability for the entire channel. The authors have performed the MD computations to construct the p_f- matrix for five members from the aquaporin family.

The osmotic transport in the cylindrical pore with the neutral wall (no interactions with the molecules) was investigated by K. S. Kim et al. [1626] and the water transfer through the lipid bilayer—by H. J. C. Berendsen and S. J. Marrink [252].

It is stressed in the latter paper that a characteristic time of the water molecule transport is rather high (about 100 psec) thus the MD should be used not for the direct simulation of the transport, but rather for the calculation (using the MD distribution function of water molecules) of the coefficients needed to close the macroscopic model of the water transfer. The authors considered the water transport through the membrane consisting of dipalmitoyl phosphatidycholine (DPPC). The membrane was assumed to be in the liquid crystalline state. The computations were performed for a rectangular periodic box at the constant pressure and the constant temperature. Never a cluster of two or more molecules inside the membrane was observed. Since the quantitative study of the water permeation process in an equilibrium simulation is impossible, the authors considered it as a diffusion in an external field described by the free energy profile, or potential of the mean force. The average velocity u_i of the ith species (here—the water) and hence its flux J_i is proportional to the gradient of the thermodynamic potential

$$u_i = -\frac{1}{\zeta_i}\nabla\mu_i, \qquad J_i = -\frac{c_i}{\zeta_i}\nabla\mu_i, \tag{10.49}$$

where the friction coefficient ζ_i is related to the diffusion coefficient $\zeta_i = R_g T/D_i$. In an ideal solution $\mu = \mu_0 + R_g T \ln c$, so the above equation is just the Fick's law $J = -D\nabla c$. In the steady state the flux is constant in every point of the membrane, thus the states on the two membrane sides $z = z_1$ and $z = z_2$ are related as

$$\Delta\mu_i = \mu_i(z_2) - \mu_i(z_1) = -J_i R_g T \int\limits_{z_1}^{z_2} \frac{dz}{c_i(z)D_i(z)}. \tag{10.50}$$

The permeation resistance (that is directly related to the permeability coefficient measured experimentally) is defined as

$$R_i^p = c_i \int\limits_{z_1}^{z_2} \frac{dz}{c_i^{\text{equil}}(z)D_i(z)}, \tag{10.51}$$

where the equilibrium water distribution $c_i^{\text{equil}}(z)$ is determined from the MD computations.

K. S. Kim et al. [1626] performed the comparison of the permeability and the diffusive drag coefficient obtained from the MD computations with the value that follow from the continuum description provided by the classical fluid dynamics, introducing two coefficients

$$G_w = \frac{p_f}{p_{f(\text{Poiseuille})}} \quad \text{and} \quad F_w = \frac{p_d}{p_{d(\text{Fick})}} \tag{10.52}$$

where the classical coefficients are expressed as

$$p_{f(\text{Poiseuille})} = \frac{\pi r_{\text{pore}}^4}{8\eta L_{\text{pore}}} \tag{10.53}$$

and

$$p_{d(\text{Fick})} = \frac{\pi r_p^2 D}{L_{\text{pore}}} \tag{10.54}$$

where the viscosity η obtained as the free viscosity η_0 corrected by the so called wall factor in the following way

$$\eta = \eta \frac{r_{\text{pore}}}{r_{\text{pore}} + k\bar{x}}. \quad (10.55)$$

Here $\bar{x}$ is the mean free path of the particle and k is the slip factor varying from 1 (no slip) up to 8 [485]. The results of the computations showed that G_w could be as low as about 0.3 and F_w smaller than 0.1.

The MD study by A. Kalra et al. [1524] of the water transport through the carbon nanotube [1524] using the hard-sphere particles to model the water molecules confirmed the existence of the continuous single file of the water (a hydrogen-bonded "water wire") and explained the paradoxical from the point of the classical fluid mechanics weak dependence of the overall conductance on the channel length. The single water file behaves as the rod that slips through the pore faster than is according to the unrestricted diffusion. The absence of the length dependence is follows from the dominance of the entry and the exit effects over the movement trough the pore itself. However, the carbon nanotube hardly could be considered as a good model of the biological channels due to the much more complex structure of the latter that could contain cavities and the extremely non-uniform distribution of the charged protein residues along the channel [1626].

The mechanisms of the proton and of the hydroxide migration along the linear water chain in the channel of the aquaglyceroporin GlpF from *Escherichia coli* was studied by M.O. Jensen et al. [1453] using the hybrid quantum-classical simulations (QM/MD) that combines the first-principle ab initio MD method of R. Car and M. Parinello (CPMD) based on the density functional theory and the classical MD method based on the force-field potentials. The system was partitioned into the QM and the MM fragments that were treated by the corresponding methods. For the QM regions interactions with the distant MM atoms were processed using a multipole expansion. The authors found that the single water file containing eight molecules is unable to support the Grotthuss-type mechanism of the proton transport.

10.1.7 Protein Denaturation

The use of the lattice models of the macromolecules [1869] reduces requirements to the computer resources due to the radical reduction of the number of the possible conformations of the macromolecule. However, these model introduce a priori the space anisotropy [2734]; thus lately the most attention is paid to the "off-lattice" models. Both the water molecules and the parts of the macromolecule are represented by the spheres of the different diameter; the interaction between the fragments i and j of the macromolecule that are not neighbours is described by the short range potential. Frequently, the cutoff Lennard-Jones potential

$$V_{ij}(r) = \begin{cases} \varepsilon\left[(\sigma_{ij}/r)^{12} - 2(\sigma_{ij}/r)^6 + v_{ij}\right], & r \leq R_{ij} \\ 0, & r > R_{ij} \end{cases} \quad (10.56)$$

is used. Here a shift v_{ij} provides the continuity of the potential at $r = R_{ij}$, while the cutoff radius depends on the properties of the elements i and j. To monitor the particles' position, the algorithm of "random lattices" could be used [2199].

The main approaches to the construction of the hybrid models are

- A representation of the space fragment of the macromolecule (for example, the phosphate or the methyl group) as a single particle or a pair of particles with some effective properties.
- Combining the discrete (atomistic) description of the macromolecule with continuum description of the solvent (water in our case) [1850]. The review of such "implicit" treatment of the water focused on the simulation of the structure and function of the biological membranes, including continuum and particle-based—tethered, non-tethered, non-tethered solvated, non-tethered implicit solvent—models has been published by G. Brannigan et al. [357].

In the HP model of the proteins [756] (as well as in its off-lattice generalization—AB model [504, 2942]) the amino acid residues are reduced to the point entities that could be either polar (P, represented by the charge or dipole) or non-polar (H) [426]. The hydrophobicity is described as the intention to reduce the area of the surface in the contact with water and is related to the reordering of the neighbour molecules into the energetically more attractive structure similar to the ice. This effect could be described formally as an attractive interaction of the non-polar amino acid residues. A hydrophobic molecule in the solution displays the water molecules that at low temperatures form an ice-like cage around the hydrophobic molecule, producing a structure that is energetically more favourable than the bulk water. The system tries to minimize the free energy by maximizing the number of the cages—exposing as many hydrophobic molecules as possible. This is the cold denaturation of the proteins [426]. At the higher temperatures the water molecules could not build the cages; the disordered water molecules around the hydrophobic amino acid residues of the protein are energetically less favourable than the bulk water; thus the proteins try to hide their hydrophobic parts inside their core. P. De Los Rios and G. Caldarelli [426] model the protein as a self-avoiding walk (SAW) on a lattice. A variable with q states (labelled from 0 to $q - 1$) is attributed to every lattice site non-occupied by the polymer. The state $q = 0$ is associated to the cage configuration which is energetically favourable when the water molecule is in the contact with the polymer and the rest states $q = 1, \ldots, q - 1$ are associated to disordered, unfavourable, configurations. The Hamiltonian of the system is written as

$$H = \sum_{i=1}^{N} \sum (-J\delta_{s_j,0} + K(1 - \delta_{s_j,0})) \quad (10.57)$$

Here the first sum run over N monomers of the polymer while the second one—over the water occupied nearest neigbours of each monomer, J and K are positive constants representing the energy of the cage and of the disordered configurations; there are no monomer-monomer interactions.

The partition function of the system is written as

$$Z_N = \sum_C Z_N(C), \tag{10.58}$$

where $Z_N(C)$ is the partition function associated to a single configuration C. Note that the maximum number of the water sites in the contact with the polymer is $M = 2(d-1)N + 2$ for the hypercube lattice in d-dimensional space; the real number of contacts is, of course, smaller than M. A particular water site could be counted more than once. The configuration partition function is expressed as

$$Z_N = q^{n_0(C)} \prod_{l=1}^{z} Y_l(C) \tag{10.59}$$

where

$$Y_l(C) = \left((q-1)e^{-\beta l K} + e^{-\beta l J}\right)_l^n (C) \tag{10.60}$$

and $n_l(C)$ is the number of water sites with l polymer contacts, z is the coordination number of the lattice, $\beta = 1/k_B T$.

G. Caldarelli and P. De Los Rios [426] analyzed the thermodynamic behaviour of the system on the two-dimensional lattice $d = 2$ using the exact enumeration technique for polymers of length up to $N = 25$. The so called Manhattan lattice was used (it is a 2D lattice on which rows (columns) are alternatively left/right (up/down) oriented). The authors state that simulations using the Manhatten lattices give results similar to those obtained on the regular Euclidian lattices. Only the bimodal case $q = 2$ was considered in [426].

The authors computed the energy of the monomer-monomer interaction ε using the partition functions for two configurations C and C' that differ in one monomer-monomer contact

$$\varepsilon = -\frac{1}{\beta} \ln \frac{Z(C')}{Z(C)} \tag{10.61}$$

and found the low and high temperature limits of this energy

$$\begin{gathered} \varepsilon \approx 2J, T \to 0 \\ \varepsilon \approx -2\left(K - \frac{J+K}{q}\right), T \to \infty \end{gathered} \tag{10.62}$$

Thus the energy of the effective monomer-monomer interaction is repulsive for low temperatures.

The explicit description of the water is necessary, in particular, for the correct simulation of the cold protein denaturation [414, 426].

The complete denaturation (cold unfolding) of the globular proteins upon decreasing temperature is similar to the abrupt swelling transition at the low temperatures of the so called thermoreversible homopolymers. C. Buzano et al. [414] used the dynamic Monte Carlo simulations to study swelling of a hydrophobic polymer in the aqueous solutions. The water was described by a 2D model on a triangular lattice using the Mercedes-Benz-type model for the interaction potentials. The polymer was represented by a self-avoiding walk on the same lattice. The authors were able to observe the low-temperature-induced swelling transition at high pressure only and formulate an important question: whether the lattice water models overestimate the stability of the ordered phase (ice)?

The simple model of the hydrophobic interaction does not reproduce the pressure effect on the cold protein denaturation; a correct simulation of this process at the high pressure requires the account of the water fluctuations [2051]. In the cited paper the non-polar heteropolymer is considered. Its interaction with the water (a partial ordering of the water molecules around the protein) is described by the Hamiltonian

$$\mathcal{H}_p = J_r n_{HB} \left(n_{max} - \sum_{i,j} n_i n_j \right) \tag{10.63}$$

where the parameter J_r defines the strength of the repulsive interaction, $n_{HB} \equiv N_{HB}/N_w$ is the numerical density of the hydrogen bonds, that are increasing the system volume, N_w is the number of water molecules, n_{max} is the maximal number of contacts of the amino acid residues with each other; $n_i = 1$ if ith node of the lattice is occupied by the residue and 0 otherwise. The expression in the round braces is the measure of the protein compactness.

The Hamiltonian of the interactions between the water molecules is written as $\mathcal{H}_{HB} = -J \sum_{i,j} \delta_{\sigma_i,\sigma_j}$; the variable σ_i specify the orientation of the water molecule—the formation of the hydrogen bond between the neighbour molecules being feasible if $\sigma_i = \sigma_j$ [2051]. The enthalpy for the system "protein in the water bath" is $\mathcal{W} = \mathcal{H}_p + \mathcal{H}_{HB} + PV$. The authors used the multicanonic algorithm of the Monte-Carlo method to compute the two-parametric (the number of the hydrogen bonds and the number of the contacts of the amino acid residues) density of states. Then the dependence of the average number of the residue-residue contacts $\bar{N}_c$ on the temperature and on the pressure could be found

$$\bar{N}_c = \sum_{N_{HB}} \sum_{N_c} N_c g(N_{HB}, N_c) \frac{\exp - \left(\frac{W(N_{HB}, N_C)}{T} \right)}{Z}, \tag{10.64}$$

where Z is the partition function. The authors performed computations from the system of a protein with 17 nonpolar residues and 383 water molecules with periodic boundary conditions.

The results have been used to construct the phase diagram of the protein-water system in the $p - T$ plane. The condition of the denaturation was formulated as $\bar{N}_c/n_{max} \geq \alpha_d$ (it was assumed $\alpha_d = 0.96$ in the paper).

The authors stress that while they were able to reproduce the experimentally determined thermodynamics of the cold denaturation the key element of which is the pressure effect on the water density, the analysis of the kinetics of this process will require more detailed models accounting for the actual number of hydrogen bonds for each water molecule and not just the average number of bonds.

The free energy of the protein denaturation and its dependence on the pH of the solution was considered by P. J. Kundrotas and A. Karshikov [1741]. The authors, in contrast to many other models that deal with a single space distribution of the protein charges, consider the protein denaturated state where titratable groups are distributed quasi randomly, obeying the protein sequence restrictions. The protein molecule was represented by a chain of N elements (titratable sites) that are located on the sphere that comprises the material of denaturated protein. The allowed positions of the elements are predetrmined by a set of uniformly distributed points on the surface forming a spherical grid. The authors have also computed the desolvation energy and the electrostatic interaction between the proteins.

There are other approaches to account for the water explicitly, for example, the model of the solvent-phantom, where the interaction of the water molecules with the macromolecules is considered while the interactions between the water molecules are neglected [2734]).

10.1.8 AFPs and Their Binding to Ice

The mechanisms of the ice growth inhibition by the AFPs and the AFGPs are difficult objects for the experimental study. In contrast to the enzymatic proteins that can often crystallized along with their ligands, there is no direct way to study the molecular interactions between the antifreeze proteins and the surface of the growing ice crystals [505]. Even more difficult is the investigation of the antifreeze protein effect on the ice nucleation [803]. Some hopes are related to new sophisticated techniques such as the "double oil layer micro-sized ice crystallization technique" that allows to minimize the influence of the container on the measurements. This approach provides a way to measure the value of the free energy barriers associated with the different dynamic steps of the nucleation [803], including the incorporation of the water molecules onto the surface of the ice nuclei at the kink sites and the inhibition of this process as a result of the adsorption of the additive or the impurity. The authors of the cited paper were able for the first time that the mechanism of the nucleation inhibition on the ice crystal and on the surface of the foreign dust particles demonstrate some common features. The authors stress that it is important to distinguish between the nucleation induction time t_{nucl} and the time t_g needed for ice crystals to grow from the critical size to the observable size in the experimental investigation of the nucleation processes. The often used the induction time is the sum of these two characteristic times and is close to the nucleation time only if $t_g \gg t_{\text{nucl}}$. Of course, this relation is often valid since the energy barrier of the nucleation is higher than that of growth, but exceptions are possible. N. Du et al. [803] also report that similarly to the surfactant molecules, the type III AFP molecules accumulate and self-organize on the surface of water that is attributed to the presence of both the hydrophobic and hydrophilic parts.

The free energy $\Delta G_{AFP-ice}$ of the AFP-ice binding could be estimated from the experimental data on the antifreeze activity. This energy proves to be rather small, being the difference of two large quantities—the energy of interaction between AFP and ice binding surfaces in vacuum and the energy of the hydration of these surfaces [1496]. The authors of the just cited paper warn that it is difficult to accurately obtain this value in computations for two reasons: (1) the uncertainty in the interaction potentials and (2) the need to include a very large number of water molecules (thousands) into the system of the AFP and ice to ensure the complete hydration. The reversible process of the AFP binding could be formally written as $AFP + ice \Leftrightarrow AFP : ice$ with k_+ and k_- being the rates of the direct and inverse reactions related through the fraction ϑ of the ice surface covered by the AFP

$$k_+(1 - \vartheta)C_{AFP} = k_-\vartheta \tag{10.65}$$

and whose ratio gives the association constant $K_a \equiv k_+/k_-$. The surface coverage could be thus expressed as

$$\vartheta = \frac{K_a C_{AFP}}{1 + K_a C_{AFP}} = \frac{C_{AFP}}{K_d + C_{AFP}}, \tag{10.66}$$

where the dissociation constant $K_d \equiv 1/K_a$ was introduced. Defining the fractional hysteresis activity ΔT_F as the ratio of the thermal hysteresis temperature at the given concentration of the AFP to the maximal thermal hysteresis temperature $\Delta T_F \equiv \Delta T_C/|DEltaT_{max}$ and assuming that it is a function of the coverage $\Delta T_F = \psi(\vartheta)$, the authors note that experimental data could be reasonably fit for the choice $\psi(x) = x$. To get the needed estimate for $\Delta G_{AFP-ice}$ it is remains to assume that it is related to K_d by the Arrhenius equation $\Delta G_{AFP-ice} = RT \ln K_d$.

It should be, however stressed, that not the energy of the AFP binding to ice matters, but the *difference* between its interaction energy with the ice and the liquid water [1659].

Still, up to date the numerical simulation is the main source of the information on the AFPs and their function.

A. Jorov et al. [1496] used the Monte Carlo for the minimization of the energy of the complexes AFP-ice in vacuum accounting for the nonbounded interactions, the hydrogen bonds and the hydration potential for proteins. They computed the MC trajectories from many different positions of the AFP at the different ice samples. If the AFP helix is considered as a rigid body, its position and orientation are characterized by six variables corresponding to three translational and three rotational degrees of freedom all of which were varied in the MC minimization. The mutual disposition of the AFP in the bounding state and ice when the helix axis of the AFP is approximately parallel to the ice surface could be described by three parameters—the distance from this axis to the surface and two angles (between the helical axis and the **a** axis on the basal plane of ice and the angle, characterizing the rotation of the AFP around the helical axis). Variation of the latter angle will cause the different AFP faces to approach the ice surface. The authors found three different binding configurations with very close values of the energy and concluded that the geometric complimentary alone does not explain the specificity of the AFP-ice association.

The consideration of the AFP-ice complex in the water using the explicit water models requires the huge execution time; A. Jorov et al. refer to the computations made by J. D. Madura et al. of the binding of the winter flounder AFP to the ice that took months at the supercomputer [1496]. The MC minimization turned out to be considerably less computationally demanding. The authors found that the convergence of the minimization process can be greatly improved by increasing the dielectric constant with only minor effect on the optimal orientation of the AFP on the ice.

The MD method was used to study the state of the antifreeze protein extracted from *Pseudopleuronectes americanus* in vacuum, in the aqueous solution and on the surface of ice [524]. The studied protein has 37 residues with 11 residue repeats. The AMBER all-atom force field for proteins [3282] and TIP3P potential for water were used. The computations were performed at the constant temperature 300 K. This choice rather than temperature 273 K was motivated by the computation complexity: it is easier to provide appropriate sampling of the phase space at the higher temperature.

The most favourable for the inhibition action of the AFP crystallographic planes of the ice were established; the relative role of the hydrogen bonds and the Van der Waals forces as well as the mutation effects (removing of one of the peptide groups) were considered. The authors identified the four distinct ice-binding regions on the AFP surface that could bond to the ice simultaneously and determined the number of the hydrogen bonds each of these regions could form with the ice surface.

This approach was also used along with the energy minimization to study the system "ice - AFP" in the vacuum and in the water. The aim was to find out what peptide groups are responsible for the suppression of the ice growth. Contrary to the existing opinion, the significant role of the hydrophobic amino acid residues was demonstrated [147, 505, 1496]. The effects of the solvent in the system "water - ice - AFP" were investigated by L. Kai et al. [1520].

G. Chen and Z. Jia [505] used a combination of the numerical approaches (the molecular docking, the energy minimization, the molecular dynamics) to find the best (i.e. the most energetically favourable) ice-binding surface of the fish type III AFP, exploiting no prior knowledge of the known ice-binding residues. They considered two arrangements of the ice: the "prism" ice representing the hexagonal ice I_h and the "random" ice, in which the water molecules are randomly positioned, still providing the same average density. The fish AFP considered consists of 66 residues; its structure is known from the X-ray crystallography and the NMR studies. The molecular dynamics simulation were performed after the energy minimization for the AFP-ice complex; no constraints were imposed at the MD stage except holding the ice oxygen atom position fixed. The authors determined the patch of the protein surface that is likely to bind to the ice; the ice-binding site of Type III AFP turned to be larger than previously supported. Participating of the peripheral hydrophobic residues support the opinion that such interactions play an important role in the AFP binding to the ice [146, 1496].

Jorov et al. presented a simple qualitative arguments while it should be so: the "rigid" water molecules incorporated into the ice hardly can provide the contacts with the AFP that are more optimal in terms of the van der Waals, the electrostatic and hydrogen bond energy in comparison with the "soft" molecules of the bulk water that leaves the hydrophobic interactions as the most probably driving force of the AFP-ice association [1496]. The water molecules around the hydrophobic groups maximize the number of the hydrogen bonds by forming the enthropically unfavourable cage-like structures. The hydrophobic groups are getting together to minimize the total surface exposed to the water. Interacting with ice, the AFP hydrophobic groups dip into the grooves on the surface of the ice crystal thus releasing the caged water to the bulk phase.

The analysis of the structure–activity relationships of the different AFPs from both the experimental data (mainly for the fish AFPs) and the numerical simulations allowed to formulate a number of the necessary conditions of the antifreeze activity [1496]:

- the AFP should be water-soluble;
- the AFP ice-binding surface should have the hydrophobic groups;
- the AFP should have the rigid backbone conformation to prevent the hydrophobic collapse of these groups in the water;
- the AFP ice-binding surface should be complimentary to the ice surface;
- the AFP hydrophobic face should not be self-complimentary to prevent the self-aggregation.

The above criteria are in a sense static ones and do not cover the important issue of the AFP molecule behaviour in respect to the crystallization front when the latter is propagating. Two points of view have been declared:

- adsorped AFP molecules are in dynamical equilibrium with the free ones and
- the absorption is permanent.

C.A. Knight and A. Wierzbicki [1659] presented strong arguments for the latter option. If a dynamic equilibrium occurs, every time when the AFP molecular is (temporary) detached, the ice front could irreversibly advance some distance into the supercooled water, since there is no energetic barrier to the new layer nucleation, except the case of growth on the basal plane [1659]. To fulfil its function, the permanently absorbed AFP molecule should resist both an engulfment by the propagating ice crystallization front and pushing forward from the interface. C.A. Knight and A. Wierzbicki suggested a quantitative measure of the AFP molecule suitability for the task, assuming it can be approximated by a spherical particle and characterized by its interfacial energies with water and ice γ_{pw} and γ_{pi} and introducing notions of the "icephobic" and "icephillic" properties that are determined by the sign of the difference $(\gamma_{pw} - \gamma_{pi})$. The authors, however, warned that this construction is based on the sharp water/ice interface that in reality could represent a transition zone of finite width. The unacceptance of the nonpermanent absorption is probably the reason why the AFP molecules are rather large: other circumstances being equal, the molecule with a smaller binding site to the ice will be more easily detached from the surface by the thermal fluctuations.

10.2 Macroscopic Models

The macroscopic models used to simulate the heat transfer *in vivo* consider the continuum media and are the same as the models (as well as the numerical methods—finite difference, finite elements, radial basis functions, spectral methods, meshfree methods, wavelet methods etc.) in other continuum mechanics spheres such as, e.g., the fluid dynamics or electromagnetics. It is the standard practice to consider only the most significant tissues, i.e. to ignore the microscopic structures such as epithelia, basal laminas, glands, nerves, etc. [262].

10.2.1 Hyperthermia

The difficulty of the numerical simulation of the hyperthermia results from the curvature of the body surface and the internal tissue composition; frequently the cubic models with the planar surfaces are used [615, 766, 1327, 3351]. The metabolic heat generation is always ignored since it is insignificant for the hyperthermia [1751]; the perfusion heat loss is also neglected in the non vascular tissues [264–266] and in some other cases, e.g., in the RF cardiac ablation [263, 437, 1434, 1435, 3097, 3098] since its effect is negligible for this procedure [1211].

10.2.1.1 RF Hyperthermia

The finite-difference schemes usually are used in the thermal analysis with the voxelized human models. I. Laakso [1750] investigated the effect of the spatial resolution on the SAR and the temperature in the head models under the RF plane-wave exposure.

Generally the error in the temperature computation is smaller than that in the SAR computation, because of the heat diffusion. A disadvantages of the FD analysis is that the model surface is discretized into small cubes, resulting in a large surface area and increasing the heat transfer from the model that leads to the in underestimation of the temperature rise, if the measured heat-transfer coefficient is directly used as the boundary condition.

The computer simulations serve as the powerful tools for the predicting the precise necrotic temperature, the ablation volume, the time of heating, the thermal history and the tissue thermal damage [182, 534, 726, 1395, 1899, 1952, 2505, 2870, 3401]. For example, M. Cavagnaro et al. [469] have proposed the different numerical models in terms of the temperature dependent thermal and dielectric properties on the temperature distribution and the ablation volume. Y. Lu et al. [1968] predicted the temperature distribution using the finite element based simulations using the water cooled MWA antenna. Y. Xu et al. [3387] computed the temperature distribution within the liver tissue using the periodically distributed tri-slot microwave antenna.

J. Kabil et al. [1518] reviewed the past and present challenges of the numerical modelling of the hypertermia. Some computational schemes have been proposed for improving the accuracy of the heat flux at the model surface [1750, 2289, 2694], in particularity, E. Neufeld et al. [2289] proposed a conformal scheme for correcting the heat flux—this method, however requires the knowledge of the normal directions for each boundary voxel.

P. Gas [1050] numerically investigated the local interstitial microwave hyperthermia of the liver tissue. The author used the finite element method to solve the wave equation for the TM wave case and the bioheat equation under the steady-state condition.

The antenna consists of the inner conductor, dielectric, the outer conductor, and the plastic catheter and two symmetric air slots in the external conductor [1388, 1886], and, thus the two-dimensional axisymmetric model was considered.

The author used the Maxwell-Ampere's and Faraday's laws in the frequency domain:

$$\nabla \times \mathbf{H} = \mathbf{J} + j\omega \mathbf{D}, \tag{10.67}$$

$$\nabla \times \mathbf{E} = -j\omega \mathbf{B} \tag{10.68}$$

where $\mathbf{E}$ and $\mathbf{H}$ are the electric and magnetic field strengths, ω is the angular frequency of the electromagnetic field and $\mathbf{J}$ is the current density given by the Ohm's law

$$\mathbf{J} = \sigma \mathbf{E}, \tag{10.69}$$

σ is the electrical conductivity, $\mathbf{D}$ and $\mathbf{B}$ are the vectors of the electric displacement density and the magnetic induction given by the material dependences

$$\mathbf{D} = \varepsilon \mathbf{E}, \qquad \mathbf{B} = \mu \mathbf{H} \tag{10.70}$$

where ε and μ are the permittivity and magnetic permeability of the medium.

The Maxwell equation could be rewritten as

$$\nabla \times \mathbf{H} = j\omega\varepsilon_0 \left(\varepsilon_r - j\frac{\sigma}{\varepsilon_0\omega}\right)\mathbf{E}, \tag{10.71}$$

$$\nabla \times \mathbf{E} = -j\omega\mu_r\mu_0\mathbf{H} \tag{10.72}$$

where ε_r and μ_r are the relative permittivity and the relative permeability of the medium, ε_0 and μ_0 are the permittivity and permeability of the vacuum, respectively. Applying the rotation operator to both sides of (10.71) and substituting (10.72) the following equation for the field distribution is obtained:

$$\nabla \times \left[\left(\varepsilon_r - j\frac{\sigma}{\varepsilon_0\omega}\right)^{-1} \nabla \times \mathbf{H}\right] - \mu_r k_0^2\mathbf{H} = 0 \tag{10.73}$$

where k_0 is the wave number in vacuum defined as $k_0 = \omega\sqrt{\varepsilon_0\mu_0}$.

Since the vector **H** (as well as **E**) in the time domain is related with the complex amplitude as

$$\mathbf{H}(\mathbf{r}, t) = Re\left[\hat{\mathbf{H}}(\mathbf{r}e^{j\omega t})\right] \tag{10.74}$$

the Eq. (10.73) in the complex domain is

$$\nabla \times \left[\hat{\varepsilon}_r^{-1} \times \hat{\mathbf{H}}\right] - \mu_r k_0^2\hat{\mathbf{H}} = 0 \tag{10.75}$$

where $\hat{\varepsilon}_r$ is the complex relative permittivity defined as

$$\hat{\varepsilon}_r(\omega) = \varepsilon_r - j\frac{\sigma}{\varepsilon_0\omega}. \tag{10.76}$$

Similarly the complex value of the vector of electric field strength is determined from the relation

$$\nabla \times \hat{\mathbf{E}} = -j\omega\mu_r\mu_0\hat{\mathbf{H}}. \tag{10.77}$$

The author considered the transverse magnetic (TM) waves, thus there are no electromagnetic field variations in the azimuthal direction. A magnetic field **H** has only the φ component

$$\mathbf{H}(\mathbf{r}, t) = H_\varphi(r, z)\mathbf{e}_\varphi e^{j\omega t} \tag{10.78}$$

and an electric field propagates in $r - z$ plane

$$\mathbf{E}(\mathbf{r}, t) = \left[E_r(r, z)\mathbf{e}_r + E_z(r, z)\mathbf{e}_z\right] e^{j\omega t}. \tag{10.79}$$

The author assumed that the liver tissue and the antenna are homogeneous media with the averaged material parameters, thus the external heat source responsible for the changing of the temperature inside the tissue is

$$Q = \frac{1}{2}\sigma |\hat{\mathbf{E}}|^2. \tag{10.80}$$

The author noted that the two-dimensional model represents a simplification of the reality but it fully reflects the treating of the pathological tissues using the interstitial microwave hyperthermia and can be used to evaluate the actual temperature distribution in the three-dimensional case. The computations showed that the temperature decreases rapidly with the distance from the microwave applicator.

T. R. Gowrishankar et al. [1141] calculated the microwave power source for the different skin layers—dermis, epidermis, subcutaneous fat—using the penetration depths for the dermis and the subcutaneous fat and the intrinsic impedances of air, dermis, and the subcutaneous fat.

P. Keangin et al. [1578] investigated the interstitial microwave ablation of the liver cancer by applying the localized heating to the tumour tissue using the microwave coaxial antenna (MCA). The goal of the MW hyperthermia is to elevate the temperature of the malignant tissue to 50° C where the cancer cells are destroyed [2098, 3323]. The tumour shrinkage has been observed as a consequence of this treatment [1313, 3323, 3423]. For example, M. J. Lopez et al. [1949] studied the effect of radiofrequency energy on the tissue shrinkage of the glenohumeral joint capsule in a sheep mode.

Keangin et al. used the cylindrical model of the liver tissue and the axisymmetric finite element method to solve the problem of the coupled system consisting of the electromagnetic wave equation, the bioheat equation and the mechanical deformation equation. The authors assumed that

- the electromagnetic wave in the MCA is characterized by the transverse electromagnetic fields;
- the electromagnetic wave in the porous liver is characterized by transverse magnetic fields;
- the dielectric properties of the porous liver are uniform and constant.

The wall of a MCA was assumed to be a perfect electric conductor $\mathbf{n} \times \mathbf{E} = 0$, at the outer surface of the porous liver a scattering boundary condition $\mathbf{n} \times \sqrt{\varepsilon}\mathbf{E} - \sqrt{\mu}H_\varphi = -2\sqrt{\mu}H_{\varphi 0}$ was specified to prevent reflection artifacts [1578].

The electric and magnetic fields in the 2D axisymmetric model are expressed as [3400]

$$\mathbf{E} = e_r \frac{C}{r} e^{j(\omega t - kz)}, \qquad \mathbf{H} = e_\varphi \frac{C}{rZ} e^{j(\omega t - kz)} \tag{10.81}$$

where

$$C = \sqrt{\frac{ZP}{\pi} \ln\left(\frac{r_{outer}}{r_{inner}}\right)}, \tag{10.82}$$

Z is the wave impedance, P is the input microwave power, r_{inner} is the dielectric inner radius, r_{outer} is the dielectric outer radius, $\omega = 2\pi f$ is the angular frequency, f is the frequency, k is the propagation constant.

The electric field is in the radial direction only inside the coaxial cable and in both the radial and the axial direction inside the tissue whereas the magnetic field is purely in the azimuth direction in the tissue. Thus for the TM wave the equation becomes scalar in H_φ

$$\nabla \times \left(\left(\varepsilon_r - \frac{j\sigma_{el}}{\omega\varepsilon_0}\right) \nabla \times H_\varphi\right) - \mu_r k_0^2 H_\varphi = 0 \tag{10.83}$$

where ε_0 is the permittivity of the free space. The authors used the linear in the temperature approximations for the dielectric properties $\varepsilon_{r,liver}$ and $\sigma_{el,liver}$.

P. Keangin et al. [1577] also studied the interstitial microwave ablation of the liver cancer using the single and double slot antenna. The double slot MCA generates a wider microwave power absorbed distribution pattern in the liver tissue than a standard single slot MCA. G. D. Guerrero López et al. [1181] used the double-slot MCA for the microwave ablation of the breast tumour.

P. Keangin and T. Rattanadecho [1576] analysed the heat transport in the porous liver during the microwave ablation. The biological tissue including the cell and the microvascular bed with the blood flow direction contains many vessels, assuming that the tissue is like a sponge, through which the blood flows and can be regarded as a porous structure [2259, 2626].

There are two models for the analysis of the heat transfer in a porous medium: the local thermal equilibrium (LTE) and the local thermal non-equilibrium (LTNE). The LTE model is based on the assumption that the tissue phase temperature is equal to the fluid phase temperature everywhere in the porous medium. The LTE assumption is not valid for the significant temperature difference between the tissue and blood phases within the porous media as in the ablation treatment. The coupled equations are solved using the axisymmetric finite element method. The influences of the blood velocities, the porosities, the input microwave powers and the positions within the porous liver on the tissue and blood temperature distributions have been investigated. The LTE assumption is found to be suitable for predicting the temperature during the MWA process for the small blood velocities.

The LTNE model in a porous medium was used by J. Fan and L. Wang [920] who developed a general bioheat transport model with both the blood and tissue macroscale temperatures. N. Afrin et al. [21] presented a model of thermal lagging in the living biological tissue based on the LTNE heat transfer model between the tissue, the arterial and venous bloods.

The Pennes equation was used for the heat transfer analysis with the heat source equal to the resistive heat generated by the electromagnetic field (the microwave power absorbed) defined as:

$$Q = \frac{1}{2}\sigma_{el,liver}|\mathbf{E}|^2. \tag{10.84}$$

Evidently, the electrical properties strongly affect the temperature increase in the tissues [1014].

Keangin et al. assumed that the liver tissues can be considered as an isotropic material. The quasi-static model for an axisymmetric geometry is described [329, 1887] by the equilibrium equations

$$\frac{\partial \sigma_{rr}}{\partial r} + \frac{\partial \sigma_{rz}}{\partial z} + \frac{\sigma_{rr} - \sigma_{\varphi\varphi}}{r} + F_r = 0, \tag{10.85}$$

$$\frac{\partial \sigma_{rz}}{\partial r} + \frac{\partial \sigma_{zz}}{\partial z} + \frac{\sigma_{rz}}{r} + F_z = 0, \tag{10.86}$$

the stress-strain relationship

$$\varepsilon_{rr} = \frac{1}{E}[\sigma_{rr} - \nu(\sigma_{\varphi\varphi} + \sigma_{zz})] + \varepsilon^{th}, \tag{10.87}$$

$$\varepsilon_{zz} = \frac{1}{E}[\sigma_{zz} - \nu(\sigma_{\varphi\varphi} + \sigma_{rr})] + \varepsilon^{th}, \tag{10.88}$$

$$\varepsilon_{\varphi\varphi} = \frac{1}{E}[\sigma_{\varphi\varphi} - \nu(\sigma_{\varphi\varphi} + \sigma_{zz})] + \varepsilon^{th}, \tag{10.89}$$

and the strain-displacement relationship

$$\varepsilon_{rr} = \frac{\partial u_r}{\partial}r, \quad \varepsilon_{zz} = \frac{\partial u_z}{\partial z}, \quad \varepsilon_{\varphi\varphi} = \frac{u}{r}, \quad \varepsilon_{rz} = \frac{1}{2}\left(\frac{\partial u_r}{\partial}z + \frac{\partial u_z}{\partial}r\right) \tag{10.90}$$

where σ is the stress, ε is the strain, F is the external body load (zero for the considered problem), E is Young's modulus, ν is the Poisson's ratio,u is the average displacement. The thermal strain

$$\varepsilon^{th} = \int_{T_{ref}}^{T} \alpha \mathrm{d}T \tag{10.91}$$

where T_{ref} is the reference temperature, α is the temperature-dependent tissue thermal expansion coefficient.

Keangin et al. studied the effect of the tissue deformation on the microwave power absorbed, the SAR distribution, the temperature distribution in the liver tissue during the MW hyperthermia. The authors found that taking into account the shape deformation gives a slightly lower maximum temperature than the model without deformation and the penetration area of the microwave of the liver tissue in the model without deformation is smaller than the penetration area of the microwave of the liver tissue in the model with the deformation.

Y. Diao et al. [747] computed the averaged absorbed power density (APD) and the temperature rise in the body models—a two-dimensional cylinder (three types of the tissues—skin, fat, and muscle) and a three-dimensional partial forearm (seven types of tissues—skin, fat, muscle, blood, tendon, cortical and cancellous bones)—for the electromagnetic exposure with frequency above 6 GHz.

The authors used the Finite-Difference Time-Domain (FDTD) method [3000] and the perfectly matched layer [2621] to truncate the simulation domain.

In the FDTD algorithm proposed by K. Yee in 1966 [3418] the electric and magnetic fields are approximated in space and time by a coupled form of Maxwell equations instead of solving a single wave equation for either the electric or the magnetic field alone. The method uses regular rectangular grids shifted in space and the half of the time step in time, thus has difficulties in handling complex geometries—the curved boundaries are approximated by staircase meshes that is evidently not the best choice for the complex shapes of the human organs. The perspective approach for such problems is the use of the unstructured grids and the local mesh refinement strategies [258, 2334, 3508]. For example, K. Kim [1621] developed the hp-finite element code for the simulation of the electromagnetic radiation and induced heat transfer in the human body. The algorithm includes three dimensional elements of all shapes (hexahedron, tetrahedron, prism, pyramid) that enable to describe the complex geometry and uses the general mesh refinement scheme, which includes isotropic and anisotropic refinements for hybrid meshes that consist of the elements of all the available shapes.

A variational formulation can be obtained by connecting the Ampére's law and the Faraday's law: one of these two equations is satisfied weakly in the distributional sense, and the other equation is imposed strongly [1621].

For the electric field formulation, K. Kim multiplied a test function **F** to the Ampére's law and integrated over the domain Ω; after integrating by parts, he obtained the following equation

$$\int_{\Omega} \mathbf{H} \cdot \nabla \times \mathbf{F} d\mathbf{x} + \int_{\partial\Omega} (\mathbf{n} \times \mathbf{H}) \cdot \mathbf{F}_t dS = \int_{\Omega} \mathbf{J}^{imp} \cdot \mathbf{F} d\mathbf{x} + \int_{\Omega} (\sigma + i\omega\varepsilon) \mathbf{E} \cdot \mathbf{F} d\mathbf{x}. \quad (10.92)$$

Excluding the magnetic field via the Faraday's law pointwise, one gets

$$\int_{\Omega} \frac{1}{\mu} \nabla \times \mathbf{E} \cdot \nabla \times \mathbf{F} \mathrm{d}\mathbf{x} - \int_{\Omega} (\omega^2 \varepsilon - i\omega\sigma) \mathbf{E} \cdot \mathbf{F} \mathrm{d}\mathbf{x}$$

$$= -i\omega \int_{\Omega} \mathbf{J}^{imp} \cdot \mathbf{F} \mathrm{d}\mathbf{x} + i\omega \int_{\partial\Omega} (\mathbf{n} \times \mathbf{H}) \cdot \mathbf{F}_t \mathrm{d}S. \tag{10.93}$$

At the interface between two media, the boundary conditions for the electric and magnetic fields are expressed as

$$\hat{\mathbf{n}} \times (\mathbf{E}_1 - \mathbf{E}_2) = 0, \quad \hat{\mathbf{n}} \cdot (\mathbf{D}_1 - \mathbf{D}_2) = \rho_s \tag{10.94}$$

and

$$\hat{\mathbf{n}} \times (\mathbf{H}_1 - \mathbf{H}_2) = \mathbf{J}_s^{imp}, \quad \hat{\mathbf{n}} \cdot (\mathbf{B}_1 - \mathbf{B}_2) = 0 \tag{10.95}$$

where $\hat{\mathbf{n}}$ is a unit normal vector from medium 1 to medium 2, ρ_s is the imposed surface charge density, $\mathbf{J}_s^{imp}$ is the imposed surface electric current density.

There are special boundary conditions [1621]:

- *Perfect electric conductor (PEC).* It cannot hold any electric fields in it, thus the tangential component of the electric field on the boundary must be zero $\hat{\mathbf{n}} \times \mathbf{E} = 0$.
- *Impressed surface current.* An antenna can be modelled by prescribing an impressed current on the surface $\hat{\mathbf{n}} \times \mathbf{H} = \mathbf{J}^{imp}$.

For the scattering problems, the Maxwell equations are formulated in the unbounded domain, and the scattered field is required to satisfy the Silver-Müller radiation condition at infinity [1621] :

$$\lim_{r \to \infty} r[\nabla \times \mathbf{E} - jk_0 \mathbf{e}_r \times (\mathbf{e}_r \times \mathbf{E})] = 0 \tag{10.96}$$

where r is a spherical coordinate centred at the scatterer.

This radiation condition means that the wave should propagate outward; the radiating electric and magnetic fields are to be transverse in the far-field region, thus the Poynting vector should be radial.

For the unbounded domain problems such as scattering or radiating waves, grid based methods such as the FDTD need to truncate the unbounded domain with the proper boundary conditions mimicking the waves at infinity. The Perfectly Matched Layer (PML) technique, which was proposed by J. P. Berenger [253] in 1994 for solving the time dependent Maxwell equations, is the most widely used one. The idea of this method is to surround the computational domain with an absorbing boundary layer of the finite thickness (the Perfectly Matched Layer) such that the waves travelling outward through this layer are changed into the evanescent ones.

This approach was re-interpreted as the complex stretch of the variables by W. C. Chew and W. H. Weedon [533].

For the 2D analysis, both the transverse-electric (TE) and the transverse-magnetic (TM) plane incident waves were considered by Diao et al. [747] as the radiation sources. For the 3D problem, a half-wave dipole antenna array was simulated. The simulated frequencies ranged from 6 to 60 GHz for the 2D problems, 28 GHz for the 3D problem.

The spatial-average APD is defined via the electromagnetic power flow as

$$S = \int\!\!\int_S Re[\mathbf{E} \times \mathbf{H}^\star] \cdot \mathrm{d}\mathbf{s}/A \tag{10.97}$$

where $\mathrm{d}\mathbf{s}$ is the integral variable vector normal to the body surface [747].

The authors found that the heating factors for the APD were almost independent of the frequency above 20 GHz and that the differences in the heating factors between the TE and TM cases are no larger than 20%.

Diao et al. [747] concluded that, since the boundary conditions significantly influence the calculation results, the discretization of the model surface with cubes results in a larger model surface area and the temperature rise is underestimated. This problem is particularly crucial for the millimetre-wave radiation due to the extremely shallow penetration depth.

For the case of the RF hyperthermia Tung et al. considered the radially symmetric problem—a spherical electrode of the radius r_0 is completely embedded and in the close contact with the biological tissue of the infinite dimension. The heat source is presented as a product of a radial and a temporal parts [868]

$$Q(r,t) = \frac{P r_0}{4\pi r^4} H(t) \tag{10.98}$$

where the total applied power outside of the electrode is

$$P = \oint_{r \geq r_0} \mathrm{d}\Omega \int_{r_0}^{\infty} r^2 Q(r) \mathrm{d}r, \tag{10.99}$$

Ω is the solid angle. Thus, the power flux per unit solid angle $\mathrm{d}P/\mathrm{d}\Omega$ varies according to the $1/r^2$ dependence.

E. Kengne et al. [1593] used the nonlinear one-dimensional heat transfer model based on the temperature-dependent blood perfusion to predict the temperature distribution in the dermis and subcutaneous tissues subjected to the point heating sources. The authors applied the Pennes type of bioheat transfer equation with the temperature-dependent blood perfusion term

$$\varrho c \frac{\partial T}{\partial t} = \lambda \frac{\partial^2 T}{\partial x^2} + c_{\mathrm{b}} \varrho_{\mathrm{b}} \omega_b(T)(T_{\mathrm{a}} - T) + \dot{q}_{\mathrm{met}} + Q^{\mathrm{ext}}, \tag{10.100}$$

the temperature-dependent blood perfusion is of the form [1777] $\omega_b(T) = \omega_0(1 + \gamma T)$ where ω_0 is the baseline perfusion and γ is the linear coefficient of the temperature dependence.

The authors used the Jacobi elliptic functions to find the initial temperature field T_0 for the basal state of dermis and subcutaneous tissues that is the solution of the boundary value problem in $[0, L]$

$$\varrho c \frac{\partial T_0}{\partial t} = \lambda \frac{\partial^2 T_0}{\partial x^2} + c_b \varrho_b \omega_b(T_0)(T_a - T_0) + \dot{q}_{met}, \tag{10.101}$$

$$\lambda \frac{dT_0}{dx}|_{x=0} = h[T_s - T_0(x)]_{x=0}, \tag{10.102}$$

where T_s is the surrounding air temperature. Because the biological body tends to keep its core temperature T_c to remain stable, the body core temperature is treated as a constant $T_0(L, t) = T_c$.

After the multiplication of (10.102) by dT_0/dx and integration, the authors got

$$\left(\frac{dT_0}{dx}\right)^2 = C$$

$$- \frac{2(\dot{q}_{met} - c_b \varrho_b \omega_0 T_a)}{\lambda} T_0 + \frac{c_b \varrho_b \omega_0 (1 - \gamma T_a)}{\lambda} T_0^2 + \frac{2\gamma c_b \varrho_b \omega_0}{3\lambda} T_0^3, \tag{10.103}$$

where C is the constant of integration.

The particular solutions of this equation are found in terms of Jacobi elliptic functions $T_0(x) = A + Bsn^2(\mu[x - L], m)$ where $sn(x, m)$ is the Jacobi elliptic sine function with modulus $m \in [0, 1]$.

Equation (10.100) with temperature-dependent blood perfusion was solved using a second-order central difference scheme in space and a Crank-Nicholson type of scheme in time. The authors considered two cases: the hypothermia and the sinusoidal point heating, which are typical the heat transfer processes involved in the soft tissues in the biological bodies, such as the heat therapy and the skin burn.

J. L. Shao et al. [2802] used the three-dimensional finite difference scheme to simulate the nano-assisted radiofrequency hyperthermia of the irregularly shaped liver tumour. The simulations were conducted to assess the temperature difference responses between the conventional RF hyperthermia and one with the inclusion of the nanoparticles on the several irregularly-shaped tumours.

The authors used the Laplace's equation $\nabla \cdot (\sigma \nabla \varphi)$ for electric field. In the absence of the experimental data on the temperature-dependence of the liver electrical conductivity [182, 482] it was assumed that the temperature-dependent behaviour of the liver tissue is similar to that of an equivalent sodium chloride solution [792] and the expression for $\sigma(T, N)$ where N is the normality derived by A. Stogryn [2945] was used (the normality of the physiological NaCl solution is $N \approx 0.105$ [182]).

The Pennes equation was used for both the healthy and the damaged tissue region with both the metabolic rate and the blood perfusion set to zero in the latter region since the perfusion in the liver carcinoma is several times less than in the healthy tissue [1433].

The effective thermal conductivity of the tissue with the injected into an intercellular space spherical particles was calculated using is the Hamilton-Crosser model [1825] as

$$\lambda_{eff}(T) = \lambda_t(T)\frac{\lambda_p + 2\lambda_t(T) - 2\eta(\lambda_t(T) - \lambda_p)}{\lambda_p + 2\lambda_t(T) + \eta(\lambda_t(T) - \lambda_p)} \tag{10.104}$$

where η is the nanoparticle volume content, the subscripts eff, t and p refer to the effective value, the liver tissue and the loaded particle, respectively.

The Eq. (10.104) could not be used to calculated the effective thermal conductivity of the tissue zone with the CNTs since the CNTs are nonspherical.

The effective thermal conductivity of the tissue with the non-spherical nanoparticles was suggested by J. Glory et al. [1100]

$$\lambda_{eff}(T) = \lambda_t(T)\frac{\lambda_t(T) + (n-1)\lambda_p - \eta(n-1)(\lambda_t(T) - \lambda_p)}{\lambda_t(T) + (n-1)\lambda_p + \eta(\lambda_t(T) - \lambda_p)} \tag{10.105}$$

where η is the nanoparticle volume content, n is the empirical shape factor $n = 3/\psi$, ψ the particle sphericity defined as the ratio of the surface area of a sphere with volume equal to that of the particle to surface area of particle.

Results have indicated that applying the selected nanoparticles with the high thermal conductivity and electrical conductivity on the targeted tissue zone promotes the heating rate while sustaining a similar ablation zone that experiences the lower maximum temperature when compared with the conventional RFA hyperthermia. Under the same nanoparticle type loading the maximum heating rate occur according to the rank (highest and lowest) of the nanoparticles' thermal conductivity. Under the same nanoparticle type loading, the region of the damaged tissue enlarges with increasing nanoparticle volume fractions.

10.2.1.2 High Intensity Focused Ultrasound (HIFU)

V. Suomi et al. [2981] used the set of the coupled first-order partial difference equations based on the conservation laws and a phenomenological loss term that accounts for the acoustic absorption

$$\frac{\partial \mathbf{u}}{\partial t} = \frac{1}{\varrho_0}\nabla p, \tag{10.106}$$

$$\frac{\partial \varrho}{\partial t} = -(2\varrho + \varrho_0)\nabla \cdot \mathbf{u} - \mathbf{u} \cdot \nabla \varrho_0, \tag{10.107}$$

$$p = c_s^2 \left(\varrho + \mathbf{d} \cdot \nabla \varrho_0 + \frac{B}{2A} \frac{\varrho^2}{\varrho_0} - L\varrho \right) \tag{10.108}$$

where p is the acoustic pressure, $\mathbf{u}$ and $\mathbf{d}$ are the acoustic particle velocity and the displacement, ϱ and ϱ_0 are the acoustic and the background density, c_s is the isentropic sound speed, B/A is the nonlinearity parameter, L is a loss operator accounting for the acoustic absorption and the dispersion that follows a frequency power law [3078].

These equations are equivalent to a generalised version of the Westervelt equation [3305] that accounts for second-order acoustic nonlinearity, the power law acoustic absorption, and a heterogeneous distribution of material properties (the sound speed, the density, the nonlinearity and the absorption coefficient).

B. E. Treeby et al. used the k-space pseudospectral method [3079] and found that the intensity loss could be roughly divided equally between the attenuation and the refraction. The authors also noted that the refraction resulted in a threefold drop in peak temperature and a 11-fold reduction in the ablated volume and also produced the large patient-to-patient variability.

T. W. H. Sheu et al. [2819, 2901] studied the high-intensity focused ultrasound thermal ablation of the liver tumour (the liver is the highly perfused organ with the blood supplied by the hepatic artery and the portal vein susceptible to the primary and metastatic tumours) using the acoustics–thermal–fluid coupling model accounting for the convective cooling in the large blood vessel and the acoustic streaming. The realistic geometry for the blood vessel and the liver was reconstructed using the MRI images—the incident finite amplitude ultrasound wave can affect the blood flow motion in the large hepatic vessels and consequently the temperature distribution in the tumour. Analytical results of the earlier studies by C. Eckar who studied the acoustic streaming profiles generated by the uniform cylindrical ultrasound beam with width much smaller than the tube diameter in an infinite circular tube [832] and by Nyborg cannot be applied to predict the acoustic streaming velocities in the blood vessels during the HIFU therapy [2823].

The authors used the Westervelt wave equation (8.16) [1225] neglecting the relaxation processes

$$\nabla^2 p - \frac{1}{c_s^2} \frac{\partial^2 p}{\partial t^2} + \frac{\delta}{c_s^4} \frac{\partial^3 p}{\partial t^3} + \frac{\beta}{\varrho_0 c_s^4} \frac{\partial^2 p^2}{\partial t^2} = 0. \tag{10.109}$$

The last term in this equation accounts for the acoustic nonlinearity which can affect the thermal and mechanical changes within the tissues [154].

The acoustic diffusivity δ that accounts for the thermal and viscous losses in a fluid is taken in the form $\delta = 2c_s^3 \alpha / \omega^2$ where α is the acoustic absorption coefficient [2819].

Under the assumption that each small element dS of the transducer surface vibrates continuously with the same velocity $u = u_0 \exp(ixt)$ normal to the surface the linear wave equation

$$\nabla^2 p - \frac{1}{c_s^2}\frac{\partial^2 p}{\partial t^2} + \frac{\delta}{c_s^4}\frac{\partial^3 p}{\partial t^3} = 0. \tag{10.110}$$

could be transformed to the diffraction integral for the velocity potential as [2339, 2459]

$$\psi_{\bar{p}} = \int\int_S \frac{u}{2\pi r} e^{-(\alpha + ikr)} \mathrm{d}S \tag{10.111}$$

where r is the distance from the source point on the transducer surface dS to a field point $\bar{p}$, k is the wave number [2815].

In a region free of the large blood vessels, the Pennes equation was employed to model the transfer of the heat in the perfused tissue region. In the region containing the large vessels, within which the flowing blood can convect heat, the biologically relevant heat source $-\varrho_b c_b \mathbf{u}$ was added to the heat equation

$$\varrho_b c_b \frac{\partial T}{\partial t} = \lambda_b \nabla^2 T - \varrho_b c_b \mathbf{u} \cdot \nabla T + Q \tag{10.112}$$

where $\mathbf{u}$ is the blood flow velocity.

The velocity of the blood flow plus the velocity resulting from the acoustic streaming due to the applied high intensity ultrasound is governed by the Navier-Stokes equation for the incompressible $(\nabla \cdot \mathbf{u})$ fluid

$$\frac{\partial \mathbf{u}}{\partial t} + (\mathbf{u} \cdot \nabla)\mathbf{u} = \frac{\mu}{\varrho_b}\nabla^2 \mathbf{u} - \frac{1}{\varrho_b}\nabla p - \frac{1}{\varrho_b}\mathbf{F} \tag{10.113}$$

where p is the static pressure, μ is the shear viscosity of the blood flow, the force vector $\mathbf{F}$ acting on the vessel blood fluid due to ultrasound is assumed to propagate along the acoustic axis $\mathbf{n}$ having the nonzero component

$$\mathbf{F} \cdot \mathbf{n} = \frac{2\alpha}{c_s} I \tag{10.114}$$

and the acoustic intensity is [2815]

$$I = \frac{(ikc\varrho_t \psi_{\bar{p}})^2}{2\varrho_t c_s}. \tag{10.115}$$

The three-dimensional problem involving the liver, the tumour and the blood vessel was analysed using the commercially available CFDRC (CFD Research Corporation, Huntsville, AL, USA) software. The convective cooling in the large blood vessel and the acoustic streaming were found to change the temperature near the blood vessel. It was shown that the acoustic streaming effect can affect the blood flow distribution in the hepatic arterial branches and leads to the mass flux redistribution.

The US simulation based on full-wave three dimensional model is a challenge: the small wavelength and the consequent need for the high resolution results in the extremely large computational domains, especially for applicators that focus the energy deeply in the body, i.e. which require modelling of a large anatomical region [2398]. This makes the application of the full-wave 3D simulations cumbersome especially for the array applicators that may have several hundreds of the transducers. Therefore the careful application-specific approximations are required for the reasonable computation time and accuracy. Alternatively, the high performance computing (HPC) techniques for the acceleration of computations, for example, the use of the computer clusters and the GPUs, may make the full-wave 3D US simulations more applicable [1332].

10.2.1.3 Laser Techniques

Frequently the laser-tissue interaction is simulated numerically using either the Lambert law or the diffusion approximation. The scattering effect hardly could be incorporated in the Lambert law modelling; the diffusion approximation is limited to the predominant scattering medium. Such treatment can introduce the significant error in the radiation flux distribution in the nonhomogeneous tissue [1188] during the short pulse laser irradiation, in which case the light suffers significant scattering in most region and the strong localized absorption in the target spot. Thus, it is necessary to consider the complete light transport through the tissue during the short pulse laser irradiation.

The intensity inside the tissue medium is governed by the transient radiative transport equation [1446, 1622, 2165]:

$$\frac{1}{c}\frac{\partial I(r,z,\Omega,t)}{\partial t}+\frac{\mu}{r}\frac{\partial}{\partial r}[rI(r,z,\Omega,t)]-\frac{1}{r}\frac{\partial}{\partial \varphi}[\eta I(r,z,\Omega,t)]+\xi\frac{\partial I(r,z,\Omega,t)}{\partial z}$$

$$+\ k_e I(r,z,\Omega,t)=\frac{k_s}{4\pi}\int\limits_{4\pi}\varphi(\Omega',\Omega)I(r,z,\Omega',t)\mathrm{d}\Omega'+S(r,z,\Omega,t) \qquad (10.116)$$

where $I(r, z, \Omega, t)$ is the scattered diffuse intensity, k_e and k are the extinction and the scattering coefficients, respectively, φ is the azimuthal angle, φ is the phase function, Ω is the solid angle, c is the velocity of light in the medium, r and z are the spatial coordinates, μ, η, ξ are the direction cosines, t is the time, S is the source term due to the laser irradiation

$$S(r,z,\Omega,t)=\frac{k_s}{4\pi}I_c(\mu^c\mu,\eta^c\eta,\xi^c\xi) \qquad (10.117)$$

where the unit vector of (μ^c, η^c, ξ^c) represents the collimated laser incident direction. The collimated intensity I_c is [1446]

$$I_c(r, z, t) = L_0 \exp -4 \ln 2[(t - z/c)t_p - 1.5]^2 \exp(2r^2/\sigma(z)^2) \exp(-k_e z) \tag{10.118}$$

where L_0 is the maximum intensity of the laser beam at the sample surface, t_p is the laser pulse width, $\sigma(z)$ is the beam radius.

The radiative heat flux can be expressed using the intensity distribution as

$$\nabla \cdot q = k_a \left(4n^2\sigma_s T^4 - \int\limits_{4\pi} \varphi(\Omega', \Omega) I(r, z, \Omega', t) \mathrm{d}\Omega' \right). \tag{10.119}$$

Most researchers have studied the thermic response of the living tissue by considering a single layer, e.g., S. Kumar and A. Srivastava [1727], S. Patidar et al. [2390], P. Kishore and S. Kumar [1644], K. C. Liu and J. C. Wang [1927], S. M. Lin et al. [1880, 1882].

In reality the biological tissues are composed of the multiple layers with different properties based on their location in the body. Few researchers have studied the thermic performance of the laser-irradiated multilayer living tissue. For example, J. H. Randrianalisoa et al. [2575] studied the effect of the short-pulsed laser irradiation with the embedded gold nanoparticles on the superficial human tissue of multiple layers. K. C. Liu et al. [1935] investigated the thermal response of the skin tissue as a triple-layered composite. J. Jiao and Z. Guo [1476] studied the thermal interaction of triple-layered skin tissue with a tumour stratified into it when the laser irradiation is focused directly on the tumour. The temperature field within the living tissue, exposed to either collimated or focussed laser irradiation, was obtained by solving the Pennes BHTE. P. Gaudarzi and A. Azimi [1137] studied the skin tissue's response, considering a single layer with laser as the heat source. The authors employed the fractional order non-Fourier heat conduction for simulating heat flow.

P. Kishore and S. Kumar [1645] computed the temperature field in the triple-layered cylindrical living tissue subjected to the laser light is determined by numerically solving the transient radiative transfer equation in conjunction with the DPL based bioheat equation.

The radiation has two components: collimated part and the diffuse part of the overall intensity make up its two components. The TRTE is used to represent the diffuse portion of the overall intensity, and it is solved using the modified discrete ordinate method (DOM). For the collimated component of the total intensity, Gaussian variation along the radial direction and the time has been considered; the variation of the collimated component along the axis is governed by the Beer–Lambert's law [1646].

The time scale corresponding to the transmission of the light falls within the range of picoseconds. In such a short pulse time, the diffusion of the thermal energy becomes insignificant. Hence, the incident pulse results in the localized temperature rise given by the equation:

$$\rho c_p \frac{\partial T}{\partial t} = -\nabla \cdot \mathbf{q}_{rad}. \tag{10.120}$$

Once the thermal energy is deposited within the tissue by the short-pulsed laser, it diffuses to the nearby normal tissues. Kishore and Kumar obtained the temperature field in the axisymmetric laser-irradiated living tissue by combining the TRTE with the DPL model of the BHTE. The authors performed the comparison of the DPL model with the Fourier and Cattaneo models. The Fourier model estimates the lower temperatures than the DPL and the hyperbolic models. This behaviour is explained by the underlying limits of the Fourier model, which assumes the infinite thermal wave speed across the biological tissue. Under this assumption, phonons spend significantly less time at any spatial position within the tissue domain, resulting in the minimal thermal energy accumulated into the tissue, consequently leading to a minimal increase in the tissue temperature. The non-Fourier models predict a more considerable temperature increase when the living tissue is exposed to the laser light than the Fourier model [1645].

The prolonged direct heating of the tumour may lead to the unfavourable protective reaction of the thermal regulation system with an increase in the blood perfusion near the tumour and release of the heat shock proteins which play a role in the process of the thermotolerance [1776]. One approach is to use the periodic heating with a single heating duration of about 10–15 s to decrease the blood flow variation [784], another is the use of the indirect heating when the tumour is not subject to collimated laser radiation.

There are two possibilities [785]:

- The absorbing heat nanoparticles are embedded not in the tumour itself but in a circular region around the tumour. Laser heating of these particles leads to the formation of the hot ring in the healthy tissue and the accumulated heat is conducted from all sides into the tumour. The difficulty of this approach is the control of the nanoparticle distribution in the circular region around the tumour.
- Alternative approach is based on the relatively weak absorption of the light by the human tissues in the therapeutic window and the higher absorption coefficient of the tumour tissue without nanoparticles.

L. A. Dombrovsky et al. [785] developed the two-dimensional model taking into account the realistic shape of the tumour, the blood perfusion through the capillaries, the volumetric heat transfer between the arterial blood and the tissue, and also the thermal conversions in the venous blood and the tumour tissue during the hyperthermia.

The authors considered the processes in the heterogeneous medium that contains the scattering (and weakly absorbing) tissue and the absorbing (and weakly scattering) nanoparticles. The radiative transfer equation (RTE) is considered in the traditional continuum theory that for the randomly polarized radiation is written as [1363, 2165]

$$\mathbf{\Omega}\nabla I_\lambda(\mathbf{r},\mathbf{\Omega}) + \beta_\lambda I_\lambda(\mathbf{r},\mathbf{\Omega}) = \frac{\sigma_\lambda}{4\pi}\int\limits_{4\pi} I_\lambda(\mathbf{r},\mathbf{\Omega}')\varphi_\lambda(\mathbf{\Omega}',\mathbf{\Omega})\mathrm{d}\mathbf{\Omega}' \tag{10.121}$$

where $I_\lambda(\mathbf{r},\mathbf{\Omega})$ is the spectral radiation intensity in the direction $\mathbf{\Omega}$, the absorption coefficient α_λ, the scattering coefficient σ_λ, and scattering phase function φ_λ depend on the coordinate $\mathbf{r}$.

The authors assumed that the medium is isotropic and the coefficients of the RTE do not depend on the direction and used the transport approximation for the scattering phase function [783]—the scattering phase function is replaced by a sum of the isotropic component and the term describing the peak of the forward scattering:

$$\varphi_\lambda = (1-\bar{\mu}_\lambda) + 2\bar{\mu}_\lambda\delta(1-\mu_s) \tag{10.122}$$

and the RTE for the isotropic scattering ($\varphi_\lambda = 1$) is simplifies to (the subscript λ is omitted for brevity)

$$\mathbf{\Omega}\nabla I(\mathbf{r},\mathbf{\Omega}) + \beta_{tr} I(\mathbf{r},\mathbf{\Omega}) = \frac{\sigma_{tr}}{4\pi}\int\limits_{4\pi} I(\mathbf{r},\mathbf{\Omega}')\mathrm{d}\mathbf{\Omega}'. \tag{10.123}$$

Dombrovsky et al. considered the axisymmetric problem for the laser induced hyperthermia assuming that

1. the body surface is optically smooth, i.e. the Fresnel's relations for reflection and refraction of the direct and scattered laser radiation at the body surface could be used;
2. the gold nanoshells are embedded in a finite circular volume around the tumour that is a small part of the computational region;
3. the tissue index of refraction is uniform in the computational region.

The authors additionally assumed that the angle between the current normal to the body surface and the normal to the flat healthy body surface is small, thus the local one dimensional solution for the directional radiation component could be used; the diffuse radiation component was determined using the well-known P1 approximation. The absorption and scattering coefficients of a composite medium which can be treated as the matrix (the host medium) containing a not too high volume fraction f_v of the small spherical particles of the same radius a was approximated as [783]

$$\alpha = \alpha_t + \frac{3}{4}\frac{Q_a}{a}, \qquad \sigma = \sigma_t + \frac{3}{4}\frac{Q_s}{a} \tag{10.124}$$

where Q_a and Q_a are the dimensionless efficiency factor of the absorption and scattering for the single particle for the gold-coated silica nanoshells taken from the paper by L. A. Dombrovsky et al. [784].

The heat transfer was computed using the two coupled energy equations for the tissue and the artery blood with the spatial and time variation of the arterial blood temperature according to the model by A. Nakayama and F. Kuwahara [2259]. The authors noted that the thermal conductivity and the radiative properties of a tissue containing the damaged cells are different from those of the healthy tissue [2451, 3488]. Particularly, the spectral absorption coefficient and the transport scattering coefficient of a damaged tissue is usually much greater than the corresponding properties of a healthy tissue [2308, 2309, 2451]. Thus the heat transfer model should include the repeating computations of the radiation field at every time step of the numerical solution for the transient problem.

M. M. Tung et al. [3096] used the hyperbolic heat transfer equation to study the pulsed laser treatment (laser thermokeratoplasty, LTK [2034]) and the radiofrequency hyperthermia (conductive keratoplasty, CK [266, 2426]) of the cornea to correct the refractive errors.

The laser heating implies the tissue heating caused by absorbing the optical energy of a high-energy laser beam. The authors considered a semi-infinite fragment of the homogeneous isotropic biological tissue in which the laser beam falls on the entire tissue surface. In the one-dimensional laser heating model the blood perfusion and the metabolic activity are neglected. The source term was obtained from the Beer-Lambert law that states that for radiation the intensity decreases exponentially with the penetration depth . Accounting for with the temporal dependence to model an energy pulse of the duration Δt Tung et al. got

$$Q(x,t) = (1 - R)bE_0 e^{-bx}[H(t) - H(t - \Delta t)] \tag{10.125}$$

where R is the dimensionless Fresnel surface reflectance, b is the absorption coefficient, E_0 is the incident energy flux at the tissue surface, $H(t)$ is the Heaviside function and the final form of the hyperbolic heat transfer equation as

$$\begin{aligned} & -\frac{\partial^2 T(x,t)}{\partial x^2} + \frac{1}{\varkappa}\left(\frac{\partial T(x,t)}{\partial t} + \tau\frac{\partial^2 T(x,t)}{\partial t^2}\right) \\ &= (1 - R)bE_0 e^{-bx}[H(t) - H(t - \Delta t) + \tau(\delta(t) - \delta(t - \Delta t))] \end{aligned} \tag{10.126}$$

where $\delta(t)$ is the Dirac delta function.

The boundary condition at the tissue surface

$$\frac{\partial T(0,t)}{\partial x} = \frac{h}{\lambda}\left(\tau\frac{\partial T(0,t)}{\partial t} + T(0,t) - T_a\right) \tag{10.127}$$

incorporates the Newton's law of cooling, T_a is the ambient temperature, $h > 0$ is the thermal convection coefficient.

10.2.1.4 Tissue Vaporization During Laser-Induced Thermotherapy

The laser-induced thermotherapy (LITT) is the minimally invasive method causing the tumour destruction due to the heat ablation and the coagulative effects [926]. The small intensity laser radiation in the Low Level Laser Therapy (LLLT) [2385] (with the power of light in the range of 1–1000 mW [1603]) has the direct effect on the molecular level of the organism [1223, 1555, 1603].

The LITT similar to other hyperthermia procedures utilizes the fact that the tumours tissue is more susceptible to the heat than the healthy tissue. The LITT is based on the local supply of the energy via an optical fibre located in a water-cooled applicator.

For the temperature of nearly 100° C the non-linear phenomena occur, such as the desiccation and the vaporization (the bubble formation) [262, 3339]. The appearance of the bubbles influence the absorption of the laser radiation: N. F. Bunkin et al. [404] showed that in the presence of the dissolved air the infrared laser irradiation heats the microbubbles of the air to the plasma temperatures that increases the absorption of the radiation, while in the absence of the dissolved gases, the water is essentially transparent due to the low probability of the multiphoton absorption.

The liver consists of about 80 % water which vaporizes if the temperatures during the treatment become sufficiently large. Usually the vaporization of the water is not included in the model of the liver ablation [1381] and there is agreement between the measured and simulated temperature until the temperature reaches approximately 100° C, then, the measured temperature stagnates while the simulated one rises further. This happens due to the phase change of water which was not included in the model [1381].

S. Blauth et al. [312] described the irradiation of the laser light by the stationary radiative transfer equation

$$s \cdot \nabla I + (\mu_a + \mu_s) I = \frac{\mu_s}{4\pi} \int\limits_{S^2} P(s \cdot s') I(s', x) \mathrm{d}s' \tag{10.128}$$

where the radiative intensity $I(s', x)$ depends on the direction $s \in S^2$ on the (unit) sphere and the position $x \in \Omega$, μ_a and μ_s are the absorption and scattering coefficients. The scattering phase function $P(s \cdot s')$ is given by the Henyey-Greenstein term [2298]

$$P(s \cdot s') = \frac{1 - g^2}{(1 + g^2 - 2g(s \cdot s'))^{3/2}} \tag{10.129}$$

where the parameter $g \in 2[-1, 1]$ is the anisotropy factor that describes the backward scattering for $g = -1$, the isotropic scattering in case $g = 0$ and the forward scattering for $g = 1$.

To solve the radiative transfer equation the authors used the so-called P1-approximation to model the radiative energy [2165]

$$I(s, x) = \varphi(x) + 3s \cdot q(x) \tag{10.130}$$

where radiative flux vector

$$q(x) = \frac{1}{4\pi} \int_{S^2} I(s, x) s \mathrm{d}s. \tag{10.131}$$

The diffusion equation for the radiative energy is [312]

$$-\nabla \cdot (D \nabla \varphi) + \mu_a \varphi = 0 \tag{10.132}$$

where the diffusion coefficient

$$D = \frac{1}{3(\mu_a + (1 - g)\mu_s)}. \tag{10.133}$$

The optical parameters μ_a, μ_s and g are sensitive to the changes of the tissue's state and once the coagulation of the cells starts, their optical parameters change and the radiation cannot enter the tissue as deeply as before. Blauth et al. [312, 313] assumed that the damage of the tissue is described by the Arrhenius law

$$w(x, t) = \int_0^t A \exp\left(-\frac{E_a}{RT(x, s)}\right) \mathrm{d}s \tag{10.134}$$

and the change of the optical parameters due to the coagulation is given by

$$\mu_a = \mu_{an}(1 - e^{-w})(\mu_{ac} - \mu_{an}), \quad \mu_s = \mu_{sn}(1 - e^{-w})(\mu_{sc} - \mu_{sn}), \tag{10.135}$$

$$g = g_n(1 - e^{-w})(g_c - g_n), \tag{10.136}$$

where the subscripts "n" and "c" indicate the properties of the native and coagulated tissue.

S. Blauth et al. [312] used two approaches to investigate the vaporization during the LITT of the liver:

- the effective specific heat (ESH) model introduced by D. Yang et al. [3400] which modifies the specific heat coefficient to account for the phase change;
- the enthalpy model which uses an additional state equation to model the phase transition.

In the effective specific heat model the phase transition is spread over a reasonably small interval around 100° C, thus the phase transition is modelled with a single PDE

$$\varrho C_p \frac{\partial T}{\partial t} - \nabla \cdot (\lambda \nabla T) + \xi_b (T - T_b) = Q_{rad} - Q_{vap} + Q_{cond} \tag{10.137}$$

where Q_{vap} is a source term for the vaporization of water and Q_{cond} is the source term for the condensation.

F. Hübner et al. [1381] suggested

$$Q_{vap} = -L \frac{\mathrm{d}W}{\mathrm{d}t} \tag{10.138}$$

where L is the latent heat of water and W is the tissue water density, thus

$$\varrho C'_p \frac{\partial T}{\partial t} - \nabla \cdot (\lambda \nabla T) + \xi_b (T - T_b) = Q_{rad} + Q_{cond} \tag{10.139}$$

where the effective specific heat capacity is given by

$$C'_p = C_p - \frac{L}{\varrho} \frac{\partial W}{\partial t}. \tag{10.140}$$

Since the water content decreases with temperature, $C'_p \geq C_p$.

S. Blauth et al. [312] used the experimental data [3401] for the water content of the bovine liver as a function of temperature.

The enthalpy model consists of two coupled equations, one for the temperature of the tissue and one for its enthalpy. S. Blauth et al. [312] formulated these equations as

$$\varrho C_p \frac{\partial T}{\partial t} = \begin{cases} \nabla \cdot (\lambda \nabla T) + \xi_b (T_b - T) + Q_{rad} + Q_{cond}, & if A \\ 0, & if B, \end{cases} \tag{10.141}$$

where A: ($T < 100°$ C or $T \geq 100°$ C and $H = \varrho L_{vap}$) and B: ($T = 100°$ C and $0 \leq H < \varrho L_{vap}$) and

$$\frac{\partial H}{\partial t} = \begin{cases} 0, & if A \\ \nabla \cdot (\lambda \nabla T) + \xi_b (T_b - T) + Q_{rad}, & if B. \end{cases} \tag{10.142}$$

In addition to the vaporization of the water, one also needs to consider the effect of the condensation of the water vapour in order to obtain an accurate model. There, it was assumed that the water vapour diffuses into the region of the lesser temperatures where it condensates and releases its latent heat obtained through the vaporization.

The authors used the experiments where the LITT was applied to the ex-vivo porcine livers from the study by F. Hübner et al. [1381] to test the vaporization models. They found that the ESH and the enthalpy model show comparable but slightly different temperature curves: the overestimation of the temperature during the middle of the experiment is usually higher for the enthalpy model.

The reason for the overestimated temperatures is probably the simplification of the condensation model: the vapour which was generated in one region can instantaneously condensate in another region, i.e., temperature can be shifted from one region to another without any delay.

The time-dependent temperature distributions produced within the biological tissues heated by a moving laser beam with the Gaussian and uniform power density profiles were studied by M. K. Loze and C. D. Wright [1965] using a time-domain method based on the Green's functions [1963].

This approach was used previously to determine the temperature distributions in the absence of the surface losses for the laser angioplasty applications by P. Loza et al. [1962] and the convective surface losses were included by M. K. Loze and C. D. Wright [1964]. The surface precooling of the skin for the port wine stain removal applications has been examined by B. Anvari et al. [95, 97] and by J. H. Torres et al. [3070].

M. K. Loze and C. D. Wright [1965] investigated the heating of a blood vessel within the skin. This situation arises in the removal of the birthmarks, such as the port-wine stains, where the laser is used to irreversibly damage the abnormal blood vessels causing the unwanted skin pigmentation. The thermal properties of the medium were assumed to be independent of temperature. Heat was produced in the medium by absorption of the optical energy from an incident laser beam. The laser beam of the wavelength Λ consisting of the plane waves and having the constant radius is incident along the positive z direction and is focused over a finite region of the medium. The light energy is absorbed within an arbitrarily shaped absorbing region causing a rise in the temperature of the medium. The temperature is governed by the 3D linear heat conduction equation

$$\varrho C \frac{\partial T(\mathbf{r}, t)}{\partial t} = \lambda \nabla^2 T(\mathbf{r}, t) + S(\mathbf{r}, t) \tag{10.143}$$

where $S(\mathbf{r}, t)$ is a heat source term.

When boundary heat losses can be modelled by the inclusion of the boundary conditions in the Green's function the temperature distribution can be obtained using the general solution [1965]

$$T(\mathbf{r}, t) = \frac{1}{\varrho C} \int_0^t \int_V G(\mathbf{r}, t|\mathbf{r}', t') S(\mathbf{r}', t') \mathrm{d}V' \mathrm{d}t' \tag{10.144}$$

where V is the volume of the region of interest and $G(\mathbf{r}, t|\mathbf{r}', t')$ is the medium Green's function

$$G(\mathbf{r}, t|\mathbf{r}', t') = G_{xy}(\mathbf{s}, t|\mathbf{s}', t')G_z(z, t|z', t') \tag{10.145}$$

where **s** is the position vector of the point (x, y) and

$$G_z(z, t|z', t') = \frac{1}{4\pi\varkappa\tau}\exp\left\{-\left[\frac{(x-x')^2+(y-y')^2}{4\varkappa\tau}\right]\right\} \tag{10.146}$$

where $\varkappa$ is the medium diffusivity $\lambda/\varrho C$,

$$G_z(z, t|z', t') = \sum_{m=0}^{\infty} A_m B_m(z) B_m(z') \exp(-\alpha_m \varkappa\tau). \tag{10.147}$$

Loze and Wright applied this model to the heating of the blood vessels within the dermis during the procedure for the port-wine stains removal. The temperature distributions obtained were in good agreement with those produced by the considerably more computationally intensive Monte Carlo optical and the finite difference thermal models. The authors found that the presence of the perfusion and the surface losses during laser heating do not appear to produce the significant reduction of the tissue temperatures.

10.2.1.5 Heat Transfer in Human Eye

Since there is no real time techniques to control the temperature distribution within the eye tissue, the numerical simulation of the interaction of the laser radiation with the human eye to determine the organ temperature is of the crucial importance—it could reduce the intraocular tissue damage induced by the heating. The ocular surface has to cope physiologically with the thermal stress of an environment that may change by more than 60° C which could dramatically affect the cellular mechanisms. For example, the thickness of the cornea depends on the endothelial cells which line the back side of that tissue and the enzyme systems within these cells that control the corneal thickness of the cornea are very temperature dependent, the blood flow, which responds to the temperature, may be affected adversely in these widely varying conditions. In the normal eye, the light is refracted first by the cornea and then by the lens to focus on the retina (the cornea contributes 60–70% of the refractive power of the eye [1109]).

The surgical methods for correcting the refractive errors of the human eye use the lasers, which are known to cause the local heating and potentially could lead to the loss of the corneal clarity as well as the opacity of the lens [1408, 3091]. The laser refractive surgery was developed to correct the refractive errors such as myopia, hyperopia and astigmatism. Two types of the laser surgical techniques—lamellar (e.g., the photorefractive keractomy, laser in situ keratomileusis or the laser sub epithelial keratomileusis) and thermal—are available to reshape the corneal curvature. The ultraviolet (UV) emitting argon fluoride (ArF) excimer laser is used to sculpt cornea in the lamellar procedures, whereas, the infrared emitting holmium yttrium aluminium garnet (Ho: YAG) laser is used to shrink cornea in the thermal procedure [1109].

The tissue heating is common in all types of the laser surgical techniques. However, the temperature monitoring of the interior of the human eye *in vivo* is not possible. Thus, the most important issue with the laser eye surgery is the estimation of the temperature rise in the eye tissues due to absorption of the high intensity laser radiation.

The laser-based treatments depend upon the light/tissue interactions that occur in the superficial layers associated with the retina and the retinal pigment epithelium (RPE); the laser radiation is primarily absorbed by the melanin granules in the RPE tissue [3525].

J. Scott [2751, 2752] was the first to use the finite element method to compute the heat transport in the human eye. A. Taflove and M. E. Brodwin [2999] probably were the first to determine the temperature rise inside the human eye due to the electromagnetic radiation. A. Emery et al. [856] computed the temperatures in the animal eye due to the microwave radiation, J. Scott [2752], K. Chua et al. [567] and M. A. Mainster [2015]—in the human eye due to the infrared radiation.

J. Scott [2751, 2752] used the two dimensional finite element model of the heat transfer in the human eye to study the cataract formation due to the infrared radiation. E. Ng and E.-H. Ooi [2293] studied the heat transfer in the human eye by applying the finite element method [2293]. A. Rafiq and M. Khanday [2561] investigated the effect of the change in the blood perfusion, porosity, evaporation, and the ambient temperature on the thermal behaviour of the human eye.

The errors in the heat treatment of the eye tissues could be dangerous, for example, the generated thermomechanical strains could cause the damage of the sensitive fibres in the optic nerve [1797, 3049] or the laser-induced thermal damage in the retina occurs [65]. The photothermal damage, as a result of absorption of radiation resulting in the localized heating, and the photomechanical damage, caused by the mechanical forces due to the rapid absorption of short pulses of light, are two of the primary damage mechanisms [304].

N. A. Alghamdi and H. M. Youssef [50] studied the interaction of the exponentially decaying laser radiation with the human eye using the dual phase-lag heat conduction model. The authors considered the eye as the spherical thermal conductive organ that consists of six layers:

- the cornea (a firm refracting surface—the transparent, dome-shaped window covering the front of the eye, supplying much of the eye's focusing power);
- the aqueous humor;
- the lens (the transparent layer, the biconvex structure, encased in a thin transparent covering);
- the vitreous humor;
- the retina;
- the sclera (the protective outer layer of the eye).

The inside of the eye body is divided by the lens into two fluid-filled sections. The vitreous humor is the most extensive section at the back of the eye that is filled with a colourless gelatinous mass; the smaller area in the front contains a transparent, water-like material called the aqueous humour [50].

Since the human eye mainly comprises the water, the metabolic heat generation plays a minor role in its temperature regulation. The sclera is the only part that has the blood vessels to helps the eye to keep its temperature. This region is modelled as a porous medium incorporating the blood circulation through the tissue.

The DPL bioheat equation with the porosity parameter β_i in the one-dimensional case is written as [2259]

$$(1-\beta_i)\left(1+\tau_{T_i}\frac{\partial}{\partial t}\right)\frac{\partial^2\vartheta_i(x,t)}{\partial x^2}=(1-\beta_i)\frac{\varrho_i C_i}{\lambda_i}\left(\frac{\partial}{\partial t}+\tau_{q_i}\frac{\partial^2}{\partial t^2}\right)\vartheta_i(x,t)$$

$$+\frac{W_{b_i}}{\lambda_i}\left(1+\tau_{q_i}\frac{\partial}{\partial t}\right)\vartheta_i(x,t)-\frac{1}{\lambda_i}\left(1+\tau_{q_i}\frac{\partial}{\partial t}\right)Q_i(x,t) \tag{10.148}$$

where $\vartheta(x,t) = T(x,t) - T_b$ is the temperature increment, $W_b = w_b C_b \varrho_b$, the porosity and the blood perfusion terms are zero in all human eye regions except the sclera and retina regions.

The heat loss from the surface S occurs through the convection and evaporation according to

$$-\lambda\left(1+\tau_T\frac{\partial}{\partial t}\right)\frac{\partial T}{\partial n}|_S = h_a(T_S - T_a) + E \tag{10.149}$$

where h_a is the ambient convection coefficient, T_a is the ambient temperature, E is the evaporative heat loss due to the tears and blinking, n is the normal direction on the surface S.

The temperature and heat flux must satisfy the continuity conditions at the boundaries between the six eye layers

$$T_i(x,t)|_{x=x_{i+1}} = T_{i+1}(x,t)|_{x=x_{i+1}}, \tag{10.150}$$

$$\lambda_i\left(1+\tau_{T_i}\frac{\partial}{\partial t}\right)\frac{\partial T_i}{\partial n}|_S = \lambda_{i+1}\left(1+\tau_{T_{i+1}}\frac{\partial}{\partial t}\right)\frac{\partial T_{i+1}}{\partial n}|_S, \tag{10.151}$$

$i = 0, 1, 2, 3, 4, 5.$

When the laser beam hits the skin surface, the laser energy is partially absorbed, scattered, and transmitted. The exponential decay of the laser power intensity along the tissue depth is governed by the Lambert–Beer law [1281, 3425, 3520]

$$I_i(x,t) = I_0 e^{-\delta x_i - \omega_i t}, \quad i = 0, 1, 2, 3, 4, 5 \tag{10.152}$$

where δ the penetration depth [50].

Alghamdi and Youssef [50] used the separation of variables

$$\vartheta(x,t) = \vartheta(x)e^{-\omega_i t} \tag{10.153}$$

and obtained the following set of the ordinary second-order homogeneous differential equations

$$\left(\frac{\mathrm{d}^2}{\mathrm{d}x^2} - \xi_i^2\right)\vartheta(x) = -\alpha_i e^{\delta_i x} \tag{10.154}$$

where

$$\xi_i^2 = \left[\frac{\varrho_i C_i(-\omega_i + \tau_{q_i}\omega_i^2)}{R_i} + \frac{W_{bi}(1-\tau_{q_i}\omega_i)}{R_i(1-\beta_i)}\right], \tag{10.155}$$

$$\alpha_i = \frac{\delta_i I_0(1-\tau_{q_i}\omega_i)}{R_i(1-\beta_i)}, \qquad R_i = \lambda_i(1-\tau_{T_i}\omega_i). \tag{10.156}$$

The set of the solutions has the form

$$\vartheta_i(x) = B_{1i}e^{\xi_i x} + B_{2i}e^{-\xi_i x} - \frac{\alpha_i}{(\delta_i^2 - \xi_i^2)}e^{-\delta_x}, \quad x_i \le x \le x_{i+1}. \tag{10.157}$$

The authors wrote out the system of the equations for the twelve coefficients $B_{10,20}, \ldots, B_{52}$.

Alghamdi and Youssef [50] performed calculations for the physiological parameters of the human eye and found, in particularity, that the perfusion rate of the blood parameter has the significant effects on the absolute temperature distribution in all the human eye layers and that the real values of the relaxation times parameters τ_q and τ_T are particularly important and have the significant effects on the heat transfer through the human eye.

B. Perrin [2433] performed the 3D boundary element method computation of the current density, the absorbed energy and the initial impedance of the human eye under the conductive keratoplasty [54] that is the non-ablative surgical technique for the treatment of the mild to moderate hyperopia (far-sightedness) [1885]. In this method a thin electrode penetrates the cornea delivering the pulsed RF energy at 350 kHz to the surrounding tissue yielding the thermally localised shrinkage and tightening of the collagen lamellae. This technique allows to correct the shape of the cornea and to treat the common types of the eye disease such as the far-sightedness or astigmatism.

The author assumed that the conductivity σ and the permittivity ε are constant. The integral formulation for the potential $\varphi(\mathbf{x}_s)$ at point $\mathbf{x}_s \in \Omega$ satisfies the boundary integral equation [364, 365]

$$c_s\varphi(x_s) + \int_\Gamma E_n^\star(\mathbf{x},\mathbf{x}_s)\varphi(x)\mathrm{d}\Gamma - \int_\Gamma \varphi^\star(\mathbf{x},\mathbf{x}_s)E_n(x)\mathrm{d}\Gamma = 0 \tag{10.158}$$

where Ω is the integration domain with boundary $\Gamma = \partial(\Omega)$, $\varphi^\star$ is the Green's function of the Laplace equation $\nabla^2\varphi^\star + \delta(\mathbf{x}, \mathbf{x}_s) = 0$, $E_n^\star$ is its normal derivative in the $\hat{\mathbf{n}}$ direction, c_s is a constant dependent on the Cauchy principal value integration of the singularity at the source point.

In the three dimensional space

$$\varphi^\star(\mathbf{x}, \mathbf{x}_s) = \frac{1}{4\pi r}, \quad E_n^\star(\mathbf{x}, \mathbf{x}_s) = \frac{\mathbf{r} \cdot \hat{\mathbf{n}}}{4\pi r^3}, \quad \mathbf{r} = \mathbf{x} - \mathbf{x}_s. \tag{10.159}$$

Using N_e flat linear triangular elements, the author got the discrete version of (10.158)

$$c_s\varphi_s \sum_{e=1}^{N_e}\sum_{\alpha=1}^{3} h(e,\alpha)\varphi_{e,\alpha} - \sum_{e=1}^{N_e} g^{(e)} E_{n,e} = 0 \tag{10.160}$$

where $\varphi_{e,\alpha}$ and $E_{n,e}$ are the potential and the normal electric field at the αth node in eth element, $g^{(e)}$ and $h(e,\alpha)$ are the BEM single and double layer integrals [364]. The assembly of this equation leads to the linear system of equations $Ax = b$, which solution provides the φ and E_n values at the boundary.

The author computed the distribution of the current density and the absorbed power within the eye including the far and near fields during the CK procedure. These results help to estimate the relative exposure level of other tissues, like the lens, the humour vitreous, the sclera, the iris anterior chamber and retina in addition to the cornea. The maximum EM power was found to be absorbed in the cornea and in the anterior chamber.

G. Sbirlea and J. P. L'Huillier [2724] used the three dimensional numerical model of the human eye to study the conduction and convection heat transfer due to an argon laser iridectomy.

The authors assumed that the aqueous and vitreous humors are viscous, incompressible fluids, the buoyancy force caused by the density variation resulting from the temperature rise is modelled using the Boussinesq approximation. The finite element method and the first order backward Euler algorithm were used.

In the laser iridectomy procedure, the laser beam strikes on the anterior corneal surface, penetrates the aqueous humor and is absorbed in the melanin layer of the iris tissue. The heat source absorbed in the iris melanin layer according to the Beer's law with a percent transmittance T_r and an absorption coefficient α is $Q(x) = T_r I_0 \alpha exp(-\alpha x)$, where x is the distance from the irradiated surface of the sample. The laser beam is assumed to be entirely transmitted by the cornea and the aqueous humor ($T_r = 1$). This assumption, evidently, overestimates the laser heat source.

Sbirlea and L'Huillier [2724] found that the convection currents induce an important modification of the temperature distribution. Along the pupillary axis the temperatures are higher on the vitreous humour, the lens and the aqueous humour in case of both the conduction and convection heat transfer then in case of the single conduction.

An overheating of the lens and the cornea endothelium by the side of the iridectomy site was observed that can be an explanation of the corneal burns and the lens opacities reported by the ophtalmoinlogists as the postoperative complications of the argon laser iridectomy [2724].

H. M. Youssef and A. A. El-Bary [3441] investigated the change in the blood perfusion, the porosity, the evaporation rate, and the ambient temperatures in the eye tissue using the non-Fourier heat conduction law.

Youssef and El-Bary [3441] considered the temperature variation on the pupillary axis where the eye is axisymmetric about and there is no need to use spherical coordinates [2270]. The authors used the Eq. (10.148); the blood perfusion and the porosity terms have been cancelled in all regions of the human eye except the sclera region, where the blood flow in the sclera is essential in maintaining the eye temperature regular with the other body organs temperature. The authors found that the lag-time, the blood perfusion, the porosity, the evaporation rate, and the ambient temperatures have significant effects on the temperature of the human eye.

P. Dai et al. [666] applied the finite element analysis to investigated the mechanical characteristics of the glaucoma in the human eye. At the high temperatures, the thermal properties of the living tissue and the blood perfusion change with temperature. The study of the non-Fourier energy transport is essential in the various biomedical applications to understand the heat transfer in the living tissues [1905, 2829, 3519].

A. Rafiq and M. Khanday [2561] formulated a mathematical model to study the thermal behaviour of the human eye in relation with the change in blood perfusion, porosity, evaporation and environmental temperatures. The thermal analysis had been performed on the multi-layered eye using Pennes bio-heat equation with appropriate boundary and interface conditions. The variational finite element method and MATLAB software were used for the solution. The thermoregulatory effect due to the blood perfusion rate, the porosity, the ambient temperature and the evaporation at various regions of human eye was considered. The main applications of this model the laser therapy and other thermoregulatory investigation on the human eye.

K. C. Gokul and D. B. Gurung [1108] investigated the thermal effects of two wheeler rider's speed in the eye when the high speed airflow into the cornea accelerates evaporation and the heat transfer.

Certain environmental conditions such as high or low temperatures, low humidity and wind flow are known factors to cause dry eye. High speed air into the eye causes significant decrease in lipid layer and tear stability. This may destroy or make thinner the lipid layer, which increases evaporation. The heat transfer between air molecules and cornea increases with air velocity. Since more air molecules would hit the corneal surface, high amount of heat is transferred by evaporation of water from precorneal tear film. The thermal boundary layer of air between cornea and environment becomes thinner. This decreases the diffusion distance between cornea and environment, causes high heat transfer.

The forced convection by hot/cold high speed airflow increases/decreases eye temperature. The eye will respond with more protective cooling/heating mechanism, such as more frequent blinking that agitates the pre-corneal tear film secreted at body core temperature. With each blink a warm lacrimal secretion is layered across the cornea giving heat both to the environment and to the cornea [1108].

There are four types of temperature effects of blinking on anterior corneal surface [1198]:

- heating/cooling due to spread of the warm tears and lipids across the surface of cornea;
- heating/cooling caused by the movement of the eyelid;
- heating/cooling via convection, radiation and tear evaporation;
- heating/cooling by the formation of the new layer of air over the surface of cornea.

In the outdoor condition, the wind accelerates the drop in temperature of the ocular surface below normal level. The eye injuries caused by cold exposure occur in individuals who try to force their eyes open in high wind or cold weather. The normal human corneal temperature ranges from 32 to 34° C [1015, 1599]. In the cold conditions and high-speed wind flow the temperature of the human eyelids and the corneal surface drops well below normal level. If the ocular surface and the eyelid temperature drop below 30° C the chances are that:

- Meibomian gland would impede its normal delivery of meibum [412];
- The blood flow rate in eyelid decreases due to vasoconstriction [1199];
- Uniform distribution of lipid layer is broken that causes greater evaporation [1665];
- Tear secretion increases to maintain eye temperature normal that causes excess tearing;
- Blink frequency increases to maintain normal eye temperature where possible [2253];
- Eye pain, blurred vision starts.

K. C. Gokul and D. B. Gurung [1108] studied the two dimensional human eye model that includes 8 components: cornea, aqueous humor, lens, vitreous humor, ciliary body, iris, retina, sclera. The human eye when eyelid is opened was considered to have 11 components: skin, orbicularis oculi, tarsal plate, cornea, aqueous humor, lens, vitreous humor, ciliary body, iris, retina and sclera.

The authors used the Pennes equation. The temperature distribution of human eye in case of eyelid opening and closure at ambient temperature 22.5 ° C and at air flow rates 20, 40 and 60 km/hr were considered. Rapid decrease in corneal temperature was observed at low atmospheric temperatures. The decreasing rate is highest in eyelid opening than blinking and lowest in closure. High amount of heat is lost to environment via convection and radiation in low ambient temperature. In open eye heat conducted from eye core to cornea is insufficient to maintain corneal temperature in normal level. Increased blinking rate is found to increase corneal temperature significantly. In our case anterior corneal temperature is increased by 2.74 ° C while blinking than in open eye at normal ambient temperature and in still air. The temperature difference at cornea during eyelid opening, blinking and closure increases as ambient temperature decreases and vice versa.

10.2.1.6 Soft Tissue Deformation

During thermal surgical procedures such as ablative treatment and hyperthermia, the soft tissue deformation due to surgical tool-tissue interaction and patient movement affect the distribution of thermal energy. Soft tissue temperature must be obtained for the deformed tissue to provide the precise delivery of thermal energy. Moreover, to enable a surgeon to visualize the simulated results immediately, the solution procedure must be fast, i.e. suitable for the real-time applications.

Various methods were reported for fast solutions of the bio-heat transfer models. Studies were focused on using parallel alternating direction explicit finite difference scheme [1277], fast Fourier transform (method [446, 757], exploiting the Graphics Processing Unit for the finite difference [1668] or finite element methods [2047], multi-grid technique [1400, 1401], the meshless point collocation method [344]. For example, M. Schwenke et al. [2749] considered the soft tissue deformation using the GPU-accelerated finite difference method. However, the finite difference method requires a regular computation grid occupying the problem domain for spatial derivatives, but human tissue/organs are irregular shapes with curvilinear boundaries, leading to difficulties in accurate computation of tissue material properties and boundary conditions.

The use of the GPU only improves the computational effort due to parallel solution. K. K. Brock et al. [380] compared the average change in prescribed dose between a static model, a model including rigid body motion and a model including rigid motion and deformation. Later the authors studied a fast explicit dynamics finite element algorithm [1831].

J. Zhang and S. Chauhan [3474, 3475] studied the tissue thermal responses on the deformed state of tissue. The authors used a transformation of the unknown deformed configuration of tissue to the known initial static non-moving configuration via a mapping function in the form of deformation gradient, which can accommodate large strains at the large deformation of the soft tissue.

10.2.2 Cryosurgery

The finite difference method for the computation of the two-dimensional temperature field around the cryoprobe was used for the first time probably by T. E. Cooper and C. J. Trezek [627]. Z. S. Deng and J.Liu [716] (see although the references in the cited paper) described the boundary element method for the computation of the temperature field in the two- and three-probes cryosurgery systems. The two-dimensional computations were performed using the finite element method in [3230] and the finite volume method on the unstructured triangular grids [843, 844]—in [3516]. An axisymmetric problem of the ice ball formation around a single cryoprobe was solved in the three-dimensional formulation in the Cartesian coordinates using the finite element method A. Fortin and Y. Belhamadia [989]. The anisotropic adaptation of the computational grid suggested by the authors in the earlier papers [238, 239] was used. The linearized equations were solved by the preconditioned (by the incomplete factorization method ILU) GMRES method. The software components from the PETSC library (http://www-unix.mcs.anl.gov/petsc/petsc-as/) were used. C. Kim et al. [1627, 1628] developed a finite element model to solve the cryosurgical problem with or without the large blood vessels. The effects of the large blood vessels were also investigated by O. I. Craciunescu and S. T. Clegg [641], Z. S. Deng and J. Liu [718] and by Z. Q. Sun et al. [2978].

M. Y. Ge et al. [1052] investigated the cryofreezing the liver tissue treated as a material having temperature-dependent thermophysical properties. The authors used the immersed boundary method (IBM) [2594, 3247] proposed by C. S. Peskin [2436] half a century ago. This method uses the Cartesian Eulerian grid points for the solution of governing equations and the Lagrangian points to represent the boundary of the immersed objects and thus avoids the tedious grid generation needed for the conventional body-fitted numerical methods. Two dimensional problem was considered under the assumptions that

- the phase transition of the biological tissue during freezing occurs in the interval between the lower and upper phase transition temperatures;
- the heat sources due to perfusion and metabolism are included for the unfrozen tissue.

The thermophysical properties of the liver tissue at the intermediate temperatures were linearly interpolated between the unfrozen and frozen values that were taken from the study by K. J. Chua [566].

M. Y. Ge et al. [1053] compared the results obtained by three different bioheat models—the Pennes model, the dual-phase-lag model and the W. Roeztel and Y. Xuan [2626, 2626] two-equation porous medium model [2626]. The authors found that at early times the tissue temperature profiles obtained by the porous model is lower than the other two models predict due to the porous parts filled with blood which increases the thermal conductivity of the tissue, and thus positively influences the freezing. However, as time elapsed, the temperature difference reduces.

The accuracy of the heat transfer computations increases if it is possible to monitor the evolution of the crystallization front [962, 2020, 2255, 3448, 3540]; in this case the boundary elements method is applicable [833, 2359]. The restoration of the domain of the cryoaction from the experimental data is a separate problem. Usually one gets a number of the two dimensional parallel image slices that should be processed to obtain the three dimensional shape of the tumour. The reconstruction of the three dimensional surface from a sequence of the two dimensional images is the ill-posed problem and the additional information should be used such as the smoothness of the transition between the adjacent input shapes and the smoothness of the reconstructed three dimensional surface; the different measures of the smoothness of the involved geometric entities could be chosen [47].

The solution of this ill-conditioned problem in the cryosurgery is greatly simplified if the *a priori* information on the shape and evolution of the frozen domain is available [2359]. Unfortunately, a considerable part of this information, for example, the assumption that the crystallization front is a simply connected convex single boundary surrounding the probe is valid for the single-probe cryosurgery systems only.

J. Alpers et al. [62] suggested a novel approach for the creation of a volumetric thermometry map without the development of a fully 3D sequence. The introduced 2.5D thermometry method utilizes any common 2D gradient-echo sequences. reference phase images are acquired for each of the eight orientations. Each newly acquired phase image will start computing the 2D thermometry map for the current orientation during the treatment. To do so, the proton resonance frequency shift (PRFS) method is used as described by V. Rieke and K. Butts Pauly [2605].

10.2.2.1 Heat Transfer in the Multiprobe Cryosurgery System

The use of the multiple cryoprobes systems allows more flexible control of the extension of the freezing domain. The success of the cryosurgery operation depends on the correct prescription of the cooling and thawing rates and on the exact localization of the domain of freezing that provide both the liquidation of the malignant tissues and the preservation of the healthy organs intact. It is not easy to accomplish due to the following reasons:

- As a rule, the cooling/thawing protocol rarely takes into account the patient specific anatomic data (the size and the geometrical shape of the tumour, the degree of the development of the capillary system, the closeness of the walls of important organs etc.).
- It is difficult to optimally place a large number of the cryoprobes (modern cryosurgery systems have up to 17 cryoprobes and, sometimes, a cryoheater) in space using the surgeon's experience alone.
- Methods of the surgery monitoring (the ultrasound diagnostics [3315] including such advanced modifications as the pulsed Doppler (PD) spectral flow analysis [1291] or the Doppler colour flow imaging (CFI) [572, 2053], the thermometric, the matrix thermovision [1418], the X-ray computer tomography [301], the NMR tomography [1791, 2035, 2998], the optical coherent tomography [553], the electric impedance tomography [834, 2027, 2554]) do not provide the complete information on the three-dimensional temperature field; the temperature distribution inside the ice ball could be determined either by the extrapolation based on the mathematical models or by the NMR tomography using the empirical correlation between the temperature and the measurements results in the Hounsfield units (HU) normalized by the contrast water-air; the accuracy of this approach is limited due to the sensitivity to the noise and the non-monotonous character of the correlation dependence. It should be also noted that the temperature of the leading front of the ice ball may be well below 0° C (e.g., – 8° C during cryoablation of the prostate [1515]).

It is relatively easy to determine the position of the boundary of the ice ball (for example, using the MR imaging due to the excellent contrast between the ice and the surrounding tissues [1243]), but not of the necrosis domain that usually has a volume equal to the one fourth of the volume of the ice ball [2022]. The required temperature should be attained with the safe margin around the tumour which is about 1 cm for the hepatic tumour and about 2–3 mm for the prostate cancer [219]. The Cryoablation Simulator (CryoSim) [1515] developed initially to support the cryosurgery of the prostate (but could be used for other organs as well) relies on the transformed ultrasound images of the ice ball and the simplified model of the heat transfer for the case of the constant thermal properties of tissues.

The value of the simulation is determined by the accuracy of its predictions. Y. Rabin [2544] listed the sources of the errors in the numerical simulation of the cryosurgery problems; one of the most important is the insufficient knowledge of the thermophysical properties of the living tissues at the low temperatures.

G. Zhao et al. [3503] used the finite element analysis implemented in the ANSYS commercial software to study the processes in the minimally invasive probe system capable of performing both the cryosurgery and the hyperthermia [1912]. The tissues were treated as the nonideal materials freezing over a temperature range, the thermophysical properties of which are temperature dependent. The enthalpy method is applied to solve the heat transfer problem with the phase change.

The heat transfer in the tissues when 3-, 6- or 18-probe cryosystems are used were simulated in [3516]. One of the aims of this study was the analysis of the sensitivity of the temperature field to the character of the temperature dependence of the effective thermal conductivity of the frozen tissues (this problem was earlier treated analytically for the one-dimensional case by Y. Rabin [2543]). The stationary problem was considered that corresponds to the exposition stage of the cryosurgery operation. In this case the temperature distribution and the position of the boundary between the frozen and unfrozen regions depend on the thermal conductivity only and do not depend on the specific heat of the tissues and on the latent heat of the phase transition. Note that the stationary two-dimensional formulation is the "worst case scenario" in respect to the preservation of the surrounding healthy tissues and organs [2546].

To derive the discrete equations, the finite volume method on the unstructured triangular grids was used [843]. The linearized equations were solved by the stabilized bi-conjugated gradient method Bi-CGSTAB [199].

10.2.2.2 Use of an Additional Heater in the Cryosurgery Systems

The use of the heater along with the cryoprobes increases the flexibility of the control over the extension of the freezing domain. The aim of using the cryoheater is to provide the protection to the important tissues and organs near the tumour that has to be destructed [2547]. The similar situation occurs when the large blood vessel is found near the tumour that prevents the effective freezing of the whole domain of the malignant tissue—the key reason for recurrence after cryoablation is the untreated tumour cells around large blood vessels [1514]. On the other hand, the possibility of unexpected complication will arise if the adjacent blood vessels are damaged [1052].

The additional cryoprobes should be placed in this case near the vessel to shield its action [259]. It should be stressed, however, that the price for the flexibility in the freezing control are the greater temperature gradients (and, hence, the greater level of the thermoelastic stresses).

An alternative to the cryoheater in the protection of the normal tissues in the vicinity of the target tumour tissues is the injection of the coolant with the nanoparticle to regulate the size and the orientation of the ice balls formed around the cryoprobe. This problem for a single cryoprobe was considered by R. Bhargava and H. Chandra [279] who used the Element Free Galerkin method (EFGM) to get the solution of the Pennes equation. The authors exploited the effective thermal conducting model of nanofluid containing the cylindrical nanoparticles proposed by S. M. S. Murshed et al. [2238].

10.2.2.3 Effect of the Tissue Thermal Conductivity

A comparison of the heat transfer computations for a single cryoprobe (an axisymmetric problem) for the cases of the constant and temperature dependent thermal conductivity was presented in [3516]. The following dependence for the effective tissue thermal conductivity (W/mK) [2281] has been used:

$$k(T) = \begin{cases} K(T), & T < T_1 \\ K(T_1) + (k_2 - K(T_1))(T - T_1)/(T_2 - T_1), & T_1 \le T \le T_2 \\ k_2, & T > T_2 \end{cases} \tag{10.161}$$

$T_1 = 260.2K, T_2 = 273.2K, k_2 = 0.49, K(T) = k_0/T^{\alpha}, k_0 = 2135, \alpha = 1.235$

It was shown that the heat transfer computations under assumption that the thermal conductivity is constant overestimate the temperature in the cryoprobe vicinity by tens of $^\circ$C and considerably underestimate the size of the cryoablation region. Earlier similar conclusions were made in [2543] on the basis of the analytical solutions of the one-dimensional equations in Cartesian, cylindrical and spherical coordinates using for the thermal conductivity the dependence $k = k_0/T^m$ (that corresponds to the temperature range $T < 260.2K$ in the relations presented above). Unfortunately, up to the present some of the heat transfer computations for the cryosurgery problems are performed for the constant or piece-wise constant thermal conductivity of the tissues [159, 2697, 3330].

In the computations performed as far as the author knows the thermal conductivity of the frozen tissue depends on the temperature only and usually is assumed to be equal to that of the ice. The accuracy of this assumption is hard to estimate *a priori*. The two cases were considered in [3516]:

- the constant thermal conductivity (note that its value does not present in the formulation of the stationary problem) and
- the temperature dependent thermal conductivity obtained by the matching of conductivity of the water and the ice.

Computations for the cryosurgery systems containing up to 18 cryoprobes showed the indeterminacy of the temperature field could be as high as several dozens of degrees Centigrade and a few centimetres in the extent of the cryoablation zone estimated by the value of the critical ("lethal") isotherm.

Development of the adequate model of the thermal conductivity of the biological tissues as the heterogeneous media that includes the ice, the extracellular matrix, the water—vitrified or bonded [1517, 1892, 3335]—is a separate problem. An approach based on the traditional methods for the effective properties of the composite materials [2454] is not sufficient since both the fraction of the unfrozen water and the state of the extracellular matrix vary during the cryoaction. It is well known that the first cycle "freezing - exposition - thawing" results in the increase of the thermal conductivity of the tissues [217] and in the considerable (dozens of per cents) increase of the volume of the ice ball for the second cycle [2021]. The destruction of the cellular membranes and the blood vessels are considered as the possible mechanisms [2350] of these alterations.

An adequate modelling of the heat transfer during the cryoaction thus should combine the macroscopic description of the heat transfer with the analysis of the media structure changes at the meso level such as the cells dehydration and the ice formation that, in turn, are mainly determined by the local cooling [167, 2350].

10.2.2.4 Mechanical Stress in Frozen Biological Objects

The mechanical stresses in the unfrozen tissues [2549, 2550, 2657] are of some interest, for example, for assessing the different possible approaches for the process of the cryoprobe insertion [47]. However, by far more important stresses in the frozen tissues that arise due to both the nonuniformity of the temperature distribution and the volumetric expansion at the phase change, since they could result in the fractures in the frozen tissues [1048, 1713, 2416] causing the irreversible organ damage and the postoperative bleeding [3366, 3506] or the stresses in the cryopreserved samples. The latter could be caused, in addition to the temperature gradients, by the nonuniformity of the tissue saturation by CPAs which are known to have the different thermal expansion coefficients [2551]. The danger related to the crack appearance in the vitrified samples is also related to the risk of the devitrification since defects could serve as the nucleation sites.

Z.-S. Deng and J. Liu [714] investigated the effects of the non-Fourier heat conduction on the thermal stress in the skin cryopreservation. The human allograft skin has been one of the most commonly used burn wound dressings [360], the fresh skin is often preferred. The only alternative for the long-term storage of skin is the cryopreservation, which was first suggested by A. Carrel [456]. The cryopreserved skin has only achieved 50–60% of the metabolic activity found in the fresh skin samples [714]. The mechanical stress that develops during the freezing of the biological tissues has been identified as one of the most important causes of the tissue damage [2548–2550, 2657].

Deng and Liu considered the simulation of the variant of the cryopreservation in which in order to avoid the film boiling when immersing the skin tissue directly into the liquid nitrogen an alternative cooling strategy for the cryopreserving is used—the skin is brought it into the contact with the copper plates at a very low temperature [1929]—under the following assumptions:

- the frozen tissue is treated as an elastic solid;
- the skin is a very thin layer, thus formulation is one dimensional;
- the skin is frozen symmetrically from its two sides;
- the relaxation times for the frozen and unfrozen phases are the same;
- the phase transition only occurs at a fixed temperature;
- thermal stress only occurs in the frozen tissue while the unfrozen tissue cannot support normal stress;

The authors found that the higher relaxation time results in the higher maximum thermal stress, and thus not considering the non-Fourier effect in the heat conduction in the biomaterial, which usually has large relaxation time, may lead to underestimation of the thermal stress.

10.2.2.5 Whole-Body Cryotherapy

The whole-body cryotherapy (WBC), or the cryosauna, [181, 347, 633, 3304] is a highly effective treatment method of a number of the diseases based on stimulating the cold receptors of the patient's skin, which provide supercooling of the skin surface to the level of $-2°$ C. To achieve such a temperature of the skin surface, it is necessary to ensure the heat removal with intensity not less than 3500 W/m2. Such a heat flux can remove gas with the temperature not higher than $-130°$ C.

D. Yerezhep et al. [3420, 3421] numerically studied the physical processes in the cryosauna using the finite element method. The mathematical model to calculate the temperature of the integumentary tissue containing the muscular layer and the subcutaneous layer during the WBC procedure included the Pennes equation and the turbulent flow equations that contain the two additional transport equations for the turbulent kinetic energy and the turbulent dissipation rate. The authors investigated the cryotherapeutic effect for patients with the different skin thickness. There is a need to study the effect that cooling has on the biological object with the different intensities of the heat removal along the length of the biological object. The authors analysed the heat flux losses from the surface of the skin of a biological object with a slight change in the geometry of the cryosauna during the whole-body cryotherapy process.

It is possible to increase the efficiency of the procedure by lowering the temperature of the cooling gas around the patient, by increasing the gas supply rate and changing the geometry of the nozzle in the cryosauna. However, the safety conditions cannot be neglected, and it is necessary to account for the fact that the gas flow distribution directly affects the uneven temperature distribution along the height of the cryosauna. The experimental determination of the temperature field distribution from the skin during the WBC process is difficult, since in a single cabin the temperature decreases once to $-130°$ C at a rate of 8° C/s [181].

10.2.2.6 Heat Transfer in Cryopreservation

The cryopreservation—the storage of cells, tissues, or other biological structure at low temperatures without injuring them—is accomplished via the two major methods

- the slow freezing [1268] with the low CPA concentration;
- the vitrification with the high concentration of CPA [2805, 2826, 3492].

Both methods aim to minimize or eliminate the cell damage during the cryopreservation by minimizing the intracellular ice crystal formation. The disadvantage of the slow freezing is the possibility of the tissue injury caused by the ice crystal formation,

The problem with the vitrification approach is the toxic effects and the osmotic shock associated with the high CPA levels. The thawing procedure could cause similar problems, such as the recrystallization and the osmotic shock. Most protocols adopt rapid thawing to prevent the intracellular ice formation by limiting time for the crystallization [1439, 3382, 3501].

Some types of cells such as, for example, the chondrocytes in the articular cartilage, are difficult for the preservation via the conventional methods [1574]. In this case the so called the liquidus tracking (LT) method is more suitable [852, 925, 2417, 3245, 3453]. The method consists in keeping the temperatures in the sample above or on the liquidus line preventing the ice crystallization and the exposition of the cryopreserved tissues to the high concentrations of the CPA at the same time [1574].

A. Piasecka-Belkhayat and A. Skorupa [2458] performed the two-dimensional numerical modelling of the heat and mass transfer in the cryopreservation of the articular cartilage sample. The authors stresses that the cryopreservation is the multiphysical field weak coupling problem that includes the modelling of the following processes:

- the heat transfer, including the phase change [1053, 1931, 2162]—the ice crystallization [1268, 2826, 3382, 3492]—and the moving boundary problem [1743, 1931, 3382];
- the mass transport at the macroscale in the extracellular matrix based on the Fick's second law [249, 1792, 2207, 2805, 2826];
- the mass transport at the microscale through the cell membrane to estimate the CPA concentration in the cells (Sect. 2.3.5).

The authors simulated the cryopreservation of the cylindrical ovine articular cartilage sample using the LT protocol proposed by D. E. Pegg et al. [2417], the chemical composition of the CPA solution is simplified to the mixture of the DMSO, KCl and the water. The finite difference method [2164] was applied.

The authors in order to account for the scatter of the experimental parameters instead of using some average values exploited the interval arithmetic [2163, 2457], thus results obtained are represented in the form of ranges that contain the correct simulation results.

The cryopreservation process following the LT protocol [2417] was simulated, the procedure includes 8 steps in a cooling phase and 7 steps in a heating phase. The history of the intracellular concentration, of the intracellular water volume in the chondrocytes and of the intracellular CPA moles number in the cell have been computed.

A. Skorupa and A. Piasecka-Belkhayat [2876] compared the heat transfer in the slow freezing and vitrification using the finite difference method with implemented directed interval arithmetic rules. The authors considered a model of a cylindrical piece of the biological material immersed a bath solution in such a way that the bath solution 'touches' the bases of the cylinder exploiting the Pennes equation with the additional source function $Q_{cr} = L\dot{f}_s$ where L is the latent heat and $\dot{f}_s = \partial f_s/\partial t$ represents the local change in volume of the frozen state in the intermediate region.

The three subdomains can be distinguished in the sample: natural, intermediate, and below freezing (melting) point. The function fs takes on a different value depending on the physical state considered in the subdomain. For the natural state and the state below the freezing (melting) point, this function is equal to 0 and 1. In the intermediate region it is assumed that $f_s = f_s(T)$.

The authors concluded that the temperature distribution for slow freezing is more regular and stabilized as a result of the time of the process. When the temperature of the bath reaches a defined minimum, the temperature in the entire area is below the freezing (melting) point. It can be assumed that it is timing that facilitates heat transport. In addition, slow freezing with an adequate cooling rate allows osmotic transport of water from the intracellular space to the extracellular matrix. In contrast, by analysing the temperature distribution in the middle of the freezing simulation, a considerable area covered by the intermediate region can be observed. This indicates a high risk of ice crystal formation.

F. Xu et al. [3382] studied the physical processes during cryopreservation that involves four steps: the CPA loading, freezing, thawing and the CPA unloading—the heat transfer at the macroscale level, crystallization, the cell volume change and the mass transport across the cell membranes at the microscale level.

The authors considered the channel-based cryopreservation methods that can be written using the cylindrical coordinate system using several heat transfer models

- the Fourier equation written by A. J. Jiao et al. [1478] as

$$\rho c \frac{\partial T}{\partial t} = \frac{1}{r}\frac{\partial}{\partial r}\left(\lambda r \frac{\partial T}{\partial r}\right) + \frac{\partial}{\partial z}\left(\lambda \frac{\partial T}{\partial z}\right) + \dot{q}_{met} \tag{10.162}$$

  simplified under several assumptions that

  - the latent heat owing to the ice crystallization is negligible;
  - the heat transfer in z direction (the straw axis) due to the geometry of a very thin and long channel is negligible;

- the properties are constant;
- the metabolic heat generation is negligible

to

$$\frac{1}{\kappa}\frac{\partial T}{\partial t} = \frac{1}{r}\frac{\partial}{\partial r}\left(r\frac{\partial T}{\partial r}\right) \tag{10.163}$$

where $\kappa = \lambda/\rho c$ is the thermal diffusivity.

- the hyperbolic (Cattaneo) heat equation— the thermal wave model of heat transfer

$$\tau_q \rho c \frac{\partial^2 T}{\partial t^2} = -\lambda \nabla^2 T + \left(\dot{q}_{met} + \tau_q \frac{\partial \dot{q}_{met}}{\partial t}\right). \tag{10.164}$$

- the simplest form of the DPL model with the first-order Taylor expansions for both q and T

$$\tau_q \rho c \frac{\partial^2 T}{\partial t^2} = -\lambda \nabla^2 T - \tau_T \lambda \nabla^2 \frac{\partial T}{\partial t} + \left(\dot{q}_{met} + \tau_q \frac{\partial \dot{q}_{met}}{\partial t}\right). \tag{10.165}$$

F. Xu et al. assumed that the ice crystallization (χ is the ratio of the total quantity of crystallized ice on cooling to the maximum crystallizable ice) is governed by the equation

$$\frac{\mathrm{d}\chi}{\mathrm{d}t} = \kappa a_1(\chi)(T_m - T)\exp\left(-\frac{Q}{RT}\right) \tag{10.166}$$

where $\kappa = L/\pi\delta^2 \nu T_m r_f$, $a_1(\chi) = \chi^{2/3}(1-\chi)$, Q is the activation energy, T_m is the final temperature of the freezing process, R is the gas constant, L is the latent heat, δ is the thickness of the transition layer between liquid and solid, r_f is the radius of the ice region, ν is the kinematic viscosity.

To load and unload the CPAs, the macroscopic flow along with the CPAs can be used—a microfluidic device can be designed and fabricated to load cells with the CPAs along a microfluidic channel, where the concentration of the CPAs varies along the channel by the diffusion [2911]. The macroscopic flow at steady state is governed by the Navier–Stokes equations. F. Xu et al. modelled the CPA transport using the convection and diffusion equation at steady state

$$\nabla \cdot (c\mathbf{v}) = \nabla \cdot (D\nabla c) \tag{10.167}$$

where c is the CPA concentration and D is the diffusion coefficient of the CPA.

10.2.3 Optimization of the Thermal Treatment

The aim of the optimization is to maximize the therapeutic effect while minimizing unwanted side effects. The suboptimal hyperthermia or the cryosurgery may leave the cells in the tumour nontreated, may lead to the damage of the healthy surrounding tissues, increase the duration of the operation and increase the risk of the post surgery complications [1671].

The threshold-temperature methods based on an isothermal contour to determine the boundary of the damaged tissue region are convenient to use, but the tissue damage depends on both temperature and the exposure time because the process is kinetic in nature [2405].

For example, D. Haemmerich et al. [1207] showed that, compared to the prediction of the Arrhenius damage model, the 50 °C-isotherm underestimated the lesion diameter by 4% for hepatic ablation and overestimated by 4.8% for cardiac ablation.

The treatment planning usually consists of the following steps [2290, 2624, 3518]:

- the acquisition of the patient-specific data with computed tomography, magnetic resonance (MR) or other medical imaging techniques;
- the segmentation of the acquired images to generate a patient representation, which includes information such as the location and shape of the targeted tumour, the surrounding vascular anatomy, etc.—the tissue segmentation is an important step of the planning since the properties, which determine the energy absorption in the tissue, vary significantly between the different tissues and organs in the human body [1670].
- the modelling and the numerical simulation of the key biophysical processes during the thermal treatment, including the energy deposition, the heat transfer and the thermal damage;
- the optimization of the steering parameters based on the simulation results;
- the post-processing and visualization.

To formulate an optimization problem one should specify

- optimization (control) parameters $\mathbf{p} = (p_1, p_2, \ldots, p_n)$
- a goal function $\mathcal{F}(\mathbf{p})$

The choice of the control parameters in, for example, the cryosurgery is evident:

- the temperature (to be more exact, its variation in time—the cooling and thawing rates and the duration of the exposition ("holding"));
- number of the cryoprobes;
- the positions of the cryoprobes [1984, 2552, 2703]

while in the MW hyperthermia the control parameters are the amplitudes and phases of the antennas of the heating system.

Sometimes (and always in the cryopreservation) the additional parameters define the saturation of the tissues by the cryoprotectant and its removal (usually a tuple of the three values—the concentration, the temperature and the time interval) for all the process stages [1551, 2771]).

A stepped cooling with a long exposition at the intermediate temperature is often used in the cryoconservation [1457, 2450] to provide the sufficient time for the extracellular ice formation and the dehydratation of the cells with a subsequent fast cooling [2444, 3066]; this procedure is referred to as the Interrupted Slow Freezing (ISF) [1911]. Note that the traditional specification of the cooling regimes by the piecewise constant cooling rate is not optimal due to the considerable non-linearity of the tissues' response [2191]. The cryoprotectant unloading in several steps is shown to increase the post-thaw fertilisation rate of the oocytes, probably due to avoiding the rupture of the fragile plasma membranes by the shearing pressure of the water rushing out of the cell to the hypotonic environment in the case of the single step unloading [2687].

It is more difficult to write the goal function [724]. In the radiation therapy a radiation dose or a probability of the destruction of the pathological tissue (based on the approximation of the experimental data) [45] are used and the corresponding goal functions are call physical and biological. Since reaction of the tissue depends on the dose that get surrounding tissues [1402], a histogram "dose-volume" is also exploited.

In the hyperthermia thermal dose quantities are optimised, which is similar to an "Arrhenius relationship" [741]. O. Szasz and A. Szasz [2994] discussed what dose definition will correspond well to the clinical protocols and listed the drawbacks of the temperature as the dose measure (the temperature does not satisfy the elementary supposition—the size of the target determining the applied dose—it does not depend on the size of the target [2995]) and advocated the absorption rate (the specific absorbed energy) criterium.

During the hyperthermia and the thermal ablation, a homogeneous tumour temperature is pursued. However, achieving this is a challenge due to the heterogeneous perfusion. An additional difficulty for the hyperthermia treatments are hot spots in the normal tissue, which limit the overall power deposition [1669].

H. W. Huang et al. [1379] developed a fast and robust temperature-based adaptive power scheme to optimise the tumour temperature, taking into account the thermal influence of the large thermally significant blood vessel in the tumour region . The optimisation becomes more computationally intensive when including vessel networks. H. W. Huang et al. [1378] demonstrated the importance of taking the thermal influence of the multiple blood vessels into account during the optimisation of power and temperature distributions for the hyperthermia treatments. A countercurrent network of straight vessels, consisting of seven generations was modelled. The optimisation was performed by continuously adjusting the power deposition in the target region, thereby aiming for the homogenous tumour temperature. Results showed that the high resolution power steering is needed to realise the homogenous temperature throughout the target region.

In the microwave hyperthermia the antenna feedings are optimized to maximize the specific absorption rate inside the tumour (the SAR-based optimization [241]—it is based on the correlation between the SAR indicators, the temperature, and the clinical outcome [434]), or to directly maximize the temperature there (the temperature-based (the T-based) optimization [1667]), involving a higher numerical cost since at every step of the optimization one has to solve the heat transfer equation [1029]. Since different tissue types have different perfusion levels, a high SAR will not always induce a high temperature. Since perfusion and other important thermal aspects like conduction and bolus cooling are not taken into account in SAR based optimisation, a more clinically relevant temperature based optimisation routine was desired: the temperature in the tumour has to be maximised, while respecting user-defined constraints to normal tissue temperatures. Therefore, SAR based optimisation does not correctly account for all temperature hot spots in normal tissue. To approximate the influence of perfusion in SAR based optimisation, the local SAR value can be divided by the local tissue perfusion [1666, 3359].

As R. Gaffoglio et al. [1029] noted that the deviations between the temperature and SAR distributions are due to the thermal boundary conditions, arising from external cooling and physiology, like the air flow in the respiratory tracts, and the major blood vessels. To asses the efficiently of the microwave power localization the target-to-hotspot SAR quotient (THQ) is introduced [1029, 2606]

$$THQ = \frac{SAR_{target}}{SAR_V}. \tag{10.168}$$

The efficiency of the SAR-based and the T-based planning is limited due to the imprecise modelling of the patient anatomy, tissue interfaces, the large uncertainty of the thermal tissue parameters [697, 2398]. For example, for the dielectric properties, variations of 30% or more have been measured in post-mortem humans [1021],and even greater variations were shown in the breast [2983, 3465] and the brain, by means of the MR measurements [3152].

The hyperthermia treatment planning (HTP) [1755, 2398, 2507] can be divided into three stages:

1. Generation of the patient model; for the improved accuracy, the patient-specific modelling [92, 1631, 2506] is performed by the segmentation the tissues using the computed tomography (CT) or the magnetic resonance (MR) images;
2. Calculation of the distribution of power deposited in the tissue—the specific absorption rate (SAR), or more generally the power density (PD) distribution is determined by the EM or the US simulation approaches; due to the highly heterogeneous and geometrically complex structure of the human body, numerical simulation techniques are essential [1666];
3. Calculation of the resulting temperature distribution in the tissue.

Due to computer limitations, the segmented CT scans have to be downscaled to a lower resolution before calculating the electric field distribution. Several methods for downscaling have been compared by J. B. Van de Kamer et al. [3141].

In order to receive clinical acceptance, the results obtained from the optimization model must be verified with real clinical data. However, the temperature distribution measurement comes with errors especially for temperature measurement of a deep-seated tumour. The significance difference in thermal conductivity of metal thermocouple and surrounding biological tissue causes errors in the temperature measurement [448]. This thermal conductivity differences issue can be overcome by applying insulation on the probe [2982]. However, the use of insulation will also cause an error due to the temperature drop across the sheath wall [1392]. The use of a magnetic field for heating nanoparticles may also heat up the metal thermocouples through the eddy currents heating. New alternative temperature mapping technique may offer a solution for obtaining higher accuracy clinical temperature reading, such as the use of magnetic resonance imaging for temperature mapping during hyperthermia treatment [459].

The optimization of the MFH should ensure that the required amount of nanoparticles is as small as possible while the nanoparticles are still able to produce high enough heating, the potential overheating may be avoided and any harm to the surrounding healthy tissue may be minimized [150, 2982]. Changing the nanoparticle size can significantly alter the ability of an the MNP to generate heat [2333, 2638] making the determination of an optimum nanoparticle size necessary for a specified set of conditions.

Planning of the hyperthermia treatment is usually based on the quadratic function defining the deviation of the computed temperature field T from the desired (optimal) one T_o:

$$\mathcal{F}(\mathbf{p}) = \int_{\Omega} (T - T_o)^2 dV; \qquad (10.169)$$

Sometimes instead of the averaged values one or several control points in the tumour region are considered [2025, 3045].

It is not necessary to restrict the tumour temperature from above. One should just specify T_d—the minimal temperature that provides the destruction of the pathological tissues. Additionally one should specify the "safe" temperature value—the maximal acceptable temperature for the healthy tissues T_{lim}.

In the solution of the optimization problem

$$\min \mathcal{F}(\mathbf{p}), \qquad \mathcal{F}(\mathbf{p}) = \int_{\Omega_{\text{tumour}}, T<T_d} (T - T_d)^2 dV \qquad (10.170)$$

restrictions on the temperature $T \leq T_{lim}$ in the region of the healthy tissue $\Omega \setminus \Omega_{\text{tumour}}$ could be either used explicitly as the box type constraints [729] or introduced into the goal function using the penalty approach [728]

$$\mathcal{F}(\mathbf{p}) = \int\limits_{\Omega_{\text{tumour}}, T<T_d} (T - T_d)^2 dV + w \int\limits_{\Omega\setminus\Omega_{\text{tumour}}, T\geq T_{\text{lim}}} (T - T_{\text{lim}})^2 dV. \quad (10.171)$$

where w is the weight factor.

H. G. Bagaria and D. T. Johnson [150] formulated for the optimization model for the MFH the objective function as $\mathcal{F} = \sqrt{E_1 + E_2}$ where $E_1 = w_1 \sum(T - 42)^2$ is the error function for the insufficient overheating in the tumour, $E_2 = w_1 \sum(T - 37)^2$ is the error function for the overheating in the healthy tissue, w_1 and w_2 are the weight factors. M. Mital and H. V. Tafreshi [2153] suggested to estimate the thermal damage value as $\int_0^{t_f} R^{(43-T)} dt$, where R is the empirical constant, t_f is the period of time.

Similar approaches are used in the cryosurgery [160, 1984]. The simplest one to specify a goal function is to define the critical ("lethal") temperature that provide the tissue destruction

$$\mathcal{F}(\mathbf{p}) = \int\limits_{\Omega} w(T) dV, \quad (10.172)$$

where the weight function $w(T)$ equals to 0 or 1 in the corresponding temperature intervals and the geometrical regions [2552, 2643]. However, this "lethal" temperature does depend on the tissue type and other factors, including the cooling/thawing protocol; the reported value differs greatly from $-2°$ C for the dog osteocytes down to $-70°$ C for the rat adenocarcinoma [3506]. W. K. Berger and J. Poledna [259], e.g., assumed the lethal temperature to be equal to $-50°$ C in their optimization analysis of the optimal probe placement in the cryosurgery of the liver [259].

R. Baissalov et al. [159] optimized the cryoaction for the six-probe system. To cope with the inertia and the non-linear response of the cryoprobes, the calibration measurements were performed and used to parameterize the temperature of the cryoprobe's tips. During the "stick" period (a clinical procedure to get the cryoprobe adhered to the tissue) $0 < t < t^{\text{st}}$ it was assumed to equal to $-10°$ C. During the freezing period (the cryosurgical operation itself) the temperature history was approximated by the two matched exponential functions

$$t^{\text{st}} < t < t^{\star} : T_i(t) = T_i^P \left(1 - b_{1i} \exp[-a_1(t - t^{\text{st}})]\right), \quad (10.173)$$

where $b_{1i} = (T_i^P - T_i^{\text{st}})/T_i^P$ and

$$t^{\text{st}} < t < t^{\text{off}} : T_i(t) = T_i^P \left(1 - b_{2i} \exp[-a_2(t - t^{\text{st}})]\right), \quad (10.174)$$

$b_{2i} = (T_i^P - T_i^{\star})/T_i^P$.

The parameters, including the temperature of dependencies matching $t^\star$, were found by fitting the experimental thermal histories. The objective (goal) function was formulated as

$$F_T = \sum_{i=1}^{n}[w(T^{\text{pr}} - T_i)^2 + \alpha S_T(T_i - U^{\text{pr}})] + \sum_{j=1}^{l}\sum_{i=1}^{m_j}\beta_j S_{N,j}(L_{ij}^{\text{pr}} - T_i), \qquad (10.175)$$

where T^{pr} is the prescribed temperature in the anatomical region of interest (ROI), l is the number of the normal tissues region near the tumour, n and m are the number of the points where the temperature is prescribed for the tumour and jth normal tissue, respectively, T_i is the actual temperature, α is the penalty factor for the exceeding the given upper temperature limit in the tumour U^{pr}, β is the penalty factor for the exceeding the given lower temperature limit in the healthy tissue, w is the weighting factor and S_T and $S_{N,j}$ are the step functions given as

$$S_T = \begin{cases} 1, & U^{\text{pr}} - T_i < 0 \\ 0, & U^{\text{pr}} - T_i > 0 \end{cases} \qquad (10.176)$$

and

$$S_{N,j} = \begin{cases} 1, & L^{\text{pr}} - T_i > 0 \\ 0, & L^{\text{pr}} - T_i < 0 \end{cases} \qquad (10.177)$$

These authors in the subsequent paper [160] had named the goal function presented above as the "quadratic temperature" form and also considered the "linear volume" form

$$F_V = \sum_{i=1}^{I_T} S_i^T(P_i^T - V_i^T) + \sum_{j=1}^{l}\sum_{i=1}^{I_N} S_{i,j}^N(V_i^{N,j} - P_i^{N,j} \qquad (10.178)$$

where I_T and I_N are the number of the isotherms considered for the tumour and the normal tissues, respectively, l is the number of the normal tissue regions, P and V with the corresponding sub- and superscripts are the prescribed and the computed volumes of the tissues completely enclosed by the ith isotherm. Weighted combination of the goal functions $F_C = w_P F_P + w_V F_V$ also was considered.

In total the twenty-four variables (the polar angle, the radius, the target temperature and the freeze time for each cryoprobe) were simultaneously optimized using the Broyden-Fletcher-Goldfarb-Shanno algorithm with an approximate reconstruction of the Hessian in the course of the iterations; the needed gradients of the goal function were determined numerically. The linear volume form was found the fastest to converge; the two other forms of the goal functions, however, provide the better solution.

In ref. [3230] the "index of the freezing exposition" is introduced as a product of the exposition time at the temperature lower than the prescribed one by the average temperature during this time interval

$$I = \int_{T_1}^{T_2} T(t)dt. \tag{10.179}$$

Unfortunately, the considered criteria are not universal ones. For example, one of the optimization goals could be minimization of the operation duration [568].

The high fidelity direct problem solvers usually are not fast enough. For example, H. P. Kok et al. [1668] stated that the very long computation time of the finite difference based DIscrete VAsculature model (DIVA) [2542] developed for the heat transfer in the system of the blood vessels makes this model impractical for the clinical applications.

Thus, one is often forced to use, at least at the early optimization stages, simplified "surrogate" models [282, 3514]. For example, S. K. Das et al. [676] used two computational schemes to accelerate the temperature optimization process: (1) the temperature distribution evaluation by the superpositioning the precomputed distributions, and (2) using the representative tissue groups (rather than every point in the domain) to evaluate the goal function.

Such low fidelity models could be physically "motivated" (do not read "justified")—e.g., reducing the spatial dimension of the problem, ignoring the "small" details, assuming the thermophysical properties to be temperature independent—or derived as the black-box models via the multivariate approximation (the regression methods, the neural networks, the kriging etc. [1072]). For example, N. Afrin and Y. Zhang [19] used the Design of Experiments (DoE) methods—the Sample Random Sampling (SRS) and the Latin Hypercube Sampling (LHS)—to build a surrogate model for the generalized dual phase lag model.

Chapter 11
Conclusions

There are several approaches for analysis of heat transfer in living tissues differing in sophistication and complexity.

The drawback of advanced models is the increased number of parameters and/or anatomy information needed to characterize the tissue. The simple Pennes bioheat equation remains a valid tool for solution of practical problems of hyperthermia, cryosurgery, thermal comfort analysis, thermal diagnostic, burn injury evaluation.

A comment should be made on the model completeness. Evidently, one has to search for a compromise between the model completeness and its tractability [3509, 3514]. The right attitude to the model development could be drawn from a letter by Blaise Pascal (14 Dec 1656): "Je n'ai fait celleci plus longue que parce que je n'ai pas eu le loisir de la faire plus courte" ("I have made this letter so long only because I did not have the leisure to make it shorter"). Thus, model development is finished not when all the relevant physical phenomena are accounted for, but when you could not simplify the model without corrupting it. This final model could be referred to as a "minimal model".

A. I. Zhmakin, *Heat Transfer in Vivo*, Springer Series in Biophysics 27,
https://doi.org/10.1007/978-3-031-95459-7_11

References

1. Aarset AV (1982) Freezing tolerance in intertidal inverterbrates (a review). Comp Biochem Physiol A 73:571–580
2. Abakali J, Ariahu CC, Nkpa NN (2006) Kinetics of osmotic dehydration of mango. J Food Process Preserv 30:597–607
3. Abazari A, Elliott Law GK, Mcann LE, Jomha JM (2009) A biomechanical triphasic approach to the transport of nondilute solutions in articular cartilage. Biophys J 97:3054–3064
4. Abbott BC, Hill AV, Howarth JV (1958) The positive and negative heat production associated with a nerve impulse. Proc R Soc Lond B Biol Sci 148:149–187
5. Abdel-Hamid B (1999) Modeling non-Fourier heat conduction with periodic thermal oscillation using the finite integral transform. Appl Math Model 23:899–914
6. Abdeljawad T (2014) On conformable fractional calculus. arXiv: 1402.6892 [math.DS]
7. Abkarian M, Lartigue C, Viallat A (2002) Tank treading and unbinding of deformable vesicles in shear flow: determination of the lift force. Phys Rev Lett 88:068103
8. Abkarian M, Faivre M, Viallat A (2007) Swinging of red blood cells under shear flow. Phys Rev Lett 98:2–5
9. Abkarian M, Faivre M, Horton R, Smistrup K, Best-Popescu CA, Stone HA (2008) Cellular-scale hydrodynamics. Biomed Mater 3:034011
10. Abouelregal AE, Abo-Dahab SM (2015) Study of the dual-phase-lag model of thermoelasticity for a half-space problem with rigidly fixed surface in the presence of a thermal shock. J Comput Theor Nanosci 12:38–45
11. Abouelregal AE, Khalil KM, Mohammed FA, Nasr ME, Zakaria A, Ahmed IE (2020) A generalized heat conduction model of higher-order time derivatives and three-phase-lags for non-simple thermoelastic material. Sci Rep 10:13625
12. Abraham JP, Sparrow EM (2007) A thermal-ablation bioheat model including liquid-to-vapor phase change, pressure- and necrosis-dependent perfusion, and moisture-dependent properties. Int J Heat Mass Transf 50:2537–2544
13. Abramowitz M, Stegun AI (1972) Handbook of Mathematical Functions. Dover, New York
14. Acker JP, Croteau IM (2004) Pre- and post-thaw assessment of intracellular ice formation. J Microsc 215:131–138
15. Acker JP, McGann L (2003) Protective effect of intracellular ice during freezing? Cryobiology 46:197–202
16. Acker JP, Lu XM, Bayley H, Fowler A, Toner M (2003) Measurement of trehalose loading of mammalian cells porated with a metal-actuated switchable pore. Biotech Bioeng 82:525–532

A. I. Zhmakin, *Heat Transfer in Vivo*, Springer Series in Biophysics 27,
https://doi.org/10.1007/978-3-031-95459-7

17. Adams JD, Scott DM, Brand NA, Suh HG, Seal AD, McDermott BP, Ganio MS, Kavouras SA (2019) Mild hypohydration impairs cycle ergometry performance in the heat: a blinded study. Scand J Med Sci Sports 29:686–695
18. Adey WR (1981) Tissue interactions with non-ionizing electromagnetic fields. Physiol Rev 61:435–514
19. Afrin N, Zhang Y (2019) Surrogate-based optimization of thermal damage to living biological tissues by laser irradiation. Front Heat Mass Transf 12:27
20. Afrin N, Zhang Y, Chen J (2014) Dual-phase lag behavior of a gas-saturated porous-medium heated by a short-pulsed laser. Int J Therm Sci 75:21–27
21. Afrin N, Zhang Y, Chen JK (2011) Thermal lagging in living biological tissue based on nonequilibrium heat transfer between tissue, arterial and venous bloods. Int J Heat Mass Transf 54:2419–2426
22. Afrin N, Zhou J, Zhang Y, Tzou DY, Chen JK (2012) Numerical simulation of thermal damage to living biological tissues indused by laser irradiation based on a generalized dual phase lag model. Numer Heat Transf A 61:483–501
23. Affortit C, Albury MS, Crichton PG, Moore AL (2002) Exploring the molecular nature of alternative oxidase regulation and catalysis. FEBS Lett. 510:121–126
24. Aghebati-Maleki A, Dolati S, Ahmadi M, Baghbanzhadeh A, Fotouhi A, Yousefi M, Aghebati-Maleki L (2020) Nanoparticles and cancer therapy: perspectives for application of nanoparticles in the treatment of cancers. J Cell Physiol 235:1962–1972
25. Agnese DM Burak WE (2005) Ablative approaches to the minimally invasive treatment of breast cancer. Cancer J 11:77–82
26. Aguilar G, Diaz SH, Lavernia EJ, Nelson JS (2002) Cryogen spray cooling efficiency: improvement of port wine stain laser therapy through multiple-intermittent cryogen spurts and laser pulses. Lasers Surg Med 31:27–35
27. Aguilar JFG (2014) Hernández: space-time farctional diffusion-advection equation with Caputo derivative. Abstr Appl Anal 2014:283019
28. Ahima RS (2020) Global warming threatens human thermoregulation and survival. J Clin Invest 130:559–561
29. Ahmadikia H, Moradi A (2012) Non-Fourier phase change heat transfer in biological tissues during solidification. Heat Mass Transf 48:1559–1568
30. Ahmadikia H, Rismanian M (2011) Analytical solution of non-Fourier heat conduction problem on a fin under periodic boundary conditions. J Mech Sci Technol 25:2919–2926
31. Ahmadikia H, Fazlali R, Moradi A (2012) Analytical solution of the parabolic and hyperbolic heat transfer equtions with constant and transient heat flux conditions on skin tissue. Int Commun Heat Mass Transf 39:121–130
32. Ahmadikia H, Moradi A, Fazlali R, Parsa A (2012) Analytical solution of non-Fourier and Fourier bioheat transfer analysis during laser irradiation of skin tissue. J Mechan Sci Technol 26:1937–1947
33. Ahmed GMY, Singh A (2011) Numerical simulation of particle migration in asymmetric bifurcation channel. J Non-Newtonian Fluid Mech 166:42–51
34. Ahmed M, Brace CL, Lee FT, Goldberg SN (2011) Principles of and advances in percutaneous ablation. Radiology 258:351–369
35. Ai H, Wu S, Gao H, Zhao L, Yang C, Zeng Y (2012) Temperature distribution analysis of tissue water vaporization during microwave ablation: experiments and simulations. Int J Hyperthermia 28:674–685
36. Aidun CK, Clausen JR (2010) Lattice-Boltzmann method for complex flows. Ann Rev Fluid Mech 42:439–472
37. Aiello LC, Wheeler P (1995) The expensive-tissue hypothesis: the brain and the digestive system in human and primate evolution. Curr Anthrop 36:199–221
38. Ainouz A (2014) On a multiscale analysis of a micro-model of heat transfer in biological tissues. arXiv:1410.7203v1 [math.AP]
39. Akbarzadeh AH, Chen Z (2013) Heat conduction in one dimensional functionally graded media based on the dual phase lag theory. Proc Inst Mech Eng C 227:744–759

40. Akbarzadeh AH, Pasini D (2014) Phase-lag heat conduction in multilayered cellular media with imperfect bonds. Int J Heat Mass Transf 75:656–667
41. Akbarzadeh AH, Fu J, Chen Z (2014) Three-phase-lag heat conduction in a functionally graded hollow cylinder. Trans Can Soc Mech Eng 38:155–171
42. Akbarzadeh AH, Cui YY, Chen ZT (2017) Thermal wave: from nonlocal continuum to molecular dynamics. RSC Adv 7:13623–13636
43. Akhmanov SA, Vysloukh VA, Chirkin AS (1988) Optics of Femtosecond Laser Pulses (in Russian). Nauka, Moscow
44. Al Nahas R, Charles A, Panicaud B, Rouhaud E, Choucair I, Saliya K, Kerner R (2020) Investigation on the use of a spacetime formalism for modeling and numerical simulations of heat phenomena. Int J Nonequil Thermodyn 45:223–246
45. Alber M, Nüsslin F (1999) An objective function for radiaion treatment optimization based on local biological measures. Phys Med Biol 44:479–493
46. Albrecht H, Fiorani F, Pieruschka R, Müller-Linow M, Jedmowski C, Schreiber L, Schurr U, Rascher U (2020) Quantitative estimation of leaf heat transfer coefficients by active thermography at varying boundary layer conditions. Front Plant Sci 10:1684
47. Albu AB, Schwartz JM, Laurendeau D, Moisan C (2003) Integrating geometric and biomechanical models of a liver tumor for cryosurgery simulations. Lect Notes Comput Sci 2673:121–131
48. Alexander RR, Griffiths JM (1993) Basic biochemical methods. Wiley, New York
49. Alexy T, Detterich J, Connes P, Toth K, Nader E, Kenyeres P, Arriola-Montenegro J, Ulker P, Simmonds MJ (2022) Physical properties of blood and their relationship to clinical conditions. Front Physiol 13:906768
50. Alghamdi NA, Youssef HM (2021) The biothermal analysis of a human eye subjected to exponentially decaying laser radiation under the dual phase-lag heat conduction law. Case Stud Thermal 25:100863
51. Ali A (1999) Statistical mechanical derivation of Cattaneo's heat flux law. J Thermophys Heat Transf 13:544–545
52. Ali YM, Zhang LC (2005) Relativistic heat conduction. Int. J. Heat Mass Transf 48:2397–2406
53. Alicki R, Miklaszewski W (2012) A resonance mechanism of efficient energy transfer mediated by Fenna-Matthews-Olson complex. arXiv:1107.0198v3 [quant-ph] (2012)
54. Alio JL, Ramzy MI, Galal A, Claramonte PJ (2005) Conductive keratoplasty for the correction of residual hyperopia after LASIK. J Refract Surg 21:698–704
55. Alizadehrad D, Imai Y, Nakaaki K, Ishikawa T, Yamaguchi T (2012) Quantification of red blood cell deformation at high-hematocrit blood flow in microvessels. J Biomech 45:2684–2689
56. Alkahtani BS, Atangana A (2017) A note on Cattaneo-Christov model with a non-singular fading memory. Thermal Sci 21:1–7
57. Allain JC, Le Lous M, Bazin S, Maroteaux P (1980) Isometric tensions developed during the hydrothermal swelling of rat skin. Connect Tissue Res 7:127–133
58. Alle H, Roth A, Geiger JRP (2009) Energy-efficient action potentials in hippocampal mossy fibers. Science 325:1405–1408
59. Allworth AE, Albertini DF (1993) Meiotic maturation in cultered bovine oocytes is accompanied by remodelling of the cumulus cell cytoskeleton. Dev Biol 158:101–112
60. Almeida MC, Steiner AA, Coimbra NC, Branco LG (2004) Thermoeffector neuronal pathways in fever: a study in rats showing a new role of the locus coeruleus. J Physiol 558:283–294
61. Almeida RD, Manadas BJ, Carvalho AP, Duarte CV (2004) Intracellular signaling mechanisms in photodynamic therapy. Biochim Biophys Acta Rev Cancer 1704:59–86
62. Alpers J, Reimert D, Rotzer M, Gerlach T, Gutberle M, Wacker F, Hensen B, Hansen C (2021) 2.5D thermometry maps for MRI-guided tumor ablation. arXiv:2108.05734
63. Altshuler GB, Zenzie HH, Erofeev AV, Smirnov MZ, Anderson RR, Dierickx C (1999) Contact cooling of the skin. Phys Med Biol 44:1003–1023

64. Alvarez FX, Jou D (2007) Memory and nonlocal effects in heat transport: from diffusive to ballistic regimes. Appl Phys Lett 90:083109
65. Amara E (1995) Numerical investigations on thermal effects of laser-ocular media interaction. Int J Heat Mass Transf 38:2479–2488
66. Amin Z, Donald JJ, Masters A, Kant R, Lees W, Brown S (1993) Interstitial laser photocoagulation therapy for liver tumours: clinical results. SPIE Laser Tissue Interact 1882:202–209
67. Amin MS, Park Y, Lue N, Dasari RR, Badizadegan K, Feld MS, Popescu G (2007) Microrheology of red blood cell membranes using dynamic scattering microscopy. Opt Exp 15:17001
68. Amin B, Shahzad A, Farina L, Parle E, McNamara L, O'Halloran M, Elahi MA (2020) Dielectric characterization of diseased human trabecular bones at microwave frequency. Med Eng Phys 78:21–28
69. Amos DE, Chen PJ (1970) Transient heat conduction with finite speeds. J Appl Mech 37:1145–1146
70. Amri A, Saidane A, Pulko S (2011) Thermal analysis of a three-dimensional breast model with embedded tumour using the transmission line matrix (TLM) method. Comput Biol Med 41:76–86
71. Amy O, Farah T, Kenneth S, Alexander W, Stephen B (2005) Magnetomotive contrast for in vivo optical coherence tomography. Opt Exp 13:6597–6614
72. An X, Lecomte MC, Chasis JA, Mohandas N, Gratzer W (2002) Shear-response of the spectrin dimer-tetramer equilibrium in the red blood cell membrane. J Biol Chem 35:31796–31800
73. Anchamparuthy V (2007) Vitrification of bovine oocytes. Ph.D Thesis, Virginia University
74. Anders J, Mysliwiec V, Preston S (1995) Low power laser irradiation alters the fluorescence intensity of rhodamine 123 labeled mitochondria in human fibroblasts. Lasers Surg Med Suppl 7:7–8
75. Andersen JS, Wilkinson CJ, Mayor T, Mortensen P, Nigg EA, Mann M (2003) Proteomic characterization of the human centrosome by protein correlation profiling. Nature 426:570–574
76. Anderson DR, Ulness DJ (2015) Properties of the Katugampola fractional derivative with potential applications in quantum mechanics. J Math Phys 56:063502
77. Anderson RR, Parrish JA (1981) The optics of human skin. J Invest Dermatol 77:13–19
78. Andorfer CA, Duman JG (2000) Isolation and characterization of cDNA clones encoding antifreeze protein of the pyrochroid beetle Dendroides canadensis. J Insect Physiol 46:365–372
79. Andrawa S, Tabrizi HB (2010) Non-Fourier effect in the presence of coupled heat and moisture transfer. Int J Heat Mass Transf 53:3080–3087
80. Andreews MD (2004) Cryosurgery for common skin conditions. Am. Family Phys 10:2355–2372
81. Andreozzi A, Brunese L, Iasiello L, Tucci C, Vanoli GP (2019) Modeling heat transfer in tumors: a review of thermal therapies. Ann. Biomed. Eng. 47:676–693
82. Andreozzi A, Iasiello M, Netti PA (2019) A thermoporoelastic model for fluid transport in tumour tissues. J R Soc Interface 16:20190030
83. Andresson B, Ekman L, Gale CC, Sundsten JW (1963) Control of thyrotrophic hormone (TSH) secretion by the "heat loss center. Acta Physiol Scand 59:12–33
84. Andrews P, Harris B, Murray G (2005) Randomized controlled trial of effects of the airflow through the upper respiratory tract of intubated brain-injured patients on brain temperature and selective brain cooling. Brit J Anaesth 94:330–335
85. Andrews ZB, Diano S, Horvath TL (2005) Mitochondrial uncoupling proteins in the CNS: in support of function and survival. Nat Rev Neurosci 6:829–840
86. Andrews CJ, Cuttle L, Simpson MJ (2016) Quantifying the role of burn temperature, burn duration and skin thickness in an in vivo animal skin model of heat conduction. Int J Heat Mass Transf 101:542–549

87. Angell CA, Sara EJ, Donnella J, MacFarlane DR (1981) Homogeneous nucleation and glass transition temperatures in solution of Li salts in D_2O and H_2O: a doubly unstable glass region. J Phys Chem 85:1461–1464
88. Angilletta MJ (2009) Thermal adaptation. Oxford University Press, Oxford
89. Anisimov SI, Kapeliovich BL, Perelman TL (1974) Electron emisson from metal surfaces exposed to ultra-short laser pulses. Sov Phys JETP 39:375–377
90. Ansari MA, Erfanzadeh M, Mohajerani E (2013) Mechanisms of laser-tissue interaction: II. Tissue thermal properties. J Lasers Med Sci 4:99–106
91. Antaki PJ (2005) New interpretation of non-Fourier heat conduction in processed meat. J Heat Transf 127:189–193
92. Antiga L (2002) Patient-specific modeling of geometry and blood flow in large arteries. PhD Thesis, Politecnico di Milano
93. Antonov VF, Smirnova EY, Shevchenko EV (1992) Lipid membranes in phase transformations (in Russian). Nauka, Moscow
94. Anvari B, Milner TE, Tanenbaum BS, Kimel S, Svaasand LO, Nelson JS (1995) Selective cooling of biological tissues: application for thermally mediated therapeutic procedures. Phys Med Biol 40:241–252
95. Anvari B, Tanenbaum BS, Milner TE, Kimel S, Svaasand LO, Nelson JS (1995) A theoretical study of the thermal response of skin to cryogen spray cooling and pulsed laser irradiation: implications for treatment of port wine stain birthmarks. Phys Med Biol 40:1451–1465
96. Anvari B, Tanenbaum BS, Milner TE, Kimel S, Svaasand LO, Nelson JS (1996) Theoretical study of the thermal response of skin to cryogen spray cooling and pulsed laser irradiation: implications for treatment of port wine stain birthmarks. Phys Med Biol 41:1245–1253
97. Anvari B, Milner TE, Tanenbaum BS, Nelson JS (1998) A comparative study of human skin thermal response to sapphire contact and cryogen spray cooling. IEEE Trans Biomed Eng 45:934–941
98. Aoki I (1989) Entropy flow and production in the human body in basal conditions. J Theor Biol 141:11–21
99. Aoki I (1990) Effects of exercise and chills on entropy production in human body. J Theor Biol 145:421–428
100. Aouane O, Thiébaud M, Benyoussef A, Wagner C, Misbah C (2014) Vesicle dynamics in a confined Poiseuille flow: from steady state to chaos. Phys Rev E 90:033011
101. Aoyagi Y, McLellan TM, Shephard RJ (1994) Effects of training and acclimation on heat tolerance in exercising men wearing protective clothing. Eur J Appl Physiol 68:234–245
102. Aoyagi Y, McLellan TM, Shephard RJ (1995) Effects of 6 versus 12 days of heat acclimation on heat tolerance in lightly exercising men wearing protective clothing. Eur J Appl Physiol 71:187–196
103. Aoyagi Y, McLellan TM, Shephard RJ (1997) Interaction of physical training and heat acclimation. Sports Med 23:173–210
104. Arav A, Natan D (2012) Directional freezing of reproductive cells and organs. Reprod Domest Anim 47(suppl 4):193–196
105. Arav A, Yavin S, Zeron Y, Natan D, Dekel I, Gacitua H (2002) New trends in gamete's cryopreservation. Mol Cell Endocrin 187:77–81
106. Arkin H, Xu LX, Holmes KR (1994) Recent developments in modeling heat transfer in blood perfused tissues. IEEE Trans Biomed Eng 41:97–107
107. Armon E, Laufer G (1987) New techniques for reducing the thermochemical damage in the course of laser surgery. Lasers Surg Med 7:162–168
108. Armstrong LE, Maresh CM (1991) The induction and decay of heat acclimatisation in trained athletes. Sports Med 12:302–312
109. Armstrong LE, Maresh CM (1998) Effects of training, environment, and host factors on the sweating response to exercise. Int J Sports Med 19:S103–S105
110. Arnoczky SP, Aksan A (2000) Thermal modification of connective tissues: basic science considerations and clinical implications. J Am Acad Orthop Surg 8:305–313

111. Arnold A, Fichera L (2021) Identification of tissue optical properties during thermal laser-tissue interactions: an ensemble Kalman filter-based approach. arXiv:2107.10340v2 [physics.med-ph]
112. Arora N, Martins D, Ruggerio D, Tousimis E, Swistel AJ, Osborne MP, Simmons RM (2008) Effectiveness of a noninvasive digital infrared thermal imaging system in the detection of breast cancer. Am J Surg 196:523–526
113. Arridge S, Schotland J (2009) Optical tomography: forward and inverse problems. Inv Probl 25:123010
114. Artaki M, Price PJ (1989) Hot phonon effects in silicon field-effect transistors. J Appl Phys 65:1317–1320
115. Artmann GM, Kelemen C, Porst D, Büldt G, Chien S (1998) Temperature transitions of protein properties in human red blood cells. Biophys J 75:3179–3183
116. Aschoff J, Wever R (1959) Anisotropy of the skin for heat transfer. Pflugers Arch Gesamte Physiol Menschen Tiere 269:130–134
117. ASHRAE Handbook Fundamentals, Atlanta (1993)
118. Ashton KG, Tracy MC, Queiroz A (2000) Is Bergmann's rule valid for mammals? Am Nat 156:390–415
119. Ashwood-Smith MJ, Friedmann GB (1979) Lethal and chromosomal effects of freezing, thawing, storage time and X-irradiation on mammalian cells preserved at $-196°$ C in dimethyl sulfoxide. Cryobiology 16:132–140
120. Ashwood-Smith MJ, Morris GJ, Fowler R, Appleton TC, Ashorn R (1988) Physical factors are involved in the destruction of embryos and oocytes during freezing and thawing procedures. Hum Reprod 3:795–802
121. Ashworth EN (1992) Formation and spread of ice in plants. Horticult Rev 13:215–255
122. Askarizadeh H, Ahmadikia H (2014) Analytical analysis of the dual-phase-lag heat transfer equation in a finite slab with periodic surface heat flux. Int J Eng 27:971–978
123. Askarizadeh H, Ahmadikia H (2014) Analytical analysis of the dual-phase-lag model of bioheat transfer equation during transient heating of skin tissue. Heat Mass Transf 50:1673–1684
124. Askarizadeh H, Ahmadikia H (2015) Periodic heat transfer in convective fins based on dual-phase-lag theory. J Thermophys Heat Transf 30:359–368
125. Askarizadeh H, Baniasadi E, Ahmadikia H (2017) Equilibrium and non-equilibrium thermodynamic analysis of high-order dual-phase-lag heat conduction. Int J Heat Mass Transf 104:301–309
126. Asllanaj F, Contassot-Vivier S, Hohmann A, Kienle A (2019) Light propagation in biological tissue. J Quant Spectr Radiat Transf 224:78–90
127. Ataie-Ashtiani B, Farhadi L (2006) A stable moving particle semi-implicit method for free surface flows. Fluid Dyn Res 38:241–256
128. Atangana A (2017) Fractal-fractional differentiation and integration: connecting fractal calculus and fractional calculus to predict complex systems. Chaos Solit Fract 102:396–406
129. Atangana A, Baleanu D (2016) New fractional derivatives with nonlocal and non-singular kernel: theory and application to heat transfer model. Thermal Sci 20:763–769
130. Atangana A, Gomez-Aguilar J (2017) A new derivative with normal distribution kernel: theory, methods and applications. Phys A Stat Mech Appl 476:1–14
131. Atangana A, Secer A (2013) A note on fractional derivatives of some special functions. Abstr Appl Anal 2013:279681
132. Ateshian GA, Likhitpanichkul M, Hung CT (2006) A mixture theory analysis for passive transport in osmotic loading of cells. J Biomech 39:464–475
133. Atkins KR, Osborne DV (1950) The velocity of second sound below 1 k. London Edinburgh Dublin Phil Mag J Sci 41:1078–1081
134. Atkinson RE (1983) Hyperthermia techniques and instrumentation. In: Storm FK (ed) Hyperthermia in cancer therapy. G. K. Hall Medical Publishers, Boston, pp 233–255
135. Attar MM, Haghpanahi M, Shahverdi H, Ilman A (2016) Thermo-mechanical analysis of soft tissue in local hyperthermia treatment. J Mech Sci Technol 30:1459–1469

136. Attwell D, Laughlin SB (2001) An energy budget for signaling in the grey matter of the brain. J Cereb Blood Flow Metab 21:1133–1145
137. Audet DM, Olbricht WL (1987) The motion of model cells at capillary bifurcations. Microvasc Res 33:377–396
138. Auer M (2000) Three-dimensional electron cryo-microscopy as a powerful structural tool in molecular medicine. J Mol Med 78:191–202
139. Aurer C, Carpenter DO (1980) Choroidal blood flow as a heat dissipating mechanism in the macula. Am J Ophthalmol 89:641–646
140. Auriault JL (2017) The paradox of Fourier heat equation: a theoretical refutation. Int J Eng Sci 118:82–88
141. Avolio AP (1980) Multi-branched model of the human arterial system. Med Biol Eng Comput 18:709–718
142. Avrami M (1940) Kinetics of Phase Change. II. Transformation-Time Relations for Random Distribution of Nuclei. J Chem Phys 8(2):212–224
143. Aweda MA, Agida M, Dada M, Awojoyogbe OB, Isah K, Faromika OP, De K, Ojambati OS (2012) Boubaker polynomials expansion scheme solution to the heat transfer equation inside laser heated biological tissues. J Heat Transf 134:064503
144. Aylott M (2006) The neonatal energy triangle. Part 2: thermoregulatory and respiratory adaptation. Paediatr Nurs 18:38–46
145. Aziz JS (2019) Analysis of biological effects of microwave energy and safe distance calculations. J Al-Rafidain Univ Coll Sci 2019:1–8
146. Baardsnes J, Kondejewski LH, Hodges RS, Chao H, Kay C, Davies PL (1999) New ice-binding face for type I antifreeze protein. FEBS Lett 463:87–91
147. Baardsnes J, Jelokhani-Niaraki M, Kondejewski LH, Kuiper MJ, Kay CM, Hodges RS, Davies PL (2001) Antifreeze protein from shorthorn sculpin: identification of the ice-binding surface. Protein Sci 10:2566–2576
148. Bächer C, Kihm A, Schrack L, Kaestner L, Laschke MW, Wagner C, Gekle S (2018) Anti-margination of microparticles and platelets in the vicinity of branching vessels. Biophys J 115(2):411–425
149. Bachiri Y, Gazeau C, Hansz J, Morisset C, Dereuddre J (1995) Succesful cryopreservation of suspension cells by encapsulation-dehydration. Plant Cell Tissue Organ Culture 43:241–248
150. Bagaria HG, Johnson DT (2005) Transient solution to the bioheat equation and optimization for magnetic fluid hyperthermia treatment. Int J Hyperth 21:57–75
151. Bagchi P (2007) Mesoscale simulation of blood flow in small vessels. Biophys J 92:1858–1877
152. Bagge U, Branemark P, Karlsson R, Skalak R (1980) Three-dimensional observations of red blood cell deformation in capillaries. Blood Cells 6:231–239
153. Bai C, Lavine A (1995) On hyperbolic heat conduction and the second law of thermodynamics. J Heat Transf 117:256–263
154. Bailey MR, Khokhlova VA, Sapozhnikov OA, Kargl SG, Crum L (2003) Physical mechanisms of the therapeutic effect of ultrasound. Acoust Phys 49:369–388
155. Baish JW (1990) Heat transport by countercurrent blood vessels in the presence of an arbitrary temperature gradient. J Biomech Eng 112:207–211
156. Baish JW (1994) Formulation of a statistical model of heat transfer in perfused tissue. J Biomech Eng 116:521–527
157. Baish JW (2000) Microvascular heat transfer. In: Bonzano JD (ed) The biomedical engineering handbook, 2nd edn, chap 98. CRC Press, Boca Raton
158. Baish JW, Ayyaswamy PS, Foster KR (1986) Heat transport mechanisms in vascular tissues: a model comparison. J Biomech Eng 108:324–331
159. Baissalov R, Sandison GA, Donnelly BJ, Saliken JC, Mckinnon JG, Muldrew K, Rewcastle JC (2000) A semi-empirical treatment planning model for optimization of multiprobe cryosurgery. Phys Med Biol 45:1085–1098
160. Baissalov R, Sandison GA, Reynolds D, Muldrew K (2001) Simultaneous optimization of cryoprobe placement and thermal protocol for cryosurgery. Phys Med Biol 46:1799–1814

161. Baker MA (1982) Brain cooling in endotherms in heat and exercise. Annu Rev Physiol 44:85–96
162. Baker MB, Baker M (2004) A new look at homogeneous freezing of water. Geophys Res Lett 31:L19102
163. Bakken AM (2006) Cryopreservation of human peripheral blood progenitor cells. Curr Stem Cell Res Ther 1:47–54
164. Bakunin OG (2003) Mysteries of diffusion and labyrinths of destiny. Phys Usp 46:309–313
165. Bal NC, Maurya SK, Sopariwala DH, Sahoo SK, Gupta SC, Shaikh SA, Pant M, Rowland LA, Bombardier E, Goonasekera SA, Tupling AR, Molkentin JD, Periasamy M (2012) Sarcolipin is a newly identified regulator of muscle-based thermogenesis in mammals. Nat Med 18:1575–1579
166. Balasubramanian SK, Coger RN (2005) Heat and mass transfer during the cryopreservation of a bioartificial liver device: a computational model. ASAIO J 51:184–193
167. Bald WB, Fraser J (1982) Cryogenic surgery. Rep Prog Phys 45:1381–1434
168. Baldwin RL (2005) Early days of studying the mechanism of protein folding. In: Buchner J, Kiefhaber T (eds) Protein folding handbook. Wiley-VCH, Weinheim, pp 3–21
169. Bale JS (1996) Insect cold hardiness: a matter of life and death. Eur J Entomol 93:369–382
170. Bale JS (2002) Insects and low temperatures: from molecular biology to distributons and abundances. Phil Trans R Soc Lond B 357:849–862
171. Baleanu D, Fernandez A (2019) On fractional operators and their classifications. Mathematics 7:830
172. Balentine DA, Wiseman SA, Bouwens LCM (1997) The chemistry of tea flavonoids. Crit Rev Food Sci Nutr 37:693–704
173. Balescu R (1995) Equilibrium and non-equilibrium statistical mechanics. Wiley, Hoboken
174. Bali R, Sharma S (2011) A model for intra-articular heat exchange in a knee joint. Tribol Lett 41:379–386
175. Ball BA, Vo A (2001) Osmotic tolerance of equine spermatozoa and the effects of soluble cryoprotectants on equine sperm motility, viability and mitochondrial membrane potential. J Androl 6:1061–1069
176. Balogh P, Bagchi P (2017) A computational approach to modeling cellular-scale blood flow in complex geometry. J Comput Phys 334:280–307
177. Balogh P, Bagchi P (2017) Direct numerical simulation of cellular-scale bloodflow in 3d microvascular networks. Biophys J 113:2815–2816
178. Banihani S, Agarwal A, Sharma R, Bayachou M (2014) Cryoprotective effect of lcarnitine on motility, vitality and dna oxidation of human spermatozoa. Andrologia 46:637–641
179. Bank H, Mazur P (1973) Visualization of freezing damage. J Cell Biol 57:729–742
180. Bannerjee A, Ogale AA, Das C, Mitra K, Subranian C (2005) Temperature distribution in different materials due to short pulse laser irradiation. Heat Trans Eng 26:41–49
181. Baranov A, Pakhomov O, Fedorov A, Ivanov V, Zaitsev A, Polyakov R (2019) Technique and technology of whole-body cryotherapy (WBC). In: Morosuk T, Sultan M (eds) Low-temperature technologies. IntechOpen, London
182. Barauskas R, Gulbinas A, Vanagas T, Barauskas G (2008) Finite element modeling of cooled-tip probe radiofrequency ablation processes in liver tissue. Comput Biol Med 38:694–708
183. Barbee K (2005) Mechanical cell injury. Ann NY Acad Sci 1066:1–18
184. Barber JO, Alberding JP, Restrepo JM, Secomb TW (2008) Simulated two-dimensional red blood cell motion, deformation, and partitioning in microvessel bifurcations. Ann Biomed Eng 36:1690–1698
185. Barber JO, Restrepo JM, Secomb TW (2011) Simulated red blood cell motion in microvessel bifurcations: effects of cell-cell interactions on cell partitioning. Cardiovasc Eng Technol 2:349–360
186. Barcroft J (1921–1922) The raison d'etre of the red corpuscle. Harvey Lectures, Series 17, 146–163

187. Barenblat GI, Entov VM, Ryzhik VM (1984) Flow of fluids and gases in porous formations (in Russian). Nedra, Moscow
188. Bargmann S, Favata A (2014) Continuum mechanical modeling of laser-pulsed heating in polycrystals: a multi-physics problem of coupling diffusion, mechanics, and thermal waves. ZAMM 94:487–498
189. Barkai E (2001) CTRW pathways to the fractional diffusion equation. arXiv:cond-mat/0108024
190. Barkai E, Klafter J (1998) Comment on subdiffusion and anomalous local viscoelasticity in actin networks. Phys Rev Lett 81:1134–1135
191. Barkai E, Metzler R, Klafter J (2000) From continuous time random walks to the fractional Fokker-Planck equation. Phys Rev E 61:132–138
192. Barker CAV, Gandlier JCC (1957) Pregnancy in a mare resulted from frozen epididymal spermatozoa. Can J Compar Med Veterin Sci 21:47–51
193. Barletta A, Zanchini E (1997) Hyperbolic heat conduction and local equilibrium: a second law analysis. Int J Heat Mass Transf 40:1007–1016
194. Barletta A, Zanchini E (1997) Unsteady heat conduction by internal-energy waves in solids. Phys Rev B 55:14208–14213
195. Barnard ACL, Lopez L, Hellums JD (1968) Basic theory of blood flow in capillaries. Microvasc Res 1:23–34
196. Barns S, Balanant MA, Sauret E, Flower R, Saha S, Gu YT (2017) Investigation of red blood cell mechanical properties using AFM indentation and coarse-grained particle method. BioMed Eng OnLine 16:140
197. Baross JA, Hoffman SE (1985) Submarine hydrothermal vents and associated gradient environments as sites for the origin and evolution of life. Orig Life Evol Bios 15:327–345
198. Barrere N, Brum J, L'Her A, Sarasüa GL, Cabeza C (2019) Vortex dynamics under pulsatile flow in axisymmetric constricted tubes. arXiv:1904.10547 [physics.flu-dyn]
199. Barrett R, Berry M, Chan TF, Demmel J, Donato J, Dongarra J, Eijkhout V, Pozo R, Romine C, der Vorst HV (1994) Templates for the solution of linear systems: building blocks for iterative methods, 2nd edn. SIAM, Philadelphia
200. Barthes-Biesel D (1991) Role of interfacial properties on the motion and deformation of capsules in shear flow. Phys A 172:103–124
201. Barun VV, Ivanov AP (2003) Thermal action of a short light pulse on biological tissues. Int J Heat Mass Transf 46:3243–3254
202. Bashkatov AN, Genina EA, Tuchin VV (2011) Optical properties of skin, subcutaneous, and muscle tissues: a review. J. Innov Opt Health Sci 4:9–38
203. Baskurt O, Meiselman H (1996) Determination of red blood cell shape recovery time constant in a Couette system by the analysis of light reflectance and ektacytometry. Biorheology 33:489–503
204. Baskurt OK, Marshall-Gradisnik S, Pyne M, Simmonds M, Brenu E, Christy R (2010) Assessment of the hemorheological profile of koala and echidna. Zoology 113:110–117
205. Batal MA, Mearawi G (2012) Conversion of absorbed light energy into diffusive thermal energy in biological tissue. Energy Proc 19:158–166
206. Batash R, Asna N, Schaffer P, Francis N, Schaffer M (2017) Glioblastoma multiforme, diagnosis and treatment. Recent literature review. Curr Med Chem 24:3002–3009
207. Batchelor GK (1970) The stress system in a suspension of force-free particles. J Fluid Mech 41:545–570
208. Batchelor GK, Green JT (1972) The hydrodynamic interaction of two small freely-moving spheres in a linear flow field. J Fluid Mech 56:375–400
209. Batchelor GK, O'Brien RW (1977) Thermal or electric conduction through a granular material. Proc R Soc Lond A 355:313–333
210. Batycky RP, Hammerstedt R, Edwards DA (1997) Osmotically driven intracellular transport phenomena. Philos Trans R Soc Lond A 355:2459–2488
211. Bauerecker S, Ulbig P, Buch V, Vrbka L, Jungwirth P (2008) Monitoring ice nucleation in pure and salty water via high-speed imaging and computer simulations. J Phys Chem C 112:7631–7636

212. Baumeister KJ, Hamill TD (1969) Hyperbolic heat conduction equation – a solution for the semi-infinite body problem. J Heat Transf 91:543–548
213. Baumgaertner A, Grudinin S, Gwan JF (2003) Molecular dynamics simulation of membrane proteins. NIC Ser 20:365–375
214. Baumgart S (2008) Iatrogenic hyperthermia and hypothermia in the neonate. Clin Perinatol 35:183–197
215. Baumler H, Neu B, Donath E, Kiesewetter H (1999) Basic phenomena of red blood cell rouleaux formation. Biorheology 36:439–442
216. Baust JG, Baust JM (Eds) (2006) Advances in Biopreservation. CRC Press, Boca Raton
217. Baust JG, Gage AA (2005) The molecular basis of cryosurgery. BJU Int 95:1187–1191
218. Baust J, Gage AA, Ma H, Zhang CM (1997) Minimally invasive cryosurgery – technological advances. Cryobiology 34:373–384
219. Baust JG, Gage AA, Clarke D, Baust JM, van Buskirk R (2004) Cryosurgery - a putative approach to molecular-based optimization. Cryobiology 48:190–204
220. Bay DC (1995) Themogenesis in the aroids. Aroideana 18:32–38
221. Bazett HC (1968) The regulation of body temperatures. In: Newburgh LH (ed) Physiology of heat regulation and the science of clothing. Hafner, New York, pp 109–192
222. Bazett HC, Mendelson ES, Love L, Libet B (1948) Precooling of blood in the arteries, effective heat capacity and evaporative cooling as factors modifying cooling of the extremities. J Appl Physiol 1:169–182
223. Beacco CM, Mordon SR, Brunetaud JM (1994) Development and experimental in vivo validation of mathematical modeling of laser coagulation. Lasers Surg Med 14:362–373
224. Beardo A, Knobloch JL, Sendra L, Bafaluy A, Frazer TD, Chao W, Hernandez-Charpak JN, Kapteyn HC, Abad B, Murnane MM, Alvarez FX, Camacho J (2021) A general and predictive understanding of thermal transport from 1D- and 2D-confined nanostructures: theory and experiment. ACS Nano 15:13019–13030
225. Beck EH, Heim R, Hansen J (2004) Plant resistance to cold stress: mechanisms and environmental signals triggering frost hardening and dehardening. J. Biosci 29:449–459
226. Beck EH, Fettig S, Knake C, Hartig K, Bhattarai T (2007) Specific and unspecific responses of plants to cold and drought stress. J Biosci 32:501–510
227. Becker SM (2012) Analytic one dimensional transient conduction into a living perfuse/non-perfuse two layer composite system. Heat Mass Transf 48:317–327
228. Becker SM (2013) One-dimensional transient heat conduction in composite living perfuse tissue. J Heat Transf 135:071002
229. Becker SM, Kuznetsov AV (2007) Thermal damage reduction associated with in vivo electroporation: a numerical investigation justifying aggressive pre-cooling. Int J Heat Mass Transf 50:105–116
230. Beek JF, van Staveren HJ, Posthumus P, Sterenborg HJC, van Gemert M (1997) The optical properties of lung as a function of respiration. Phys Med Biol 42:2263–2272
231. Beerling DJ, Terry AC, Mitchell PL, Callaghan TV, Gwynn-Jones D, Lee JA (2001) Time to chill: effects of simulated global climate change on leaf ice nucleation temperatures of subarctic vegetation. Am J Botany 88:628–633
232. Behura AK, Prasad BN, Prasad L (2013) Burn depth prediction using analytical and numerical solution of Pennes bioheat equation. Int J Innov Appl Stud 3:215–220
233. Beignon F, Gueguen N, Tricoire-Leignel H, Mattei C, Lenaers G (2022) The multiple facets of mitochondrial regulations controlling cellular thermogenesis. Cell Mol Life Sci 79:525
234. Bejan A (1997) Constructal-theory network of conducting paths for cooling a heat generating volume. Int J Heat Mass Transf 40:799–816
235. Bejan A (2001) The tree of convective heat streams: its thermal insulation function and the predicted 3/4-power relation between the body heat loss and body size. Int J Heat Mass Transf 44:699–704
236. Belayev AM, Prokhorov GG (2017) History of cryomedicine (in Russian). In: Prokhorov GG, Belyaev AM, Prokhorov DG (eds) Fundamentals of clinical cryomedicine. Kniga Po Trebovaniyu, Saint-Petersburg, pp 21–28

237. Belding H, Hatch T (1955) Index for evaluating heat stress in terms of resulting physiological strain. Heat Pip Air Cond 27:129–136
238. Belhamadia Y, Fortin A, Chamberland E (2004) Anisotropic mesh adaptation for the solution of the Stefan problem. J Comput Phys 194:233–255
239. Belhamadia Y, Fortin A, Chamberland E (2004) Three-dimensional anisotropic mesh adaptation for phase change problem. J Comput Phys 201:753–770
240. Bell DG, Tikusis P, Jacobs I (1992) Relative intensityof muscular contraction during shivering. J Appl Physiol 72:2336–2342
241. Bellizzi GG, Drizdal T, van Rhoon GC, Crocco L, Isernia T, Paulides MM (2018) The potential of constrained SAR focusing for hyperthermia treatment planning: analysis for the head & neck region. Phys Med Biol 64:015013
242. Ben Adda F (1997) Geometric interpretation of the differentiability and gradient of real order. Compt Rend Ser I Math 326:931–934
243. Ben Adda F (1997) Geometric interpretation of the fractional derivative. J Fract Calc 11:21–51
244. Bennett CH (1982) The thermodynamics of computation - a review. Int J Theor Phys 21:905–940
245. Bennett KM, Hyde JS, Schmainda KM (2006) Water diffusion heterogeneity index in the human brain is insensitive to the orientation of applied magnetic field gradient. Magn Resonan Med 56:235–239
246. Benson EE, Lynch PT, Jones J (1992) Detection of lipid peroxidation products in cryoprotected and frozen rice cells: consequences for post-thaw survival. Plant Sci 85:107–114
247. Benson DA, Meerschaert MM, Revielle J (2013) Fractional calculus in hydrological modelling: a numerical perspective. Adv Water Resour 51:479–497
248. Benson JD, Benson CT, Critser JK (2014) Mathematical model formulation and validation of water and solute transport in whole hamster pancreatic islets. Math Biosci 254:64–75
249. Benson JD, Higgins AZ, Desai K, Eroglu A (2018) A toxicity cost function approach to optimal CPA equilibration in tissues. Cryobiology 80:144–155
250. Bento D, Rodrigues RO, Faustino V, Pinho D, Fernandes CS, Pereira AI, Garcia V (2018) Lima: deformation of red blood cells, air bubbles, and droplets in microfluidic devices: flow visualizations and measurements. Micromachines 9:151
251. Benzinger TH (1969) Heat regulation: homeostasis of central temperature in man. Physiol Rev 49:671–759
252. Berendsen HJC, Marrink SJ (1993) Molecular dynamics of water transport through membranes: water from solvent to solute. Pure Appl Chem 65:2513–2520
253. Berenger JP (1994) A perfectly matched layer for the absorption of electromagnetic waves. J Comput Phys 114:185–200
254. Berezhkovskii A, Hummer G (2002) Single-file transport of water molecules through a carbon nanotube. Phys Rev Lett 89:064503
255. Berg BA (2000) Introduction to multicanonical Monte Carlo simulations. Fields Inst Commun 26:1–24
256. Berg N, Fuchs L, Wittberg L (2019) Flow characteristics and coherent structures in a centrifugal blood pump. Flow Turbul Combust 102:469–483
257. Bergamasco L, Alberhini M, Fasano M, Cardellini A, Chiavazzo E, Asinari P (2018) Mesoscopic moment equations for heat conduction: characteristic features and slow-fast mode decomposition. Entropy 20:126
258. Berger MJ, LeVeque RJ (1998) Adaptive mesh refinement using wave-propagation algorithms for hyperbolic systems. SIAM J Numer Anal 35:2298–2316
259. Berger WK, Poledna J (2001) New strategies for the placement of cryoprobes in malignant tumors of the liver for reducing the probability of recurrences after hepatic cryosurgery. Int J Colorectal Dis 16:331–339
260. Berglund LG (2002) Simulation of human termoregulatory regulatory responses in microcooling in hot environments. In: 32nd International Conference on Environment Systems, 2002-01-2412

261. Bering EA (1974) Effects of profound hypothermia and circulatory arrest on cerebral oxygen metabolism and cerebrospinal fluid electrolyte composition in dogs. J Neurosurg 39:199–204
262. Berjano EJ (2006) Theoretical modeling for radiofrequency ablation: state-of-the-art and challenges for the future. BioMed Eng OnLine 5:24
263. Berjano EJ, Hornero F (2004) Thermal-electrical modeling for epicardial atrial radiofrequency ablation. IEEE Trans Biomed Eng 51:1348–1357
264. Berjano EJ, Saiz J, Ferrero JM (2002) Radio-frequency heating of the cornea: theoretical model and in vitro experiments. IEEE Trans Biomed Eng 49:196–205
265. Berjano EJ, Alió JL, Saiz J, Ferrero JM (2003) Ring electrode for radio-frequency heating of the cornea: modelling and in vitro experiments. Med Biol Eng Comput 41:30–639
266. Berjano EJ, Alió JL, Saiz J (2005) Modeling for radio-frequency conductive keratoplasty: implications for the maximum temperature reached in the cornea. Physiol Meas 26:157–172
267. Berkelmann L, Bader A, Meshksar S, Dierks A, Hatipoglu G, Krauss JK, Schwabe K, Manteuffel D, Ngezahayo A (2019) Tumour-treating fields (TTFields): investigations on the mechanism of action by electromagnetic exposure of cells in telophase/cytokinesis. Sci Rep 9:7362
268. Bernal JD, Fowler RH (1933) A theory of water and ionic solution, with particular reference to hydrogen and hydroxyl ions. J Chem Phys 1:515–548
269. Bernard A, Fuller BJ (1996) Cryopreservation of human oocytes: a review of current problems and perspectives. Hum Reprod Update 2:193–207
270. Bernard S, Buist M, Monteiro O, Smith K (2003) Induced hypothermia using large volume, ice-cold intravenous fluid in comatose survivors of out-of-hospital cardiac arrest: a preliminary report. Resuscitation 56:9–13
271. Bertoloni G, Sacchetto R, Baro E, Ceccherelli F, Jori G (1993) Biochemical and morphological changes in escherichia coli irradiated by coherent and non-coherent 632.8 nm light. J Photochem Photobiol B Biol 18:191–196
272. Bertot LC, Sato M, Tateishi R, Yoshida H, Koike K (2011) Mortality and complication rates of percutaneous ablative techniques for the treatment of liver tumors: a systematic review. Eur Radiol 21:2584–2596
273. Bertsova YV, Popov VN, Bogachev AV (2004) NADH oxidation by mitochondria from the thermogenic plant arum orientale. Biochemistry (Moscow) 69:580–584
274. Bessonov N, Sequeira A, Simakov S, Vassilevskii Y, Volpert V (2016) Methods of blood flow modelling. Math Model Nat Phenom 11:1–25
275. Betti L, Lycett SJ, von Craramo-Taubade N, Pearson OM (2015) Are human hands and feet affected by climate? A test of Allen's rule. Am J Phys Anthropol 158:132–140
276. Betz T, Lenz M, Joanny J, Sykes C (2009) ATP-dependent mechanics of red blood cells. Proc Natl Acad Sci USA 106:15320–15325
277. Beybalaev VD (2010) Mathematical model of heat conduction in fractal medium. Math Models Comput Simul 2:91–97
278. Beybalaev VD, Aliverdiev AA, Magomedov RA, Meilanov RR, Akhmedov EN (2017) Modeling of freezing processes by an one-dimensional thermal conductivity equation with fractional differention operator. Vestn Samar Gos Techn Univ 21:376–387
279. Bhargava R, Chandra H (2017) Study of bioheat transfer phase change during cryosurgery for an irregular tumor tissue using EFGM. arXiv:1711.02994v1 [q-bio.QM]
280. Bhattacharya A, Mahajan RL (2003) Temperature dependence of thermal conductivity of biological tissues. Physiol Meas 24:769–783
281. Bhavaraju NC, Valvano JW (1999) Thermophysical properties of swine myocardium. Int J Thermophys 20:665–676
282. Bhosekar A, Ierapetritou M (2018) Advances in surrogate-based modeling, feasibility analysis, and optimization: a review. Comput Chem Eng 108:250–267
283. Bhowmick S, Bischof JC (1998) Supraphysiological thermal injury in dunning AT-1 prostate tumor cells. Adv Heat Mass Transf Biotechnol ASME HTD 362:77–78

284. Bhowmik A, Singh R, Repaka R, Mishra S (2013) Conventional and newly developed bioheat transport models in vascularized tissues: a review J Thermal Biol 38:107–125
285. Bianchi L, Cavarzan F, Ciampitti L, Cremonesi M, Grilli F, Saccomandi P (2022) Thermophysical and mechanical properties of biological tissues as a function of temperature: a systematic literature review. Int J Hyperthermia 39:297–340
286. Bianko P, Robey PG (2001) Stem cells in tissue engineering. Nature 414:118–121
287. Biben T, Misbah C (2003) Tumbling of vesicles under shear flow within an advected-field approach. Phys Rev E 67:031908
288. Bicher HI, Mitagvaria N (1981) Circulatory responses of malignant tumors during hyperthermia. Microvasc Res 21:19–26
289. Bienertová-Vaškău J, Nečesánek I, Novák J, Vinklárek J, Zlámal F (2014) Stress entropic load as a transgenerational epigenetic response trigger. Med Hypotheses 82:271–274
290. Biesty S, Platt R (1998) Incredible body. Dorling Kindersley Limited, London
291. Bilgili M, Simsek E, Sahin A, Ozbek A (2015) Examination of human heat loss in five Mediterranen regions. Physiol Behav 149:61–68
292. Bilgili M, Sahin B, Simsek E, Ozbek A, Yasar A (2019) Heat loss and exergy flow through respiration of the human body under different meteorological conditions. Eur Mech Sci 3:45–51
293. Bingi VN, Savin AV (2003) Effects of weak magnetic fields on biological systems: physical aspects. Phys Usp 46:259–292
294. Biot MA (1956) Thermoelasticity and irreversible thermodynamics. J Appl Phys 27:240–253
295. Bird H (1949) James Arnott, MD (Aberdeen), 1797–1883, a pioneer in refrigeration. Anaesthesia 4:10–17
296. Biro PA, Stamps JA (2008) Are animal personality traits linked to life-history productivity? Trends Ecol Evol 23:361–368
297. Bischof JC (2006) Micro and nanoscale phenomenon in bioheat transfer. Heat Mass Transf 42:955–966
298. Bischof JC, Diller KR (2018) From nanowarming to thermoregulation: new multiscale applications of bioheat transfer. Annu Rev Biomed Eng 20:301–327
299. Bischof J, He X (2005) Thermal stability of proteins. Ann NY Acad Sci 1066:1–22
300. Bischof JC, Wolkers WF, Tsvetkova NM, Oliver AE, Crowe JH (2002) Lipid and protein changes due to freezing in Dunning AT-1 cells. Cryobiology 45:22–32
301. Bischof JC, Mahr B, Choi JH, Behling M, Mewes D (2007) Use of X-ray tomography to map crystalline and amorphous phases in frozen biomaterials. Ann Biomed Eng 35:292–304
302. Bisht S, Faiq M, Tolahunase M, Dada R (2017) Oxidative stress and male infertility. Nat Rev Urol 14:470–485
303. Bisschops LLA, Hoedemaekers CWE, Simons KS, van der Hoeven JG (2010) Preserved metabolic coupling and cerebrovascular reactivity during mild hypothermia after cardiac arrest. Crit Care Med 38:1542–1547
304. Bixler1 JN, Hokr BH, Oian CA, Noojin GD, Thomas RJ, Yakovlev VV (2015) Assessment of tissue heating under tunable laser radiation from 1100 nm to 1550 nm. arXiv:1509.08022v1 [physics.bio-ph]
305. Bizunok SN, Sventitskii EN (1983) Study of aqueous solutions of cryoprotectors using NMR (in Russian). In: Water in biological systems and their components, Leningrad University
306. Blackburn HD (2004) Development of national animal genetic resource programs. Reprod Fertil Develop 16:27–32
307. Blana A, Walter B, Rogenhofer S, Wieland NF (2003) High intensity focused ultrasound for the treatment of localized prostate cancer: 5-year experience. Urology 63:297–300
308. Blanco JM, Gee G, Wildt DE, Donoghue AM (2000) Species variation in osmotic, cryoprotectant and cooling rate tolerance in poultry, eagle and pelegrine falcon spermatozoa. Biol Reprod 63:1164–1171

309. Blanco JM, Long JA, Gee G, Donoghue AM, Wildt DE (2008) Osmotic tolerance of avian spermatozoa: influence of time, temperature, cryoprotectant and membrane ion pump function on sperm viability. Cryobiology 56:8–14
310. Blank M (2008) Protein and DNA interactions with electromagnetic fields. Electromagn Biol Med 28:3–23
311. Blank M, Goodman R (2009) Electromagnetic fields stress living cells. Pathophysiology 16:71–78
312. Blauth S, Hübner F, Leithäuser C, Siedow N, Vogl TJ (2019) Mathematical modeling of vaporization during laser-induced thermotherapy in liver tissue. arXiv:1910.12515v1 [math.AP]
313. Blauth S, Hübner F, Leithäuser C, Siedow N, Vogl TJ (2020) Mathematical modeling of vaporization during laser-induced thermotherapy in liver tissue. J Math Industry 10:16
314. Blázquez-Castro A (2017) Direct $^1O_2$ optical excitation: a tool for redox biology. Redox Biol 13:39–59
315. Bleehen NM, Howard GCW (1988) Clinical hyperthermia: methods and results. In: Bleehen NM (ed) Radiobiology in radiotherapy, pp 209–231
316. Bligh J (2006) A theoretical consideration of the means whereby the mammalian core temperature is defended at a null zone. J Appl Physiol 100:1332–1337
317. Bligh J, Johnson KG (1973) Glossary of terms for thermal physiology. J Appl Physiol 35:941–961
318. Blouin-Demers G, Nadeau P (2005) The cost-benefit model of thermoregulation does not predict lizard thermoregulatory behavior. Ecology 86:560–566
319. Blowers S (2018) Modelling brain temperatures in healthy patients and those with induced hypothermia. PhD Thesis, The University of Edinburgh
320. Bluestein D, Rambod E, Gharib M (2000) Vortex shedding as a mechanism for free emboli formation in mechanical heart valves. J Biomech Eng 122:125–134
321. Blum NT, Yildirim A, Chattaraj R, Goodwin AR (2017) Nanoparticles formed by acoustic destruction of microbubbles and their utilization for imaging and effects on therapy by high intensity focused ultrasound. Theranostics 7:694–702
322. Bocquet B, Van de Velde JC, Mamouni A, Leroy Y, Giaux G, Delannoy J, Delvalee D (1990) Microwave radiometric imaging at 3 GHz for the exploration of breast tumours. IEEE Trans Microw Theory Techn 38:791–793
323. Bodo E, Jens L, Martin S (1998) Optimization of temperature distributions for regional hyperthermia based on a nonlinear heat transfer model. Ann NY Acad Sci 858:36–46
324. Bogerd CP, Rossi RM, Brühwiler PA (2011) Thermal perception of ventilation changes in full-face motorcycle helmets: subject and manikin study. Ann Occup Hyg 55:192–201
325. Bogerd CP, Walker I, Brühwiler PA, Rossi RM (2014) The effect of a helmet on cognitive performance is, at worst, marginal: a controlled laboratory study. Appl Ergon 45:671–676
326. Bogerd CP, Aerts JM, Annaheim S, Bröde P, de Bruyne G, Flouris A, Kuklane K, Sotto Mayor T, Rossi RM (2015) A review on ergonomics of headgear: thermal effects. Int J Ind Ergon 45:1–12
327. Bogin B, Hermanussen M, Scheffler C (2022) Bergmann'srule is a "just so" story of human body size. J Physiol Anthropol 41:15
328. Boldyrev AA, Kyaivyaryainen EI, Ilyukha VA (2006) Biomembranology (in Russian). Institute of Biology of Karelian Research Centre, Russian Acad Science, Petrzavodsk (2006)
329. Boley BA, Weiner JH (1960) Theory of thermal stress. Wiley, New York
330. Bommadevara M, Zhu L (2002) Temperature difference between the body core and arterial blood supplied to the brain during hyperthermia or hypothermia in humans. Biomechan Model Mechanobiol 1:137–149
331. Bordin-Aykroyd S, Dias RB, Lynch E (2019) Laser-tissue interaction. EC Dental Sci 18:2303–2308
332. Boregowda SC (1998) Thermodynamic modeling and analysis of human stress responses. PhD thesis. Old Dominion University

333. Boregowda SC, Tiwari SN, Chaturvedi SK, Hou GJW (1998) Thermodynamic analysis of an unsteady 3D human thermal model. AIAA Paper No. 98-0841
334. Both S, Czel B, Fülöp T, Grof G, Gyenis A, Kovács R, Ván P (2016) Verhas: deviation from the Fourier law in room-temperature heat pulse experiments. J. Non-Equil Thermodyn 41:41–48
335. Bottin E, Lacroix Z, Leger S (2016) In-vivo heat retention comparison of eyelid warming masks. Contact Lens Anterior Eye 39:311–315
336. Bouchama A, Knochel JP (2002) Heat stroke. N Engl J Med 346:1978–1988
337. Boughton RS, Spencer SK (1987) Electrosurgical fundamentals. J Am Acad Dermatol 16:862–867
338. Boukai AI, Bunimovich Y, Tahir-Kheli J, Goddard WA III, Heath JR (2008) Silicon nanowires as efficient thermoelectric materials. Nature 451:168–171
339. Boulant JA (2006) Neuronal basis of Hammel's model for setpoint thermoregulation. J Appl Physiol 100:1347–1354
340. Boulant JA, Bignall KE (1973) Hypothalamic neuronal responses to peripheral and deep-body temperatures. Am J Physiol 225:1371–1374
341. Boulant JA, Hardy JD (1974) The effect of spinal and skin temperatures on the firing rate and thermosensitivity of preoptic neurons. Am J Physiol 240:639–660
342. Boules DJ, Lillford PJ, Rees DA, Shanks IA (2002) Preface. Philos Trans R Soc Lond B 357:829
343. Boulnois JL (1986) Photophysical processes in recent medical laser developments: a review. Lasers Med Sci 1:47–66
344. Bourantas GC, Ghommem M, Kagadis GC, Katsanos K, Loukopoulos VC, Burganos VN, Nikiforidis GC (2014) Real-time tumor ablation simulation based on the dynamic mode decomposition method. Med Phys 41:053301
345. Boutron P (1983) Glass-forming tendency, stability of the amorphous state, and cryoprotection of red blood cells. J Phys Chem 87:4273–4276
346. Bouvet V, Ben RE (2003) Antifeeze glycoproteins. structure, conformation, and biological applications. Cell Biochem Biophys 39:133–144
347. Bouzigon R, Grappe F, Ravier G, Dugue B (2016) Whole- and partial-body cryostimulation/cryotherapy: current technologies and practical applications. J Thermal Biol 61:67–81
348. Bower RE, Carbo BG, Pace MM, Graham EF (1973) Effects of dilution and glycerol on the release of glutaminic-oxaloacetic transaminase (GOT) from boar spematozoa. J Anim Sci 36:319–324
349. Bowler K, Laudien H, Laudien I (1983) Cellular heat injury. J Therm Biol 8:426–430
350. Bowman HF (1981) Heat transfer and thermal dosimetry. J. Microw Power 16:121–133
351. Bown S (1983) Phototherapy in tumors. World J Surg 7:700–709
352. Box FM, van der Geest RJ, Rutten M, Reiber JH (2005) The influence of flow, vessel diameter, and nonnewtonian blood viscosity on the wall shear stress in a carotid bifurcation model for unsteady flow. Invest Radiol 40:277–294
353. Brace CL (2010) Microwave tissue ablation: biophysics, technology and applications. Crit Rev Biomed Eng 38:65–78
354. Bradshaw WE, Holzapfel CM (2008) Genetic response to rapid climate change: it's seasonal timing that matters. Mol Ecol 17:157–166
355. Braitenberg V, Schüz A (1998) Cortex: statistics and geometry of neuronal connectivity. Springer, Berlin
356. Branca C, Magazu S, Maisano G, Migliardo P, Mineo PG (2001) Comparison of thermophysical properties in disaccharide aqueous solutions. Phys Script 64:390–397
357. Brannigan G, Lin LCL, Brown FLH (2006) Implicit solvent simulation models for biomembranes. Eur Biophys J 35:104–124
358. Braslavsky I, Lipson SG (1999) The double-pyramid structure of dendritic ice growing from supercooled water. J Cryst Growth 198(199):56–61
359. Braunmüller S, Schmid L, Sackmann E, Franke T (2012) Hydrodynamic deformation reveals two coupled modes/time scales of red blood cell relaxation. Soft Matter 8:11240–11248

360. Bravo D, Rigley TH, Gibran N, Strong DM, Newman-Gage H (2000) Effect of storage and preservation methods on viability in transplantable human skin allografts. Burns 26:367–378
361. Brazhnikov AM, Karpychev VA, Luikova AV (1975) One engineering method of calculating heat conduction processes. Inzh-Fiz Zh 28:677–680
362. Brdiczka D, Wallimann T (1994) The importance of the outer mitochondrial compartment in regulation of energy metabolism. Mol Cell Biochem 133–134:69–83
363. Breasted J (1930) The edwin smith surgical papyrus. The University of Chicago Oriental Institute Publications, Chicago
364. Brebbia C, Dominguez J (1992) Boundary elements, an introductory course, 2nd edn. McGraw-Hill, New York
365. Brebbia C, Telles J, Wrobel L (1984) Boundary elements techniques. Springer, Berlin
366. Breman JW (2006) Leaf-tissue freeze-tolerance mechanisms in bahiagrass (Paspalum notatum Flügge). Ph.D Thesis, University of Florida
367. Bremmell KE, Evans A, Prestidge CA (2006) Deformation and nano-rheology of red blood cells: an AFM investigation. Colloids Surf B 50:43–48
368. Brengelmann GL (1993) Specialized brain cooling in humans. FASEB J 7:1148–1153
369. Breton G, Danyluk J, Ouellet F, Sarhan F (2000) Biotechnological applications of plant freezing associated proteins. Biotechnol Annu Rev 6:57–99
370. Brettel K, Leibl W (2001) Electron transfer in photosystem I. Biochim Biophys Acta 1507:100–114
371. Brey EM, King TW, Johnson C, McIntire LV, Reece GP, Patric CW (2002) A technique for quntitative three-dimensional analysis of microvascular structure. Microvasc Res 63:279–294
372. Briels WJ, Tepper HL (1997) Crystal growth of the Lennard-Jones (100) surface by means of equilibrium and nonequilibrium Molecular Dynamics. Phys Rev Lett 79:5074–5077
373. Brigham PA, McLoughlin E (1996) Burn incidence and medical care use in the United States: estimates, trends, and data sources. J Burn Care Rehabil 17:95–107
374. Brinck H, Werner J (1994) Estimation of the thermal effect of blood flow in a branching countercurrent network using a three-dimensional vascular model. J Biomech Eng 116:324–330
375. Brinck H, Werner J (1995) Use of vascular and non-vascular models for the assessment of temperature distribution during induced hyperthermia. Int J Hyperthermia 11:615–626
376. Brinkkoetter PT, Song H, Losel R, Schnetzke U, Gottmann U, Feng Y, Hanusch C, Beck GC, Schnuelle P, Wehling M, van der Woude FJ, Yard BA (2008) Hypothermic injury: the mitochondrial calcium, ATP and ROS love-hate triangle out of balance. Cell Physiol Biochem 22:195–204
377. Brissot P, Ropert M, le Lan C, Loreal O (2012) Non-transferrin bound iron: a key role in iron overload and iron toxicity. Biochim Biophys Acta 1820:403–410
378. Brix G, Seebass M, Hellwig G, Griebel J (2002) Estimation of heat transfer and temperature rise in partial-body regions during MR procedures: an analytical approach with respect to safety considerations. Magn Reson Imag 20:65–76
379. Brochard F, Lennon JF (1975) Frequency spectrum of the flicker phenomenon in erythrocytes. J Phys 36:1035–1047
380. Brock KK, McShan DL, Ten Haken RK, Hollister SJ, Dawson LA, Balter JM (2003) Inclusion of organ deformation in dose calculations. Med Phys 30:290–295
381. Brockbank KGM, Smith KM (1993) Synergistic interaction of low-molecular-weight polyvinylpyrrolidones with dimethylsolfoxide during cell preservation. Transplant Proc 25:3185
382. Brockbank KGM, Walsh JR, Song YC, Taylor MJ (2003) Vitrification: preservation of cellular implants. In: Ashammakhi N, Ferretti P (eds) Topics in Tissue Engineering, chap.12. https://www.researchgate.net/publication/52002243_Vitrification_Preservation_of_Cellular_Implants
383. Bron AJ, Tiffany JM (2004) The contribution of meibomian disease to dry eye. Ocular Surf 2:149–164

384. Bronstein VL, Steponkus PL (1995) Nucleation and growth of ice in concentrated solutions of ethylene glycol. Cryobiology 32:1–22
385. Brooks SM, Brooks NP (1980) The human body. Structure and function in health and disease. The C.N. Mosby Company, St. Louis
386. Brown VJ (1989) Development of computer modelling techniques for microwave thermography. PhD Thesis, University of Glasgow
387. Brown WS, Dewey WA, Jacobs HR (1970) Thermal properties of teeth. J Dent Res 49:752–755
388. Brown SL, Hunt JW, Hill RP (1992) Differential thermal sensitivity of tumour and normal tissue microvascular response during hyperthermia. Int J Hyperth 8:501–514
389. Brown CL, Bale JS, Walters KFA (2004) Freezing induces a loss of freeze tolerance in an overwintering insect. Proc R Soc Lond B 271:1507–1511
390. Brush RA, Griffith M, Mlynarz A (1994) Characterization and quantification of intrinsic ice nucleators in winter rye (Secale cereale) leaves. Plant Physiol 104:725–735
391. Bryant G, Wolfe J (1989) Can hydration forces induce lateral phase separations in lamellar phases? Eur Biophys J 16:369–374
392. Bryant G, Wolfe J (1992) Interfacial forces in cryobiology and anhydrobiology. Cryo-Lett 13:23–36
393. Bryant G, Koster KL, Wolfe J (2001) Membrane behaviour in seeds and other systems at low water content: the various effects of solutes. Seed Sci Res 11:17–25
394. Bryngelson SH, Freund JB (2016) Capsule-train stability. Phys Rev Fluids 1:033201
395. Bryngelson SH, Freund JB (2018) Global stability of flowing red blood cell trains. Phys Rev Fluids 3:073101
396. Bryngelson SH, Gueniat F, Freund JB (2019) Irregular dynamics of cellular blood flow in a model microvessel. Phys Rev E 100:012203
397. Bubnovskaya L, Belous A, Solopan S, Kovelskaya A, Bovkun L, Podoltsev A, Kondtratenko I, Osinsky S (2014) Magnetic fluid hyperthermia of rodent tumors using manganese perovskite nanoparticles. J Nanoparticles 2014:278761
398. Bucharan S, Cross S, Acker JP, Toner M, Carpenter JF, Pyatt D (2002) A novel technique for the long-term cryopreservation of hematopoietic progenitor cells using intracellular trehalose. Exp Hemotol 30:131
399. Bucherer C, Lelievre JC, Lacombe C (1988) Theoretical and experimental study of the time dependent flow of red blood cell suspension through narrow pores. Biorheology 25:639–649
400. Buckley LB (2008) Linking traits to energetics and population dynamics to predict lizard ranges in changing environments. Am Nat 171:E1–E19
401. Buckley LB, Nufio CR, Kingsolver JG (2013) Phenotypic clines, energy balances and ecological responses to climate change. J Animal Ecol 83:41–50
402. Bulte JWM, Kraitchman DL (2004) Iron oxide MR contrast agents for molecular and cellular imaging. NMR Biomed 17:484–499
403. Bunkin FV, Lyakhov GA, Shipilov KF (1995) Thermal self-action of acoustic wave packets in a liquid. Phys Usp 38:1099–1118
404. Bunkin NF, Ninham BW, Babenko VA, Suyazov NV, Sychev AA (2010) Role of dissolved gas in optical breakdown of water: differences between effects due to helium and other gases. J Phys Chem B 114:7743–7752
405. Buono MJ, Sjoholm NT (1988) Effect of physical traning on peripheral sweat production. J Appl Physiol 65:811–814
406. Burch NJ, Lehouck RB (2010) The nonlocal Cattaneo-Vernotte equations. SAND2010-8783P
407. Burton A (1934) The application of the theory of heat flow to the study of energy metabolism. J Nurition 7:497–533
408. Buscarini E, Buscarini L (2004) Radiofrequency thermal ablation with expandable needle of focal liver malignancies: complication report. Eur J Radiol 14:31–37

409. Buscher BA, McMeekin TO, Goodwin D (2000) Treatment of leg telangiectasia by using a long-pulse dye laser at 595 nm with and without dynamic cooling device. Lasers Surg Med 27:171–175
410. Bushev YG, Davletbaeva SV, Muguet F (2003) The 3-attractor water model: Monte-Carlo simulations with a new, effective 2-body potential (BMW). Molecules 8:226–242
411. Busija DW, Leffler CW (1987) Hypothermia reduces cerebral metabolic rate and cerebral blood flow in new born pigs. Am J Physiol 253:H869–H873
412. Butovich IA, Arciniega JC, Wojtowicz JC (2010) Meibomian lipid films and the impact of temperature. Investig Ophthalm Vis Sci 51:5508–5518
413. Buyukdagli S, Joyeux M (2007) Theoretical investigation of finite size effects at DNA. arXiv:physics/0703082v1 [physics.bio-ph]
414. Buzano C, De Stefanis E, Pretti M (2007) Low-temperature-induced swelling of a hydrophobic polymer; a lattice approach. J Chem Phys 126:074904
415. Byakov VM, Stepanov SV (2006) The mechanism for the primary biological effects of ionizing radiation. Phys Usp 49:467–487
416. Bynum GD, Pandolf KB, Schuette WH, Goldman RF, Lees DE, Whang-Peng J, Atkinson ER, Bull JM (1978) Induced hyperthermia in sedated humans and the concept of critical thermal maximum. Am J Physiol 235:R228–R236
417. Byrd PF, Friedman MD (1971) Handbook of elliptic integrals for engineers and scientists, 2nd edn. Springer-Verlag, Berlin
418. Byrne C, Lee J, Chew SAN, Lim CL, Tan YM (2006) Continuous thermoregulatory responses to mass-participation distance running in heat. Med Sci Sports Exercise 38:803–810
419. Cabanac M (1986) Keeping a cool head. News Physiol Sci 1:41–44
420. Cabanac M (2006) Adjustable set point: to honor Harold T. Hammel. J Appl Physiol 100:1338–1346
421. Cabanac M, Brinnel H (1985) Blood flow in the emissary veins of the human head during hyperthermia. Eur J Appl Physiol 54:172–176
422. Cabanac M, Caputa M (1979) Natural selective cooling of the human brain: evidence of its occurrence and magnitude. J Physiol (Lond) 286:255–264
423. Cahill DG, Ford WK, Goodson KE, Mahan GD, Majumdar A, Maris HJ, Merlin R, Phillpot SR (2003) Nanoscale thermal transport. J Appl Phys 93:793–818
424. Cahill DG, Braun PV, Chen G, Clakr DR, Fan S, Goodson KE, Keblinski P, King WP, Mahan GD, Majumdar A, Maris HJ, Phillpot SR, Pop E, Shi L (2014) Nanoscale thermal transport. II. 2003–2012. Appl Phys Rev 1:011305
425. Cairns CJ, Andrews PJ (2002) Management of hyperthermia in traumatic brain injury. Curr Opin Crit Care 8:106–110
426. Caldarelli G, de los Rios P (2001) Cold and warm denaturation of proteins. J Biol Phys 27:229–241
427. Campelo F, Hernandez-Machado A (2006) Dynamic model and stationary shapes of fluid vesicles. Eur Phys J E 20:37–45
428. Canham PB (1970) The minimum energy of bending as a possible explanation of the biconcave shape of the human red blood cell. J Theor Biol 26:61–81
429. Canhamp B, Burtona C (1968) Distribution of size and shape in populations of normal human red blood cells. Circul Res 22:405–422
430. Canney MS, Khokhlova VA, Bessonova OV, Bailey MR, Crum LA (2010) Shock-induced heating and millisecond boilin in gels and tissues due to high intensity focused ultrasound. Ultrasound Med. Biol. 36:250–267
431. Cannon WB (1929) Organization for Physiological Homeostasis. Physiol. Rev. 9:399–431
432. Cannon B, Nedergaard J (2004) Brown adipose tissue: functionand physiological significance. Physiol Rev 84:277–359
433. Cannon B, Nedergaard J (2011) Nonshivering thermogenesis and its adequate measurement in metabolic studies. J Exp Biol 214:242–253

434. Canters RA, Wust P, Bakker JF, Van Rhoon GC (2009) A literature surveyon indicators for characterisation and optimisation of sardistributions in deep hyperthermia, a plea for standardisation. Int J Hyperthermia 25:593–608
435. Cantwell CP, Obyrne J, Eustace S (2004) Current trends in treatment of osteoid osteoma with an emphasis on radiofrequency ablation. Eur Radiol 14:607–617
436. Cao BY, Guo ZY (2007) Equation of motion of phonon gas and non-Fourier heat conduction. J Appl Phys 102:053503
437. Cao H, Vorperian VR, Tungjitkusolmun S, Tsai JZ, Haemmerich D, Choy YB, Webster JG (2001) Flow effect on lesion formation in RF cardiac catheter ablation. IEEE Trans Biomed Eng 48:425–433
438. Caputo M, Fabrizio M (2015) A new definition of fractional derivative without singular kernel. Progr Fract Differ Appl 1:73–85
439. Caputo M, Fabrizio M (2016) Applications of new time and spatial fractional derivatives with exponential kernels. Progr Fract Differ Appl 2:1–11
440. Carboni EJ (2017) The margination and transport of particles in blood flow. PhD Thesis, University of Connecticut
441. Carboni EJ, Bognet BH, Bouchillon GM, Kadilak AL, Shor LM, Ward MD, Ma AWK (2016) Direct tracking of particles and quantification of margination in blood flow. Biophys J 111:1487–1495
442. Careau V, Bininda-Emonds ORP, Thomas DW, Réale D, Humphries MM (2009) Exploration strategies map along fast-slow metabolic and life-history continua in muroid rodents. Funct Ecol 23:150–156
443. Careau V, Montiglio PO, Garant D, Pelletier F, Speakman JR, Humphries MM, Réale D (2015) Energy expenditure and personality in wild chipmunks. Behav Ecol Sociobiol 69:653–661
444. Carignano MA, Shepson PB, Szleifer I (2005) Molecular dynamics simulations of ice growth from supercooled water. Mol Phys 103:2957–2967
445. Carlisle HJ, Ingram DL (1973) The influence of body core temperature and peripheral temperatures on oxygen consumption in the pig. J Physiol 231:341–352
446. Carluccio G, Erricolo D, Oh S, Collins CM (2013) An approach to rapid calculation of temperature change in tissue using spatial filters to approximate effects of thermal conduction. IEEE Trans Biomed Eng 60:1735–1741
447. Carnevale KA (2004) Finite-difference model of cell dehydratation during cryopreservation. M.S. Thesis, Georgia Institute of Technology
448. Carnochan P, Dickinson RJ, Joiner MC (1986) The practical use of thermocouples for temperature measurement in clinical hyperthermia. Int J Hyperth 2:1–19
449. Caro CG (2012) The mechanics of the circulation, 2nd edn. Cambridge University Press, Cambridge
450. Caro CG, Fitz-Gerald JM, Schroter RC (1971) Atheroma and arterial wall shear. Observation, correlation and proposal of a shear dependent mass transfer mechanism for atherogenesis. Proc R Soc Lond B Biol Sci 177:109–159
451. Carpenter JF, Hansen TN (1992) Antifreeze protein modulates cell survival during cryopreservation: mediation through influence on ice crystal growth. Proc Natl Acad Sci USA 89:8953–8957
452. Carpin A, Mainardi F (1997) Fractals and fractional calcuus in continuum mechanics. Springer, Wien
453. Carpinteri A, Cornetti P, Sapora A, Di Paola M, Zingales M (2009) Fractional calculis in solid mechanics: local versus non-local approach. Phys Scr T136:014003
454. Carr RT, Wickham LL (1991) Influence of vessel diameter on red-cell distribution at microvascular bifurcations. Microvasc Res 41:184–196
455. Carr RT, Geddes JB, Wu F (2005) Oscillations in a simple microvascular network. Ann Biomed Eng 33:764–771
456. Carrel A (1912) The preservation of tissue and its application in surgery. J Am Medical Assn 59:523–530

457. Carrey J, Mehdaoui B, Respaud M (2011) Simple models for dynamic hysteresis loop calculations of magnetic single-domain nanoparticles: application to magnetic hyperthermia optimization. J Appl Phys 109:083921
458. Carslaw HS, Jaeger JC (1959) Conduction of heat in solids. Clarendon Press, Oxford
459. Carter DL, MacFall JR, Clegg ST, Wan X, Prescott DM, Charles HC, Samulski TV (1998) Magnetic resonance thermometry during hyperthermia for human high-grade sarcoma. Int J Radiat Oncol Biol Phys 40:815–822
460. Carter DC, Lim K, Ho JX, Wright BS, Twigg PD, Miller TY, Chapman J, Keeling K, Ruble J, Vekilov PG, Thomas BR, Rosenberger F, Chernov AA (1999) Lower dimer impurity incorporation may result in higher perfection of HEWL crystals grown in microgravity: a case study. J Cryst Growth 196:623–637
461. Casas-Vázquez J, Jou D (1997) Lagrange multipliers in extended irreversible thermodynamics and in informational statistical thermodynamics. Brazilian J Phys 27:547–559
462. Casas-Vázquez J, Jou D (2003) Temperature in nonequilibrium states: a review of open problems and current proposals. Rep Progr Phys 66:1937–2023
463. Castellani JW, Young AJ, Ducharme MB, Giesbrecht GG, Glickman E, Sallis RE (2006) Prevention of cold injuries during exercise. Med Sci Sports Exercise 38:2012–2029
464. Castellani MP, Rioux TP, Castellani JW, Potter AW, Xu X (2021) A geometrically accurate 3 dimensional model of human thermoregulation for transient cold and hot environments. Comput Biol Med 138:104892
465. Castro MA, Rodríguez F, Escolano J, Martín JA (2013) Exact and analytic-numerical solutions of lagging models of heat transfer in a semi-infinite medium. Abstr Appl Anal 2013:397053
466. Castro MA, Martín JA, Rodríguez F (2017) Unconditional stability of a numerical method for the dual-phase-lag equation. Math Probl Eng 2017:1650380
467. Catttaneo C (1958) A form of heat conduction equation which eliminated the paradox of instantaneous propagation. Comp Rend 247:431–433
468. Cattaneo L, Zunino P (2014) Computational models for fluid exchange between microcirculation and tissue interstitium. Netw Heterogeneous Media 9:135–159
469. Cavagnaro M, Pinto R, Lopresto V (2015) Numerical models to evaluate the temperature increase induced by ex vivo microwave thermal ablation. Phys Med Biol 60:3287–3311
470. Cavender-Bares J (2005) Impacts of freezing on long-distance transport in woody plants. In: Holbrook NM, Zwieniecki M, Melcher P (eds) Vascular transport in plant. Elsevier, Oxford, pp 401–424
471. Cebremedhin KG, Wu B, Perano K (2016) Modeling conductive cooling for thermally stressed dairy cows. J Thermal Biol 56:91–99
472. Cerda E, Mahadevan L (2003) Geometry and physics of wrinkling. Phys Rev Lett 90:074302
473. Cervadoro A, Giverso C, Pande R, Sarangi S, Preziosi L, Wosik J, Brazdeikis A, Decuzzi P (2013) Design maps for the hyperthermic treatment of tumors with superparamagnetic nanoparticles. PLoS One 8:e57332
474. Cetingül MP, Herman C (2011) Quantification of the thermal signature of a melanoma lesion. Int J Thermal Sci 50:421–431
475. Cevc G (1995) Material transport across permeability barriers by means of lipid vesicles. In: Lipowsky R, Sackmann E (eds) Handbook of biological physics. Elsevier, Amsterdam, pp 465–490
476. Chai CY, Lin MT (1972) Effects of heating and cooling the spinal cord and medulla oblongata on thermoregulation in monkeys. J Physiol 225:297–308
477. Chakrabarty J, Banerjee D, Pal D, De J, Ghosh A, Majumder GC (2007) Shedding off specific lipid constituents from sperm cell membrane during cryopreservation. Cryobiology 54:27–35
478. Chambers R, Halle P (1932) The formation of ice in protoplasm. Proc R Soc B110:336–352
479. Chan WL, Averback RS, Cahill DG, Lagoutchev A (2008) Dynamics of femtosecond laser-induced melting of silver. Phys Rev B 78:214107
480. Chandrasekharan K (1985) Elliptic functions. Springer, Berlin

481. Chang HY, Li X, Karniadakis GE (2017) Modeling of biomechanics and biorheology of red blood cells in type 2 diabetes mellitus. Biophys J 113:481–490
482. Chang I (2003) Finite element analysis of hepatic radiofrequency ablation probes using temperature-dependent electrical conductivity. Biomed Eng Online 2:12
483. Chang L, Liu X, Fan D, Miao Y, Zhang H, Ma H, Liu QY, Ma P, Xue WM, Luo YE, Fan HM (2016) The efficiency of magnetic hyperthermia and in vivo histocompatibility for human-like collagen protein-coated magnetic nanoparticles. Int J Nanomed 11:1175–1185
484. Chao NH, Chiang CP, Hsu HW, Tsai CT, Lin TT (1994) Toxicity tolerance of oyster embryos to selected cryoprotectants. Aquat Living Resour 7:99–104
485. Chapman S, Cowling TG (1970) The mathematical theory of non-uniform gases, 3rd edn. Cambridge University Press, Cambridge
486. Chapuisat G, Dronne M, Grenier E, Hommel M, Gilquin H, Boissel J (2008) A global phenomenological model of ischemic stroke with stress on spreading depressions. Prog Biophys Mol Biol 97:4–27
487. Charkoudian N (2003) Skin blood flow in adult human thermoregulation: how it works, when it does not, and why. Mayo Clin Proc 78:603–612
488. Charny CK (1992) Mathematical models of bioheat transfer. Adv Heat Transf 22:19–155
489. Charny CK, Weinbaum S, Lewin RL (1990) An evaluation of the Weinbaum-Jiji bioheat equation for normal and hyperthermic conditions. J Biomech Eng 112:80–87
490. Chaseling GK, Crandall CG, Gagnon D (2020) Skin blood flow measurements during heat stress: technical and analytical considerations. Am J Physiol Regul Integr Comp Physiol 318:R57–R69
491. Chato JC (1980) Heat transfer to blood vessels. J Biomech Eng 102:110–118
492. Chato JC (1981) Reflection on the history of heat and mass transfer in bioengineering. J Biomech Eng 103:97–101
493. Chato JC (1989) Fundamentals of bioheat transfer. Springer, Berlin
494. Chato JC (1992) A view of the history of heat transfer in bioengineering. Adv Heat Transf 22:1–18
495. Chechkin A, Metzler R, Klafter J, Gonchar VY (2008) Introduction to the theory of Lévy flights. In: Klages R, Radons G, Sokolov IM (eds) Anomalous transport. Foundations and applications, pp 129–162. Wiley, Hoboken
496. Chefranov SG (2013) On conditions of negativity of friction resistance for non-stationary modes of blood flow and possible mechanism of affecting of environmental factors on energy effectiveness of cardio-vascular system functioning. arXiv:1301.6603 [physics.flu-dyn]
497. Chen G (2005) Nanoscale energy transport and conversion: a parallel treatment of electrons, molecules, phonons, and photons. OUP, Oxford
498. Chen W (2006) Time-space fabric underlying anomalous diffusion. Chaos Soltons Fract 28:923–929
499. Chen G (2021) Non-Fourier phonon heat conduction at the microscale and nanoscale. Nat Rev Phys 3:555–569
500. Chen J, Beraun JE (2001) Numerical study of ultrashort laser pulse interaction with metal films. Numer Heat Trnsf A 40:1–20
501. Chen PJ, Gurtin ME (1968) On a theory of heat conduction involving two temperatures. Z Angew Math Phys 19:614–627
502. Chen W, Holm S (2004) Fractional Laplacian time-space models for linear and nonlinear lossy medium exhibiting arbitrary frequency power-law dependencies. J Acoust Soc Am 115:924–930
503. Chen MM, Holmes KR (1980) Microvascular contribution to tissue heat transfer. Ann NY Acad Sci 335:137–150
504. Chen M, Huang WQ (2006) Heuristic algorithm for off-lattice protein folding problem. J Zhejiang Univ Sci B 7:7–12
505. Chen G, Jia Z (1999) Ice-binding surface of fish III antifreeze. Biophys J 77:1602–1608
506. Chen W, Liang Y (2017) New methodologies in fractional and fractal derivatives modeling. Chaos Solitons Fract 102:72–77

507. Chen W, Pang G (2016) A new definition of fractional Laplacian with application to modelling three-dimensional nonlocal heat conduction. J Comput Phys 309:350–367
508. Chen G, Shakouri A (2002) Heat transfer in nanostructures for solid-state energy conversion. J Heat Transf 124:242–252
509. Chen PJ, Gurtin ME, Williams WO (1968) A note on nonsimple heat conduction. Z Angew Math Phys 19:969–970
510. Chen PJ, Gurtin ME, Williams WO (1969) On the thermodynamics of non-simple elastic materials with two temperatures. Z Angew Math Phys 20:107–112
511. Chen SS, Wright NT, Humphrey JD (1998) Heat-induced changes in the mechanics of a collagenous tissue: isothermal, isotonic shrinkage. J Biomech Eng 120:382–388
512. Chen SS, Wright NT, Humphrey JD (1998) Phenomenological evolution equations for heat-induced shrinkage of a collagenous tissue. IEEE Trans Biomed Eng 45:1234–1240
513. Chen SY, Doolen GD (1998) Lattice boltzmann method for fluid flow. Annu Rev Fluid Mech 30:329–364
514. Chen X, Saidel GM (2009) Mathematical modeling of thermal ablation in tissue surrounding a large vessel. J Biomech Eng 131:011001
515. Chen T, Acker JP, Cheley S, Bayley H, Fowler A, Toner M (2001) Beneficial effect of intracellular trehalose on the membrane integrity of dried mammalian cells. Cryobiology 43:168–181
516. Chen HT, Song JP, Liu KC (2004) Study of hyperbolic heat conduction problem in IC chip. Jap J Appl Phys 43:4404–4410
517. Chen JK, Tzou DY, Beraun JE (2006) A semiclassical two-temperature model for ultrafast laser heating. Int J Heat Mass Transf 49:307–316
518. Chen W, Sun H, Zhang X, Korošak D (2010) Anomalous diffusion by fractal and fractional derivatives. Comput Math Appl 59:1754–1758
519. Chen Y, Yan Y, Zhang K (2010) On the local fractional derivative. J Math Anal Appl 362:17–33
520. Chen AC, Arany PR, Huang YY, Tomkinson EM, Sharma SK, Kharkwal GB, Saleem T, Mooney D, Yull FE, Blackwell TS (2011) Low-level laser therapy activates NF-kB via generation of reactive oxygen species in mouse embryonic broblasts. PloS One 6:e22453
521. Chen H, Adam A, Cheng Y, Tang S, Hatung J, Bao E (2015) Localization and expression of heat shock protein 70 with rat myocardial cell damage induced by heat stress in vitro and in vivo. Molecul Med Rep 11:2276–2284
522. Chen J, Liu X, Hu Y, Chen X, Tan S (2023) Cryopreservation of tissues and organs: present, bottlenecks, and future. Front Vet Sci 10:1201794
523. Cheng CHC (2003) Freezing avoidance in polar fishes. In: Gerday C (ed) Encyclopedia of life support systems (EOLSS) - theme 6.73 extremophiles. Eolss Publishers, London
524. Cheng A, Merz KM (1997) Ice-binding mechanism of winter flounder antifreeze proteins. Biophys J 73:2851–2873
525. Cheng H, Plewes D (2002) Tissue thermal conductivity by magnetic resonanse thermometry and focused ultrasound heating. J Magn Reson Imag 16:598–609
526. Cheng K, Sun S (2010) Recent advances in syntheses and therapeutic applications of multifunctional porous hollow nanoparticles. Nano Today 5:183–196
527. Cheng L, Xu M, Wang L (2008) From Boltzmann transport equation to single-phase-lagging heat conduction. Int J Heat Mass Transf 51:6018–6023
528. Cheong WF, Prahl SA, Welch AJ (1990) A review of the optical properties of biological tissues. IEEE J Quant Electron 26:2166–2185
529. Chernov AA (1997) Protein versus conventional crystals: creation of defects. J Cryst Growth 174:354–361
530. Cherry EM, Eaton JK (2013) Shear thinning effects on blood flow in straight and curved tubes. Phys Fluids 25:073104
531. Cheung A, Neyzari A (1984) Deep local hyperthermia for cancer therapy: external electromagnetic and ultrasound techniques. Cancer Res 44:4736–4744

532. Cheuvront SN, Kenefick RW (2014) Dehydration: physiology, assessment, and performance effects. Compr Physiol 4:257–285
533. Chew WC, Weedon WH (1994) A 3D perfectly matched medium from modified Maxwell's equations with stretched coordinates. Microw Opt Technol 7:599–604
534. Chiang J, Wang P, Brace C (2013) Computational modelling of microwave tumour ablations. Int J Hyperthermia 29:308–325
535. Chichindaev AV, Khromova IV (2018) Heat exchange in the "human-environment" system in conditions of low temperatures (in Russian). NSTU Publisher, Novosibirsk
536. Chichindaev AV, Khromova IV, Dyachenko YV (2018) Influence of heat transfer between calculated elements on thermal processes in the human body – environment system (in Russian). Proceedings of the Russian Higher School Academy of Sciences, pp 106–113
537. Chichindaev AV, Dyachenko YV, Sparin VA, Khromova IV (2019) Life support systems and equipment for aircraft (in Russian). NSTU Publisher, Novosibirsk
538. Chidamaram R, Sikka SK (2003) Neutron crystallography – then and now. Curr Sci 85:871–880
539. Chien S (1970) Shear dependence of effective cell volume as a determinant of blood viscosity. Science 168:977–979
540. Chien S (1987) Red cell deformability and its relevance to blood flow. Rev Physiol 49:177–192
541. Chien S, Jan K (1973) Ultrastructural basis of the mechanism of rouleaux formation. Microvasc Res 5:155–166
542. Chien S, Usami S, Bertles JF (1970) Abnormal rheology of oxygenated blood in sickle cell anemia. J Clin Invest 49:623–634
543. Chiriță S (2019) On high-order approximations for describing the lagging behavior of heat conduction. Math Mech Solids 24:1648–1667
544. Chiriță S, Ciarlettac M, Tibullo V (2015) On the wave propagation in the time differential dual-phase-lag thermoelastic model. Proc R Soc A 471:20150400
545. Chiriță S, Ciarlettac M, Tibullo V (2021) Qualitative properties of solutions in the time differentialdual-phase-lag model of heat conduction. arXiv:2103.10083v1 [math.AP]
546. Chiu CH, Chan HH, Ho WS, Yeung CK, Nelson JS (2003) Prospective study of pulsed dye laser in conjunction with cryogen spray cooling for treatment of port wine stains in chinese patients. Dermatol Surg 29:909–915
547. Chiu VH, Deshpande VV, Postma HC, Lau CN, Miko C, Forro L, Bockrath M (2005) Ballistic phonon thermal transport in multiwalled carbon nanotubes. Phys Rev Lett 95:226101
548. Chiu-Lam A, Staples E, Pepine CJ, Rinaldi C (2021) Perfusion, cryopreservation, and nanowarming of whole hearts using colloidally stable magnetic cryopreservation agent solutions. Sci Adv 7:eabe3005
549. Chizhov AV, Amakhin D, Smirnova EY, Zaitsev AV (2022) Ictal wavefront propagation in slices and simulations with conductance-based refractory density model. PLoS Comput Biol 18:e1009782
550. Chklovskii DB, Schikorski T, Stevens CF (2002) Wiring optimization in cortical circuits. Neuron 43:341–347
551. Cho S, Shin MH, Kim YK, Seo JE, Lee YM, Park CH, Chung JH (2009) Effects of infrared radiation and heat on human skin aging in vivo. J Investig Dermatol Symp Proc 14:15–19
552. Choi JH, Han B, Bishof JC (2004) Effects of a cryoprotective agent on thermal properties of solutions at subzero temperatures. ASME Int Mech Eng Congr 2004:1–2
553. Choi B, Milner TE, Kim J, Goodman JN, Vargas G, Aguilar G, Nelson JS (2005) Use of optical coherence tomography to monitor biological tissue freezing during cryosurgery. J Biomed Opt 9:282–286
554. Choi JH, Marshall R, Neimark M, Konstas A, Lin E, Chiang Y, Mast H, Rundek T, Mohr J, Pile-Spellman J (2010) Selective brain cooling with endovascular intracarotid infusion of cold saline: a pilot feasibility study. Am J Neuroradiol 31:928–934

555. Choi J, Morrissey M, Bischof JC (2013) Thermal processing of biological tissue at high temperatures: impact of protein denaturation and water loss on the thermal properties of human and porcine liver in the range 25–80 C. J Heat Transf 135:1–7
556. Choi JH, Yoon SH, Park SG, Choi SH (2016) Analytical solution of the Cattaneo-Vernotte equation for non-Foirier heat conduction. J Korean Soc Marine Eng 40:389–396
557. Choi J, Zhou H, Landig R, Wu HY, Yu X, Stetina S, Kucsko G, Mango S, Needleman D, Samuel ADT, Maurer P, Park H, Lukin MD (2020) Probing and manipulating embryogenesis via nanoscale thermometry and temperature control. arXiv:2001.02664v1 [physics.bio-ph]
558. Chou Y, Yang RJ (2008) Application of CESE method to simulate non-Fourier heat conduction in finite medium with pulse surface heating. Int J Heat Mass Transf 51:3525–3534
559. Chou Y, Yang RJ (2009) Two-dimensional dual-phase-lag thermal behavior in single-/multi-layer structures using CESE method. Int J Heat Mass Transf 52:239–249
560. Choudhuri S (2007) On a thermoelastic three-phase-lag model. J Therm Sci 30:231–238
561. Chretien D, Benit P, Ha HH, Keipert S, El-Khoury R, Chang YT, Jastroch M, Jacobs HT, Rustin P, Rak M (2018) Mitochondria are physiologically maintained at close to 50 degrees c. PLoS Biol 16:e2003992
562. Christov CI (2009) On frame indiffent formulation of the Maxwell-Cattaneo model of finie-speed heat conduction. Mech Res Comm 36:481–486
563. Christov CI (2019) Comments on "Scattering cancellation-based cloaking for the Maxwell-Cattaneo heat waves". arXiv: 1908.02188
564. Christov CI, Jordan PM (2005) Heat conduction paradox involving second-sound propagation in moving media. Phys Rev Lett 94:154301
565. Chu KF, Dupuy DE (2014) Thermal ablation of tumours: biological mechanisms and advances in therapy. Nat Rev Cancer 14:199–208
566. Chua KJ (2013) Fundamental experiments and numerical investigation of cryofreezing incorporating vascular network with enhanced nano-freezing. Int J Therm Sci 70:17–31
567. Chua K, Ho J, Chou S, Islam M (2005) On the study of the temperature distribution within a human eye subjected to a laser source. Int Commun Heat Mass Transf 32:1057–1065
568. Chua KJ, Chou SK, Ho JC (2007) An analytical study on the thermal effects of cryosurgery on selective cell destruction. J Biomech 40:100–116
569. Chung SH, Mazur E (2009) Surgical applications of femtosecond lasers. J Biophotonics 2:557–572
570. Chung S, Vafai K (2014) Mechanobiology of low-density lipoprotein transport within an arterial wall–impact of hyperthermia and coupling effects. J Biomech 47:137–147
571. Chung H, Dai T, Sharma SK, Huang YY, Carroll JD, Hamblin MR (2012) The nuts and bolts of low-level laser (light) therapy. Ann Biomed Eng 40:516–533
572. Church CC, Barnett SB (2012) Ultrasound-induced heating and its biological consequences. In: ter Haar G (ed) The safe use of ultrasound in medical diagnosis, pp 46–68
573. Church CC, Carstensen EL, Nyborg WL, Carson PL, Frizzell LA, Bailey MR (2008) The risk of exposure to diagnostic ultrasound in postnatal subjects: nonthermal mechanisms. J Ultrasound Med 27:565–592
574. Churchill TA, Storey KB (1994) Effects of dehydration on organ metabolism in the frog Pseudacris crucifer: hyperglycemic responses to dehydration mimic freezing-induced cryoprotectant production. J Comput Physiol B 164:492–498
575. Chvapil M, Jensovsky L (1963) The shrinkage temperature of collagen fibres isolated from the tail tendons of rats of various ages and from different places of the same tendon. Gerontologia 1:18–29
576. Cicekli U (2003) Computational model for heat transfer in the human eye using the finite elemet method. MS Thesis, Lousiana State University
577. Ciesielski M, Duda M, Mochnacki B (2016) Comparison of bio-heat transfer numerical models based on the Pennes and Cattaneo-Vernotte equations. J Appl Math Comput Mech 15:33–38

578. Ciesielski M, Siedlecki J, Janik MK (2020) Mathematical modelling of thermal and electrical processes during electrosurgical resection of colorectal polyps. Int J Eng Sci 154:103351
579. Cifra M, Fields JZ, Farhadi A (2011) Electromagnetic cellular interactions. Progr Biophys Mol Biol 105:223–246
580. Ciliberto S (2017) Experiments in stochastic thermodynamics: short history and perspectives. Phys Rev X 7:021051
581. Cimmelli VA (2007) An extension of Liu procedure in weakly nonlocal thermodynamics. J Math Phys 48:113510
582. Cimmelli VA (2009) Different thermodynamics theories and different heat conduction laws. J Non-Equilib Thermodyn 34:299–333
583. Cimmelli VA (2021) Local versus nonlocal constitutive theories of nonequilibrium thermodynamics: the Guyer-Krumhansl equation as an example. Zeit Angewandte Math Phys 72:195
584. Cimmelli VA, Jou D, Ruggeri T, Ván P (2014) Entropy principle and recent results in non-equilibrium theories. Entropy 16:1756–1807
585. Cimmelli VA, Kosińsky W, Saxton K (1991) A new approach to the theory of heat conduction with finite wave speeds. Le Matem XLVI:95–105
586. Cimmelli VA, Sellitto A, Jou D (2010) Nonequilibrium temperatures, heat waves, and nonlinear heat transport equations. Phys Rev B 81:054301
587. Cimmelli VA, Sellitto A, Jou D (2010) Nonlinear evolution and stability of the heat flow in nanosystems: beyond linear phonon hydrodynamics. Phys Rev B 82:184302
588. Cimmelli VA, Sellitto A, Triani V (2010) A generalized Coleman-Noll procedure for the exploitation of the entropy principle. Proc R Soc A 466:911–925
589. Cimmelli VA, Sellitto A, Triani V (2010) A new perspective on the form of the first and second laws in rational thermodynamics: Korteweg fluids as an example. J Non-Equilib Thermodyn 35:251–265
590. Cimmelli VA, Ván P (2005) The effect of non-locality on the evolution of higher order fluxes in nonequilibrium thermodynamics. J Math Phys 46:112901
591. Cioc R (2016) Physical and geometrical interpretation of Grunwald-Letnikov differintegrals: measurement of path and acceleration. Fract Calcul Appl Anal 19:161–172
592. Clark RS, Kochanek PM, Marion DW, Schiding JK, White M, Palmer AM, DeKosky ST (1996) Mild posttraumatic hypothermia reduces mortality after severe controlled cortical impact in rats. J Cereb Blood Flow Metabol 16:253–261
593. Clarke DM, Baust JM, Buskirk RGV, Baust JG (2001) Chemo-cryo combination therapy: an ajunctive model for the treatment of prostate cancer. Cryobiology 42:274–285
594. Clarke DM, Baust JM, Buskirk RGV, Baust JG (2004) Addition of anticancer agents enhances freezing-induced prostate cancer cell death: implications of mitochondrial involvement. Cryobiology 49:45–61
595. Clarkson DP, Schatte CL, Jordan JP (1972) Thermal neutral temperature of rats in helium-oxygen, argon-oxygen, and air. Am J Physiol 222:1494–1498
596. Clas SD, Dalton CR, Hancock BC (1999) Differential scanning calorimetry: applications in drug development. Pharm. Sci Technol Today 2:311–320
597. Clausen JR, Reasor DA Jr, Aidun CK (2011) The rheology and microstructure of concentrated non-colloidal suspensions of deformable capsules. J Fluid Mech 685:202–234
598. Cobb JP, Hotchkiss RS, Karl IE, Buchman TG (1996) Mechanisms of cell injury and death. Brit J Anaesth 77:3–10
599. Cognié Y, Barill G, Poulin N, Mermillord P (2003) Current status of embryo technologies in sheep and goat. Theriogenology 59:171–188
600. Cohen ML (1977) Measurements of the thermal properties of human skin. A review. J Invest Derma 69:333–338
601. Cokelet G, Goldsmith H (1991) Decreased hydrodynamic resistance in the 2- phase flow of blood through small vertical tubes at low flow-rates. Circul Res 68:1–17
602. Coleman BD (1961) Foundations of linear viscoelesticity. Rev Mod Phys 33:239–249

603. Coleman BD (1964) Thermodynamics of materials with memory. Arch Rational Mech Anal 17:1–46
604. Coleman BD, Gurtin ME (1967) Equipresence and constitutive equations for rigid heat conductors. Zeitsch Angewandte Math Phys 18:199–208
605. Coleman BD, Mizel VJ (1967) A general theory of dissipation in materials with memory. Arch Rational Mech Anal 27:255–274
606. Coleman BD, Noll W (1963) The thermodynamics of elastic materials with heat conduction and viscosity. Arch Ration Mech Anal 13:167–178
607. Coleman BD, Owen DR (1974) A mathematical foundation for thermodynamics. Arch Ration Mech Anal 54:1–104
608. Coleman BD, Fabrizio M, Owen DR (1982) On the thermodynamics of second sound in dielectric crystals. Arch Rat Mech Anal 80:135–158
609. Coleman BD, Fabrizio M, Owen DR (1985) Thermodynamics and the constitutive relations for second sound in crystals. In: Grioli G (ed) Thermodynamics and constitutive equations, vol 228. Springer, Berlin, pp 20–43
610. Collins CA, Rojas E (1982) Temperature dependence of the sodium channel gating kinetics in the node of ranvier. Q J Exp Physiol 67:41–55
611. Commission on Genetic Resources for Food and Agriculture (2007) FAO of the UN: the state of the world's animal genetic resouces for food and agriculture. FAO, Rome
612. Committee on the Mathematics and Physics of Emerging Dynamic Biomedical Imaging (1996) Mathematics and Physics of Emerging Biomedical Imaging. NAP
613. Compte A, Metzler R (1997) The generalized Cattaneo equation for the description of anomalous transport processes. J Phys A Math Gen 30:7277–7289
614. Conejero JA, Peris A, Trujillo M (2010) Chaotic asymptotic behavior of the hyperbolic heat transfer equation solutions. Int J Bifurc Chaos 20:2943–2947
615. Conil E, Hadjem A, Gati A, Wong M, Wiart J (2011) Influence of plane-wave incidence angle on whole body and local exposure at 2100 MHz. IEEE Trans Electromagn Compat 53:48–52
616. Connor CW, Hynynen K (2002) Bio-acoustic thermal lensing and nonlinear propagation in focused ultrasound surgery using large focal spots: a parametric study. Phys Med Biol 47:1911–1928
617. Conrad MC, Anderson JL, Garrett JB (1971) Chronic collateral growth after femoral artery occlusion in the dog. J Appl Physiol 31:550–555
618. Considine MJ, Daley DO, Whelan J (2001) The expression of alternate oxidase and uncoupling protein during fruit ripening in mango. Plant Physiol 126:1619–1629
619. Consiglieri L, dos Santos I, Haemmerich D (2003) Theoretical analysis of the heat convection coefficient in large vessels and the significance for thermal ablative therapies. Phys Med Biol 48:4125–4134
620. Consolazio CF, Matoush LR, Nelson RA, Torres JB, Isaac GJ (1963) Environmental temperature and energy expenditures. J Appl Physiol 18:65–68
621. Conway J, Hawley MS, Seagar AD, Brown BH, Barber DC (1985) Applied potential tomography (ART) for noninvasive thermal imaging during hyperthermia treatment. Electron Lett 21:836–838
622. Conzemius M, Brown T, Zhang Y, Robinson R (2002) A new animal model of femoral head osteonecrosis: one that progresses to human-like mechanical failure. J Orthoped Res 20:303–309
623. Cookson AN, Lee J, Michler C, Chabiniok R, Hyde E, Nordsletten DA, Sinclair M, Siebes M, Smith NP (2012) A novel porous mechanical framework for modelling the interaction between coronary perfusion and myocardial mechanics. J Biomech 45:850–855
624. Cooper SM, Dawber RPR (2001) The history of cryosurgery. J R Soc Med 94:196–201
625. Cooper IS, Lee A (1961) Cryostatic congelation: a system for producing a limited controlled region of cooling or freezing of biological tissues. J Nerve Mental Dis 33:259–263
626. Cooper IS, Lee A (1961) Cryothalamectomy – hypothermic coagulation: a technical advance in basal ganglia surgery. Preliminary report. J Am Geriatr Soc 9:714–718

627. Cooper TE, Trezek GJ (1970) Analytical prediction of the temperature field emanating from a cryogenic surgical cannula. Cryobiology 7:79–93
628. Cooper TE, Trezek GJ (1971) Correlation of thermal properties of some human tissue with water content. Aerospace Med 42:24–27
629. Cordasco D, Bagchi P (2014) Orbital drift of capsules and red blood cells in shear flow. Phys Fluids 25:091902
630. Coris EE, Ramirez AM, Van Durme DJ (2004) Heat illness in athletes. Sports Med 34:9–16
631. Costanzo JP, Grenot C, Lee RE (1995) Supercooling, ice inoculation and freeze tolerance in the European common lizard. Lacerta vivipara J Comp Physiol B 165:238–244
632. Costanzo JP, Baker PJ, Lee RE (2006) Physiological responses to freezing in hatchlings of freeze-tolerant and -intolerant turtles. J Comp Physiol B 176:697–707
633. Costello JT, Donnelly AE, Karki A, Selfe J (2014) Effects of whole body cryotherapy and cold water immersion on knee skin temperature. Int J Sports Medi 35:35–40
634. Coticchio G, De Santis L, Rossi G, Borini A, Albertini D, Scaravelli G, Alecci C, Bianchi V, Nottola S, Cecconi S (2006) Sucrose concentration influences the rate of human oocytes with normal spindle and chromosome configurations after slow-cooling cryopreservation. Hum Reprod 21:1771–1776
635. Cotta RM (1993) Integral transforms in computational heat and fluid flow. CRC Press, Boca Raton
636. Cotta RM, Cotta BP, Naveira-Cotta CP, Cotta-Pereira G (2010) Hybrid integral transforms analysis of the bioheat equation with variable properties. Int J Thermal Sci 49:1510–1516
637. Coussios CC, Farng CH, Ter Haar G, Roy RA (2007) Role of acoustic cavitation in the delivery and monitoring of cancer treatment by high-intensity focused ultrasound (HIFU). Int J Hyperthermia 23:105–120
638. Coutant CC (1987) Thermal preference: when does an asset become a liability? Environm Biol Fish 18:161–172
639. Crabo BG (2001) Physiological aspects of stallion semen cryopreservation. AAEP Proc 47:291–295
640. Craciunescu O, Clegg TS (1997) Perturbations of large vessels on induced temperature distributions. Part a: three-dimensional simulation study. Adv Heat Mass Transf Biotechnol 335:193–198
641. Craciunescu OI, Clegg ST (2001) Pulsatile blood flow effects on temperature distribution and heat transfer in rigid vessels. J Biomech Eng 123:500–505
642. Craciunescu O, Das SK, Dewhirst MK (1999) Three-dimensional microvascular networks fractal structure: a potential for tissue characterization? Advan. Heat Mass Transf Biotechnol 363:9–13
643. Craciunescua OI, Stauffer PR, Soher BJ, Wyatt CR, Arabe O, Maccarini P, Das SK, Cheng KS, Wong TZ, Jones EL, Dewhirst M, Vujaskovic Z, MacFall J (2009) Accuracy of real time noninvasive temperature measurements using magnetic resonance thermal imaging in patients treated for high grade extremity soft tissue sarcomas. Med Phys 36:4848–4858
644. Craig AB, Dvorak M (1966) Thermal regulation during water immersion. J Appl Physiol 21:1577–1585
645. Craig AD, Chen K, Bandy D, Reiman EM (2000) Thermosensory activation of insular cortex. Nat Neurosci 3:184–190
646. Cramer WA, Knaff DB (1991) Energy transduction in biological membranes. A textbook of bioenergetics. Springer, New York
647. Crezee J, Lagendijk JJ (1990) Experimental verification of bioheat transfer theories: measurement of temperature profiles around large artificial vessels in perfused tissue. Phys Med Biol 35:905–923
648. Crezee J, Lagendjik JJW (1992) Temperature uniformity during hyperthermia: impact of large vessels. Phys Med Biol 37:1321–1337
649. Criado-Sancho M, Llebot J (1993) Behavior of entropy in hyperbolic heat conduction. Phys Rev E 47:4104–4107

650. Criado-Sancho JM, Jou D, Casas-Vázquez J (2006) Nonequilibrium kinetic temperatures in flowing gases. Phys Lett A 350:339–341
651. Cross MM (1965) Rheology of non-Newtonian fluids: a new flow equation for pseudoplastic systems. J Colloid Sci 20:417–437
652. Crowe JH, Crowe LM, Chapman D (1984) Preservation of membranes in anhydrobiotic organisms: the role of trehalose. Science 223:701–703
653. Crowe JH, Crowe LM, Carpenter JF, Rudolph AS, Wistrom CA, Spargo BJ, Anchordoguy TJ (1989) Interactions of sugars with membranes. Biochim Biophys Acta 947:367–384
654. Crozier L, Dwyer G (2006) Combining population-dynamic and ecophysiological models to predict climate-induced insect range shifts. Am Nat 167:853–866
655. Cui ZF, Barbene JC (1991) The influence of model parameter values on the prediction of skin surface temperature: II. Contact problems. Phys Med Biol 36:1607–1620
656. Cummins L, Nauenberg M (1983) Thermal effects of laser radiation in biological tissue. Biophys J 42:99–102
657. Curley S (2003) Radiofrequency ablation of malignant liver tumors. Ann Surg Oncol 10:338–347
658. Curra FP, Mourad PD, Khokhlova VA, Cleveland RO, Crum LA (2000) Numerical simulations of heating patterns and tissue temperature response due to high intensity focused ultrasound. IEEE Trans Ultrason Ferroelectr Freq Control 47:1077–1089
659. Curry MR (1995) Cryopreservation of semen from domestic livestock. In: Day JG, McLellan MR (eds) Methods in molecular biology, vol 38. Humana Press, Totowa, pp 189–197
660. Curry MR, Millar JD, Watson PF (1994) Calculated optimal cooling rates for ram and human sperm cryopreservation fail to conform with empirical observations. Biol Reprod 51:1014–1021
661. Cvetkovic M, Polyak D, Hirata A (2016) The electromagnetic-thermal dosimetry for the homogeneous human brain model. Eng Anal Bound Elem 63:61–73
662. Dabak N, Tomak Y, Piskin A, Gulman B, Ozcan H (2003) Early results of a modified techniques of cryosurgery. Int Orthopaedics 27:249–253
663. Dae MW, Gao DW, Ursell PC, Stillson CA, Sessler DI (2003) Safety and efficacy of endovascular cooling and re-warming for induction and reversal of hypothermia in human-sized pigs. Stroke 34:734–738
664. Dagan Z, Weinbaum S, Jiji LM (1986) Parametric study of the three layer microcirculatory model for surface tissue energy exchange. J Biomech Eng 108:89–96
665. Dai W, Nassar N (2000) A domain decomposition method for solving three-dimensional heat transfer equations in a doubly-layered thin film with microscale thickness. Numer Heat Transf 127:243–255
666. Dai P, Han H, Zhao Y, Fan M (2016) Finite element analysis of the mechanical characteristics of glaucoma. J Mech Med Biol 16:1650060
667. D'Amico S, Claverie P, Collins T, Georlette D, Gratia E, Hoyoux A, Meuwis MA, Feller G, Gerday C (2002) Molecular basis of cold adaptation. Phil Trans R Soc Lond B 357:917–925
668. Damor RS, Kumar S, Shukla AK (2013) Numerical solution of fractional bioheat equation with constant and sinusoidal heat flux condition on skin tissue. Am J Math Anal 1:20–24
669. Damor RS, Kumar S, Shukla AK (2014) Numerical simulation of fractional bioheat equation in hyperthermia treatment. J Mech Med Biol 14:1450018
670. Damor RS, Kumar S, Shukla AK (2015) Parametric study of fractional bioheat equation in skin tissue with sinusoidal heat flux. Fract Differ Calculus 5:43–53
671. Damseh RA, Al-Odat MQ, Al-Azar TA, Shannak BA, Sal-Hussein FM (2006) Numerical investigation and validation of hyperbolic heat conduction model applied to fast precooling of a slab food product. J Ind Inst Sci 86:695–703
672. D'Angelo C, Quarteroni A (2008) On the coupling of 1d and 3d diffusion-reaction equations: application to tissue perfusion problems. Math Models Meth Appl Sci 18:1481–1504
673. Danker G, Verdier C, Misbah C (2008) Rheology and dynamics of vesicle suspension in comparison with droplet emulsion. J Non-Newtonian Fluid Mech 152:156–167

674. Dao M, Lim CT, Suresh S (2003) Mechanics of the human red blood cell deformed by optical tweezers. J Mech Phys Solids 51:2259–2280
675. Das K, Mishra SC (2014) Study of thermal behavior of a biological tissue: an equivalence of Pennes bioheat equation of Wulff continuum model. J Thermal Biol 45:103–109
676. Das SK, Clegg CT, Samulski TV (1999) Computational techniques for fast hyperthermia temperature optimization. Med Phys 26:319–328
677. Datta NR, Ordóonez SG, Gaipl U, Paulides MM, Crezee H, Gellermann J, Marder D, Puric E, Bodis S (2015) Local hyperthermia combined with radiotherapy and-/or chemotherapy: recent advances and promises for the future. Cancer Treat Rev 41:742–753
678. Datta NR, Rogers S, Klingbiel D, Gomez S, Puric E, Bodis S (2016) Hyperthermia and radiotherapy with or without chemotherapy in locally advanced cervical cancer: a systematic review with conventional and network meta analyses. Int J Hyperthermia 32:809–821
679. Davalos RV, Mir LM, Rubinsky B (2005) Tissue ablation with irreversible electroporation. Ann Biomed Eng 33:223–231
680. DaVee TS, Reineberg EJ (1980) Extreme hypothermia and ventricular fibrillation. Ann Emerg Med 9:100–107
681. Davidson SRH, James DF (2003) Drilling in bone: modeling heat generation and temperature distribution. J Biomech Eng 125:305–314
682. Davies CR, Saidel GM, Harasaki H (1997) Sensitivity analysis of 1-D heat transfer in tissue with temperature-dependent perfusion. J Biomechem Eng 119:77–80
683. Davies CR, Saidel GM, Harasaki H (1997) Sensitivity analysis of one-dimensional heat transfer in tissue with temperature-dependent perfusion. ASME J Biomech Eng 119:77–80
684. Davies PL, Baardsnes J, Kuiper MJ, Walker VK (2002) Structure and function of antifreeze proteins. Philos Trans R Soc Lond B 357:927–935
685. Davies AM, Weinberg U, Palti Y (2013) Tumor treating fields: a new frontier in cancer therapy. Ann New York Acad Sci 1291:86–95
686. Davis TRA (1963) Non-shivering thermogenesis. Fed Proc 22:777–782
687. Davit Y, Bell CG, Byrne H, Chapman LAC, Kimpton LS, Lang GE, Leonard KHL, Oliver JM, Pearson NC, Shipley RJ, Waters SL, Whiteley JP, Wood BD, Quintard M (2013) Homogenization via formal multiscale asymptotics and volume averaging: how do the two techniques compare? Adv Water Resour 62 B:178–206
688. Davydov EV, Lubashevsky IA, Milyaev VA, Musin RF (2001) Nondiffusive heat transfer in muscle tissue. Preliminary results. arXiv: cond-mat/0102006
689. Day WA (1997) On rates of propagation of heat according to Fourier theory. Quart Appl Math 55:127–138
690. Day SH, Nicoll-Griffith DA (1999) Cryopreservation of rat and human liver slices by rapid freezing. Cryobiology 38:154–159
691. Day JG, Stacey GN (eds) (2007) Cryopreservation and freeze-drying protocols (methods in molecular biology). Humana Press, Totowa
692. de Boer JF, Srinivas SM, Malekafzali A (1998) Imaging thermally damaged tissue by polarization sensitive optical coherence tomography. Opt Exp 3:212–218
693. de Dear RJ, Arens E, Zhang H, Ouro M (1997) Convective and radiative heat transfer coefficients for individual human body segments. Int J Meteorol 40:141–156
694. de Gennes PJG (1985) Wetting: statics and dynamics. Rev Mod Phys 57:827–863
695. de Graaf IAM, Koster HJ (2003) Cryopreservation of precision-cut tissue slices for application in drug metabolism research. Toxicol Vitro 17:1–17
696. de Graaf IAM, Draaisma AL, Schoeman O, Fahy GM, Groothuis GMM, Koster HJ (2007) Cryopreservation of rat precision-cut liver and kidney slices by rapid freezing and vitrification. Cryobiology 54:1–12
697. de Greef M, Kok HP, Correia D, Bel A, Crezee J (2010) Optimization in hyperthermia treatment planning: the impact of tissue perfusion uncertainty. Med Phys 37:4540–4550
698. de Kanter R, Olinga P, Hof I, de Jager M, Verwillegen W, Sloof M, Koster H, Meijer D, Groothuis G (1998) A rapid and simple method for cryppreservation of human liver slices. Xenobiotica 28:225–234

699. De la Rosette J, D'Ancona FC, Debruyne F (1997) Current status of thermotherapy of the prostate. J Urol 157:430–438
700. De Mendoza AM, Michlíková S, Berger J, Karschau J, Kunz-Schughart LA, McLeod DD (2020) Mathematical model for the thermal enhancement of radiation response: thermodynamic approach. arXiv:2007.00317v2 [physics.bio-ph]
701. de Oliveira EC, Teneiro Machado JA (2014) A review of definitions for fractional derivatives and integral. Math Prob Eng 2014:238459
702. de Santos RM, Barreta MH, Frajblat M, Cucco DC, Mezzaliza JC, Bunn S, Cruz FB, Viera AD, Mezzaliza A (2006) Vacuum-cooled liquid nitrogen increases the developmental ability of vitrified-warmed bovine oocytes. Ciêcia Rural 36:1501–1506
703. del Castillo-Negrette D (2008) Fractional diffusion models of anomalous transport. In Klages R, Radons G, Sokolov IM (eds) Anomalous transport. Foundations and applications. Wiley-VCH, Weinheim, pp 163–192
704. Debenetti PG (2003) Supercooled and glassy water. J Phys Condens Matter 15:R1669–R1726
705. Decuzzi P, Ferrari M (2006) The adhesive strength of non-spherical particles mediated by specific interactions. Biomaterials 27:5307–5314
706. DeGroot DW, Gallimore RP, Thompson SM, Kenefick RW (2013) Extremity cooling for heat stress mitigation in military and occupational settings. J Ther Biol 38:305–310
707. Dehkordi FJ, Shakeri-Zadeh A, Khoei S, Ghadiri H, Shiran MB (2013) Thermal distribution of ultrasound waves in prostate tumor: comparison of computational modeling with in vivo experiments. Biomathematics 2013:428659
708. Delkhosh M (2013) Introduction of derivatives and integrals of fractional order and its applications. Appl Math Phys 1:103–119
709. Dellimore JW, Dunlop MJ, Canham PB (1983) Ratio of cells and plasma in blood flowing past branches in small plastic channels. Am J Physiol 244:H635–H643
710. Denet AR, Vanbever R, Preat V (2004) Skin electroporation for transdermal and topical drug delivery. Adv Drug Deliv Res 56:659–674
711. Deng SX, Ge XX (2019) Local fractional Helmholtz simulation for heat conduction in fractal media. Thermal Sci 23:1671–1675
712. Deng ZS, Liu J (2001) Blood perfusion-based model for characterizing the temperature fluctuations in living tissues. Phys A 300:521–530
713. Deng ZS, Liu J (2002) Analytical study on bioheat transfer problems with spatial or transient heating on skin surface or inside biological bodies. J Biomech Eng 124:638–649
714. Deng ZS, Liu J (2003) Non-Fourier heat conduction effect on prediction of temperature transients and thermal stress in skin cryopreservation. J Thermal Stresses 26:779–798
715. Deng ZS, Liu J (2004) Mathematical modeling of temperature mapping over skin surface and its implementation in thermal disease diagnostics. Comput Biol Med 34:495–521
716. Deng ZS, Liu J (2004) Modeling of multidimensional freezing problem during cryosurgery by the dual reciprocity boundary element method. Eng Anal Bound Elem 28:97–108
717. Deng ZS, Liu J (2004) Monte Carlo simulation of the effects of large blood vessels during hyperthermia. Lect Notes Comput Sci 3314:437–442
718. Deng ZS, Liu J (2005) Effect of configuration between cryoprobe and large blood vessels on the tissue freezing during cryosurgery. Conf Proc IEEE Eng Med Biol Soc 1–7:490–493
719. Deng ZS, Liu J (2005) Numerical simulation of selective freezing of target biological tissues following injection of solutions with specific thermal properties. Cryobiology 50:183–192
720. Deng ZS, Liu J (2006) Numerical study of the effects of large blood vessels on three-dimensional tissue temperature profiles during cryosurgery. Num Heat Transf A 49:47–67
721. Deng G, Andrews DW, Laursen RA (2003) Amino acid sequence of a new type of antifreeze protein, from the longhorn sculpin Myoxocephalus octodecimspinosis. FEBS Lett 402:17–20
722. Denicol GS, Kodama T, Koide T, Mota P (2007) Causal theory of relativistic dissipative hydrodynamics. Brazil J Phys 37:1047–1054

723. Denlinger DL, Lee RE (1998) Physiology of cold sensitivity. In: Hallman GJ, Denlinger DL (eds) Temperature sensitivity in insects and application in integrated pest management, pp 55–95. Westview Press, Boulder
724. Dennis B, Dulikravich G, Rabin Y (2000) Optimization of organ freezing protocols with specified allowable thermal stress levels. IMECE 2000, Orlando, FL, Nov. 5–10, HTD-Vol. 368/BED-Vol. 47, pp 33-48
725. Derby J, Chelikowsky J, Sinno T, Dai B, Kwon YI, Lun L, Pandy A, Yeckel A (2007) Large-scale numerical modeling of melt and solution crystal growth. In: Sowronski M, DeYoreo J, Wang C (eds) Perspective on inorganic, organic and biological crystal growth: from fundamentals to applications, pp 139–158. American Institute of Physics, College Park
726. Deshazer G, Prakash P, Merck D, Haemmerich D (2017) Experimental measurement of microwave ablation heating pattern and comparison to computer simulations. Int J Hyperthermia 33:74–82
727. Deuflhard P, Hochmuth R (2002) Multiscale analysis of thermoregulation in the human microvascular system. ZIB–Report 02–31
728. Deuflhard P, Mehrmann V (2002) Numerical analysis and scientific computing in key technologies. In: DFG research center mathematics for key technologies. Berliner Mathematische Gesellschaft, Berlin, pp 9–21
729. Deuflhard P, Seebass M, Stalling D, Beck R, Hege HC (1997) Hyperthermia treatment planning in clinical cancer therapy: modelling, simulation, and visualization. Konrad-Zuse-Zentrum, Berlin, Preprint SC 97-26
730. Deuflhard P, Schiela A, Weiser M (2012) Mathematical cancer therapy planning in deep regional hyperthermia. Acta Numer 21:307–378
731. Devarakonda SB, Myers MR, Lanier M, Dumoulin C, Banerjee RK (2017) Assessment of gold nano-particle mediated enhanced hyperthermia using MR guided high-intensity focused ultrasound ablation procedure. Nano Lett 17:2532–2538
732. Devasagayam T, Kamat JP (2002) Biological significance of singlet oxygen. Ind J Exp Biol 40:680–692
733. Devireddy RV, Swanlund DJ, Roberts KP, Bischof JC (1999) Subzero water permeability of mouse spermatozoa in the presence of extracellular ice and cryoprotective agents. Biol Reprod 61:764–775
734. Devireddy RV, Swanlund DJ, Roberts KP, Pryor JL, Bischof JC (2000) The effect of extracellular ice and cryoprotective agents on the water permeability parameters of human sperm plasma membrane during freezing. Hum Reprod 15:1125–1135
735. Devireddy RV, Leo PH, Lowengrub JS, Bischof JC (2002) Measurement and numerical analysis of freezing in solutions enclosed in a small container. Int J Heat Mass Transf 45:1915–1931
736. Devireddy RV, Swanlund DJ, Alghamdi AS, Duoos LA, Troedsson MHT, Bischof JC, Roberts KP (2002) Measured effect of collection and cooling conditions on the motility and the water transport parameters at subzero temperatures of equine spermatozoa. Reproduction 124:643–648
737. Devor A, Sakadzic S, Srinivasan VJ, Yaseen MA, Nizar K, Saisan PA, Tian P, Dale AM, Vinogradov SA, Franceschini MA, Boas DA (2012) Frontiers in optical imaging of cerebral blood flow and metabolism. J Cerebr Blood Flow Metabol 32:1259–1276
738. DeVries A (1971) Glycoproteins as biological antifreeze agents in Antarctic fishes. Science 172:1152–1155
739. DeVries A (1984) Role of glycopeptides and peptides in inhibition of crystallization of water in polar fishes. Philos Trans R Soc Lond B 304:575–588
740. DeVries A, Wohlshlag D (1969) Freezing resistance in some Antarctic fishes. Science 163:1073–1075
741. Dewey WC (2009) Arrhenius relationships from the molecule and cell to the clinic. Int J Hyperth 25:3–20
742. Dewhirst MW, Vujaskovic Z, Jones E, Thrall D (2005) Re-setting the biologic rationale for thermal therapy. Int J Hyperth 21:779–790

743. Dewhirst MW, Abraham J, Viglianti B (2015) Evolution of thermal dosimetry for application of hyperthermia to treat cancer. Adv Heat Transf 47:397–416
744. Dhar A, Saito K, Derrida B (2013) Exact solution of a Lévy walk model for anomalous heat transport. Phys Rev E 87:010103
745. Dhar P, Paul A, Narasimhan A, Das SK (2016) Analytical prediction of sub-surface thermal history in translucent tissue phantoms during plasmonic photo-thermotherapy (PPTT). J Therm Biol 62B:143–149
746. Dhar A, Kundu A, Kundu A (2019) Anomalous heat transport in one dimensional systems: a descripton using non-local fractional-type diffusion equation. arXiv: 1911.04457
747. Diao Y, Rashed EA, Hirata A (2020) Assessment of absorbed power density and temperature rise for nonplanar body model under electromagnetic exposure above 6 GHz. Phys Med Biol 65:224001
748. Dickerson E, Dreaden E, Huang X, El-Sayed IH, Chu H, Pushpanketh S, McDonald J, El-Sayed M (2008) Gold nanorod assisted near-infrared plasmonic photothermal therapy (PPTT) of squamous cell carcinoma in mice. Cancer Lett 269:57–66
749. Dickinson MB (2002) Heat transfer and vascular cambium necrosis in the boles of trees during surface fires. In: Viegas DX (ed) Forest fire research & wildland fire safety. Millpress, Rotterdam, pp 1–13
750. Dickinson MB, Johnson EA (2001) Fire effects on trees. Fire effects on trees. Forest fires: behavior and ecological effects. Academic Press, New York, pp 477–525
751. Dickson JA, Calderwood SK (1980) Temperature range and selective sensitivity of tumors to hyperthermia: a critical review. Ann NY Acad Sci 335:180–205
752. Diederich CJ, Hynynen K (1999) Ultrasound technology for hyperthermia. Ultrasound Med Biol 25:871–887
753. Diethelm K (2010) The analysis of fractional differential equations. An application-oriented exposition using differential operators of caputo type. Springer, Berlin
754. Dietrich WD, Bramlett HM (2010) The evidence for hypothermia as a neuroprotectant in traumatic brain injury. Neurotherapeutics 7:43–50
755. Dietrich WD, Atkins CM, Bramlett HM (2009) Protection in animal models of brain and spinal cord injury with mild to moderate hypothermia. J Neurotrauma 26:301–312
756. Dill KA (1985) Theory for the folding and stability of globular proteins. Biochemistry 24:1501–1509
757. Dillenseger JL, Esneault S (2010) Fast FFT-based bioheat transfer equation computation. Comput Biol Med 40:119–123
758. Diller KR (1992) Modeling of bioheat transfer processes at high and low temperatures. Adv Heat Transf 22:157–357
759. Diller KR (1996) Pioneers in cryobiology: Julius von Sachs (1832–1897). Cryo-Lett 17:201–212
760. Diller KR (1997) Engineering-based contributions in cryobiology. Cryobiology 34:304–314
761. Diller KR (2005) Bioheat and mass transfer as viewed through a microscope. J Biomech Eng 127:67–84
762. Diller KR (2015) Heat transfer in health and healing. J Heat Transf 137:103001
763. Diller KR (2015) Therapeutic recruitment of thermoregulation in humans by selective thermal stimulation along the spine. Adv Heat Transf 47:341–396
764. Diller KR, Pearce JA (1999) Issues in modeling thermal alterations in tissues. Ann N Y Acad Sci 888:153–164
765. Dimbylow PJ (1993) FDTD calculations of SAR for a dipole closely coupled to the head at 900 MHz and 1.9 GHz. Phys Med Biol 38:361–368
766. Dimbylow PJ, Hirata A, Nagaoka T (2008) Intercomparison of whole-body averaged SAR in European and Japanese voxel phantoms. Phys Med Biol 53:5883–5897
767. Dimova R (2014) Recent developments in the field of bending rigidity measurements on membranes. Adv Colloid Interface Sci 208:225–234
768. Ding JP, Pickard B (1993) Modulation of mechanosensitive cation channels by temperature. Plant J 3:713–720

769. Dintenfass L (1968) Internal viscosity of the red cell and a blood viscosity equation. Nature 219:956–958
770. Dinnetes A, Dai S, Jiang S, Yang X (2000) Somatic cell nuclear transfer with vitrified recipient oocytes in cattle. Theriogenology 52:215
771. Discher DE (2000) New insights into erythrocyte membrane organization and microelasticity. Curr Opin Hematol 7:117–122
772. Discher DE, Mohandas N, Evans EA (1994) Molecular maps of red cell deformation: hidden elasticity and in situ connectivity. Science 266:1032–1035
773. Ditchfield R, Olbricht WL (1996) Effects of particle concentration on the partitioning of suspensions at small divergent bifurcations. J Biomech Eng 118:287–294
774. Ditkin VA, Prudnikov AP (1965) Integral transforms and operational calculus. Pergamon Press, Oxford
775. Dixit A, Gade U (2015) A case study on human bio-heat transfer and thermal comfort within CFD. Build Environ 94:122–130
776. Dixon JA (1982) Surgical applications of lasers. Proc. IEEE 70:579–588
777. Dixon M, Grace J (1983) Natural convection from leaves at realistic Grashof numbers. Plant Cell Environ 6:665–670
778. Dmitriev AS (2015) Inroduction to nano thermophysics (in Russian). BINOM, Moscow
779. Dodd GD III, Frank MS, Aribandi M, Chopra S, Chintapalli KN (2001) Radiofrequency thermal ablation: computer analysis of the size of the thermal injury created by overlapping ablations. Am J Roentgenol 177:777–782
780. Dodds PS, Rothman DH, Weitz JS (2001) Re-examination of the "3/4-law" of metabolism. J Theor Biol 209:9–27
781. Dodson W, Dimitrakopoulos P (2010) Tank-treading of erythrocytes in strong shear flows via a nonstiff cytoskeleton-based continuum computational modeling. Biophys J 99:2906–2916
782. Dombrovsky LA (2012) The use of transport approximation and diffusion-based models in radiative transfer calculations. Comput Thermal Sci 4:297–315
783. Dombrovsky LA, Baillis D (2010) Thermal radiation in disperse systems: an engineering approach. Begell House, New York
784. Dombrovsky LA, Timchenko V, Jackson M, Yeoh GH (2011) A combined transient thermal model for laser hyperthermia of tumors with embedded gold nanoshells. Int J Heat Mass Transf 54:5459–5469
785. Dombrovsky LA, Timchenko V, Jackson M (2012) Indirect heating strategy for laser induced hyperthermia: an advanced thermal model. Int J Heat Mass Transf 55:4688–4700
786. Donaldson GC, Keating WR, Saunders RD (2003) Cardiovascular responses to heat stress and their adverse consequencies in healthy and vulnerable human populations. Int J Hyperthermia 19:225–235
787. Donikier V, VanLaethem JL, Ickx B, VanGansbeke D, Goldman S, Gelin M (2003) Local ablative treatments for liver metastases: the current situation. Acta Chir Belg 103:452–457
788. Donner J, Thompson S, Kreuzer M, Baffou G, Quidant R (2012) Mapping intracellular temperature using green fluorescent protein. Nano Lett 12:2107–2111
789. Donoghue AM, Wishart GJ (2000) Storage of poultry semen. Animal Reprod Sci 62:213–232
790. Doost SN, Zhong L, Su B, Morsi YS (2016) The numerical analysis of non-Newtonian blood flow in human patient-specific left ventricle. Comput Meth Progr Biomed 127:232–247
791. Dorr LN, Hynynen K (1992) The effects of tissue heterogeneities and large blood vessels on the thermal exposure induced by short high-power ultrasound pulses. Int J Hyperth 8:45–59
792. Dos Santos I, Haemmerich D, Schutt D, da Rocha AF, Menezes LR (2009) Probabilistic finite element analysis of radiofrequency liver ablation using the unscented transform. Phys Med Biol 54:627
793. Doss JD (1982) Calculation of electric fields in conductive media. Med Phys 9:566–573

794. Doufas AG, Akca O, Barry A, Petrusca DA, Suleman MI, Morioka N, Guarnaschelli JJ, Sessler DI (2002) Initial experience with a novel heat-exchanging catheter in neurosurgical patients. Anesthes Analges 95:1752–1756
795. Dougherty TJ (1998) Photodynamic therapy. J Nat Cancer Inst 90:889–905
796. Doxsey S (2001) Re-evaluating centrosome function. Nat Rev Mol Cell Biol 2:688–698
797. Doyeux V, Podgorski T, Peponas S, Ismail S, Coupier G (2010) Behavior of flowing bodies near a bifurcation: elucidating the Zweifach-Fung effect. In: The 2nd European conference on microfluidics, Toulouse, pp 1–17
798. Doyeux V, Podgorski T, Peponas S, Ismail M, Coupier G (2011) Spheres in the vicinity of a bifurcation: elucidating the Zweifach-Fung effect. J Fluid Mech 674:359–388
799. Dremin V, Novokova I, Rafailov E (2022) Simulation of thermal field distribution in biological tissue and cell culture media irradiated with infrared wavelengths. Opt Exp 30:23078
800. Dreyer W, Struchtrup H (1993) Heat pulse experiments revisited. Contin Mech Thermodyn 5:3–50
801. Driscoll TA, Maki KL (2007) Searching for rare growth factors using Milticanonical Monte Carlo methods. SIAM Rev 49:673–692
802. Drizdal T, Paulides MM, van Holthe N, van Rhoon GC (2018) Hyperthermia treatment planning guided applicator selection for subsuperficial head and neck tumors heating. Int J Hyperth 34:704–713
803. Du N, Liu XY, Hew CL (2003) Ice nucleation inhibition. J Biol Chem 278:36000–36004
804. Du Q, Liu C, Wang X (2004) A phase field approach in the numerical study of the elastic bending energy for vesicle membranes. J Comput Phys 198:450–468
805. Du Bois EF (1921) The basal metabolism in fever. J Am Med Assoc 77:352–355
806. Dube PA, Imbeau D, Dubeau D, Auger I (2019) Worker heat stress prevention and work metabolism estimation: comparing two assessment methods of the heart rate thermal component. Ergonomics 62:1066–1085
807. Dubko TS (2006) On application of the fluorescent analysis in cryobiology. In: Visnik Kharkivskogo natsion. Univ. im. V.N. Karazina, Serie: Biology, pp 221–231
808. Duck FA (1990) Physical property of tissues - a comprehensive reference book. Academic, London
809. Duck FA (2012) The propagation of ultrasound through tissue. In: ter Haar G (ed) The safe use of ultrasound in medical diagnosis, pp 4–17
810. Dudar TE, Jain RK (1984) Differential response of normal and tumor microcirculation to hyperthermia. Cancer Res 44:605–612
811. Dulińska I, Targosz M, Strojny W, Lekka M, Czuba P, Balwierz W, Szymoński M (2006) Stiffness of normal and pathological erythrocytes studied by means of atomic force microscopy. J Biochem Biophys Methods 66:1–11
812. Duman JG (2002) The inhibition of ice nucleators by insect antifreeze proteins is enhanced by glycerol and citrate. J Comput Physiol B 172:163–168
813. Duman JG, Serianni AS (2002) The role of endogeneous antifreeze protein enhancers in the hemolymph thermal hysteresis activity in the beetle Dendroides canadensis. J Insect Physiol 48:103–111
814. Dumet D, Engelmann F, Chabrillange N, Duval Y (1993) Cryopreservation of oil palm (Elaeis guineensis) somatic embryos involving a disiccation step. Plant Cell Rep 12:352–355
815. Dumont F, Marechal PA, Gervais P (2003) Influence of cooling rate on Saccharomyces cerevisiae destruction during freezing: unexpected viability at ultra-rapid cooling rates. Cryobiology 46:33–42
816. Dumont F, Marechal PA, Gervais P (2004) Cell size and water permeability as determining factors for cell viability after freezing at different cooling rates. Appl Environ Microbiol 70:268–272

817. Dumont F, Marechal PA, Gervais P (2006) Involvement of two specific causes of cell mortality in freeze-thaw cycles with freezing to - 196 ° C. Appl Environ Microbiol 72:21330–21335
818. Duong VTT, Phan AD, Lien NTH, Hue DT, Hoa DQ, Nga DT, Nhung TH, Viet NA (2007) Near-infrared photothermal response of plasmonic gold-coated nanoparticles in tissues. arXiv:1710.04764v2 [cond-mat.mes-hall]
819. Dupire J, Abkarian M, Viallat A (2010) Chaotic dynamics of red blood cells in a sinusoidal flow. Phys Rev Lett 104:168101
820. Dupire J, Socol M, Viallat A (2012) Full dynamics of a red blood cell in shear flow. Proc Natl Acad Sci 109:20808–20813
821. Durkee JW, Antich PP (1991) Characterization of bioheat transport using an exact solution to the cylindrical geometry, multi-region, time-dependent bioheat equation. Phys Med Biol 36:1377–1406
822. Durkee JW, Antich PP (1991) Exact solutions to the multi-region time-dependent bioheat equation with transient heat sources and boundary conditions. Phys Med Biol 36:345–368
823. Durkee JW, Antich PP, Lee CE (1990) Exact solutions to the multiregion time-dependent bioheat equation. I: solution development. Phys Med Biol 35:847–867
824. Durkee JW, Antich PP, Lee CE (1990) Exact solutions to the multiregion time-dependent bioheat equation. II: Numerical evaluation of the solutions. Phys Med Biol 35:869–889
825. Dutta J, Kundu B (2021) Hybrid analytical models to estimate non-equilibrium temperatures in live-tissues based on appropriate initial thermal-field and non-invasive therapeutic heating. J Appl Phys 129:105102
826. Dutz S, Hergt R (1985) Magnetic nanoparticle heating and heat transfer on a microscale: basic principles, realities and physical limitations of hyperthermia for tumour therapy. Int J Hyperth 29:790–800
827. Dutz S, Hergt R (2014) Magnetic particle hyperthermia–a promising tumour therapy? Nanotechnology 25:452001
828. Duvernoy HM, Delon S, Vannson JL (1981) Cortical blood vessels of the human brain. Brain Res Bull 7:519–579
829. Dzherbashyan MM, Nersesian AD (1968) Fractional derivatives and the Cauchy problem for differential equations of fractional order (in Russian). Izv Acad Nauk Armjanskoy SSR Mat 3:3–29
830. Dzwinel W, Boryczko K, Yuen DA (2003) A discrete-particle model of blood dynamics in capillary vessels. J Colloid Interf Sci 258:163–173
831. Ebrahimi S, Bagchi P (2022) A computational study of red blood cell deformability effect on hemodynamic alteration in capillary vessel networks. Sci Rep 12:4304
832. Eckart C (1948) Vortices and streams caused by sound waves. Phys Rev 73:68–76
833. Edd JF, Rubinsky B (2006) Detecting cryoablation with EIT and the benefit of including ice front imaging data. Physiol Meas 27:S175–S185
834. Edd JF, Horowitz L, Rubinsky B (2005) Temperature dependence of tissue impedivity in electrical impedance tomogrphy of cryosurgery. IEEE Tras Biomech Eng 52:695–701
835. Edd JF, Horowitz L, Davalos RV, Mir LM, Rubinsky B (2006) In vivo results of a new focal tissue ablation technique: irreversible electroporation. IEEE Trans Biomed Eng 53:1409–1415
836. Eddin M (2003) Timeline: lipids on the frontier: a century of cell-membrane bilayers. Nat Rev Mol Cell Biol 4:414–418
837. Edelhauser HE, van Horn DL, Gallum AB, Schultz RO (1971) Experimental rehydration of cryopreserved corneal tissue. Invest Ophthalmol 10:100–107
838. Edwards EA, Finkelstein NA, Dunlley SQ (1951) Speclrophotometery of living human skin in ultraviolet range. J Invest Dermatol 16:311–321
839. Edwards AM, Philips RA, Watkins NW, Freeman MP, Murphy EJ, Afanasyev V, Buldyrev SV, da Luz MGE, Rapaso EP, Stanley HE, Viswanathan M (2007) Revisiting L'evy flight search patterns of wandering albatrosses, bumblebees and deer. Nature 449:1044–1048

840. Egan GF, Johnson J, Farrell M, Mcallen R, Zamarripa F, Mckinley MJ, Lancaster J, Denton D, Fox PT (2005) Cortical, thalamic, and hypothalamic responses to cooling and warmingthe skin in awake humans: a positron-emission tomographystudy. Proc Natl Acad Sci USA 102:5262–5267
841. Egelmann F (2004) Plant cryopreservation: progress and prospects. In Vitro Cell Dev Biol Plant 40:427–433
842. Eggleton CD, Popel AS (1998) Large deformation of red blood cell ghosts in a simple shear flow. Phys Fluids 10:1834–1845
843. Egorov YE, Zhmakin AI (1998) Numerical simulation of low-Mach number gas mixture flows with heat and mass transfer using unstructured grid. Comp Mat Sci 11:204–220
844. Egorov YE, Galyukov AO, Zhmakin AI (1999) Computation of flow and deposition in epitaxial reactors using 3D adaptive unstructured grids. Lect Notes Comput Sci 8:267–277
845. Ehrlich CP, Papanicolaou N, Treves S, Hurwitz RA, Richard P (1982) Splenic scintigraphy using Tc-99m-labeled heat-denatured red blood cells in pediatric patients: concise communication. J Nucl. Med. 23:209–213
846. Eisenberg DP, Steif PS, Rabin Y (2014) On the effects of thermal history on the development and relaxation of thermo-mechanical stress in cryopreservation. Cryogenics 64:86–94
847. Ekpenyong AE, Whyte G, Chalut K, Pagliara S, Lautenschläger F, Fiddler C, Paschke S, Keyser UF, Chilvers ER, Guck J (2012) Viscoelastic properties of differentiating blood cells are fate- and function-dependent. PLoS One 7:e45237
848. Ekstrand V (2005) Therapeutic applications of acoustic and electromagnetic energy. PhD Thesis, Karolinska Institutet, Stockholm
849. El-Danasouri I, Selman H (2005) Vitrification versus conventional cryopreservation technique. Middle East Fertil Soc J 10:205–206
850. El-Kebir H, Lee Y, Bentsman J (2022) Physics-based learning of parameterized thermodynamics from real-time thermography. arXiv:2203.13148v2 [cs.CV]
851. el Sayed SO, Dyson M (1990) Comparison of the effect of multi-wavelength light produced by a cluster of semiconductor diodes and of each individual diode on mast cell number and degranulation in intact and injured skin. Lasers Surg Med 10:559–568
852. Elford BC, Walter CA (1972) Effects of electrolyte composition and pH on the structure and function of smooth muscle cooled to 79°C in unfrozen media. Cryobiology 9:82–100
853. Elias H, Schwartz D (1969) Surface areas of the cerebral cortex of mammals determined by stereological methods. Science 166:111–113
854. Eliazar II, Schlesinger MF (2013) Fractional motions. Phys Rep 527:10–129
855. Elliott GD, Wang S, Fuller BJ (2017) Cryoprotectants: a review of the actions and applications of cryoprotective solutes that modulate cell recovery from ultra-low temperatures. Cryobiology 76:74–91
856. Emery A, Kramar P, Guy A, Lin J (1975) Microwave induced temperature rises in rabbit eyes in cataract research. J Heat Transf 97:123–128
857. Emmanuel S, Berkowitz B (2007) Continuous time random walks and heat transfer in porous media. Transf Porous Med 67:413–430
858. Emsminger I, Busch F, Hunter NPA (2006) Photostasis and cold acclimation: sensing low temperature through photosynthesis. Physiol Plant 126:28–44
859. Enden G, Popel AS (1994) A numerical study of plasma skimming in small vascular bifurcations. J Biomech Eng 116:79–88
860. Engelhardt H, Gaub H, Sackmann E (1984) Viscoelastic properties of erythrocyte membranes in high frequency electric fields. Nature 307:378–380
861. Engelmann F (2000) Importance of cryopreservation for the conservation of plant genetic resources. In: Engelmann F, Takagi H (eds) Cryopreservation of tropical plant germplasm - current research progress and applications. IPGRI, pp 8–20
862. Engelmann F, Takagi H (eds) (2000) Cryopreservation of tropical plant germplasm - current research progress and applications. Intern, Plant Genetic Resources Institute
863. English MJM (2001) Physical principles of heat transfer. Curr. Anasth Crit Care 12:66–71

864. English NJ, Mooney DA (2007) Denaturation of hen egg white lysozyme in electromagnetic fields: a molecular dynamics study. J Chem Phys 126:091105
865. Engstrom KG, Meiselman HJ (1995) Analysis of red blood cell membrane area and volume regulation using micropipette aspiration and perfusion. Biorheology 32:115–116
866. Enthel P, Adeagbo WA, Sugihara M, Rollmann G, Zayak AT, Kreth M, Kadau K (2004) Molecular dynamics simulations in biology, chemistry and physics. Lect Notes Phys 642:177–206
867. Epstein Y, Moran DS (2006) Thermal comfort and the heat stress indices. Ind Health 44:388–398
868. Erez A, Shitzer A (1980) Controlled destruction and temperature distributions in biological tissues subjected to monoactive electrocoagulation. J Biomech Eng 102:42–49
869. Eriksson S, Hurme R, Rhen M (2002) Low-temperature sensors in bacteria. Philos Trans R Soc Lond B 357:887–893
870. Eroglu A, Russo MJ, Bieganski R, Fowler A, Cheley S, Bayley H, Toner M (2004) Intracellular trehalose improves the survival of cryopreserved mammalian cells. Nat Biotechnol 18:163–167
871. Escolano J, Rodriguez F, Castro MA, Vives F, Martin JA (2011) Exact and analytic-numerical solutions of bidimensional lagging models of heat conduction. Math Comput Model 54:1841–1845
872. Esmailie F, Francoeur M, Ameel T (2020) A three-dimensional thermal model of the human cochlea for magnetic cochlear implant surgery. arXiv:2011.10183v1 [physics.app-ph]
873. Espanol P (1998) Fluid particle model. Phys Rev E 57:2930–2948
874. Etheridge ML, Bischof JC (2013) Optimizing magnetic nanoparticle based thermal therapies within the physical limits of heating. Ann Biomed Eng 41:78–88
875. Etheridge ML, Xu Y, Rott L, Choi J, Glasmacher B, Bischof JC (2014) RF heating of magnetic nanoparticles improves the thawing of cryopreserved biomaterials. Technology 2:229–242
876. Etienne RS, Apol ME, Olff HA (2006) Demystifying the West, Brown & Enquist model of the allometry of metabolism. Funct Ecol 20:394–399
877. Evans EA (1983) Bending elastic modulus of red blood cell membrane derived from buckling instability in micropipet aspiration tests. Biophys J 43:27–30
878. Evans EA (1989) Structure and deformation properties of red blood cells: concepts and quantitative methods. Meth Enzymol 173:3–35
879. Evans EA, Hochmuth RM (1976) Membrane viscoelasticity. Biophys J 16:1–11
880. Evans AA, Levine AJ (2013) Reflection and refraction of flexural waves at geometric boundaries. Phys Rev Lett 111:038101
881. Evans E, Skalak R (1980) Mechanics and thermodynamics of biomembranes. CRC Press, Boca Raton
882. Evans W, Fish J, Keblinski P (2006) Thermal conductivity of ordered molecular water. Lockheed Martin Corp. Rep. LM-06K008
883. Evans DW, Moran EC, Baptista PM, Soker S, Sparks JL (2013) Scale-dependent mechanical properties of native and decellularized liver tissue. Biomech Model Mechanobiol 12:569–580
884. Evans AA, Bhaduri B, Popescuc G, Levine AJ (2017) Geometric localization of thermal fluctuations in red blood cells. Proc Natl Acad Sci U S A 114:2865–2870
885. Evensen G (2009) The ensemble Kalman filter for combined state and parameter estimation. IEEE Control Syst Mag 29:83–104
886. Everett MA, Yeargers E, Sayre RM, Olson RL (1966) Penetration of epidermis by ultraviolet rays. Photochem Photobiol 5:533–542
887. Ezzat M, Al-Sowayan N, Al-Muhiameed Z, Ezzat S (2014) Fractional modelling of Pennes bioheat transfer equation. Heat Mass Transf 50:907–914
888. Ezzat MA (2020) The effects of thermal and mechanical material properties on tumorous tissue during hyperthermia treatment. J Thermal Biol 92:102649

889. Ezzat MA, El-Karamany AS (2002) The uniqueness and reciprocity theorems for generalized thermo-viscoelasticity with two relaxation times. Int J Eng Sci 40:1275–1284
890. Ezzat MA, El Karamany AS (2003) On uniqueness and reciprocity theorems for generalized thermoviscoelasticity with thermal relaxation. Can J Phys 81:823–833
891. Ezzat MA, El-Karamany AS, Samaan AA (2001) State space formulation to generalized thermo-viscoelasticity with thermal relaxation. J Thermal Stresses 24:823–846
892. Ezzat MA, El Karamany AS, El-Bary AA (2009) State space approach to one dimensional magneto-thermoelasticity under the Green-Naghdi theories. Can J Phys 87:867–878
893. Ezzat MA, El Karamany AS, Fayik MA (2012) Fractitional order theory in thermoelastic solid with three-phase-lag heat transfer. Arch Appl Math 82:557–572
894. Ezzat MA, El-Bary AA, Fayik MA (2013) Fractional Fourier law with three-phase lag of thermoelasticity. Mech Adv Mater Struct 20:593–602
895. Ezzat MA, El-Karamany AS, El-Bary AA (2015) Thermo-viscoelastic materials with fractional relaxation operators. Appl Math Model 39:7499–7512
896. Ezzat MA, El-Bary AA, Al-Sowayan NS (2016) Tissue responses to fractional transient heating with sinusoidal heat flux condition on skin surface. Anim Sci J 87:1304–1311
897. Fa KS (2005) Fokker-Planck description and diffusive phonon transport. arXiv:cond-mat/0510673
898. Fabrizio M (2016) Fractional rheological models for thermodynamical systems. Dissipation and free energies. Fract Calcul Appl Anal 17:206–223
899. Fabrizio M, Franchi F (2014) Delayed thermal models: stability and thermodynamics. J Thermal Stresses 37:160–173
900. Fabrizio M, Lazzari B (2014) Stability and second law of thermodynamics in dual-phase-lag heat conduction. Int J Heat Mass Trans 74:484–489
901. Fabrizio M, Morro A (2003) Thermodynamics and second sound in a two-fluid model of helium II. Revisited. J Non-Equilibr Thermodyn 28:69–84
902. Fabrizio M, Lazzari B, Tibullo V (2017) Stability and thermodynamic restrictions for a dual-phase-lag thermal model. J Non-Equil Thermodyn 42:243–252
903. Fagrell B (1984) Microcirculation of the skin. In: Fagrell B (ed) The physiology and pharmacolcogy of the microcirculation. Elsevier, London, pp 133–180
904. Fahmy AM (2019) A new LRBFCM-GBEM modeling algorithm for general solution of time fractional-order dual phase lag transfer problems in functionally graded tissues. Num Heat Transf A Appl 75:616–626
905. Fahnestock GR, Hare RC (1964) Heating of tree trunks in surface fires. J Forestry 62:799–805
906. Fahraus R, Lindqvist T (1931) The viscosity of .the blood in narrow capillary tubes. Am J Physiol-Leg Content 96:562–568
907. Fahy G (2020) The history of organ cryopreservation research. Cryobiology 97:268
908. Fahy GM, Wowk B (2015) Principles of cryopreservation by vitrification. Methods Mol. Biol. 1257:21–82
909. Fahy GM, MacFarlane DR, Angell CA, Meryman HT (1984) Vitrification as an approach to cryopreservation. Cryobiology 21:407–426
910. Fahy GM, Wowk B, Wu J, Paynter S (2004) Improved vitrification solutions based on the predictability of vitrification solution toxicity. Cryobiology 48:22–35
911. Fahy GM, Wowk B, Wu J (2006) Cryopreservation of complex systems: the missing link in the regenerative medicine supply chain. Rejuvenation Res 9:279–291
912. Fajardo LF (1984) Pathological effects of hyperthermia in normal tissues. Cancer Res 44:4826s–4835s
913. Falahatkar S, Nouri-Borujerdi A, Najafi M, Mohammadzadeh A (2017) Numerical solution of non-Fourier heat transfer during laser irradiation on tooth layers. J Mech Sci Technol 31:6085–6092
914. Falahatkar S, Nouri-Borujerdi A, Mohammadzadeh A, Najafi M (2018) Evaluation of heat conduction in a laser irradiated tooth with the three-phase-lag bio-heat transfer model. Therm Sci Eng Prog 7:203–212

915. Falcini F, Garra R, Voller V (2019) Modeling anomalous heat diffusion: comparing fractional derivative and non-linear diffusivity treatments. Int J Thermal Sci 137:584–588
916. Falk D (1990) Brain evolution in homo: the "radiator" theory. Behav Brain Sci 13:333–381
917. Falk MH, Issels RD (2001) Hyperthermia in oncology. Int J Hyperthermia 17:1–18
918. Famá A, Restuccia L, Ván P (2020) Generalized ballistic-conductive heat transport laws inthree-dimensional isotropic materialsa. arXiv:1902.10980 [cond-mat.stat-mech]
919. Fan J, Wang L (2011) Analytical theory of bioheat transport. J Appl Phys 109:104202
920. Fan J, Wang L (2011) A general bioheat model at macroscale. Int J Heat Mass Transf 54:722–726
921. Fanger PO (1970) Thermal comfort, analysis and applications in environmental engineering. McGraw-Hill, New York
922. Farina L, Weiss N, Nissenbaum Y, Cavagnaro M, Lopresto V, Pinto R, Tosoratti N, Amabile C, Cassarino S, Goldberg SN (2014) Characterisation of tissue shrinkage during microwave thermal ablation. Int J Hyperth 30:419–428
923. Farina L, Nissenbaum Y, Cavagnaro M, Goldberg SN (2018) Tissue shrinkage in microwave thermal ablation: comparison of three commercial devices. Int J Hyperth 34:382–391
924. Farnell S, Maxwell L, Tan S, Rhodes A (2005) Temperature measurement: comparison of non-invasive methods used in adult critical care. J Clin Nurs 14:632–639
925. Farrant J (1965) Mechanism of cell damage during freezing and thawing and its prevention. Nature 205:1284–1287
926. Fasano A, Hömberg D, Naumov D (2010) On a mathematical model for laser-induced thermotherapy. Appl Math Model 34:3831–3840
927. Fasla B, Benmouna R, Benmouna M (2014) On the hyper thermal therapy of tumor tissues by direct laser heating and gold nano particles. J Biomater Nanobiotechnol 5:44–51
928. Fazlali R, Ahmadikia H (2013) Analytical soltion of thermal wave models on skin tissue under arbitrary periodic boundary conditions. Int J Thermophys 34:139–159
929. Fearnhead P, Kunsch HR (2018) Particle filters and data assimilation. Ann Rev Stat Appl 5:421–449
930. Febbraio MA (2001) Alterations in energy metabolism during exercise and heat stress. Sports Med 31:47–59
931. Feder ME, Hofmann GE (1999) Heat-shock proteins, molecular chaperones, and the stress response: evolutionary and ecological physiology. Annu Rev Physiol 61:243–282
932. Fedorenko A, Lishko P, Kirichok Y (2012) Mechanism of fatty-acid dependent UCP1 uncoupling in brown fat mitochondria. Cell 151:400–413
933. Fedosov DA (2010) Multiscale modeling of blood flow and soft matter. Ph.D. Thesis, Brown University
934. Fedosov DA, Caswell B, Karniadakis GE (2010) Dissipative particle dynamics modeling of red blood cells. In: Pozrikidis C (ed) Computational hydrodynamics of capsules and cells. CRC Press, Boca Raton, p 36
935. Fedosov DA, Caswell B, Karniadakis GE (2010) A multiscale red blood cell model with accurate mechanics, rheology, and dynamics. Biophys J 98:2215–2225
936. Fedosov DA, Caswell B, Karniadakis GE (2010) Systematic coarse-graining of spectrin-level red blood cell models. Comput Methods Appl Mech Eng 199:1937–1948
937. Fedosov DA, Caswell B, Popel AS, Karnidakis GE (2010) Blood flow and cell-free layer in microvessels. Microcirculation 17:615–628
938. Fedosov DA, Pan W, Caswell B, Gompper G, Karniadakis GE (2011) Predicting human blood viscosity in silico. Proc Natl Acad Sci USA 108:11772–11777
939. Fedosov DA, Noguchi H, Gompper G (2013) Multiscale modeling of blood flow: from single cells to blood rheology. Biomech Model Mechanobiol 13:239–258
940. Fedosov DA, Peltomäki M, Gompper G (2014) Deformation and dynamics of red blood cells in flow through cylindrical microchannels. Soft Matter 10:4258
941. Fehér A, Kovács R (2021) On the evaluation of non-Fourier effects in heat pulse experiments. Int J Eng Sci 169:103577

942. Fehér A, Kovács R (2022) Analytical evaluation of non-Fourier heat pulse experiments on room temperature. IFAC PapersOnLine 55(18):87–92
943. Fehér A, Lukacs N, Somlai L, Fodor T, Szücs M, Fülöp T, Ván P, Kovács R (2021) Size effects and beyond-Fourier heat conduction in room-temperature experiments. J Non-Equil Thermodyn 46:403–411
944. Fehling PC, Haller JM, Lefferts WK, Hultquist EM, Wharton M, Rowland TW, Smith DL (2015) Effect of exercise, heat stress and dehydration on myocardial performance. Occup Med 65:317–323
945. Fei YB, Wei LB, Gao SQ, Lu MC, Wang BH, Li ZF, Zhang YM, Shu NH, Jiang Y, Wang WX (2001) Isolation, purification and characterization of secondary structure of antifreeze protein from Ammopiptantus mongolicus. Chin Sci Bull 46:495–498
946. Feldman BJ (2018) Thermoregulation in mice, rats and humans: an insight into the evolution of human hairlessness. arXiv:1812.02003 [q-bio.OT]
947. Feng Y, Oden JT, Rylander MN (2008) A two-state cell damage model under hyperthermic conditions: theory and in vitro experiments. J Biomed Eng 130:041016
948. Fenton BM, Carr RT, Cokelet GR (1985) Nonuniform red cell distribution in 20 to 100 micrometers bifurcations. Microvasc Res 29:103–126
949. Fernandez A, Özarslan MA, Baleanu D (2019) On fractional calculus with general analytic kernels. Appl Math Comput 354:248–265
950. Fernández RG, Abascal JLF, Vega C (2006) The melting point of ice i_h for common water models calculated from direct coexistence of the solid-liquid interface. J Chem Phys 124:144506
951. Ferng AS, Schipper D, Connell AM, Marsh KM, Knapp S, Khalpey Z (2017) Novel vs clinical organ preservation solutions: improved cardiac mitochondrial protection. J Cardiothorac Surg 12:7
952. Ferras LL, Ford NJ, Morgado ML, Nobrea JM, Rebelo MS (2015) Fractional Pennes bioheat equation: theoretical and numerical studies. Fract Calculus Appl Anal 18:1080–1106
953. Ferreira MS, Yanagihara JI (2009) A transient three-dimensional heat transfer model of the human body. Int Commun Heat Mass Transf 36:718–724
954. Ferrer M, Jou D (1991) Higher-order fluxes and the speed of thermal waves. Int J Heat Mass Transf 34:3055–3060
955. Fhuong NL, Yamashita M, Yoo S, Ito K (2016) Prediction of convective heat transfer coefficient of human upper and lower airway surfaces in steady and unsteady breathing conditions. Build Environ 100:172–185
956. Fiala D, Lomas KJ, Stohrer M (1999) A computer model of human ther-moregulation for a wide range of environmental conditions: the passive system. J Appl Physiol 87:1957–1972
957. Fiala D, Lomas KJ, Stohrer M (2001) Computer prediction of human thermoregulatory and temperature responses to a wide range of environmental conditions. Int J Biometeorol 45:143–159
958. Fiala D, Havenith G, Bröde P, Kampmann B, Jendritzky G (2012) UTCI-Fiala multi-node model of human heat transfer and temperature regulation. Int J Biometeorol 56:429–441
959. Fichera G (1992) Is the Fourier theory of heat propagation paradoxical? Int Rend Circ Matem Palermo Ser II XLI 41:5–28
960. Field SB (1987) Hyperthermia in the treatment of cancer. Phys Med Biol 32:789–811
961. Field SB, Morris CC (1983) The relationship between heating time and temperature: its relevance to clinical hyperthermia. Radiother Oncol 1:179–186
962. Filippov YP, Vasilkov AP (1979) A method for simulation of the thermal regime of biological tissue in local cryoaction (in Russian). Eng Phys J 36:1100–1107
963. Finkelstein AV, Ptitsyn OB (2002) Physics of protein (in Russian). Institute of Protein RAS, Moscow
964. Finney JL (2004) Water? What's so special about it? Philos Trans R Soc Lond B 359:1145–1165
965. Finnikov KA, Korzun AM, Kolesnichenko LV (2011) Mathematial modelling of heat transfer. J Stress Physiol Biochem 7:424–436

966. Firsov N, Priezzhev A, Klimova N, Tyurina A (2006) Fundamental laws of the deformational behavior of erythrocytes in shear flow. J Eng Phys Thermophys 79:118–124
967. Fischer TM (2004) Shape memory of human red blood cells. Biophys J 86:3304–3313
968. Fischer TM (2007) Tank-tread frequency of the red cell membrane: dependence on the viscosity of the suspending medium. Biophys J 93:2553–2561
969. Fischer TM, Stöhr-Liesen M, Schmid-Schönbein H (1978) The red cell as a fluid droplet: tank tread-like motion of the human erythrocyte membrane in shear flow. Science 202:894–896
970. Fischer JP, Dams J, Gotz MH, Kerker E, Loesel FH, Messer C, Suhm MH, Billie JF (1994) Plasma-mediated ablation of brain tissue with picosecond laser pulses. Appl Phys B 58:493–499
971. Fiskerstran EJ, Ryggen K, Norvang LT, Svaasand LO (1997) Clinical effects of dynamic cooling during pulsed laser treatment of port-wine stains. Lasers Med Sci 12:320–327
972. Fitz-Gerald JM (1969) Mechanics of red-cell motion through very narrow capillaries. Proc R Soc Lond B 14:151–165
973. Fletcher DA, Mullins RD (2010) Cell mechanics and the cytoskeleton. Nature 463:485–492
974. Fletcher GL, Hew CL, Davies PL (2001) Antifreeze proteins of teleost fishes. Annu Rev Physiol 63:359–390
975. Flory PJ, Garrett RR (1958) Phase transition in collagen and gelatin systems. J Am Chem Soc 80:4836–4845
976. Flyckt V, Raaymakers B, Lagendijk J (2006) Modelling the impact of blood flow on the temperature distribution in the human eye and the orbit: fixed heat transfer coeffcients versus the Pennes bioheat model versus discrete blood vessels. Phys Med Biol 51:5007
977. Foda E, Almesri I, Awbi HB, Sirén K (2011) Models of human thermoregulation and the prediction of local and overall thermal sensations. Build Environ 46:2023–2032
978. Fonseca F, Martin M, Morris GI (2006) Stabilization of frozen Lactobacillus delibrueckii subsp. bulgaricus in glycerol suspension: freezing kinetics and storage temperature effects. Appl Environ Microbiol 72:6472–6482
979. Foo JJ, Liu KK, Chan V (2003) Thermal effect on a viscously deformed liposome in a laser trap. Ann Biomed Eng 31:354–362
980. Foo JJ, Liu KK, Chan V (2004) Viscous drag of deformed vesicles in an optical trap: experiments and simulations. AIChE J 50:249–254
981. Foo JJ, Chan V, Feng ZQ, Liu KK (2006) Human red blood cell deformed under thermal fluid flow. Biomed Mater 1:1–7
982. Foote RH (2002) The history of artificial insemination: selected notes and notables. J Anim Sci 80:1–10
983. Föppl A (1907) Vorlesungen uber technische Mechanik. B G Teubner, Bd 5, Leipzig, p 132
984. Forest V, Peoćh M, Campos L, Guyotat D, Vergnon JM (2005) Effect of cryotherapy or chemotherapy on apoptosis in a non-small-cell lung cancer xenografted into SCID mice. Cryobiology 50:29–37
985. Fork R, Brito Cruz CH, Becker PC, Shank CV (1987) Compression of optical pulses to six femtoseconds by using cubic phase compression. Opt Lett 12:483–485
986. Formosa LE, Ryan MT (2018) Mitochondrial OXPHOS complex assembly lines. Nat Cell Biol 20:511–513
987. Forsyth AM, Wan J, Ristenpart WD, Stone HA (2010) The dynamic behavior of chemically "stiffened" red blood cells in microchannel flows. Microvasc Res 80:37–43
988. Fort J, Pujol T, Cukrowski AS (2000) Several-temperature systems: extended irreversible thermodynamics and thermal wave propagation. J Phys A 33:6953–6973
989. Fortin A, Belhamadia Y (2005) Numerical prediction of freezing fronts in cryosurgery: comparison with experimental results. Comput Meth Biomech Biomed Eng 8:241–249
990. Foster KR (2000) Thermal and nonthermal mechanisms of interaction of radio-frequency energy with biological systems. IEEE Trans Plasma Sci 28:15–23

991. Foster KR, Cheever E, Leonard JB, Blum FD (1984) Transport properties of polymer solutions. Biophys J 45:95–984
992. Foster KR, Ziskin MC, Balzano Q (2016) Thermal response of human skin to microwave energy: a critical review. Health Phys 111:528–541
993. Foulks GN, Bron AJ (2003) Meibomian gland dysfunction: aclinicals cheme for description, diagnosis, classification, and grading. Ocular Surf 1:107–126
994. Fournier JB, Lacoste D, Raphael E (2004) Fluctuation spectrum of fluid membranes coupled to an elastic meshwork: Jump of the effective surface tension at the mesh size. Phys Rev Lett 92:018102
995. Fox SW, Dose K (1972) Molecular evolution and the origins of life. Freeman, San Francisco
996. Franca MV, Orlande HRB (2018) Estimation of parameters of the dual-phase-lag model for heat conduction in metal-oxide-semiconductor field-effect transistors. Int Commun Heat Mass Transf 92:107–111
997. Francis C, Frederic L, Sylvie L, Prasanna P, Henri D (2009) Scaling laws for branching vessels of human cerebral cortex. Microcirculation 16:331–344
998. Franco W, Wang GX, Nelson JS, Aguilar G (2004) Radial heat transfer dynamics during cryogen spray cooling. ASME Int Mech Eng Congr Exhib ASME2004-59609 (2004)
999. Franconi C, Vrba J, Montecchia F (1993) 27 MH hybrid evanescent-mode applicators. Int J Hyperth 9:655–673
1000. Frangioni JV (2003) In vivo near-infrared fluorescence imaging. Curr Opin Chem Biol 7:626–634
1001. Franke T, Hoppe RHW, Linsenmann C, Schmid L, Willbold C, Wixforth A (2011) Numerical simulation of the motion of red blood cells and vesicles in microfluidic flows. Comput Visual Sci 14:167–180
1002. Frankel JI, Vick B, Ozisik MN (1985) Flux formulation of hyperbolic heat conduction. J Appl Phys 58:3340–3345
1003. Franks F (2003) Nucleation of ice and its management in ecosystems. Philos Trans R Soc Lond A 361:557–574
1004. Frazer TD, Knobloch JL, Hoogeboom-Pot KM, Nardi D, Chao W, Falcone RW, Murnane MM, Kapteyn HC, Hernandez-Charpak JN (2019) Engineering nanoscale thermal transport: size- and spacing-dependent cooling of nanostructures. Phys Rev Appl 11:024042
1005. Fredrickson KL (2004) Optimization of cryosurgical probes for cancer treatment. Ph.D. Thesis, University of Wisconsin, Madison
1006. Freund JB (2013) The flow of red blood cells through a narrow spleen-like slit. Phys Fluids 25:110807
1007. Freund JB (2014) Numerical simulation of flowing blood cells. Ann Rev Fluid Mech 46:67–95
1008. Freund JB, Orescanin MM (2011) Cellular flow in a small blood vessel. J Fluid Mech 671:466–490
1009. Frey B, Weiss EM, Rubner Y, Wunderlich R, Ott OJ, Sauer R, Fietkau R, Gaipl US (2012) Old and new facts about hyperthermia-induced modulations of the immune system. Int J Hyperth 28:528–542
1010. Friedrich RE, Quade M, Jowe N, Kroetz P, Amling M, Kohlrusch FK, Zusting J, Gosau M, Schlüter H, Miller RJD (2020) Ablation precision and thermal effects of a picosecond infrared laser (PIRL) on roots of human teeth: a pilot study ex vivo. In Vivo 34:2325–2336
1011. Friend PJ (2020) Strategies in organ preservation-a new golden age. Transplantation 104:1753–1755
1012. Fry FJ, Johnson LK (1978) Tumor irradiation with intense ultrasound. Ultrasound Med Biol 4:337–341
1013. Fujikawa S (1981) The effect of various cooling rates on the membrane ultrastructure of frozen human erythrocites and its relation to the extent of haemolysis after thawing. J Cell Sci 49:369–382
1014. Fujimoto M, Hirata A, Wang J, Fujiwara O, Shiozawa T (2006) FDTD-derived correlation of maximum temperature increase and peak SAR in child and adult head models due to dipole antenna. IEEE Trans Electromagn Compat 48:240–247

1015. Fujishima H, Toda I, Yamada M, Sato N, Tsubota K (1996) Corneal temperature in patients with dry eye evaluated by infrared radiation thermometry. Brit J Ophthalm 80:29–32
1016. Fujiwara H, Ishikawa T, Lima R, Matsuki N, Imai Y, Kaji H, Nishizawa M, Yamaguchi T (2009) Red blood cell motions in high-hematocrit blood flowing through a stenosed microchanne. J Biomech 42:838–846
1017. Fülöp T, Kovács R, Lovas A, Rieth A, Fodor T, Szücs M, Ván P, Gróf G (2018) Emergence of non-Fourier hierarchies. Entropy 20:832
1018. Fung YC (1973) Stochastic flow in capillary blood vessels. Microvasc Res 5:34–48
1019. Fung YC (1997) Biomechanics: circulation. Springer, Berlin
1020. Fuxe K, Rivera A, Jacobsen KX, Hoistad M, Leo G, Horvath TL, Staines W, De la Calle A, Agnati LF (2005) Dynamics of volume transmission in the brain. Focus on catecholamine and opioid peptide communication and the role of uncoupling protein 2. J Neural Trans 112:65–76
1021. Gabriel C, Gabriel S, Corthout E (1996) The dielectric properties of biological tissues: I. Literature survey. Phys Med Biol 41:2231–2249
1022. Gabriel S, Lau RW, Gabriel C (1996) The dielectric properties of biological tissues: II. Measurements in the frequency range 10 Hz to 20 GHz. Phys Med Biol 41:2251–2269
1023. Gabriel S, Lau RW, Gabriel C (1996) The dielectric properties of biological tissues: III. Parametric models for the dielectric spectrum of tissues. Phys Med Biol 41:2271–2293
1024. Gabris E, Rybaczuk M, Kedzia A (2005) Fractal models of circulatory system. Symmetrical and asymmetrical approach comparison. Chaos Solit Fract 24:707–715
1025. Gabris E, Rybaczuk M, Kedzia A (2006) Blood flow simulation through fractal models of circulatory system. Chaos Solit Fract 27:1–7
1026. Gaehtens P (1980) Flow of blood through narrow capillaries: rheological mechanisms determining capillary hematocrit and apparent viscosity. Biorheology 17:183–189
1027. Gaehtgens P, Dührssen C, Albrecht K (1980) Motion, deformation, and interaction of blood cells and plasma during flow through narrow capillary tubes. Blood Cells 6:799–817
1028. Gaffin L (2001) Pathophysiology of heatstrokes. In: Lounsbury DE, Randolh KB, Burr RE (eds) Medical aspects of harsh environments. Office of the Surgeon General, Department of the Army, vol 1, pp. 161–208
1029. Gaffoglio R, Righero M, Giordanengo G, Zucchi M, Vecchi G (2020) Fast optimization of temperature focusing in hyperthermia treatment of sub-superficial tumors. arXiv:2006.13261v2 [cs.CE]
1030. Gafiychuk VV, Lubashevsky IA (1999) Mathematical description of heat transfer in living tissue. arXiv:adap-org/9911001v1
1031. Gafiychuk VV, Lubashevsky IA (2021) Mathematical description of heat transfer in living tissue (Part II). arXiv:adap-org/9911002v1
1032. Gage AA (1998) History of cryosurgery. Semin Surg Oncol 14:99–109
1033. Gage AA, Baust J (1998) Mechanisms of tissue injury in cryosurgery. Cryobiology 37:171–186
1034. Gage AA, Baust JG (2002) Cryosurgery – a review of recent advances and current issues. Cryo Lett 23:69–78
1035. Gagge AP, Stolwijk JAJ, Nishi Y (1971) An effective temperature scale based on a simple model of human physiological regulator response. ASHRAE Trans 77:247–262
1036. Gagge AP, Fobelets AP, Berglund LG (1986) A standard predictive index of human response to the thermal environment. ASHRAE Trans 92:709–731
1037. Gahwiler BH, Mamoon AM, Schlapfer WT, Tobias CA (1972) Effects of temperature on spontaneous bioelectricactivity of cultured nerve cells. Brain Res 40:527–533
1038. Gallardo M, Paulini F, Corral A, Balcerzyk M, Lucci CM, Ambroise J, Merola M, Fernandez-Maza L, Risco R, Dolmans M, Amorim A (2018) Evaluation of a new freezing protocol containing 20% dimethyl sulphoxide concentration to cryopreserve human ovarian tissue. Reprod Biomed Online 37:653–665
1039. Gallegos AD, Malaga C (2017) Natural convection in eccentric spherical annuli. Eur J Mech B 65:464–471

1040. Galligan M (2006) Proposed guidelines for skin-to-skin treatment of neonatal hypothermia. Am J Matern/Child Nurs 31:298–304
1041. Galloway SDR (1999) Cardiovascular and metabolic responses to heat stress. J Physiol 52:10S
1042. Gambassi G, Cerbai E, Pahor M, Capograssi MC, Carbonin P, Mugelli A (1994) Temperature modulates calcium homeostatis and ventricular arrhythmias in myocardial prepaartions. Cardiovasc Res 28:391–399
1043. Gandhi NS, Dupuy DE (2005) Image-guided radiofrequency ablation as a new treatment option for patients with lung cancer. Semin Roentgenol 40:171–181
1044. Gandolfi M, Celardo GL, Borgonovi F, Ferrini G, Avella A, Banfi F, Giannetti C (2016) Emergent ultrafast phenomena in correlated oxides and heterostructures. Phys Script 92:034004
1045. Ganeshan K, Chawla A (2017) Warming the mouse to model human diseases. Nat Rev Endocrinol 13:458–465
1046. Ganguly M, Miller S, Mitra K (2015) Model development and experimental validation for analyzing initial transients of irradiation of tissues during thermal therapy using short pulse lasers. Las Surg Med 47:711–722
1047. Gao B, Langer S, Corry PM (1995) Application of the time-dependent Green's function and Fourier transforms to the solution of the bioheat equation. Int J Hyperth 11:267–285
1048. Gao DY, Lin S, Watson PF, Critser LK (1995) Fracture phenomena in an isotonic salt solution during freezing and their elimination using glycerol. Cryobilogy 32:270–284
1049. Gas P (2011) Essential facts on the history of hyperthermia and their connections with electromedicine. Przegl Electrotech 87:37–40
1050. Gas P (2012) Temperature distribution of human tissue in interstitial microwave hyperthermia. Przegl Electrotech 88:144–146
1051. Gates DM (1968) Transpiration and leaf temperature. Annu Rev Plant Physiol 19:211–238
1052. Ge MY, Chua KJ, Shu C, Yang WM (2015) Analytical and numerical study of tissue cryofreezing via the immersed boundary method. Int J Heat Mass Transf 83:1–10
1053. Ge MY, Shu C, Yang WM, Chua KJ (2017) Incorporating an immersed boundary method to study thermal effects of vascular systems during tissue cryo-freezing. J Therm Biol 64:92–99
1054. Gebremedhin KG, Wu B (2001) A model of evaporative cooling of wet skin surface and fur layer. J Therm Biol 26:537–545
1055. Gebremedhin KG, Wu B (2002) Simulation of sensible and latent heat losses from wet-skin surface and fur layer. J Thermal Biol 277:291–297
1056. Geerling G, Tauber J, Baudouin C, Goto E, Matsumoto Y, O'Brien T (2011) The international workshop on meibomian gland dysfunction: report of the subcommittee on management and treatment of meibomian gland dysfunction. Investig Ophthalmol Vis Sci 52:2050–2064
1057. Gehl J (2003) Electroporation: theory and methods, perspectives for drug delivery, gene therapy and research. Acta Physiol Scand 177:437–447
1058. Geiger S, Emmanuel S (2010) Non-Fourier thermal transport in fractured geological media. Water Resour Res 46. https://agupubs.onlinelibrary.wiley.com/doi/full/10.1029/2009WR008671
1059. Gellermann J, Wlodarczyk W, Feussner A, Fahling H, Nadobny J, Hildebrandt B, Felix R, Wust P (2005) Methods and potentials of magnetic resonance imaging for monitoring radiofrequency hyperthermia in a hybrid system. Int J Hyperth 21:497–513
1060. Gembarovich J, Gembarovich JJ (2004) Non-Fourier heat conduction modeling in a finite medium. Int J Thermophys 25:1261–1272
1061. Gembarovich J, Majernik V (1988) Non-Fourier propagation of heat pulse in finite medium. Int J Heat Mass Transf 31:1073–1080
1062. Gera N, Yang A, Holtzman TS, Lee SX, Wong ET, Swanson KD (2015) Tumor treating fields perturb the localization of septins and cause aberrant mitotic exit. PLOS ONE 10:e0125269
1063. Gerasimov AN (1948) Generalization of the linear deformation laws and applications to the problems of inner friction (in Russian). Appl Math Mech 12:529–539

1064. Gerner EW (1987) Thermal dose and time–temperature factors for biological responses to heat shock. Int J Hyperth 3:319–327
1065. Gerweck LE (1977) Modification of cell lethality at elevated temperatures the pH effect. Radiat Res 70:224–235
1066. Ghani S, ElBialy EMAA, Bakochristou F, Gamaledin SMA, Rashwan MM (2017) The effect of forced convection and PCM on helmets' thermal performance in hot and arid environments. Appl Therm Eng 111:624–637
1067. Ghazanfarian J, Shomali Z, Abbassi A (2015) Macro- to nanoscale heat and mass transfer: the lagging behavior. Int J Therm 36:1416–1467
1068. Ghazizadeh HR, Azimi A, Maerefat M (2012) An inverse problem to estimate relaxation parameter and order of fractionality in fractional single-phase-lag heat equation. Int J Heat Mass Transf 55:2095–2101
1069. Ghazizadeh HR, Maerefat M (2010) Modeling diffusion to thermal wave heat propagation by using fractional heat conduction constitutive model. Iran J Mech Eng 11:66–76
1070. Ghobrial IM, Witzig TE, Adjei AA (2005) Targeting apoptosis pathways in cancer therapy. CA Cancer J Clin 55:178–194
1071. Giammaria F, Andras S (2006) Hyperthermia today: electric energy, a new opportunity in cancer treatment. J Cancer Res Ther 2:41–46
1072. Giesing JP, Barthelemy JFM (1998) A summary of industry MDO applications and needs. AIAA-98-4737
1073. Gifford ME, Clay TA, Careau V (2014) Individual (co) variation in standard metabolic rate, feeding rate, and exploratory behavior in wild-caught semiaquatic salamanders. Physiol Biochem Zool 87:384–396
1074. Giladi M, Munster M, Schneiderman RS, Voloshin T, Porat Y, Blat R, Zielinska-Chomej K, Haag P, Bomzon Z, Kirson ED, Weinberg U, Viktorsson K, Lewensohn R, Palti Y (2017) Tumor treating fields (TTFields) delay DA damage repair following radiation treatment of glioma cells. Radiat Oncol 12:206–206
1075. Gilbert JA, Hill PJ, Laybourn-Parry J (2004) Demonstration of antifreeze protein activity in Antarctic lake bacteria. Microbiology 150:171–180
1076. Gilchrist R, Medal R, Shorey W, Hanselman R, Parrott J, Taylor C (1957) Selective inductive heating of lymph. Ann Surg 146:596–606
1077. Gill P, Moghadam TT, Ranjbar B (2010) Differential scanning calorimetry techniques: applications in biology and nanoscience. J Biomol Tech 21:167–193
1078. Gillespie DT (1976) A general method for numerically simulating the stochastic time evolution of coupled chemical reactions. J Comput Phys 22:403–434
1079. Gilliland GL, Tung M, Ladner J (1996) The biological macromolecule crystallization database and NASA protein crystal growth archive. J Res Natl Inst Stand 101:309–320
1080. Gilmor A, Liu J, Peter AT, Critser JK (1998) Determination of plasma membrane characteristics of boar spermatozoa and their relevance to cryopreservation. Biol Reprod 58:28–36
1081. Gilmore JA, Liu J, Woods EJ, Peter AT, Critser JK (2000) Cryoprotective agent and temperature effects on human sperm membrane permeabilities: convergence of theoretical and empirical approaches for optimal cryopreservation methods. Hum Reprod 15:335–343
1082. Giordano MA, Gutierrez G, Rinaldi C (2010) Fundamental solutions to the bioheat equation and their application to magnetic fluid hyperthermia. Int J Hyperth 26:475–484
1083. Giovambattista N, Angell CA, Sciortino F, Stanley HE (2004) Glass-transition temperature of water: a simulation study. Phys Rev Lett 93:047801
1084. Giovambattista N, Rossky PJ, Debenedetti PD (2006) Effect of pressure on the phase behavior and structure of water confined between nanoscale hydrophobic and hydrophilic plates. Phys Rev E 73:041604
1085. Girasole M, Cricenti A, Generosi R, Congiu-Castellano A, Boumis G, Amiconi G (2001) Artificially induced unusual shape of erythrocytes: an atomic force microscopy study. J Microsc 204:46–52
1086. Girasole M, Dinarelli S, Boumis G (2012) Structure and function in native and pathological erythrocytes: a quantitative view from the nanoscale. Micron 43:1273–1286

1087. Giraud MN, Motta C, Boucher D, Grizard G (2000) Membrane fluidity predicts the outcome of cryopreservation of human spermatozoa. Hum Reprod 15:2160–2164
1088. Gisolfi CV (1996) Fluid balance and optimal performance. Nutri Rev 54:S159–S168
1089. Gisolfi CV, Mora F (2000) The hot brain: survival. Temperature and the human body. MIT Press, Cambridge
1090. Gisolfi CV, Robinson S (1969) Relations between physical training, acclimatization, and heat tolerance. J Appl Physiol 26:530–534
1091. Gisolfi CV, Wenger CB (1984) Temperature regulation during exercise: old concepts, new ideas. Exerc Sport Sci Rev 12:339–372
1092. Givoni B, Goldman RF (1972) Predicting rectal temperature response to work, environment, and clothing. J Appl Physiol 32:812–822
1093. Givoni B, Goldman RF (1973) Predicting heart rate response to work, environment, and clothing. J Appl Physiol 34:201–204
1094. Glass DE, Ozisik MN, McRae DS, Vick B (1986) Hyperbolic heat conduction with temperature-dependent thermal conductivity. J Appl Phys 59:1861–1865
1095. Glass DE, Ozisik MN, Vick B (1987) Non-Fourier effects on transient temperature resulting from periodic on-off heat flux. Int J Heat Mass Transf 30:1623–1631
1096. Glazer ES, Curley SA (2011) Non-invasive radiofrequency ablation of malignancies mediated by quantum dots, gold nanoparticles and carbon nanotubes. Ther Deliv 2:1325–1330
1097. Glazier DS (2005) Beyond the '3/4-power law': variation in the intra- and interspecific scaling of metabolic rate in animals. Biol Rev Cambridge Philos Soc 80:611–662
1098. Gleich B, Weizenecker J (2005) Tomographic imaging using the nonlinear response of magnetic particles. Nature 435:1217–1217
1099. Glenn TN, Rastregar S, Jacques SL (1996) Finite element analysis of temperature controlled coagulation in laser irradiated tissue. IEEE Trans. Biomed. Eng. 43:79–86
1100. Glory J, Bonetti M, Helezen M, Mayne-L'Hermite M, Reynaud C (2008) Thermal and electrical conductivities of water-based nanofluids prepared with long multiwalled carbon nanotubes. J Appl Phys 103:094309
1101. Glossary of verification and validation terms. http://www.grc.nasa.gov/WWW/wind/valid/tutorial/glossary.html
1102. Glowinski R, Pan TW, Periaux J (1994) A fictitious domain method for Dirichlet problem and applications. Comput. Methods Appl. Mech. Eng. 111:283–303
1103. Glowinski R, Pan TW, Periaux J (1994) A fictitious domain method for external incompressible viscous flow modeled by Navier- Stokes equations. Comput. Methods Appl. Mech. Eng. 112:133–148
1104. Gluckman PD, Wyatt JS, Azzopardi D, Ballard R, Edwards AD, Ferriero DM, Polin RA, Robertson CM, Thoresen M, Whitelaw A, Gunn AJ (2005) Selective head cooling with mild systemic hypothermia after neonatal encephalopathy: multicentre randomised trial. Lancet 365:663–670
1105. Go MR, Dundon JM, Karlowicz DJ, Domingo CB, Muscarella P, Melvin WS (2004) Delivery of radiofrequency energy to the lower esophageal sphincter improves symptoms of gastroesophageal reflux. Surgery 136:786–794
1106. Gobin MA, O'Neal P, Watkins DM, Halas NJ, Drezek RA, West JL (2005) Near infrared laser-tissue welding using nanoshells an exogenous absorber. Lasers Surg Med 37:123–129
1107. Godoy S, García-Colín LS (1997) Nonvalidity of the telegrapher's diffusion equation in two and three dimensions for crystalline solids. Phys Rev E 55:2127–2131
1108. Gokul KC, Gurung DB (2019) Mathematical model: thermal effects of two wheeler rider's speed in his/her eye. J Math Sci Model 2:143–154
1109. Gokul KC, Dil BG, Pushpa RA (2015) Mathematical model: comparative study of thermal effects of laser in corneal refractive surgeries. Appl Appl Math 10:620–633
1110. Goldman RF (2001) Introduction to heat-related problems in military operations. In: Lounsbury DE, Randolh KB, Burr RE (eds) Medical aspects of harsh environments, vol 1. Office the Surgeon General, Department of Army, Falls Church, pp 3–42

1111. Goldman N, Leforestier C, Saykally RJ (2005) A "first principles" potential energy surface for liquid water from VRT spectroscopy of water clusters. Philos Trans R Soc A 363:493–508
1112. Goldstein RE, van de Meent JW (2015) A physical perspective on cytoplasmic streaming. Interface Focus 5:20150030
1113. Goldstein RE, Tuval I, van de Meent JW (2008) Microfluidics of cytoplasmic streaming and its implications for intracellular transport. Proc Natl Acad Sci USA 105:3663–3667
1114. Goloviznin VM, Kondratenko PS, Matveev LV, Korotkin IA, Dranikov IL (2010) Anomalous diffusion of radionuclides in strongly nonuniform geological formations (in Russian). Nauka, Moscow
1115. Goltsev A, Kozlova A, Dubrava Y, Ostankov T, Sirous M (2007) Cryopreservation as a method for regulating the functional potential of bone marrow hemopoietic stem cells. Cell Preserv Tech 5:33–40
1116. Gomez H, van der Zee KG (2018) Computational phase-field modeling. In: Encyclopedia of computational Mechanics, pp. 1–35
1117. Gómez-Aguilar JF, Razo-Hernandez R, Granados-Lieberma D (2014) A physical interpretation of fractional calculus in observables terms: analysis of fracional time constant and the transtory response. Rev Mexicana Fisica 60:32–38
1118. Gompper G, Fedosov DA (2015) Modeling microcirculatory blood flow: current state and future perspectives. WIREs Syst Biol Med 8:157–168
1119. Gonzalez-Alonso J, Mora-Rodriguez R, Below PR, Coyle EF (1995) Dehydration reduces cardiac output and increases systemic and cutaneous vascular resistance during exercise. J Appl Physiol 79:1487–1496
1120. Gonzalez-Alonso J, Mora-Rodriguez R, Below PR, Coyle EF (1997) Dehydration markedly impairs cardiovascular function in hyperthermic endurance athletes during exercise. J Appl Physiol 82:1229–1236
1121. González-Suárez A, Gutierrez-Herrera E, Berjano E, Jimenez-Lozano JN, Franco W (2015) Thermal and elastic response of subcutaneous tissue with different fibrous septa architectures to RF heating: numerical study. Lasers Surg Med 47:183–195
1122. Gonzalez-Narvaez E, López de Haro M, Vázquez F (2021) Internal structure and heat conduction in rigid solids: a two temperature approach. arXiv:2106.01323
1123. Goodarzi-Ardakani V, Taebi-Rahni M, Salimi MR, Ahmadi G (2016) Computational simulation of temperature and velocity distribution in human upper respiratory airway during inhalin of hot air. Respir Physiol Neurobiol 223:49–58
1124. Goodman R, Bassett CAL, Henderson A (1983) Pulsing electromagnetic fields induce cellular transcription. Science 220:1283–1285
1125. Gopalakrishnan J (2002) A mathematical model for irrigated epicardial radiofrequency ablation. Ann Biomed Eng 30:884–893
1126. Gopalakrishnan SS, Pier B, Biesheuvel A (2014) Dynamics of pulsatile flow through model abdominal aortic aneurysms. J Fluid Mech 758:150–179
1127. Gorbach AM, Heiss J, Kufta C, Sato S, Fedio P, Kammerer WA, Solomon J, Oldfield EH (2003) Intraoperative infrared functional imaging of human brain. Ann Neurol 54:297–309
1128. Gordon CJ (1985) Relationship between autonomic and behavioral thermoregulation in the mouse. Physiol Behav 34:687–690
1129. Gordon CJ (2012) Thermal physiology of laboratory mice: defining thermoneutrality. J Therm Biol 37:654–685
1130. Gordon SA, Surrey K (1960) Red and far-red action on oxidative phosphorylation. Radiat Res 12:325–339
1131. Gorenflo R, Mainardi F (1997) Integral and differential equations of fractional order. In: Carpinteri A, Mainardi F (eds) Fractals and fractional calculus in continuum mechanics. Springer, Wien
1132. Gorenflo R, Mainardi F (2003) Fractional diffusion processes: probability distributions and continuos time random walk. In: Rangarajan G, Ding M (eds) Processes with long range correlations. Springer, Berlin, pp 148–166

1133. Gorenflo R, Mainardi F (2008) Continuous time random walk, Mittag-L waiting time and fractional diffusion: mathematical aspects. In Klages R, Radons G, Sokolov IM (eds) Anomalous transport. Foundations and applications. Wiley-VCH, Weinheim, pp 93–116
1134. Gorenflo R, Mainardi F, Moretti D, Pagnini G, Paradisi P (2002) Fractional diffusion: probability disributions and random walk models. Phys A 305:106–112
1135. Gorenflo R, Vivoli A, Mainardi F (2004) Discrete and continuous random walk models for space-time fractional diffusion. Nonlinear Dyn 38:101–116
1136. Gorton AJ (1996) Measurements and analysis of the microwave dielectric properties of human and animal tissues. Ph.D. Thesis, University of Glasgow
1137. Goudarzi P, Azimi A (2019) Numerical simulation of fractional non-Fourier heat conduction in skin tissue. J Therm Biol 84:274–284
1138. Gould IG, Tsai P, Kleinfeld D, Linninger A (2017) The capillary bed offers the largest hemodynamic resistance to the cortical blood supply. J Cerebral Blood Flow Metabol 37:52–68
1139. Goulet CT, Thompson MB, Michelangeli M, Wong BBM, Chapple DG (2017) Thermal physiology: a new dimension of the pace-of-life syndrome. J Anim Ecol 86:1269–1280
1140. Gov N, Zilman AG, Safran S (2003) Cytoskeleton confinement and tension of red blood cell membranes. Phys Rev Lett 90:228101
1141. Gowrishankar TR, Stewart DA, Martin GT, Weaver JC (2004) Transport lattice models of heat transport in skin with spatially heterogeneous, temperature-dependent perfusion. Biomed Eng Online 3:1–17
1142. Goya GF, Grazu V, Ibarra MR (2008) Magnetic nanoparticles for cancer therapy. Curr Nanosci 4:1–16
1143. Grabski GK, Lesnic D, Johansson BT (2016) Identification of a time-dependent bio-heat blood perfusion coefficient. Int Commun Heat Mass Transf 75:218–222
1144. Grad H (1958) Principles of the kinetic theory of gases. In: Flugge s (ed) Thermodynamik der gase [Thermodynamics of gases]. Handbuch der physic [Encyclopedia of physics], vol 3–12. Springer, Berlin
1145. Graether SP, Kulper MJ, Gagné SM, Walker VK, Jia Z, Sykes BD, Davies PL (2000) β-helix structure and ice-binding properties of a hyperactive antifreeze protein from an insect. Nature 406:325–328
1146. Graham LA, Davies PL (2005) Glycine-rich antifreeze proteins from snow fleas. Science 310:461
1147. Graham LA, Liow YC, Walker VK, Davies PL (1977) Hyperactive antifreeze proteins from beetles. Nature 388:727–728
1148. Grahn D, Brock-Utne JG, Watenpaugh DE, Heller HC (1998) Recovery from mild hypothermia can be accelerated by mechanically distending blood vessels in the hand. J Appl Physiol 85:1643–1648
1149. Gránásy L (1999) Cahn-Hillard-type density functional calculations for homogeneous uce nucleation in undercooled water. J Mol Struct 485–486:523–526
1150. Grant NM (2010) Thermogenesis in plants: the mode of heating and regulation in hot flowers. PhD Thesis, University of Wollongong
1151. Grassman A, Peters F (1999) Experimental investigation of heat conduction in wet sand. Heat Mass Transf 35:289–294
1152. Gratzl T, Dohr G, Schmidt-Kloiber H, Reichel E (1991) Histological distinction of mechanical and thermal defects produced by nanosecond laser pulses in striated muscle at 1064 nm. Proc SPIE 1427:55–62
1153. Gray JJ (2004) The interaction of proteins with solid surfaces. Curr Op Struct Biol 14:110–115
1154. Green DM, Diller KR (1978) Measurement of burn induced leakage of macromolecules in living tissue. J Biomech Eng 100:153–158
1155. Green A, Naghdi P (1991) A re-examnation of the basic postulates of thermomechanics. Proc Roy Soc Lond 357:171–194

1156. Green A, Naghdi P (1992) On undamped heat waves in an elastic solid. J Thermal Stress 15:253–264
1157. Green AE, Rivlin RS (1957) The mechanics of non-liner materials with memory (Part I). Arch Rational Mech Anal 1:1–21
1158. Green AE, Rivlin RS (1959) The mechanics of non-liner materials with memory (Part III). Arch Rational Mech Anal 4:387–404
1159. Green AE, Rivlin RS, Spencer AJM (1959) The mechanics of non-liner materials with memory (Part II). Arch Rational Mech Anal 3:82–90
1160. Gresits I, Thuroszy G, Sagi O, Gyure-Garami B, Markus G, Simon F (2018) Non-calorimetric determination of absorbed power during magnetic based hyperthermia. arXiv:1803.09077v2 [physics.med-ph]
1161. Greve B, Hammes S, Raulin C (2001) The effect of cold air cooling on 585 nm pulsed dye laser treatment of port-wine stains. Dermatol Surg 27:633–636
1162. Grier B, Alyanak E, White M, Camberos J, Figliola R (2014) Numerical integration techniques for discontinuous manufactured solutions. J Comput Phys 278:193–203
1163. Griffith M, Ewart KW (1995) Antifreeze proteins and their potential use in frozen foods. Biotech Adv 13:375–402
1164. Griffith M, Yaish MW (2004) Antifreeze proteins in overwintering plants: a tale of two activities. Trends Plant Sci 9:399–405
1165. Griffith M, Antikainen M, Hon WC, Pihakaski-Maunsbach K, Yu XM, Chun JU, Yang DSC (1997) Antifreeze proteins in winter rye. Plant Physiol 100:327–332
1166. Griffith M, Lumb C, Wiseman SB, Wisniewski M, Johnson RW, Marangoni AG (2005) Antifreeze proteins modify the freezing process in planta. Plant Physiol 138:330–340
1167. Grigoletto EC, de Oliveira EC (2013) Fractional versions of the fundamental theorem of calculus. Appl Math 4:23–33
1168. Grinstein SV, Kost OA (2001) Structural and functional feature of membrane proteins (in Russian). Adv Biol Chem 41:77–104
1169. Grmela M, Lebon G (1998) Finite-speed propagation of heat: a nonlocal and nonlinear approach. Phys A 248:428–441
1170. Grmela M, Lebon G, Dubois C (2011) Multiscale thermodynamics and mechanics of heat. Phys Rev E 83:061134
1171. Groot RD, Warren PB (1997) Dissipative particle dynamics: bridging the gap between atomistic and mesoscopic simulation. J Chem Phys 107:4423–4435
1172. Grosberg AY, Khokhlov AR (1986) Phase transitions in polymer and biopolymer systems. Sov Phys Usp 29:797–799
1173. Gross DC, Proebsting EL, MacCrindle-Zimmerman H (1988) Development, distribution and characterization of intrinsic, non-bacterial ice nuclei in Prunus wood. Plant Physiol 88:915–922
1174. Grysa K, Maciag A, Adamczyk-Krasa J (2014) Trefftz functions applied to direct and inverse non-Fourier heat conduction problems. J Heat Transf 136:091302
1175. Gu X, Wei Y, Yin X, Li B, Yang R (2018) Phononic thermal properties of two-dimensional materials. Rev Mod Phys 90:041002
1176. Guatteo E, Chung KK, Bowala TK, Bernardi G, Mercuri NB, Lipski J (2005) Temperature sensitivity of dopaminergic neurons of the substantia nigra pars compacta: involvement of transient receptor potential channels. J Neurophysiol 94:3069–3080
1177. Guckenberger A, Kihm A, John T, Wagner C, Gekle S (2018) Numerical-experimental observation of shape bistability of red blood cells flowing in a microchannel. Soft Matter 14:2032–2043
1178. Guedes JC (2016) Mathematical modelling of human thermoregulation in hot environments. Ph.D. Thesis, University of Porto
1179. Güémez J, Fiolhais C, Fiolhais M (2002) Reproducing Black's experiments: freezing point depression and supercooling of water. Eur J Phys 23:83–91

1180. Guenther JF, Seki S, Kleinhans FW, Edashige K, Roberts DM, Mazur P (2006) Extra- and intra-cellular ice formation in stage I and II Xenopus laevis oocytes. Cryobiology 52:401–416
1181. Guerrero López GD, Cepeda Rubio MFJ, Hernández Jácquez JI, Hernandez A, Salas L, Perezgasga F, Flores García F (2017) Computational FEM model, phantom and ex vivo swine breast validation of an optimized double-slot microcoaxial antenna designed for minimally invasive breast tumor ablation: theoretical and experimental comparison of temperature, size of lesion, and SWR, preliminary data. Comput Math Meth Med 2017:1562869
1182. Guglietta F, Behr M, Biferale L, Falcucci G, Sbragaglia M (2020) On the effects of membrane viscosity on transient red blood cell dynamics. arXiv:2004.04109v1 [physics.flu-dyn]
1183. Guide for the verification and validation of computational fluid dynamics simulations. AIAA G-077-1998 (1998)
1184. Guido S, Tomaiuolo G (2009) Microconfined flow behavior of red blood cells in vitro. Compt Rendus Phys 10:751–763
1185. Guieu JD, Hardy JD (1970) Effects of preoptic and spinal cord temperature in control of thermal polypnea. J Appl Physiol 28:540–542
1186. Guntur SR, Lee KI, Paeng DG, Coleman AJ, Choi MJ (2013) Temperature-dependent thermal properties of ex vivo liver undergoing thermal ablation. Ultrasound Med Biol 39:1771–1784
1187. Guo P, Hillyard SD, Fu BM (2003) A two-barrier compartment model for volume flow across amphibian skin. Am J Physiol Regul Integr Comp Physiol 285:R1384–R1394
1188. Guo Z, Wan SK, Kim KH, Kosaruja C (2003) Comparing diffusion approximation with radiation transfer analysis for light transport in tissues. Opt Rev 10:415–421
1189. Guo Q, Duffy SP, Matthews K, Santoso AT, Scott MD, Ma H (2014) Microfluidic analysis of red blood cell deformability. J Biomech 47:1767–1776
1190. Guo R, Zheng J, Chen J (2017) MRI RF-induced heating in heterogeneous human body with implantable medical device. In: Halefoglu AM (ed) High-resolution neuroimaging - basic physical principles and clinical applications. InTech, Rijeka, p 71384
1191. Gupta P, Srivastava A (2019) Non-Fourier transient thermal analysis of biological tissue phantoms subjected to high intensity focused ultrasound. Int J Heat Mass Transf 136:1052–1063
1192. Gupta PK, Singh J, Rai KN (2010) Numerical simulation for heat transfer in tissues during thermal therapy. J Thermal Biol 35:295–301
1193. Gupta A, Dai T, Hamblin MR (2014) Effect of red and near-infrared wavelengths on low-level laser (light) therapy-induced healing of partial-thickness dermal abrasion in mice. Lasers Med Sci 29:257–265
1194. Gupta KJ, Mur LAJ, Neelwarne B (eds) (2015) Alternative respiratory pathways in higher plants. Wiley, Hoboken
1195. Gurtin ME, Pipkin AC (1969) A general theory of heat conduction with finite wave speed. Arch Rat Mech Anal 31:113–126
1196. Gurtin ME, Williams WO (1966) On the Clausius-Duhem inequality. Z Angew Math Phys 17:626–633
1197. Gurtin ME, Williams WO (1967) An axiomatic foundation for continuum thermodynamics. Arch Ration Mech Anal 26:83–117
1198. Gurung DB, Gokul KC (2016) Mathematical model of thermal effects of blinking in human eye. Int J Biomath 9:1650006
1199. Gurung DB, Saxena VP (2010) Pheripheral temperature distribution in human subjects exposed to wind flow. Int J Math Eng 2:728–740
1200. Guy AW (1971) Electromagnetic fields and relative heating patterns due to a rectangular aperture source in direct contact with bilayered biological tissue. IEEE Trans. Microw Theory Tech 19:214–223

1201. Gyarmati I (1977) The wave approach of thermodynamics and some problems of non-linear theories. J Non-Equil Thermodyn 2:233–260
1202. Habash RWY, Bansal R, Krewski D, Alhafid HT (2006) Thermal therapy, Part 1: an introduction to thermal therapy. Crit Rev Biomed Eng 34:459–489
1203. Habash RWY, Bansal R, Krewski D, Alhafid HT (2006) Thermal therapy, part 2: hyperthermia techniques. Crit Rev Biomed Eng 34:491–542
1204. Hackman P (2001) Recognizing and understanding the cold-stressed term infant. Neonat Netw 20:35–41
1205. Haden RL (1940) The human red blood cell. https://www.ccjm.org/content/ccjom/8/2/111.full.pdf
1206. Haemmerich D, Laeseke PF (2005) Thermal tumour ablation: devices, clinical applications and future directions. Int J Hyperth 21:755–760
1207. Haemmerich D, Webster J, Mahvi D (2003) Thermal dose versus isotherm as lesion boundary estimator for cardiac and hepatic radiofrequency ablation. In: Proceedings of the 25th annual international conference of the IEEE engineering in medicine and biology society. IEEE, Piscataway, pp 134–137
1208. Haemmerich D, Santos ID, Schutt DJ, Webster JG, Mahvi DM (2006) In vitro measurements of temperature-dependent specific heat of liver tissue. Med Eng Phys 28:194–197
1209. Hagedorn M, Kleinhans FW, Artemov D, Pilatus U (1998) Characterization of a major permeability barrier in the zebrafish embryo. Biol Reprod 59:1240–1250
1210. Hainer B (1991) Fundamentals of electrosurgery. J Am Board Family Pract 4:419–426
1211. Haines DE, Watson DD (1989) Tissue heating during radiofrequency catheter ablation: a thermodynamic model and observations in isolated perfused and superfused canine right ventricular free wall. Pacing Clin Electrophysiol 12:962–976
1212. Hajnová V, Zlámal F, Lenárt P, Bienertova-Vasku J (2021) Homeostatic model of human thermoregulation with bistability. Sci Rep 11:17327
1213. Hale JP, Marcelli G, Parker KH, Winlovea CP, Petrov PG (2009) Red blood cell thermal fluctuations: comparison between experiment and molecular dynamics simulations. Soft Matter 5:3603–3606
1214. Hales JR, Iriki M, Tsuchiya K, Kozawa E (1978) Thermally-induced cutaneous sympathetic activity related to blood flow through capillaries and arteriovenous anastomoses. Pflüg Arch 375:17–24
1215. Hales JRS, Fawcett AA, Bennett JW, Needham AD (1978) Thermal control of blood flow through capillaries and arteriovenous anastomoses in skin of sheep. Pflüg Arch 378:55–63
1216. Halimi MT, Dhahri H, Khedher NB, Hassen MB, Sakli F (2009) Thermal properties of industrial safety helmets. Build Environ 5:833–844
1217. Hall CM, Moeendarbary E, Sheridan GK (2021) Mechanobiology of the brain in ageing and Alzheimer's disease. Eur J Neurosci 53:3851–3878
1218. Hallaj I, Cleveland R, Hynynen K (2001) Simulations of the thermo-acoustic lens effect during focused ultrasound surgery. J Acoust Soc Am 109:2245–2253
1219. Halle B (2004) Biomolecular crystallography: structural changes during flash-cooling. Proc Natl Acad Sci USA 101:4793–4798
1220. Halle B (2004) Protein hydration dynamics in solution: a critical survey. Philos Trans R Soc Lond B 359:1207–1224
1221. Halpern D, Secomb TW (1991) The squeezing of red blood cells through capillaries with near-minimal diameters. J Fluid Mech 203:381–400
1222. Halpern D, Secomb TW (1991) Viscous motion of disc-shaped particles through parallel-sided channels with near minimal widths. J Fluid Mech 231:545–560
1223. Hamblin MR, Demidova TN (2006) Mechanisms of low level light therapy. Proc SPIE 6140:614001
1224. Hamilton G (1998) Investigations of the thermal properties of human and animal tissues. PhD Thesis, University of Glasgow
1225. Hamilton MF, Blackstock DT (1998) Nonlinear acoustics. Academic Press, Boston

1226. Hammel HT, Jackson DC, Stolwijk JA, Hardy J, Stromme SB (1963) Temperature regulation by hypothalamic proportional control with an adjustable set point. J Appl Physiol 18:1146–1154
1227. Hammel HT, Maggert J, Kaul R (1976) Effects of altering spinal cord temperature on temperature regulation in the adélie penguin, pygoscelis adeliae. Pflüg Arch 362:1–6
1228. Han B, Bishof JC (2004) Direct cell injury associated with eutectic crystallization during freezing. Cryobiology 48:8–21
1229. Han B, Bishof JC (2004) Thermodynamic nonequilibrium phase change behavior and thermal properties of biological solutions for cryobioligy applications. J Biomech Eng 126:196–203
1230. Han Z, Bischof JC (2020) Perspective: critical cooling and warming rates as a function of CPA concentration. Cryo Lett 41:185–193
1231. Han DW, Hyon SH, Park JC, Park KD, Park YH, Park HK (2006) Non-frozen preservation of mammalian tissue using green tea polyphenolic compounds. Biomed Mater 1:R18–R29
1232. Han J, Sydykov B, Yang H, Sieme H, Oldehof H, Wolkers WF (2019) Spectroscopic monitoring of transport processes during loading of ovarian tissue with cryoprotective solutions. Sci Rep 9:15577
1233. Hand JW (1986) Physical techniques in clinical hyperthermia (Electronic & electrical engineering research studies). Wiley, Hoboken
1234. Hand JW (2008) Modelling the interaction of electromagnetic fields (10 MHz–10 GHz) with the human body: methods and applications. Phys Med Biol 53:243–286
1235. Hand JW, Ledda JL, Evans NTS (1982) Temperature distribution in tissues subjected to local hyperthermia by RF induction heating. Brit J Cancer 45(suppl V):31–35
1236. Hand JW, Law RW, Lagendijk JW, Lin X, Burl M, Young IR (1999) Electromagnetic and thermal modeling of SAR and temperature fields in tissue due to an RF decoupling coil. Magn Reson Med 42:183–192
1237. Hang SC (1995) The method of space-time conservation element and solution element: a new approach for the Navier-Stokes and Euler equations. J Comput Phys 119:295–324
1238. Hanna EG, Tait PW (2015) Limitations to thermoregulation and acclimatization challenge human adaptation to global warming. Int J Environ Res Publ Health 12:8034–8074
1239. Hansen E (2003) Modelling heat transfer in a bone-cement-prothesis system. J Biomech 36:787–795
1240. Hansson T, Oostenbrink C, van Gunsteren WF (2002) Molecular dynamics simulations. Curr Opin Struct Biol 12:190–196
1241. Hanyu Y, Ichikawa M, Matoumoto G (1992) An improved cryofixation method: cryoquenching of small tissue blocks during microwave irradiation. J Microsc 165:225–235
1242. Haque N, Ludri A, Hossain SA, Ashutosh M (2012) Comparative studies on temperature threshold for heat shock protein 70 induction in young and adult murrah buffaloes. J Anim Physiol Anim Nutr (Berl) 96:920–929
1243. Harada J, Dohi M, Mogami T, Fukuda K, Miki K, Kishimoto NFK, Simizu T, Miyasaka K (2001) Initial experience of percutaneous renal cryosurgery under the guidance of a horizontal open MRI system. Rad Med 19:291–296
1244. Harari C, Magagna M, Bedoya M, Lee F, Lubner M, Hinshaw J, Ziemlewicz T, Brace C (2016) Microwave ablation: comparison of simultaneous and sequential activation of multiple antennas in liver model systems. Radiology 278:95–103
1245. Harden JL, MacKintosh JL, Olmstead PD (2005) Budding and domain shape transformations in mixed lipid films and bilayer membranes. Phys Rev E 72:011903
1246. Hardy JD, Du Bois EF (1940) Differences between men and women in their response to heat and cold. Proc Nat Acad Sci USA 26:389–398
1247. Hardy JD, Dubois EF (1937) Basal metabolism, radiation, convection and vaporization at temperatures of 22 to 35c. J Nutrition 15:477–497
1248. Hardy JD, Dubois EF (1937) Regulation of heat loss from the human body. Proc Nat Acad Sci USA 23:624–631

1249. Hariharan P, Myers MR, Banerjee RK (2007) HIFU procedures at moderate intensities - effect of large blood vessels. Phys Med Biol 52:3493–3513
1250. Harkness J, Whittington RB (1970) Blood-plasma viscosity: an approximate temperature-invariant arising from generalised concepts. Biorheology 6:169–187
1251. Harrap MJM, Rands SA, de Ibarra NH, Whitney HM (2017) The diversity of floral temperature patterns, and their use by pollinators. eLife 6:e31262
1252. Harris B, Andrews P, Marshall I, Robinson T, Murray G (2008) Forced convective head cooling device reduces human cross-sectional brain temperature measured by magnetic resonance: a non-randomized healthy volunteer pilot study. Brit J Anaesth 100:365–372
1253. Harris JJ, Jolivet R, Attwell D (2012) Synaptic energy use and supply. Neuron 75:762–777
1254. Harris GR, Church CC, Dalecki D, Ziskin MC, Bagley JE (2016) Comparison of thermal safety practice guidelinesfor diagnostic ultrasound exposures. Ultasound Med Biol 42:345–357
1255. Harrison SM, Bers DM (1989) Influence of the tempeartuer on calcium sensitivity of myofilaments of skinned ventricular muscle from the rabbit. J Gen Physiol 93:411–428
1256. Hashido M, Kidera AK, Ikeguchi M (2007) Water transport in aquaporins: osmotic permeability matrix analysis of molecular dynamics simulations. Biophys J 93:373–385
1257. Hassanpour S, Saboonchi A (2014) Interstitial hyperthermia treatment of countercurrent vascular tissue: a comparison of Pennes, WJ and porous media bioheat models. J Ther Biol 46:47–55
1258. Hassanpour S, Saboonchi A (2015) The numerical assessment of volume averaging method in heat transfer modeling of tissue-like porous media. Int Commun Heat Mass Transf 63:41–48
1259. Hatch TF (1963) Assessment of heat stress. In Hardy JD (ed) Temperature - its measurement and control in science and industry, vol. 3, part 3. Reinhold, New York, pp 307–318
1260. Hatfield HS (1953) Measurement of the thermal conductivity of animal tissue. J Physiol 120:35–38
1261. Hatfield HS, Pugh LGC (1951) Thermal conductivity of human fat and muscle. Nature 168:918–919
1262. Hatton L (1997) The T experiments: errors in scientific software. IEEE Comput Sci Eng 4:27–38
1263. Hatwar R (2017) Applications of heat and mass transfer analysis in bio-medicine and materials. PhD Thsis, Johns Hopkins University
1264. Havenith G, Fiala D (2015) Thermal indices and thermophysiological modeling for heat stress. Compr Physiol 6:255–302
1265. Havlin S, Buldyrev SV, Goldberger AL, Mantegna RN, Ossadnik SM, Peng CK, Simons M, Stanley HE (1995) Fractals in biology and medicine. Chaos Solitons Fract 6:171–201
1266. Hawkins D, Houreld N, Abrahamse H (2005) Low level laser therapy (LLLT) as an effective therapeutic modality for delayed wound healing. Ann NY Acad Sci 1056:486–493
1267. Hayashi T, Fukuda N, Uchiyama S, Inada N (2015) A cell-permeable fluorescent polymeric thermometer for intracellular temperature mapping in mammalian cell lines. PLoS One 10:e0117677
1268. Hayashi Y, Horiguchi I, Kinooka M, Sugiyama H (2020) Slow freezing process design for human induced pluripotent stem cells by modeling intracontainer variation. Comput Chem Eng 132:106597
1269. Haymes EM, Wells CL (1986) Environment and human performance. Human Kinetics, Champaign
1270. Haynie DT (2008) Biological thermodynamics, 2nd edn. Cambridge University Press, Cambridge
1271. Hayward JN, Baker MA (1968) Role of cerebral arterial blood in the regulation of brain temperature in the monkey. Am J Physiol 215:389–403
1272. Hayward JN, Baker MA (1969) A comparative study of the role of the cerebral blood in the regulation of brain temperature in five mammals. Brain Res 16:417–440

1273. He X (2011) Thermostability of biological systems: fundamentals, challenges, and quantification. Open Biomed Eng J 5:47–73
1274. He JH (2014) A tutorial feview on fractal spacetime and fractional calculus. Int J Theor Phys 53:3698–3718
1275. He JH (2018) Fractal calculus and its geometrical explanation. Results Phys 10:272–276
1276. He JH, Liu F (2013) Local fractional iterative method for fractal heat transfer in silk cocoon hierarchy. Nonlinear Sci Lett 4:15–20
1277. He ZZ, Liu J (2016) An efficient parallel numerical modeling of bioheat transfer in realistic tissue structure. Int J Heat Mass Transf 95:843–852
1278. He ZZ, Liu J (2017) A coupled continuum-discrete bioheat transfer model for vascularized tissue. Int J Heat Mass Transf 107:544–556
1279. He Q, Zhu L, Weinbaum S, Lemons DE (2002) Experimental measurements of temperature variations along paired vessels from 200 to 1000 μm in diameter in rat hind leg. J Biomed Eng 124:656–661
1280. He Q, Zhu L, Weinbaum S (2003) Effect of blood flow on thermal equilibration and venous rewarming. Ann Biomed Eng 31:659–666
1281. He Y, Shirazaki M, Liu H, Himeno R, Sun Z (2006) A numerical coupling model to analyze the blood flow, temperature, and oxygen transport in human breast tumor under laser irradiation. Comput Biol Med 36:1336–1350
1282. He JH, Wang Q, Sun J (2011) Can polar bear hairs absorb environmental energy? Thermal Sci 15:911–913
1283. He JH, Li ZB, Wang Q (2016) A new fractional derivative and its application to explanation of polar bear hairs. J King Saud Univ Sci 28:190–192
1284. He Y, Laugersen K, Kamp D, Sultan SA, Oddershede LB, Jauffred L (2019) Effects and side-effects of plasmonic photothermal therapy in brain tissue. arXiv:1902.06578v1 [physics.med-ph]
1285. He S, Zhang Y, Huang Z, Zhang G, Gao Y (2020) Influence of internal structure and composition on head's local thermal sensation and temperature distribution. Atmosphere 11:218
1286. Hein A, Daanen M, Racinais S, Periard JD (2018) Heat acclimation decay and re-induction: a systematic review and meta analysis. Sports Med 48:409–430
1287. Heinrich V, Ritchie K, Mohandas N, Evans E (2020) Elastic thickness compressibilty of the red cell membrane. Biophys J 81:1452–1463
1288. Heldin CH, Rubin K, Pietras K, Östman A (2004) High interstitial fluid pressure - an obstacle in cancer therapy. Nat Rev Cancer 4:806–813
1289. Helfmann W, Jannes P, Jessen C (1981) Total body thermosensitivity and its spinal and supraspinal fractions in the conscious goose. Pflüg Arch 391:60–67
1290. Helfrich W (1973) Elastic properties of lipid bilayers: theory and possible experiments. Zeitschr Naturforsch C 28:693–703
1291. Helmy S, Bader Y, Koch M, Tiringer D, Kollmann C (2015) Measurement of thermal effects of doppler ultrasound: an in vitro study. PLoS ONE 10:e0135717
1292. Henderson PW, Lewis GK, Shaikh N, Sohn A, Weinstein AL, Olbricht WL, Spector JA (2010) A portable high-intensity focused ultrasound device for noninvasive venous ablation. J Vascular Surg 51:707–711
1293. Hendijanifard M, Willis DA (2013) Nanosecond time-resolved measurements of transient hole opening during laser micromachining of an aluminum film. J Heat Transf 135:091202
1294. Henriques FC (1947) Studies of thermal injuries. V. The predictability and the significance of thermally induced rate process leading to irreversible epidermal injury. Arch Pathol 43:489–502
1295. Henriques FC, Moritz AR (1947) Studies of thermal injuries. I. The conduction of heat to and through skin and the temperatures attained therein. A theoretical and an experimental investigation. Am J Pathol 23:530–549
1296. Henry PSH (1948) The diffusion of moisture and heat through textiles. Discuss Faraday Soc 3:243–257

1297. Henry J (2005) Basal metabolic rate studies in humans: measurement and development of new equations. Publ Health Nutrition 8:1133–1152
1298. Henry SD, Guarrera JV (2012) Protective effects of hypothermic ex vivo perfusion on ischemia/reperfusion injury and transplant outcomes. Transplant Rev 26:163–175
1299. Henyey LG, Greenstein JL (1941) Diffuse radiation in the galaxy. Astrophys J 93:70–83
1300. Herbert K, Bhusate L, Scott D, Diamantopolos C, Perrett D (1989) Effect of laser light at 820 nm on adenosine nucleotide levels in human lymphocytes. Las Life Sci 3:37
1301. Heremans JP, Dresselhaus MS (2013) When thermoelectrics reached the nanoscale. Nat Nanotechnol 8:471–473
1302. Hergt R, Andra W, d'Ambly CG, Hilger I, Kaiser W, Richter U, Schmidt HG (1998) Physical limits of hyperthermia using magnetite fine particles. IEEE Trans Magn 34:3745–3754
1303. Hergt R, Dutz S, Müller R, Zeisberger M (2006) Magnetic particle hyperthermia: nanoparticle magnetism and materials development for cancer therapy. J Phys Condens Matter 18:S2919
1304. Herlach D, Galenko P, Holland-Moritz D (2007) Metastable solids from undercooled melts. Elsevier, Amsterdam
1305. Hernandez RM, Orive G, Murua A, Pedraz JL (2010) Microcapsules and microcarriers for in situ cell delivery. Adv Drug Deliv Rev 62:711–730
1306. Herrera L (2019) Casual heat conduction contravening the fading memory paradigm. Entropy 21:950
1307. Herrera L, Pavón D (2001) Hyperbolic theories of dissipation: why and when do we need them? arXiv:gr-qc/0111112
1308. Hertig BA, Belding HS (1963) Evaluation and control of heat hazards. In Hardy JD (ed) Temperature - its measurement and control in science and industry, vol 3, part 3. Reinhold, New York, pp 347–355
1309. Hertz PE, Huey RB, Stevenson RD (1993) Evaluating temperature regulation by field-active ectotherms: the fallacy of the inappropriate question. Am Natural 142:796–818
1310. Hess M, Hildebrandt MD, Müller F, Kruber S, Kroetz P, Schumacher U, Reimer R, Kammal M, Püschel K, Wöllmer W, Miller D (2013) Picosecond infrared laser (PIRL): an ideal phonomicrosurgical laser? Eur Arch Otorhinol 270:2927–2937
1311. Hey EN, Katz G (1970) The optimum thermal environment for naked babies. Arch Dis Child 45:328–334
1312. Heymans N, Podlubny I (2006) Physical interpretation of initial conditions for fractional differential equations with Riemann-Liouville fractional derivatives. Rheolog Acta 45:765–772
1313. Hildebrandt B, Wust P, Ahlers O, Dieing A, Sreenivasa G, Kerner T, Felix R, Riess H (2002) The cellular and molecular basis of hyperthermia. Crit Rev Oncol Hematol 43:33–56
1314. Hilfer R (2002) Experimental evidence for fractional time evolution in glass forming materials. Chem Phys 284:399–408
1315. Hilfer R (2008) Threefold introduction to fractional derivatives. In: Klages R, Radons G, Sokolov IM (eds) Anomalous transport. Foundations and applications. Wiley-VCH, Weinheim, pp 17–74
1316. Hilfer R (2019) Mathematical and physical interpretations of fractional derivatives and integrals. In: Kochubei A, Luchko Y (eds) Handbook of fractional calculus with applications, vol 1. de Gruyter, Berlin, pp 47–85
1317. Hilger I, Hergt R, Kaiser WA (2005) Towards breast cancer treatment by magnetic heating. J Magnet Magnet Mater 293:314–319
1318. Hilger I, Hergt R, Kaiser WA (2005) Use of magnetic nanoparticle heating in the treatment of breast cancer. IEEE Proc Nanobiotechnol 152:33–39
1319. Hill AV (1929) The heat production and recovery of crustacean nerve. Proc R Soc Lond B 105:133–176
1320. Hill KK (2021) Oxygen binding thermodynamics of human hemoglobin in the red blood cell. Ph.D. Thesis, University of Nebraska

1321. Hillig WB (1998) Measurement of interfacial free energy for ice/water system. J Crystal Growth 183:463–468
1322. Hilt M, Pesch W, Zimmermann W (2003) Two paradigms of spatiotemporal chaos: spiral convection and waves on biomembranes. NIC Ser 20:453–463
1323. Hincha DH (2002) Cryoprotectin: a plant lipid-transfer protein homologue that stabilizes membranes during freezing. Philos Trans R Soc Lond B 357:909–916
1324. Hindmarsh JP, Russel AB, Chen XD (2003) Experimental and numerical analysis of the temperature transition of a suspended freezing water droplet. Int J Heat Mass Transf 46:1199–1213
1325. Hines-Peralta A, Hollander Y, Solazzo S, Horkan C (2004) Hybrid radiofrequency and cryoablation device: preliminary results in an animal model. J Vasc Interv Radiol 15:1111–1120
1326. Hirashima Y, Takaba M, Endo S, Hayashi N, Yamashita K, Takaku A (1998) Intracerebral temperature in patients with hydrocephalus of varying aetiology. J Neurol Neurosurg Psychiatr 64:792–794
1327. Hirata A, Watanabe H, Shiozawa T (2002) SAR and temperature increase in the human eye induced by obliquely incident plane waves. IEEE Trans Electromagn Compat 44:592–594
1328. Hirata A, Watanabe S, Fujiwara O, Kojima M, Sasaki K, Shiozawa T (2007) Temperature elevation in the eye of anatomically based human head models for plane-wave exposures. Phys Med Biol 52:6389–6399
1329. Hirata A, Funahashi D, Kodera S (2019) Setting exposure guidelines and product safety standards for radio-frequency exposure at frequencies above 6 GHz: brief review. Ann Telecommun 74:17–24
1330. Hirsch LR, Stafford RJ, Bankson JA, Sershen SR, Rivera B, Price RE, Hazle JD, Halas NJ, West JL (2003) Nanoshell-mediated near-infrared thermal therapy of tumors under magnetic resonance guidance. Proc Nat Acad Sci 100:13549–13554
1331. Hiser C, McIntosh L (1990) Altemative oxidase of potato is an integral membrane protein synthesized de novo during aging of tuber slices. Plant Physiol 93:312–318
1332. Hlawitschka M, McGough RJ, Ferrara KW, Kruse DE (2011) Fastultrasound beam prediction for linear and regular two-dimensional arrays. IEEE Trans Ultrason Ferroelectr Freq Control 58:2001–2012
1333. Ho JR, Kuo CP, Jiaung WS (2013) Study of heat transfer in multilayered structure within the framework of dual-phase-lag heat conduction model using lattice Boltzmann method. Int J Heat Mass Transf 46:55–69
1334. Hochmuth R, Deuflhard P (2004) Multiscale analysis of thermoregulation in the human microvascular system. Math Methods Appl Sci 27:971–989
1335. Hochmuth R, Worthy PR, Evans EA (1979) Red cell extensional recovery and the determination of membrane viscosity. Biophys J 26:101–114
1336. Hodgkin AL, Huxley AF (1952) Movement of sodium and potassium ions during nervous activity. ColdSpringHarb Symp Quant Biol 17:43–52
1337. Hodson DA, Eason G, Barbenel JC (1986) Modeling transient heat transfer through the skin and superficial tissues–1: surface insulation. ASME J Biomech Eng 108:183–188
1338. Hodson DA, Barbenel JC, Eason G (1989) Modeling transient heat-transfer through the skin and a contact material. Phys Med Biol 34:1493–1507
1339. Hoffmann NE, Bishof JC (2002) The cryobiology of cryosurgical injury. Urology 60(suppl 2A):40–49
1340. Hoffman A, Sgró C (2011) Climate change and evolutionary adaptation. Nature 470:479–485
1341. Hollerbach R (1996) On the theory of the geodynamo. Phys Earth Planet Interior 98:163–185
1342. Holmer I (2006) Protective clothing in hot environments. Industr Health 44:401–413
1343. Holopainen R (2012) A human thermal model for improved thermal comfort. Ph.D. Thesis, Aalto University

1344. Holt RD (2009) Bringing the hutchinsonian niche into the 21st century: ecological and evolutionary perspectives. Proc Nat Acad Sci USA 106:19659–19665
1345. Holzapfel GA (1996) On large strain viscoelasticity: continuum formulation and finite element applications to elastomeric structures. Int J Numer Meth Eng 39:3903–3926
1346. Hong K, Georgiades C (2010) Radiofrequency ablation: Mechanism of action and devices. J Vascul Intervent Radiol 21:S179–S186
1347. Hong J, Rubinsky B (1994) Patterns of ice formation in normal and malignant breast tissues. Cryobiology 31:109–120
1348. Honnor F, Thomas MA (1969) Packaging and cooling problems associated with microelectronic equipment. Microelectron Reliab 8:331–337
1349. Hoogerbrugge P, Koelman J (1992) Simulating microscopic hydrodynamic phenomena with dissipative particle dynamics. Europhys Lett 19:155
1350. Hooshmand P, Moradi A, Khezry B (2015) Bioheat transfer analysis of biological tissues induced by laser irradiation. Int J Thermal Sci 90:214–223
1351. Horgan CO, Quintanilla R (2005) Spatial behaviour of solutions of the dual-phase-lag heat equation. Math Methods Appl Sci 28:43–57
1352. Horvath TL, Warden CH, Hajos M, Lombardi A, Goglia F, Diano S (1999) Brain uncoupling protein 2: uncoupled neuronal mitochondria predict thermal synapses in homeostatic centers. J Neurosci 19:10417–10427
1353. Hoshino T, Tronsmo AM, Matsumoto N, Araki T, Georges F, Goda T, Ohgiya S, Ishizaki K (1998) Freezing resistance among isolates of a psychrophillic fungus, Typhula Ishikariensis, from Norway. Proc NIPR Symp Polar Biol 11:112–118
1354. Hossain MM, Gu M (2016) Radiative cooling: principles, progress, and potentials. Adv Sci 3:1500360
1355. Hossain SS, Hughes TJR, Decuzzi P (2014) Vascular deposition patterns for nanoparticles in an inflamed patient-specific arterial tree. Biomech Model Mechanobiol 13:585–597
1356. Hou HJM (2011) Enthalpy, entropy, and volume changes of electron transfer reactions in photosynthetic proteins. In: Mizutani T (ed) Application of thermodynamics to biological and materials science. InTech, Rijeka, pp 93–110
1357. Hou SM, Hsueh YS, Lin J, Wu HC, Lin FH (2009) The in vivo performance of biomagnetic hydroxyapatite nanoparticles in cancer hyperthermia therapy. Biomaterials 30:3956–3960
1358. Hou Y, Sun Z, Raw W, Liu J (2018) Nanoparticle-mediated cryosurgery for tumor therapy. Nanomed Nanotechnol Biol Med 14:493–506
1359. Houdas Y, Ring E (2013) Human body temperature: its measurement and regulation. Springer, Berlin
1360. Houston AI (2010) Evolutionary models of metabolism, behaviour and personality. Philos Trans R Soc 365:3969–3975
1361. Howard R, Scheiner A, Cunningham J, Gatenby R (2019) Cytoplasmic convection currents and intracellular temperature gradients. PLoS Comput Biol 15:e1007372
1362. Howarth C, Gleeson P, Attwell D (2012) Updated energy budgets for neural computation in the neocortex and cerebellum. J Cereb Blood Flow Metab 32:1222–1232
1363. Howell JR, Siegel R, Mengüc MP (2010) Thermal radiation heat transfer. CRC Press, New York
1364. Howell JR, Siegel R, Menguc MP (2011) Thermal Radiation Heat Transfer, 5th edn. CRC Press, London
1365. Howells EB (2015) Measuring temperature. Anasth Intes Care Med 16:358–362
1366. Hristov J (2013) A note on the integral approach to non-linear heat conduction with Jeffrey's fading memory. Thermal Sci 17:733–737
1367. Hristov J (2016) Transient heat diffusion with a non-singular fading memory: from the Cattaneo constitutive equation with Jeffrey's kernel to the Caputo-Fabrizio time-fractional derivative. Thermal Sci 20:757–762
1368. Hristov J (2017) Steady-state heat conduction in a medium with spatial non-singular fading memory: derivation of Caputo-Fabrizio space-fractional derivative from Cattaneo concept with Jeffrey's kernel and analytical solutions. Thermal Sci 21:827–839

1369. Hristov J (2019) Bio-heat models revisited: concepts, derivations, nondimensalization and fractionalization approaches. Front Phys 7:189
1370. Hsu R, Secomb TW (1989) Motion of non-axisymmetric red blood cells in cylindrical capillaries. J Biomech Eng 111:147–151
1371. Hu Y, Zhang XY, Li XF (2022) Thermoelastic analysis of biological tissue during hyperthermia treatment for moving laser heating using fractional dual-phase-lag bioheat conduction. Int J Thermal Sci 182:107806
1372. Hua ZZ, Xu HY, Zhou GY, Liu JF, Huang H, Ding WX (2001) Analysis of thermal stress and fracture during cryopreservation of blood vessel. Sci China Ser E 44:159–163
1373. Huang Z (2005) A review of progress in clinical photodynamic therapy. Technol Cancer Res Treat 4:283–293
1374. Huang HW (2013) Influence of blood vessel on the thermal lesion formation during radiofrequency ablation for liver tumors. Med Phys 40:073303
1375. Huang HW (2015) Confluent thermal lesion formation in liver with radio frequency ablation by using internally cooled multiple-electrode technique: computational results. J Appl Sci Eng 18:275–288
1376. Huang Z, Ruan X (2017) Nanoparticle embedded double-layer coating for daytime radiative cooling. Int J Heat Mass Transf 104:890–896
1377. Huang J, Holt RG, Cleveland RO, Roy RA (2004) Experimental validation of a tractable numerical model for focused ultrasound heating in flow-through tissue phantoms. J Acoust Soc Am 116:2451–2458
1378. Huang HW, Liauh C, Shih TC, Horng TL, Lin WL (2010) Significance of blood vessels in optimization of absorbed power and temperature distributions during hyperthermia. Int J Heat Mass Transf 53:5651–5662
1379. Huang HW, Liauh C, Chou CY, Shih TC, Lin WL (2012) A fast adaptive power scheme based on temperature distribution and convergence value for optimal hyperthermia treatment. Appl Therm Eng 37:103–111
1380. Hubel A, Norman J, Darr TB (1999) Cryobiology characteristics of genetically modified hematpoietic progenitor cells. Cryobiology 38:140–153
1381. Hübner F, Leithäuser C, Bazrafshan B, Siedow N, Vogl TJ (2017) Validation of a mathematical model for laser-induced thermotherapy in liver tissue. Lasers Med Sci 32:1399–1409
1382. Huey RB, Kingsolver JG (1989) Evolution of thermal sensitivity of ectotherm performance. Trends Ecol Evol 4:131–135
1383. Huey RB, Kingsolver JG (2011) Variation in universal temperature dependence of biological rates. Proc Nat Acad Sci USA 108:10377–10378
1384. Huey RB, Slatkin M (1976) Costs and benefits of lizard thermoregulation. Quart Rev Biol 51:363–384
1385. Huges MA, Dunn MA (1996) The molecular biology of plant acclimation to low temperatures. J Exp Bot 47:291–305
1386. Huizenga C, Hui Z, Arens E (2001) A model of human physiology and comfort for assessing complex thermal environments. Build Environ 36:691–699
1387. Hulbert AJ, Else PL (2004) Basal metabolic rate: history, composition, regulation, and usefulness. Physiol Biochem Zool 77:869–876
1388. Hurter W, Reinbold F, Lorenz WJ (1991) A dipole antenna for interstitial microwave hyperthermia. IEEE Trans Microw Theory Tech 39:1048–1054
1389. Huttunen JMJ, Huttunen T, Malinen M, Kaipio JP (2006) Determination of heterogeneous thermal parameters using ultrasound induced heating and MR thermal mapping. Phys Med Biol 51:1011–1032
1390. Hyder F, Rothman DL, Bennett MR (2013) Cortical energy demands of signaling and nonsignaling components in brain are conserved across mammalian species and activity levels. Proc Natl Acad Sci USA 110:3549–3554
1391. Hynynen K (2010) MRI-guided focused ultrasound treatments. Ultrasonics 50:221–229

1392. Hynynen K, Martin CJ, Watmough DJ, Mallard JR (1983) Errors in temperature measurement by thermocouple probes during ultrasound induced hyperthermia. Brit J Radiol 56:969–670
1393. Hynynen K, Colucci V, Chung A, Jolesz F (1996) Noninvasive arterial occlusion using MRI-guided focused ultrasound. Ultrasound Med Biol 22:1071–1077
1394. Iasiello M, Vafai K, Andreozzi A, Bianco N (2016) Low-density lipoprotein transport through an arterial wall under hyperthermia and hypertension conditions - an analytical solution. J Biomech 49:193–204
1395. Idir M, Youcef F, Aziz O, Emmanuel K, Ahmed L (2018) Improved thermal ablation of tumors using a real-time local data measurement system. Biomed J Sci Tech Res 11:8754–8764
1396. Ijima E, Li K, Hikage T, Nagai A, Murakami Y, Arima T, Ishitake T, Masuda H (2023) Intensity-dependent temperature rise induced by local exposure to 26.5 GHz quasi-millimeter-wave in rat. In Vivo 37:2092–2099
1397. Ilyin V, Procaccia I, Zagorodny A (2010) Stochastic processes crossing from ballistic to fractional diffusion with memory: exact results. Phys Rev E 81:030105(R)
1398. Ilyin V, Procaccia I, Zagorodny A (2013) Fokker-Planck equation with memory the crossover from ballistic to diffusive processes in many-particle systems and incompressible media. Phys Rev E 16:13004
1399. Inderbitzen B, Yon S, Lasheras J, Dobak J, Perl J, Steinberg GK (2002) Safety and performance of a novel intravascular catheter for inducing and reversal of hypothermia in a porcine model. Neurosurgery 50:364–370
1400. Indik RA, Indik JH (1994) A new computer method to quickly and accurately compute steady-state temperatures from ferromagnetic seed heating. Med Phys 21:1135–1144
1401. Indik JH, Indik RA, Cetas TC (1994) Fast and efficient computer modeling of ferromagnetic seed arrays of arbitrary orientation for hyperthermia treatment planning. Int J Radiat Oncol Biol Phys 30:653–662
1402. Intensity Modulated Radiation Therapy Collaborative Working Group (2001) Intensity-modulated radiotherapy: current status and issues of interest. Int J Radiat Oncol Biol Phys 51:880–914
1403. Iordana A, Alexandru S (2017) Advanced thermo-mechanical analysis in the magnetic hyperthermia. J Appl Phys 122:1–13
1404. Irdani T, Carletti B, Ambrogioni L, Roversi P (2006) Rapid-cooling and storage of plant nematodes at $-140°$ C. Cryobiology 52:319–322
1405. Irimia D, Karlsson OM (2002) Kinetics and mechanism of intercellular ice propagation in a micropatterned tissue construct. Biophys J 82:1858–1868
1406. Irimia D, Karlsson OM (2005) Kinetics of intracellular ice formation in one-dimensional arrays of interacting biological cells. Biophys J 88:647–660
1407. Isbert C, Ritz JP, Roggan A, Schuppan D, Ruhl M, Buhr HJ, Germer CT (2004) Enhancement of the immune response to residual intrahepatic tumor tissue by laser-induced thermotherapy (LITT) compared to hepatic resection. Lasers Surg Med 35:284–292
1408. Ishihara M, Morimoto Y, Sato S, Kikuchi M, Arai T, Obara M (2001) Temperature measurement for energy-efficient ablation by thermal radiation with a microsecond time constant from the corneal surface during ArF excimer laser ablation. Front Med Biol Eng 11:167–175
1409. Ishikawa T, Pedley T (2007) Diffusion of swimming model micro-organisms in a semi-dilute suspensions. J Fluid Mech 588:437–462
1410. Isichenko MV (1992) Percolation, statistical topography an transport in random media. Rev Mod Phys 64:961–1043
1411. Iszály Z, Gresits I, Márián IG, Thuróczy G, Sági O, Márkus BG, Simon F, Náandori I (2022) Polarised superlocalization in magnetic nanoparticle hyperthermia. arXiv:2112.08925v2 [cond-mat.soft]
1412. Ito A, Shinkai M, Honda H, Kobayashi T (2005) Medical applications of functionalized magnetic nanoparticles. J Biosci Bioeng 100:1–11

1413. Ito K (2004) Thermogenic plants and plant hormones. Regul Plant Growth Develop 39:167–173
1414. Ito K, Seymour RS (2005) Expression of uncoupling protein and alternative oxidase depends on lipid or carbohydrate substrates in thermogenic plants. Biol Lett 1:427–430
1415. Ito K, Ito T, Onda Y, Uemura M (2004) Temperature–triggered periodical thermogenic oscillations in skunk cabbage (symplocarpus foetidus). Plant Cell Physiol 45:257–264
1416. Ito-Inaba Y, Sato M, Masuko H, Hida Y, Toyooka K, Watanabe M, Inaba T (2009) Developmental changes and organelle biogenesis in the reproductive organs of thermogenic skunk cabbage (Symplocarpus renifolius). J Exp Bot 60:3909–3922
1417. Ivancic A, Lebot V, Roupsard O, Garcia JQ, Okpul T (2011) Thermogenic flowering of taro (colocasia esculenta, araceae). Canad J Bot 82:1557–1565
1418. Ivanitskii GR (2006) Modern matrix thermovision in biomedicine. Phys Usp 49:1263
1419. Iyiola OS, Nwaeze ER (2016) Some new results on the new conformable fractional calculus with application using D'Alambert approach. Prog Fract Differ Appl 2:1–7
1420. Iyoho AE, Jang TS, Nair SS (2004) Human thermal model with extremities for asymmetric environments. In: Proceedings of the 2004 American control conference, Boston, FrA14.1
1421. Iyoho AE, Ng LJ, MacFadden L (2017) Modeling of gender differences in thermoregulation. Milit Med 182:295–303
1422. Izhar LI, Petrou M (2012) Thermal imaging in medicine. Adv Imag Elect Phys 171:41–114
1423. Izzo F, Granata V, Grassi R, Fusco R, Palaia R, Delrio P, Carrafiello G, Azoulay D, Petrillo A, Curley SA (2019) Radiofrequency ablation and microwave ablation in liver tumors: an update. Oncologist 24:e990–e1005
1424. Jacobus WE (1985) Respiratory control and the integration of heart high-energy phosphate metabolism by mitochondrial creatine kinase. Annu Rev Physiol 47:707–725
1425. Jacques SL (1992) Laser-tissue interactions: photochemical, photothermal, and photomechanical. Surg Clin N Am 72:531–558
1426. Jacques SL (2006) Ratio of entropy to enthalpy in thermal transitions in biological tissues. J Biomed Opt 11:041108
1427. Jacques SL (2013) Optical properties of biological tissues: a review. Phys Med Biol 58:R37–R61
1428. Jacques SL, Nelson JS, Wright WH, Milner TE (1993) Pulsed photothermal radiometry of port-wine stain lesions. Appl Opt 32:2439–2446
1429. Jafri MS, Dudycha SJ, O'Rourke, B (2001) Cardiac energy metabolism: models of cellular respiration. Annu Rev Biomed Eng 3:57–81
1430. Jaglo-Ottosen KR, Gilmour SR, Zarka DG, Schabenberger O, Thomashow MF (1998) Arabidopis CBF1 overexpression induces cor genes and enhances freezing tolerance. Science 280:104–106
1431. Jain RE (1983) Bioheat transfer: Mathematical models of thermal systems. In: Storm FK (ed) Hyperthermia in Cancer Therapy. G. K. Hall Medical Publishers, Boston, pp 9–46
1432. Jain J, Paulson R (2006) Oocyte cryopreservation. Fertil Steril 86:1037–1046
1433. Jain RK, Ward-Hartley KA (1984) Tumor blood flow-characterization, modifications, and role in hyperthermia. IEEE Trans Sonics Ultrason 31:504–525
1434. Jain MK, Wolf PD (1999) Temperature-controlled and constant power radio-frequency ablation: what affects lesion growth? IEEE Trans Biomed Eng 46:1405–1412
1435. Jain MK, Wolf PD (2000) A three-dimensional finite element model of radiofrequency ablation with blood flow and its experimental validation. Ann Biomed Eng 28:1075–1084
1436. Jain MK, Wolf PD, Henriquez C (1995) Chilled-tip electrode radio frequency ablation of the endocardium: a finite element study. In: Proceedings of 17th international conference of the engineering in medicine and biology society
1437. Jamieson NV, Sundberg R, Lindell S, Southard JH, Belzer FO (1988) A comparison of cold storage solutions for hepatic preservation using the isolated perfused rabbit liver. Cryobiology 25:300–310
1438. Jamieson SW, Kapelanski DP, Sakakibara N, Manecke GR, Thistlethwaite PA, Kerr KM, Channick RN, Fedullo PF, Auger WR (2003) Pulmonary endarterectomy: Experience and lessons learned in 1,500 cases. Ann Thoracic Surg 76:1457–1462

1439. Jang TH, Park SC, Yang JH, Kim JY, Seok JH, Park US, Choi CW, Lee SR, Han J (2017) Cryopreservation and its clinical applications. Integr Med Res 6:12–18
1440. Janik M, Kleinhans FW, Hageborn M (2000) Overcoming a permeability barrier by microinjecting cryoprotectants into zebrafish embryos (Brachydanio rerio). Cryobiology 41:25–34
1441. Janke W (2007) Introduction to simulation techniques. Lect Notes Phys 716:207–260
1442. Janssen I, Heymsfield SB, Wang ZM, Ross R (1985) Skeletal muscle mass and distribution in 468 men and women aged 18–88 yr. J Appl Physiol 89:81–88
1443. Jasinski M (2018) Modelling of thermal damage in soft tissue subjected to laser irradiation. J Appl Math Comput Mech 17:29–41
1444. Jasinsky M, Majhrzak E, Turchan L (2016) Numerical analysis of the interactions between laser and soft tissues using generalized dual-phase lag equation. Appl Math Model 40:750–762
1445. Jastroch M, Wuertz S, Kloas W, Klingenspor M (2005) Uncoupling protein 1 in fish uncovers an ancient evolutionary history of mammalian nonshivering thermogenesis. Physiol Genom 22:150–156
1446. Jaunich M, Raje S, Kim K, Mitra K, Guo Z (2008) Bio-heat transfer analysis during short pulse laser irradiation of tissues. Int J Heat Mass Transf 51:5511–5521
1447. Jayalakshmy MS, Philip J (2010) Thermophysical properties of plant leaves and their influence on the environment temperature. Int J Thermophys 31:2295–2304
1448. Jeffree CE, Reed ND, Smith JAC, Dale JE (1987) Water droplets and ice deposits in leaf intercellular spaces: redistribution of water during cryofixation for scanning electron microscopy. Planta 172:20–37
1449. Jeffrey DJ (1973) Conduction through a random suspension of spheres. Proc R Soc Lond A 335:355–367
1450. Jeffreys H (1929) The earth, its origin. history and physical constitution. CUP, Cambridge
1451. Jenne JW, Preusser T, Gunther M (2012) High-intensity focused ultrasound: principles, therapy guidance, simulations and applications. Z Med Phys 22:311–322
1452. Jensen MO, Mouritsen OG (2006) Single-channel water permeability of Escherichia coli aquaporins AqpZ and GlpF. Biophys J 90:2270–2284
1453. Jensen MO, Rüthlisberger U, Rovira C (2005) Hydroxide and proton migration in aquaporins. Biophys J 89:1744–1759
1454. Jeon S, Hurley KR, Bischof JC, Haynes CL, Hogan CJ (2016) Quantifying intra- and extracellular aggregation of iron oxide nanoparticles and its influence on specific absorption rate. Nanoscale 8:16053–16064
1455. Jeon S, Oberreit DR, Van Schooneveld G, Gao Z, Bischof JC, Haynes CL, Hogan CJ (2016) Ion-mobility-based quantification of surface-coating-dependent binding of serum albumin to superparamagnetic iron oxide nanoparticles. ACS Appl Mater Interfaces 8:24482–24490
1456. Jeong SW, Liu H, Chen WR (2003) Temperature control in deep tissue ablation. Proc SPIE 5068:216–226
1457. Jeremias E, Bedaiwy MA, Nelson D, Biscotti CV, Falcone T (2003) Assessment of tissue injury in cryopreserved ovarian tissue. Fertil Steril 79:651–653
1458. Jessen C (2001) Selective brain cooling in mammals and birds. Jap J Physiol 51:291–301
1459. Jessen C (2001) Temperature regulation in humans and other mammals. Springer, Berlin
1460. Jessen C, Ludwig O (1971) Spinal cord and hypothalamus as core sensors of temperature in the conscious dog. II. Addition of signals. Pflüg Arch 324:205–216
1461. Jessen C, Mayer ET (1971) Spinal cord and hypothalamus as core sensors of temperature in the conscious dog. I. Equivalence of responses. Pflüg Arch 324:189–204
1462. Jessen C, Simon E (1971) Spinal cord and hypothalamus as core sensors of temperature in the conscious dog. III. Identity of functions. Pflüg Arch 324:217–226
1463. Jessen C, McLean JA, Calvert DT, Findlay JD (1972) Balanced and unbalanced temperature signals generated in the spinal cord of the ox. Am J Physiol 222:1343–1347
1464. Jessen C, Felde D, Volk P, Kuhnen G (1990) Effects of spinal cord temperature on the generation and transmission of temperature signals in the goat. Pflüg Arch 416:428–433

1465. Jeun M, Bae S, Tomitaka A, Takemura Y, Park KH, Paek SH, Chung KW (2009) Effects of particle dipole interaction on the ac magnetically induced heating characteristics of ferrite nanoparticles for hyperthermia. Appl Phys Lett 95:082501
1466. Ji L (2006) Time parallelization studies in bio-molecular dynamics simulations. MS Thesis, Florida St. University
1467. Ji CC, Dai W, Sun ZZ (2018) Numerical method for solving the time-fractional dual-phase-lagging heat conduction equations the tempearture-jump boundary conditions. J Sci Comput 75:1307–1336
1468. Jia N, Yu L, Yang KX, Wang RM, Luo XN, Xu QZ (2016) A novel exercise thermophysiology comfort prediction model with fuzzy logic. Mobile Inform Syst 2016:8586493
1469. Jiang F (2006) Solution and analysis of hyperbolic heat propagation in hollow spherical objects. Heat Mass Transf 42:1083–1091
1470. Jiang X, Chen S (2015) Analytical and numerical solutions of time fractional anomalous thermal diffusion equation in composite medium. Z Angew Math Mech 95:156–164
1471. Jiang X, Qi H (2012) Thermal wave model of bioheat transfer with modified Riemann-Liouville fractional derivative. J Phys A Math Theor 45:485101
1472. Jiang L, Tsai HL (2007) Modeling of ultra short laser pulse-train processing of metal thin film. Int J Heat Mass Transf 50:3461–3470
1473. Jiang XY, Xu MY (2010) The time fractional heat conduction equation in general orthogonal curvilinear coordinate and cylindrical coordinate systems. Phys A Stat Math Appl 389:3368–3374
1474. Jiang F, Liu D, Zhou J (2002) Non-Fourier heat conduction phenomena in porous material heated by microsecond laser pulse. Microsc Therm Eng 6:331–346
1475. Jiang SC, Ma N, Li HJ, Zhang XX (2002) Effects of thermal properties and geometrical dimensions on skin burn injuries. Burns 28:713–717
1476. Jiao J, Guo Z (2009) Thermal interaction of short-pulsed laser focused beams with skin tissues. Phys Med Biol 54:4225–4241
1477. Jiao A, Riegler R, Ma H, Petersen GP (2005) Thin film evaporation effect on heat transport capability in a grooved heat pipe. J Microfluidics Nanofluidics 1:227–233
1478. Jiao A, Han X, Critser JK, Ma H (2006) Numerical investigations of transient heat transfer characteristics and vitrification tendencies in ultra-fast cell cooling processes. Cryobiology 52:386–392
1479. Jiao AJ, Ma HB, Critser JK (2009) Experimental investigation of cryogenic oscillating heat pipes. Int J Heat Mass Transf 52:3504–3509
1480. Jin B, Mazur P (2015) High survival of mouse oocytes/embryos after vitrification without permeating cryoprotectants followed by ultra-rapid warming with an IR laser pulse. Sci Rep 5:9271
1481. Jin B, Kleinhans F, Mazur P (2014) Survivals of mouse oocytes approach 100% after vitrification in 3-fold diluted media and ultra-rapid warming by an IR laser pulse. Cryobiology 68:419–430
1482. Jing L, Yao L, Zhao M, Peng LP, Liu M (2018) Organ preservation: from the past to the future. Acta Pharmacol Sin 39:845–857
1483. Johannsen M, Gneveckow U, Eckelt L, Feussner A, Waldöfner N, Scholz, Deger S, Wust P, Loening SA, Jordan A (2005) Clinical hyperthermia of prostate cancer using magnetic nanoparticles: presentation of a new interstitial technique. Int J Hyperth 21:637–647
1484. Johnsen HK, Folkow LP (1988) Vascular control of brain cooling in reindeer. Am J Physiol 254:R730–R739
1485. Johnson CC, Guy AW (1972) Nonionizing electromagnetic wave effects in biological materials and systems. Proc IEEE 60:692–718
1486. Johnson PC, Saidel GM (2002) Thermal model for fast simulation during magnetic resonance imaging guidance of radio frequency tumor ablation. Ann Biomed Eng 30:1152–1161
1487. Johnson JM, Minson CT, Kellogg DL (2014) Cutaneous vasodilator and vasoconstrictor mechanisms in temperature regulation. Compr Physiol 4:33–89

1488. Johnston KA, Bennett AF (eds) (1996) Animals and temperature: phenotypic and evolutionary adaptation. Cambridge University Press, Cambridge
1489. Johnson WA, Mehl RF (1939) Reaction kinetics in processes of nucleation and growth, Trans. AIME, 135, p. 416
1490. Jones CH (1987) Medical thermography. IEE Proc 134 A:225–236
1491. Jones EL, Oleson JR, Prosnitz LR, Samulski TV, Vujaskovic Z, Yu D, Sanders LL, Dewhirst MW (2005) Randomized trial of hyperthermia and radiation for superficial tumors. J Clin Oncol 23:3079–3085
1492. Jordan A, Wust P, Fähling HP, John W, Hinz A, Felix R (1993) Inductive heating of ferrimagnetic particles and magneticfluids: physical evaluation of their potential for hyperthermia. Int J Hyperth 9:51–68
1493. Jordan A, Scholz R, Maier-Hauff K, Johannsen M, Wust P, Nodobny J, Schirra H, Schmidt H, Deger S, Loening S, Lanksch W, Felix R (2001) Presentation of a new magnetic field therapy system for the treatment of human solid tumors with magnetic fluid hyperthermia. J Magn Magn Mat 225:118–126
1494. Jordan A, Wust P, Fähling H, John W, Hinz A, Felix R (2009) Inductive heating of ferrimagnetic particles and magnetic fluids: physical evaluation of their potential for hyperthermia. Int J Hyperth 25:499–511
1495. Jorgensen WL, Chandrasekhar J, Madura JD, Impey RW, Klein ML (1983) Comparison of simple potential functions for simulating liquid water. J Chem Phys 79:926–935
1496. Jorov A, Zhorov BS, Yang DSC (2004) Theoretical study of interaction of winter flounder antifreeze protein with ice. Protein Sci 13:1524–1537
1497. Joseph DD, Presiosi L (1989) Heat waves. Rev Mod Phys 61:41–73
1498. Joseph DD, Presiosi L (1990) Addendum to the paper Heat waves. Rev Mod Phys 62:375–391
1499. Joseph A, Guevara-Torres A, Schallek J (2019) Imaging single-cell blood flow in the smallest to largest vessels in the living retina. eLife 8:e45077
1500. Jou D, Cimmelli VA (2016) Constitutive equations for heat conduction in nanosystems and nonequilibrium processes: an overview. Commun Appl Ind Math 7:196–222
1501. Jou D, Camacho J, Grmela M (1991) On the nonequilibrium thermodynamics of non-Fickian diffusion. Macromolecules 24:3597–3602
1502. Jou D, Casa-Vázquez J, Lebon G (1988) Extended irreversible thermodynamics. Rep Prog Phys 51:1105–1179
1503. Jou D, Casa-Vázquez J, Lebon G (1999) Extended irreversible thermodynamics revisited (1988–1998). Rep Prog Phys 62:1035–1142
1504. Jou D, Casa-Vázquez J, Lebon G (2008) Extended irreversible thermodynamics of heat transport. A brief introduction. Proc Eston Acad Sci 57:118–126
1505. Jou D, Cimmelli VA, Sellitto A (2011) Dynamical temperature and renormalized flux variable in extended thermodynamics of rigid heat conductors. J Non-Equil Thermodyn 36:373–392
1506. Jowett N, Wöllmer W, Reimer R, Zustin J, Schumacher U, Wiseman PW, Mlynarek AM, Böttcher A, Dalchow CV, Lörincz BB, Knecht R, Miller RJ (2014) Bone ablation without thermal or acoustic mechanical injury via a novel picosecond infrared laser (PIRL). Otolaryngol Head Neck Surg 150:385–393
1507. Joyner MJ, Dietz NM (1997) Nitric oxide and vasodilation in human limbs. J Appl Physiol 83:1785–1796
1508. Jumarie G (2006) Modified Riemann-Liouville derivative and fractional Taylor series of nondifferentiable functions further results. Comput Math Appl 51:1376–1376
1509. Jumarie G (2007) Lagrangian mechanics of farctional order, Hamilton-Jacobi fractional PDE and Taylor's series of nondifferentialable functions. Chaos Soltons Fract 32:969–987
1510. Jumarie G (2007) The Minlowski's space-time is consistent with differential geometry of fractional order. Phys Lett 363:5–11
1511. Jumarie G (2013) The Leibniz rule for fractional derivatives holds with non-differentiable functions. Math Stat 1:50–52

1512. Jumarie G (2013) On the derivative chain-rules in fractional calculus via fractional difference and their application to systems modelling. Open Phys 11:617–633
1513. Jung O, Lindner C, Pantermehl S, Barbeck M (2021) Heat development during medical drilling: influencing factors and examination methods - overview and first results. In Vivo 35:3011–3017
1514. Jungraithmayr W, Szarzynski M, Neeff H, Haberstroh J, Kirste G, Schmitt-Graeff A, Farthmann EH, Eggstein S (2004) Significance of total vascular exclusion for hepatic cryotherapy: an experimental study. J Surg Res 116:32–41
1515. Junkun M, Kelly TJ, Zaim A, Young K, Keck RW, Selman S, Jankun J (1999) Computer model for cryosurgery of the prostate. Comput Aid Surg 4:193–199
1516. Jussofie A, Kirsch M, de Groot H (1998) ca^{2+}-dependent cytotoxicity of H_2O_2 in L929 cells: the role of H_2O_2-induced Na^+ influx. Free Radic Biol Med 25:712–719
1517. Kaatze U (1990) On the existence of bound water in biological systems as probed by dielectric spectroscopy. Phys Med Biol 35:1663–1681
1518. Kabil J, Belguerras L, Trattnig S, Pasquier C, Felblinger J, Missoffe A (2016) A review of numerical simulation and analytical modeling for medical devices safety in MRI. IMIA Yearbook Medical Inform 10:152–158
1519. Kaganov MI, Lifshitz IM, Tanatarov MV (1957) Relaxation between electrons and crystalline lattices. Sov Phys JETP 4:173–178
1520. Kai L, Yan W, Hongwei T, Guangju C, Zhenhe T (2007) Computational simulations on the fish-type-II antifreeze protein-ice-solvent system. Front Biol China 2:180–183
1521. Kalambur VS, Han B, Hammer BE, Shield TW, Bischof JC (2005) In vitro characterization of movement, heating and visualization of magnetic nanoparticles for biomedical applications. Nanotechnology 16:1221–1233
1522. Kalkowsky B, Kampmann B (2006) Physiological strain of miners at hot working places in German coal mines. Ind Health 44:465–473
1523. Kalmbach AS, Waters J (2012) Brain surface temperature under a craniotomy. J Neurophysiol 108:3138–3146
1524. Kalra A, Garde S, Hummer G (2003) Osmotic water transport through carbon nanotube membranes. Proc Natl Acad Sci USA 100:10175–10180
1525. Kaminski W (1990) Hyperbolic heat conduction equation for materials with a nonhomogeneous inner structure. J Heat Transf 112:555–560
1526. Kampinga HH (2006) Cell biological effects of hyperthermia alone or combined with radiation or drugs: a short introduction to newcomers in the field. Int J Hyperth 22:191–196
1527. Kampinga HH, Dikomey E (2001) Hyperthermic radiosensitization: mode of action and clinical relevance. Int J Radiat Biol 77:399–408
1528. Kanatdzikis MG (2009) Nanostructured thermoelectrics: the new paradigm? Chem Mater 22:648–659
1529. Kandel ER, Schwartz JH, Jessell TM (1991) Principles of neural science, 3rd edn. Appleton and Lange, Norwalk
1530. Kandra D (2004) Tissue interactions with lasers and liquid nitrogen — a novel cryopreseravtion method. M.S. Thesis, Bangalore University
1531. Kandra D, Devireddy R (2008) Numerical simulation of local temperature distorsions during ice nucleation of cells in suspension. Int J Heat Mass Transf 51:5655–5661
1532. Kang Z, Wang F (2019) An advanced three-dimensional thermoregulation model of the human body: development and validation. Int Commun Heat Mass Transf 107:34–43
1533. Kanno H, Speedy RJ, Angell CA (1975) Supercooling of water to $-96°$ C underr pressure. Science 189:880–881
1534. Kannojiya V, Das AK, Das PK (2021) Simulation of blood as fluid: a review from rheological aspects. IEEE Rev Biomed Eng 14:327–341
1535. Kanosue K, Hosono T, Yanase-Fujiwara M (1994) Hypothalamic network for thermoregulatory vasomotoractivity. Am J Physiol 267:R283–R288

1536. Kanosue K, Hosono T, Zhang YH, Chen XM (1998) Neuronal networks controlling thermoregulatory efectors. Prog Brain Res 115:49–62
1537. Kaoui B, Biros G, Misbah C (2009) Why do red blood cells have asymmetric shapes even in a symmetric flow? Phys Rev Lett 103:188101
1538. Kaoui B, Tahiri N, Biben T, Ez-Zahraouy H, Benyoussef A, Biros G, Misbah C (2011) Complexity of vesicle microcirculation. Phys Rev E 84:041906
1539. Kappiyoor R, Liangruksa M, Ganguly R, Ishwar K (2010) The effects of magnetic nanoparticle properties on magnetic fluid hyperthermia. J Appl Phys 108:094702
1540. Karaki W, Ghaddar N, Ghali K, Kuklane K, Holmér I, Vanggaard L (2013) Human thermal response with improved AVA modeling of the digits. Int J Thermal Sci 67:41–52
1541. Karakoyun R, Romano A, Nordström J, Ericzon BG, Nowak G (2019) Type of preservation solution, UW or HTK, has an impact on the incidence of biliary stricture following liver transplantation: a retrospective study. J Transplant 2019:8150736
1542. Karbowski J (2009) Thermodynamic constraints on neural dimensions, firing rates, brain temperature and size. J Comput Neurosci 27:415–436
1543. Karch R, Neumann F, Neumann M, Schreiner W (1999) A three-dimensional model for arterial tree representation, generated by constrained constructive optimization. Comput Biol Med 29:19–38
1544. Kargol M (2002) Mechanistic approach to membrane mass transport processes. Cell Mol Biol Lett 7:983–993
1545. Kargol M, Kargol A (2003) Mechanistic formalism for membrane transport generated by osmotic and mechanical pressure. Gen Physiol Biophys 22:51–68
1546. Kargol M, Kargol A, Przestalski M, Siedlecki J, Karpińska M, Rogowski M (2005) Human red blood cells' physiological water exchange with the plasma. Ann Acad Medicae Biol 50:237–240
1547. Karlsson JOM (2001) A theoretical model of intracellular devitrification. Cryobiology 42:154–169
1548. Karlsson JOM (2002) Cryopreservation: freezing and vitrification. Science 296:655–656
1549. Karlsson OM (2004) Theoretical analysis of unidirectional intercellular ice propagation in stratified cell clusters. Cryobiology 48:357–361
1550. Karlsson JO, Toner M (1996) Long-term storage of tissues by cryopreservation: critical issues. Biomaterials 17:243–256
1551. Karlsson JOM, Toner M (2000) Cryopreservation. In: Lanza RP, Langer R, Vacanti JP (eds) Principles of tissue engineering, 2nd edn. Academic Press, San Diego, pp 180–194
1552. Karma A, Rappel W (1998) Quantitative phase-field modeling of dendritic growth in two and three dimensions. Phys Rev E 57:4323–4349
1553. Karnatovskaia LV, Wartenberg KE, Freeman WD (2014) Therapeutic hypothermia for neuroprotection history, mechanisms, risks, and clinical applications. The Neurohospitalist 4:153–163
1554. Karnaukhov VN, Gakhova EN, Uteshov VK (eds) (1996) Conservation of genetic resources (in Russian). Puschino Sci Center, RAS
1555. Karu TI (1988) Molecular mechanism of the therapeutic effect of low-intensity laser radiation. Las Life Sci 2:53–74
1556. Karu T, Kalendo G, Letokhov V, Lobko V (1982) Biostimulation of hela cells by low-intensity visible light. Il Nuovo Cimento D 1:828–840
1557. Karu T, Kalendo G, Letokhov V, Lobko V (1983) Stimulation of e. coli growth by laser and incoherent red light. Il Nuovo Cimento D 2:1138–1144
1558. Karu T, Pyatibrat L, Kalendo G (1995) Irradiation with He Ne laser increases ATP level in cells cultivated in vitro. J Photochem Photobiol B Biol 27:219–223
1559. Kaspar C (2011) Thermoregulation of gene expression in encapsulated cells by magnetic field-directed, nanoparticle-mediated heat induction. Ph.D. Thesis, Ludwig-Maximilians-Universität, München
1560. Kasza KE, Rowat AC, Liu J, Angelini TE, Brangwynne CP, Koenderink GH, Weitz DA (2007) The cell as a material. Curr Opin Cell Biol 19:101–107

1561. Katić K, Li R, Zeiler W (2016) Thermophysiological models and their applications: a review. Build Environ 106:286–300
1562. Katika K, Pilon L (2006) Steady-state directional diffuse reflectance and fluorescence of human skin. Appl Opt 45:4174–4183
1563. Katkov II (ed) (2012) Current frontiers in cryobiology. InTech, Rijeka
1564. Katkov II (2014) Stopping biological clocks. The science and art of biopreservation. BioProc Int 12:42–52
1565. Katkov II, Bolyukh VF (2017) Krioblast ^{TM}: a promising cryogenic platform for kinetic vitrification by hyperfast cooling. In: Proceedings of the 4th international workshop heat transfer, IWHT2017, Las Vegas, TFEC-IWHT2017-17397
1566. Katkov II, Levine F (2004) Prediction of the glass transition temperature of water solutions: comparison of different models. Cryobiology 49:62–82
1567. Kato M, Shinzawa K, Yoshikawa S (1981) Cytochrome-oxidase is a possible photoreceptor in mitochondria. Photochem Photobiol 2:4–5
1568. Katschinski D (2004) On heat and cells and proteins. News Physiol Sci 19:11–15
1569. Katugampola UN (2014) New approach to a generalized fractional derivative. Math Anal Appl 6:1–15
1570. Katugampola UN (2014) A new fractional derivative with classical properties. arXiv: 1410.6535 [math.CA]
1571. Katzfuss M, Stroud JR, Wikle CK (2016) Understanding the ensemble Kalman filter. Am Stat 70:350–357
1572. Kaur P, Hurwitz MD, Krishnan S, Asea A (2011) Combined hyperthermia and radiotherapy for the treatment of cancer. Cancers 3:3799–3823
1573. Kay JE, Tsemekhman V, Larson B, Baker M, Swanson B (2003) Comment on evidence for surface-initiated homogeneous nucleation. Atmos Chem Phys 3:1439–1443
1574. Kay AG, Hoyland JA, Rooney P, Kearney JN, Pegg DE (2015) A liquidus tracking approach to the cryopreservation of human cartilage allografts. Cryobiology 71:77–84
1575. Kazimi MS, Erdman CF (1989) On the interface of two suddenly contacting materials. J Heat Transf 97:615–617
1576. Keangin P, Rattanadecho T (2013) Analysis of heat transport on local thermal non-equilibrium in porous liver during microwave ablation. Int J Heat Mass Transf 67:46–60
1577. Keangin P, Rattanadecho T, Wessapan P (2011) An analysis of heat transfer in liver tissue during microwave ablation using single and double slot antenna. Int Commun Heat Mass Transf 38:757–766
1578. Keangin P, Wessapan P, Rattanadecho T (2011) Analysis of heat transfer in deformed liver cancer modeling treated using a microwave coaxial antenna. Appl Therm Eng 31:3243–3254
1579. Keangin P, Vafai K, Rattanadecho P (2013) Electromagnetic field effects on biological materials. Int J Heat Mass Transf 65:389–399
1580. Kearney M, Porter W (2009) Mechanistic niche modelling: combining physiological and spatial data to predict species ranges. Ecol Lett 12:334–350
1581. Kearns DR (1971) Physical and chemical properties of singlet molecular oxygen. Chem Rev 71:395–427
1582. Keatinge WR, Cannon P (1960) Freezing-point of human skin. Lancet 1:11–14
1583. Kedem O, Katchalsky A (1958) Thermodynamic analysis of the permeability of biological membranes to non-electrolytes. Biochim Biophys Acta 27:229–246
1584. Keijer J, Li M, Speakman JR (2019) What is the best housing temperature to translate mouse experiments to humans? Mol Metabol 25:168–176
1585. Kelemen C, Chien S, Artmann GM (2001) Temperature transition of human hemoglobin at body temperature: effect of calcium. Biophys J 80:2622–2630
1586. Keller KH, Seiler L (1971) Ananalys is of peripheral heat transfer in man. J Appl Physiol 30:779–786
1587. Keller SR, Skalak R (1982) Motion of a tank-treading ellipsoidal particle in a shear flow. J Fluid Mech 120:27–47

1588. Kellogg DL (2006) In vivo mechanisms of cutaneous vasodilation and vasoconstriction in humans during thermoregulatory challenges. J Appl Physiol 100:1709–1718
1589. Kelly KM, Nanda VS, Nelson JS (2002) Treatment of port-wine stain birthmarks using the 1.5-msec pulsed dye laser at high fluences in conjunction with cryogen spray cooling. Dermatol Surg 28:309–313
1590. Kelso MB (1995) A study of the use of combined thermal and microwave modelling of body regions for microwave thermography. Ph.D. Thesis, University of Glasgow
1591. Kengne E, Lakhssassi A (2015) Bioheat transfer problem for one-dimensional spherical biological tissues. Math Biosci 269:1–9
1592. Kengne E, Liu WM (2006) Exact solutions of the derivative nonlinear Schrödinger equation for a nonlinear transmission line. Phys Rev E 73:1–8
1593. Kengne E, Lakhssassi A, Vaillancourt R (2012) Temperature distributions for regional hypothermia based on nonlinear bioheat equation of Pennes type: dermis and subcutaneous tissues. Appl Math 3:217–224
1594. Kengne E, Saydá M, Lakhssassi A (2013) Effect of convective term on temperature distribution in biological tissue. Eur Phys J Plus 128:1–10
1595. Kennedy JE, Wu F, ter Haar GR, Gleeson FV, Phillips RR, Middleton MR, Canston D (2004) High-intensity focused ultrasound for the treatment of liver tumors. Ultrasonics 42:931–935
1596. Keren K, Yam PT, Kinkhabwala A, Mogilner A, Theriot JA (2009) Intracellular fluid moving cells. Nat Cell Biol 11:1219–1224
1597. Kerrigan L (2006) Cryopreservation of bacteria. ATCC Connect 26:3–4
1598. Kerse C, Yavas S, Kalacioglu H, Asik MD, Akcaalan O, Idlay O (2014) arXiv:1410.6730v1 [physics.optics]
1599. Kessel L, Johnson L, Aridsson H, Larsen M (2010) The relationship between body and ambient temperature and corneal temperature. Investig Ophthalm Vis Sci 51:6593–6597
1600. Kevan PG, Nunes-Silva P, Sudarsan R (2018) Thermal regimes in hollow stems of herbaceous plants - concepts and models. Int J Biomet 62:2057–2062
1601. Khairy P, Chauvet P, Lehmann J, Lambert J, Macle L, Tanguay JF, Sirois MG, Santoianni D, Dubuc M (2003) Lower incidence of thrombus formation with cryoenergy versus radiofrequency catheter ablation. Circulation 107:2045–2050
1602. Khaled ARA, Vafai K (2003) The role of porous media in modeling flow and heat transfer in biological tissues. Int J Heat Mass Transf 46:4989–5003
1603. Khalid MZ (2016) Mechanism of laser/light beam interaction at cellular and tissue level and study of the influential factors for the application of low level laser therapy. arXiv:1606.04800v1 [physics.med-ph]
1604. Khalil R, Al Horani M, Yousef A, Sababheh M (2014) A new definition of fractional derivative. J Comput Appl Math 264:65–70
1605. Khan AA, Vincent JFV (1996) Mechanical damage induced by controlled freezing in apple and potato. J Texture Stud 27:143–157
1606. Khan MH, Sink RK, Manstein D, Eimerl D, Anderson RR (2005) Intradermally focused infrared laser pulses: thermal effects at defined tissue depths. Lasers Surg Med 36:270–280
1607. Khan MAM, Ahrné L, Oliveira JC, Oliveira FAR (2008) Prediction of water and soluble solids concentration during osmotic dehydration of mango. Food Bioproduct Process 86:7–13
1608. Khanafer K, Vafai K (2000) Synthesis of mathematical models representing bioheat transport. Adv Numer Heat Transf 3:1–28
1609. Khanafer K, Vafai K (2006) The role of porous media in biomedical engineering as related to magnetic resonance imaging and drug delivery. Heat Mass Transf 42:939–953
1610. Kharakoz DP (2001) On the possibility of the physiological role of the transition "liquid-solid" in biological membranes (in Russian). Adv Biol Chem 41:333–364
1611. Khayat RE, Ostoja-Starzewski M (2011) On objective rate of heat and stress fluxes. Connection with micro/nano-scale heat convection. Discr Cont Dyn Syst Ser B 15:991–998
1612. Khayat RE, de Bruyn J, Niknami M, Stranges DF, Khorasani RMH (2015) Non-Fourier effects in macro- and microscale non-isothermal flow of liquids and gases. Rev Int J Thermal Sci 997:163–177

1613. Kheifets L, Repacholi A, Saunders R (2003) Thermal stress and radiation protection principles. Int J Hyperth 19:215–224
1614. Khintchine A, Lévy P (1936) Sur leslois stables. Compt Rend 202:374–376
1615. Khlebtzov NG, Dykhman LA (2010) Optical properties and biomedical applications of plasmonic nanoparticles. J Quant Spectr Radiat Transf 111:1–35
1616. Khosla K, Wang Y, Hagedorn M, Qin Z, Bischof JC (2017) Gold nanorod induced warming of cryogenically stabilized embryos enhances viability. ACS Nano 11:7869–7878
1617. Khromova IV, Chichindaev AV (2018) Heat exchange processes modeling in "human body - thermal protection - environment" system. Adv Eng Res 158:181–185
1618. Khvesyuk VI, Skryabin AS (2017) Heat conductivity of nanostructures. High Temp 55:428–450
1619. Kilbride P, Meneghel J, Fonseca F, Morris J (2021) The transfer temperature from slow cooling to cryogenic storage is critical for optimal recovery of cryopreserved mammalian cells. PLoS ONE 16:e025957
1620. Kilbride P, Meneghel J, Chawda MM, Ross S, Crompton T (2022) Scaling up cryopreservation from cell suspensions to tissues: challenges and successes. In: Quain M (ed) Cryopreservation - applications and challenges. Intech Open, London
1621. Kim K (2013) Finite element modeling of electromagnetic radiation and induced heat transfer in the human body, PhD Thesis, The University of Texas at Austin
1622. Kim KH, Guo Z (2004) Ultrafast radiation heat transfer in laser tissue welding and soldering. Numer Heat Transf A 46:23–46
1623. Kim K, Guo Z (2007) Multi-time-scale heat transfer modeling of turbid tissues exposed to short-pulsed irradiations. Comput Meth Prog Med 86:112–123
1624. Kim A, Wilson BC (2010) Measurement of ex vivo and in vivo tissue optical properties: methods and theories. In: Welch A, van Gemert M (eds) Optical-thermal response of laser-irradiated tissue. Springer, Berlin, pp 267–319
1625. Kim JH, Hahn EW, Tokita N, Nisce LZ (1977) Local tumor hyperthermia in combination with radiation therapy. I. Malignant cutaneous lesions. Cancer 40:161–169
1626. Kim KS, Davis IS, Macpherson PA, Pedley TJ, Hill AE (2005) Osmosis in small pores: a molecular dynamics study of the mechanism of solvent transport. Proc R Soc A 461:273–296
1627. Kim C, O'Rourke AP, Mahvi DM, Webster JG (2007) Finite-element analysis of ex vivo and in vivo hepatic cryoablation. IEEE Trans Bio-Med Eng 54:1177–1185
1628. Kim C, O'Rourke AP, Will JL, Mahvi DM, Webster JG (2008) Finite-element analysis of hepatic cryoablation around a large blood vessel. IEEE Trans Bio-Med Eng 55:2087–2093
1629. Kim S, Ong PK, Yalcin O, Intagietta M, Johnon PC (2009) The cellfree layer in microvascular blood flow. Biorheol 46:181–189
1630. Kim S, Ong PK, Yalcin O, Intaglietta M, Johnson PC (2009) The cell-free layer in microvascular blood flow. Biorheology 46:181–189
1631. Kim HJ, Vignon-Clementel IE, Coogan JS, Figueroa CA, Jansen KE, Taylor CE (2010) Patient-specific modeling of blood flow and pressure in human coronary arteries. Ann Biomed Eng 38:3195–3209
1632. Kim JM, Jeong YJ, Chung HJ, Lee C, Oh CH (2010) Real-time interactive magnetic resonance (MR) temperature imaging in both aqueous and adipose tissues using cascaded deep neural networks for MR-guided focused ultrasound surgery (MRgFUS). arXiv:1908.10995 [eess.IV]
1633. Kim KH, Do W, Park S (2018) Improving resolution of MR images with an adversarial network incorporating images with different contrast. Med Phys 45:3120–3131
1634. Kim J, Choi W, Park EY, Kang Y, Lee KJ, Kim HH, Kim WJ, Kim C (2019) Real-time photoacoustic thermometry combined with clinical ultrasound imaging and high intensity focused ultrasound. IEEE Trans Biomed Eng 66:3330–3338
1635. Kimizuka N, Suzuki T (2007) Supercooling behavior in aqueous solution. J Phys Chem 111:2268–2273

1636. Kimizuka N, Viriyarattanasak C, Suzuki T (2008) Ice nucleation and supercooling behavior of polymer aqueous solution. Cryobiology 56:80–87
1637. Kingma BRM (2012) Human thermoregulation; a synergy between physiology and mathematical modeling. Doctoral Thesis, Maastricht University
1638. Kingma B, Frijns A, van Marken Lichtenbelt W (2012) The thermoneutral zone: implications for metabolic studies. Front Biosci 4:1975–1985
1639. Kinosita K, Tsong TY (1977) Formation and resealing of pores of controlled sizes in human erythrocite membrane. Nature 268:438–441
1640. Kirkpatrick S, Gelatt CD, Vecchi MP (1953) Optimization by simulated annealing. Science 220:671–680
1641. Kirson ED, Gurvich Z, Schneiderman R, Dekel E, Itzhaki A, Wasserman Y, Schatzberger R, Palti Y (2004) Disruption of cancer cell replication by alternating electric fields. Cancer Res 64:3288–3295
1642. Kirson ED, Dbaly V, Tovarys F, Vymazal J, Soustiel JF, Itzhaki A, Mordechovich D, Steinberg-Shapira S, Gurvich Z, Schneiderman R, Wasserman Y, Salzberg M, Ryffel B, Goldsher D, Dekel E, Palti Y (2007) Alternating electric fields arrest cell proliferation in animal tumor models and human brain tumors. Proc Nat Acad Sci USA 104:10152–10157
1643. Kiryakova V (1994) Generalized fractional calculus and applications. Longman, Harlow
1644. Kishore P, Kumar S (2021) Analytical investigation of non-Fourier bioheat transfer in the axisymmetric living tissue exposed to pulsed laser heating using finite integral transform technique. J Heat Transf 143:121201
1645. Kishore P, Kumar S (2023) Thermodynamic stability analysis of non-Fourier model-based bio-heat transfer in laser-irradiated cylindrical-shaped multilayered biological tissue. Heat Transf 52:4857–4879
1646. Kishore P, Kumar S, Patel VM (2022) Conjugate heat transfer analysis of laser-irradiated cylindrical-shaped biological tissue embedded with the optical inhomogeneity. Int Commun Heat Mass Transf 137:106302
1647. Kiyatkin EA (2007) Brain temperature fluctuations during physiological and pathological conditions. Eur J Appl Physiol 101:3–17
1648. Kiyatkin EA (2010) Brain temperature homeostasis: physiological fluctuations and pathological shifts. Front Biosci 15:73–92
1649. Kiyatkin EA, Brown PL, Wise RA (2002) Brain temperature fluctuation: a reflection of functional neural activation. Eur J Neurosci 16:164–168
1650. Klafter J, Sokolov IM (2011) First steps in random walks: from tools to applications. OUP, Oxford
1651. Klafter J, Zumofen G (1993) Scale-invariant motion in intermittent chaotic systems. Phys Rev E 47:851–863
1652. Kleiber M (1947) Body size and metabolic rate. Physiol Rev 27:511–541
1653. Kleinhans FW (1998) Membrane permeability modeling: Kedem- Katchalsky vs a two-parameter formalism. Cryobiology 37:271–289
1654. Klinger HG (1974) Heat transfer in perfused biological tissue I: general theory. Bull Math Biol 36:403–415
1655. Klinger HG (1978) Heat transfer in perfused biological tissue II: the macroscopic temperature distribution in tissue. Bull Math Biol 40:183–199
1656. Klinger HG (1980) The description of heat transfer in biological tissue. Ann NY Acad Sci 335:133–136
1657. Klostergaard J, Barta M, Tomasovic SP (1989) Hyperthermic modulation of tumor necro-sis factor-dependent monocyte/macrophage tumor cytotoxicity in vitro. J Biol Response Mod 8:262–277
1658. Knight MC (2002) Signal transduction leading to low-temperature tolerance in Arabidopsis thaliana. Philos Trans R Soc Lond B 357:871–875
1659. Knight CA, Wierzbicki A (2001) Adsorption of biomolecules to ice and their effects upon ice growth. 2. A discussion of the basic mechanism of "antifreeze" phenomena. Crystal Growth Des 1:439–436

1660. Knight CA, DeVries AL, Oolman LD (1984) Fish antifreeze protein and the freezing and recrystallization of ice. Nature 308:295–296
1661. Knochel JP, Dotin LN, Hamburger RC (1974) Heat stress, exercise, and muscle: effects on urate metabolism an drenal function. Am J Intern Med 81:321–328
1662. Knutson RM (1974) Heat production and temperature regulation in eastern skunk cabbage. Science 186:746–747
1663. Kobayashi Y, Ito Y, Ostapenko V, Sakai M, Matsushita N, Imai K, Shimizu K, Aruga A, Tanigawa K (2014) Whole-body heat treatment stimulates antigen-specific T cell responses in humans. Immunol Lett 162(1):256–261
1664. Koch C (1998) Biophysics of computation. Oxford University Press, Oxford
1665. Koh S, Tung C, Kottaiyan R, Zavislam J, Yoon G, Aquavella J (2012) Effect of airlfow exposure on the tear mensiscus. J Ophthalmol 2012:983182
1666. Kok HP (2007) Treatment planning for locoregional and intraluminal hyperthermia. Ph.D. Thesis, Universiteit van Amsterdam
1667. Kok HP, van Haaren PM, van de Kamer JB, Wiersma J, van Dijk JD, Crezee J (2005) High-resolution temperature-based optimization for hyperthermia treatment planning. Phys Med Biol 50:3127–3141
1668. Kok HP, van den Berg CA, Bel A, Crezee J (2013) Fast thermal simulations and temperature optimization for hyperthermia treatment planning, including realistic 3D vessel networks. Med Phys 40:103303
1669. Kok HP, Gellermann J, van den Berg CAT, Crezee H, Stauffer P, Hand J (2013) Thermal modelling using discrete vasculature for thermal therapy: a review. Int J Hyperth 29:336–345
1670. Kok HP, Wust P, Stauffer PR, Bardati F, van Rhoon GC, Crezee J (2015) Current state of the art of regional hyperthermia treatment planning: a review. Radiat Oncol 10:196
1671. Kok HP, Kotte A, Crezee J (2017) Planning, optimisation and evaluation of hyperthermia treatments. Int J Hyperth 33:593–607
1672. Kok HP, Cessman ENK, Ceelen W, Brace CL, Ivkov R, Grull H, Terr Haar G, Wust P, Creeze J (2020) Heating technology for malignant tumors: a review. Int J Hyperth 37:711–741
1673. Kolios MC, Sherar MD, Hunt JW (1995) Large blood vessel cooling in heated tissue: a numerical study. Phys Med Biol 40:477–494
1674. Kolios MC, Worthington AE, Sherar MD, Hunt JW (1998) Experimental evaluation of two simple thermal models using transient temperature analysis. Phys Med Biol 43:3325–3340
1675. Kolios MC, Worthington AE, Holdsworth DW, Sherar MD, Hunt JW (1999) An investigation of the flow dependence of temperature gradients near large vessels during steady state and transient tissue heating. Phys Med Biol 44:1479–1497
1676. Kolpashchikov VL, Schnip AI (1978) Constitutive heat transfer equations for materials with memory. Int J Heat Mass Transf 21:155–161
1677. Kolwankar KM (2013) Local fractional calculus: a review. arXiv:1307.0739
1678. Kolwankar KM, Gangal AD (1997) Hölder exponents of irregular functions and local fractional derivatives. Pramana J Phys 48:49–68
1679. Kolwankar KM, Gangal AD (1998) Local fractional derivatives and fractal functions of several variables. arXiv: physics/9801010 [math-ph]
1680. Kolwankar KM, Gangal AD (1998) Local fractional Fokker-Plank equation. Phys Rev Lett 80:214–217
1681. Konc J, Cseh S, Varga E, Kriston R, Kanyó K (2005) Cryopreservation of oocytes and embryos in human assisted reproduction. J Reprod Endokrinol 2:251–258
1682. Kondo N (2001) The control of sweating rate and skin blood flow during exercise. In: Gisolfi CV (ed) Exercise, nutrition, and environmental stress, vol 1. Cooper Publishing Group, Traverse City, pp 153–178
1683. Kondo E, Yasuda K, Kitamura N, Kudoh T, Minami A, Tohyama H (2005) The effect of electrothermal shrinkage on the biomechanical properties of the anterior cruciate ligament: an experimental study. Arthroscopy 21:448–456

1684. Kondrashina AV, Ogurtsov VI, Papkovsky DB (2015) Comparison of the three optical platforms for measurement of cellular respiration. Anal Biochem 468:1–3
1685. Kondratiev TV, Wold R, Aasum E, Tveita T (2008) Myocardial mechanical dysfunction and calcium overload following rewarming from experimental hypothermia in vivo. Cryobiology 56:15–21
1686. König K (2000) Laser tweezers and multiphoton on life science. Histochem Cell Biol 114:79–92
1687. König K, Svaasand L, Liu Y, Sonek G, Patrizio P, Tadir Y, Berns MW, Tromberg BJ (1996) Determination of motility forces human spermatozoa using an 800 nm optical trap. Cell Mol Biol 42:501–509
1688. König K, Tadir Y, Patrizio P, Berns MW, Tromberg BJ (1996) Effect of ultraviolet exposure and near-infrared laser tweezers on human spermatozoa. Hum Reprod 11:2162–2164
1689. Kononenko VL (2009) Flicker in erythrocytes. I. Theoretical models and registration techniques. Biol Membr 26:352–369
1690. Konstas AA, Neimark MA, Laine AF, Pile-Spellman J (2007) A theoretical model of selective cooling using intracarotid cold saline infusion in the human brain. J Appl Physiol 102:1329–1340
1691. Körber CS, Englich S, Rau G (1990) Intracellular ice formation: cryomicroscopical observation and calorimetric measurement. J Microsc 161:313–325
1692. Korn CB, Schwarz US (2007) Mean first passage times for bond formation for a Brownian particlein linear shear flow above a wall. J Chem Phys 126:09103
1693. Korn CB, Schwarz US (2008) Dynamic states of cells adhering in shear flow: from slipping to rolling. Phys Rev E 77:041904
1694. Körner C, Bergmann HW (1998) The physical defects of the hyperbolic heat conduction equation. Appl Phys A 67:397–401
1695. Korotkova N, Barthlott W (2009) On the thermogenesis of the titan arum (Amorphophallus titanum). Plant Signal Behav 4:1096–1098
1696. Korpan NN (2007) A history of cryosurgery: its development and future. J Am College Surg 204:314–324
1697. Koshimoto C, Mazur P (2002) Effects of cooling and warming rate to and from $-70°$ C, and effect of further cooling from -70 to -196 ° C on the motility of mouse spermatozoa. Biol Reprod 66:1477–1484
1698. Kostädt P, Liu M (2000) On the causality and stability of the relativistic diffusion equation. Phys Rev D 62:023003
1699. Kotnik T, Miklavcic D (2000) Theoretical evaluation of the distributed power dissipation in biological cells exposed to electric field. Bioelectromagnetics 21:385–394
1700. Kotte A, van Leeuwen G, de Bree J, van der Koijk J, Crezee H, Lagendjik J (1996) A description of discrete vessel segments in thermal modelling of tissues. Phys Med Biol 41:865–884
1701. Kotte ANTJ, Van Leeuwen GMJ, Lagendijk JJW (1999) Modelling the thermal impact of a discrete vessel tree. Phys Med Biol 44:57–74
1702. Kou HS, Shih TC, Lin WL (2003) Effect of the directional blood flow on thermal dose distribution during thermal therapy: an application of a Green function based on the porous model. Phys Med Biol 48:1577–1589
1703. Kovács R (2017) Heat conduction beyond Fourier's law: theoretical predictions and experimental validation. Ph.D. Thesis, Budapest University of Technology and Economics
1704. Kovács R (2019) On the rarefied gas experiments. Entropy 21:718
1705. Kovács R, Ván P (2015) Generalzed heat conduction in heat pulse experiments. Int J Heat Mass Transf 83:613–620
1706. Kovács R, Ván P (2016) Models of ballistic propagation of heat at low temperatures. Int J Thermophys 37:95
1707. Kovács R, Ván P (2017) Thermodynamical consistency of the dual phase lag heat conduction equation. Contin Mech Thermodyn 30:223–230

1708. Kovács R, Ván P (2018) Second sound and ballistic heat conduction: NaF experiments revisited. Int J Heat Mass Transf 117:682–690
1709. Krams I, Kivleniece I, Kuusik A, Krama T, Freeberg TM, Mänd R, Vrublevska J, Rantala MJ, Mänd M (2013) Predation selects for low resting metabolic rate and consistent individual differences in anti-predator behavior in a beetle. Acta Etholog 16:163–172
1710. Kratochvil MJ, Seymour AJ, Li TL, Paşca SP, Kuo CJ, Heilshorn SC (2019) Engineered materials for organoid systems. Nat Rev Mater 4:606–622
1711. Kregel KC (2002) Heat shock proteins: modifying factors in physiological stress responses and acquired thermotolerance. J Appl Physiol 92:2177–2186
1712. Kristiansen E, Pedersen S, Ramlov H, Zachariassen KE (1999) Antifreeze activity in the cerambycid beetle Rhagium inquisitor. J Comp Physiol B 169:55–60
1713. Kroener C, Luyet B (1966) Formation of cracks during vitrification of glycerol solutions and disappearance of cracks during rewarming. Biodynamica 10:47–52
1714. Kroll MH, Hellums JD, Larry VM, Schafer AI, Moake JL (1996) Platelets and shear stress. Blood 88:1525–1541
1715. Kucsko G, Maurer PC, Yao NY, Kubo M, Noh HJ, Lo PK, Park H, Lukin MD (2013) Nanometre-scale thermometry in a living cell. Nature 500:54–58
1716. Kucsko G, Maurer PG, Yao NY, Kubo M, Noh H, Lo P, Park H, Lukin MD (2013) Nanometer scale thermometry in a living cell. Nature 500:54–58
1717. Kücük T, Baser I (2007) Cryopreswervation of female fertility: a review on the basis of cryobiology for obstetrics and gynecology residents. Eur Clinics Obstet Gynecol 3:97–102
1718. Kudinov IV, Kudinov VA (2014) Investigation of the pressure distribution in a flow of a viscous fluid in a pipeline under hydraulic-shock conditions with account for the relaxation properties of the fluid. J Eng Phys Thermophys 87:344–354
1719. Kudinov IV, Eremin AV, Kudinov VA, Dovallo AI, Zhukov VV (2020) Mathematical model of damped elastic rod oscillations with dual-phase-lag. Int J Solids Struct 200:231–241
1720. Kuennen MR, Gillum TL, Amorim FT, Kwon YS, Schneider SM (2010) Palm cooling to reduce heat strain in subjects during simulated armored vehicle transport. Eur J Appl Physiol 108:1217–1223
1721. Kuflic EG (1994) Cryosurgery updated. J Am Acad Dermatol 31:925–944
1722. Kulish VV, Lage JL (2000) Fractional-diffusion solution for transient local temperature and heat flux. Trans ASME 122:372–376
1723. Kulish VV, Novozhilov VB (2004) An integral equation for the dual-lag model of heat transfer. J Heat Transf 126:805–808
1724. Kumar D, Rai KN (2016) A study on thermal damage during hyperthermia treatment based on DPL model for multilayer tissues using finite element Legendre wavelet Galerkin approach. J Therm Biol 62:170–180
1725. Kumar D, Rai KN (2017) Numerical simulation of time fractional dual-phase-lag model of heat transfer within skin tissue during thermal therapy. J Thermal Biol 67:49–58
1726. Kumar S, Srivastava A (2015) Thermal analysis of laser-irradiated tissue phantoms using dual phase lag model coupled with transient radiative transfer equation. Int J Heat Mass Transf 90:166–179
1727. Kumar S, Srivastava A (2015) Thermal analysis of laser-irradiated tissue phantoms using dual phase lag model coupled with transient radiative transfer equation. Int J Heat Mass Transf 90:466–479
1728. Kumar S, Mitra K, Yamada Y (1998) Hyperbolic damped-wave models fot transient light-pulse propagation in scattering media. Appl Opt 35:3372–3378
1729. Kumar PBS, Gompper G, Lipowsky R (1999) Modulated phases in multicomponent fluid membranes. Phys Rev E 60:4610–4618
1730. Kumar P, Buldyrev SV, Starr FW, Giovambattista N, Stanley HE (2005) Thermodynamics, structure, and dynamics of water confined between hydrophobic plates. Phys Rev E 72:051503
1731. Kumar A, Henríquez-Rivera RG, Graham MD (2014) Flow-induced segregation in confined multicomponent suspensions: effects of particle size and rigidity. J Fluid Mech 738:423–462

1732. Kumar P, Kumar D, Rai KN (2015) A mathematical model for hyperbolic space-fractional bioheat transfer during thermal therapy. Proc Eng 127:56–62
1733. Kumar P, Kumar D, Rai KN (2015) A numerical study on dual-phase-lag model of bio-heat transfer during hyperthermia treatment. J Thermal Biol 49–50:98–105
1734. Kumar P, Kumar D, Rai KN (2016) A study on DPL model of heat transfer in bi-layer tissues during MFH treatment. Comput Biol Med 75:160–172
1735. Kumar D, Singh S, Rai KN (2016) Analysis of classical Fourier, SPL and DPL heat transfer model in biological tissues in presence of metabolic and external heat source. Heat Mass Transf 52:1089–1107
1736. Kumar D, Kumar P, Rai KN (2017) Numerical solution of non-linear dual-phase-lag bioheat transfer equation within skin tissues. Math Biosci 293:56–63
1737. Kumar R, Vashishth AK, Ghangas S (2019) Non-local heat conduction approach in a bi-layer tissue during magnetic fluid hyperthermia with dual phase lag model. Biomed Mater Eng 30:387–402
1738. Kumar R, Vashishth AK, Ghangas S (2019) Phase-lag effects in skin tissue during transient heating. Int J Appl Mech Eng 24:603–623
1739. Kumar D, Dhar P, Paul A (2020) Thermal response of dielectric nanoparticle infused tissue phantoms during microwave assisted hyperthermia. arXiv:2006.13568
1740. Kumar M, Upadhyay S, Singh S, Rai KN (2023) Mathematical modelling and simulation of three phase lag bio-heat transfer model during cancer treatment. Int J Thermal Sci 184:108002
1741. Kundrotas PJ, Karshikoff A (2002) Modeling of denaturated state for calculation of the electrostatic contribution to protein stability. Protein Sci 11:1681–1686
1742. Kundu B, Dewanjee D (2015) A new method for non-Fourier thermal response in a single layer skin tissue. Case Stud Therm Eng 5:79–88
1743. Kundu P, Sukumar S, Kar SP (2018) Numerical modeling for freezing and cryogenic preservation for viability of biological tissue. Mater Today Proc 5:18823–18832
1744. Kurz A (2008) Physiology of thermoregulation. Best Pract Res Clin Anaesth 22:627–644
1745. Kushlan JA (1985) The vestiary hypothesis of human hair reduction. J Hum Evol 14:29–32
1746. Kusnezov D, Bulgac A, Dang GD (1999) Quantum Lévy processes and fractional kinetics. Phys Rev Lett 82:1136–1139
1747. Kusuoka H, Ikoma Y, Futaki S, Suga H, Kitabatake A, Kamada T, Inoue M (1991) Positive inotropism in hypothermia partially depends on an increase in maximal Ca(2+)-activated force. Am J Physiol 261:H1005–H1010
1748. Kvadsheim PH, Folkow LP, Blix SP (1966) Thermal conductivity of minke whale blubber. J Thermal Biol 21:123–128
1749. Kwak HS, Im HG, Shim EB (2016) A model for allometric scaling of mammalian metabolism with ambient heat loss. Integrat Med Res 5:30–36
1750. Laakso I (2009) Assessment of the computational uncertainty of temperature rise and SAR in the eyes and brain under far-field exposure from 1 to 10 GHz. Phys Med Biol 54:3393–3404
1751. Labonte S (1994) A computer simulation of radiofrequency ablation of the endocardium. IEEE Trans Biomed Eng 41:883–890
1752. Lac E, Morel A, Barthes-Biesel D (2007) Hydrodynamic interaction between two identical capsules in simple shear flow. J Fluid Mech 573:149–169
1753. Laeseke P, Lee F, Weide DVD, Brace C (2009) Multiple-antenna microwave ablation: spatially distributing power improves thermal profiles and reduces invasiveness. J Interv Oncol 2:65–72
1754. Lagendijk J (1982) The influence of blood flow in large vessels on the temperature distribution in hyperthermia. Phys Med Biol 27:17–23
1755. Lagendijk JJ (2000) Hyperthermia treatment planning. Phys Med Biol 45:R61
1756. Lagendijk JJ, Hofman P, Schipper J (1988) Perfusion analyses in advanced breast carcinoma during hyperthermia. Int J Hyperth 4:479–495

1757. Lagendijk JJW (1982) A mathematical model to calculate temperature distributions in human and rabbit eyes during hyperthermic treatment. Phys Med Biol 27:1301–1311
1758. Lahiri BB, Bagavathiappan S, Jayakumar T, Philip J (2012) Medical applications of infrared thermoraphy: a review. Infrared Phys Technol 55:221–235
1759. Lai D, Chen Q (2016) A two-dimensional model for calculating heat transfer in the human body in a transient and non-uniform thermal environment. Energy Build 118:114–122
1760. Lai H (1992) Research on the neurological effects of nonionizing radiation at the University of Washington. Bioelectromagnetics 13:513–526
1761. Lakes RS (2009) Viscoelastic materials. Cambridge University Press, Cambridge
1762. Lakhssassi A, Kengne E, Semmaoui H (2010) Investigation of nonlinear temperature distribution in biological tissues by using bioheat transfer equation of Pennes' type. Nat Sci 2:131–138
1763. Lakhssassi A, Kengne E, Semmaoui H (2010) Modified Pennes' equation modelling bio-heat transfer in living tissues: analytical and numerical analysis. Nat Sci 2:1375–1385
1764. Laloi M, Klein M, Riesmeier JW, Müller-Röber B, Fleury C, Bouillaud F, Ricquier D (1997) A plant cold-induced uncoupling protein. Nature 389:135–136
1765. Lam TS (1986) Laser stimulation of collagen synthesis in human skin fibroblast cultures. Lasers Life Sci 1:61–77
1766. Lam TT (2010) Thermal propagation in solids due to surface laser pulsation and oscillation. Int J Thermal Sci 49:1639–1648
1767. Lamond AI, Earnshaw WC (1998) Structure and function in the nucleus. Science 280:547–553
1768. Lamprecht I, Seymour RS, Schultze-Motel P (1998) Direct and indirect calorimetry of thermogenic flowers of the sacred lotus. Nelumbo nucifera. Thermochim Acta 309:5–16
1769. Lamprecht I, Schmolz E, Blanco L, Romero C (2002) Flower ovens: thermal investigations on heat producing plants. Thermochim Acta 391:107–118
1770. Lamzin IM, Khayrullin RM (2014) The quality assessment of stored red blood cells probed using atomic-force microscopy. Anat Res Int 2014:1–5
1771. Lan X, Wang Y, Peng J, Si Y, Ren J, Ding B, Li B (2021) Designing heat transfer pathways for advanced thermoregulatory textiles. arXiv:2103.04763v1 [physics.app-ph]
1772. Landau LD, Lifshitz EM (1984) Theory of elasticity, 3rd edn. Elsevier, Amsterdam
1773. Landau LD, Lifshitz EM (1987) Fluid mechanics. Pergamon, Oxford
1774. Landau LD, Lifsic EM, Pitaevskii LP, Kosevich AM (1986) Theory of elasticity. Course of theoretical physics, vol 7. Pergamon, Oxford
1775. Landauer R (1961) Irreversibility and heat generation in the computing process. IBM J Res Dev 5:183–191
1776. Landry J, Chréttien P, Bernier D, Nicole LM, Marceau N, Tanguay RM (1982) Thermotolerance and heat shock proteins induced by hyperthermia in rat liver cells. Int J Radiat Oncol Biol Phys 8:59–62
1777. Lang J, Erdmann B, Seebass M (1999) Impact of nonlin-ear heat transfer on temperature control in regional hyperthermia. IEEE Trans Biomed Eng 46:1129–1138
1778. Lanier WL (1995) Cerebral metabolic rate and hypothermia: their relationship with ischemic neurologic injury. J Neurosurg Anesth 7:216–221
1779. Lanier WL, Iaizzo PA, Murray MJ (1992) The effects of forced-air cooling and rewarming on systemic and central nervous physiology in isoflurane-anesthetized dogs. Resuscitation 23:121–136
1780. Lanotte L, Mauer J, Mendez S, Fedosov DA, Fromenta JM, Claeria V, Gompper G, Abkarian M (2016) A new look at blood shear-thinning. arXiv:1608.03730v1 [physics.bio-ph]
1781. Lanotte L, Mauer J, Mendez S, Fedosov DA, Fromental JM, Claveria V, Nicoul F, Gompper G, Abkarian M (2016) Red cells' dynamic morphologies govern blood shear thinning under microcirculatory flow conditions. Proc Natl Acad Sci USA 113:13289–13294
1782. Lanotte L, Mauer J, Mendez S, Fedosov DA, Fromental JM, Claveria V, Nicoud F, Gompper G, Abkarian M (2016) Correction for Lanotte et al. Red cells' dynamic morphologies govern blood shear thinning under microcirculatory flow conditions. Proc Natl Acad Sci USA 113:E8207–E8207

1783. Lapshin A (1954) Heat and temperature conduction of raw fat material and molten fat (in Russian). Myasnaya Indus 25:55–56
1784. Laptook AR, Shalak L, Corbett RJ (2001) Differences in brain temperature and cerebral blood flow during selective head versus whole-body cooling. Pediatrics 108:1103–1110
1785. Laraoui A, Aycock-Rizzo H, Gao Y, Lu X, Riedo E, Meriles CA (2015) Imaging thermal conductivity with nanoscale resolution using a scanning spin probe. Nat Commun 6:8954
1786. Lassen NA (1985) Normal average value of cerebral blood flow in younger adults is 50ml/100g/min. J Cereb Blood Flow Metab 5:347–349
1787. Lau CP, Tai YT, Lee PW (1995) The effects of radio frequency ablation versus medical therapy on the quality-of-life and exercise capacity in patients with accessory pathway-mediated supraventricular tachycardia: a treatment comparison study. Pacing Clin Electrophysiol 18:424–432
1788. Laurent S, Bridot JL, Elst LV, Muller RN (2010) Magnetic iron oxide nanoparticles for biomedical applications. Future Med Chem 2:427–449
1789. Laurent S, Dutz S, Häfeli UO, Mahmoudi M (2011) Magnetic fluid hyperthermia: focus on superparamagnetic iron oxide nanoparticles. Adv Colloid Interface Sci 166:8–23
1790. Lauri J, Bykov A, Fabritius T (2016) Quantification of cell-free layer thickness and cell distribution of blood by optical coherence tomography. J Biomed Opt 21:040501
1791. Lauterbur PC (2003) All science is interdisciplinary - from magnetic moments to molecules to men. http://nobelprize.org/nobel-prizes/medicine/laureates/2003/lauterbur-lecture.pdf
1792. Lawson A, Mukherjee IN, Sambanis A (2012) Mathematical modeling of cryoprotectant addition and removal for the cryopreservation of engineered or natural tissues. Cryobiology 64:1–11
1793. Layne JR, Jones AL (2001) Freeze tolerance in the gray treefrog: cryoprotectant mobilization and organ dehydration. J Exper Zool 290:1–5
1794. Layne JR, Kuharski DK (2000) Effects of prolonged freezing and super cooling on body composition, puparation, and adult emergence of Eurosta solidaginis. Environ Entomol 30:12–16
1795. Layne JR, Lee RE (1995) Adaptation of frogs to survive freezing. Clim Res 5:53–59
1796. Lázaro GR, Hernandez-Machadoa A, Pagonabarraga I (2014) Rheology of red blood cells under flow in highly confined microchannels: I. Effect of elasticity. Soft Matter 10:7195
1797. Le Grand Y, El Hage SG (1980) Physiological optics. Springer, New York
1798. Leber B, Mayrhauser U, Leopold B, Koestenbauer S, Tscheliessnigg K, Stadlbauer V, Stiegler P (2012) Impact of temperature on cell death in a cell-culture model of hepatocellular carcinoma. Anticancer Res 32:915–922
1799. Leble V, Lima R, Dias R, Fernandes C, Ishikawa T, Imai Y, Yamaguchi T (2011) Asymmetry of red blood cell motions in a microchannel with a diverging and converging bifurcation. Biomicofluidics 5:044120
1800. Lebon G (2014) Heat conduction at micro and nanoscales: a review through the prism of extended irreversible thermodynamics. J Non-Equilib Thermodyn 39:35–59
1801. Lebon G, Jou D, Casas-Vázquez J (1980) An extension of the local equilibrium hypothesis. J Phys A Math Gen 13:275–290
1802. Lebon G, Ruggieri M, Valenti A (2008) Extended thermodynamics revisited: renormalized flux variables and second sound in rigid solids. J Phys Condens Matter 20:025223
1803. Lebon J, Machraft H, Grmela M, Debois C (2011) An extended thermodynamic model of transient heat conduction at sub-continuum scales. Proc R Soc A 467:3245–3256
1804. Lee RE, Costanzo JP (1998) Biological ice nucleation and ice distribution in cold-hardy ectothermic animals. Annu Rev Physiol 60:55–72
1805. Lee RE, McGrath JJ, Morason RT, Taddeo RM (1993) Survival of intracellular freezing, lipid coalescence and osmotic fragility in fat body cells of the freeze-tolerant gall fly Eurosta solidaginis. J Insect Physiol 39:445–450
1806. Lee JC, Callaway JC, Foehring RC (2005) Effects of temperature on calcium transients and ca^{2+}-dependent afterhyperpolarizations in neocortical pyramidal neurons. J Neurophysiol 93:2012–2020

1807. Lee BW, Faller R, Sum AK, Vattulainen I, Patra M, Karttunen M (2006) Structural effects of small molecules on phospholipid bilayers investigated by molecular simulations. Fluid Phase Equilibria 225(1):63. physics/0407083
1808. Lee HS, Lee DH, Cho S, Chung JH (2006) Minimal heating dose: a novel biological unit to measure infrared irradiation. Photodermatol Photoimmunol Photomed 22:148–152
1809. Lee JKW, Nio AQX, Lim CL, Teo YNE, Byrne C (2010) Thermoregulation, pacing and fluid balance during mass participation distance running in a warm and humid environment. Eur J Appl Physiol 109:887–898
1810. Lee TR, Yoo SS, Yang J (2016) Generalized plasma skimming model for cells and drug carriers in the microvasculature. arXiv:1609.04973v3 [q-bio.TO]
1811. Lee CH, Hong CH, Liao WT, Yu HS (2017) Differential immunological effects of infrared irradiation and its associated heat in vivo. J Dermatol Sci 86:e80–e81
1812. Leeuwen GMJV, Kotte ANT, de Bree J, der Koijk JFV, Crezee H, Lagendjik JJW (1997) Accuracy of geometrical modelling of heat transfer from tissue to blood vessels. Phys Med Biol 42:1451–1460
1813. Leftov V, Regirer S, Shadrina N (1982) Blood rheology (in Russian). Meditsina, Moskow
1814. Lei H, Karniadakis GE (2012) Quantifying the rheological and hemodynamic characteristics of sickle cell anemia. Biophys J 102:185–194
1815. Lei H, Karniadakis GE (2013) Probing vasoocclusion phenomena in sickle cell anemia via mesoscopic simulations. Proc Natl Acad Sci USA 110:11326–11330
1816. Leibo SP, McGrath JJ, Cravalho EG (1978) Microscopic observations of intracellular ice formation in unfertilized mouse ova as a function of cooling rate. Cryobiology 15:257–271
1817. Leibo SP, Picton HM, Godsen RG (2002) Cryopreservation of human spermatozoa. In: Vayena E, Rowe PJ, Griffin PD (eds) Current practicies and controversies in assisted reproduction. World Health Organization, Geneva, pp 152–165
1818. Leighton TG (2007) Rapporteur report: mechanisms and interactions. Progr Biophys Molec Biol 93:280–294
1819. Leis AP, Beck M, Gruska M, Best C, Hegerl R, Baumeister W, Leis JW (2006) Cryo-electron tomography of biological specimens. IEEE Signal Proc Mag 95:95–103
1820. Lemons DE, Chien S, Crawshaw LI, Weinbaum S, Jiji LM (1987) The significance of vessel size and type in vascular heat transfer. Am J Physiol 253:R128–R135
1821. Len JS, Koh WSD, Tan SX (2010) The roles of reactive oxygen species and antioxidants in cryopreservation. Biosci Rep 39:BSR20191601
1822. Lencioni R, Della PC, Bartolozzi C (2005) Percutaneous image-guided radiofrequency ablation in the therapeutic management of hepatocellular carcinoma. Abdom Imag 30:401–408
1823. Lenné T, Bryany G, Koster KL, Holcomb R (2006) Quantitative analysis of the effect of sugars on membrane phase transition and interlamellar hydration forces. Cryobiology 53(3):443–443
1824. Lennie P (2003) The cost of cortical computation. Curr Biol 13:493–497
1825. Leong KC, Yang C, Murshed SMS (2006) A model for the thermal conductivity of nanofluids – the effect of interfacial layer. J Nanopart Res 8:245–254
1826. Lepock JR (2003) Cellular effects of hyperthermia: relevance to the minimum dose for thermal damage. Int J Hyperth 19:252–266
1827. Lerbret A, Bordat P, Affouard F, Descamps M, Migliardo F (2005) How homogeneous are the trehalose, maltose and sucrose water solutions? An insight from molecular dynamics simulations. J Phys Chem B 109:11046–11057. cond-mat/0503579
1828. Letnikov AV (1868) Theory of differentiation of fractional order. Math Sb 3:1–7
1829. Levant M, Steinberg V (2016) Intermediate regime and a phase diagram of red blood cell dynamics in a linear flow. Phys Rev E 94:062412
1830. Leverett L, Lynch E, Alfrey C, Hellums J (1972) Red blood-cell damage by shear-stress. Biophys J 12:257–273
1831. Levermore C, Pomraning G (1981) A flux-limited diffusion theory. Astrophys J 248:321–334

1832. Levin EI, Zhmakin AI (1996) Load balancing for computations of steady viscous flows using adaptive unstructured grids. Lect Notes Comput Sci 1156:272–277
1833. Levin RL, Cravalho EG, Higgins CG (1978) The concentration-polarization effect in a multicomponent electrolyte solution – the human erythrocyte. J Theor Biol 71:225–254
1834. Levitt M, Park BH (1993) Water: now you see it, now you don't. Structure 1:223–226
1835. Levitt M, Hirshberg M, Sharon R, Laidig KE, Daggett V (1997) Calibration and testing of a water model for simulation of the molecular dynamics of proteins and nucleic acids in solution. J Phys Chem B 25:5051–5061
1836. Levy Y, Onuchic JN (2006) Water mediation in protein folding and molecular recognition. Annu Rev Biophys Biomol Struct 35:389–415
1837. Lewandowska M, Malinowski L (2006) An analytical soluton of the hyperbolic heat conduction equation for the case of a finite medium symmetrically heated on both sides. Int Commun Heat Mass Transf 33:61–69
1838. Lewis JK, Bischof JC, Braslavsky I, Brockbank KGM, Fahy GM, Fuller BJ, Rabin Y, Tocchio A, Woods EJ, Wowk BG, Acker JA, Giwa S (2016) The grand challenges of organ banking: proceedings from the first global summit on complex tissue cryopreservation. Cryobiology 72:169–182
1839. Li X (2014) Heat conduction in nanoscale materials: a statistical-mechanics derivation of the local heat flux. Phys Rev E 90:032112
1840. Li KC, Chen HT (2009) Analysis for the dual-phase-lag bio-heat transfer during magnetic hyperthermia treatment. Int J Heat Mass Transf 52:1185–1192
1841. Li SN, Cao BY (2016) Lorentz covariance of heat conduction laws and a Lorentz-covariant heat conduction model. Appl Math Model 40:5532–5541
1842. Li SN, Cao BY (2016) On defects of Taylor series approximation in heat conduction models. Int J Heat Mass Transf 98:824–832
1843. Li SN, Cao BY (2016) On thermodynamics problems in the single-phase-lagging heat conduction models. Entrophy 18:391
1844. Li SN, Cao BY (2019) Anomalous heat conduction equations based on non-Brownian descriptions. Phys A 526:121141
1845. Li W, Fan S (2018) Nanophotonic control of thermal radiation for energy applications. Opt Exp 26:15995–16021
1846. Li C, Deng W (2007) Remarks on fractional derivatives. Appl Math Comput 187:777–784
1847. Li C, Ito K (2014) Numerical and experimental estimation of convective heat transfer coefficient of human body under strong forced convective flow. J Wind Eng Indus Aerodyn 126:107–117
1848. Li N, Andorfer CA, Duman JG (1998) Enhancement of insect antifreeze protein activity by solutes of low molecular mass. J Exp Biol 201:2243–2251
1849. Li J, Dao M, Lim C, Suresh S (2005) Spectrin-level modeling of the cytoskeleton and optical tweezers stretching of the erythrocyte. Biophys J 88:3707–3719
1850. Li X, Hassan SA, Mehler EL (2005) Long dynamics simulations of proteins using atomistic fields and a continuum representation of solvent effects: calculation of structural and dynamic properties. Poteins Struct Funct Bioinf 60:464–484
1851. Li G, Saenz J, Godke RA, Devireddy RV (2006) Effect of glycerol and cholesterol-loaded cyclodextrin on freezing-induced water loss in bovine spermatozoa. Reproduction 131:875–886
1852. Li C, Qian D, Chen YQ (2011) On Riemann-Liouville and Caputo derivatives. Discr Dyn Nat Soc 2011:562494
1853. Li Y, Fu ZY, Su BL (2012) Hierarchically structured porous materials for energy conversion and storage. Adv Funct Mater 22:4634–4667
1854. Li L, Tan H, Gu Z, Liu Z, Geng Y, Liu Y, Tong H, Tang Y, Qiu J, Su L (2013) Heat stress induces apoptosis through a ca2+-mediated mitochondrial apoptotic pathway in human umbilical vein endothelial cells. PLoS One 9:e111083

1855. Li l, Yu B, Liang M, Yang S, Zou M (2014) A comprehensive study of the effective thermal conductivity of living biological tissue with randomly distributed vascular trees. Int J Heat Mass Transf 72:616–621
1856. Li L, Zhou L, Zhang Y (2014) Thermal wave superposition and reflection phenomena during femtosecond laser interaction with thin gold film. Numer Heat Transf A 65:1139–1153
1857. Li X, Zhong Y, Jazar R, Subic A (2014) Thermal-mechanical deformation modelling of soft tissues for thermal ablation. Bio-Med Mater Eng 24:2299–2310
1858. Li X, Zhong Y, Smith J, Gu C (2017) Non-Fourier based thermal-mechanical tissue damage prediction for thermal ablation. Bioengineered 8:71–77
1859. Li X, Li C, Xue Z, Tian X (2018) Analytical study of transient thermo-mechanical responses of dual-layer skin tissue with variable thermal material properties. Int J Thermal Sci 124:459–466
1860. Li D, Liao W, Rigden AJ, Liu X, Wang D, Malyshev S, Shevliakova E (2019) Urban heat island: aerodynamics or imperviousness? Sci Adv 5:eeau4299
1861. Li CY, Lin SM, Wan YY (2020) Prediction of temperature field and thermal damage of multilayer skin tissues subjected to time-varying laser heating and fluid cooling by a semianalytical method. Math Problems Eng 2020:5074280
1862. Li X, Qin QH, Tian X (2020) Thermo-viscoelastic analysis of biological tissue during hyperthermia treatment. Appl Math Model 79:881–895
1863. Li XY, Qin QH, Tian XG (2020) Thermo-viscoelastic analysis of biological tissue during hyperthermia treatment. Appl Math Model 79:881–895
1864. Li X, Li Y, Luo P, Tian X (2021) Relationship between the nonlocal effect and lagging behavior in bioheat transfer. J Heat Transf 143:051201
1865. Liang XG, Ge XS, Zhang YP, Wang GJ (1991) A convenient method of measuring the thermal conductivity of biological tissue. Phys Med Biol 36:1599–1605
1866. Licina D, Melikov AK, Sekhar C, Tham KW (2015) Air temperature investigation microenvironment around a human body. Build Environ 92:39–47
1867. Lidell ME, Betz MJ, Leinhard OD, Heglind M, Elander L, Slawik M, Mussack T, Nilsson D, Romu T, Nuutila P, Virtanen KA, Beuschlein F, Persson A, Borga M, Enerbäck S (2013) Evidence for two types of brown adipose tissue in humans. Nat Med 19:631–634
1868. Liebermann J, Nawroth F, Isachenko V, Isachenko E, Rahimi G, Tucker MJ (2002) Potential importance of vitrification in reproductive medicine. Biol Reprod 67:1671–1680
1869. Lifshits IM, Grossberg AY, Khokhlov AR (1979) Volume interactions in the statistical physics of a polymer macromolecule. Sov Phys Usp 22:123–142
1870. Lighthill MJ (1968) Pressure-forcing of tightly fitting pellets along fluid-filled elastic tubes. J Fluid Mech 34:113–143
1871. Lillford PJ, Holt CB (2002) In vitro use of biological cryoprotectants. Philos Trans R Soc Lond B 357:945–951
1872. Lillywhite HB, Albert JS, Sheehy CM, Seymour RS (2012) Gravity and the evoluttion of cardiopulmonary morphology in snakes. Comp Biochem Physiol 161:230–242
1873. Lim CL (2020) Fundamental concepts of human thermoregulation and adaptation to heat: a review in the context of global warming. Int J Environ Res Publ Health 17:7795
1874. Lim CL, Ng KKC, Lee LKH (1997) The effects of prolonged passive heat exposure and basic military training on thermoregulatory and cardiovascular responses in recruits from a tropical country. Mil Med 162:623–627
1875. Lim CL, Byrne C, Lee J (2008) Human thermoregulation and measurement of body temperature in exercise and clinical settings. Ann Acad Med Singap 37:347–353
1876. Lima R, Ishikawa T, Imai Y, Takeda M, Wada S, Yamaguchi T (2008) Radial dispersion of red blood cells in blood flowing through glass capillaries: the role of hematocrit and geometry. J Biomech 41:2188–2197
1877. Lima R, Nakamura M, Omori T, Ishikawa T, Wada S, Yamaguchi T (2011) Microscale flow dynamics of red blood cells in microchannels: an experimental and numerical analysis. Comput Meth Appl Sci 19:297–309

1878. Lima E, De Biasi E, Zysler RD, Mansilla MV, Mojica-Pisciotti ML, Torres E, Calatayud MP, Marquina C, Ricardo Ibarra M, Goya GF (2014) Relaxation time diagram for identifying heat generation mechanisms in magnetic fluid hyperthermia. J Nanopart Res 16:2791
1879. Lin DTW (2011) A molecular dynamics study of the heat transfer phenomena in the bio-tissue. Int J Sci Eng 1:17–21
1880. Lin SM (2013) Analytical solutions of bio-heat conduction on skin in Foiurier and non-Fourier models. J Mech Med Biol 13:1350063
1881. Lin JT (2016) Analysis of thermal therapy using infrared diode lasers operated in continuous wave and pulsed mode. Med Dev Diagn Eng 2:29–35
1882. Lin S, Li C (2009) Analytical solutions of non-fourier bio-heat conductions for skin subjected to pulsed laser heating. Int J Therm Sci 26:352–369
1883. Lin SM, Li CY (2016) Analytical solutions of non-Fourier bio-heat conductions for skin subjected to pulsed laser heating. Int J Therm Sci 110:146–158
1884. Lin JC, Lin MF (1982) Microwave hyperthermia-induced bloodbrain barrier alterations. Radiat Res 89:77–87
1885. Lin D, Manche E (2003) Two-year results of conductive keratoplasty for the correction of low to moderate hyperopia. J Cataract Refract Surg 29:2339–2350
1886. Lin JC, Wang YJ (1987) Interstitial microwave antennas for thermal therapy. Int J Hyperth 3:37–47
1887. Lin S, Gao DY, Yu XC (1990) Thermal stress induced by water solidification in a cylindrical tube. J Heat Transf 112:1079–1082
1888. Lin S, Lo W, Tan H, Chan JY, Chen WL, Wang SH, Sun Y, Lin WC, Chen JS, Hsu CJ, Tjiu JW, Yu HS, Jee SH, Dong CY (2006) Prediction of heat-induced collagen shrinkage by use of second harmonic generation microscopy. J Biomed Opt 11:34020
1889. Lin M, Xu F, Lu TJ, Bai BF (2010) A review of heat transfer in human tooth - experimental characterization and mathematical modeling. Dental Mater 26:501–513
1890. Lin Z, Ruan X, Zou J, Fu X (2014) Experimental study of cavitation phenomenon in a centrifugal blood pump induced by the failure of inlet cannula. Chin J Mech Eng 27:165–170
1891. Linacre E (1972) Leaf temperatures, diffusion resistances, and transpiration. Agric Meteorol 10:365–382
1892. Ling G (2004) What determines water content of a living cell? Physiol Chem Phys Med NMR 36:1–19
1893. Linninger AA, Gould IG, Marinnan T, Hsu CY, Chojecki M, Alaraj A (2013) Cerebral microcirculation and oxygen tension in the human secondary cortex. Ann Biomed Eng 41:2264–2284
1894. Liou YC, Tocilj A, Davies PL, Jia Z (2000) Mimicry of ice structure by surface hydroxyls and water of a beta-helix antifreeze protein. Nature 406:322–324
1895. Lipowsky HH (2005) Microvascular rheology and hemodynamics. Microcirculation 12:5–15
1896. Lipton JM (1985) Thermoregulation in pathological states. In: Shitzer A, Eberhart RC (eds) Heat transfer in medicine and biology, vol 1. Plenum Press, New York, pp 9–105
1897. Liu IS (1972) Method of Lagrange multipliers for exploitation of the entropy principle. Arch Ration Mech Anal 46:131–148
1898. Liu J (2000) Preliminary survey on the mechanisms of the wave-like behaviors of heat transfer in living tissues. Forsch Ingenieur 66:1–10
1899. Liu D, Brace C (2017) Numerical simulation of microwave ablation incorporating tissue contraction based on thermal dose. Phys Med Biol 62:2070–2086
1900. Liu D, Pan F (2016) Advances in cryopreservation of organs. J Huazhong Univ Sci Technol Med Sci 36:153–161
1901. Liu B, Wang LCH, Belke DD (1991) Effect of low-temperature on the cytolic free Ca^{2+} in rat ventricular myocytes. Cell Calcium 12:11–18
1902. Liu J, Xu LX (1999) Estimation of blood perfusion using phase shift in temperature response to sinusoidal heating the skin surface. IEEE Trans Biomed Eng 46:1037–1043

1903. Liu C, Mi CC, Li BQ (2009) Transient temperature response of pulsed-laser-induced heating for nanoshell-based hyperthermia treatment. IEEE Trans Nanotechnol 8:69–706
1904. Liu FJ, Li ZB, Zhang S, Liu HY (2015) He's fractional derivative for heat conduction in a fractal medium arising in silkworm coocon hierarchy. Thermal Sci 19:1155–1159
1905. Liu J, Chen X, Xu LX (1999) New thermal wave aspects on burn evaluation of skin subjected to instantaneous heating. IEEE Trans Biomed Eng 46:420–428
1906. Liu J, Deng ZS (2009) Nano-cryosurgery: advances and challenges. J Nanosci Nanotech 9:4521–4542
1907. Liu J, Jia W, Vu H, Aguilar G (2006) Heat transfer between single cryogenic droplet and epoxy skin phantom. In: Proceedings of 9th Annual Conference on Liquid Atomization and Spray Systems, Toronto
1908. Liu J, Ren ZP, Wang CC (1995) Interpretation of living tissue's temperature oscillations by thermal wave theory. Chin Sci Bull 40:1493–1495
1909. Liu J, Ren ZP, Wang CC (1996) Simulating experimental study on the mechanisms of temperature oscillation in living tissue. J Basic Sci Eng 4:173–182
1910. Liu J, Wang CC, Ren ZP (1997) Theory and experiments on temperature oscillations effects in living tissue. J Tsinghua Univ Sci Tech 2:91–95
1911. Liu J, Woods EJ, Agca Y, Critser ES, Critser JK (2000) Cryobiology of rat embryos II: a theoretical model for the development of interrupted slow freezing procedures. Biol Reprod 63:1303–1312
1912. Liu J, Zhou Y, Yu T, Gui L, Deng Z, Lv Y (2004) Minimally invasive probe system capable of performing both cryosurgery and hyperthermia treatment on target tumor in deep tissues. Min Invas Ther Allied Technol 13:47–57
1913. Liu J, Zhu JK (1998) A calcium sensor homolog required for plant salt tolerance. Science 280:1943–1945
1914. Liu KC (2008) Thermal propagation analysis for living tissues with surface heating. Int J Thermal Sci 47:507–513
1915. Liu KC (2011) Thermal lagging in concentric living tissues with Gaussian distribution source. Int J Heat Mass Transf 54:2829–2836
1916. Liu KC (2015) Analysis for high-order effects in thermal lagging to thermal responses in bioloical tissue. Int. J. Heat Mass Transf 81:347–354
1917. Liu W, Balandin AA (2005) Thermal conduction in $Al_xGa_{1-x}N$ alloys and thin films. J Appl Phys 97:073710
1918. Liu Z, Bilston L (2000) On the viscoelastic character of liver tissue: experiments and modeling of the linear behavior. Biorheology 37:191–201
1919. Liu KC, Chang PC (2007) Analysis of dual-phase-lag heat conduction in cylindrical system with a hybrid method. Appl Math Model 31:369–380
1920. Liu KC, Chen HT (2010) Investigation for the dual phase lag behavior of bio-heat transfer. Int J Thermal Sci 49:1138–1146
1921. Liu KC, Chen HT (2015) Analysis of the bioheat transfer problem with pulse boundary heat flux using a generalized dual-phase-lag model. Int Commun Heat Mass Transf 65:31–36
1922. Liu KC, Chen PJ (2008) Finite propagation of heat transfer in multilayer tissue. J Thermophys Heat Transf 22:775–782
1923. Liu KC, Chen TM (2018) Analysis of the thermal response and requirement for power dissipation in magnetic hyperthermia with the effect of blood temperature. Int J Heat Mass Transf 126:1048–1056
1924. Liu KC, Chen YS (2016) Analysis of heat transfer and burn damage in a laser irradiated living tissue with the generalized dual-phase-lag model. Int J Thermal Sci 103:1–9
1925. Liu KC, Cheng PJ (2019) Numerical analysis of power dissipation requirement in magnetic hyperthermia problems. J Therm Biol 86:102430
1926. Liu Y, Liu WK (2006) Rheology of red blood cell aggregation by computer simulation. J Comput Phys 220:139–154
1927. Liu KC, Wang JC (2014) Analysis of thermal damage to laser irradited tissue based on the dual-phase-lag model. Int J Heat Mass Transf 70:621–628

1928. Liu KC, Yang YC (2020) Numerical analysis of local nonequilibrium heat transfer in layered spherical tissue during magnetic hyperthermia. Comput Meth Biomech Biomed Eng 23:968–980
1929. Liu L, Zhou YX (2002) Analytical study on the freezing and thawing processes of biological skin with finite thickness. Heat Mass Transf 38:319–326
1930. Liu SC, Derick LH, Palek J (1987) Visualization of the hexagonal lattice in the erythrocyte membrane skeleton. J Cell Biol 104:527–536
1931. Liu Z, Wan RG, Muldrew K, Sawchuk S, Rewcastle J (2004) A level set variational formulation for coupled phase change/mass transfer problems: application to freezing of biological systems. Finite Elem Anal Des 40:1641–1663
1932. Liu KC, Cheng PJ, Wang JC (2014) Analysis of thermal damage in a laser-irradiated based on the non-Fourier model. Int J Eng Technol 6:132–135
1933. Liu KC, Cheng PJ, Wang YN (2011) Analysis of non-Fourier thermal behaviour for multi-layer skin model. Thermal Sci 15:561–567
1934. Liu Z, Sun X, Tang J, Tang Y, Tong H, Wen Q, Liu Y, Su L (2011) Intestinal inflammation and tissue injury in response to heat stress and cooling treatment in mice. Mol Med Rep 4:437–443
1935. Liu KC, Wang YN, Chen YS (2012) Investigation on the bio-heat transfer with the dual-phase-lag effect. Int J Therm Sci 58:29–35
1936. Liu P, Liu J, Duan H (2012) Thermal modelling for endocardiac radiofrequency ablation: comparison of hyperbolic bioheat equation and Pennes bioheat equation with finite element method. arXiv:1204.0098 [math.AP]
1937. Liu KC, Wang CC, Cheng PJ (2013) Analysis of non-Fourier thermal behavior in layered tissue with pulse train heating. Appl Mech Mater 479–480:496–500
1938. Liu X, Zhu Y, Zhang F, Gong XF (2013) Estimation of temperature elevation generated by ultrasonic irradiation in biological tissues using the thermal wave model. Chin Phys B 22:024301
1939. Liu L, Zheng LC, Liu FW, Zhang XX (2016) An improved heat conduction model with Riesz fractional Cattaneo-Christov flux. Int J Heat Mass Transf 103:1191–1197
1940. Liu L, Zheng L, Liu F (2018) Research on macroscopic and microscopic heat transfer mechanisms based on non-Fourier constitutive models. Int J Heat Mass Transf 127:165–172
1941. Liu ZL, Li H, Qian Y, Buffet P, Dao M, Karniadakis GE (2021) Computational modeling of biomechanics and biorheology of heated red blood cells. Biophys J 120:4663–4671
1942. Lloyd TD, Orr S, Skett P, Berry DP, Dennison AR (2003) Cryopreservation of hepatocytes: a review of current methods for banking. Cell Tissue Bank 4:3–15
1943. Loesel FH, Fisher FP, Suhan H, Bille JF (1998) Non-thermal ablation of neural tissue with femtosecond laser pulses. Appl Phys B 66:121–128
1944. Loh CY, Hultgren SC, Chang SC (2001) Wave computation in compressible flow using space-time conservation element and solution element method. AIAA J 39:794–801
1945. Loitsanskii LG (1970) Mechanics of gases and liquids, 3rd edn (in Russian). Nauka, Moscow
1946. Lomolino MV (2005) Body size evolution in insular vertebrates: generality of the island rule. J Biogeogr 32:1683–1699
1947. London RA, Glinsky ME, Zimmerman GB, Bailey DS, Eder DC, Jacques SL (1997) Laser-tissue interaction modeling with LATIS. Appl Opt 36:9068–9074
1948. Long Q, Xu XY, Ramnarine KV, Hoskins P (2001) Numerical investigation of physiologically realistic pulsatile flow through arterial stenosis. J Biomech 34:1229–1242
1949. Lopez MJ, Hayashi K, Fanton GS, Thabit G, Markel MD (1998) The effect of radiofrequency energy on the ultrastructure of joint capsular collagen. Arthroscopy 14:495–501
1950. Lopéz Molina JAL, Rivera MJ, Berjano E (2012) Analytical model based on a cylindrical geometry to study RF ablation with needle-like internally cooled electrode. Math Probl Eng 2012:834807
1951. Lopresto V, Pinto R, Farina L, Cavagnaro M (2017) Microwave thermal ablation: effects of tissue properties variations on predictive models for treatment planning. Med Eng Phys 46:63–70

1952. Lopresto V, Pinto R, Farina L, Cavagnaro M (2017) Treatment planning in microwave thermal ablation: clinical gaps and recent research advances. Int J Hyperth 33:83–100
1953. Lopresto V, Strigari L, Farina L, Minosse S, Pinto R, D'Alessio D, Cassano B, Cavagnaro M (2018) CT-based investigation of the contraction of ex vivo tissue undergoing microwave thermal ablation. Phys Med Biol 63:055019
1954. Lopresto V, Argentieri A, Pinto R, Cavagnaro M (2019) Temperature dependence of thermal properties of ex vivo liver tissue up to ablative temperatures. Phys Med Biol 64:1–13
1955. Lord HW, Shulman Y (1967) A generalized dynamical theory of thermoelasticity. J Mech Phys Solids 15:299–309
1956. Lorenzo CF, Hartley TT (1998) Initialization, conceptualization, and application in the generalized fractional calculus. NASA/TP-1998-208415
1957. Lorthois S, Cassot F (2010) Fractal analysis of vascular networks: insights from morphogenesis. J Theor Biol 262:614–633
1958. Losada J, Nieto JJ (2015) Properties of a new fractionalmderivative withou singular kernel. Prog Fract Differ Appl 1:87–92
1959. Lovelock JE (1953) The haemolysis of human red blood cells by freezing and thawing. Biochim Biophys Acta 10:414–436
1960. Lovelock JE, Bishop MWH (1959) Prevention of freezing damage to living cells by dimethyl sulphoxide. Nature 183:1394–1395
1961. Loyd D, Karlsson M, Erlandsson BE, Sjödin JG, Ask P (1997) Computer analysis of hyperthermia treatment of the prostate. Adv Eng Softw 28:347–351
1962. Loza P, Kouznetsov D, Ortega R (1994) Temperature distribution in a uniform medium heated by absorption of a gaussian light beam. Appl Opt 33:3831–3836
1963. Loze MK, Wright CD (1997) Temperature distributions in semiinfinite and finite-thickness media as a result of absorption of laser light. Appl Opt 36:494–507
1964. Loze MK, Wright CD (1998) Temperature distributions in laser heated semi-infinite and finite-thickness media with convective surface losses. Appl Opt 37:6822–6832
1965. Loze MK, Wright CD (2001) Temperature distributions in laser-heated biological tissue with application to birthmark removal. J Biomed Opt 6:74–85
1966. Lozina-Lozinski LK (1974) Studies in cryobiology. Coronet Books, London
1967. Lu DS, Raman SS, Limanond P, Aziz D, Economou J, Busuttil R, Sayre J (2003) Influence of large peritumoral vessels on outcome of radiofrequency ablation of liver tumors. J Vasc Interv Radiol 14:1267–1274
1968. Lu Y, Nan Q, Li L, Liu Y (2009) Numerical study on thermal field of microwave ablation with water-cooled antenna. Int J Hyperth 25:108–115
1969. Lu L, Rahimian A, Zorin D (2017) Contact-aware simulations of particulate Stokesian suspensions. J Comput Phys 347:160–182
1970. Lu L, Morse MJ, Rahimian A, Stadler G, Zorin D (2019) Scalable simulation of realistic volume fraction red blood cell flows through vascular networks. In: The international conference for high performance computing, networking, storage, and analysis (SC'19), November 17–22, 2019, Denver
1971. Lubashevsky I, Gafiychuk V (1999) Mathematical description of heat transfer in living tissue. arXiv:adap-org/991101
1972. Lubashevsky IA, Gafiychuk VV (2000) Cooperative mechanism of self-regulation in hierarchical living systems. SIAM J Appl Math 60:633–663
1973. Lubashevsky IA, Priezzhev AV, Gafiychuk VV, Cadjan MG (1996) Free boundary model for local thermal coagulation. Proc SPIE 2681:81–91
1974. Lubashevsky IA, Priezzhev VV, Gafiychuk VV, Cadjan MG (1996) Free-boundary model for local thermal coagulation. Proc SPIE 2681:81–91
1975. Lubashevsky IA, Priezzhev AV, Gafiychuk VV, Cadjan MG (1997) Local thermal coagulation due to laser - tissue interaction as irreversible phase transition. J Biomed Opt 2:95–105
1976. Lubashevsky IA, Gafiychuk VV, Datsko BY (2002) Anomalous properties of heat diffusion in living tissue caused by branching artery network. Qualitative description. arXiv:cond-mat/0201057v1 [cond-mat.soft]

1977. Luby-Phelps K (2000) Cytoarchitecture and physical properties of cytoplasm: volume, viscosity, diffusion, intracellular surface area. Int Rev Cytol 192:189–221
1978. Luchakov YI, Arokina NK, Nesmeyanov AA (2018) Convective and conducive heat transfer in various arras of organism. J New Med Technol e-Edition
1979. Luchko Y (2004) Operational rules for a mixed operator of the Erdélye-Kober type. Frac Calc Appl Anal 7:339–364
1980. Luchko Y, Punzi A (2011) Modeling anomaluos heat transfer in geothermal reservoirs via fractional diffusion equations. Int J Geomath 1:257–276
1981. Ludwig N, Trecroci A, Caumo A, Alberti G, Michielon G, Formenti D, Gargano M (2016) Dynamics of thermographic skin temperature response during squat exercise at two different speeds. J Therm Biol 59:58–63
1982. Luikov AV (1975) System of differential equations of heat and mass transfer in capillary-porous bodies (Review). Int J Heat Mass Transf 9:139–152
1983. Lundheim R (2004) Physiological and ecological significance of biological ice nucleators. Philos Trans R Soc Lond B 357:937–943
1984. Lung DC, Stahovich TF, Rabin Y (2004) Computerized planning for multiprobe cryosurgery using a force-field analogy. Comp Meth Biomech Biomed Eng 7:101–110
1985. Lunin VY, Afonin PV, Urzhumtsev AG (2006) Statistical modeling and the principle of the maximal likelihood in crystallography of macromolecules (in Russian). Math Biol Bioinf 1:17–26
1986. Luo P, Li X, Tian X (2021) Nonlocal thermoelasticity and its application in thermoelastic problem with temperature-dependent thermal conductivity. Eur J Mech A Solids 87:104204
1987. Luo W, Kramer R, de Kort Y, Rense P, van Marken Lichtenbelt W (2022) The effects of a novel personal comfort system on thermal comfort, physiology and perceived indoor environmental quality, and its health implications - stimulating human thermoregulation without compromising thermal comfort. Indoor Air 32:e12951
1988. Lupetti P (2005) Cryotechniques for electron microscopy: a minireview. In: Evangelista V, Barsanti L, Passarelli V, Gualtieri P (eds) From cells to proteins: imaging nature across dimensions. Springer, Berlin, pp 53–70
1989. Lusk G (1928) The elements of the science of nutrition. Johnson Reprint Corporation, London
1990. Luyet BJ (1937) The vitrification of organic colloids and protoplasm. Biodynamica 1:1–14
1991. Luyet BJ, Hodapp R (1938) Revival of frog spermatozoa vitrified in liquid air. Proc Soc Exp Biol 39:433–434
1992. Luzzi R, Vasconcellos AR, Casas-Vázquez J, Jou D (1997) Characterization and measurement of a nonequilibrium temperature-like variable in irreversible thermodynamics. Phys A 234:699–714
1993. Lv YG, Liu J (2007) Effect of transient temperature on thermoreceptor response and thermal sensation. Build Environ 42:656–664
1994. Lv YG, Deng ZS, Liu J (2005) 3-D numerical study on the induced heating effects of embedded micro/nanoparticles on human body subject to external medical electromagnetic field. IEEE Trans Nanobiosci 4:284–294
1995. Lv YG, Deng ZS, Liu J (2005) Study on the induced heating effects of embedded micro/nano particles on human body subject to extternal medical electromagnetic field. IEEE Trans Nanobiosci 4:284–294
1996. Lyubarev AE, Kurganov BI (2000) Study of the irreversible thermal protein denaturation using differential scanning calorimetry (in Russian). Adv Biol Chem 40:43–84
1997. Ma N, Gao X, Zhang XX (2004) Two-layer simulation model of laser-induced interstitial thermo-therapy. Lasers Med Sci 18:184–189
1998. Ma W, Liu W, Li M (2015) Modeling heat transfer from warm water to foot; analytical solution and experimental validation. Int J Thermal Sci 98:364–371
1999. MacFarlane DR, Fragoulis M (1986) Theory of devitrification in multicomponent glass forming systems under diffusion control. Phys Chem Glasses 27:228–234

2000. MacFarlane DR, Scheirer J, Smedley SI (1986) Pressure coefficient of conductance and of glass transition temperatures in concentrated aqueous LiCl, LiI and $AlCl_3$ solutions. J Phys Chem 90:2168–2173
2001. Machlenkin A, Goldberger O, Tirosh B, Paz A, Volovitz I, Bar-Haim E, Lee SH, Vadai E, Tzehoval E, Eisenbach L (2005) Combined dendritic cell cryotherapy of tumor induced systemic antimetastatic immunity. Clin Cancer Res 11:4955–4961
2002. Maciaszek JL, Lykotrafitis G (2011) Sickle cell trait human erythrocytes are significantly stiffer than normal. J Biomech 44:657–61
2003. Madan A, Dehaan R, Mudra D, Carrol K, Lecluyse E, Parkinson A (1999) Effect of cryopreservation on cytochrome P-450 enzyme induction in cultured rat hepatocytes. Drug Metab Dispos 27:327–335
2004. Madden CJ, Morrison SF (2019) Central nervous system circuits that control body temperature. Neurosci Lett 696:225–232
2005. Madsen SJ, Wilson BC (2012) Optical properties of brain tissue. In: Madsen SJ (ed.) Optical methods and instrumentation in brain imaging and therapy. Sprinher, Berlin, pp 1–22
2006. Madsen PL, Sperling BK, Warming T, Schmidt JF, Secher NH, Wildschiodtz G (1993) Middle cerebral artery blood velocity and cerebral blood flow and o_2 uptake during dynamic exercise. J Appl Physiol 74:245–250
2007. Maggs AC (2006) Multiscale Monte Carlo algorithm for simple fluids. Phys Rev Lett 97:197802
2008. Magin RL, Ingo C, Colon-Perez L, Triplett W, Mareci TH (2013) Characterization of anomalous diffusion in porous biological tissues using fractional derivatives and entropy. Micropor Mesopor Mat 178:39–43
2009. Mahjoob S, Vafai K (2009) Analytical characterization of heat transport through biological media incorporating hyperthermia treatment. Int J Heat Mass Transf 52:1608–1618
2010. Mahjoob S, Vafai K (2010) Analysis of bioheat transport through a dual layer biological media. J Heat Transf 132:031101
2011. Mainardi F (2004) Application of integral transforms in fractional diffusion processes. Integral Transf Special Funct 15:477–484
2012. Mainardi F, Gorenflo R (2000) On Mag-Leffler function in fractional evaluation processes. J Comput Appl Math 118:283–299
2013. Mainardi F, Gorenflo R (2007) Time-fractional derivatives in relaxation processes: a tutorial survey. Int J Theor Appl 10:269–308
2014. Mainardi F, Luchko Y, Pagnini G (2001) The fundamental solution of the space-time fractional diffusion equation. Fract Calc Appl Anal 4:153–192
2015. Mainster MA (1979) Ophthalmic applications of infrared lasers thermal considerations. Invest. Ophthalmol Vis Sci 18:414–420
2016. Majchrzak E, Mochnacki B (2019) Numerical model of biological tissue heating using the models of bio-heat transfer with delays. E3S Web of Conferences 128:02002
2017. Majchrzak E, Dziatkiewicz G, Paruch M (2008) The modelling of heating a tissue subjected to external electromagnetic field. Acta Bioeng Biomech 10:29–37
2018. Majeed A, Zeeshan A, Alamri SZ, Ellahi R (2018) Heat transfer analysis in ferromagnetic viscoelastic fluid flow over a stretching sheet with suction. Neural Comput Appl 30:1947–1955
2019. Majumdar A (1993) Microscale heat conduction in dielectric thin films. J Heat Transf 115:7–18
2020. Mala T (2003) Cryoablation of liver tissue. MD Thesis, University of Oslo
2021. Mala T (2006) Cryoablation of liver tumors - a review of mechanisms, techniques and clinical outcome. Min Invas Ther 15:9–17
2022. Mala T, Samset E, Aurdal L, Edwin B, Gladhaug I, Soreide O (2001) Magnetic resonance imaging estimated three-dimensional temperature distribution in liver cryolesions. A study of cryolesion characteristics assumed necessary for tumour ablation. Cryobiology 43:268–275

2023. Malek A, Abbasi G (2015) Heat treatment modelling using strongly continuous semigroups. Comput. Biol. Med. 62:65–75
2024. Malek AM, Alper SL, Izumo S (1999) Hemodynamic shear stress and its role in atherosclerosis. JAMA 282:2035–2042
2025. Malinen M, Huttunen T, Kaipio JP (2003) Thermal dose optimization method for ultrasound surgery. Phys Med Biol 48:745–762
2026. Mallamace F, Chen SH, Broccio M, Corsaro C, Crupi V, Baglioni P, Fratini E, Vannucci C, Stanley HE (2007) Role of solvent in the dynamical transitions of proteins: the case of the lysozyme-water system. J Chem Phys 127:045104
2027. Malmivuo J, Plonsey R (1995) Bioelectromagnetism - principles and applications of bioelectric and biomagnetic fields. Oxford University Press, New York
2028. Maloney SK, Mitchell D, Mitchell G, Fuller A (2007) Absence of selectivebrain cooling in unrestrained baboons exposed to heat. Am J Physiol Regul Integr Comp Physiol 292:R2059–R2067
2029. Mamonov AB, Coalson RD, Zeidel ML, Mathai JC (2007) Water and deuterium oxide permeability through Aquaporin 1: MD predictions and experimental verification. J Gen Physiol 130:111–116
2030. Mamantov MA, Gibson-Reinemer DK, Linck EB, Sheldon KS (2021) Climate-driven range shifts of montane species vary with elevation. Global Ecol Biogeogr 30:784–794
2031. Mandal BB (2000) Cryopreservation of yam apices: a comarative study with three different techniques. In: Engelmann F, Takagi H (eds) Cryopreservation of tropical plant germplasm - current research progress and applications. IPGRI, Rome, pp 233–237
2032. Mandal BB, Chandel KPS, Diwvedi S (1996) Cryopreservation of yam (*Dioscorea spp.*) shoot apices by encapsulation-dehydration. Cryo-Lett 17:165–174
2033. Manns F, Milne PJ, Gonzalez-Cirre X, Denham DB, Parel JM, Robinson DS (1998) In-situ temperature measurements with thermocouple probes during laser interstitial thermometry (LITT): quantification and correction of a measurement artifact. Lasers Surg Med 23:94–103
2034. Manns F, Borja D, Parel JM, Smiddy W, Culbertson W (2003) Semianalytical thermal model for subablative laser heating of homogeneous nonperfused biological tissue: application to laser thermokeratoplasty. J Biomed Optics 8:288–297
2035. Mansfield P (2003) Snap-shot MRI. http://nobelprize.org/nobel-prizes/medicine/laureates/2003/mansfield-lecture.pdf
2036. Manstein D, Herron GS, Sink RK, Tanner H, Anderson RR (2004) Fractional photothermolysis: a new concept for cutaneous remodeling using microscopic patterns thermal injury. Lasers Surg Med 34:426–438
2037. Mantegna RN, Stanley HE (1994) Stochastic process with ultraslow convergence to a Gaussian: the truncated Lévy flight. Phys Rev Lett 73:2946–2949
2038. Manteifel V, Bakeeva L, Karu T (1997) Ultrastructural changes in chondriome of human lymphocytes after irradiation with HeNe laser: appearance of giant mitochondria. J Photochem Photobiol B Biol 38:25–30
2039. Manuchehrabadi N, Gao Z, Zhang J, Ring HL, Shao Q, Liu F, McDermott M, Fok A, Rabin Y, Brockbank KGM, Garwood M, Haynes CL, Bischof JC (2017) Improved tissue cryopreservation using inductive heating of magnetic nanoparticles. Sci Transl Med 9:eaah4586
2040. Mao Y, Xu M (2015) Lattice Boltzmann numerical analysis of heat transfer in nano-scale silicon films induced by ultra-fast laser heating. Int J Therm Sci 89:210–221
2041. Mao YD, Xu MT (2015) Non-Fourier heat conduction in thin gold film heated by an ultra-fast-laser. Sci China Technol Sci 58:1–12
2042. Marcelli G, Parker KH, Winlove CP (2005) Thermal fluctuations of red blood cell membrane via a constant-area particle-dynamics model. Biophys J 89:2473–2480
2043. Marchant TR (1993) Thermal waves for nonlinear hyperbolic heat conduction. Math Comput Model 10:111–121
2044. Marciak-Kozlowska J, Kozlowski M (2010) Heat analysis of biological tissue exposed to laser pulses. Lasers Eng 20:279–295

2045. Marcove R, Miller T (1969) The treatment of primary and metastatic bone tumors by cryosurgery. Surg Clin Am 49:421–430
2046. Margesin R, Neuner G, Storey KB (2007) Cold-loving microbes, plants, and animals – fundamental and applied aspects. Naturwissenschaften 94:77–99
2047. Mariappan P, Weir P, Flanagan R, Voglreiter P, Alhonnoro T, Pollari M, Moche M, Busse H, Futterer J, Portugaller HR, Sequeiros RB, Kolesnik M (2017) GPU-based RFA simulation for minimally invasive cancer treatment of liver tumours. Int J Comput Assist Radioyl Surg 12:59–68
2048. Marion DW, Leonov Y, Ginsberg M, Katz LM, Kochanek PM, Lechleuthner A, Nemoto EM (1996) Resuscitative hypothermia. Crit Care Med 24:S81–S89
2049. Marion DW, Penrod LE, Kelsey SF, Obrist WD, Kochanek PM, Palmer AM, Wisniewski SR, DeKosky ST (1997) Treatment of traumatic brain injury with moderate hypothermia. N Engl J Med 336:540–546
2050. Márkus F (2011) Can a Lorentz invariant equation describe thermal energy propagation problems? In: Vikhrenko VS (ed) Heat conduction - basic research. BoD - Books on Demand, Norderstedt, pp 155–176
2051. Marqués MI, Borreguero JM, Stanley HE, Dokholyan NV (2003) Possible mechanism for cold denaturation of proteins at high pressure. Phys Rev Lett 91:138101
2052. Marth W, Aland S, Voigt A (2016) Margination of white blood cells: acomputational approach by a hydrodynamic phase field model. J Fluid Mech 790:389–406
2053. Martin K (2010) The acoustic safety of new ultrasound technologies. Ultrasound 18:110–118
2054. Mascio CE, Myers JA, Edmonds HL, Austin EH (2009) Near-infrared spectroscopy as a guide for an intermittent cerebral perfusion strategy during neonatal circulatory arrest. ASAIO J 55:287–290
2055. Mascio PD, Martinez GR, Miyamoto S, Ronsein GE, Medeiros MHG, Cadet J (2019) Singlet molecular oxygen reactions with nucleic acids, lipids, and proteins. Chem Rev 119:2043–2086
2056. Maslianitsyna AI, Ermolinsky PB, Lugovtsov AE, Priezzhev AV (2020) Study by optical techniques of the dependence of aggregation parameters of human red blood cells on their deformability. J Biomed Photon Eng 6:3371
2057. Massarweh NN, Cosgriff N, Slakey DP (2006) Electrosurgery: history, principles, and current and future uses. J Am Coll Surg 202:520–530
2058. Massarweh NN, Slakey D (2006) Electrosurgery: history, principles, and current and future uses. J Am Coll Surg 202:520–530
2059. Massie I, Selden C, Hodgson H, Fuller B, Gibbons S, Morris GJ (2014) GMP cryopreservation of large volumes of cells for regenerative medicine: active control of the freezing process. Tissue Eng 20:693–702
2060. Massip A (2001) Cryopreservation of embryos of farm animals. Reprod Domest Anim 36:49–55
2061. Masuda H, Ushiyama A, Hirota S, Wake K, Wattanabe S, Yamanaka Y, Taki M, Ohkubo C (2007) Effects of acute exposure to a 1439 MHz electromagnetic field on the microcirculatory parameters in rat brain. In vivo 21:555–562
2062. Masuda H, Ushiyama A, Hirota S, Wake K, Wattanabe S, Yamanaka Y, Taki M, Ohkubo C (2007) Effects of subchronic exposure to a 1439 MHz electromagnetic field on the microcirculatory parameters in rat brain. In vivo 21:563–570
2063. Mathot KJ, Nicolaus M, Araya-Ajoy YG, Dingemanse NJ, Kempenaers B, Grémillet D (2014) Does metabolic rate predict risk-taking behaviour? A field experiment in a wild passerine bird. Funct Ecol 29:239–249
2064. Matsumoto M, Saito S, Ohmine I (2002) Molecular dynamics simulation of the ice nucleation and growth process leading to water freezing. Nature 416:409–413
2065. Matsumoto Y, Allen JS, Yoshizawa S, Ikeda T, Kaneko Y (2005) Medical ultrasound with microbubbles. Exp Thermal Fluid Sci 29:255

2066. Matsumoto S, Matsusita M, Morita T, Kamachi H, Tsukiyama S, Furukawa Y, Koshida S, Tachibana Y, Nishimura S, Todo S (2006) Effects of synthetic antifreeze glycoprotein analogue on islet cell survival and function during cryopreservation. Cryobiology 52:90–98
2067. Matsunaga D, Imai Y, Yamaguchi T, Ishikawa T (2016) Rheology of a dense suspension of spherical capsules under simple shear flow. J Fluid Mech 786:110–127
2068. Matthias A, Ohlson KB, Fredriksson JM, Jacobsson A, Nedergaard J, Cannon B (2000) Thermogenic responses in brown fat cells are fully UCP1-dependent. J Biol Chem 275:25073–25081
2069. Matzarakis A, Mayer H, Iziomon MG (1999) Application of a universal thermal index: physioloical effective temperature. Int J Biometeorol 43:76–84
2070. Mauer J (2016) Shapes and dynamics of blood cells in Poiseuille and shear flows. Ph.D. Thesis, Universität Köln
2071. Mauer J, Mendez S, Lanotte L, Nicoud F, Abkarian M, Gompper G, Fedosov DA (2018) Flow-induced transitions of red blood cell shapes under shear. Phys Rev Lett 121:118103
2072. Maugin GA, Muschik W (1994) Thermodynamics with internal variables Part I. General concepts. J Non-Equilib Thermodyn 19:217–249
2073. Maugin GA, Muschik W (1994) Thermodynamics with internal variables Part II. Applications. J Non-Equilib Thermodyn 19:250–289
2074. Maurer MJ, Thompson HA (1973) Non-Fourier effects at high heat flux. J Heat Transf 95:284–286
2075. Maxfield FR, van Meer G (2010) Cholesterol, the central lipid of mammalian cells. Curr Opin Cell Biol 10:422–429
2076. Mayer B, Oberbauer R (2003) Mitochondrial regulation of apoptosis. News Physiol Sci 18:89–94
2077. Mayhan WG (2000) Leukocyte adherence contributes to disruption of the blood-brain barrier during activation of mast cells. Brain Res 869:112–120
2078. Mayr E (1956) Geographical character gradients and climatic adaptation. Evolution 10:105–108
2079. Mazur P (1956) Studies on the effects of subzero temperatures on the viability of spores of aspergillus flavus. I. The effect of rate of warming. J Gen Physiol 39:869–888
2080. Mazur P (1963) Kinetics of water loss from cells at subzero temperatures and the likelihood of intracellular freezing. J Gen Physiol 47:347–369
2081. Mazur P (1970) Cryobiology: the freezing of biological systems. Science 168:939–949
2082. Mazur P (1984) Freezing of living cells: mechanisms and implications. Am J Physiol 247:125–142
2083. Mazur P (1997) Biophysical and biological factors determining the ability to achieve long-term cryobiological preservation. ORNL/CP - 95102
2084. Mazur P, Kleinhans FW (2008) Relantionship between intracellular ice formation in oocytes of the mouse and Xenopus and the physical state of the external medium – a revisit. Cryobiology 56:22–27
2085. Mazur P, Koshimoto C (2002) Is intracellular ice formation the cause of death of mouse sperm frozen at high cooling rates? Biol Reprod 66:1485–1490
2086. Mazur P, Seki S, Pinn IL, Kleinhans FW, Edashige K (2005) Extra- and intra-cellular ice formation in mouse oocytes. Cryobiology 51:29–53
2087. Mazur P, Pinn IL, Kleinhans FW (2007) The temperature of intracellular ice formation in mouse oocytes vs. the unfrozen fraction at that temperature. Cryobiology 54:223–233
2088. Mazza S, Battaglia LF, Miller MW, Dewey WC, Edwards MJ, Abramowicz JS (2004) The ∇t thermal dose concept 2: in vitro cellular effects. J Therm Biol 29:151–156
2089. Mazza G, Gandol M, Capone M, Ban F, Giannetti C (2021) Thermal dynamics and electronic temperature waves in layered correlated materials. arXiv:2102.07685
2090. McAchran SE, Lesani OA, Resnick M (2005) Radiofrequency ablation of renal tumors: past, present, and future. Urology 66:15–22
2091. McAllen RM, McKinley MJ (2018) Efferent thermoregulatory pathways regulating cutaneous blood flow and sweating. Handb Clin Neurol 156:305–316

2092. McBride A, Bargmann S, Pond D, Limbert G (2016) Thermoelastic modelling of skin at finite deformations. J Thermal Biol 62:201–209
2093. McBridea D, Bargmannb S, Pondd D, Limberte A (2016) Thermoelastic modelling of the skin at finite deformations. J. Thermal Biol 62(part B):201–209
2094. McCabe KM, Hernandez M (2010) Molecular thermometry. Pediatr Res 67:469–475
2095. McCormick EJ, Sanders MS (1982) Human factors in engineering and design, 5th edn. McGraw-Hill, New York
2096. McDonough JM, Kunadian I, Kumar RR, Yang T (2006) An alternative discretization and solution procedure for the dual phase-lag equation. J Computat Phys 219:163–171
2097. McGahan J, Dodd J (2001) Radiofrequency ablation of the liver: current status. Am J Roentgenol 176:3–16
2098. McGahan JP, Brock JM, Tesluk H, Gu WZ, Schneider P, Browing PD (1992) Hepatic ablation with use of radio-frequency electrocautery in the animal model. J Vasc Intervenet Radiol 3:291–297
2099. Mchedlishvili GI (1989) Microcirculation of blood. General principles of control and disturbances (in Russian). Nauka, Leningrad (1989)
2100. Mchedlishvili G, Maeda N (2001) Blood flow structure related to red cell flow: a determination of blood fluidity in narrow microvessels. Jap J Physiol 51:19–30
2101. McIntosh JR (2001) Electron microscopy of cells: a new beginning for a new century. J Cell Biol 153:F25–F32
2102. McIntosh R, Nicastro D, Mastronarde D (2005) New views of cells in 3D: an introduction to electron tomography. Trends Cell Biol 15:43–51
2103. McKenzie JM, Voss IC, Siegel DI (2007) Groundwater flow with energy transport and water-ice phase change: numerical simulations, benchmarks, and application to freezing in peat bogs. Adv Wat Res 30:966–983
2104. McLellan TM, Selkirk GA (2006) The management of heat stress for the firefighter: a review of work conducted on behalf of the Toronto fire service. Ind Health 44:414–426
2105. McIntosh L (1994) Molecular biology of the alternative oxidase. Plant Physiol 105:781–786
2106. McWhirter JL, Noguchi H, Gompper G (2009) Flow-induced clustering and alignment of vesicles and red blood cells in microcapillaries. Proc Natl Acad Sci USA 106:6039–6043
2107. Meeuse B (1966) The voodoo lily. Sci Am. 215:80–88
2108. Meilanov R, Shabanova M, Akhmedov E (2011) A research note om a solution of Stefan problem with fractional time and space derivatives. Int Rev Chem Eng 3:810–813
2109. Meilanov RP, Shabanova MR (2007) Heat conduction equation for the media with the fractal structure (in Russian). Top Technol 8:84–85
2110. Meilanov RP, Shabanova MR (2011) Peculiarities of solutions to the heat conduction equation in fractional derivatives. Tech Phys 56:903–908
2111. Meiri S, Dayan T (2003) On the validity of Bergmann's rule. J Biogeograph 30:331–351
2112. Mekjavic IB, Sundberg CJ, Linnarsson D (1991) Core temperature 'null zone'. J Appl Physiol 71:1289–1295
2113. Mellergard P (1995) Intracerebral temperature in neurosurgical patients: intracerebral temperature gradients and relationships to consciousness level. Surg Neurol 43:91–95
2114. Mercer GN, Gill AM, Weber RO (1994) A time-dependent model of fire impact on seed survival in woody fruits. Austral J Botany 42:71–81
2115. Merrill EW, Gilliland ER, Lee TS, Salzman EW (1966) Blood rheology: effect of fibrinogen deduced by addition. Circ Res 18:437–446
2116. Meryman HT (1971) Cryoprotective agents. Cryobiology 8:173–183
2117. Mesquita MV, Vasconcellos AR, Luzzi R, Mascarenhas S (2004) Systems biology: an information-theoretic-based thermo-statistical approach. Brazil J Phys 34:459–488
2118. Metropolis N, Rosenbluth AW, Rosenbluth MN, Tellera AH, Teller E (1953) Equations of state calculations by fast computing machines. J Chem Phys 21:1087–1092
2119. Metzler R, Barkai E, Klafter J (1999) Anomalous diffusion and relaxation close to thermal equilbrium: a fractional Fokker-Planck equation approach. Phys Rev Lett 82:3563–3567

2120. Metzler R, Barkai E, Klafter J (1999) Deriving fractional Fokker-Planck equations from a generalised master equation. Erophys Lett 46:431–436
2121. Metzler R, Compte A (1999) Stochastic foundation of normal and anomalous Cattaneo-type transport. Phys A Stat Mech Appl 268:454–468
2122. Metzler R, Klafter J (2000) The random walk's guide to anomalous diffusion: a fractional dynamics approach. Phys Rep 339:1–77
2123. Metzler R, Klafter J (2004) The restaurant at the end of the random walk: recent developments in the description of anomalous transport by fractional dynamics. J Phys A Math Gen 37:R161–R208
2124. Michenfelder JD, Milde JH (1991) The relationship among canine brain temperature, metabolism, and function during hypothermia. Anesthesiology 75:130–136
2125. Michenfelder JD, Milde JH (1992) The effect of profound levels of hypothermia (below 14°c) on canine cerebral metabolism. J Cereb Blood Flow Metab 12:877–880
2126. Mignon T, Mendez S (2021) A theoretical investigation of the frisbee motion of red blood cells in shear flow. Math Model Nat Phenom 16:23
2127. Mikic BB (1967) A model rate equation for transient thermal conduction. Int J Heat Mass Transf 10:1899–1904
2128. Milani V, Noessner E (2006) Effects of thermal stress on tumor antigenicity and recognition by immune effector cells. Cancer Immunol Immunother 55:312–319
2129. Miles CA (1993) Kinetics of collagen denaturation inmammalian lens capsules studied by differential scanning calorimetry. Int J Biol Macromol 15:265–271
2130. Miles CA, Bailey AJ (2001) Thermally labile domains in the collagen molecule. Micron 32:325–332
2131. Milhorat TH (1975) The third circulation revisited. J Neurosurg 42:628–645
2132. Millar C (2003) 3D Simulation techniques for biological ion channels. Ph.D. Thesis, Glasgow
2133. Millar C, Asenov A, Roy S (2005) Self-consistent particle simulation of ion channels. J Comput Theor Nanoscience 2:1–12
2134. Millenbaugh NJ, Roth C, Sypniewska R, Chan V, Eggers JS, Kiel JL, Blystone RV, Mason PA (2008) Gene expression changes in the skin of rats induced by prolonged 35 GHz millimeter-wave exposure. Radiat Res 169:288–300
2135. Miller LK (1982) Cold-hardiness strategies of some adult and immature insects oberwintering in interior Alaska. Comp Biochem Physiol A 73:595–604
2136. Miller DL (2012) Induction of pulmonary hemorrhage in rats during diagnostic ultrasound. Ultrasound Med Biol 38:1476–1482
2137. Miller K, Ross B (1993) An introduction to the fractional calculus and fractional differential equations. Wiley, New York
2138. Miller DL, Smith NB, Bailey MR, Czarnota GJ, Hynynen K, Makin RS (2012) Overview of therapeutic ultrasound applications and safety considerations. J Ultrasound Med 31:623–634
2139. Miller DL, Dou C, Raghavendran K (2015) Anesthetic techniques influence the induction of pulmonary capillary hemorrhage during diagnostic ultrasound scanning in rats. J Ultrasound Med 34:289–297
2140. Millian AJ (1984) Whole-body hyperthermia induction techniques. Cancer Res 44(Suppl):4869s–4872s
2141. Millien V, Lyons SK, Olson L, Smith FA, Wilson AB, Yom-Tov Y (2006) Ecotypic variation in the context of global climate change: revisiting the rules. Ecol Lett 9:853–869
2142. Mills WJ (2001) Clinical aspects of freezing cold injury. In: Lounsbury DE, Randolh KB, Burr RE (eds) Medical aspects of harsh environments, vol 1. Office The Surgeon General, Department of the Army USA, Falls Church, pp 429–466
2143. Milner TE, Goodman DM, Tanenbaum BS, Nelson JS (1995) Depth profiling of laser-heated chromophores in biological tissues by pulsed photothermal radiometry. J Opt Soc Am A 12:1479–1488
2144. Milner TE, Smithies DJ, Goodman DM, Lau A, Nelson JS (1996) Depth determination of chromophores in human skin by pulsed photothermal radiometry. Appl Opt 35:3379–3385

2145. Milovanov VK (1934) Principles of artificial insemination (in Russian). State Publishing House, Moscow
2146. Minnich AJ, Dresselhaus M, Ren Z, Chen G (2009) Bulk nanostructured thermoelectric materials: current research and future prospects. Energy Env Sci 2:466–479
2147. Minorsky PV (1989) Temperature sensing in plants: a review and an hypothesis. Plant Cell Environ 12:195–135
2148. Mir LM, Rubinsky B (2002) Treatment of cancer with cryochemotherapy. Br J Cancer 86:1658–1660
2149. Mir LM, Belehradek M, Domenge C, Liboniski B, Orlowski S, Belehradek J, Schwaab B, Paoletti C (1991) Electrochemotherapy. A novel antitumor treatment: first clinical trial. C R Acad Sci Ser III 313:613–618
2150. Mishima O, Stanley HE (1998) The relationship between liquid, supercooled and glassy water. Nature 396:329–335
2151. Mishra SC, Chungh P, Kumar P, Mitra K (2006) Development and comparison of the DTM, the DOM and the FVM formulations for the short-pulse laser transport through a participating medium. Int J Heat Mass Transf 49:1820–1832
2152. Misra JC, Shit GC, Rath HJ (2010) Flow and heat transfer of a MHD viscoelastic fluid in a channel with stretching walls: some applications to haemodynamics. arXiv:1007.1228v1 [physics.flu-dyn]
2153. Mital M, Tafreshi HV (2012) A methodology for determining optimal thermal damage in magnetic nanoparticle hyperthermia cancer treatment. Int J Numer Methods Biomed Eng 28:205–213
2154. Mitchell P (1961) Coupling of phosphorylation to electron and proton transfer by a chemiosmotic type of physiology. Nature 191:144–148
2155. Mitchell P (1979) Keilin's respiratory chain concept and its chemiosmotic consequences. Science 206:1148–1159
2156. Mitchell JW, Myers GE (1968) An analytical model of the countercurrent heat exchange phenomena. Biophys J 8:897–911
2157. Mitra K, Kumar S, Vedavarz A, Moallemi MK (1995) Experimental evidence of hyperbolic heat conduction in processed meat. J Heat Transf 117:568–573
2158. Mitry RR, Hughes RD, Dhawan A (2002) Progress in human hepatocytes: isolation, culture & cryopreservation. Semin Cell Dev Bio 13:463–467
2159. Moares EM (2011) Time varying heat conduction in solids. In: Vikrenko V (ed) Heat conduction - basic research. InTech, Rijeka, pp 178–202
2160. Mochalin V, Shenderova O, Ho D, Gogotsi Y (2012) The properties and applications of nanodiamonds. Nat Nanotech 7:11–23
2161. Moche M, Busse H, Futterer JJ, Hinestrosa CA, Seider D, Brandmaier P, Kolesnik M, Jenniskens S, Blanco Sequeiros R, Komar G, Pollari M, Eibisberger M, Portugaller HR, Voglreiter P, Flanagan R, Mariappan P, Reinhardt M (2020) Clinical evaluation of in silico planning and real-time simulation of hepatic radiofrequency ablation. Eur Radiol 30:934–942
2162. Mochnacki B, Majchrzak E (2017) Numerical model of thermal interactions between cylindrical cryoprobe and biological tissue using the dual-phase lag equation. Int J Heat Mass Transf 108:1–10
2163. Mochnacki B, Piasecka-Belkhayat A (2013) Numerical modeling of skin tissue heating using the interval finite difference method. Mol Cell Biomech 10:233–244
2164. Mochnacki B, Suchy J (1993) Numerical methods in computations of foundry processes. Polish Scientific Publishers PWN, Cracow
2165. Modest MF (2003) Radiative heat transfer, 2nd edn. McGraw-Hill, New York
2166. Mofrad M (2009) Rheology of the cytoskeleton. Annu Rev Fluid Mech 41:433–453
2167. Moghadam MN, Abdel-Sayed P, Camine VM, Pioletti DP (2015) Impact of synovial fluid flow on temperature reulation in knee cartilage. J Biomech 48:370–374
2168. Mohajer M, Ayani MB, Tabrizi HB (2016) Numerical study of non-Fourier heat conduction in a biolayer spherical living tissue during hyperthermia. J Therm Biol 62:181–188

2169. Mohammed Y, Verhey JF (2005) A finite element method model to simulate laser interstitial thermo therapy in anatomical inhomogeneous regions. BioMed Eng Online 4:1–16
2170. Mohandas N, Chasis J (1993) Red blood cell deformability, membrane material properties and shape: regulation by transmembrane, skeletal and cytosolic proteins and lipids. Semin Hematol 30:171–192
2171. Mohandas N, Evans E (1994) Mechanical properties of the red cell membrane in relation to molecular structure and genetic defeats. Ann Rev Biophys Biomol Struct 23:787–818
2172. Mohandas N, Gallagher PG (2008) Red cell membrane: past, present, and future. Blood 112:3939–3948
2173. Molecular Dynamics Machine. http://atlas.riken.go.jp/mdm/index.html
2174. Molina JAL, Rivera MJ, Berjano E (2014) Fourier, hyperbolic and relativistic heat equations: a compative analytic study. Proc Roy Soc A 470:20140547
2175. Monroy AF, Dhindsa RS (1995) Low-temperature signal transduction: induction of cold acclimation-specific genes of alfaalfa by calcium at 25° C. Plaant Cell 7:321–331
2176. Mont M, Einhorn T, Sponseller P, Hungerford D (1998) The trapdoor procedure using autogeneous and cancellous bone grafts for osteonecrosis of the femoral head. J Bone Joint Surg 80B:56–62
2177. Montroll EW, Weiss GH (1965) Random walks on lattices. II. J Math Phys 6:167
2178. Moore Y, Sofer P, Ilovich M (2001) The science and technology behind cryosurgery. Technical notes, Galil Medical
2179. Moradi A, Ahmadikia H (2012) Numerical study of solidification in biological tissue with blood flow and metabolism effects by the dual phase lag model. J Eng Med 226:406–416
2180. Moran DS, Epstein Y (2006) Evaluation of the envrionmental stress index (ESI) for hot/dry and hot/wet climates. Indus Health 46:399–403
2181. Moran DS, Eliyahu U, Heled Y, Rabinovitz S, Hoffman J, Margaliot M (2004) Core temperature measurement by microwave radiometry. J Thermal Biol 29:539–542
2182. Moreno AJ, Buldyrev SV, Nave EL, Saika-Voivod I, Sciortino F, Tartaglia P, Zaccarelli E (2005) Energy landscape of a simple model for strong liquids. Phys Rev Lett 95:157802
2183. Morgan K (1998) Thermoneutral zone and critical temperatures of horses. J Thermal Biol 23:59–61
2184. Morimoto RI, Tissieres A, Georgopoulos C (1990) Stress proteins in biology and medicine. Cold Spring Harbor, New York
2185. Morimoto R, Laakso I, Santis V, Hirata A (2016) Relationship between peak spatial-averaged specific absorption rate and peak temperature elevation in human head in frequency range of 1–30 GHz. Phys Med Biol 61:5406–5425
2186. Moritz AR, Henriques FC (1947) Studies of thermal injury, II: the relative importance of time and surface temperature in the causation of cutaneous burns. Am J Pathol 23:695–720
2187. Moriyama E, Salcman M, Broadwell RD (1991) Blood-brain barrier alteration after microwave-induced hyperthermia is purely a thermal effect: I. Temperature and power measurements. Surg Neurol 35:177–182
2188. Morley MJ (1966) Thermal conductivities of muscles, fat and bones. J Food Techn 1:303–311
2189. Mornet S, Vasseur S, Grasset F, Duguet E (2004) Magnetic nanoparticle design for medical diagnosis and therapy. J Mater Chem 14:2161–2175
2190. Morris GJ (2006) Rapidly cooled human sperm: no evidence of intracellular ice formation. Hum Reprod 21:2075–2083
2191. Morris GJ, Acton E, Avery S (1999) A novel approach to sperm cryopreservation. Hum Reprod 14:1013–1021
2192. Morris JM, Goodrich M, Acton E, Fonesca F (2006) The high viscosity encountered during freezing in glycerol solutions: effect on cryopreservation. Cryobiology 52:323–334
2193. Morrison PR (1946) Temperature regulation in three central American mammals. J Cell Comp Physiol 27:125–137
2194. Morrison SF, Nakamura K (2019) Central mechanisms for thermoregulation. Annu Rev Physiol 81:285–308

2195. Morse PM, Feshbach H (1953) Methods of theoretical physics. McGraw-Hill, New York
2196. Moseley PL (1997) Heat shock proteins and heat adaptation of the whole organism. J Appl Physiol 83:1413–1417
2197. Moshrefi-Torbati M, Hammond JK (1998) Physical and geometrical interpretation of fractional operators. J Frank Inst 335:1077–1086
2198. Moskowitz MJ, Paulsen KD, Ryan TP, Pang D (1994) Temperature field estimation using electrical impedance profiling methods. Int J Hyperth 10:229–245
2199. Mouritsen OG, Bloom M, Ipsen JH, Miao L, Nielsen M, Polson J, Thewalt J, Zhu H, Zuckermann MJ (2004) Random-lattice simulations of biomembranes. NIC Ser 23:347–374
2200. Moyers-Gonzalez MA, Owens RG (2010) Mathematical modelling of the cell-depleted peripheral layer in the steady flow of blood in a tube. Biorheology 47:39–71
2201. Moyes CD, Schulte PM (2005) Principles of animal physiology. Pearson/Benjamin Cummings, San Francisco
2202. Mrevlishvili GM (1979) Low-temperature calorimetry of biological macromolecules. Sov Phys Usp 22:433–455
2203. Mrozek S, Vardon F, Geeraerts T (2012) Brain temperature: physiology and pathophysiology after brain injury. Anesth Res Pract 2012:989487
2204. Mugnano J, Lee R, Taylor R (1996) Fat body cells and calcium phosphate spherules induce ice nucleation in the freeze-tolerant larvae of the gall fly Eurosta solidaginis (diptera, tephritidae). J Exp Biol 199:465–471
2205. Muguruma C, Okamoto Y, Mikami M (2002) An application of the canonical Monte Carlo method to the bulk water system. Int El J Mol Des 1:583–592
2206. Muhsin K, Gökhan S (2008) Modelling airfow, heat transfer and moisture transport around a standing human body by computational fluid dynamics. Int Commun Heat Mass Transf 35:1159–1164
2207. Mukherjee IN, Li Y, Song YC, Long RC, Sambanis A (2008) Cryoprotectant transport through articular cartilage for long-term storage: experimental and modeling studies. Osteoarthr Cartil 16:1379–1386
2208. Mukherjee PK, Bhattacharya J (2007) Phenomenological theory of the nematic to lamellar phase transition in lyotropic liquid crystals. J Chem Phys 126:024901
2209. Mukhopadhyay S, Prasad R, Kumar R (2011) On the theory of two-temperature thermoelasticity with two phase-lags. J Therm Stress 34:352–365
2210. Muldrew K, Rewcastle J, Donnelly BJ, Saliken JC, Liang S, Goldie S, Olson M, Baissalov R, Sandison G (2001) Flounder antifreeze peptides increase the efficacy of cryosurgery. Cryobiology 42:182–189
2211. Mulier S, Jiang Y, Jamart J, Wang C, Feng Y, Marchal G, Ni Y (2015) Bipolar radiofrequency ablation with 2x2 electrodes as a building block for matrix radiofrequency ablation: ex vivo liver experiments and finite element method modelling. Int J Hyperth 31:649–665
2212. Mullen SF (2007) Advances in fundamental cryobiology of mammalian oocytes. Ph.D. Thesis, University of Missouri-Columbia
2213. Mullen SF, Agka Y, Critser JK (2004) Modeling of probability of MII spindle disruption in bivine oocytes as a function of total osmolality using logistic regression and its application toward improved CPA addition and removal procedures. Cell Pres Tech 2:145–155
2214. Mullen SF, Rosenbaum M, Critser JK (2007) The effect of osmotic stress on the cell volume, metaphase II spindle and developmental potential of in vitro matured porcine oocytes. Cryobiology 54:281–289
2215. Muller AW (1995) Were the first organisms heat engines? A new model for biogenesis and the early evolution of biological energy conversion. Prog Biophys Mol Biol 63:193–231
2216. Muller AWJ (1996) The thermosynthesis model for the origin of life and the emergence of regulation by ca^{2+}. Essays Biochem 31:103–119
2217. Muller AWJ (2005) Thermosynthesis as energy source for the RNA world: a model for the bioenergetics of the origin of life. BioSystems 82:93–102
2218. Muller AWJ (2009) Emergence of animals from heat engines Part 1. Before the snowball Earths. Entropy 11:463–512

2219. Muller AWJ (2012) Major transitions in evolution linked to thermal gradients above hydrothermal vents. arXiv:1212.2271 [physics.gen-ph]
2220. Muller AWJ, Schulze-Makuch D (2006) Sorption heat engines: simple inanimate negative entropy generators. Phys A 362:369–381
2221. Muller AWJ, Schulze-Makuch D (2006) Thermal energy and the origin of life. Orig Life Evol Biosph 36:177–189
2222. Müller I (1972) On the frame dependence of stress and heat flux. Arch Rat Mech Anal 45:241–250
2223. Müller I (1999) Speed of propagation in classical and relativistic extended thermodynamics. Living Rev Relat 2:1–32
2224. Müller GJ, Sliney DH (1989) Dosimetry of laser radiation in medicine and biology. SPIE Press, Bellingham
2225. Müller I, Ruggeri T (1998) Rational extended thermodynamics. Springer, New York
2226. Müller-Reichert T, Hohenebrg H, O'Toole ET, Mcdonald K (2003) Cryoimmobilization and three-dimensional visualization of C. eleans ultrastructure. J Microscopy 212:71–80
2227. Mullis AM (1997) Rapid solidification and a finite velocity for the propagation of heat. Mater Sci Eng 226–228:28–32
2228. Mullis AM (1997) Rapid solidification within the framework of a hyperbolic conduction model. Int J Heat Mass Transf 40:4085–4094
2229. Murakami A, Nagao K, Sakaguchi R, Kida K, Hara Y, Mori Y, Okabe K, Harada Y, Umeda M (2022) Cell-autonomous control of intracellular temperature by unsaturation of phospholipid acyl chains. Cell Rep 38:110487
2230. Murakami S, Zeng D, Hayashi T (1999) CFD analysis of wind environment around a human body. J Wind Eng Ind Aerodyn 83:393–408
2231. Murase K (2016) An integral-transform approach to the bioheat transfer problems in magnetic hyperthermia. arXiv:1604.07138 [cs.CE]
2232. Murase K, Oonoki J, Takata H, Song R, Angraini A, Ausanai P, Matsushita T (2011) Simulation and experimental studies on magnetic hyperthermia with use of superparamagnetic iron oxide nanoparticles. Radiol Phys Technol 4:194–202
2233. Murata N, Ishizaki-Nishigawa O, Higashi S, Hayashi H, Tasaka Y, Nishida I (1992) Genetically engineered alteration in the chilling sensitivity of plants. Nature 356:710–713
2234. Murdoch AI (1983) On material frame-indifference, intrinsic spin, and certain constitutive relations motivated by the kinetic theory of gases. Arch Rat Mech Anal 83:185–194
2235. Murphy DJ (1977) Metabolic and tissue solute changes associated with changes in the freezing tolerance of the bivalve mollusc *Modiolus Demissus*. J Exp Biol 69:1–12
2236. Murray CD (1926) The physiological principle of minimum work applied to the angle of branching of arteries. J Gen Physiol 9:835
2237. Murray HM, Hew CL, Kao KR, Fletcher GL (2002) Localisation of cells from winter flounder gill expressing a skin type antifreeze protein gene. Can J Zool 80:110–119
2238. Murshed SMS, Leong KC, Yang KC (2008) Investigations of thermal conductivity and viscosity of nanofluids. Int J Therm Sci 47:560–568
2239. Muschik W (1998) Objectivity and frame indifference, revisited. Arch Mech 50:541–547
2240. Musielak M (2009) Red blood cell-deformability measurement: review of techniques. Clin Hemorheol Microcirc 42:47–64
2241. Muschik W, Restuccia L (2008) Systematic remarks on objectivity ad frame-indifference, liquid crystal theory as an example. Arch Appl Mech 78:837–854
2242. Mustafa A, Abhilash RH, Chauhan S (2017) Organ movement and targeting during non-invasive therapy and surgery. SM J Biomed Eng 3:2014
2243. Mutsenko V, Knaack S, Lauterboeck L, Tarusin D, Sydykov B, Cabiscol R, Ivnev J, Belikan J, Beck A, Dipresa D, Lode L, Khassawna T, Kampschulte M, Scharf R, Petrenko AY, Korossis S, Wolkers WF, Gelinsky M, Glasmacher B, Gryshkov O (2020) Effect of 'in air'freezing on post-thaw recovery of Callithrix jacchus mesenchymal stromal cells and properties of 3d collagen-hydroxyapatite scaffolds. Cryobiology 92:215–230

2244. Nabil M, Zunino P (2016) A computational study of cancer hyperthermia based on vascular magnetic nanoconstructs. R Soc Open Sci 3:160287
2245. Nabil M, Decuzz P, Zunino P (2015) Modelling mass and heat transfer in nano-based cancer hyperthermia. R Soc Open Sci 2:150447
2246. Nada H, van der Eerden JPJM (2003) An intermolecular potential model for the simulation of ice and water near the melting point: a six-site model of H_2O. J Chem Phys 118:7401
2247. Nadel ER, Pandolf KB, Roberts MF, Stolwijk JAJ (1974) Mechanisms of thermal acclimation to exercise and heat. J Appl Physiol 37:515–520
2248. Nadel ER, Mack GW, Nose H, Tripathi A (1987) Tolerance to severe heat and exercise: perophera vascular responses to body fluid changes. In: Hales RJS, Richards DAB (eds) Heat stress: physical exertion and environment. Elsevier, Amsterdam, pp 117–131
2249. Nagashima K (2006) Central mechanisms for thermoregulationin a hot environment. Industr Health 44:359–367
2250. Nagashima H, Kashiwasaki N, Ashman RJ, Grupen CG, Seamark RF, Nottle MB (1994) Recent advances in cryopreservation of porcine embryos. Theriogenelogy 41:113–118
2251. Nagashima H, Kashiwasaki N, Ashman RJ, Grupen CG, Nottle MB (1995) Cryopreservation of porcine embryos. Nature 374:416
2252. Nagy KA, Odell DK, Seymour SR (1972) Temperature regulation by the inflorescence of Philodendron. Science 178:1195–1197
2253. Nakamori K, Odawara M, Nakajima T, Mizutani T, Tsubota K (1997) Blinking is controlled primarily by ocular surface conditions. Am J Ophthalmol 124:23–30
2254. Nakamura T, Hasegawa S, Shio H, Uyesaka N (1994) Rheologic and pathologic significance of red cell passage through narrow pores. Blood Cells 20:151–165
2255. Nakamura R, Tuncali K, Morton PR, Hata N, Silverman SG, Kikinis R, Jolesz FA, Zientara GP (2004) Control system for mr-guided cryotherapy – short-term prediction of therapy boundary using automatic segmentation and 3D optical flow. Lect Notes Comput Sci 3217:542–550
2256. Nakano T, Kikugawa G, Ohara T (2013) Molecular heat transfer in lipid bilayers with symmetric and asymmetric tail chains. J Heat Transf 135(661301):1–8
2257. Nakasako M (2004) Water-protein interactions from high-resolution protein crystallography. Philos Trans R Soc Lond A 359:1191–1206
2258. Nakashima K, Yamaguchi-Shinozaki K (2006) Regulons involved in osmotic stress-responsive and cold-responsive gene expression in plants. Physiol Plant 126:62–71
2259. Nakayama A, Kuwahara F (2008) A general bioheat transfer model based on the theory of porous media. Int J Heat Mass Transf 51:3190–3199
2260. Nakayama A, Kuwahara F, Umemoto T, Hayashi T (2002) Heat and fluid flow within anisotropic porous medium. J Heat Transf 124:746–753
2261. Nakayama A, Kuwahara F, Kodama Y (2006) A thermal dispersion flux transport equation and its mathematical modelling for heat and fluid flow in a porous medium. J Fluid Mech 563:81–96
2262. Nakayama A, Sano Y, Yoshikawa K (2010) A rigorous derivation of the bioheat equation for local tissue heat transfer based on a volume averaging theory. Heat Mass Transf 46:739–746
2263. Nakhushev AM (2003) Fractional calculus and applications. Fismatlit, Moscow
2264. Namakshenas P, Mojra A (2019) Numerical study of non-Fourier thermal ablation of benign thyroid tumor by focused ultrasound (FU). Biocybern Biomed Eng 39:571–585
2265. Namakshenas P, Mojra A (2020) Microstructure-based non-Fourier heat transfer modeling of HIFU treatment for thyroid cancer. Comput Methods Program Biomed 197:105698
2266. Nanev CN (2007) Protein crystal nucleation: recent notions. Cryst Res Technol 42:4–12
2267. Nanopoulos DV (1995) Theory of brain function, quantum mechanics and superstrings. CERN-TH/95-128
2268. Naqvi KR, Waldenstream S (2005) Brownian motion description of heat conduction by phonons. Phys Rev Lett 95:065901
2269. Narasimhan TN (1999) Fourier's heat conduction equation: history, influences and connections. Proc Indian Acad Sci 108:117–148

2270. Narasimhan A, Sadasivam S (2013) Non-Fourier bio heat transfer modelling of thermal damage during retinal laser irradiation. Int J Heat Mass Transf 60:591–597
2271. Narasimhan A, Jha KK, Gopal L (2010) Transient simulations of heat transfer in human eye undergoing laser surgery. Int J Heat Mass Transf 53:482–490
2272. Narayanamurti V, Dynes RC, Andres K (1975) Propagation of sound and second sound using heat pulses. Phys Rev B 11:2500–2524
2273. Nash GB, Meiselman HJ (1985) Alteration of red cell membrane viscoelasticity by heat treatment: effect on cell deformability and suspension viscosity. Biorheology 22:73–84
2274. Natarajan S, Raman S, Priester AM, Garritano J, Margolis DJA, Lieu P, Macairan ML, Huang J, Grundfest W, Marks LS (2016) Focal laser ablation of prostate cancer: phase I clinical trial. J Urol 196:68–75
2275. Nath S (2016) The thermodynamic efficiency of ATP synthesis in oxidative phosphorylation. Biophys Chem 219:69–74
2276. Naveira-Cotta CP, Cotta RM, Orlande HRB, Fudym O (2009) Eigenfunction expansions for transient diffusion in heterogeneous media. Int J Heat Mass Transf 52:5029–5039
2277. Nayanajith PGH, Saha SC, Gu YT (2014) Deformation of a single red blood cell in a microvessel. ANZIAM J 55:C64–C79
2278. Nealson KH, Conrad PG (1999) Life: past, present and future. Philos Trans R Soc Lond B 354:1923–1939
2279. Neimark MA, Konstas AA, Choi JH, Laine AF, Pile-Spellman J (2008) Brain cooling maintenance with cooling cap following induction with intracarotid cold saline infusion: a quantitative model. J Theor Biol 253:333–344
2280. Nekhaeva TL (2017) Cryoconservation (in Russian). In: Prohorov GG, Belyaev AM, Prokhorov DG (eds) Fundamentals of clinical cryomedicine. Kniga Po Trebovaniyu, Saint-Petersburg, pp 131–147
2281. Nellis GF, Klein SA, Keppler F, Fredrickson K (2008) Design data for optimization of cryosurgical probes ASHRAE 1246-TRP
2282. Nelson DA (1990) The biology of myelopoiesis. Clin Lab Med 10:649–659
2283. Nelson JS, Millner TE, Anvari B, Tanelbaum BS, Kimel BS, Svaasand LO, Jacques SL (1995) Dynamic epidermal cooling during pulsed-laser treatment of port-wine stain – a new methodology with preliminary clinical evaluation. Acta Dermatol 131:695–700
2284. Nelson JS, Milner TE, Lars OS, Kimel S (1995) Laser pulse duration must match the estimated thermal relaxation time for successful photothermolysis of blood vessels. Lasers Med Sci 10:9–12
2285. Nelson JS, Milner TE, Anvari B, Tanenbaum BS, Svaasand LO, Kimel S (1996) Dynamic epidermal cooling in conjunction with laser-induced photothermolysis of port wine stain blood vessels. Lasers Surg Med 19:224–229
2286. Neu B, Meiselman HJ (2002) Depletion-mediated red blood cell aggregation in polymer solutions. Biophysical J 83:2482–2490
2287. Neuberger T, Schopf B, Hofmann H, Hofmann M, von Rechenberga B (2005) Superparamagnetic nanoparticles for biomedical applications: possibilities and limitations of a new drugdelivery system. J Magn Magn Mat 293:483–496
2288. Neueder A, Gipson TA, Batterton S, Lazell HJ, Farshim PP, Paganetti P, Housman DE, Bates GP (2017) HSF1-dependent and -independent regulation of the mammalian in vivo heat shock response and its impairment in Huntington's disease mouse models. Sci Rep 7:12556
2289. Neufeld E, Chavannes N, Samaras T, Kuster N (2007) Novel conformal technique to reduce staircasing artifacts at material boundaries for FDTD modeling of the bioheat equation. Phys Med Biol 52:4371–4381
2290. Neufeld E, Paulides MM, van Rhoon GC, Kuster N (2012) Numerical modeling for simulation and treatment planning of thermal therapy. In: Moros EG (ed) Physics of thermal therapy: fundamentals and clinical applications. CRC Press, Boca Raton, pp 119–136

2291. Neves EB, Salamunes ACC, de Oliveira RM, Stadnik AMW (2017) Effect of body fat and gender on body temperature distribution. J Therm Biol 70:1–8
2292. Newman WH, Lele PP, Bowman HF (1990) Limitations and significance of thermal washout data obtained during microwave and ultrasound hyperthermia. Int J Hyperth 6:771–784
2293. Ng EYK, Ooi EH (2006) FEM simulation of the eye structure with bio-heat analysis. Comput Methods Programs Biomed 82:268–276
2294. Ng EYK, Sudharsan NM (2001) Effect of blood flow, tumour and cold stress in a female breast: a novel time-accurate computer simulation. Proc Inst Mech Eng H J Eng Med 215:393–404
2295. Nguyen DT, Colvin ME, Yeh Y, Feeney RE, Fink WH (2004) Intermolecular interaction studies of winter flounder antifreeze protein reveal the existence of thermally accessible binding state. Biopolymers 75:109–117
2296. Ni D, Shi HJ, Yin YJ, Niu LS (2006) Morphology and phase behavior of two-component lipid membranes. J Biol Phys 32:369–381
2297. Nicolis G, Prigogine I (1989) Exploring complexity. Freeman, New York
2298. Niemz MH (2007) Laser-tissue interactions: fundamentals and applications. Springer, Berlin
2299. Nigmatullin RR (1992) A fractional integral and its physical interpretation. Theor Math Phys 90:242–251
2300. Nigmatullin RR, Le Mehaute A (2005) Is there geometrical/physical meaning of the fractional integral with complex exponent. J Non-Crystalline Solids 351:2888–2899
2301. Nilkolić V, Said-Houari B (2021) Local well-posedness of a coupled Westervelt-Pennes model of nonlinear ultrasonic heating. arXiv:2108.06999v1 [math.AP]
2302. Nilkolić V, Said-Houari B (2022) The Westervelt-Pennes model of nonlinear thermoacoustics: global solvability and asymptotic behavior. J Differ Equ 336:628–653
2303. Nijhout HF, Best J, Reed MC (2014) Escape from homeostasis. Math Biosci 257:104–110
2304. Niklas KJ (2004) Plant allometry: is there a grand unifying theory? Biol Rev Cambridge Philos Soc 79:871–889
2305. Niklas KJ (2006) Scaling the paths of resistance. New Phytol 169:219–221
2306. Niklas KJ, Enquist BJ (2001) Invariant scaling relationships for interspecific plant biomass production rates and body size. Proc Nat Acad Sci USA 98:2922–2927
2307. Niklas KJ, Kutschera U (2015) Kleiber's law: how the fire of life ignited debate, fueled theory, and neglected plants as model organisms. Plant Signal Behav 10:e1036216
2308. Nilsson AMK, Lucassen GW, Verkruysse W, Anderson-Engels S, van Gemert MJS (1997) Changes in optical properties of human whole blood in vitro due to slow heating. Photochem Photobiol 65:366–373
2309. Nilsson AMK, Sturesson C, Liu DL, Anderson-Engels S (1998) Changes in spectral shape of tissue optical properties in conjunction with laser-induced thermotherapy. Appl Opt 37:1256–1267
2310. Nobel PS (2009) Physicochemical and environmental plant physiology. Elsevier Academic Press, Oxford
2311. Noguchi H (2009) Swinging and synchronized rotations of red blood cells in simple shear flow. arXiv:0905.0180v2 [cond-mat.soft]
2312. Noguchi H, Gompper G (2005) Shape transitions of fluid vesicles and red blood cells in capillary flows. Proc Natl Acad Sci USA 102:14159–14164
2313. Nollen EAA, Brunsting JF, Roelofsen H, Weber LA, Kampinga HH (1999) In vivo chaperone activity of heat shock protein 70 and thermotolerance. Mol Cell Biol 19:2069–2079
2314. Noroozi MJ, Saedodin S, Gangi DD (2016) A new solution for nonlinear dual phase lag heat conduction problem. Alexandria Eng J 55:1745–1752
2315. Noroozi MJ, Saedodin S, Gangi DD (2016) Nonlinear solution to a non-Fourier heat conduction problem in a slab heated by laser source. Arch Mech Eng LXIII:129–144
2316. Ntziachristos V (2006) Fluorescence molecular imaging. Annu Rev Biomed Eng 8:1–33
2317. Nunziato JW (1971) On heat conduction in materials with memory. Q Appl Math 29:187–204

2318. Nussmeier NA (2002) A review of risk factors for adverse neurological outcome after cardiac surgery. J Extracorp Technol 34:4–10
2319. Nybo L, Secher NH, Nielsen B (2002) Inadequate heat release from the human brain during prolonged exercise with hyperthermia. J Physiol 545:697–704
2320. Nygård K, Hakala M, Manninen S, Andrejczuk A, Itou M, Sakurai Y, Pettersson L, Hämäläinen K (2006) Compton scattering study of water versus ice I_h: intra- and intermolecular structure. Phys Rev E 74:031503
2321. Nyíri B (1991) On the entropy current. J Non-Equil Thermodyn 16:179–186
2322. Obana A, Gohto Y (2002) Scanning laser system for photodynamic therapy of choroidal neovascularization. Lasers Surg Med 30:170–175
2323. Oberkampf WL, Trucano TG (2002) Verification and validation in computational fluid dynamics. Prog Aerospace Sci 38:209–272
2324. O'Brien WD (2007) Ultrasound - biophysics mechanisms. Prog Biophys Mol Biol 93:212–255
2325. Ocal S, Ozcan MU, Basdogan I, Basdogan C (2010) Effect of preservation period on the viscoelastic material properties of soft tissue with implication for liver transplantation. J Biomech Eng 132:1–7
2326. Oden JT (2002) The promise of computational engineering and science: will it be kept? IACM Exp 12:12–15
2327. Odibat ZM, Shawagfeh NT (2007) Generized Taylor's formula. Appl Math Comput 186:285–294
2328. Ohno N, Terada N, Murata, SI, Katoh R, Ohno S (2005) Application of cryotechniques with freeze-substitution for the immunohistochemical demonstration of intranuclear pcreb and chromosome territory. J Histoch Cytochem 53:55–62
2329. Ohshiro T, Calderhead RG (1989) Low level laser therapy: a practical introduction. Brit J Surg 76:11–18
2330. Okabe K, Inada N, Gota C, Harada Y, Funatsu T, Uchiyama S (2012) Intracellular temperature mapping with a fluorescent polymeric thermometer and fluorescence lifetime imaging microscopy. Nat Commun 3:705
2331. Okabe K, Sakaguchi R, Shi B, Kiyonaka S (2018) Intracellular thermometry with fluorescent sensors for thermal biology. Pflügers Arch 470:717–731
2332. Okajima J, Maruyama S, Takeda H, Atsuki K (2009) Dimensionless solutions and general characteristics of bioheat transfer during thermal therapy. J Thermal Biol 34:377–384
2333. Okawa K, Sekine M, Maeda M, Tada M, Abe M, Matsushita N, Nishio K, Handa H (2006) Heating ability of magnetite nanobeads with various sizes for magnetic hyperthermia at 120 kHz, a noninvasive frequency. J Appl Phys 99:08H102
2334. Okoniewski M, Okoniewska E, Stuchly MA (1997) Three-dimensional subgridding algorithm for FDT. IEEE Trans Antennas Propag 45:422–429
2335. Olalla-Tárraga MA, Rodríuez MA, Hawkins BA (2006) Broad-scale patterns of body size in squamate reptiles of Europe and North America. J Biogeograph 33:781–793
2336. Oldham K, Spanier J (1974) The fractional calculus; theory and applications of differentiation and integration to arbitrary order. Academic Press, New York
2337. Onda Y, Kato Y, Abe Y, Ito T, Morohashi M, Ito Y, Ichikawa M, Matsukawa K, Kakizaki Y, Koiwa H, Ito K (2008) Functional coexpression of the mitochondrial alternative oxidase and uncoupling protein underlies thermoregulation in the thermogenic florets of skunk cabbage. Plant Physiol 146:636–645
2338. O'Neal D, Hirsch L, Halas N, Payne J, West J (2004) Photo-thermal tumor ablation in mice using near infrared-absorbing nanoparticles. Cancer Lett 209:171–176
2339. O'Neil HT (1949) Theory of focusing radiators. J Acoust Soc Am 21:516–526
2340. O'Neill DP, Peng T, Stieglar P, Mayrhauser U, Koestenbauer S, Tscheliessnigg K, Payne S (2011) A three-state mathematical model of hyperthermic cell death. Ann Biomed Eng 39:570–579
2341. Onsager L (1931) Reciprocal relations in irreversible processes. Phys Rev 37:405–426

2342. Ontololan J, Borres M, Patac A, Maglasang G (2013) Review of fractal and fractal derivatives in relation to the physics of fractals. UV J Res 219–228. https://www.researchgate.net/publication/326148157_Review_of_Fractal_and_Fractal_Derivatives_in_relation_to_the_Physics_of_Fractals
2343. Ooi EH, Ang WT, Ng EYK (2009) A boundary element model for investigating the effects of eye tumor on the temperature distribution inside the human eye. Comput Biol Med 39:667–677
2344. Ordonez-Miranda J, Alvarado-Gil JJ (2009) Thermal wave oscillations and thermal relaxation time determination in a hyperbolic heat transport model. Int J Thermal Sci 48:2053–2062
2345. Ordonez-Miranda J, Alvarado-Gill JJ (2010) Thermal characterization of granular materials using a thermal-wave resonant cavity under the dual-phase lag model of heat conduction. Granul Mater 12:569–577
2346. Ordonez-Miranda J, Alvarado-Gill J (2011) On the stability of the exact solutions of the dual-phase lagging model of heat conduction. Nanosc Res Lett 6:327
2347. Orel D, Rozman J (2003) A computer simulation of ultrasound thermal bio-effect in embrionic models. Radioengineering 12:26–30
2348. Orief Y, Dafopoulos K (2005) Vitrification: will it replace the conventional gamete cryopreservation techniques? Middle East Fert Soc J 10:171–184
2349. O'Rourke A, Haemmerich D, Prakash PJ, Converse MC, Mahvi DM, Webster J (2007) Current status of liver tumor ablation devices. Expert Rev Med Dev 4:523–537
2350. Orpwood RD (1981) Biophysical and engineering aspects of cryosurgery. Phys Med Biol 26:555–575
2351. Ortega-Palacios R, Trujillo-Romero CJ, Cepeda Rubio MFJ, Vera A, Leija L, Reyes JL, Ramires-Estudillo MC, Morales-Alvarez F, Vega-Lppez MA (2018) Feasibility of using a novel 2.45GHz double short distance slot coaxial antenna for minimally invasive cancer breast microwave ablation therapy: computational model, phantom, and in vivo swine experimentation. J Healthcare Eng 2018:5806753
2352. Ortega-Julia J, Ortega D, Leliaert J (2022) Estimating the heating of complex nanoparticle aggregates for magnetic hyperthermia. arXiv:2207.14551v1 [cond-mat.mes-hall]
2353. Ortner V, Kaspar C, Halter C, Töllner L, Mykhaylyk O, Walzer J, Günzburg WH, Dangerfield JA, Hohenadl C, Czerny T (2012) Magnetic field-controlled gene expression in encapsulated cells. J Control Release 158:424–432
2354. Orvar BL, Sangwan V, Omann F, Dhindsa RS (2000) Early steps in cold sensing by plant cells: the role of actin cytoskeleton and membrane fluidity. Plant J 23:785–794
2355. Osler TJ (1970) Leibniz rule for fractional derivatives generalized and application to infinite series. SIAM J Appl Math 18:658–674
2356. Osler TJ (1973) A correction to Leibniz rule for fractional derivatives. SIAM J Math Anal 4:456–459
2357. Otanicar TP, Phelan PE, Prasher RS, Rosengarten G, Taylor RA (2010) Nanofluid-based direct absorption solar collector. J Renew Sustain Energy 2:033102
2358. Othman MIA, Ali MGS, Farouk RM (2011) The effect of relaxation time on the heat transfer and temperature distribution in tissues. World J.Mech 1:283–287
2359. Otten DM, Rubinsky B (2005) Front-tracking image reconstruction algorithm for EIT-monitored cryosurgery using boundary element method. Physiol Meas 26:503–516
2360. Oura S, Tamaki T, Hirai I, Yoshimasu T, Ohta F, Nakamura R, Okamura Y (2007) Radiofrequency ablation therapy in patients with breast cancers two centimeters or less in size. Breast Cancer 14:48–54
2361. Overgaard J (1977) Effect of hyperthermia on malignant cells in vivo: a review and a hypothesis. Cancer 39:2637–2646
2362. Overgaard K, Overgaard J (1972) Investigations on the possibility of a thermic tumour therapy–i: short-wave treatment of a transplanted isologous mouse mammary carcinoma. Eur J Cancer 8:65–78

2363. Oxtoby DW (2002) Density functional methods in the statistical mechanics of materials. Annu Rev Mater Res 32:39–52
2364. Oxtoby DW (2003) Crystal nucleation in simple amd complex fluids. Philos Trans R Soc Lond A 361:419–427
2365. Özen S, Helhel S, Cerezci O (2008) Heat analysis of biological tissue exposed to microwave by using thermal wave model of bio-heat transfer (TWMBT). Burns 34:45–49
2366. Özisik MN, Tzou DY (1994) On the wave theory in heat conduction. ASME J Heat Transf 116:526–535
2367. Özisik MN, Vick B (1984) Propagation and reflection of thermal wave in a finite medium. Int J Heat Mass Transf 27:1845–1854
2368. Özkavukcu S, Erdemli E (2002) Cryopreservation: basic knowledge and biophysical effects. J Ankara Med School 24:187–196
2369. Pagnini G (2012) Erdélye-Kober fractional diffusion. Frac Calc Appl Anal 15:117–127
2370. Pagnini G, Mura A, Mainardi F (2012) General fractional master equation for self-similar stochastic processesling anomalous diffusion. Int J Stoch Anal 2012:427383
2371. Pakdel E, Naebe M, Sun L, Wang X (2019) Advanced functional fibrous materials for enhanced thermoregulating performance. ACS Appl Mater Interfaces 11:13039–13057
2372. Pan W, Caswell B, Karniadakis GE (2010) A low-dimensional model for the red blood cell. Soft Matter 6:4366–4376
2373. Panasenko OO, Kim MV, Gusev NB (2003) Structure and properties of small heat shock proteins (in Russian). Adv Biol Chem 43:59–98
2374. Pandolf KB, Stroschein LA, Drolet LL, Gonzalez RR, Sawka MN (1986) Prediction modeling of physiological responses and human performance in the heat. Comput Biol Med 16:319–329
2375. Panicaud B, Rouhaud E (2014) A frame-indifferent model for a thermo-elastic material beyond the three-dimensional Eulerian and Lagrangian descriptions. Contin Mech Thermodyn 26:79–93
2376. Panjehpour M, Wilke A, Frazier DL, Overholt BF (1991) Hyperthermia treatment using a computer controlled Nd:YAG laser system in combination with surface cooling. Proc SPIE 1427:307–315
2377. Pankhurst QA, Thanh NTK, Jones SK, Dobson J (2009) Progress in applications of magnetic nanoparticles in biomedicine. J Phys D Appl Phys 42:224001
2378. Pankhurst QA, Thanh NKT, Jones SK, Dobson J (2009) Progress in applications of magnetic nanoparticles in biomedicine. J Phys D Appl Phys 42:224001
2379. Papadjopoulos D, Hui S, Vail WJ, Poste G (1976) Studies on the membrane fusion. I. Interactions of pure phospholipid membranes and the effect of myristic acid, lysoletichin, proteins and dimetylsulfoxide. Biochim Biophys Acta 448:254–264
2380. Papkovsky DB, Dmitriev RI (2013) Biological detection by optical oxygen sensing. Chem Soc Rev 42:8700–8732
2381. Parker EJ, Jenkins RJ, Butler CP, Abbott GL (1961) Flash method of determining thermal diffusivity, heat capacity, and thermal conductivity. J Appl Phys 32:1679–1684
2382. Parker DL, Smith V, Sheldon P, Crooks LE, Fussell L (1982) Temperature distribution measurements in two-dimensional NMR imaging. Med Phys 10:321–325
2383. Parmesan C, Yohe G (2003) A globally coherent fingerprint of climate change impacts across natural systems. Nature 421:37–42
2384. Parsons K (2014) Human thermal environments: the effects of hot, moderate, and cold environments on human health, comfort, and performance, 3rd edn. CRC Press, Boca Raton
2385. Pascu ML (2000) Laser physics elements to consider for low level laser therapy. Laser Ther 13:114–125
2386. Pasenkiewicz-Gierula M, Murzyn K, Rog T, Czaplewski C (2000) Molecular dynamics simulation studies of lipid bilayer systems. Acta Biochim Polonica 47:601–611
2387. Passarella S, Perlino E, Quagliariello E, Baldassarre L, Catalano I, Cingolani A (1983) Evidence of changes, induced by He Ne laser irradiation, in the optical and biochemical properties of rat liver mitochondria. Bioelectrochem Bioenerget 10:185–198

2388. Passarella S, Casamassima E, Molinari S, Pastore D, Quagliariello E, Catalano I, Cingolani A (1984) Increase of proton electrochemical potential and ATP synthesis in rat liver mitochondria irradiated in vitro by helium-neon laser. FEBS Lett 175:95–99
2389. Patel D, Franklin KA (2014) Temperature-regulation of plant architecture. Plant Signal Behav 4:577–579
2390. Patidar S, Kumar S, Srivastava A, Singh S (2017) Lattice Boltzmann method-based solution of radiative transfer equation for investigating light propagation through laser-irradiated tissue phantoms. Int Commun Heat Mass Transf 84:144–149
2391. Paton BC (2001) Cold, casualties, and conquests: the effects of cold on warfare. In: Lounsbury DE, Randolh KB, Burr RE (eds) Medical aspects of harsh environments, vol 1. Office The Surgeon General, Department of Army, USA, Falls Church, pp 313–349
2392. Paul A, Paul A (2018) Computational study of photo-thermal ablation of large blood vessel embedded tumor using localized injection of gold nanoshells. J Therm Biol 78:329–342
2393. Paul A, Paul A (2020) Thermomechanical analysis of a triple layered skin structure in presence of nanoparticle embedding multi-level blood vessels. Int J Heat Mass Transf 148:119076
2394. Paul A, Paul A (2020) Thermomechanical assessment of breast tumor subjected to focused ultrasound and interstitial laser heating. arXiv:2007.14053 [physics.med-ph]
2395. Paul A, Narasimhan A, Kahlen F, Das S (2014) Temperature evolution in tissues embedded with large blood vessels during photo-thermal heating. J Thermal Biol 41:77–87
2396. Paulides M, Bakker J, Neufeld E, Zee J, Jansen P, Levendag P, Rhoon G (2007) The HYPERcollar: a novel for hyperthermia in the head and neck. Int J Hyperth 23:567–576
2397. Paulides MM, Bakker JF, Chavannes N, van Rhoon GC (2007) A patch antenna design for application in a phased-array head and neck hyperthermia applicator. IEEE Trans Biomed Eng 54:2057–2063
2398. Paulides MM, Stauffer PR, Neufeld E, Maccarini PF, Kyriakou A, Canters RAM, Diederich CJ, Bakker JF, Van Rhoon GC (2013) Simulation techniques in hyperthermia treatment planning. Int J Hyperth 29:346–357
2399. Paulides MM, Verduijn GM, Van Holthe N (2016) Status quo and directions in deep head and neck hyperthermia. Radiat Oncol 11:809–821
2400. Pauly H, Schwan HP (1971) Mechanism of absorption of ultrasound in liver tissue. J Acoust Soc Am 50:692–698
2401. Pavel M, Gradinariu G, Stancu A (2008) Study of the optimum dose of ferromagnetic nanoparticles suitable for cancer therapy using MFH. IEEE Trans Magn 44:3205–3208
2402. Pavŝelj N, Miklavĉiĉ D (2011) Resistive heating and electropermeabilization of skin tissue during in vivo electroporation: a coupled nonlinear finite element model. Int J Heat Mass Transf 54:2294–2302
2403. Payne SR, Young OA (1995) Effects of pre-slaughter administration of antifreeze proteins on frozen meat quality. Meat Sci 41:147–155
2404. Peaceman DW (1978) Interpretation of well-block pressures in numerical reservoir simulation. Soc Petroleum Eng J 18:183–194
2405. Pearce JA (2010) Models for thermal damage in tissues: processes and applications. Crit Rev Biomed Eng 38:1
2406. Pearce RS (1999) Molecular analysis of acclimation to cold. Plant Growth Reg 29:47–76
2407. Pearce RS, Willison JHM (1985) Wheat tissues freeze-etched during exposure to extracellular freezing: distribution of ice. Planta 163:295–303
2408. Pearce SC, Mani V, Boddicker RL, Johnson JS, Weber TE, Ross JW, Rhoads RP, Baumgard LH, Gabler NK (2013) Heat stress reduces intestinal barrier integrity and favors intestinal glucose transport in growing pigs. PLoS One 8:e70215
2409. Pech M, Wieners G, Fischbach F, Hengst S, Beck A, Warschewske G, Hanninen EL, Wust P, Ricke J (2004) Synchronous CT-guided brachytherapy in patients at risk for incomplete interstitial laser ablation of liver malignancies. Med Laser Appl 19:73–82

2410. Peco C, Rosolen A, Arroyo M (2013) An adaptive meshfree method for phase-field models of biomembranes. Part II: a lagrangian approach for membranes in viscous fluids. J Comput Phys 249:320–336
2411. Pedersen TH, Videm V, Svennevig JL, Karlsen H, Ostbakk RW, Jensen O, Mollnes TE (1997) Extracorporeal membrane oxygenation using a centrifugal pump and a servo regulator to prevent negative inlet pressure. Ann Thorac Surg 63:1333–1339
2412. Peek MCL, Ahmed M, Napoli A, Usiskin S, Baker R, Douek M (2017) Minimally invasive ablative techniques in the treatment of breast cancer: a systematic review and meta-analysis. Int J Hypertherm 33:191–202
2413. Pegg DE (1966) Cryobiology. Phys Med Biol 11:209–224
2414. Pegg DE (2005) The role of vitrification techniques of cryopreservation in reproductive medicine. Hum Fertil 8:231–239
2415. Pegg DE, Harris AR (1970) A platinum resistance thermomemter for use in cryobiology. Phys Med Biol 15:409–416
2416. Pegg DE, Wusteman MC, Boylan S (1996) Fractures in cryopreserved elastic arteries: mechanism and prevention. Cryobiology 33:658–659
2417. Pegg DE, Wang L, Vaughan D (2006) Cryopreservation of articular cartilage. Part 3: the liquidus-tracking method. Cryobiology 52:360–368
2418. Pendergraft SS, Sadri-Ardekani H, Atala A, Bishop CE (2017) Three-dimensional testicular organoid: a novel tool for the study of human spermatogenesis and gonadotoxicity in vitro. Biol Reprod 96:720–732
2419. Peng Y, Cui Y (2020) Advanced textiles for personal thermal management and energy. Joule 4:724–742
2420. Peng T, O'Neill DP, Payne SJ (2011) A two-equation coupled system for determination of liver tissue temperature during thermal ablation. Int J Heat Mass Transf 54:2100–2109
2421. Peng L, Su B, Yu A, Jiang X (2019) Review of clothing for thermal management with advanced materials. Cellulose 26:1–34
2422. Pennes HH (1948) Analysis of tissue and arterial blood flow temperatures in the resting human forearm. J Appl Physiol 1:93–122; reprinted: Ibid, 1998, 85, 5–34
2423. Penrose R (1994) Shadows of the mind. Oxford University Press, Oxford
2424. Penta R, Ambrosi D (2015) The role of the microvascular tortuosity in tumor transport phenomena. J Theor Biol 364:80–97
2425. Penta R, Ambrosi D, Quarteroni A (2015) Multiscale homogenization for fluid and drug transport in vascularized malignant tissues. Math Models Methods Appl Sci 25:79–108
2426. Peratta A (2007) Electromagnetic modelling of a human eye exposed to conductive keratoplasty. WIT Trans Biomed Health 12:223–232
2427. Percy A, Widman S, Rizzo JA, Tranquilli M, Elefteriades JA (2009) Deep hypothermic circulatory arrest in patients with high cognitiveneeds: full preservation of cognitive abilities. Ann Thorac Surg 87:117–123
2428. Perdikaris P, Grinberg L, Karniadakis GE (2016) Multiscale modeling and simulation of brain blood flow. Phys Fluids 28:021304
2429. Periasamy M, Herrera JL, Reis FCG (2017) Skeletal muscle thermogenesis and its role in whole body energy metabolism. Diabetes Metab J 41:327–336
2430. Periasamyorcid M, Herrera JL, Reis FCG (2017) Skeletal muscle thermogenesis and its role in whole body energy metabolism. Diab Metabol J 41:327–336
2431. Perić M, Schäfer M, Schreck E (1992) Numerical simulation of complex fluid flows on MIMD computers. In: Pelz RB, Ecer A, Häuser J (eds) Parallel computational fluid dynamics'92, pp 311–324
2432. Perl W, Hirsh RL (1966) Local blood flow in kidney tissue by heat clearance measurements. J Theor Biol 19:251–280
2433. Perrin B (2007) Investigation of short-time heat transfer effects by an optical pump-probe method. In: Volz S (ed) Microscale and nanoscale heat transfer. Springer, Berlin, pp 333–359

2434. Pertaya N, Celik Y, DiPrinzio CL, Wettlaufer JS, Davies PL, Braslavsky I (2007) Growth-melt asymmetry in ice crystals under the influence of spruce budworm antifreeze protein. J Phys Condens Matter 19:412101
2435. Peshkov V (1944) Second sound in helium II. J Phys-USSR 8:381–381
2436. Peskin CS (1972) Flow patterns around heart valves - numerical method. J Comput Phys 10:252–271
2437. Peskin CS (2002) The immersed boundary method. Acta Numer 11:479–517
2438. Petcu D, Vizman D, Paprzycki M (2006) Heuristic load balancing for CFD codes executed in heterogeneous computing environments. Scalable Comput Pract Exp 7:15–23
2439. Petersen A, Heschel I, Rau G, Glasmacher B (2004) Optimization of cryopreservation procedures by controlled nucleation. Biomed Tech 49:632–633
2440. Petersen A, Schneider H, Rau G, Glasmacher B (2006) A new approach for freezing aqueous solutions under active control of the nucleation temperature. Cryobiology 53:248–257
2441. Pethick WA, Murray HJ, McFadyen P, Brodie R, Gaul CA, Stellingwer T (2019) Effects of hydration status during heat acclimation on plasma volume and performance. Scand J Med Sci Sports 29:189–199
2442. Petrat F, Li T, Dehne N, de Groot H, Rauen U (2006) Sodium as a major mediator of no-induced cell death in cultured hepatocytes. Life Sci 79:1606–1615
2443. Petrikaite V, Matulis D (2011) Thermodynamics of natural and synthetic inhibitor binding to human Hsp90. In: Mizutani T (ed) Application of thermodynamics to biological and materials science. InTech, Rijeka, pp 77–92
2444. Petrunkina AM (2007) Fundamental aspects of gamete cryobiology. J Reproduktionsmed Endokrinol 4:78–91
2445. Petryk AA, Giustini AJ, Gottesman RE, Kaufman PA, Hoopes P (2013) Magnetic nanoparticle hyperthermia enhancement of cisplatin chemotherapy cancer treatment. Int J Hyperth 29:845–851
2446. Petsev DN, Thomas BR, Yau ST, Tsekova D, Nanev C, Wilson WW, Vekilov PG (2001) Temperature-independent solubility and interactions between apoferritin monomers and dimers in solution. J Cryst Growth 232:21–29
2447. Pettigrew RT, Galt JM, Ludgate CM, Horn DB, Smith AN (1974) Circulatory and biochemical effects of whole body hyperthermia. Brit J Surg 61:727–730
2448. Peyrounette M, Davit Y, Quintard M, Lorthois S (2018) Multiscale modelling of blood flow in cerebral microcirculation: Details at capillary scale control accuracy at the level of the cortex. PLoS ONE 13:e0189474
2449. Pfaff RT, Liu J, Gao D, Li TK, Critser JK (1998) Water and DMSO membrane permeability characteristics of in-vivo and in-vitro-derived and cultured oocytes and embryos. Mol Hum Reprod 4:51–59
2450. Pfaff RT, Agca Y, Liu J, Woods EJ, Peter AT, Critser JK (2000) Cryobiology of rat embryos I. Determination of zygote membrane permeability coefficients for water and cryoprotectants, their activation energies, and the development of improved cryopreservation methods. Biol Reprod 63:1294–1302
2451. Pfefer TJ, Choi B, Vargas G, McNally KM, Welsh AJ (2000) Pulsed laser-induced thermal damage in whole blood. J Biomech Eng 122:196–202
2452. Pfefer TJ, Smithies DJ, Milner TE, van Gemert MJ, Nelson JS, Welch AJ (2000) Bioheat transfer analysis of cryogen spray cooling during laser treatment of port wine stains. Lasers Surg Med 26:145–157
2453. Phadnis A, Kumar S, Srivastava A (2016) Numerical investigation of thermal response of laser-irradiated biological tissue phantoms embedded with gold nanoshells. J Thermal Biol 61:16–28
2454. Pham DC, Torquato S (2003) Strong-contrst expansions and approximations for the effective conductivity of multiphase composites. J Appl Phys 94:6591
2455. Pham L, Dahiya R, Rubinsky B (1999) An in vivo study of antifreeze protein adjuvant cryosurgery. Cryobiology 38:169–175

2456. Phillips JC, Braun R, Wang W, Gumbart J, Tajkhorshid E, Villa E, Chipot C, Skeel RD, Kale L, Schulten K (2005) Scalable molecular dynamics with NAMD. J Comput Chem 26:1781–1802
2457. Piasecka-Belkhayat A (2011) Interval boundary element method for 2D transient diffusion problem using the directed interval arithmetic. Eng Anal Bound Elem 35:259–263
2458. Piasecka-Belkhayat A, Skorupa A (2021) Numerical study of heat and mass transfer during cryopreservation process with application of directed interval arithmetic. Materials 14:2966
2459. Pierce AD (1981) Acoustics. An introduction to its physical principles and applications. McGraw-Hill, New York
2460. Pietruszka M, Lewicka S (2007) Effect of temperature on plant elongation and cell wall extensibility. Gen Physiol Biophys 26:40–47
2461. Pikkula BM, Torres JH, Tunnell JW, Anvari B (2001) Cryogen spray cooling: effects of droplet size and spray density on heat removal. Lasers Surg Med 28:103–112
2462. Pinard MA, Huffman J (1997) Fire resistance and bark properties of trees in a seasonally dry forest in eastern Bolivia. J Trop Ecol 13:727–740
2463. Pinho D, Pereira AI, Lima R (2010) Red blood cells motion in a glass microchannel. AIP Conf Proc 1281:963–966
2464. Pinisetty D (2005) Molecular dynamics simulations of biological membranes in the presence of cryoprotectants. MS Thesis, Louisiana State University
2465. Piper HM, Siegmund B, Ladilov YV, Schlüter KD (1993) Calcium and sodium control in hypoxic-reoxygenated cardiomyocytes. Basic Res Cardiol 88:471–482
2466. Pivkin IV, Karniadakis GE (2008) Accurate coarse-grained modeling of red blood cells. Phys Rev Lett 101:118105
2467. Plieth C (1999) Temperature sensing by plants: calcium-permeable channels as primary sensor – a model. J Membr Biol 172:121–127
2468. Plonsey R, Heppner DB (1967) Considerations of quasi-stationarity in electrophysiological systems. Bull Math Biophys 29:657–664
2469. Podio-Guidugli P (2016) For a statistical interpretation of Helmholtz/thermal displacement. Continuum Mech Thermodyn 28:1–5
2470. Podio-Guidulgi P (2012) Dissipative entropy makes the heat equation hyperbolic. Atti della Accademia Peloritana dei Pericolanti 90:1–8
2471. Podlubny I (1998) Fractional differential equations. Academic Press, Cambridge
2472. Podlubny I (2000) Matrix approach to discrete fractional calaculus. Frac Calc Appl Anal 3:359–386
2473. Podlubny I (2002) Geometric and physical interpretation of fractional integration and fractional differentiation. Frac Calc Appl Anal 5:36–386
2474. Podlubny I, Despotovic V, Skovranek T, McNaughton BH (2007) Shadows on the walls: geometric interpretation of fractional integration. J Online Math Appl 7:1664
2475. Podlubny I, Chechkin A, Chen YQ, Vinagre BMJ (2009) Matrix approach to discrete fractional calaculus II: partial differential equations. J Comput Phys 228:3137–3153
2476. Podorvanov VV (2014) Thermogenesis in plants. Ukr Bot J 71:96–103
2477. Polderman KH (2004) Application of therapeutic hypothermia in the intensive care unit. Intens Care Med 30:757–769
2478. Polge C, Smith AU, Parkers AS (1949) Revival of spermatozoa after vitrification and dehydration at low temperatures. Nature 164:666–676
2479. Pollacco DA, Farina L, Wismayer PS, Farrugia L, Sammut CV (2018) Characterization of the dielectric properties of biological tissues and their correlation to tissue hydration. IEEE Trans Dielectr Electr Insul 25:2191–2197
2480. Polwaththe Gallage H, Saha S, Sauret E, Flower R, Gu Y (2015) Numerical investigation of motion and deformation of a single red blood cell in a stenosed capillary. Int J Comput Methods 12:1540003
2481. Ponder E (1962) The coefficient of thermal conductivity of blood and of various tissues. J Gen Physiol 45:545–551

2482. Ponti RD (2005) Cryothermal energy ablation of cardiac arrhythmias 2005: state of the art. Indian Pacing Electrophysiol J 5:12–24
2483. Poole RJ (2012) The deborah and Weissenberg numbers. Rheol Bull 53:32–39
2484. Poor HZ, Moosavi H, Moradi A, Menghari HG, Parastarfeizabadi M (2014) Investigation of the dual-phase lag effects in biological tissues during laser irradiation. Int J Mech Syst Eng 4:33–46
2485. Pop E, Goodson KE (2006) Thermal phenomena in nanoscale transistors. J Electron Packag 128:102
2486. Pop E, Sinha S, Goodson KE (2006) Heat generation and transport in nanometer-scale transistors. Proc IEEE 94:1587–1601
2487. Popel AS, Johnson PC (2005) Microcirculation and hemorheology. Ann Rev Fluid Mech 37:43–69
2488. Popovic D, McCartney L, Beasley C, Lazebnik M, Okoniewski M, Hagness S, Booske JH (2005) Precision open-ended coaxial probes for in vivo and ex vivo dielectric spectroscopy of biological tissues at microwave frequencies. IEEE Trans Microw Theory Techn 53:1713–1722
2489. Pörtner HO, Knust R (2007) Climate change affects marine fishes through the oxygen limitation of thermal tolerance. Science 315:95–97
2490. Potten CS, Wilson JW, Booth C (1997) Regulation and significance of apoptosis in the stem cells of the gastrointestinal epithelium. Stem Cells 15:82–93
2491. Potter AW (2019) Modeling individualized thermoregulatoryresponses to clothing and activity, in hot and cold environments. Ph.D. Thesis, The State University of New Jersey
2492. Poulou L, Botsa E, Thanou I, Ziakas P, Thanos L (2015) Percutaneous microwave ablation vs radiofrequency ablation in the treatment of hepatocellular carcinoma. World J Hepatol 7:1054–1063
2493. Pournaras CJ, Rungger-Brandle E, Riva CE, Hardarson SH, Stefansson E (2008) Regulation of retinal blood flow in health and disease. Prog Retinal Eye Res 27:284–330
2494. Povstenko YZ (2005) Fractional heat conduction equation and assotiated thermal stress. J Thermal Stress 28:83–102
2495. Povstenko YZ (2009) Theory of thermoelasticity based on the space-time fractional heat conduction equation. Phys Scr 136:014017
2496. Povstenko YZ (2011) Fractional Cattaneo-type equations and generalized thermoelasicity. J. Thermal Stress 34:97–114
2497. Povstenko Y, Povstenko Y (2013) Fractional heat conduction in an infinite medium with a spherical inclusion. Entropy 137:4122–4133
2498. Powell SL (2002) Contact cooling and it's effects on manual dexterity. Ph.D. Thesis, Loughborough University
2499. Pozoz RS, Danzl DF (2001) Human physiological responses to cold stress and hypothermia. In: Lounsbury DE, Randolh KB, Burr RE (eds) Medical aspects of harsh environments, vol 1. Office The Surgeon General, Department Army, USA, Falls Church, pp 351–382
2500. Pozrikidis C (1992) Boundary integral and singularity methods for linearized viscous flow. Cambridge University Press, Cambridge
2501. Pozrikidis C (2001) Effect of membrane bending stiffness on the deformation of capsules in simple shear flow. J Fluid Mech 440:269–291
2502. Pozrikidis C (2003) Numerical simulation of the flow-induced deformation of red blood cells. Ann Biomed Eng 31:1194–1205
2503. Pozrikidis C (2005) Axisymmetric motion of a file of red blood cells through capillaries. Phys Fluids 17:031503
2504. Prado G, Farutin A, Misbah C, Bureau L (2015) Viscoelastic transient of confined red blood cells. Biophys J 108:2126–2136
2505. Prakash P (2010) Theoretical modeling for hepatic microwave ablation. Open Biomed Eng J 4:27–38

2506. Prasad B, Ha YH, Lee SK, Kim JK (2016) Patient-specificsimulation for selective liver tumor treatment with non-invasive radiofrequency hyperthermia. J Mech Sci Technol 30:5837–5845
2507. Prasad B, Kim JK, Kim S (2019) Role of simulations in the treatment planning of radiofrequency hyperthermia therapy in clinics. J Oncol 2019:9685476
2508. Preda L, Mihailescu M, Preda A (2009) Application of fractional derivative to the relaxation of laser target. UPB Sci Bull Ser A 71:11–20
2509. Preusser T, Weihusen A, Peitgen HO (2005) On the modelling of perfusion in the simulation of RF-ablation. Proceedings of Simulation und Visualisierung (SimVis), Magdeburg, pp 259–268
2510. Pries AR, Ley K (1989) Red cell distribution at microvascular bifurcations. Microvasc Res 38:81–101
2511. Pries AR, Secomb TW (2008) Blood flow in microvascular networks. In: Microcirculation, pp 3–36. Elsevier, Amsterdam
2512. Pries AR, Ley K, Gaehtgens P (1986) Generalization of the Fahraeus principle for microvessel networks. Am J Physiol 251:H1324–H1332
2513. Pries AR, Ley K, Claassen M, Gaehtgens P (1989) Red cell distribution at microvascular bifurcations. Microvasc Res 38:81–101
2514. Pries AR, Neuhaus D, Gaehtgens P (1992) Blood viscosity in tube flow: dependence on diameter and hematocrit. Am J Physiol 263:H1770–H1778
2515. Pries AR, Secomb TW, Gaehtgens P (1996) Biophysical aspects of blood flow in the microvasculature. Cardiovasc Res 32:654–667
2516. Prokhorov GG (Ed) (2001) Advances in cryomedicine (in Russian). Saint Petersburg
2517. Prokhorov GG (2012) Cryosurgery in Russia. In: Xu K, Korpan NN, Niu L (eds) Modern cryosurgery for cancer. World Scientific, Singapore, pp 55–68
2518. Prokhorov DG, Zhmakin AI (2017) Cryoaction on biological tissues (in Russian). In: Prohorov GG, Belyaev AM, Prokhorov DG (eds) Fundamentals of clinical cryomedicine. Kniga Po Trebovaniyu, Saint-Petersburg, pp 50–79
2519. Prohorov GG, Belyaev AM, Prokhorov DG (Eds) (2017) Fundamentals of clinical cryomedicine (in Russian). Kniga Po Trebovaniyu, Saint-Petersburg
2520. Prosser CL, Bishop D, Brown F, Jahn T, Wulff VJ (1952) Comparative animal physiology. Saunders Company, Centennial
2521. Provatas N, Elder K (2011) Phase-field methods in materials science and engineering. Wiley, Hoboken
2522. Provenzale A, Smith LA, Vio R, Murante G (1992) Distinguishing between low-dimensional dynamics and randomness in measured time series. Phys D Nonlinear Phenom 58:31–49
2523. Pu Z (2005) A dynamic model of the human/cooling system/clothing/a dynamic model of the human/cooling system/clothing/environment system environment system. Ph.D. Thesis, University of Central Florida
2524. Pugh LGCE, Corbett JL, Johnson RH (1967) Rectal temperatures, weight losses and sweat rates in marathon running. J Appl Physiol 23:347–352
2525. Puglisi JL, Bassani RA, Bassani JW, Amin JN, Bers DM (1996) Temperature and relative contributions of Ca transport systems in cardiac myocyte relaxation. Am J Physiol 270:H1772–H1778
2526. Pugsley MK, Tabrizchi R (2000) The vascular system. An overview of structure and function. J Pharmacol Toxicol Methods 44:333–340
2527. Puig-de Morales M, Turner KT, Butler JP, Fredberg JJ, Suresh S (2007) Viscoelasticity of the human red blood cell. J Appl Physiol 293:597–605
2528. Pujo-Menjouet L (2016) Blood cell dynamics: half of a century of modelling. Math Model Nat Phenom 11:92–115
2529. Pushkar NS, Itkin YA, Bronstein VL, Gordiyenko EA, Kozmin YV (1976) On the problem of dehydration and intracellular crystallization during freezing of cell suspensions. Cryobiology 13:147–152

2530. Pushkar NS, Shrago MI, Belous AM, Kalugin YV (1978) Cryoprotectors (in Russian). Naukova Dumka, Kyiv
2531. Qi HT, Xu HY, Guo XW (2007) The generalized Taylor's formula. Appl Math Comput 186:286–293
2532. Qin Z, Bischof JC (2012) Thermophysical and biological responses of gold nanoparticle laser heating. Chem Soc Rev 41:1191–1217
2533. Qin Z, Durand LG, Allard L, Cloutier G (1998) Effects of a sudden flow reduction on red blood cell rouleau formation and orientation using RF backscattered power. Ultrasound Med Biol 24:503–511
2534. Qiu TQ, Tien CL (1993) Heat transfer mechanisms during short-pulse laser heating of metals. J Heat Transf 115:835–841
2535. Qiu Y, Wu W, Yuan X, Xiao B, Ren Z (2012) Human thermoregulation model for estimating thermal protection performance of anti-G garments. Chin J Mech Eng 25:47–54
2536. Quintanilla R (2001) Exponential stability in the dual-phase-lag heat conduction theory. J Non-Equilib Thermodyn 27:217–227
2537. Quintanilla R, Racke R (2006) A note on stability in dual-phase-lag heat conduction. Int J Heat Mass Transf 49:1209–1213
2538. Quintanilla R, Racke R (2015) Spatial behavior in phase-lag heat conduction. Differ Integral Equ 28:291–308
2539. Raabe D (2007) A texture-component avrami model for predicting recrystallization textures, kinetics and grain size. Modell Simul Mater Sci Eng 15:39–63
2540. Raaymakers B, Crezee J, Lagendijk JJ (2000) Modelling individual temperature profiles from an isolated perfused bovine tongue. Phys Med Biol 45:765–780
2541. Raaymakers BW, Kotte AN, Lagendijk J (2000) How to apply a discrete vessel model in thermal simulations when only incomplete vessel data are available. Phys Med Biol 45:3385–3401
2542. Raaymakers BW, Kotte ANTJ, Lagendijk J (2009) Discrete vasculature (DIVA) model simulating the thermal impact of individual blood vessels for in vivo heat transfer. In: Minkowycz WJ (ed) Advances in numerical heat transfer, vol 3, pp 121–148
2543. Rabin Y (2000) The effect of temperature-dependent thermophysical properties in heat transfer simulations of biomaterials in cryogenic temperatures. Cryo-Lett 21:163–170
2544. Rabin Y (2003) A general model for the propagation of uncertainty in measurements into heat transfer simulations and its application to cryobiology. Cryobiology 46(2):109–120
2545. Rabin Y, Shitzer A (1996) A new cryosurgical device for controlled freezing, part I: setup and validation test. Cryobiology 33:82–92
2546. Rabin Y, Stahovich TF (2002) The thermal effect of urethral warming during cryosurgery. Cryo-Lett 23:361–374
2547. Rabin Y, Stahovich TF (2003) Cryoheater as a means of cryosurgery control. Phys Med Biol 48:619–632
2548. Rabin Y, Steif PS (1996) Analysis of thermal stresses around a cryosurgical probe. Cryobiology 33:276–290
2549. Rabin Y, Steif PS (1998) Thermal stresses in a freezing sphere and application to cryosurgery. J Appl Mech 65:328–333
2550. Rabin Y, Steif PS (2000) Thermal stresses modelling in cryosurgery. Int J Solids Struct 37:2363–2375
2551. Rabin Y, Taylor MJ, Wolmark N (1998) Thermal expansion measurements of frozen biological tissues at cryogenic temperatures. J Biomech Eng 120:259–266
2552. Rabin Y, Lung DC, Stahovich TF (2004) Computerized planning of cryosurgery using cryoprobes and cryoheaters. Techn Cancer Res Treatment 3:229–243
2553. Racinais S, Moussay S, Nichols D, Travers G, Belfekih T, Schumacher YO, Periard JD (2019) Core temperature up to 41.5°c during UCI road cycling world championships in the heat. Brit J Sports Med 53:426–429
2554. Radai MM, Abboud S, Rubinsky B (1999) Evaluation of the impedance technique for cryosurgery in a theoretical model of the head. Cryobiology 38:51–59

2555. Radebaugh R (2005) Heat transfer in cryogenic catheters. In: Kakaç S, Vasiliev LL, Bayazitoglu Y, Yener Y (eds) Microscale heat transfer. US Government, pp 445–464
2556. Radhakrishnan R, Trout BL (2003) Nucleation of crystalline phases of water in homogeneous and inhomogeneous environments. Phys Rev Lett 90:158301
2557. Radhakrishnan R, Trout BL (2003) Nucleation of hexagonal (I_h) ice in liquid water. J Am Chem Soc 125:7743–7747
2558. Radhakrishnan R, Trout BL (2005) Order parameter approach to understanding and quatifying the physico-chemical behavior of complex systems. In: Yip S (ed) Handbook of materials modeling. Volume 1: methods and models. Springer, Berlin, pp 1–14
2559. Radio frequency/microwave radiation. PADS, 10-18-2000 (2000)
2560. Raff M (1998) Cell suicide for beginners. Nature 396:119–122
2561. Rafiq A, Khanday M (2016) Thermal behavior of human eye in relation with change in blood perfusion, porosity, evaporation and ambient temperature. J Therm Biol 62:138–142
2562. Rahideh H, Malekzadeh P, Golbahar Haghighi MR (2011) Non-Fourier heat conduction analysis with temperature-dependent thermal conductivity. ISRN Mech Eng 2011:321605
2563. Rahimian A, Veerapaneni SK, Zorin D, Biros G (2015) Boundary integral method for the flow of vesicles with viscosity contrast in three dimensions. J Comput Phys 298:766–786
2564. Rahimy M (2010) Applications of fractional diferential equations. Appl Math Sci 4:2453–2461
2565. Raichle ME (1983) Neurogenic control of blood-brain barrier permeability. Acta Neuropathol Suppl 8:75–79
2566. Raichle ME (2003) Functional brain imaging and human brain function. J Neurosci 23:3959–3962
2567. Raleva K, Vasileska D, Goodnick SM, Nedjalkov M (2008) Modeling thermal effects in nanodevices. IEEE Trans Electron Dev 55:1306–1313
2568. Rallison JM, Acrivos A (1978) A numerical study of thedeformation and burst of a viscous drop in an extensional flow. J Fluid Mech 89:191–200
2569. Ramachandran T (1996) Water vaporization from heated tissue: an in vitro study by differential scanning calorimetry. Lasers Surg Med 19:413–415
2570. Ramachandran T, Screenivasan K, Sivakumar R (1996) Water vaporization from heated tissue: an in vitro study by differential scanning calorimetry. Lasers Surg Med 19:413–415
2571. Ramanathan NL (1964) A new weighting system for mean surface temperature of the human body. J Appl Physiol 19:531–533
2572. Ramsay J (1964) The rectal complex of the mealworm *Tenebrio molitor L. (Coleoptera, Tenebrionadae)*. Philos Trans R Soc Lond B 248:279–314
2573. Rand R (1964) Mechanical properties of red cell membrane. II. Viscoelastic breakdown of membrane. Biophys J 4:303–316
2574. Rand RW, Dashe AM, Paglia DE, Conway LW, Solomon DH (1964) Stereotactic cryohypophysectomy. JAMA 189:255–259
2575. Randrianalisoa JH, Dombrovsky LA, Lipiński W, Timchenko V (2014) Effects of short-pulsed laser radiation on transient heating of superficial human tissues. Int J Heat Mass Transf 78:488–497
2576. Raouf I, Khalid S, Khan A, Lee J, Kim HS, Kim MH (2020) A review on numerical modeling for magnetic nanoparticle hyperthermia: progress and challenges. J Therm Biol 91:102644
2577. Rasaiah JC, Lynden-Bell RM (2001) Computer simulation studies of the structure and dynamics of ions and non-polar solutes in water. Philos Trans R Soc Lond A 359:1545–1574
2578. Raschke K (1960) Heat transfer between the plant and the environment. Annu Rev Plant Physiol 11:111–126
2579. Raszkowski T, Samson A (2017) Numeical approaches to the heat transfer in modern electronic structures. Comput Sci 18:71–93
2580. Rauen U, de Groot H (2008) Inherent toxicity of organ preservation solutions to cultured hepatocytes. Cryobiology 56:88–92

2581. Rautenberg W, Necker R, May B (1972) Thermoregulatory responses of the pigeon to changes of the brain and the spinal cord. Pflüg Arch 338:31–42
2582. Ravin R, Cai TX, Pursley RH, Garmendia-Cedillos M, Pohida T, Freidlin RZ, Wang H, Zhuang Z, Giles AJ, Williamson AH, Gilbert MR, Basser PJ (2020) A novel in vitro device to deliver induced electromagnetic fields to cell and tissue cultures. arXiv:2011.09698v1 [physics.bio-ph]
2583. Reasor DA Jr, Clausen JR, Aidun CK (2013) Rheological characterization of cellular blood in shear. J Fluid Mech 726:497–516
2584. Recktenwald SM, Graessel K, Maurer FM, John T, Gekle S, Wagner C (2021) Red blood cell shape transitions and dynamics in time-dependent capillary flows. arXiv:2112.10442v1 [physics.bio-ph]
2585. Reed BM (ed) (2007) Plant cryopreservation: a practical guide. Springer, Berlin
2586. Reed KL, Brown TD, Conzemius MG (2003) Focal cryogenic insults for inducing segmental osteonecrosis: computational and experimental assessments of thermal fields. J Biomech 36:1317–1326
2587. Rees AR, Sternberg MJE (1984) From cells to atoms: an illustrated introduction to molecular biology. Blackwell Science, Oxford (1984)
2588. Regirer SA (Ed) (1971) Blood circulation hydro-dynamics (in Russian). Mir, Moscow
2589. Reichel F, Mauer J, Nawaz AA, Gompper G, Guck J, Fedosov DA (2019) High-throughput microfluidic characterization of erythrocyte shapes and mechanical variability. Biophys J 117:14–4
2590. Reichold J, Stampanoni M, Keller AL, Buck A, Jenny P, Weber B (2009) Vascular graph model to simulate the cerebral blood flow in realistic vascular networks. J Cerebral Blood Flow Metabol 29:1429–1443
2591. Reinhound P, Iren FV, Kijne JW (2000) Cryopreservation of undifferentiated plant cells. In: Engelmann F, Takagi H (eds) Cryopreservation of tropical plant germplasm - current research progress and applications. IPGRI, Rome, pp 91–100
2592. Reith J, Jorgensen HS, Pedersen PM, Nakayama H, Raaschou HO, Jeppesen LL, Olsen TS (1996) Body temperature in acute stroke: relation to stroke severity, infarct size, mortality, and outcome. Lancet 347:422–425
2593. Reitman ML (2018) Of mice and men - environmental temperature, body temperature, and treatment of obesity. FEBS Lett 592:2098–2107
2594. Ren WW, Shu C, Wu J, Yang WM (2012) Boundary condition-enforced immersed boundary method for thermal flow problems with dirichlet temperature condition and its applications. Comput Fluids 57:40–51
2595. Rethfeld B, Ivanov DS, Garcia ME, Anisimov SI (2017) Modelling ultrafast laser ablation. J Phys D Appl Phys 50:193001
2596. Reverberi AP, Bagnerini P, Maga L, Bruzzone AG (2008) On the non-linear Maxwell-Cattaneo equation with nonconstant diffusivity: shock and discontinuity waves. Int J Heat Mass Transf 51:5327–5332
2597. Rheistädter MC, Seydel T, Demmel F (2005) Molecular motion in lipid bilayers studied by the neutron backscattering technique. Phys Rev E 71:061908
2598. Ricci AD, Rizzo A, Bonucci C, Tavolari S, Palloni A, Frega E, Mollica V, Tober N, Mazzotta E, Felicani C, Serra C, Brandi G (2020) The (eternal) debate on microwave ablation versus radiofrequency ablation in BCLA-A hepatocellular carcinoma. In Vivo 34:3421–3429
2599. Richardson RB, Mailloux RJ (2023) Mitochondria need their sleep: redox, bioenergetics, and temperature regulation of circadian rhythms and the role of cysteine-mediated redox signaling, uncoupling proteins, and substrate cycles. Antioxidants 12:674
2600. Richardson S, Swietach P (2016) Red blood cell thickness is evolutionarily constrained by slow, hemoglobin-restricted diffusion in cytoplasm. Sci Rep 6:36018
2601. Richtmyer RD, Morton KW (1967) Difference methods for initial-value problems. Interscience, New York
2602. Ricklefs RE, Wikelski M (2002) The physiology/life-history nexus. Trends Ecol Evol 17:462–468

2603. Rida M, Karaki W, Ghaddar N, Ghali K (2013) Robust modeling of human thermal response to hand and forearm cooling for improved performance and energy efficiency. J Appl Mech Eng 2:129
2604. Riedl V, Utza L, Castrillóna G, Grimmerc T, Rauscheckerc JP, Plonerc M, Fristoni KJ, Drzezgaj A, Sorga C (2016) Metabolic connectivity mapping reveals effective connectivity in the resting human brain. Proc Nat Acad Sci USA 113:428–433
2605. Rieke V, Butts Pauly K (2008) MR thermometry. J Magn Reson Imag 27:376–390
2606. Rijnen Z, Togni P, Roskam R, van de Geer RSG, Goossens RH, Paulides MM (2015) Quality and comfort in head and neck hyperthermia: a redesign according to clinical experience and simulation studies. Int J Hyperth 31:823–830
2607. Ring EFJ (1990) Quantitative thermal imaging. Clin Phys Physiol Meas 11(suppl A):87–95
2608. Ring EFJ (1995) Quantitative thermal imaging in rheumatology. Biomed Thermol 15:69–71
2609. Rinker B, Cui XD, Cibull ML, Fink B, Gao DY, Vasconez HC (2008) Cryopreservation of composite tissue transplants. HAND 3:17–23
2610. Ritchie KP, Keller BM, Syed KM, Lepock JR (1994) Hyperthermia (heat shock)-induced protein denaturation in liver, muscle and lens tissue as determined by differential scanning calorimetry. Int J Hyperth 10:605–618
2611. Ritossa F (1962) A new puffing pattern induced by temperature shock and dnp in drosophila. Cell Mol Life Sci 18:571–573
2612. Roach PJ (1998) Verification and validation in computational science and engineering. Hermosa Publishers, Albuquerque
2613. Roache PJ (2002) Code verification by the method of manufactured solutions. J Fluids Eng 124:4–10
2614. Robert CP, Casella G (2004) Monte carlo statistical methods. Springer, New York
2615. Roberts PH, Soward AM (1992) Dynamo theory. Annu Rev Fluid Mech 24:459–512
2616. Roberts DA, Detre JA, Bolinger L, Insko EK, Lenkinski RE, Pentecost MJ, Leigh JS (1995) Renal perfusion in humans: MR imaging with spin tagging of arterial water. Radiology 196:281–286
2617. Robins HI, Hugander A, Cohen JD (1989) Whole body hyperthermia in the treatment of neoplastic disease. Radiol Clin North Am 27:603–610
2618. Robinson CH (2001) Cold adaptation in Arctic and Antarctic fungi. New Phytol 151:341–353
2619. Robinson MP, Wusteman MC, Wang L, Pegg DE (2002) Electromagnetic re-warming of cryopreserved tissues: effect of choice of cryoprotectant and sample shape on uniformity of heating. Phys Med Biol 47:2311–2325
2620. Robles V, Barbosa V, Herráez MP, Martinez-Piramo S, Cancela ML (2007) The antifreeze protein type I (AFP I) increases seabream (Sparus aurata) embryos tolerance to low temperatures. Theriogenology 68:284–289
2621. Roden JA, Gedney SD (2000) Convolutional PML (CPML): an efficient FDTD implementation of the CFS-PML for arbitrary media. Microw Opt Technol Lett 27:334–338
2622. Rodrigues DB, Pereira PJS, Limao-Vieira P, Stauffer PR, Maccarini PF (2013) Study of the one dimensional and transient bioheat transfer equation: multi-layer solution development and applications. Int J Heat Mass Transf 62:153–162
2623. Roeder F (1934) Die messung der warmeleitzahl der menschlichen haut und ihre veranderlichkeit. Z Biol 95:164–168
2624. Roemer RB (1999) Engineering aspects of hyperthermia therapy. Ann Rev Biomed Eng 1:347–376
2625. Roemer RB, Oleson JR, Cetas TC (1985) Oscillatory temperature response to constant power applied to canine muscle. Am J Physiol 249:R153–R158
2626. Roetzel W, Xuan Y (1998) Transient response of the human limb to an external stimulus. Int J Heat Mass Transf 41:229–239
2627. Rogers K (Ed) (2011) The human body. Blood: physiology and circulation. Britannica Educational Publishing, Chicago

2628. Rogolino P, Kovács R, Ván P, Cimmelli VA (2018) Generalized heat conduction equations: parabolic and hyperbolic models. Cont Mech Thermodyn 30:1245–1258
2629. Rogov AG, Sukhanova EI, Uralskaya LA, Aliverdieva DA, Zvyagilskaya RA (2014) Review: alternative oxidase: distribution, induction, properties, structure, regulation, and functions. Biochemistry 79:1615–1634
2630. Roman S, Lorthois S, Duru P, Risso F (2013) Flow of concentrated red blood cell suspensions at micro-bifurcations: an in vitro experimental study. Comput Methods Biomech Biomed Eng 16:33–35
2631. Romanovsky AA (2014) Skin temperature: its role in thermoregulation. Acta Physiol 210:498–507
2632. Romanovsky AA, Ivanov AI, Shimansky YP (2002) Selected contribution: ambient temperature for experiments in rats: a new method for determining the zone of thermal neutrality. J Appl Physiol 92:2667–2679
2633. Romero-Méndez R, Chu K, Vu H, Franko W, Aguilar G (2005) Confinement of freezing front by laser irradiation during cryosurgery. In: Proceedings of HT2005, ASME Summer Heat Transfer Conference, San Francisco, pp 1–7
2634. Romero-Méndez R, Franko W, Aguilar G (2007) Laser-assisted cryosurgery of prostate: numerical study. Phys Med Biol 52:463–468
2635. Rose GD, Wolfenden R (1993) Hydrogen bonding, hydrophobicity, packing, and protein folding. Annu Rev Biophys Biomol Struct 22:381–415
2636. Rosen AD (2001) Nonlinear temperature modulation of sodium channel kinetics in GH(3)cells. Biochim Biophys Acta 1511:391–396
2637. Rosen A, Stuchly MA, Van der Vorst A (2002) Applications of RF/microwaves in medicine. IEEE Trans Microw Theory Tech 50:963–974
2638. Rosensweig RE (2002) Heating magnetic fluid with alternating magnetic field. J Magn Magn Mater 252:370–374
2639. Rosolen A, Peco C, Arroyo M (2013) An adaptive meshfree method for phase-field models of biomembranes. Part I: approximation with maximum-entropy basis functions. J Comput Phys 249:303–319
2640. Rosomoff HL, Holaday DA (1954) Cerebral blood flow and cerebral oxygen consumption during hypothermia. Am J Physiol 179:85–88
2641. Ross B (1977) The development of fractional calculus 1695–1990. Historia Math 4:75–89
2642. Rossi S, Zanier ER, Mauri I, Columbo A, Stocchetti N (2001) Brain temperature, body core temperature, and intracranial pressure in acute cerebral damage. J Neurol Neurosurg Psychiatr 71:448–454
2643. Rossi MR, Tanaka D, Shimada K, Rabin Y (2007) An efficient numerical technique for bioheat simulations and its application to computerized cryosurgery planning. Comput Methods Prog Biomed 85:41–50
2644. Rossmann C, Haemmerich D (2014) Review of temperature dependence of thermal properties, dielectric properties, and perfusion of biological tissues at hyperthermic and ablation temperatures. Crit Rev Biomed Eng 42:467–492
2645. Rossmann C, Garrett-Mayer E, Rattay F, Haemmerich D (2014) Dynamics of tissue shrinkage during ablative temperature exposures. Physiol Meas 35:55–67
2646. Rosti ME, Brandt L (2018) Suspensions of deformable particles in a Couette flow. J Non-Newton Fluid Mech 262C:3–11
2647. Rosti ME, Brandt L, Mitra D (2018) Rheology of suspensions of viscoelastic spheres: deformability as an effective volume fraction. Phys Rev Fluid 3:012301
2648. Rostovtseva T, Colombini M (1996) ATP flux is controlled by a voltage-gated channel from the mitochondrial outer membrane. J Biol Chem 271:28006–28008
2649. Roti JL (2008) Cellular responses to hyperthermia (40–46°C): cell killing and molecular events. Int J Hyperth 24:3–15
2650. Roversi PF, Cosi E, Irdani T (2008) Chill sensitivity and cryopreservation of eggs of the greater wax moth Galleria mellonella (Lepidoptera: Pyralidae). Cryobiology 56:1–7

2651. Rowland LA, Bal NC, Periasamy M (2015) The role of skeletal-muscle-based thermogenic mechanisms in vertebrate endothermy. Biol Rev Camb Philos Soc 90:1279–1297
2652. Royauté R, Anderson M, Berdal A, Garrison CR, Dochtermann NA (2018) Paceless life? A meta-analysis of the pace-of-life syndrome hypothesis. Behav Ecol Sociobiol 72:64
2653. Ruane J, Sonnio A (2006) The role of biotechnology in exploring and protecting agricultural genetic resources. Food and Agriculture Organization of the UN, Rome, 187 pp
2654. Rubin MB (1992) Hyperbolic heat conduction and the second law. Int J Eng Sci 30:1665–166
2655. Rubinsky B (2000) Cryosurgery. Annu Rev Biomed Eng 2:157–187
2656. Rubinsky B (2003) Principles of low temperature cell preservation. Heart Fail Rev 8:277–284
2657. Rubinsky B, Cravalho EG, Mikic B (1980) Thermal stresses in frozen organs. Cryobiology 17:66–73
2658. Rubinsky B, Lee CJ, Bastacky J, Onik J (1987) The process of freezing and the mechanism of damage during hepatic cryosurgery. Cryobilogy 27:85–97
2659. Rubinsky B, Onik G, Mikus P (2007) Irreversible electroporation: a new ablation modality – clinical implications. Technol Cancer Res Treat 6:1–12
2660. Rubner M (1912) The laws of energy consumption in nutrition. Elsevier, Amsterdam
2661. Rucker RB (2007) Allometric scaling, metabolic body size and interspecies comparisons of basal nutritional requirements. Anim Physiol Anim Nutr 91:148–56
2662. Ruggeri T, Sugiyama M (2015) Rational extended thermodynamics beyond the monatomic gas. Springer, Berlin
2663. Rugh JP, Farrington RB, Bharathan D, Vhalinos A, Burke R, Huizenga C, Zhang H (2004) Predicting human thermal comfort in a transient nonuniform thermal environment. Eur J Appl Physiol 92:721–727
2664. Rukolaine SA (2014) Unphysical effects of the dual-phase-lag model of heat conduction. Int J Heat Mass Transf 78:58–63
2665. Rukolaine SA (2017) Unphysical effects of the dual-phase-lag model of heat conduction: higher-order approximations. Int J Heat Mass Transf 113:83–88
2666. Rukolaine SA (2024) Hyperbolicity of the ballistic-conductive model of heat conduction: the reverse side of the coin. Z Angew Math Phys 75:50
2667. Rukolaine S, Samsonov A (2012) The delayed uncoupled continuos-time random walks do not provide a model for the telegraph equation. arXiv: 1202.6468
2668. Rukolaine S, Samsonov A (2013) Local immobilization of particles in mass transfer described by a Jeffreys-type equation. Phys Rev E 88:062116
2669. Rupp ME, Heermann J, Upho ME (2004) Need for a reliable system to measure body temperature. Am J Infect Control 32:184
2670. Rutlant J, Pommer AC, Meyers SA (2003) Osmotic tolerance limits and properies of rhesus monkey (macaca mulatta) spermatozoa. J Androl 24:534–541
2671. Rutman RS (1995) On physical interpretations of fractional integration and differentiation. Theor Math Phys 105:1509–1519
2672. Ryabov YE, Puzenko A, Feldman Y (2004) Nonmonotonic relaxation kinetics of confined systems. Phys Rev B 69:014204
2673. Rylander MN, Feng Y, Bass J, Diller KR (2005) Thermally induced injury and heat-shock protein expression in cells and tissues. Ann NY Acad Sci 1066:222–242
2674. Saccomandi P, Schena E, Silvestri S (2013) Techniques for temperature monitoring during laser-induced thermotherapy: an overview. Int J Hyperth 29:609–619
2675. Saedodin S, Torabi M, Eskandar H (2010) Thermal analysis of workpiece under Electrical Discharge Mashining (EDM), using hyperbolic heat conduction model. Int J Des Manufact Technol 3:17–24
2676. Sahoo N, Ghosh S, Narasimhan A, Das SK (2014) Investigation of non-Fourier effects in bio-tissues during laser-assisted photothermal therapy. Int J Thermal Sci 76:208–220

2677. Sahoo N, Narasimhan A, Dhar P, Das SK (2018) Non-Fourier thermal transport induced structural hierarchy and damage to collagen ultrastructure subjected to laser irradiation. Int J Hyperth 34:229–242
2678. Sahuquillo J, Vilalta A (2007) Cooling the injured brain: how does moderate hypothermia influence the pathophysiology of traumatic brain injury. Curr Pharmaceut Des 13:2310–2322
2679. Sajjadi AY, Mitra K, Grace M (2011) Ablation of subsurface tumors using an ultrashort pulse laser. Opt Lasers Eng 49:451–456
2680. Saka MN, Pandolf KB (2001) Physical exercise in hot climates: physiology, performance, and biomedical issues. In: Lounsbury DE, Randolh KB, Burr RE (eds) Medical aspects of harsh environments, vol 1, pp 87–133. Office of the Surgeon General, Department of the Army, Falls Church
2681. Sakai A (1960) Survival of the twig of woody plants at -196 ° C. Nature 185:393–394
2682. Salahuddin S, La Gioia A, Shahzad A, Elahi M, Kumar A, Kilroy D, Porter E, O'Halloran M (2018) An anatomically accurate dielectric profile of the porcine kidney. Biomed Phys Eng Exp 4:25–42
2683. Salari K, Knupp P (2000) Code verification by the method of manufactured solutions. Sandia Report SAND2000-1444, Sandia National Laboratories
2684. Salazar A (2006) Energy propagation of thermal waves. Eur J Phys 27:1349–1355
2685. Salehyar S, Zhu Q (2016) Deformation and internal stress in a red blood cell as it is driven through a slit by an incoming flow. Soft Matter 12:3156–3164
2686. Saliken JC, Donnelly BJ, Rewcastle JC (2002) The evolution and state of modern technology for prostate cryosurgery. Urology 60(suppl 2A):26–33
2687. Salinas-Flores L, Adams SL, Lim MH (2008) Determination of the membrane permeability characteristics of pacific oyster, Crassotera gigas, oocytes and development of optimal methods to add and remove ethylene glycol. Cryobiology 56:43–52
2688. Salinas-Flores L, Adams SL, Wharton DA, Downes MF, Lim MH (2008) Survival of pacific oyster, Crassotera gigas, oocytes in relation to the intracellular ice formation. Cryobiology 56:28–35
2689. Salloum M, Ghaddar N, Ghali K (2007) A new transient bioheat model of the human body and its integration to clothing model. Int J Thermal Sci 46:371–384
2690. Salloum M, Ma R, Zhu L (2008) An in-vivo experimental study of the temperature elevation in animal tissue during magnetic nanoparticle hyperthermia. Int J Hyperth 24:589–601
2691. Salomao M, Zhang X, Yang Y, Lee S, Hartwig JH, Chasis JA, Mohandas M, An X (2008) Protein 4.1 R-dependent multiprotein complex: new insights into the structural organization of the red blood cell membrane. Proc Natl Acad Sci USA 105:8026–8031
2692. Salt RW (1959) Survival of frozen fat body cells in insects. Nature 184:1426
2693. Salt RW (1962) Intracellular freezing in insects. Nature 193:1207–1208
2694. Samaras T, Christ A, Kuster N (2006) Effects of geometry discretization aspects on the numerical solution of the bioheat transfer equation with the FDTD technique. Phys Med Biol 51:N221–N229
2695. Samko SG, Kilbas AA, Marichev OI (1993) Fractional integrals and derivatives. Theory and applications. Gordon and Breach Science Publishers, London
2696. Samsel RW, Perelson AS (1984) Kinetics of rouleau formation. II. Reversible reactions. Biophys J 45:805–824
2697. Samset E, Mala T, Edwin B, Gladhaug I, Soreide O, Fosse E (2001) Validation of estimated 3D temperature maps during hepatic cryo surgery. Magn Res Imag 19:715–721
2698. Samsonov A, Popov SV (2013) The effect of a 94 GHz electromagnetic field on neuronal microtubules. Bioelectromagnetics 34:133–144
2699. Sanbonmatsu KY, Tung CS (2007) High performance computing in biology: multimillion atom simulations of nanoscale systems. J Struct Biol 157:470–480
2700. Sánchez R, Carreras BA, van Milligen BP (2005) Fluid limit of nonintegrable continuos-time random walks in termes of fractional differential equations. Phys Rev E 71:011111
2701. Sandell JL, Zhu TC (2011) A review of in-vivo optical properties of human tissues and its impact on PDT. J Biophoton 12:773–787

2702. Sanderson T, Ume C, Jarzinski J (1995) Hyperbolic heat conduction caused by temporally modulated laser pulses. Ultrasonics 33:423–427
2703. Sandison GA (2002) Future directions for cryosurgery computer planning. Urology 60(suppl 2A):50–55
2704. Sands D (2011) Pulsed laser heating and melting. In: Vikhrenko V (ed) Heat transfer - engineering applications. InTech, Rijeka, pp 47–70
2705. Sangwan V, Foulds I, Sinh J, Dhindsa RS (2001) Cold-activation of Brassica napus BN115 promoter is mediated by structural changes in membranes and cytoskeleton, and requires Ca^{2+} influx. Plant J 27:1–12
2706. Sankaran V, Walsh JT, Maitland DJ (2002) Comparative study of polarized light propagation in biologic tissues. J Biomed Opt 7:300–306
2707. Sano F, Washio T, Matsmae M (2019) Measurements of specific heat capacities required to build computer simulation models for laser thermotherapy of brain lesions. Tokai J Exp Clin Med 44:80–84
2708. SantaLucia J, Hicks D (2004) Thethermodynamics of DNA structural motifs. Annu Rev Biophys Biomol Struct 33:415–440
2709. Sanyal DC, Maji NK (2001) Thermoregulation through skin under variable atmospheric and physiological conditions. J Theor Biol 208:451–456
2710. Sanz B, Calatayud MP, De Biasi E, Lima E, Mansilla MV, Zysler RD, Ibarra MR, Goya GF (2016) In silico before in vivo: how to predict the heating efficiency of magnetic nanoparticles within the intracellular space. Sci Rep 6:38733
2711. Sanz PD, de Elvira C, Martino M, Zaritzky N, Otero L, Carrasco JA (1999) Freezing rate simulation as an aid to reducing crystallization damage in foods. Meat Sci 52:275–278
2712. Sapareto SA, Dewey WC (1984) Thermal dose determination in cancer therapy. Int J Radiat Oncol Biol Phys 10:787–800
2713. Saparov SM, Kozono D, Rothe U, Agre P, Pohl P (2001) Water and ion permeation of aquaporin-1 in planar lipid bilayers. J Biol Chem 276:31515–31520
2714. Saper CB, Lu J, Chou TC, Gooley J (2005) The hypothalamic integrator for circadian rhythms. Trends Neurosci 28:152–157
2715. Saragusty J (2015) Directional freezing for large volume cryopreservation. Methods Mol Biol 1257:381–397
2716. Sarkisov GN (2006) Structural models of water. Phys Usp 49:809–820
2717. Sasaki K, Mizuno M, Wake K, Watanabe S (2017) Monte Carlo simulations of skin exposure to electromagnetic field from 10 GHz to 1 THz. Phys Med Biol 62:6993–7010
2718. Savage MV, Brengelmann GL (1996) Control of skin blood flow in the neutral zone of human body temperature regulation. J Appl Physiol 80:1249–1257
2719. Savage VM, West GB (2007) A quantitative, theoretical framework for understanding mammalian sleep. Proc Natl Acad Sci USA 104:1051–1056
2720. Sawka MN (1992) Human tolerance and heat strain during exercise: influence of hydration. J Appl Physiol 73:368–375
2721. Sawka MN, Latzka WA, Mattot RP, Montain SJ (1998) Hydration effects on temperature regulation. Int J Sports Med 19:S108–S110
2722. Sawka MN, Montain SJ, Latzka WA (2001) Hydration effects on thermoregulation and performance in the heat. Comp Biochem Physiol Part A 128:679–690
2723. Sawyer WG, Hamilton MA, Fregly BJ, Banks SA (2003) Temperature modeling in a total knee joint replacement using patient-specific kinematics. Tribol Lett 15:343–351
2724. Sbirlea G, L'Huillier JP (1997) A powerful finite element for analysis of argon laser iridectomy: influence of natural convection on the human eye. WIT Trans Biomed Health 4:67–79
2725. Scala A, Starr FW, Nave EL, Stanley HE, Sciortino F (2000) Free energy surface of supercooled water. Phys Rev E 62:8016
2726. Scalas E, Gorenflo R, Mainardi F, Raberto M (2002) Revisiting the derivation of the fractional diffusion equation. arXiv: 0210166 [cond-mat.dis-nn]

2727. Schäfer AT, Kauffmann JD (1999) What happens in freezing bodies? Experimental study of histological tissue change caused by freezing injuries. Forensic Sci Intern 102:149–158
2728. Scheffer L, Bitler A, Ben-Jacob E, Korenstein R (2001) Atomic force pulling: probing the local elasticity of the cell membrane. Eur Biophys J 30:83–90
2729. Scher H, Lax M (1973) Stochastic transport in a disordered solid. I. Theor Phys Rev B 7:4491
2730. Schieke SM, Schroeder P, Krutmann J (2003) Cutaneous effects of infrared radiation: from clinical observations to molecular response mechanisms. Photodermatol Photoimmunol Photomed 19:228–234
2731. Schlader ZJ, Johnson BD, Pryor RR, Stooks J, Clemency BM, Hostler D (2020) Human thermoregulation during prolonged exposure to warm and extremely humid environments expected to occur in disabled submarine scenarios. Am J Physiol Regul Integr Comp Physiol 318:R950–R960
2732. Schleeh J, Mateos J, Íñiguez-de La-Torre I, Wadefalk N, Nilsson P, Grahn J, Minnich A (2015) Phonon black-body radiation limit for heat dissipation in electronics. Nat Mater 14:187–192
2733. Schlick T, Skeel RD, Brunger AT, Kalé LV, Board JA, Hermans J, Schulten K (1999) Algorithmic challenges in computational molecular biophysics. J Comput Phys 151:9–48
2734. Schmid F, Dúchs D, Lenz O (2004) Amphiphiles at interfaces: simulation of structure and phase behavior. NIC Ser 23:323–346
2735. Schmid-Schöbein GW, Diller KR (2005) Transport properties in biomedical systems: a roadmap for future research directions. Ann Biomed Eng 33:1136–1141
2736. Schmid-Schönbein H, Wells R (1969) Fluid drop-like transition of erythrocytes under shear. Science 165:288–291
2737. Schmid-Schönbein H, Wells RE (1971) Rheological properties of human erythrocytes and their influence upon the "anomalous" viscosity of blood. Ergeb Physiol 63:146–219
2738. Schmidt JD, Doyle J, Larison S (1998) Prostate cryoablation: update 1998. CA Cancer J Clin 48:239–253
2739. Schmidt MH, Swang TW, Hamilton IM, Best JA (2017) State-dependent metabolic partitioning and energy conservation: a theoretical framework for understanding the function of sleep. PLoS ONE 12:e0185746
2740. Schmitt JM, Kumar G (1996) Turbulent nature of refractive-index variations in biological tissue. Opt Lett 21:1310–1312
2741. Schnider U, Mazur P (1987) Relative influence of unfrozen fraction and salt concentration on the survival of slowly frozen eight-cell mouse embryos. Cryobilogy 24:17–41
2742. Schober R, Bettag M, Sabel M, Ulrich F, Hessel S (1993) Fine structure of zonal changes in experimental Nd:YAG laser-induced intersti-tial hyperthermia. Lasers Surg Med 13:234–241
2743. Scholander PF (1955) Evolution of climatic adaptation in homeotherms. Evolution 9:15–26
2744. Scholander PF, Hock R, Walters V, Johnson F, Irving L (1950) Heat regulation in some arctic and tropicalm ammals and birds. Biol Bull 99:237–258
2745. Schreuder MDJ, Brewer CA, Heine C (2001) Modelled influences of non-exchanging trichomes on leaf boundary layers and gas exchange. J Theor Biol 210:23–32
2746. Schuepp PH (1993) Leaf boundary layers. New Phytol 59:477–507
2747. Schürmann HW, Serov SV (2004) Traveling wave solutions of a generalized modified. Kadomtsev-Petviashvili equation. J Math Phys 45:2181–2187
2748. Schwartz GG, Rosenblum RA (1981) Allometry of primate hair density and the evolution of human hairlessness. Am J Phys Anthropol 55:9–12
2749. Schwenke M, Georgii J, Preusser T (2017) Fast numerical simulation of focused ultrasound treatments during respiratory motion with discontinuous motion boundaries. IEEE Trans Biomed Eng 64:1455–1468
2750. Sciortino F, Nave EL, Scala A, Stanley HE, Starr FW (2002) Water and its energy landscape. Eur Phys J E 9:233–237

2751. Scott J (1988) The computation of temperature rises in the human eye induced by infrared radiation. Phys Med Biol 33:243–257
2752. Scott J (1988) A finite element model of heat transport in the human eye. Phys Med Biol 33:227–241
2753. Scott M (2005) The modeling of blood rheology in small vessels. Ph.D. Thesis, University of Waterloo
2754. Scott K, Acker JP (2003) Red blood cells beyond the 10-year storage limit: time for a review? Cell Preserv Technol 1:227
2755. Scott KL, Lecak J, Acker JP (2005) Biopreservation of red blood cells: past, present and future. Transfus Med Rev 19:127–142
2756. Scott AA, Waugh DW, Zaitchik BF (2018) Reduced urban heat island intensity under warmer conditions. Environ Res Lett 13:064003
2757. Searles JA, Carpenter JF, Randolph TW (2001) The ice nucleation temperature determines the primary drying rate of lyophillization for samples frozen on a temperature-controlled shelf. J Pharm Sci 90:860–871
2758. Sears MW, Angilletta MJ Jr (2015) Costs and benefits of thermoregulation revisited: both the heterogeneity and spatial structure of temperature drive energetic costs. Am Nat 185:E94–E102
2759. Secomb TW (1995) Mechanics of blood flow in the microcirculation. Symp Soc Exp Biol 49:305–321
2760. Secomb TW (2011) Mechanics and computational simulation of blood flow in microvessels. Med Eng Phys 33:800–804
2761. Secomb TW (2017) Blood flow in the microcirculation. Annu Rev Fluid Mech 49:443–461
2762. Secomb TW, Hsu R (1996) Analysis of red blood cell motion through cylindrical micropores: effects of cell properties. Biophys J 71:963–966
2763. Secomb TW, Pries AR (2011) The microcirculation: physiology at the mesoscale. J Physiol 589:1047–1052
2764. Secomb TW, Pries AR (2013) Blood viscosity in microvessels: experiment and theory. C R Phys 14:470–478
2765. Secomb TW, Skalak R, Özkaya N, Gross JE (1986) Flow of axisymmetric red blood cells in narrow capillaries. J Fluid Mech 163:405–423
2766. Secomb T, Hsu R, Beamer N, Coull BM (2000) Theoretical simulation of oxygen transport to brain by networks of microvessels: effects of oxygen supply and demand on tissue hypoxia. Microcirculation 7:237–247
2767. Secomb TW, Hsu R, Park EY, Dewhirst MW (2004) Green's function methods for analysis of oxygen delivery to tissue by microvascular networks. Ann Biomed Eng 32:1519–1529
2768. Secomb TW, Styp-Rekowska B, Pries AR (2007) Two-dimensional simulation of red blood cell deformation and lateral migration in microvessels. Ann Biomed Eng 35:755–765
2769. Sedunov BI, Frank-Kamenetskii DA (1963) Dielectric constants of biological. Sov Phys Usp 7:1066–1071
2770. Seegenschmiedt MH, Vernon CC (1996) A historical perspective on hyperthermia in oncology. In: Seegenschmiedt MH, Fessenden P, Vernon CC (eds) Thermo-radiotherapy and thermo-chemiotherapy. Springer, Berlin, pp 3–46
2771. Segino M, Ikeda M, Hirahara F, Sato K (2005) In vitro follicular development of cryopreserved mouse ovarian tissue. Reproduction 130:187–192
2772. Seifert U (1997) Configurations of fluid membranes and vesicles. Adv Phys 46:13–137
2773. Seip R, Ebbini ES (1995) Studies on the three-dimensional temperature response to heating fields using diagnostic ultrasound. IEEE Trans Biomed Eng 42:828–839
2774. Sellitto A, Cimmelli VA (2012) A continuum approah to thermomass theory. J Heat Transf 134:112402
2775. Sellitto A, Cimmelli VA, Jou D (2012) Analysis of three nonlinear effects in a continuum approach to heat transport in nanosysytems. Phys D 241:1344–1350
2776. Sellitto A, Cimmelli VA, Jou D (2013) Entropy flux and anomalous axial heat transport at the nanoscale. Phys Rev B 87:054302

2777. Sellitto A, Cimmelli VA, Jou D (2013) Thermoelectric effects and size dependency of the figure-of-merit in cylindrical nanowires. Int J Heat Mass Transf 57:109–116
2778. Sellitto A, Cimmelli VA, Jou D (2016) Mesoscopic theories of heat transfer in nanosystems. Springer, Berlin
2779. Sellitto A, Rogolino P, Carlomagno I (2016) Heat-pulse propagation along nonequilibrium nanowires in thermomass theory. Commun Appl Ind Math 7:39–55
2780. Sellitto A, Carlomagno I, Di Domenico M (2021) Nonlocal and nonlinear effects in hyperbolic heat transfer in a two-temperature model. Z Angew Math Phys 72:7
2781. Selye H (1956) The stress of life. The McGraw-Hill, New York
2782. Selye H (1976) Forty years of stress research: principal remaining problems and misconceptions. Can Med Assoc J 115:53–56
2783. Sen S, Subramanian S, Discher DE (2005) Indentation and adhesive probing of a cell membrane with AFM: theoretical model and experiments. Biophys J 89:3203–3213
2784. Seo BM, Miura M, Sonoyama W, Coppe C, Stanyon R, Shi S (2005) Recovery of stem cells from cryopreserved peridontal ligament. J Dent Res 84:907–912
2785. Seppänen O, Fisk W, Mendell M (1999) Association of ventilation rates and CO_2 concentrations with health and other human responses in commercial and institutional buildings. Indoor Air 9:226–252
2786. Serdyuk IN (2002) Physical methods in structural molecular biology in the beginning of 21 century (in Russian). Adv Biol Chem 42:3–28
2787. Serdyukov SI (2001) A new version of extended irreversible thermodynamics and dual-phase-lag model in heat transfer. Phys Lett A 281:16–20
2788. Serdyukov SI (2013) Higher order heat and mass transfer equations and their justification in extended irreversible thermodynamics. Theor Found Chem Eng 47:89–103
2789. Serota HM, Gerard RW (1938) Localized temperature changes in the cat brain. J Neurophysiol 1:115–124
2790. Serrat MA, King D, Lovejoy CO (2008) Temperature regulates limb length in homeothermsby directly modulating cartilage growth. Proc Nat Acad Sci USA 105:19348–19353
2791. Serrone J, Kocaeli H, Douglas Mast T, Burgess MT, Zuccarello M (2012) The potential applications of high-intensity focused ultrasound (HIFU) in vascular neurosurgery. J Clin Neurosci 19:214–221
2792. Seymour RS (2001) Biophysics and physiology of temperature regulation in thermogenic flowers. Biosci Rep 21:223–236
2793. Seymour RS, Schultze-Motel P (1996) Thermoregulating lotus flowers. Nature 383:305
2794. Seymour RS, Schultze-Motel P (1997) Heat-producing flowers. Endeavour 21:125–129
2795. Seymour RS, White CR, Gibernau M (2003) Heat reward for insect pollinators. Nature 426:243–244
2796. Shafranov VV, Borkhunova EN, Korotkii NG, Vissarionov VA, Stenko AG (2003) Keloid scars: etiology, clinical, morphological, physical diagnostics and RF-cryo therapy (in Russian). CPR Delovaya kniga
2797. Shahidi V, Savard P (1994) A finite element model for radiofrequency ablation of the myocardium. IEEE Trans Biomed Eng 41:963–968
2798. Shahzad A, Khan S, Jones M, Dwyer RM, O'Halloran M (2017) Investigation of the effect of dehydration on tissue dielectric properties in ex vivo measurements. Biomed Phys Eng Exp 3:1–9
2799. Shaitan KB, Saraikin SS (1999) Molecular dynamics method. http://www.moldyn.ru/library/md/default.htm
2800. Shalygin AN, Krotov KA (1990) Magnetic capture of individual biological cells and model aggregates of cell membranes. Phys Usp 33:541–553
2801. Shankaran S, Laptook AR, Ehrenkranz RA, Tyson JE, Mcdonald SA, Donovan EF, Fanaroff AA, Poole WK, Wright LL, Higgins RD, Finer NN, Carlo WA, Duara S, Oh W, Cotten CM, Stevenson DK, Stoll BJ, Lemons JL, Guillet R, Jobe AH (2005) Whole-body hypothermia for neonates with hypoxic-ischemic encephalopathy. N Engl J Med 353:1574–1584

2802. Shao JL, Arjun B, Leo HC, Chua KJ (2017) Nano-assisted radiofrequency ablation of clinically extracted irregularly shaped liver tumors. J Thermal Biol 66:101–113
2803. Shapiro Y, Seidman DS (1990) Field and clinical observations of exertional heat stroke patients. Med Sci Sports Exerc 22:6–14
2804. Sharan M, Popel AS (2001) A two-phase model for flow of blood in narrow tubes with increased effective viscosity near the wall. Biorheology 38:415–428
2805. Shardt N, Al-Abbasi KK, Yu H, Jomha NM, McGann LE, Elliott JAW (2016) Cryoprotectant kinetic analysis of a human articular cartilage vitrification protocol. Cryobiology 73:80–92
2806. Shariat SF, Raptidis G, Masatoschi M, Bergamaschi F, Slawin KM (2005) Pilot study of radiofrequency interstitial tumor ablation (RITA) for the treatment of radio-recurrent prostate cancer. Prostate 65:260–267
2807. Sharma R, Hooyberghs H, Lauwaet D, De Ridder K (2019) Urban heat island and future climate change–implications for Delhi's heat. J Urban Health 96:235–251
2808. Shaw A, Martin K (2012) The acoustic output of diagnostic ultrasound scanners. In ter Haar G (ed) The safe use of ultrasound in medical diagnosis, pp 18–45
2809. Shaw JM, Jones GM (2003) Terminology associated with vitrification and other cryopreservation procedures for oocytes and embryos. Hum Reprod Update 9:583–605
2810. Shelby JP, White J, Ganesan K, Rathod PK, Chiu DT (2003) A microfluidic model for single-cell capillary obstruction by Plasmodium falciparum-infected erythrocytes. Proc Natl Acad Sci USA 100:14618–14622
2811. Shen W, Zhang J, Yang F (2005) Modeling and numerical simulation of bioheat transfer and biomechanics in soft tissue. Math Comput Model 41:1251–1265
2812. Shen WS, Zhang J, Yang FQ (2005) Skin thermal injury prediction with strain energy. Int J Nonlinear Sci Numer Simul 6:2317–2328
2813. Shen X, Ding G, Wei J, Zhao L, Zhou Y, Deng H, Lao L (2006) An infrared radiation study of the biophysical characteristics of traditional moxibustion. Compl Ther Med 14:213–219
2814. Shen Z, Coupier G, Kaoui B, Polack B, Harting J, Misbah C, Podgorski T (2016) Inversion of hematocrit partition at microfluidic bifurcations. Microvasc Res 105:40–46
2815. Shenoi MM, Shah NB, Griffin RJ, Vercellotti GM, Bischof JC (2011) Nanoparticle preconditioning for enhanced thermal therapies in cancer. Nanomedicine 6:545–563
2816. Sherief HH, El-Sayed AMA, Abd El-Latief AM (2010) Fractional order theory of thermoelasicity. Int J Solid Struct 47:269–273
2817. Sherwin RP, Richters A, Richters V (1967) The occurrence of a cyclosis-like phenomenon in human lung cancer cells in vitro. Cancer Res 27:152–158
2818. Sherwood JM, Dusting J, Kaliviotis E, Balabani S (2012) The effect of red blood cell aggregation on velocity and cell-depleted layer characteristics of blood in a bifurcating microchannel. Biomicrofluidics 6:024119
2819. Sheu TWH, Solovchuk MA, Chen AWJ, Thiriet M (2011) On an acoustics-thermal-fluid coupling model for the prediction of temperature elevation in liver tumor. Int J Heat Mass Transf 54:4117–4126
2820. Shi X, Datta AK (1998) Thermal stresses from large volumetric expansion during freezing of biomaterials. J Biomech Eng 120:720–726
2821. Shi T, Reeves RH, Gilinsky DA, Friedmann EI (1997) Characterization of viable bacteria from Siberian permafrost by 16sr DNA sequencing. Microbiol Ecol 33:169–179
2822. Shi X, Datta AK, Mukharjee S (1999) Thermal fracture in a biomaterial during rapid freezing. J Thermal Stress 22:275–292
2823. Shi X, Martin RW, Vaezy S, Crum LA (2002) Quantitative investigation of acoustic streaming in blood. J Acoust Soc Am 111:1110–1121
2824. Shi J, Chen Z, Shi M (2009) Simulation of heat transfer of biological tissue during cryosurery based on vascular trees. Appl Thermal Eng 29:1792–1798
2825. Shi L, Pan TW, Glowinski R (2012) Numerical simulation of lateral migration of red blood cells in Poiseuille flows. Int J Numer Methods Fluids 68:1393–1408
2826. Shi M, Feng S, Zhang X, Ji C, Xu F, Lu TJ (2018) Droplet based vitrification for cell aggregates: numerical analysis. J Mech Behav Biomed Mater 82:383–393

2827. Shih TC, Kou HS, Liauch CT, Lin WL (2002) Thermal models of bioheat transfer equations in living tissue and thermal dose equivalence due to hypertermia. Biomed Eng Appl Basis Commun 14:40–56
2828. Shih TC, Kou HS, Lin WL (2002) Effect of effective tissue conductivity onthermal dose distributions of living tissues with directional blood flow during thermal therapy. Int Commun Heat Mass Transf 29:115–126
2829. Shih TC, Kou HS, Liauh CT, Lin WL (2005) The impact of thermal wave characteristics on thermal dose distribution during thermal therapy: a numerical study. Med Phys 32:3029–3036
2830. Shih TC, Yuan P, Lin WL, Kou HS (2007) Analytical analysis of the Pennes bioheat transfer equation with sinusoidal heat flux condition on skin surface. Med Eng Phys 29:946–953
2831. Shimada K, Asahina E (1975) Visualization of intracellular ice crystals in very rapidly frozen cells at − 27° C. Cryobiology 12:209–218
2832. Shimmen T, Yokota E (2004) Cytoplasmic streaming in plants. Curr Opin Cell Biol 16:68–72
2833. Shingleton WW (1962) Selective heating and cooling of tissue in cancer chemotherapy. Ann Surg 156:408–414
2834. Shinozaki K, Yamaguchi-Shinozaki K (2000) Molecular responses to dehydration and low temperatures: differences and cross-talk between two stress signaling pathways. Curr Opin Plant Biol 3:217–233
2835. Shiota K (1982) Neural tube defects and maternal hyperthermia in early pregnancy: epidemiology in a human embryo population. Am J Med Genet 12:281–288
2836. Shipley RJ, Chapman SJ (2010) Multiscale modelling of fluid and drug transport in vascular tumours. Bull Math Biol 72:1464–1491
2837. Shirkavand A, Nazif HR (2019) Numerical study on the effects of blood perfusion and body metabolism on the temperature profile of human forearm in hyperthermia conditions. J Therm Biol 84:339–350
2838. Shirmohammadi R (2008) Temperature transients in spherical medium irradiated by laser pulse. Int Commun Heat Mass Transf 35:1017–1023
2839. Shitzer A, Chato JC (1978) Analytical solutions to tie problem of transient heat transfer in living tissue. J Biomech Eng 100:202–210
2840. Shitzer A, Eberhart RC (1985) Heat transfer in medicine biology: analysis and applications. Plenum, New York
2841. Shiu YT, McIntire LV (2003) In vitro studies of erythrocyte-vascular endothelium interactions. Annu Rev Biomed Eng 31:1299–1313
2842. Shlesinger MF (2008) Random processes with infinite moments. In: Klages R, Radons G, Sokolov IM (eds) Anomalous transport. Foundations and applications. Wiley, Hoboken, pp 75–90
2843. Shnoll EE, Grivtsov AG (1996) Molecular dynamics method in physical chemistry (in Russian). Nauka, Moscow
2844. Shomalia Z, Kovács R, Ván P, Kudinov IV, Ghazanfariane J (2022) Recent progresses and future directions of lagging heat models in thermodynamics and bioheat transfer. Contin Mech Thermodyn 34:637–679
2845. Shrivastava D (Ed) (2018) Theory and applications of heat transfer in humans. Wiley, Hoboken
2846. Shrivastava D, Roemer R (2005) An analitycal study of heat transfer in a finite region with two blood vessels and general dirichlet boundary conditions. Int J Heat Mass Transf 48:4090–4102
2847. Shrivastava D, Roemer R (2005) A general analitycal derivation of a new, source term dependent 2-D Poisson conduction shape factors. Int J Heat Mass Transf 47:4293–4300
2848. Shrivastava D, Roemer RB (2005) An analytical study of "Poisson conduction shape factors" for two thermally significant vessels in a finite, heated tissue. Phys Med Biol 50:3627–3641
2849. Shrivastava D, Roemer RB (2006) Readdressing the issue of thermally significant blood vessels using a countercurrent vessel network. Phys Med Biol 128:210–216

2850. Shrivastava D, Vaughan J (2009) A generic bioheat transfer thermal model for a perfused tissue. J Biomech Eng 131:1–12
2851. Sidebottom S, Buckley S, Pudney P, Twigg S, Jarman C, Holt C, Telford J, McArthur A, Worall D, Hubbard R, Lillford P (2000) Heat-stable antifreeze proteins from grass. Nature 406:256
2852. Siedow JN, Umbach AL (1995) Plant mitochondrial electron transfer and molecular biology. Plant Cell 7:821
2853. Sierociuk D, Dzielinski A, Sarwas G, Petra I, Podlubny I, Skovranek T (2013) Modeling heat transfer in heterogeneous media using fractional calculus. Philos Trans Ser A 371:20120146
2854. Sierociuk D, Skovranek T, Macias M, Podlubny I, Petra I, Dzielinski A, Ziubinsky P (2015) Diffusion process modeling by using fractional-order models. Appl Math Comput 257:2–11
2855. Siesjo B (1978) Brain energy metabolism. Wiley, New York
2856. Silva JM (2009) From the discovery of the circulation of the blood to the first steps in hemorheology: part 2. Portuguese J Cardiol 28:1405–1439
2857. Silva FRGB, Goncalves G, Lenzi MK, Lenzi EK (2013) An extension of the Luikov system equations of heat and mass transfer. Int J Heat Mass Transf 63:233–238
2858. Silva NP, Bottoglieri A, Conceicao RC, O'Halloran M (2020) Characterisation of ex vivo liver thermal properties for electromagnetic-based hyperthermic therapies. Sensors 20:3004
2859. Simankov D (2021) Classification of the factors of functioning of the individual human thermoregulation system and prediction of its state (local thermophysical aspects). arXiv:2111.01818 [physics.med-ph]
2860. Simanovskii DM, Mackanos MA, Irani R, O'Connell-Rodwell CH, Contag CH, Schwettman HA, Palanker DV (2006) Cellular tolerance to pulsed hyperthermia. Phys Rev E 74:011915
2861. Simo E, Zambe JIC, Tabekoueng FW, Fotso PR, Kuichew DK, Tetsa FT (2020) Numerical analysis of heat transfer in biological tissues irradiated by pulsed laser: treatment of angioma, condyloma and trachea tumors. Indian J Eng 17:384–395
2862. Simon E (1974) Temperature regulation: the spinal cord as a site of extrahypothalamic thermoregulatory functions. Rev Physiol Biochem Pharmacol 71:1–76
2863. Simon E, Pierau FK, Taylor DCM (1986) Central and peripheral thermal control of effectors in homeothermic temperature regulation. Physiol Rev 66:235–309
2864. Singer SJ, Nicolson GL (1972) The fluid mosaic model of the structure of cell membranes. Science 175:720–731
2865. Singh S, Kumar S (2015) Freezing of biological tissues during cryosurgery using hyperbolic heat conduction model. Math Model Anal 20:443–456
2866. Singh S, Melnik R (2019) Coupled thermo-electro-mechanical models for thermal ablation of biological tissues and heat relaxation time effects. Phys Med Biol 64:245008
2867. Singh S, Melnik R (2020) Thermal ablation of biological tissues in disease treatment: a review of computational models and future directions. Electromagn Biol Med 39:49–88
2868. Singh G, Paul A, Shekhar H, Paul A (2020) Pulsed ultrasound assisted thermo-therapy for subsurface tumor ablation: a numerical investigation. J Therm Sci Eng Appl 13:1–44
2869. Singh J, Gupta PK, Rai KN (2011) Solution of fractional bioheat equations by finite difference method and HPM. Math Comput Model 54:2316–2325
2870. Singh S, Repaka R, Al-Jumaily A (2018) Sensitivity analysis of critical parameters affecting the efficacy of microwave ablation using taguchi method. Int J RF Microw Comput-Aid Eng 29:2011581
2871. Skalak R, Branemark PI (1969) Deformation of red blood cells in capillaries. Science 164:717–719
2872. Skalak R, Tozeren A, Zarda RP, Chien S (1973) Strain energy function of red blood cell membranes. Biophys J 13:245–264
2873. Skalak R, Keller SR, Secomb TW (1981) Mechanics of blood flow. J Biomech Eng 103:102–115
2874. Ŝkop V, Guo J, Liu N, Xiao C, Hall KD, Gavrilova O, Reitman ML (2020) Mouse thermoregulation: introducing the concept of the thermoneutral point. Cell Rep 31:107501

2875. Ŝkop V, Xiao C, Liu N, Gavrilova O, Reitman ML (2021) The effects of housing density on mouse thermal physiology depend on sex and ambient temperature. Mol Metabol 53:101332
2876. Skorupa A, Piasecka-Belkhayat A (2023) Comparison of heat transfer phenomena for two different cryopreservation methods: slow freezing and vitrification. J Appl Math Comput Mech 22:57–69
2877. Skotheim JM, Secomb TW (2007) Red blood cells and other nonspherical capsules in shear flow: oscillatory dynamics and the tank-treading-to-tumbling transition. Phys Rev Lett 98:078301
2878. Skovsen E, Snyder JW, Lambert JDC, Ogilby PR (2005) Lifetime and diffusion of singlet oxygen in a cell. J Phys Chem B 109:8570–8573
2879. Skubatz H (2014) Thermoregulation in the appendix of the sauromatum guttatum(Schott) inflorescence. Bot Stud 55:68
2880. Skubatz H, Nelson TA, Meeuse BJD, Bendich AJ (1991) Heat production in thevoodoo lily (sauromatum guttatum) as monitored by infrared thermography. Plant Physiol 95:1084–1088
2881. Small A (1959) Skunk cabbages, symplocarpus foetidus. Bull Torrey Bot Club 86:413–421
2882. Smallwood M, Bowles DJ (2002) Plants in cold climate. Philos Trans R Soc Lond B 357:831–847
2883. Smallwood M, Worall D, Byass L, Elias L, Ashford D, Doucet CJ, Holt C, Telford J, Lillford P, Bowles DJ (1999) Isolation and characterization of a novel antifreeze protein from carrot (Daucus carota). Biochem J 340:385–391
2884. Smiley ST, Reers M, Mottola-Hartshorn C, Lin M, Chen A, Smith T, Steele GD Jr, Chen LB (1991) Intracellular heterogeneity in mitochondrial membrane potentials revealed by a J-aggregate-forming lipophilic cation JC-1. Proc Natl Acad Sci USA 88:3671–3675
2885. Smith JE (2005) Cooling methods used in the treatment of exertional heat illness. Br J Sports Med 39:503–507
2886. Snedden WA, Fromm H (2001) Calmodulin as a versatile calcium signal transducer in plants. New Phytol 151:35–66
2887. Snyder GI, Toberer ES (2008) Complex thermoelectric materials. Nat Mater 7:105–114
2888. Sobban CB, Thomas S, Peterson GP (2017) Microscale transport phenomena for bioengineering applications: recent advances. J Adv Nanomater 2:41–59
2889. Sobol EN (1999) Heating and structural alterations in cartilage under laser radiation. IEEE J Quant Electron 35:532–539
2890. Sobolev SL (1991) Transport properties and travelling waves in systems with local nonequilibrium. Sov Phys Usp 34:217–229
2891. Sobolev SL (1993) Two-temperature discrete model for nonlocal heat conduction. J Phys III 3:2261–2269
2892. Sobolev SL (1994) Equations of transfer in non-local media. Int J Heat Mass Transf 37:2175–2182
2893. Sobolev SL (1994) Heat conduction equation for systems with inhomogeneous internal structure. J Eng Phys Thermophys 66:436–440
2894. Sobolev SL (1994) Space-time nonlocal model for heat conduction. Phys Rev E 50:3255–3258
2895. Sobolev SL (1995) Two-temperature Stefan problem. Phys Lett A 197:243–246
2896. Sobolev SL (1997) Local non-equilibrium transport models. Phys Usp 40:1042–1053
2897. Sobolev SL (2018) Hyperbolic heat conduction, effective temperature, and third law for nonequilibrium systems with heat flux. Phys Rev E 97:022122
2898. Sobolev SL (2018) On hyperbolic heat-mass transfer equation. Int J Heat Mass Transf 122:629–630
2899. Sokolov IM, Klafter J, Blumen A (2002) Fractional kinetics. Phys Todsay 55:48–54
2900. Solon LR, Aronson R, Gould G (1961) Physiological implications of laser beams the very high radiation flux densities of optical masers point to important biomedical applications. Science 134:1506–1508
2901. Solovchuk MA, Sheu TWH, Thiriet M, Lin WL (2011) Effect of acoustic streaming on tissue heating due to high-intensity focused ultrasound. arXiv:1111.2908 [physics.flu-dyn]

2902. Solovchuk MA, Sheu TWH, Thiriet M (2013) Simulation of nonlinear westervelt equation for the investigation of acoustic streaming and nonlinear propagation effects. J Acoust Soc Am 134:3931–3942
2903. Solovchuk MA, Sheu TWH, Thiriet M, Lin WL (2013) On a computational study for investigating acoustic streaming and heating during focused ultrasound ablation of liver tumor. J Appl Therm Eng 56:62–76
2904. Solovchuk MA, Thiriet M, Sheu TWH (2014) Computational study for investigating acoustic streaming and heating during acoustic hemostasis. arXiv:1411.6343v1 [physics.flu-dyn]
2905. Somia N, Verma IM (2000) Gene therapy: trials and tribulations. Nat Rev Genet 1:91–99
2906. Song CW (1984) Effect of local hyperthermia on blood flow and microenvironment: a review. Cancer Res 44:4721S–4730S
2907. Song CW, Lokshina A, Rhee JG, Patten M, Levitt SH (1984) Implication of blood flow in hyperthermia treatment of tumors. IEEE Trans Biomed Eng 31:9–16
2908. Song WJ, Weinbaum S, Jiji LM, Lemons LD (1988) A combined macro and microvascular model for whole limb heat trans. J Biom Eng 110:259–268
2909. Song CW, Choi IB, Nah BS, Sahu SK, Osborn JL (1995) Microvasculature and perfusion in normal tissues and tumours. In: Seegenschmiedt MH, Fessenden P, Vernon CC (eds) Thermoradiometry and Thermochemotherapy, vol 1. Springer, Berlin, pp 139–156
2910. Song S, Zhang Y, Fong CC, Tsang CH, Yang Z, Yang M (2003) CDNA microarray analysis of gene expression profiles in human broblast cells irradiated with red light. J Investig Dermatol 120:849–857
2911. Song YS, Moon S, Hulli L, Hasan SK, Kayaalp E, Demirci U (2009) Microfluidics for cryopreservation. Lab Chip 9:1874–1881
2912. Song P, Gao H, Zhang M, Yang F, Li SS, Kang B, Xu JJ, Chen HY (2019) Transient microscopy for measuring heat transfer in single cells. arXiv:1901.00141 [physics.bio-ph]
2913. Song X, Teng A, Wei J, Song L (2020) Research on the laser-brain tissue interaction by finite element analysis. arXiv:2009.13907 [physics.med-ph]
2914. Sonin AS (1987) Lyothropic nematics. Sov Phys Usp 30:875–896
2915. Sonksen J, Craggs J (1999) Circulation of the skin. Curr Anaesth Crit Care 10:58–63
2916. Sonnichsen FD, DeLuca IC, Davies PL, Sykes BD (1996) Refind solution structure of type III antifreeze protein: hydrophobic groups may be involved in the energetics of the protein-ice interacion. Structure 4:1325–1327
2917. Sonoda MT, Moreira NH, Martínez L, Favero FW, Vechi SM, Martins LR, Skaf MS (2004) A review on the dynamics of water. Brazil J Phys 34:3–16
2918. Soo ETL, Yip GWC, Lwin ZM, Kumar SD, Bay BH (2008) Heat shock proteins as novel therapeutic targets in cancer. In Vivo 22:311–316
2919. Sordillo LA, Pu Y, Pratavieira SA, Budansky Y, Alfano RR (2014) Deep optical imaging of tissue using the second and third near-infrared spectral windows. J Biomed Opt 19:56004
2920. Sorensen DN, Voigt LK (2003) Modelling flow and heat transfer around a seated human body by computational fluid dynamics. Build Environ 38:753–762
2921. Sottile ML, Nadin SB (2018) Heat shock proteins and DNA repair mechanisms: an updated overview. Cell Stress Chaperones 23:303–315
2922. Souzu H, Mazur P (1978) Temperature dependence of the survival of human erythrocytes frozen slowly in various concentration of glycerol. Biophys J 23:89–100
2923. Sparks JP, Campbell GS, Black RA (2000) Liquid water content of wood tissue at temperatueres below 0° C. Can J For Res 30:624–630
2924. Stachecki J (2004) An overview of oocytes preservation. Reprod Biomed Online 9:152–163
2925. Stacy R, Eroglu A, Fowler A, Biggers J, Toner M (2006) Thermal characterisation of Nakagata's mouse sperm freezing protocol. Cryobiology 52:99–117
2926. Stam S, Gardel ML (2014) Cutting through the noise: the mechanics of intracellular transport. Dev Cell 30:365–366
2927. Stamenovic D (2008) Rheological behavior of mammalian cells. Cell Mol Life Sci 65:3592–3605

2928. Stanczyk M, Telega JJ (2002) Modelling of heat transfer in biomechanics - a review. Part I. Soft tissues. Acta Bioengineer Biomech 4:32–61
2929. Stanczyk M, Van Leeuwen GMJ, Van Steenhoven AA (2007) Discrete vessel heat transfer in perfused tissue–model comparison. Phys Med Biol 52:2379–2391
2930. Stanislavsky AA (2004) Probabilistic interpretation of the integral of fractional order. Theor Math Phys 138:418–431
2931. Stanley HE, Buldyrev SV, Goldberger AL, Goldberger ZD, Havlin S, Mantegna RN, Ossadnik SM, Peng CK, Simons M (1994) Statistical mechanics in biology: how ubiquitous are long-range correlations? Phys A 205:214–253
2932. Starr FW, Angell CA, Stanley HE (2003) Prediction of entropy and dynamical properties of water below the homogeneous nucleation temperature. Phys A 323:51–66
2933. Steck LN, Sparrow EM, Abraham JP (2011) Non-invasive measurement of the hhuman core temperature. Int J Heat Mass Transf 54:975–982
2934. Steen PA, Soule EH, Michenfelder JD (1979) Detrimental effect of prolonged hypothermia in cats and monkeys with and without regional cerebral ischemia. Stroke 10:522–529
2935. Stefanoic S, Sütterlin M, Gaiser T, Schaff C, Neumann M, Berger L, Froemmel M, Tuschy B, Berlit S (2023) Microscopic, macroscopic and thermal impact of argon plasma, diode laser, and electrocoagulation on ovarian tissue. In Vivo 37:531–538
2936. Steif PS, Palastro MC, Rabin Y (2007) Analysis of the effect of partial vitrification on stress development in cryopreserved blood vessels. Med Eng Phys 29:661–670
2937. Steiner AA, Branco LGS (2002) Hypoxia-induced anapyrexia: implications and putative mediators. Annu Rev Physiol 64:263–288
2938. Steponkus PL (1993) Effects of freezing and cold acclimation on the plasma membrane of isolated protoplasts. US Dept. Energy. Progress Rep. DE-FG02-84ER13214, 7pp
2939. Steward DL (2004) Effectiveness of multilevel (tongue and palate) radiofrequency tissue ablation for patients with obstructive sleep apnea syndrome. Laryngoscope 114:2073–2084
2940. Stier JS, Filiault DL, Wisniewski M, Palta JP (2003) Visualiszation of freezing progression in turfgrasses using infrared video thermography. Crop Sci 43:415–420
2941. Stigliano RV, Shubitidze F, Kekalo K, Baker I, Giustini AJ, Hoopes PJ (2013) Understanding mNP hyperthermia for cancer treatment at the cellular scale. Proc SPIE Int Soc Opt Eng 8584:85840E
2942. Stillinger FH, Head-Gordon T (1995) Collective aspects of protein folding illustrated by a toy model. Phys Rev E 52:2872–2877
2943. Stocks JM, Taylor NAS, Tipton M, Greenleaf JE (2004) Human physiological responses to cold exposure. Aviat Space Environ Med 75:444–457
2944. Stoffner R, Kremser C, Schullian P, Haidu M, Widmann G, Bale RJ (2012) Multipolar radiofrequency ablation using 4–6 applicators simultaneously: a study in the ex vivo bovine liver. Eur J Radiol 81:2568–2575
2945. Stogryn A (1971) Equations for calculating the dielectric constant of saline water. IEEE Trans Microw Tech 19:733–736
2946. Stoll AM (1967) Heat transfer in biotechnology. Adv Heat Transf 4:65–141
2947. Stoll AM (1967) Heat transfer in biotechnology. In: Hartnett JP, Irvine TF (eds) Advacnes in heat transfer. Academic Press, New York, pp 65–141
2948. Stoll AM, Chianta MA, Piergallini JR (1979) Thermal conduction effects in human skin. Aviation Space Environ Med 50:778–787
2949. Stolwijk JAJ, Hardy JD (1966) Temperature regulation in man – a theoretical study. Pflug Arch 291:129–162
2950. Storey KB (2006) Reptile freeze tolerance: metabolism and gene expression. Cryobiology 52:1–16
2951. Storey KB, Storey JM (1986) Freeze tolerant frogs: cryoprotectants and tissue metabolism during freeze/thaw cycles. Can J Zool 64:49–56
2952. Storey KB, Storey JM (1996) Natural freezing survival in animals. Annu Rev Ecol Syst 27:365–386

2953. Storey KB, Storey JM (2004) Physiology, biochemistry and molecular biology of vertebrate freeze tolerance: the wood frog. In: Benson E, Fuller B, Lane N (eds) Life in the frozen state. CRC Press, Boca Raton, pp 243–274
2954. Storey KB, Bischof J, Rubinsky B (1992) Cryomicroscopic analysis of freezing in liver of the freeze-tolerant wood frog. A J Physiol Regul Integr Comp Physiol 263:R185–R194
2955. Storm FK (1983) Hyperthermia in cancer therapy. Hall, Boston
2956. Storm FK (1989) Clinical hyperthermia and chemotherapy. Radiol Clin North Am 27:621–627
2957. Stott SL, Karlsson JOM (2003) Simulating the effects of intercellular ice propagation during cryosurgery. In: Summer Bioengineering Conference, Sonesta Beach Resort
2958. Stott SL, Irimia D, Karlsson JOM (2004) Parametric analysis of intercellular ice propagation during cryosurgery, simulated using Monte Carlo techniques. Technol. Cancer Res Treat 3:113–123
2959. Strambini GB, Gabellini E (1996) Proteins in frozen solutions: evidence of ice-induced partial unfolding. Biophys J 70:971–976
2960. Stranges DF, Khayat RE, Albaalbaki B (2013) Thermal convection in non-Fourier fluids. Linear Stabil Int J Thermal Sci 74:14–23
2961. Strasberg S, Linehan D (2003) Radiofrequency ablation of liver tumors. Curr Probl Surg 40:459–498
2962. Stratigos AJ, Dover JS (2000) Overview of lasers and their properties. Dermatol Ther 13:2–16
2963. Strey H, Peterson M, Sackmann E (1995) Measurement of erythrocyte membrane elasticity by flicker eigenmode decomposition. Biophys J 69:478–488
2964. Strezer F (2002) Microwave medical devices. IEEE Microw Mag 3:65–70
2965. Strohbehn JW (1983) Temperature distributions from interstitial RF electrode hyperthermia systems: theoretical predictions. Int J Radiat Oncol Biol Phys 9:1655–1667
2966. Strom CS, Liu XY, Jia Z (2004) Antifreeze protein-induced morphological modification mechanisms linked to ice binding surface. J Biol Chem 279:32407–32417
2967. Studholme CV (1997) Modeling heat and mass transport in biological tissues during freezing. M.S. Thesis, University of Alberta
2968. Sturesson C (1998) Medical laser-induced thermotherapy. Models and applications. Lund Report on Atomic Physics LRAP-235 (1998)
2969. Sturesson C, Andersson-Engels S (1996) Mathematical modelling of dynamic cooling and pre-heating, used to increase the depth of selective damage to blood vessels in laser treatment of port wine stains. Phys Med Biol 41:413–428
2970. Sudár A, Futaki G, Kovács R (2021) Continuum modeling perspectives of non-Fourier heat conduction in biological systems. J Non-Equil Thermodyn 46:371–381
2971. Sukstanskii AL, Yablonskiy DA (2006) Theoretical model of temperature regulation in the brain during changes in functional activity. Proc Nat Acad Sci USA 103:12144–12149
2972. Sumida S (2006) Recovery rate, osmotic fragility, and oxygen dissociation curve of red blood cells preserved for 30 years or longer. Cryobiology 53(3):385–385
2973. Sumida S (2006) Transfusion and transplantation of cryopreserved cells and tissues. Cell Tissue Bank 7:265–365
2974. Sumimoto R, Kamada N, Jameson NV, Fukuda Y, Dohi K (1991) A comparison of a new solution combining histidine and lactobionate with UW solution and Eurocollins for rat liver preservation. Transplantation 51:589–593
2975. Sumpter ML (2004) Jonson-Mehl-Avrami kinetics of intracellular ice formation in confluent tissue constructs. M.S. Thesis, Georgia Institute of Technology
2976. Sun C, Munn LL (2008) Lattice-Boltzmann simulation of blood flow in digitized vessel networks. Comput Math Appl 55:1594–1600
2977. Sun F, Wang GX, Kelly KM, Aguitar G (2005) Numerical modeling of the tissue freeze-thaw cycle during cutaneous cryosurgery using liquid nitrogen spray. In: Proceedings of the institute of mechanical engineers congress and exhibition IMECE2005, Orlando, pp 1–8

2978. Sun ZQ, Yang Y, Liu J (2012) In vivo experiments and numerical investigations on nanocryosurgical freezing of target tissues with large blood vessels. J Biomed Nanotechnol 8:10–18
2979. Sun XG, Hao X, Zhang Y, Baleanu D (2017) Relaxation and diffusion models with non-singular kernels. Phys A Stat Mech Appl 468:590–596
2980. Sun Y, Rahbani JF, Jedrychowski MP, Riley CL, Vidoni S, Bogoslavski D, Hu B, Dumesic PA, Zeng X, Wang AB, Knudsen NH, Kim CR, Marasciullo A, Millan AL, Chouchani ET, Kazak L, Spiegelman BM (2021) Mitochondrial TNAP controls thermogenesis by hydrolysis of phosphocreatine. Nature 593:580–585
2981. Suomi V, Jaros J, Treeby B, Cleveland RO (2018) Full modelling of high-intensity focused ultrasound and thermal heating in the kidney using realistic patient models. IEEE Trans Biomed Eng 65:2660–2670
2982. Suriyanto, Ng EYK, Kumar SD (2017) Physical mechanism and modeling of heat generation and transfer in magnetic fluid hyperthermia through Néelian and Brownian relaxation: a review. BioMed Eng OnLine 16:36
2983. Surowiec AJ, Stuchly SS, Barr JB, Swarup A (1988) Dielectric propertiesof breast carcinoma and the surrounding tissues. IEEE Trans Biomed Eng 35:257–263
2984. Sutera SP, Skalak P (1993) The history poiseuille's law. Annu Rev Fluid Mech 25:1–19
2985. Sutherland JC (2002) Biological effects of polychromatic light. Photochem Photobiol 76:164–170
2986. Suzuki Y, Tateishi N, Soutani M, Maeda N (1996) Deformation of erythrocytes in microvessels and glass capillaries: effects of erythrocyte deformability. Microcirculation 3:49–57
2987. Suzuki A, Fomin SA, Chugunov VA, Niibori Y, Hashida T (2016) Fractional diffusion modeling of heat transfer in porous and fractured media. Int J Heat Mass Transf 103:611–618
2988. Svanes K, Zweifach BW (1968) Variations in small blood vessel hematocrits produced in hypothermic ratsby micro-occlusion. Microvasc Res 1:210–220
2989. Sverdrup PG, Sinha S, Asheghi M, Uma S, Goodson KE (2001) Measurement of ballistic phonon conduction near hotspots in silicon. Appl Phys Lett 78:3331–3333
2990. Sverdrup PG, Sungtaek Y, Goodson KE (2001) Sub-continuum simulations of heat conduction in silicon-on-insulater transistors. Trans ASME 123:130–137
2991. Swietach P, Tiffert T, Mauritz J, M, Seear R, Esposito A, Kaminski, CF, Lew VL, Vaughan-Jones RD (2010) Hydrogen ion dynamics in human red blood cells. J Physiol 588:4995–5014
2992. Swishchev IM, Kusalik PG (1994) Crystallization of liquid water in a molecular dynamics simulation. Phys Rev Lett 73:975–978
2993. Szasz A (2019) Thermal and nonthermal effects of radiofrequency on living state and applications as an adjuvant with radiation therapy. J Radiat Cancer Res 10:1–17
2994. Szasz O, Szasz A (2016) Heating, efficacy and dose of local hyperthermia. Open J Biophys 6:10–18
2995. Szasz O, Szasz A (2021) Approaching complexity: hyperthermia dose and its possible measurement in oncology. Open J Biophys 11:68–132
2996. Szasz A, Vincze GY, Szasz O, Szasz N (2003) An energy analysis of extracellular hyperthermia. Electromag Biol Med 22:103–115
2997. Szücz M, Pavelka M, Kovács R, Fülöp T, Vän P, Grmela M (2021) A case study of non-Fourier heat conduction using internal variables and GENERIC. arXiv:2103.03326v4 [physics.class-ph]
2998. Tacke J, Adam G, Sellhaus B, Glowinski A, Heschel I, Schäffter T, Schorn R, Grosskortenhaus S, Rau G, Günter RW (2001) Experimental MR imaging-guided interstitial cryotherapy of the brain. Am J Neuroradiol 22:431–440
2999. Taflove A, Brodwin ME (1975) Computation of the electromagnetic fields and induced temperatures within a model of the microwave-irradiated human eye. IEEE Trans Microw Theor Tech 23:888–896

3000. Taflove A, Hagness SC (2005) Computational electrodynamics: the finite-difference time-domain method, 3rd edn. Artech House, Boston
3001. Tahiri N, Biben T, Ez-Zahraouy H, Benyoussef A, Misbah C (2013) On the problem of slipper shapes of red blood cells in the microvasculature. Microvasc Res 85:40–45
3002. Taifel Y (1972) On the parabolic, hyperbolic and discrete formulation for the heat conduction equation. Int J Heat Mass Transf 15:369–372
3003. Tajkhorshid E, Zhu F, Schulten K (2005) Kinetic theory and simulation of single-channel water transport. In Yip S (ed), Handbook of materials modeling, vol. I: methods and models. Springer, Amsterdam, pp 1797–1822
3004. Takahashi K (1997) Mesomechanics of continua and revised field equations. JSME Int J Ser A 40:99–107
3005. Takeishi N, Rosti ME, Imai Y, Wada S, Brandt L (2019) Hemorheology in dilute, semi-dilute and dense suspensions of red blood cells. J Fluid Mech 872:818–848
3006. Takemori T (1995) A fundamental model of the human thermal system for prediction of thermal comfort. Heat Transfer-Jap Res 24:147–165
3007. Talaee MR, Kabiri A (2017) Analytical solution of hyperbolic bioheat equation in spherical coordinates applied in radiofrequency heating. J Mech Med Biol 17:1750072
3008. Tamames I, King C, Huang CY, Telischi FF, Hoer ME, Rajguru SM (2018) Theoretical evaluation and experimental validation of localized therapeutic hypothermia application to preserve residual hearing after cochlear implantation. Ear Hear 39:712–719
3009. Tan CL, Knight ZA (2018) Regulation of body temperature by the nervous system. Neuron 98:31–48
3010. Tang DW, Araki N (1996) Non-Fourier heat conduction in a finite medium under periodic surface thermal disturbance. Int J Heat Mass Transf 39:1585–1590
3011. Tang DW, Araki N (1997) On non-Fourier temperature wave and thermal relaxation time. Int J Thermophys 18:493–504
3012. Tang X, Pikal MJ (2005) The effect of stabilizers and denaturants on the cold denaturation temperatures of proteins and implications for freeze-drying. Pharm Res 22:1167–1175
3013. Tanghe A, Dijck PV, Colavizza D, Thevelin JM (2004) Aquaporin-mediated improvement of freeze tolerance of saccharomyces cerevisiae is restricted to rapid freezing conditions. Appl Environ Microbiol 70:3377–3382
3014. Tangwongsan C, Will JA, Webster JG, Meredith KL, Mahvi DM (2004) In vivo measurement of swine endocardial convective heat transfer coefficient. IEEE Trans Biomed Eng 51:1478–1486
3015. Tanimoto R, Takumi T, Yuichiro H, Nakai Y, Shindo Y, Oka K, Horoi N, Funahashi A (2016) Detection of temperature difference in neuronal cells. Sci Rep 6:22071
3016. Tansey EA, Johnson CD (2015) Recent advances in thermoregulation. Adv Physiol Educ 39:139–148
3017. Tap CY, Chua CK, Dong ZL, Liu ZH, Zhang DO, Loh LE, Sing SL (2015) Review of selective laser melting: materials and applications. Appl Phys Rev 2:041101
3018. Tappel AL (1966) Effects of low temperatures and freezing on enzymes and enzymes systems. In: Meryman HT (ed) Cryobiology. AP, London, pp 163–177
3019. Tarasov VE (2011) Fractional dynamics: applications of fractional calculus to dynamics of particles, fields and media. Nonlinear physical science. Springer, Berlin
3020. Tarasov VE (2013) No violation of the Lebniz rule. Commun Nonlinear Sci Numer Simul 18:2945–2948
3021. Tarasov VE (2015) Vector calculus in non-integer dimensional space and its applications to fractal media. Commun Nonlinear Sci Numer Simul 20:360–374
3022. Tarasov VE (2016) Heat transfer in fractal materials. Int J Heat Mass Transf 93:427–430
3023. Tarasov VE (2016) Leibniz rule and fractional derivative of power functions. J Comput Nonlinear Dyn 11:031014
3024. Tarasov VE (2016) On chain rule for fractional derivatives. Commun Nonlinear Sci Numer Simul 30:1–4

3025. Tarasov VE (2017) Interpretation of fractional derivatives as reconstruction from sequence of integer derivatives. Fundam Inf 151:431–442
3026. Tarasov V (2018) No nonlocality, no farctional derivative. Nonlinear Sci Numer Simul 62:157–163
3027. Tateishi AA, Ribeiro HV, Lenzi EK (2017) The role of fractional time-derivative operators on anomalous diffusion. Front Phys 5:52
3028. Taylor MJ, Campbell RN, Rutledge RN, Brockbank KGM (2001) Comparison of UNISOL with EuroCollins solution as a vehicle solution for cryoprotectants. Trans Proc 33:667–679
3029. Taylor NAS, Machado-Moreira DA, van den Heuel AMJ, Caldwell AN (2014) Hands and feet: physiological insulators, radiators and evaporators. Eur J Appl Physiol 114:2037–2060
3030. Taylor MJ, Weegman BP, Baicu SC, Giwa SE (2019) New approaches to cryopreservation of cells, tissues, and organs. Trans Med Hemothe 46:197–215
3031. Telenkov SA, Tanenbaum BS, Goodman DM, Nelson JS, Milner TE (1999) In vivo infrared tomographic imaging of laser-heated blood vessels. IEEE J Sel Top Quant Electron 5:1193–1199
3032. Teneiro Machado JA (2009) Fractional derivatives: probability interpretation and frequency response of rational approximations. Commun. Nonlinear Sci Num Simul 14:3492–3497
3033. Teneiro Machado JA (2009) A probabilistic interpretation of the fractional-order differentiation. Fract Calcul Appl Anal 6:73–80
3034. Teneiro Machado JA, Kiryakova V, Mainardi F (2010) A poster about the recent history of fractional calculis. Fract Calculu Appl Anal 13:329–334
3035. Teneiro Machado JA, Kiryakova V, Mainardi F (2011) Recent history of fractional calculis. Commun Nonlinear Sci Numer Simul 16:1140–1153
3036. ter Haar G (ed) (2012) The safe use of ultrasound in medical diagnosis, 3rd edn. The British Institute of Radiology, London
3037. ter Haar G, Coussios C (2007) High intensity focused ultrasound: physical principles and devices. Int J Hyperth 23:89–104
3038. ter Haar G, Rivens I, Chen L, Riddler S (1991) High intensity focused ultrasound for the treatment of rat tumours. Phys Med Biol 36:1495–1501
3039. Ternov IM (1995) Synchrotron radiation. Phys Usp 38:409–434
3040. Terry LI, Roemer RB, Walter GH, Booth D (2014) Thrips' responses to thermogenic associated signals in a cycad pollination system: the interplay of temperature, light, humidity and cone volatiles. Funct Ecol 28:857–867
3041. Texter J, Tirrell M (2001) Chemical processing by self-assembly. AIChE J 47:1706–1710
3042. Tharasanit T, Colenbrander B, Stout TAE (2005) Effect of cryopreservation on the cellular integrity of equine embryos. Reproduction 129:789–798
3043. Tharp HS, Zhang W (1993) Analytical study of temperature oscillations in living tissue. IEEE Trans Biomed Eng 40:108–110
3044. Thellier F, Monchoux F, Spagnol S, Sassi M (2009) Measurement of ambient air temperature for evaluation of human heat convective losses. Measurement 42:62–70
3045. Thiebaut C, Lemonnier D (2002) Three-dimensional modelling and optimization of thermal fields induced in a human body during hyperthermia. Int J Thermal Sci 41:500–508
3046. Thiesen B, Jordan A (2008) Clinical applications of magneticnanoparticles for hyperthermia. Int J Hyperth 24:467–474
3047. Thirumala S, Forman JM, Monroe WT, Devireddy RV (2007) Freezing and post-thaw apoptotic behaviour of cells in the presence of palmitoyl nanogold particles. Nanotechnology 18:195104
3048. Thomsen S (1991) Pathologic analysis of photothermal and photomechanical effects of laser-tissue interactions. Photochem Photobiol 53:825–835
3049. Thompson KP, Ren QS, Parel JM (1992) Therapeutic and diagnostic application of lasers in ophtalmology. Proc IEEE 80:838–860
3050. Thompson KA, Richa J, Liebhaber SA, Storey BT (2001) Dialysis addition of trehalose/glycerol cryoprotectant allows recovery of cryopreserved mouse spermatozoa with satisfactory fertilizing ability as assessed by yield of live young. J Androl 22:339–344

3051. Thurston GB (1972) Viscoelasticity of human blood. Biophys J 12:1205–1217
3052. Thurston GY (1979) Rheological parameters for the viscosity, visco-elasticity and thixotropy of blood. Biorheology 16:149–162
3053. Tikuisis P, Bell DG, Jacobs I (1990) Shivering onset, metabolic response, and convective heat transfer during cold air exposure. J Appl Physiol 70:1996–2002
3054. Tilahun M, Scott EP, Vick B (1999) The question of thermal waves in heterogeneous and biological materials. In: Proceedings of the ASME 1999 international mechanical engineering congress and exposition. Advances in heat and mass transfer in biotechnology, Nashville, TN, USA, November 14–19, pp145–152
3055. Timofeev SF (2001) Bergmann's principle and deep-water gigantism in marine crustaceans. Biol Bull Russ Acad Sci 28:646–650
3056. Tingle SJ, Figueiredo RS, Moir JAG, Goodfellow M, Talbot D, Wilson CH (2019) Machine perfusion preservation versus static cold storage for deceased donor kidney transplantation. Cochrane Database Syst Rev 3:CD011671
3057. Titchmarsh ES (1937) Introduction to the theory of fourier transforms. OUP, Oxford
3058. Tiwari R, Kumar R, Abouelregal AE (2021) Analysis of a magneto-thermoelastic problem in a piezoelastic medium using the non-local memory-dependent heat conduction theory involving three phase lags. Mech Time-Depend Mater 26: 271–287
3059. Todorova T, Seitsonen AP, Hutter J (2005) Molecular dynamics simulation of liquid water: hybrid density functionals. UCRL-JRNL-215365
3060. Togni P, Drizdal T, Vrba J, Vannucci L (2009) Slot-line applicator for microwave hyperthermia. J Microw Power Electromag Energy ONLINE 43:24–30
3061. Tokunaga Y, Wicomb WN, Concepcion W, Nakazato P, Collins GM, Esquivel CO (1991) Successful 20-hour rat liver preservation with chlorpromazine in sodium lactobionate sucrose solution. Surgery 110:80–86
3062. Tomaiuolo G, Guido S (2011) Start-up shape dynamics of red blood cells in microcapillary flow. Microvasc Res 82:35–41
3063. Tomaiuolo G, Simeone M, Martinelli V, Rotoli B, Guido S (2009) Red blood cell deformation in microconfined flow. Soft Matter 5:3736–3740
3064. Tomashow MF (1999) Plant cold acclimation: freezing tolerance genes and regulatory mechanisms. Rev Plant Physiol Plant Mol Biol 50:571–599
3065. Toner M, Cravalho EG, Karel M (1990) Thermodynamics and kinetics of intracellular ice formation during freezing of biological cells. J Appl Phys 67:1582–1593; Erratum: 70, 4653, 1991
3066. Toner M, Cravalho EG, Stachechki J, Fitzgerald T, Tompkins RG, Yamush ML, Armant DR (1993) Nonequilibrium freezing of one-cell mouse embryos. Biophys J 64:1906–1921
3067. Tong JK, Huang X, Boriskina SV, Loomis J, Xu Y, Chen G (2015) Infrared-transparent visible-opaque fabrics for wearable personal thermal management. arXiv:1507.04269 [physics.optics]
3068. Torell LM, Nilsson SK (1980) Calculation of skin temperature increase caused by subcutaneous veins perpendicular to the skin surface. Phys Med Biol 25:85–91
3069. Torre D (1970) Cutaneous cryosurgery. Y State J Med 70:2551–2554
3070. Torres JH, Nelson JS, Tanenbaum BS, Milner TE, Goodman DM, Anvari B (1999) Estimation of internal skin temperatures in response to cryogen spray cooling: implications for laser therapy of port wine stains. IEEE J Sel Top Quant El 5:1058–1066
3071. Torres JD, Talens P, Escriche I, Chiralt A (2006) Influence of process conditions on mechanical properties of osmotically dehydrated mango. J Food Eng 74:240–246
3072. Torres P, Ziabari A, Torello A, Bafaluy J, Camacho J, Cartoixa X, Shakouri A, Alvarez F (2018) Emergence of hydrodynamic heat transport in semiconductors at the nanoscale. Phys Rev Mater 2:076001
3073. Tortora GJ, Derrickson B (2012) The cardiovascular system: blood vessels and hemodynamics. In: Principles of anatomy & physiology, 13th edn. Wiley, Hoboken
3074. Torvi CL, Dale JD (1994) A finite element model of skin subjected to a flash fire. J Biomed Eng 116:250–255

3075. Tran-Son-Tay R, Sutera S, Rao P (1984) Determination of red blood cell membrane viscosity from rheoscopic observations of tank-treading motion. Biophys J 46:65–72
3076. Trad FS, Toner M, Biggers JD (1998) Effects of cryoprotectants and ice-seeding temperature on intracellular freezing and survival of human oocytes. Hum Reprod 14:1569–1577
3077. Tran-Son-Tay R, Kan HC, Udaykumar HS, Damay E, Shyy W (1998) Rheological modelling of leukocytes. Med Biol Eng Comput 36:246–250
3078. Treeby BE, Cox B (2010) Modeling power law absorption and dispersion for acoustic propagation using the fractional Laplacian. J Acoust Soc Am 127:2741–2748
3079. Treeby BE, Jaros J, Rendell AP, Cox B (2012) Modeling nonlinear ultrasound propagation in heterogeneous media with power law absorption using a k-space pseudospectral method. J Acoust Soc Am 131:4324–4336
3080. Trejo-Soto C, Lázaro GR, Pagonabarraga I, Hernández-Machado A (2022) Microfluidics approach to the mechanical properties of red blood cell membrane and their effect on blood rheology. Membranes 12:217
3081. Trepat X, Deng L, An SS, Navajas D, Tschumperlin DJ, Gerthoffer WT, Butler JP, Fredberg JJ (2007) Universal physical responses to stretch in the living cell. Nature 447:592–595
3082. Trinkaus E (1981) Neanderthal limb proportions and cold adaptation. In: Stringer CB (ed) Aspects of Human Evolution. Taylor and Francis, London, pp 187–224
3083. Trounson AO, Mohr L (1983) Human pregnancy following cryopreservation, thawing and transfer of eight-cell embryos. Nature 305:707–709
3084. Trucu D, Ingham DB, Lesnic D (2011) Reconstruction of the space- and time-dependent blood perfusion coefficient in bio-heat transfer. Heat Transf Eng 32:800–810
3085. Truesdell C, Noll W (2003) The non-linear field theories of mechanics. Springer, Berlin
3086. Trueta J (1953) The influence of the blood supply in controlling bone growth. Bull Hosp Joint Disease 14:147–157
3087. Trujillo M, Berjano E (2013) Review of the mathematical functions used to model the temperature dependence of electrical and thermal conductivities of biological tissue in radiofrequency ablation. Int J Hyperth 29:590–597
3088. Tryba AK, Ramirez JM (2004) Hyperthermia modulates respiratory pacemaker bursting properties. J Neurophysiol 92:2844–2852
3089. Tsubota KI, Wada S (2010) Elastic force of red blood cell membrane during tank-treading motion: consideration of the membrane's natural state. Int J Mech Sci 52:356–364
3090. Tsubota K, Wada S, Yamaguchi T (2006) Simulation study on effects of hematocrit on blood flow properties using particle method. J Biomech Sci Eng 1:159–170
3091. Tsutomu O (1991) Thermal effect of infra-red radiation on the eye: a study based on a model. Ann Occup Hyg 35:1–12
3092. Tu J, Hwang JH, Chen T, Fan T, Guo X, Crum LA, Zhang D (2012) Controllable in vivo hyperthermia effect induced by pulsed high intensity focused ultrasound with low duty cycles. Appl Phys Lett 101:124102
3093. Tuchin VV (1997) Light scattering study of tissues. Phys Usp 40:495–515
3094. Tuchin VV (2009) Optical spectroscopy of biologicalmaterials. In: Andrews DL (ed) Encyclopedia of applied spectroscopy. Wiley-VCH, Weinheim, pp 555–626
3095. Tuchin VV (2016) Tissue optics and photonics: light-tissue interaction. J Biomed Photon Eng 2:030201
3096. Tung MM, Trujillo M, Lopéz Molina JAL, Rivera MJ, Berjamo EJ (2009) Modeling the heating of biological tissue based on the hyperbolic heat transfer equation. Math Comput Model 50:665–672
3097. Tungjitkusolmun S, Woo EJ, Cao H, Tsai JZ, Vorperian VR, Webster JG (2000) Thermal-electrical finite element modelling for radio frequency cardiac ablation: effects of changes in myocardial properties. Med Biol Eng Comput 38:562–568
3098. Tungjitkusolmun S, Vorperian VR, Bhavaraju N, Cao H, Tsai JZ, Webster JG (2001) Guidelines for predicting lesion size at common endocardial locations during radio-frequency ablation. IEEE Trans Biomed Eng 48:194–201

3099. Tungjikusolmun S, Tyler ST, Haemmerich D, Tsai JZ, Cao H, Webster JG, Lee FT, Mahvi DM, Vorperian VR (2002) Three-dimensional finite-element analysis for radio-frequency hepatic tissue ablation. IEEE Trans Med Eng 49:3–8
3100. Turk JR, Laughin MH (2004) Physical activity and atherosclerosis: which animal model? Can J Appl Physiol 29:657–683
3101. Tuszynski JA, Wenger C, Friesen D, Preto J (2016) An overview of sub-cellularmechanisms involved in the action of TTFields. Int J Env Res Publ Health 13:1128
3102. Tyshenko MG, Doucet D, Davies PL, Walker VK (1997) The antifreeze potential of the spruce budworm thermal hysteresis protein. Nat Biotechnol 15:887–890
3103. Tzou D (2011) Nonlocal behavior in phonon transport. Int J Heat Mass Transf 54:475–481
3104. Tzou DY (1989) On the thermal shock wave induced by moving heat source. J Heat Transf 111:232–238
3105. Tzou DY (1993) An engineering assessment to the relaxation time in thermal wave propagation. Int J Heat Mass Transf 36:1845–1851
3106. Tzou DY (1995) Experimental support for the lagging behavior in heat propagation. J Thermophys Heat Transf 9:686–693
3107. Tzou DY (1995) The generalized lagging responce in small-scale and high-rate heating. Int J Heat Transf 38:3231–3240
3108. Tzou DY (2001) Ultrafast laser heating on metal films: effect of microvoids. J Thermophys Heat Transf 16:30–35
3109. Tzou DY (2012) Lagging behavior in biological systems. J Heat Transf 134:051006
3110. Tzou DY (2015) Macro- to microscale heat transfer: the lagging behavior, 2nd edn. Wiley, New York
3111. Tzou DY, Dai W (2009) Thermal lagging in multi-carrier systems. Int J Heat Mass Transf 52:1206–1213
3112. Tzou D, Guo ZY (2010) Nonlocal behavior in thermal lagging. Int J Therm Sci 49:1133–1137
3113. Tzou DY, Özisik MN, Chiffelle RJ (1994) The lattice temperature in the microscopic two-step model. ASME J Heat Transf 116:1034–1038
3114. Uchaikin VV (2003) Self-similar anomalous diffusion and Lévy-stable laws. Phys Usp 46:821–849
3115. Uchiyama S, Tsuji T, Ikado K, Yoshida A, Kawamoto K, Hayashi T, Inada N (2015) A cationic fluorescent polymeric thermometer for the ratiometric sensing of intracellular temperature. Analyst 140:4498–4506
3116. Uddin MN, Uddin MM, Alam M (2020) Comparative mathematical study of blood flow through stenotic and aneurysmatic artery with the presence and absence of blood clots. Malays J Math Sci 16(3):599–623
3117. Underwood L, Vircondelet C, Jermy M (2018) Thermal comfort and drag of a streamlined cycling helmet as a function of ventilation hole placement. Proc Inst Mech Eng Part P J Sports Eng Technol 232:15–21
3118. US Department of Energy, Office of Science (2004) A science-based case for large-scale simulation, v. 2. US Department of Energy Office of Science Laboratory, Washington
3119. Ushiyama A, Ohkubo C (2004) Acute effects of low-frequency electromagnetic fields on leukocyte-endothelial interactions in vivo. In Vivo 18:125–132
3120. Usov NA, Liubimov BY (2012) Dynamics of magnetic nanoparticle in a viscous liquid: application to magnetic nanoparticle hyperthermia. J Appl Phys 112:023901
3121. Utoh J, Harasaki H (1992) Damage to erythrocytes from long-term heat stress. Clin Sci (Lond). 82(1):9–11
3122. Uzoigwe C (2006) The human erythrocyte has developed the biconcave disc shape to optimise the flow properties of the blood in the large vessels. Med Hypotheses 67:1159–1163
3123. Vaezy S, Martin R, Yazij H, Kaczkowski P, Keilman G, Carter S, Caps M, Chi EY, Bailey M, Crum L (1998) Hemostasis of punctured blood vessels using high-intensity focused ultrasound. Ultrasound Med Biol 24:903–910

3124. Vaezy S, Martin R, Kaczkowski P, Keilman G, Goldman B, Yaziji H, Carter S, Caps M, Crum L (1999) Use of high-intensity focused ultrasound to control bleeding. J Vascul Surg 29:533–542
3125. Vaezy S, Martin R, Crum L (2001) High intensity focused ultrasound: a method of hemostasis. Echocardiography 18:309–315
3126. Vaezy S, Noble ML, Keshavarzi A, Paun M, Prokop AF, Cornejo C, Sharar S, Chi EY, Crum LM, Martin RW (2004) Liver hemostasis with high-intensity ultrasound: repair and healing. Ultrasound Med Biol 23:217–225
3127. Väinölä FA (2000) Genetic and physiological aspects of cold hardness in rhododendron. Ph.D. Thesis, University of Helsinki
3128. Vainstein BK (1966) X ray structure analysis of globular proteins. Phys Usp 9:251–275
3129. Vajda T (1999) Cryo-bioorganic chemistry: molecular interactions at low temperature. Cell Mol Life Sci 56:398–414
3130. Valerio D, Machado JT, Kiryakova V (2011) Some pioneers of the applicatoons of fractional calculus. Fract Calculus Appl Anal 17:552–578
3131. Vali G (1996) Ice nucleation— a review. In: Kulmula M, Wagner P (eds) Nucleation and atmospheric aerosols. Pergamon Press, Oxford, pp 271–279
3132. Valvano JW, Allen JT, Bowman HF (1984) The simultaneous measurement of thermal conductivity, thermal diffusivity, and perfusion in small volumes of tissue. J Biomech Eng 106:192–197
3133. Valvano JW, Cochran JR, Diller K (1985) Thermal conductivity and diffusivity of biomaterials with self-heated thermistors. Int J Thermophys 6:301–311
3134. Ván P (2001) Weakly nonlocal irreversible thermodynamics - the Guyer-Krumhansl and the Cahn-Hilliard equations. Phys Lett A 290:88–92
3135. Ván P (2015) Theories and heat pulse experiments of non-Fourier heat conduction. arXiv:1501.04234v1 [cond-mat.stat-mech]
3136. Ván P, Bíró TS (2008) Relativistic hydrodynamics - causality and stability. Eur Phys J Spec Top 155:201–212
3137. Ván P, Fülöp T (2012) Universality in heat conduction theory: weakly nonlocal thermodynamics. Ann Phys 524:470–478
3138. Ván P, Czel B, Fülöp T, Grof G, Gyenis A, Verhas J (2013) Expeimental aspects of heat conduction beyond Fourier. ArXiv: 1305.3583 [cond-mat.mtrl-sci]
3139. Ván P, Kovach R, Fülöp T (2014) Thermodynamic hierarchy of evolution equations. arXiv: 1412.4490 [cond-mat.stat-mech]
3140. Ván P, Berezovski A, Fülöp T, Gróf G, Lovas A, Verhás J (2017) Guyer-Krumhansl-type heat conduction at room temperature. Eur Phys Lett 118:50005
3141. Van de Kamer JB, Kroeze H, De Leeuw AAC, Lagendijk JJW (2001) Quasistatic zooming of FDTD E-field computations: the impact of down-scaling techniques. Phys Med Biol 46:1539–1551
3142. Van der Kooi C, Kevan PG, Koski M (2019) The thermal ecology of flowers. Ann Botany 124:343–353
3143. Van Der Zee J (2002) Heating the patient: a promising approach? Ann Oncol 13:1173–1184
3144. van der Zee J, González D, van Rhoon GC, van Dijk JDP, van Putten WLJ, Hart AAM (2000) Comparison of radiotherapy alone with radiotherapy plus hyperthermia in locally advanced pelvic tumours: a prospective, randomised, multicentre trial. Lancet 355:1119–1125
3145. Van Gemert MJC, Welch AJ (1989) Clinical use of laser - tissue interaction. IEEE Eng Med Biol Mag 8:10–13
3146. Van-Gemert MJC, Jacques SL, Sterenborg HJCM, Star WM (1989) Skin optics. IEEE Trans Biomed Eng 36:1146–1153
3147. van Hemmen JL, Peterson MA (2005) Crumpling of curved sheets: generalizing the föppl–von kármán equations. arXiv:cond-mat/0503158v2 [cond-mat.soft]
3148. van Leeuwen GMJ, Lagendijk JJW, Leersum BJAMV, Zwamborn APM, Hornsleth SN, Kotte ANTJ (1999) Calculation of change in brain temperature due to exposure to a mobile phone. Phys Med Biol 44:2367–2379

3149. Van Leeuwen CMJ, Van Steenhoen A (2005) Heat tansfer in humans - loal and whole-body. In: Advanced course and workshop on blood flow, Warsaw
3150. Van Leeuwen GM, Kotte AN, Lagendijk JJ (1998) A flexible algorithm for construction of 3-D vessel networks for use in thermal modeling. IEEE Trans Biomed Eng 45:596–604
3151. Van Leeuwen GM, Hand JW, Lagendijk JJ, Azzopardi DV, Edwards AD (2000) Numerical modeling of temperature distributions within the neonatal head. Pediatr Res 48:351–356
3152. van Lier AL, Brunner DO, Pruessmann KP, Klomp DW, Luijten PR, Lagendijk JJ (2012) B1 (+) phase mapping at 7 T and its application for in vivo electrical conductivity mapping. Magn Reson Med 67:552–561
3153. van Lier AL, Kotte AN, Raaymakers BW, Lagendijk JJ, Berg CA (2012) Radiofrequency heating induced by 7T head MRI: thermal assessment using discrete vasculature or Pennes' bioheat equation. J Magn Res Imag 35:795–803
3154. Van Meer G, Voelker DR, Feigenson GW (2008) Membrane lipids: where they are and how they behave. Nat Rev Mol Cell Biol 9:112
3155. van Milligen BP, Carreras BA, Sánchez R (2005) The foundations of diffusion revisited. Plasma Phys Control Fusion 47:B743–B754
3156. van Ooijen AM, van Marken Lichtenbelt W, van Steenhoven AA, van Westerterp KR (2005) Cold-induced heat production preceding shivering. Brit J Nutr 93:387–391
3157. van Rhoon GC (2013) External electromagnetic methods and devices. In: Moos EG (ed) Physics of thermal therapy. Fundamentals and clinical applications. CRC Press, Boca Raton
3158. Van Wagner CE (1973) Height of crown scorch in forest fires. Canad J Forest Res 3:373–378
3159. VanBavel E (2007) Effects of shear stress on endothelian cells: possible relevance for ultrasound applications. Prog Biophys Mol Biol 93:374–383
3160. Vanderzwalmen P, Bertin G, Debauche C, Standaert V, van Rosendaal E, Vandervorst M, Bollen N, Zech H, Mukaida T, Takahashi K, Schoysman R (2002) Births after vitrification at morula and blastocyst stages: effect of artificial reduction of the blastocoelic cavity before vitrification. Hum Reprod 17:744–751
3161. Vanlerberghe GC, McIntosh L (1997) Alternative oxidase: from gene to function. Ann Rev Plant Physiol Plant Mol Biol 48:703–734
3162. Vanne A, Hynynen K (2003) MRI feedback temperature control for focused ultrasound surgery. Phys Med Biol 48:31–43
3163. Varchanis S, Dimakopoulos Y, Wagner C, Tsamopoulos J (2018) How viscoelastic is human blood plasma? Soft Matter 14:4238–4251
3164. Vasquez-Rivera A, Sommer KK, Oldenhof H, Higgins Z, Brockbank KGM, Hilfiker A, Wolkers WF (2018) Simultaneous monitoring of different vitrification solution components permeating into tissues. Analyst 143:420–428
3165. Vatamanu J, Kusalik PG (2007) Molecular dynamics methodology to investigate steady-state heterogeneous crystal growth. J Chem Phys 126:124703
3166. Vaziri A, Mahadevan L (2008) Localized and extended deformations of elastic shells. Proc Natl Acad Sci USA 105:7913–7918
3167. Vaziri S, Yalon E, Rojo MM, Suryavanshi SV, Zhang H, McClellan CJ, Bailey CS, Smithe KKH, Gabourie AJ, Chen V, Deshmukh S, Bendersky L, Davydov AV, Pop E (2019) Ultrahigh thermal isolation across heterogeneously layered two-dimensional materials. Sci Adv 5:eaax1325
3168. Vázquez F, Ván P, Kovács R (2020) Ballistic-diffusive model for heat transport in superlattices and the minimum effective heat conductivity. Entropy 22:167
3169. Vedavarz A, Kumar S, Moallemi MK (1994) Significance on non-Fourier heat waves in conduction. J Heat Transf 116:221–224
3170. Vedenov AA, Levchenko EB (1983) Supramolecular liquid crystal structures in solutions of amphiphilic molecules. Sov Phys Usp 9:747–774
3171. Vega C, Abascal JLF, Sanz E, MacDowell LG, McBride C (2005) Can simple models describe the phase diagram of water? J Phys Condens Matter 17:S3283–S3288
3172. Velizhanin KA, Chien CC, Dubi Y, Zwolak M (2011) Driving denaturation: nanoscale heating as a probe of DNA melting. Phys Rev E 83:050906(R)

3173. Venkatasubramanian R, Siivola E, Colpitts T, O'Quinn B (2001) Thin-film thermoelectric devices with high room-temperature figures of merit. Nature 413:597–602
3174. Venkatasubramanian RI, Grassl ED, Barocas VH, Lafontaine D, Bischof JC (2006) Effects of freezing and cryopreswervation on the mechanical proerties of arteries. Ann Biomed Eng 34:823–832
3175. Vera C, Skelton R, Bossens F, Sung LA (2005) 3D nanomechanics of an erythrocyte junctional complex in equibiaxial and anisotropic deformations. Ann Biomed Eng 33:1387–1404
3176. Vercesi AE, Borecký J, Maia I, Arruda P, Cuccovia IM, Chaimovich H (2006) Plant uncoupling mitochondrial proteins. Annu Rev Plant Biol 57:383–404
3177. Verdaasdonk RM, Borst C, van Gemert MJC (1990) Explosive onset of continuous wave laser tissue ablation. Phys Med Biol 35:1129–1144
3178. Verhás J (1997) Thermodynamics and rheology. Kluwer Academic, Cambridge
3179. Verhey JF, Mohammed Y, Ludwig A, Giese K (2003) Implementation of a practical model for light and heat distribution using laser-induced thermoterapy near to a large vessel. Phys Med Biol 48:3595–3610
3180. Verkman AS (2002) Solute and macromolecule diffusion in cellular aqueous compartments. Trends Biochem Sci 27:27–33
3181. Verkruysse W, Majaron B, Tanenbaum BS, Nelson JS (2000) Optimal cryogen spray cooling parameters for pulsed laser treatment of port wine stains. Lasers Surg Med 27:165–170
3182. Vermeersch B, De May G (2008) Non-Fourier heat conduction in nano-scaled electronic structures. Anal Integr Circ Sig Proc 55:197–204
3183. Vermeersch B, Shakouri A (2014) Spatiotemporal flux memory to nondiffusive transport. arXiv: 1412.8571 [cond-mat.stat-mech]
3184. Vermeersch B, Shakouri A (2015) Nonlocality in microscale heat conduction. arXiv:1412.6555v2 [cond-mat.mes-hall]
3185. Vernotte P (1958) Les paradoxes de la theorie de l' equation de la chaleur. Comput Rendus 246:3145–3155
3186. Vernotte P (1961) Some possible complications in the phenomena of thermal conduction. Comput Rendus 252:2190–2191
3187. Vigh L, Los DA, Horwath I, Murata N (1993) The primary signal in the biological perception of temperature; Pd-catalyzed hydrogenation of membrane lopids stimulated the expression of the desA gene in Synechocystis. Proc Natl Acad Sci USA 90:9090–9094
3188. Vigh L, Mareska B, Hartwood JL (1998) Does the membrane's physical state control the expression of heat shock and other genes? Trends Biochem Sci 23:369–374
3189. Vigh L, Escribá P, Sonnleitner A, Sonnleitner M, Piotto S, Maresca B, Horväth I, Harwood JL (2005) The significance of lipid composition for membrane activity: new concepts and ways of assessing function. Prog Lipid Res 44:303–344
3190. Vigier G, Thollet G, Vassoille R (1987) Cubic and hexagonal ice formation in water-glycerol mixtures (50% w/w). J Cryst Growth 84:309–315
3191. Vineis CJ, Shakouri A, Majumdar A, Kanatzidis MG (2010) Nanostructured thermoelectrics: big efficiency gains from small features. Adv Mater 22:3970–3980
3192. Vines RG (1968) Heat transfer through bark, and the resistance of trees to fires. Austral J Botany 16:499–514
3193. Viswanathan C, Zhu JK (2002) Molecular genetic analysis of cold-regulated gene transcription. Philos Trans R Soc Lond B 357:877–886
3194. Viswanathan C, Zhu J, Zhu JK (2006) Molecular genetic analysis of cold-regulated gene transcription. Physiol Plant 126:52–61
3195. Vitkova V, Mader MA, Misbah C, Podgorski T (2007) Rheology of dilute suspensions of vesicles and red blood cells. https://www.researchgate.net/publication/1757874_Rheology_of_dilute_suspensions_of_vesicles_and_red_blood_cells
3196. Vo A, Ekpenyong A (2022) Fractional calculus modeling of cell viscoelasticity quantifies drug response and maturation more robustly than integer order models. arXiv:2201.02589 [physics.bio-ph]

3197. Vogel S (2006) Living in a physical world. IX. Making and maintaining liquid water. J Biosci 31:525–536
3198. Vogel S (2009) Leaves in the lowest and highest winds: temperature, force and shape. New Phytol 183:13–26
3199. Vogel A, Venugopalan V (2003) Mechanisms of pulsed laser ablation of biological tissues. Chem Rev 103:577–644
3200. Vogl TJ, Farshid P, Naguib NNN, Zangos S (2013) Thermal ablation therapies in patients with breast cancer liver metastases: a review. Eur Radiol 23:797–804
3201. Vogt WC, Barton JK, Agrawal A, Pfefer TJ (2018) Measurement and thermal dependence of biological tissue optical properties. Theor Appl Heat Transf Hum 1:355–378
3202. Voituron Y (2002) To freeze or not to freeze? An evolutionary perspective on the cold-hardiness stratigies of overwintering ectotherms. Am Natural 160:255–270
3203. Volgushev M, Vidyasagar TR, Chistiakova M, Eysel UT (2000) Synaptic transmission in the neocortex during reversible cooling. Neuroscience 98:9–22
3204. Voller VR, Falcini F, Garcia R (2013) Fractional Stefan problem exhibiting lumped and distributed memory effects. Phys Rev E 87:042401
3205. Volosevich P, Zmitrenko N, Levanov E, Severina E (2009) Dynamic and heating of plasma subject to heat flux relaxation. Math Models Comput Simul 1:189–199
3206. Volterra V (1930) Theory of functional and of integral and integro-differential equations. Blackie and Son, London
3207. Volz S (Ed) (2007) Microscale and nanoscale heat transfer. Springer, Berlin
3208. von Kármán T (1910) "Festigkeitsproblem im Maschinenbau," Encyk. D. Math. Wiss. IV:311–385
3209. Vonnegut B (1947) The nucleation of ice formation by silver iodide. J Appl Phys 18:593–595
3210. Vorlícek J, Vrbova B, Vrba J (2011) Prospective applications of microwaves in medicine. In: Gali-Muhtasib H (ed) Advances in cancer therapy, pp 507–532. http://dx.doi.org/10.5772/1780
3211. Vrba J, Franconi C, Montecchia F, Vannucci I (1993) Evanescent-mode applicators for hyperthermia. IEEE Trans Biomed Eng 40:397–407
3212. Vrba J, Franconi C, Lapes M (1996) Theoretical limits for the penetration depth of intracavitary applicators. Int J Hyperth 12:737–742
3213. Vrba J, Lape M, Oppl L (1999) Technical aspects of microwave thermotherapy. Bioelectrochem Bioenerg 48:305–309
3214. Vrbka L, Jungwirth P (2005) Brine rejection from freezing salt solutions: a molecular dynamics study. Phys Rev Lett 95:148501
3215. Vrbka L, Jungwirth P (2007) Molecular dynamics simulations of freezing of water and salt solutions. J Mol Liq 134:54–70
3216. Vujaskovic Z, Song CW (2004) Physiological mechanisms underlying heat-induced radiosensitization. Int J Hyperth 20:163–174
3217. Vyas R, Rustgi ML (1992) Green's function solution to the tissue bioheat equation. Med Phys 19:1319–1324
3218. Vyas DCM, Kumar S, Srivastava A (2016) Porous media based bio-heat transfer analysis on counter-current artery vein tissue phantoms: applications in photo thermal therapy. Int J Heat Mass Transf 99:122–140
3219. Wada H (2003) Dynamics of phase separation in confined two-component fluid membranes. J Phys Soc Jap 72:3142–3150
3220. Wagner AM, Moore AL (1997) Structure and function of the plant alternative oxidase: its putative role in the oxygen defence mechanism. Biosci Rep 17:319–333
3221. Wagner KR, Zuccarello M (2005) Local brain hypothermia for neuroprotection in stroke treatment and aneurysm repair. Neurol Res 27:238–245
3222. Wagner C, Steffen P, Svetina S (2013) Aggregation of red blood cells: from rouleaux to clot formation. C R Phys 14:459–469

3223. Wagshul ME, McAllister JP, Limbrick DD, Yang S, Mowrey W, Goodrich J, Meiri A, Morales DM, Kobets A, Abbott R (2021) MR elastography demonstrates reduced white matter shear stiffness in early-onset hydrocephalus. Neuroimage Clin 30:102579
3224. Wahlberg J, Southhard JH, Belzer FO (1986) Development of a cold storage solution for pancreas preservation. Cryobiology 23:477–482
3225. Wakaki M (1979) Thermal effect of the co_2 laser and our technique in dealing with it. In: Kaplan I (ed) Laser surgery. Academic, Jerusalem, pp 27–42
3226. Walburn FJ, Schneck D (1976) A constitutive equation for whole human blood. Biorheology 1:201–210
3227. Waldorf HA, Alster TS, McMillan K, Kauvar AN, Geronemus RG, Nelson JS (1997) Effect of dynamic cooling on 585-nm pulsed dye laser treatment of port-wine stain birthmarks. Dermatol Surg 23:657–662
3228. Waldron HA (1980) Investigation into the temperature field produced during endometrial cryosurgery of the perfused human uterus. Phys Med Biol 25:323–331
3229. Wall MS, Deng XH, Torzilli PA, Doty SB, O'Brien SJ, Warren RF (1999) Thermal modification of collagen. J Shoulder Elbow Surg 8:339–344
3230. Wan R, Liu Z, Muldrew K, Rewcastle J (2003) A finite element model for ice ball evolution in a multi-probe cryosurgery. Comput Methods Biomech Biomed Eng 6:197–208
3231. Wang LQ (1994) Generalized Fourier law. Int J Heat Mass Transf 37:2627–2634
3232. Wang L (2005) Identification and characterization of protein enhancers of antifreeze proteins from overwintering beetle larva *Dendroides canadensis*. Ph.D. Thesis, University of Notre Dame
3233. Wang LV (2008) Prospects of photoacoustic tomography. Med Phys 35:5758–5767
3234. Wang HD, Guo ZY (2019) Physical heat transfer. Front Heat Mass Transf 13:20
3235. Wang LV, Hu S (2012) Photoacoustic tomography: in vivo imaging from organelles to organs. Science 335:1458–1462
3236. Wang P, Li T (2019) Which wavelength is optimal for transcranial low-level laser stimulation? J Biophoton 12:e201800173
3237. Wang GX, Prasad V (2000) Microscale heat and mass transfer and non-equilibrium phase change in rapid solidification. Mater Sci Eng A 292:142–148
3238. Wang T, Xing ZW (2013) Erythrocyte hemodynamics in stenotic microvessels: a numerical investigation. Phys Rev E 88:042711
3239. Wang L, Xu M (2002) Well-posedness of dual-phase-lagging heat conduction equation: higher dimensions. Int J Heat Mass Transf 45:1165–1171
3240. Wang SQ, Zhou ZQ (1999) Alpha-stat calibration of indo-1 fluorescence and measurement of intracellular free calcium in rat ventricular cells at different temperatures. Life Sci 65:871–877
3241. Wang L, Xu M, Zhou X (2001) Well-posedness and solution structure of dual-phase-lagging heat conduction. Int J Heat Mass Trans 44:1650–1669
3242. Wang SS, Chen J, Keltner L, Christophersen J, Zheng F, Krouse M, Singhal A (2002) New technology for deep light distribution in tissue for phototherapy. Cancer J 8:154–163
3243. Wang J, Kalinichev AG, Kirkpatrick RJ (2004) Molecular modeling of water structure in nano-pores between brucite (001) surfaces. Geochim Cosmochim Acta 68:3351–3365
3244. Wang H, Dai W, Bejan A (2007) Optimal temperature distribution in a 3D triple-layered skin structure embedded with artery and vein vasculature and induced by electromagnetic radiation. Int J Heat Mass Transf 50:1843–1854
3245. Wang L, Pegg DE, Lorrison J, Vaughan D, Rooney P (2007) Further work on the cryopreservation of articular cartilage with particular reference to the liquidus tracking (LT) method. Cryobiology 55:138–147
3246. Wang T, Pan TW, Xing ZW, Glowinski R (2009) Numerical simulation of rheology of red blood cell rouleaux in microchannels. Phys Rev E 79:041916
3247. Wang ZL, Fan JR, Luo K, Cen K (2009) Immersed boundary method for the simulation of flows with heat transfer. Int J Heat Mass Transf 52:4510–4518

3248. Wang M, Yang N, Guo ZY (2011) Non-Foirier heat conductions in nanomaterials. J Appl Phys 110:064310
3249. Wang QL, He JH, Li ZB (2012) Fractional model for heat conduction in polar bear hairs. Thermal Sci 16:339–342
3250. Wang SY, Huang S, Borca-Tasciuc DA (2013) Potential sources of errors in measuring and evaluating the specific loss power of magnetic nanoparticles in an alternating magnetic field. IEEE Trans Magnet 49:255–262
3251. Wang T, Xing ZW, Xing DY (2013) Structure-induced dynamics of erythrocyte aggregates by microscale simulations. J Appl Math 2013:409387
3252. Wang H, Wang B, Normoyle KP, Jackson K, Spitler K, Sharrock MF, Miller CM, Best C, Llano D, Du R (2014) Brain temperature and its fundamental properties: a review for clinical neuroscientists. Front Neurosci 8:307
3253. Wang K, Tavakkoli F, Wang S, Vafai K (2015) Analysis and analytical characterization of bioheat transfer during radiofrequency ablation. J Biomech 48:930–940
3254. Wang Z, Zhao G, Wang T, Yu Q, Su M, He X (2015) Three-dimensional numerical simulation of the effects of fractal vascular tree on tissue temperature and intracelluar ice formation during combined canser therapy of cryosurgery and hyperthermia. Appl Thermal Eng 90:296–304
3255. Wang G, Zhang L, Wang X, Tai BL (2016) An inverse method to reconstruct the heat flux produced by bone grinding tools. Int J Thermal Sci 101:85–92
3256. Wang T, Rongin U, Xing Z (2016) A micro-scale simulation of red blood cell passage through symmetric and asymmetric bifurcated vessels. Sci Rep 6:20262
3257. Wang BX, Tao MJ, Ai Q, Xin T, Lambert N, Ruan D, Cheng YC, Nori F, Deng FG, Long GL (2018) Quantum simulation of photosynthetic energy transfer. arXiv:1801.09475v2 [quant-ph]
3258. Wang Y, Wang Q, Luo Y, Jiang L, Zeng Z, Gan L, Chen J, Han H, Zou J (2020) Comparative study of pulsed versus continuous high-intensity focused ultrasound ablation using in vitro and in vivo models. J Ultrasound Med 39:259–271
3259. Wang R, Chen L, Jia Y, Liu L, Sun L, Liu Y, Li Y (2022) Heat production and volatile biosynthesis are linked via alternative respiration in magnolia denudata during floral thermogenesis. Front Plant Sci 13:955665
3260. Wang X, Wang E, Zhao G (2023) Advanced cryopreservation engineering strategies: the critical step to utilize stem cell products. Cell Regen 12:28
3261. Ward JC, Wilks J (1951) The velocity of second sound liquid helium near the absolute zero. Lond Edinburgh Dublin Philos Mag J Sci 42:314–316
3262. Wargocki P, Wyon DP, Sundell J, Clausen G, Fanger PO (2000) The effects of outdoor air supply rate in an office on perceived air quality, sick building syndrome (SBS) symptoms and productivity. Indoor Air 10:222–236
3263. Warner RM, Shuttleworth R, Benson JD, Eroglu A, Higgins AZ (2021) General tissue mass transfer model for cryopreservation applications. Biophys J 120:4980–4991
3264. Warren WE, Chen PJ (1973) Wave propagation in the two temperature theory of thermoelasticity. Acta Mech 16:21–33
3265. Wass CT, Lanier WL, Hofer RE, Scheithauer BW, Andrews AG (1995) Temperature changes of 1°c alter functional neurological outcome and histopathology in a canine model of complete cerebral ischemia. Anesthesia 83:325–335
3266. Watanabe H, Kobayashi Y, Hashizume M, Fujie M (2009) Modeling the temperature dependence of thermophysical properties: study on the efffect of temperature dependence for RFA. In Engineering in medicine and biology society. Proceedings of the annual international conference of the IEEE, pp 5100–5105
3267. Watanabe H, Yamazaki Kobayashi Y, Hashizume M, Isobe Y, Lu X, Miyashita T, Ohdaira T, Fujie M (2012) Validation of accuracy of liver model with temperature-dependent thermal conductivity by comparing the simulation and in vitro rf ablation experiment. In Engineering in medicine and biology society. Proceedings of the annual international conference of the IEEE, pp 5712–5717

3268. Watling JR, Grant NM, Miller RE, Robinson SA (2008) Mechanisms of thermoregulation in plants. Plant Signal Behav 3:595–597
3269. Waugh R, Evans EA (1979) Thermoelasticity of red blood cell membrane. Biophys J 26:115–132
3270. Waynant RE (2011) Lasers in medicine. CRC Press, Boca Raton
3271. Weaver JC (2000) Electroporation of cells and tissues. IEEE Trans Plasma Sci 28:24–33
3272. Weaver ME, Ingram DL (1969) Morphological changes in swine associated with environmental temperature. Ecology 50:710–713
3273. Webb RC, Bonifas AP, Behnaz A, Zhang YH, Yu KJ (2013) Ultrathin conformal devices for precise and continuous thermal characterization of human skin. Nat Mater 12:938–944
3274. Webb RC, Pielak RM, Bastien P, Ayers J, Niittynen J, Kurniawan J, Manco M, Lin A, Cho NH, Malyrchuk V, Balooch G, Rogers JA (2015) Thermal transport characteristics of human skin measured in vivo using ultrathin conformal arrays of thermal sensors and actuators. PLoS ONE 10:e0118131
3275. Webster WS, Edwards MJ (1984) Hyperthermia and the induction of neural tube defects in mice. Teratology 29:417–425
3276. Wei Y, Ullah G, Ingram J, Schiff SJ (2014) Oxygen and seizure dynamics: II. Computational modeling. J Neurophysiol 112:213–223
3277. Weinbaum S, Jiji LM (1985) A new simplified bioheat equation for the effect of blood flow on average tissue temperature. J Biomech Eng 107:131–139
3278. Weinberg M, Kapral R (1989) Phase transformation kinetics in finite inhomogeneously nucleated systems. J Chem Phys 91:7146–7152
3279. Weinbaum S, Jiji L, Lemons DE (1984) Theory and experiment for the effect of vascular temperature on surface tissue heat transfer–part 1: anatomical founation and model conceptualization. J Biomech Eng 106:246–251
3280. Weinbaum S, Jiji L, Lemons DE (1984) Theory and experiment for the effect of vascular temperature on surface tissue heat transfer–Part 2: model formulation and solution. J Biomech Eng 106:331–341
3281. Weinbaum S, Xu LX, Zhu L, Ekpene A (1997) A new fundamental bioheat equation for muscle tissue: part I. Blood perfusion term. J Biomech Eng 119:278–288
3282. Weiner SJ, Kollman PA, Case DA, Singh UC, Ghio C, Alagona G, Profeta JS, Weiner P (1984) A new force field for molecular dynamics simulation of nucleic acid and proteins. J Am Chem Soc 106:764–784
3283. Weinstein KJ (2005) Body proportions in ancient andeans from high and low altitudes. Am J Phys Anthropol 128:569–585
3284. Weiss GH (1994) Aspects and applications of the random walk. North Holland, Amsterdam
3285. Weiss RA, Sadick NS (2000) Epidermal cooling crystal collar device for improved results and reduced side effects on leg telangiectasias using intense pulsed light. Dermatol Surg 26:1015–1018
3286. Weiss JA, Maker BN, Govindjee S (1996) Finite element implementation of incompressible, transversely isotropic hyperelasticity. Comput Methods Appl Mech Eng 135:107–128
3287. Welch AJ (1984) The thermal response of laser irradiated tissue. IEEE J Quant Electron QE-20:1471–1481
3288. Welch AJ, van Gemert MJC (Eds) (2011) Optical-thermal response of laser irradiated tissue. Springer, Berlin
3289. Weller AB (1999) A review of the physiological responses to exercise and cold stress, and the factors that may influence body cooling in man. J Physiol 52:12S–13S
3290. Weller S (1950) Thermo-osmosis of gases through a membrane. Nature 165:199–200
3291. Wells RE, Merrill EW (1961) Shear rate dependence of the viscosity of whole blood and plasma. Science 133:763–764
3292. Wells R, Schmid-Schönbein H (1969) Red celldeformation and fluidity of concentrated cell suspensions. J Appl Physiol 27:213–217

3293. Wenger CB (2001) Human adaptation to hot environments. In Lounsbury DE, Randolh KB, Burr RE (eds) Medical aspects of harsh environments, vol 1, pp 51–86. Office of the Surgeon General, Department of Army, Falls Church
3294. Wenger C, Salvador R, Basser PJ, Miranda C (2015) The electric field distribution in the brain during TTFields therapy and its dependence on tissue dielectric properties and anatomy: a computational study. Phys Med Biol 60:7339–7357
3295. Wenger C, Miranda R, Salvador P, Thielscher A, Bomzon Z, Giladi M, Mrugala MM, Korshoej AR (2018) A review on tumor-treating fields (TTFields): clinical implications inferred from computational modeling. IEEE Rev Biomed Eng 11:195–207
3296. Werner J (1989) Thermoregulatory models. Recent research, current applications and future development. Scand J Work Environ Health 15:34–46
3297. Werner J, Buse M (1988) Temperature profiles with respect to inhomogeneity and geometry of human body. J Appl Physiol 65:1110–1118
3298. Wessapan T, Rattanadecho P (2014) Influence of ambient temperature on heat transfer in the human eye during exposure to electromagnetic field at 900 MHz. Int J Heat Mass Transf 70:378–388
3299. Wessapan T, Rattanadecho P (2016) Flow and heat transfer in biological tissue due to electromagnetic near-field exposure effects. Int J Heat Mass Transf 97:174–184
3300. West GB, Brown JH (2005) The origin of allometric scaling laws in biology from genomes to ecosystems: towards a quantitative unifying theory of biological structure and organization. J Exp Biol 208:1575–1592
3301. West GB, Brown JH, Enquist BJ (1997) A general model for the origin of allometric scaling laws in biology. Science 276:122–126
3302. West GB, Brown JH, Enquist BJ (1999) The fourth dimension of life: fractal geometry and allometric scaling of organisms. Science 284:1677–1679
3303. West GB, Woodruff WH, Brown JH (2002) Allometric scaling of metabolic rate from molecules and mitochondria to cells and mammals. Proc Nat Acad Sci USA 99(suppl 1):2473–2478
3304. Westerlund T, Oksa J, Smolander J, Mikkelsson M (2003) Thermal responses during and after whole-body cryotherapy. J Thermal Biol 28:601–608
3305. Westervelt PJ (1963) Parametric acoustic array. J Acoust Soc Am 35:535–537
3306. Westin J (2008) An improved thermoregulatory model for cooling garment applications with transient metabolic rates. Ph.D. Thesis, University of Central Florida
3307. Weymann HD (1967) Finite speed of propagation in heat conduction, diffusion, and viscous shear motion. Am J Phys 35:488–496
3308. Wharton DA, Ferns DJ (1995) Survival of intracellular freezing by the Antarctic nematode Panagrolaimus Davidi. J Exp Biol 198:1381–1387
3309. Wheeler P (1984) The evolution of bipedality and loss of functional body hair in hominoids. J Hum Evol 13:91–98
3310. Whitaker S (1986) Flow in porous media I: a theoretical derivation of Darcy's law. Transp Porous Med 1:3–25
3311. White MD, Cabanac M (1995) Respiratory heat loss and core temperatures during submaximal exercise. J Thermal Biol 20:489–496
3312. White CR, Seymour RS (2003) Mammalian basal metabolic rate is proportional to body mass$^{2/3}$. Proc Nat Acad Sci USA 100:4046–4049
3313. White CR, Seymour RS (2005) Allometric scaling of mammalian metabolism. J Exp Biol 208:1611–1619
3314. Whittaker ET, Watson GN (1927) A course of modern analysis. Cambridge University Press, Cambridge
3315. Whittingham TA (2007) Medical diagnostic applications and sources. Prog Biophys Mol Biol 93:84–110
3316. Whittingham DG, Lyon MF, Gleniser PH (1977) Long-term storage of mouse embryos at -196° C. Genet Res 29:177–181

3317. Widehem P, Cochet N (2003) Pseudomonas syringae as an ice nucleator – application to freeze-concentration. Process Biochem 39:405–410
3318. Wilke M (2022) $l_p - l_q$-theory for a quasilinear non-isothermal Westervelt equation. arXiv:2204.09630v2 [math.AP]
3319. Williams RJP (1961) Possible functions of chains of catalysts. J Theor Biol 1:1–17
3320. Wilson BC, Jacques SL (1990) Optical reflectance and transmittance of tissues: principles and applications. IEEE J Quant Electron 26:2186–2199
3321. Wilson SB, Spence VA (1988) A tissue heat transfer model for relating dynamic skin temperature changes to physiological parameters. Phys Med Biol 33:895–912
3322. Wilson O, Goldman RF, Molnar GW (1976) Freezing temperature of finger skin. J Appl Physiol 41:551–558
3323. Winter A, Laing J, Paglione R, Sterzer F (1985) Microwave hyperthermia for brain tumors. Neurosurgery 17:387–399
3324. Winter L, Seifert F, Zilberti L, Murbach M, Ittermann B (2021) MRI-related heating of implants and devices: a review. J Magn Reson Imag 53:1646–1665
3325. Wisniewski M, Fuller M, Glenn DM, Palta J, Carter J, Gusta L, Griffith M, Duman J (2001) Factors involved in ice nucleation and propagation in plants: an overview based on new insights gained from the use of infrared thermography. Búïsindi Icel Agr Sci 14:41–47
3326. Wissler EH (1964) A mathematical model of the human thermal system. Bull Math Biophys 26:147–166
3327. Wissler EH (1987) Comments on the new bioheat transfer equation proposed by Weinbaum and Jiji. Trans ASME J Biom Eng 109:226–233
3328. Wissler EH (1998) Pennes' 1948 paper revisited. J Appl Physiol 85:35–41
3329. Withers LA, Davey MR (1978) A fine-structural study of the freeze-preservation of plant tissue cultures I. The frozen state. Protoplasma 94:207–219
3330. Woitowicz A, Stelmann S, Jankun J (2003) Computer simulation of prostate cryoablation - fast and accurate approximation of exact solutions. Comput Aid Surg 8:91–98
3331. Woldhuis B, Tangelder G, Slaaf DW, Reneman RS (1992) Concentration profile of blood platelets differs in arterioles and venules. Am J Physiol Heart Circul Physiol 262:H1217–H1223
3332. Wolfe J (2002) Cellular thermodynamics. In: Encyclopedia of life sciences. Macmilllan, New York, pp 1–17
3333. Wolfe J, Steponkus PL (1983) Mechanical properties of the plasma membrane of isolated plant protoplasts. Plant Physiol 71:276–285
3334. Wolfe J, Dowgert MF, Steponkus PL (1985) Dynamics of membrane exchange of the plasme membrane and the lysis of isolated protoplasts during rapid expansion in area. J Membr Biol 86:127–138
3335. Wolfe J, Bryant G, Koster KL (2002) What is 'unfreezable water', how unfreezable is it and how much is there? Cryo Lett 23:157–166
3336. Wolkers WF, Balasubramanian SK, Ongstad EL, Zec HC, Bischof JC (2007) Effects of freezing on membranes and proteins in LNCaP prostate tumor cells. Biochim Biophys Acta 1768:728–736
3337. Womersley JR (1955) Method for the calculation of velocity, rate of flow and viscous drag in arteries when the pressure gradient is known. J Physiol 127:553–563
3338. Womersley JR (1957) Oscillatory flow in arteries: the constrained elastic tube as a model of arterial flow and pulse transmision. Phys Med Biol 2:178–187
3339. Wood MA, Shaffer KM, Ellenbogen AL, Ownby ED (2005) Microbubbles during radiofrequency catheter ablation: composition and formation. Heart Rhythm 2:397–403
3340. Worall D, Elias L, Ashford D, Smallwood M, Sidebottom C, Lillford P, Tellford J, Holt C, Bowles D (1998) A carrot leucine-rich-repeat protein that inhibits ice recrystallization. Science 282:115–117
3341. Workmaster BA, Palta J, Wisniewski M (1999) Ice nucleation and propagation in cranberry uprights and fruit using infrared thermography. J Am Soc Horticult Sci 124:619–625

3342. Wowk B, Leitl E, Rasch CM, Mesbah-Karimi N, Harris SB, Fahy GM (2000) Vitrification enhancement by synthetic ice blocking agents. Cryobiology 40:228–236
3343. Wright NT, Humphrey JD (2002) Denaturation of collagen via heating: an irreversible rate process. Ann Rev Biomed Eng 4:109–128
3344. Wright SN, Kochunov P, Mut F, Bergamino M, Brown KM, Mazziotta JC, Toga AW, Cebral JR, Ascoli GA (2013) Digital reconstruction and morphometric analysis of human brain arterial vasculature from magnetic resonance angiography. Neuroimage 82:170–181
3345. Wu J (2007) Shear stress in cells generated by ultrasound. Prog Biophys Mol Biol 93:363–373
3346. Wu J (2018) Acoustic streaming and its applications. Fluids 3:108
3347. Wu J, Nyborg WL (2008) Ultrasound, cavitation bubbles and their interaction with cells. Adv Drug Deliv Rev 60:1103–1116
3348. Wu YL, Weinbaum S, Jiji L (1993) A new analytic technique for 3-D heat transfer from a countercurrent blood vessels. Int J Heat Mass Transf 36:1073–1083
3349. Wu HL, Ma Y, Peng XF (2004) Freezing-thawing characteristics of botanical tissue and influence of water morphology. Chin Phys Lett 21:345–347
3350. Wu J, Long Q, Xu SX, Anwar R, Padhani AR (2009) Study of tumor blood perfusion and its variation due to vascular normalization by anti-angiogenic therapy based on 3D angio genic microvasculature. J Biomech 42:712–721
3351. Wu T, Tan L, Shao Q, Zhang C, Zhao C, Li Y, Conil E, Hadjem A, Wiart J, Lu B, Xiao L, Wang N, Xie Y, Zhang S (2011) Chinese adult anatomical models and the application in evaluation of RF exposures. Phys Med Biol 56:2075–2089
3352. Wu C, Chen X, Zhou X (2016) Performance of novel solar assisted bian stone thermal therapy. Int J Heat Mass Transf 100:445–450
3353. Wu Y, Jelezko F, Plenio MB, Weil T (2016) Diamond quantum devices in biology. Angewandte Chem Int Ed 55:6586–6598
3354. Wu J, Fu M, Tong X, Qin Y (2017) Heat stress evaluation at the working face in hot coal mines using an improved thermophysiological model. Int J Heat Technol 35:67–74
3355. Wu K, Liu J, Saha R, Ma B, Su D, Peng C, Sun J, Wang JP (2019) Irregularly shaped $\gamma' - Fe_4N$ nanoparticles for hyperthermia treatment and t2 contrast-enhanced magnetic resonance imaging with minimum dose. arXiv:1910.06842 [physics.app-ph]
3356. Wuitchik SJS, Mogensen S, Barry TN, Paccard A, Jamniczky HA, Barrett RDH, Rogers SM (2022) Evolution of thermal physiology alters the projected range of threespine stickleback under climate change. Mol Ecol 31:2312–2326
3357. Wulff W (1974) The energy conservation equation for living tissue. IEEE Trans Biomed Eng 21:494–495
3358. Wunsch C, Ferrari R (2004) Vertical mixing, energy, and the general circulation of the oceans. Annu Rev Fluid Mech 36:281–314
3359. Wust P, Seebass M, Nadobny J, Deuflhard P, Monich G, Felix R (1996) Simulation studies promote technological development of radiofrequency phased array hyperthermia. Int J Hyperth 12:477–494
3360. Wust P, Hildebrandt B, Sreenivase G, Rau B, Gellermann J, Riess H, Felix R, Schlag P (2002) Hyperthermia in combined treatment of cancer. Lancet Oncol 3:487–497
3361. Wust P, Gneveckow U, Johannsen M, Böhmer D, Henkel T, Kahmann F, Sehouli J, Felix R, Ricke J, Jordan A (2006) Magnetic nanoparticles for interstitial thermotherapy-feasibility, tolerance and achieved temperatures. Int J Hyperth 22:673–685
3362. Wusterman MC, Rauen U, Simmonds J, Hunds N, Pegg DE (2008) Reduction of cryoprotectant toxicity in cells in suspension by use of a sodium-free vehicle solution. Cryobiology 56:72–79
3363. Wusterman MC, Simmonds J, Vaughan D, Pegg DE (2008) Vitrification of rabbit tissues with propylene glycol and trehalose. Cryobiology 56:62–71
3364. Wyndham CH (1973) The physiology of exercise under heat stress. Ann Rev Physiol 35:193–220

3365. Xantheas SS, Dunning THJ (1993) *Ab initio* studies of cyclic water clusters $(H_2O)_n$, n = 1–6. I. Optimal structures and vibrational spectra. J Chem Phys 99:8774–8792
3366. Xe X, Bischof J (2005) Analysis of thermal stress in cryosurgery of kidneys. J Biomech Eng 127:656–661
3367. Xu M (2011) Themodynamic basis of dual-phase-lagging of heat conduction. J Heat Transf 133:041401
3368. Xu K (2012) History of cryosurgery. In: Xu K, Korpan NN, Niu L (eds) Modern cryosurgery for cancer. World Scientific, Singapore, pp 3–28
3369. Xu M, Hu H (2011) A ballistic-diffusive heat conduction model extracted from Boltzmann transport equation. Proc R Soc A 467:1851–1864
3370. Xu M, Wang L (2002) Thermal oscillation and resonance in dual-phase-lagging heat conduction. Int J Heat Mass Transf 45:1055–1061
3371. Xu M, Wang L (2005) Dual-phase-lagging heat conduction based on Boltzmann transport equation. Int J Heat Mass Transf 48:5616–5624
3372. Xu X, Tikuisis P (2014) Thermoregulatory modeling for cold stress. Compr Physiol 4:1057–1081
3373. Xu HY, Jiang XY (2015) Time fractional dual-phase-lag conduction equation. Chin Phys B 24:034401
3374. Xu F, Lu T (2011) Introduction to skin biothermomechanics and thermal pain. Springer, Berlin
3375. Xu J, Zou BM, Xu P (2006) A new model for heat conduction of nanofluids based on fractal distribution of nanoparticles. J Phys D Appl Phys 39:4486–4490
3376. Xu F, Lu TJ, Seffen KA (2008) Biothermomechanical behavior of skin tissue. Acta Mech Sin 24:1–23
3377. Xu F, Seffen KA, Lu TJ (2008) Non-Fourier analysis of skin biothermomechanics. Int J Heat Mass Transf 51:2237–2259
3378. Xu F, Wen T, Lu TJ (2008) Skin biothermomechanics for medical treatments. J Mech Behav Biomed Mater 1:172–187
3379. Xu KC, Niu LZ, He WB, He YS, Li YF, Zuo JS (2008) A pilot study on the combination of cryosurgery and ^{125}iodine seed implantation for treatment of locally advanced pancreatic cancer. World J Gastroenterol 14:1603–1611
3380. Xu F, Lu TJ (2009) Skin biothermomechanics: modeling and experimental characterization. Adv Appl Math 43:147–248
3381. Xu F, Lu TJ, Seffen KA, Ng EYK (2009) Mathematical modeling of skin bioheat transfer. Appl Mech Rev 62:050801
3382. Xu F, Moon S, Zhang X, Shao L, Song YS, Demirci U (2010) Multi-scale heat and mass transfer modelling of cell and tissue cryopreservation. Philos Trans R Soc A Math Phys Eng Sci 368:561–583
3383. Xu M, Guo J, Wang L, Cheng L (2011) Thermal wave interference as the origin of the overshooting phenomena in dual-phase-lagging heat conduction. Int J Heat Sci 50:825–830
3384. Xu F, Wang PF, Lin M, Lu TJ, Ng EYK (2010) Quantifying the underlying mechanism of skin thermal damage: a review. J Mech Med Biol 10:373–400
3385. Xu K, Chiu D, Shiu CTW (2012) Overview of cryosurgery for tumors. In: Xu K, Korpan NN, Niu L (eds) Modern cryosurgery for cancer. World Scientific, Singapore, pp 103–122
3386. Xu H, Jing X, Qi H (2013) Fractional Cattaneo heat equation in a semi-infinite medium. Chin Phys B 22:014401
3387. Xu Y, Moser MA, Zhang E, Zhang W, Zhang B (2019) Large and round ablation zones with microwave ablation: a preliminary study of an optimal aperiodic tri-slot coaxial antenna with the matching network section. Int J Thermal Sci 140:539–548
3388. Xu J, Psikuta A, Lia J, Annaheim S, Rossi RM (2021) Evaluation of the convective heat transfer coefficient of human body and its effect on the human thermoregulation predictions. Build Environ 196:107778
3389. Xua F, Lub T, Seffen KA (2008) Biothermomechanics of skin tissues. J Mech Phys Solids 56:1852–1884

3390. Xuan Y, Roetzel W (1997) Bioheat equation of the human thermal system. Chem Eng Technol 20:268–276
3391. Yablonskiy DA, Ackerman JJH, Raichle ME (2000) Coupling between changes in human brain temperature and oxidative metabolism during prolonged visual stimulation. Proc Nat Acad Sci 97:7603–7608
3392. Yamada K, Aoki T, Enami Y, Tashiro Y, Zenaou Z, Koizumi T, Kusano T, Wada Y, Shibata H, Tomiaka K, Siriratsivawong K, Hoffman RM, Murakami M (2020) An improved encapsulation method for cryopreserving hepatocytes for functional transplantation using a thermo-reversible gelation polymer. In Vivo 34:2309–2316
3393. Yamanaka S (2020) Pluripotent stem cell-based cell therapy – promise and challenges. Cell Stem Cell 20:523–531
3394. Yan JF, Liu J (2008) Nanocryosurgery and its mechanisms for enhaning freezing efficiency of tumor tissues. Nanomed Nanotechnol Biol Med 4:79–87
3395. Yan Z, Buldyrev SV, Giovambattista N, Stanley HE (2005) Structural order for one-scale and two-scale potentials. Phys Rev Lett 95:130604
3396. Yan Z, Buldyrev SV, Kumar P, Giovambattista N, Debenetti PG, Stanley HE (2007) Structure of the first- and second-neigbor shells of simulated water: quantitative relation to translational and orientational order. Phys Rev E 76:051201
3397. Yang XJ (2012) Generalized local fractional Taylor's formula with local farctional derivative. J Exp Syst 1:1–5
3398. Yang WI, Addona T, Nair DG, Qi L, Ravikumar TS (2003) Apoptosis induced by cryo-injury in human colorectal cancer cells is associated with mitochondrial dysfunction. Int J Cancer 103:360–369
3399. Yang KT, Pan SF, Chen CI, Hsu SM, Tseng YZ, Yang SM, Wu ML (2004) Mitochondrial Na^+ overload is caused by oxidative stress and leads to activation of the caspase 3-dependent apoptotic mashinery. FASEB J 18:1442–1444
3400. Yang D, Converse M, Mahvi DM, Webster JG (2007) Expanding the bioheat equation to include tissue internal water evaporation during heating. IEEE Trans Biomad Eng 54:1382–1388
3401. Yang D, Converse M, Mahvi M, Webster J (2007) Measurement and analysis of tissue temperature during microwave liver ablation. IEEE Trans Biomed Eng 54:150–155
3402. Yang X, Forouzan O, Burns JM, Shevkoplyas SS (2011) Traffic of leukocytes in microfluidic channels with rectangular and rounded cross-sections. Lab Chip 11:3231–3240
3403. Yang N, Xu X, Zhang G, Li B (2012) Thermal transport in nanostructures. AIP Adv 2:041410
3404. Yang XJ, Srivastava HM, Teneiro Machado JA (2015) A new fractional derivative withput singular kernel: application to the modelling of the steady heat flow. arXiv: 1601.01623
3405. Yang Y, Yao R, Li B, Liu H, Jiang L (2015) Modelling of the microclimate saturation inside a safety helmet. Build Environ 87:1–9
3406. Yang XJ, Han Y, Li J, Liu WX (2016) On steady heat flow problem involving Yang-Srivastava-Machado fractional derivative without singular kernel. Thermal Sci S717–S721
3407. Yang XJ, Zhang ZZ, Teneiro Machado JA, Baleanu D (2016) On local fractional operators view of computational complexity. Diffusion and relaxation defined on Cantor sets. Thermal Sci 20:S755–S767
3408. Yang B, Cheng X, Dai D, Olofsson T, Li H, Meier A (2019) Real-time and contactless measurements of thermal discomfort based on human poses for energy efficient control of buildings. Build Environ 162:106284
3409. Yang B, Li X, Hou Y, Meier A, Cheng X, Choi JH, Wang F, Wang H, Wagner A, Yan D, Li A, Olofsson T, Li H (2020) Non-invasive (non-contact) measurements of human thermal physiology signals and thermal comfort/discomfort poses - a review. Energy Build 224:110261
3410. Yazdi M, Sheikhzadeh M (2014) Personal cooling garments: a review. J Text Inst 105:1231–1250

3411. Ye H, De S (2017) Thermal injury of skin and subcutaneous tissues: a review of experimental approaches and numerical models. Burns 43:909–932
3412. Ye T, Li H, Lam KY (2010) Modeling and simulation of microfluid effects on deformation behavior of a red blood cell in a capillary. Microvasc Res 80:453–463
3413. Ye T, Phan-Thien N, Khoo BC, Lim CT (2014) Dissipative particle dynamics simulations of deformation and aggregation of healthy and diseased red blood cells in a tube flow. Phys Fluids 26:111902
3414. Ye T, Phan-Thien N, Lim CT (2016) Particle-based simulations of red blood cells. J Biomech 49:2255–2266
3415. Ye T, Phan-Thien N, Lim CT, Peng L, Shi H (2017) Hybrid smoothed dissipative particle dynamics and immersed boundary method for simulation of red blood cells in flows. Phys Rev E 95:063314
3416. Ye T, Shi H, Peng L, Li Y (2017) Numerical studies of a red blood cell in rectangular microchannels. J Appl Phys 122:084701
3417. Yeagle PL (2016) The membranes of cells. Academic Press, Cambridge
3418. Yee K (1966) Numerical solution of initial boundary value problems involving Maxwell's equations in isotropic media. IEEE Trans Antennas Propagt 14:302–307
3419. Yeo TP (2004) Heat stroke: a comprehensive review. AACN Clin Issues 15:280–293
3420. Yerezhep D, Yemelyanov A, Kuznetsov Y, Perminov A, Pronin V, Minikaev A, Yerezhep A (2019) Consideration of the effectiveness of cryotherapy for different thicknesses of the skin using numerical modelling. J Phys Conf Ser 1333:032093
3421. Yerezhep D, Yemelyanov A, Kuznetsov Y, Perminov A, Pronin V, Minikaev A, Yerezhep A (2020) Numerical analysis of the heat flux density of the whole-body cryotherapy (WBC) object coming from the skin. IOP Conf Ser Mat Sci Eng 826:012019
3422. Yetkinler DN, Greenleaf JE, Sherman OH (2002) Histologic analysis of radiofrequency energy chondroplasty. Clin Sports Med 21:649–661
3423. Yi GQ, Gu B, Chen LK (2014) The safety and efficacy of magnetic nano-iron hyperthermia therapy on rat brain glioma. Tumor Biol 35:2445–2449
3424. Yi G, Li L, Luo M, He X, Zou Z, Gu Z, Su L (2017) Heat stress induces intestinal injury through lysosome- and mitochondria-dependent pathway in vivo and in vitro. Oncotarget 8:40741–40755
3425. Yilbas B, Al-Aqeeli N (2009) Analytical investigation into laser pulse heating and thermal stresses. Opt Laser Technol 41:132–139
3426. Yildrim ED, Ozerdam B (2008) A numerical simulation study for human passive thermal system. Energy Build 40:1117–1123
3427. Yin X, Thomas T, Zhang J (2013) Multiple red blood cell flows through microvascular bifurcations: cell free layer, cell trajectory, and hematocrit separation. Microvasc Res 89:47–56
3428. Yin Y, Ren Y, Li H, Qi H (2020) Characteristic analysis of light and heat transfer in photothermal therapy using multiplelight-source heating strategy. Int J Thermal Sci 158:106533
3429. Ying L, Biros G, Zorin D (2016) A high-order 3D boundary integral equation solver for elliptic PDEs in smooth domains. J Comput Phys 219:247–275
3430. Yong YH, Pope JM, Wolfe J (1998) The effects of solutes on the freezing properties of and hydration forces in lipid lamellar phases. Biophys J 74:1949–1965
3431. Yong KW, Pingguan-Murphy B, Xu F, Wan Abas WAB, Choi JR, Omar SZ, Noor Azmi MA, Chua KH, Safwani WKZ (2015) Phenotypic and functional characterization of long-term cryopreserved human adipose-derived stem cells. Sci Rep 5:9596
3432. Yoo S, Ito K (2020) Multi-stage optimization of local environmental quality by comprehensive computer simulated person as a sensor for HVAC control. Adv Build Energy Res 14:171–188
3433. Yoshida S, Yoshida A, Kinoshita S (2020) Evaluation methods of adaptation cities. In: Takebayashi H, Moriyama M (eds) Adaptation measures for urban heat islands. Academic Press, Cambridge, pp 115–159

3434. Yoshino H, Minari O (1987) Heat-induced dissociation of human erythrocyte spectrin dimer into monomers. Biochim Biophys Acta 905:100–108
3435. Young GS (1998) Thermodynamic characterization of skin, hide and similar materials composed of fibrous collagen. Stud Conserv 43:65–79
3436. Young S, Bolton P, Dyson M, Harvey W, Diamantopoulos C (1989) Macrophage responsiveness to light therapy. Lasers Surg Med 9:497–505
3437. Young VR, Yu YM, Fikagawa NK (1991) Protein and energy interactions throughout life. Metabolic basis and nutritional implications. Acta Paediatr Scand Suppl 373:5–24
3438. Youssef HM (2006) Two-dimensional generalized thermoelasticity problem for a half-space subjected to ramp-type heating. Eur J Mech A/Solids 25:745–763
3439. Youssef HM (2010) Theory of fractional order generalized thermoelasticity. J Heat Transf 132:1–7
3440. Youssef HM, Alghamdi NA (2020) Modeling of one-dimensional thermoelastic dual-phase-lag skin tissue subjected to different types of thermal loading. Sci Rep 10:3399
3441. Youssef HM, El-Bary AA (2020) The thermal behavior analysis of the human eye under the heat conduction law with one relaxation time. Alexandria Eng J 59:5263–5271
3442. Yu ZW, Quinn PJ (2000) The effect of dimethyl sulphoxide on the structure and phase behaviour of palmitoleoylphospatidylethanolamine. Biochim Biophys Acta 1509:440–450
3443. Yu H, van Gunsteren WF (2004) Charge-on-spring polarizable water models revisited: from water clusters to liquid water to ice. J Chem Phys 121:9549–9564
3444. Yu Y, Hill AP, Mccormick DA (2012) Warm body temperature facilitates energy efficient cortical action potentials. PLoS Comput Biol 8:e1002456
3445. Yu B, Jiang X, Wan C (2016) Numerical alorithms to estimate relaxation parameters and Caputo fractional derivative for a fractional thermal wave model in spherical composite medium. Appl Math Comput 274:106–118
3446. Yu FC, Lu KC, Lin SH, Chen GS, Chu P, Gao GW, Lin YF (1997) Energy metabolism in exertional heat stroke with acute renal failure. Nephrol Dial Transplant 12:2087–2092
3447. Yu HS, Wu CS, Kao YH, Chiou MH, Yu CL (2003) Helium-neon laser irradiation stimulates migration and proliferation in melanocytes and induces repigmentation in segmental-type vitiligo. J Investig Dermatol 120:56–64
3448. Yu TH, Zhou YX, Liu J (2003) Dynamic low-frequency electrical impedance of biological materials subject to freezing and its implementation in cryosurgical monitoring. Int J Thermophys 24:513–531
3449. Yu TH, Liu J, Zhou YX (2004) Using electrical impedance detection to evaluate the viability of biomaterials subject to freezing or thermal injury. Anal Bioanal Chem 378:1793–1800
3450. Yu TH, Ling J, Zhou YX (2005) Selective freezing of target biological tissues after injection of solutions with specific thermal properties. Cryobiology 50:174–182
3451. Yu YJ, Hu W, Tian XG (2014) A generalized thermoelasicity model based on memory-dependent derivative. Int J Eng Sci 81:123–134
3452. Yu YJ, Li CL, Xue Z, Tian XG (2016) The dilemma of hyperbolic heat conduction and its settlement byincorporating spatially nonlocal effect at nanoscale. Phs Lett A 380:255–261
3453. Yu X, Zhang S, Chen G (2019) Modeling the addition/removal of dimethyl sulfoxide into/from articular cartilage treated with the liquidus-tracking method. Int J Heat Mass Transf 141:719–730
3454. Yue K, Yu C, Lei Q, Luo Y, Zhang X (2014) Numerical simulation of effect of vessel bifurcation on heat transfer in the magnetic fluid hyperthermia. Appl Thermal Eng 69:11–18
3455. Zabardaev V, Denisov S, Klafter J (2015) Lévy walks. Rev Mod Phys 87:483–530
3456. Zachariassen KE, Hammel HT (1976) Nucleating agents in the haemolymph of insects tolerant to freezing. Nature 262:285–287
3457. Zachariassen KE, Husly JA (1982) Antifreeze effect of thermal hysteresis agents protects highly supercooled insects. Nature 298:865–867
3458. Zachariassen KE, Kristiansen E (2000) Ice nucleation and antinucleation in nature. Cryobiology 41:257–279

3459. Zamir M (2001) Fractal dimensions and multifractility in vascular branching. J Theor Biol 212:183–190
3460. Zanchini E (1999) Hyperbolic heat-conduction theories and nondecreasing entrophy. Phys Rev B 60:991–997
3461. Zangi R (2004) Water confined to a slab geometry: a review of recent computer simulation studies. J Phys Condens Matter 16:S5371–S5388
3462. Zangos E, Mark MG, Balzer JO, Engelmann BK, Straub R, Eichler K, Hergoz C, Lehnert T, Sollner O, Heller M, Thalhammer A, Vogl TJ (2004) Neoadjuvant transarterial chemoembolization (TAGE) before percutaneous MR-guided laser-induced thermotherapy (LITT): results in large-size primary and secondary liver tumors. Med Laser Appl 19:98–108
3463. Zarda P, Chien S, Skalak R (1977) Elastic deformations of red blood cells. J Biomech 10:211–221
3464. Zaslavsky GM (2008) Hamiltonian chaos and fractional dynamics. OUP, Oxford
3465. Zastrow E, Davis SK, Lazebnik M, Kelcz F, Van Veen BD, Hagness SC (2008) Development of anatomically realistic numerical breast phantoms with accurate dielectric properties for modeling microwave interactions with the human breast. IEEE Trans Biomed Eng 55:2792–2800
3466. Zavos PM, Graham EF (1983) Effects of various degrees of supercooling and nucleation temperatures on fertility of frozen turkey spermatozoa. Cryobiology 20:553–559
3467. Zecova M, Terpak J (2015) Fractional heat conduction models and thermal diffusivity determination. Math Problems Eng 2015:753936
3468. Zenker W, Kubik S (1996) Brain cooling in humans – anatomical considerations. Anatom Embryol 193:1–13
3469. Zezell DM, Ana PA, Pereira TM, Correa PR, Velloso W (2011) Heat generation and transfer on biological tissues due to high-intensity laser irradiation. In: dos Santos Bernardes MA (ed) Developments in heat transfer. IntechOpen, London
3470. Zhai LN, Li J (2015) Prediction methods of skin burn for performance evaluation of thermal protective clothing. Burns 41:1385–1396
3471. Zhang ZM (2007) Nano/microscale heat transfer. McGraw-Hill, New York
3472. Zhang Y (2009) Generalized dual-phase lag bioheat equations based on nonequilibrium heat transfer in living biological tissues. Int J Heat Mass Transf 52:4829–4834
3473. Zhang J, Chauhan S (2019) Fast computation of soft tissue thermal response under deformation based on fast explicit dynamics finite element algorithm for surgical simulation. arXiv:1911.11845 [cs.CE]
3474. Zhang J, Chauhan S (2019) Real-time computation of bio-heat transfer in the fast explicit dynamics finite element algorithm (FED-FEM) framework. Numer Heat Transf B 75:217–238
3475. Zhang J, Chauhan S (2020) Fast computation of soft tissue thermal response under deformation based on fast explicit dynamics finite element algorithm for surgical simulation. Comput Methods Program Biomed 187:105244
3476. Zhang G, Li B (2006) Thermal conductivity of nanotubes revisited: Effects of chirality, isotope impurity, tube length, and temperature. arXiv:cond-mat/0501194 [cond-mat.mtrl-sci]
3477. Zhang K, Schubert G (2000) Magnetohydrodynamics in rapidly rotating spherical shells. Annu Rev Fluid Mech 32:411–445
3478. Zhang K, Sejnowski TJ (2000) A universal scaling law between gray matter and white matter of cerebral cortex. Proc Natl Acad Sci USA 97:5621–5626
3479. Zhang Q, Jackson TH, Ungan A, Gao D (1999) Numerical modeling of continuous hybrid heating of cryopreserved tissue. Int J Heat Mass Transf 42:395–403
3480. Zhang YT, Liu J, Zhou YX (2002) Pilot study on cryogenic heat transfer in biological tissues embedded with large blood vessels. Forsch Ingenier 67:188–197
3481. Zhang A, Xu LX, Sandison GA, Zhang J (2003) A microscale model for prediction of breast cancer cell damage during cryosurgery. Cryobiology 47:143–154

3482. Zhang DM, Li L, Zhihua L, Li G, Xinyu T (2005) Non-Fourier conduction model with thermal source term of ultra short high power pulsed laser ablation and temperature involvement before melting. Phys B 364:285–293
3483. Zhang J, Sandison GA, Murthy JY, Xu LX (2005) Numerical simulation for heat transfer in prostate cancer cryosurgery. J Biomech Eng 127:279–294
3484. Zhang X, Gu H, Fujii M (2006) Experimental study on the effective thermal conductivity and thermal diffusivity of nanofluids. Int J Thermophys 27:569–580
3485. Zhang Q, Cong Y, Qu S, Luo S, Yang G (2008) Cryopreservation of gametophytes of Laminaria japonica (phaeophyta) using encapsulation-dehydration with two-step cooling method. J Ocean Univ Chin 7:65–71
3486. Zhang H, Arens E, Huizenga C, Han T (2010) Thermal sensation and comfort models for non-uniform and transient environments: part I: local sensation of individual body parts. Build Environ 45:380–388
3487. Zhang H, Arens E, Huizenga C, Han T (2010) Thermal sensation and comfort models for non-uniform and transient environments: part II: local comfort of individual body parts. Build Environ 45:389–398
3488. Zhang JZ, Zhang XX, Audette M (2011) A photothermal model of selective photothermolysis with dynamically changing vaporization temperature. Lasers Med Sci 26:630–640
3489. Zhang W, Cai X, Holm S (2014) Time-fractional heat equations and negative absolute temperatures. Comput Math Appl 67:164–171
3490. Zhang J, Zhong Y, Gu C (2017) Ellipsoid bounding region-based chainmail algorithm for soft tissue deformation in surgical simulation. Int J Inter Des Manuf 12:903–918
3491. Zhang Y, Chen B, Li D (2017) Non-Fourier effect of lasermediated thermal behaviors in biotissues: a numerical study by the dual-phase-lag model. Int J Heat Mass Transf 108:1428–1438
3492. Zhang Y, Zhao G, Chapal Hossain SM, He X (2017) Modeling and experimental studies of enhanced cooling by medical gauze for cell cryopreservation by vitrification. Int J Heat Mass Transf 114:1–7
3493. Zhang HH, Liu Y, Xiong XYZ, Shi GM, Wang CY, Sha WEI (2019) Investigating thermal cooling mechanisms of human body under exposure to electromagnetic radiation. arXiv:1901.04917 [physics.med-ph]
3494. Zhang J, Hills J, Zhong Y, Shirinzadeh B, Smith J, Gu C (2019) Temperature-dependent thermomechanical modeling of soft tissue deformation. J Mech Med Biol 18:1840021
3495. Zhang Y, Wang H, Stewart S, Jiang B, Ou W, Zhao G, He X (2019) Cold-responsive nanoparticle enables intracellular delivery and rapid release of trehalose for organic-solvent-free cryopreservation. Nano Lett 19:9051–9061
3496. Zhang X, Zheng L, Liu L, Zhang X (2020) Modeling and simulation on heat transfer in blood vessels subject to a transient laser irradiation. J Heat Transf 142:4045669
3497. Zhang Y, Shi X, You W, Tao Z, Zhong Y, Kabeer FC, Maldonado P, Oppeneer PM, Bauer M, Rossnagel K, Kapteyn H, Murnane M (2020) Coherent modulation of the electron temperature and electron-phonon couplings in a 2D material. Proc Nat Acad Sci 117:8788–8793
3498. Zhang J, Lay J, Roberts SK, Chauhan S (2021) Towards real-time finite-strain anisotropic thermo-visco-elastodynamic analysis of soft tissues for thermal ablative therapy. Comput Methods Program Biomed 198:105789
3499. Zhang Q, Sun Y, Yang J, Soh AK, Wang X (2021) Theoretical analysis of thermal response in biological skin tissue subjected to multiple laser beams. Case Stud Thermal Eng 24:100853
3500. Zhao X, Chua KJ (2012) Studying the thermal effects of a clinically-extracted vascular tissue during cryo-freezing. J Therm Biol 37:556–563
3501. Zhao G, Fu J (2017) Microfluidics for cryopreservation. Biotechnol Adv 35:323–336
3502. Zhao D, Luo M (2019) Representations of acting processes and memory effects: general fractional derivative and its application to theory of heat conduction with finite wave speeds. Appl Math Comput 346:531–544

3503. Zhao G, Bai X, Luo D, Gao D (2006) Modeling the heat transfer problem for the novel combined cryosurgery and hyperthermia system. Cryo Lett 27(2):115–126
3504. Zhao R, Antaki J, Naik T, Bachman TN, Kameneva MV, Wu ZJ (2006) Microscopic of erythrocyte deformation dynamics. Biorheology 43:747–765
3505. Zhao G, Luo DW, Liu ZF, Gao DY (2007) Comparative study of the cryosurgical processes with two different cryosurgical systems: the Endocare cryoprobe system versus the novel combined cryosurgery and hyperthermia system. Latin Am Appl Res 37:215–222
3506. Zhao G, Zhang H, Guo XJ, Luo D, Gao DY (2007) Effect of blood flow and metabolism on multidimensional heat transfer during cryosurgery. Med Eng Phys 29:205
3507. Zhao Y, Cai YG, Yang XJ (2016) A local fractional derivative with applications to fractal relaxation and diffusion phenomena. Thermal Sci 20:S723–S727
3508. Zhmakin AI (1997) A memory-efficient unstructured grid refinement algorithm for computation of 3D steady viscous flows. Commun Numer Methods Eng 13:219–228
3509. Zhmakin AI (2004) Heat transfer problems in crystal growth. Keynote lecture. ICHMT international symposium on advances in computational heat transfer, Norway. CD-ROM Proc, Begell House, p 24
3510. Zhmakin AI (2008) Physical aspects of cryobiology. Phys Usp 51:231–252
3511. Zhmakin AI (2009) Fundamentals of cryobiology: physical phenomena and mathematical models (biological and medical physics, biomedical engineering). Springer, Berlin
3512. Zhmakin AI (2017) Action of low temperatures on biological objects (in Russian). In: Prohorov GG, Belyaev AM, Prokhorov DG (eds) Fundamentals of clinical cryomedicine. Kniga Po Trebovaniyu, Saint-Petersburg, pp 29–49
3513. Zhmakin AI (2017) Heat transfer in vivo: phenomena & models. In: Kulacki F (ed) Handbook of thermal science and engineering. Springer, Berlin, pp 2333–2379
3514. Zhmakin AI (2020) Modelling of mass and heat transfer in bulk single crystal growth. LAM Lambert Academic, Saarbrücken
3515. Zhmakin AI (2023) Non-fourier heat conduction. From phase-lag models to relativistic and quantum transport, Springer, Berlin
3516. Zhmakin AI, Ofengeim DK (2006) Stationary heat transfer in a multiprobe cryosurgery system. Tech Phys Lett 32:765–767
3517. Zhong-Can OY, Helfrich W (1989) Bending energy of vesicle membranes: general expressions for the first, second, and third variation of the shape energy and applications to spheres and cylinders. Phys Rev A Gen Phys 39:5280–5288
3518. Zhou J (2015) Modeling and simulation of thermal ablation in vascularized tissues. Ph.D. Thesis, The University of Texas at Austin
3519. Zhou J, Zhang Y, Chen J (2008) Non-Fourier heat conduction effect on laser-induced thermal damage in biological tissues. Numer Heat Transf A Appl 54:1–19
3520. Zhou J, Chen J, Zhang Y (2009) Dual-phase lag effects on thermal damage to biological tissues caused by laser irradiation. Comput Biol Med 39:286–293
3521. Zhou J, Zhang Y, Chen JK (2009) An axisymmetric dual phase-lag bioheat model for laser heating of living tissues. Int J Thermal Sci 48:1477–1485
3522. Zhou B, Xu F, Chen CQ, Lu TJ (2010) Strain rate sensitivity of skin tissue under thermomechanical loading. Philos Trans R Soc A 368:679–690
3523. Zhou X, Xiong J, Lian Z (2017) Predication of skin temperature and thermal comfort under two-way transient environments. J Therm Biol 70:15–20
3524. Zhou Y, Yu Z, Ma Q, Guo G, Tu Z, Zhang D (2019) HIFU therapy based on magneto-acousto-electrical tomography. IEEE Trans Biomed Eng 66:666–674
3525. Zhu L (2009) Heat transfer applications in biological systems. In: Kutz M (ed) Biomedical engineering & design handbook. McGraw-Hill Professional, New York, pp 2.33–2.67
3526. Zhu L, Weinbaum S (1995) A model for heat transfer from embedded blood vessels in 2-D tissue preparations. J Biomech Eng 117:64–73
3527. Zhu L, Weinbaum S, Lemons DE (1996) Microvascular thermal equilibration in rat cremaster muscle. Ann Biomed Eng 24:109–123

3528. Zhu L, Xu LX (1999) Evaluation of the effectiveness of transurethral radio frequency hyperthermia in canine prostate: temperature distribution analysis. J Biomed Eng 121:584–590
3529. Zhu L, Lemons DE, Weinbaum S (1995) A new approach for prediction the enhancement in the effective conductivity of perfused muscle tissue due to hyperthermia. Ann Biomed Eng 23:1–12
3530. Zhu C, Bao G, Wang N (2000) Cell mechanics: mechanical response, cell adhesion, and molecular deformation. Ann Rev Biomed Eng 2:189–226
3531. Zhu F, Tajkhorshid E, Shulten K (2004) Theory and simulation of water permeation in Aquaporin-1. Biophys J 86:50–57
3532. Zhu M, Ackerman JJ, Sukstanskii AL, Yablonskiy DA (2006) How the body controls brain temperature: the temperature shielding effect of cerebral blood flow. J Physiol 101:1481–1488
3533. Zhu L, Schappeler T, Cordero-Tumangday C, Rosengart AJ (2009) Thermal interactions between blood and tissue: development of a theoretical approach in predicting body temperature during blood cooling/rewarming. Adv Numer Heat Transf 3:197–219
3534. Zhu M, Ackerman JJ, Yablonskiy DA (2009) Body and brain temperature coupling: the critical role of cerebral blood flow. J Compr Physiol 179:701–710
3535. Zhu H, Wang H, Li D, Xiao Z, Su H, Kuang X (2019) Evaluation of the human thermal comfort under simulated weightlessness: an experimental study based on the power spectrum analysis of the heart rate variability. Micrograv Sci Technol 31:73–83
3536. Zhu Y, Sharaf S, Ali BJ (2021) Sports health quantification method and system implementation based on multiple thermal physiology simulation. Appl Math Nonlinear Sci 7:1–10
3537. Zhukovski K, Oskolkov D, Gubina N (2018) Some exact solutions to non-Fourier heat equations with substantial derivative. Axioms 7:48
3538. Ziskin MC, Alekseev SI, Foster KR, Balzano Q (2018) Tissue models for RF exposure evaluation at frequencies above 6 GHz. Bioelectromagnetic 39:173–189
3539. Zitting KM, Vujovic N, Yuan RK, Isherwood CM, Medina JE, Wang W, Buxton OM, Williams JS, Czeisler CA, Duffy JF (2018) Human resting energy expenditure varies with circadian phase. Curr Biol 28:3685–3690
3540. Zlochiver S, Rosenfeld M, Abboud S (2005) Contactless bio-impedance monitoring technique for brain cryosurgery in a 3D head model. Ann Biomed Eng 33:616–625
3541. Zolfaghari A, Maerefat M (2011) Bioheat transfer. In: Dos Santos Bernardes MA (ed.) Developments in heat transfer. InTech, Rijeka, pp 153–170
3542. Zubair M (2018) Nonlinear mathematical models involving non-Fourier heat flux. Ph.D. Thesis, Quaid-i-Azam Univercity, Islamabad

Index

A. I. Zhmakin, *Heat Transfer in Vivo*, Springer Series in Biophysics 27,
https://doi.org/10.1007/978-3-031-95459-7

The manufacturer's authorised representative in the EU is Springer Nature Customer Service Centre GmbH, Europaplatz 3, 69115 Heidelberg, Germany. If you have any concerns regarding our products, please contact ProductSafety@springernature.com

Printed and bound by CPI Group (UK) Ltd, Croydon, CR0 4YY
30/09/2026
02225589-0019